W9-AMW-677

Seventh Edition

OCEANOGRAPHY

A View of Earth

M. Grant Gross

Chesapeake Research Consortium
and University of Maryland

Elizabeth Gross

Scientific Committee on Oceanic Research
and The Johns Hopkins University

Prentice Hall
Upper Saddle River, New Jersey 07458

Contents

8 Currents 199

9 Waves 229

10 Tides 255

13 Benthos 339

14 Coasts, Coastal Oceans, and Large Lakes 369

15 The Ocean on a Changing Earth 403

PREFACE

The Seventh Edition, like its predecessors, introduces students to the ocean—how it works, how it influences our lives, and how it may affect our future. The level of difficulty remains unchanged; a high-school-level background in science is assumed.

A second objective is to introduce students to science as a process of continual inquiry and open testing of ideas. We show how scientists approach problems, using a variety of tools, including measurements of the ocean made by instruments on Earth-orbiting satellites and powerful computer-modeling techniques that predict ocean behavior.

In modern society, science provides an important framework for structuring our understanding of the world. Oceanography is an example of a modern science that is involved with many aspects of our rapidly changing environment and society.

We wrote this book to present oceanography to non-scientists. The study of the ocean introduces students to the solid Earth sciences, the atmosphere, the life sciences and the physical sciences—in short, all of the basic sciences. Thus, it should help students acquire an understanding of how science works. Further, an understanding of ocean science provides a useful background to individuals dealing with a variety of problems and conflicts in industry, government, and the private sector.

Finally, oceanography is an example of a science that continually expands our view of how the Earth works. Almost weekly, newspaper articles and TV or radio programs describe some new, unfamiliar and fascinating aspect of our oceans. In our careers we have discovered that oceanography is an endlessly exciting field of science. In this edition, we seek to share that excitement with our readers.

Features of the Seventh Edition

Updated Material

Throughout the book we have included examples of recent developments in ocean research and technology that have broadened our view of Earth and the ocean. Particular attention is paid to applications of ocean science that benefit society. Sections dealing with global climate change, the coastal ocean, aquaculture, and new technologies have been added or updated. Many new illustrations have been added, and the references have been expanded to include recent publications.

This edition has been revised to broaden the readers' perspective on the role of ocean science in their daily lives. Each chapter includes a "media box" in which developments in oceanography published in recent newspaper articles are used to expand on a topic in the chapter. Special topics highlight the achievements of individual marine scientists that produced major advances in our understanding of the ocean. Particular attention is given to applications of ocean science and their importance to solving immediate problems of the environment and of humanity. The human aspects of ocean sciences, such as the impacts of rapid development and population growth in coastal areas, are emphasized.

New Features

This edition has been extensively revised for easy, independent student use. More examples are provided, and more bridges between sections are provided to emphasize the many links among oceanic processes. Each chapter has been rewritten and reorganized in a sequence that emphasizes how the various aspects of ocean sciences are integrated.

The artwork has been revised and expanded. Many new graphics and new color photographs have been added to illustrate major features in the text. Explanatory drawings and labels have been added to help readers interpret photographs and other diagrams.

Various study tools have been improved to help students assess their progress. Review questions have been revised and expanded. New "critical thinking" questions have been added to each chapter to help students integrate materials between chapters. Chapter outlines and study objectives have also been revised. Lists of key terms have been added to each chapter.

A new Prologue has been added to emphasize major themes that extend through more than one chapter. Two new chapters have been added. One deals with the coastal ocean and large lakes where many conflicts are arising because of increasing population pressure and limitations on resource availability. The other new chapter deals with the ocean's role in global climate change. Two new appendixes were added—one provides brief biographical information about individuals important in ocean science, and the other discusses employment possibilities in oceanography and other environmental sciences.

Clarity

This text has been thoroughly revised, sentence by sentence. We worked closely with a developmental editor to reorganize the entire text and to make illustrations more easily understood. For example, mathematical equations are replaced, where possible, by other formulations. Technical words and phrases are defined when they are first introduced.

The Glossary has been updated and expanded to include the many terms and concepts arising from new instruments and techniques. It will be a useful reference tool both for students new to science and the more advanced reader. All terms in the Glossary are shown in boldface in the text when they are first introduced. Other important terms are shown in italics to help students in reviewing the material in the chapter.

Balanced Coverage

The Seventh Edition retains a balanced coverage of all aspects of ocean sciences. The coverage of biology has been expanded and reorganized. The chapter on sediments has been extensively revised and integrated with chapters on related topics.

Acknowledgments

We are deeply indebted to our many colleagues and students who taught us as we have all worked and learned together. We want especially to express our appreciation for the hard work and encouragement from Bob McConnin, editor; Irene Nunes, developmental editor; Peg Gluntz, copy editor; Tim Flem, production editor; Barbara Salz and Sabina Dowell, photo researchers; Julie Hotchkiss, indexer; and finally, Ray Mullaney. All have worked hard to help us meet the many deadlines, and we are grateful for their assistance and encouragement.

Over many years, our friends and colleagues have also contributed to the book through conversations, questions, criticisms, and through helping us locate photos and other artwork.

University of Washington: P. Jumars, K. Banse, K. Aagard, J. Delaney
University of Alaska: V. Alexander, T. Royce
MBARI: P. Brewer, B. Robison
University of Hawaii: R. Stroup, D. Karl, S. Smith
University of Delaware: J. Sharp, F. Webster
University of Maryland: K. Tenore, T. Malone, W. Boicourt, E. Houde, W. Boynton
Virginia Institute of Marine Sciences: J. Milliman, H. Ducklow
National Science Foundation: B. Haq, C. Sancetta, M. Reeve, C. Dybas, P. Penhale, P. Dauphine
Anne Arundel Community College: D. Lear
University of North Carolina: D. Frankenburg, H. Pearl, C. Neumann
Old Dominion University: W. Dunstan, L. Atkins
Harvard University: J. McCarthy, A. Robinson
Texas A&M University: W. Merrell, G. Fryxell, G. Rowe
University of California, Los Angeles: W. Hamner
Scripps Institution of Oceanography: F. Azam
University of Miami: G. Ostlund, G. Brass, O. Brown
University of California, Santa Barbara: K. MacDonald, K. Bruland; A. Alldredge
Dalhousie University, Canada: R. Fournier, E. Mills
Bedford Institute of Oceanography, Canada: J. Lauzier, K. Mann
University of Southern California: R. Douglas
Duke University: R. Barber, T. Johnson
Harbor Branch Oceanographic Institution: M. Youngbluth, P. Blades-Ecklebarger
University of East Anglia, United Kingdom: P. Williamson
Université de Québec, Canada: K. Juniper
University of Kiel, Germany: W. Ham
The Johns Hopkins University: T. Osborn, P. Olson, S. Stanley
Universität Bern, Switzerland: A. Fuchs

M. Grant Gross
Elizabeth Gross

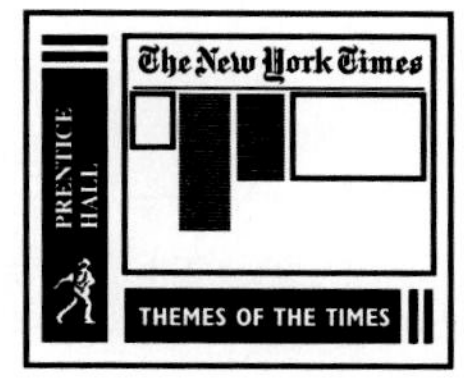

PROLOGUE

In order to examine Earth and its ocean, we need to understand several scientific themes. These themes are briefly stated here in the prologue. We start with the modern approach to studying Earth and its ocean: considering Earth as a set of interacting systems.

The Earth System

Because satellites allow us to observe Earth globally, scientists can now treat our planet as a single system. This field of study, called Earth System Science, involves three activities:

Characterizing the component parts of the system
Specifying the interactions among these component parts
Modeling the behavior of the total system

The components chosen for each model depend on the processes to be modeled and the time and space scales of the model.

To begin, we define a **system** as a set of interacting components and reservoirs of mass or energy. These components and reservoirs are affected by factors acting on the system, such as temperature. For Earth, the atmosphere, cryosphere, hydrosphere, lithosphere, and the biosphere are the major components making up what is called the **Earth system** (Fig. P–1).

We represent systems by diagrams in which boxes stand for the various components and arrows show interactions among them, as Figure P–1 shows. (The components of a system are usually referred to as **subsystems.** In this text, we use the terms component and subsystem.) Scientists combine various representations, then use the combinations to investigate how systems respond to changes in some factor, such as variations in incoming solar radiation or the composition of the atmosphere.

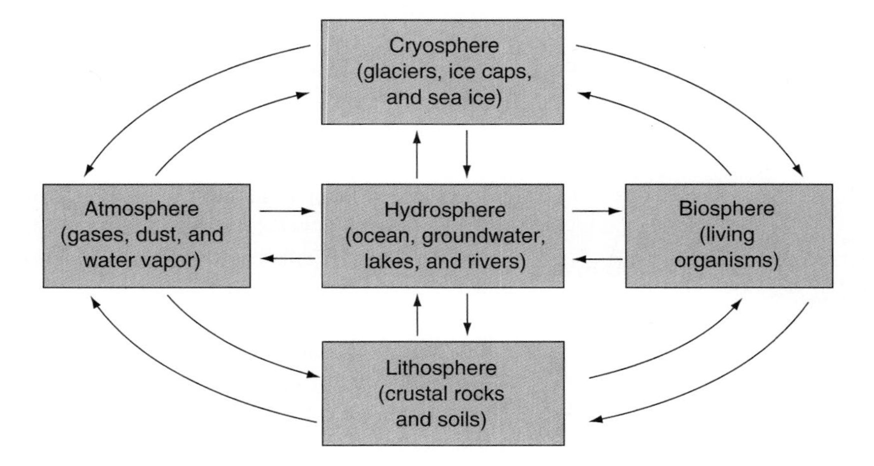

Figure P–1
Simplified representation of the Earth System and its major subsystems.

In analyzing any system, the most difficult part of the process is determining the nature of the interactions of the system. One of the most important processes affecting these interactions is **feedback,** a process that affects the flow of information from one part of the system to another. Feedback can be either positive or negative (Fig. P–2). **Positive feedback** indicates that a change in some component results in the same sense of change in some other part of the system. For example, if an increase in temperature causes an increase in water evaporation, we say there is positive feedback. **Negative feedback** is the opposite: an increase in some part of the system causes a decrease elsewhere. An example is when an increase in atmospheric dust causes lower atmospheric temperatures.

Time and Space

In our study of the ocean and related parts of Earth, we shall consider events and processes that span enormous scales of time and space. In many cases, these ranges are impossible for humans to observe or experience or, in some cases, even to comprehend easily.

Time Scales

The longest time scales involve the formation of the universe and the Solar System, including Earth. As we see in Figure P–3, these processes took place over many billions of years—exactly how many billions of years is still hotly disputed by astronomers. Humans study these ancient processes via optical and radio telescopes that can peer back into the very early stages of formation of the Solar System. On the other extreme of the time scale are the rapid movements of molecules in liquid water as they continually form and reform in various aggregates. Some of these processes can be studied using instruments that detect the subtle changes in the substances as their atoms and molecules rearrange themselves.

Direct human observations of Earth and ocean processes are limited to the time interval of a human lifetime, which means that the span of time involved runs from fewer than 100 years down to a fraction of second. Beyond these limits, we are forced to use other techniques involving various kinds of instruments, such as the optical and radio telescopes just mentioned above.

Written human history extends back about 5,000 years. We can use the records of our ancestors to study Earth processes, especially those involving the land. Such records have been useful in studying climate change in areas inhabited by literate human societies. For instance, there are records of especially good or especially bad harvests, including famines. For the Nile, there are records of river floods for thousands of years. The fact that we know nothing about uninhabited areas or areas inhabited by preliterate peoples obviously limits the information available to us.

Figure P–2
The complex role of water in climatic processes illustrates how feedback may be either positive or negative.

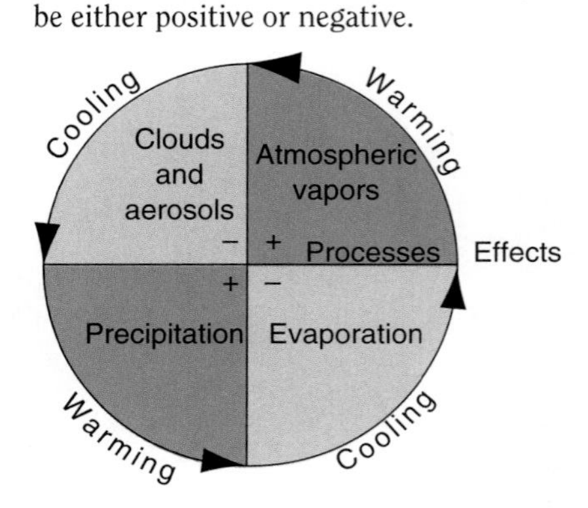

Beyond the limits of human history, we can use various proxy records to study climate. Perhaps the most familiar proxy record is annual tree rings, which provide us with information about temperature and rainfall, recorded in the variations of the growth rings laid down during the year. Long-lived trees cover much the same time span as written history, but provide much wider coverage of areas. Some marine corals also deposit annual bands, and so they, too, preserve a record of the environment they experience in the ocean. Such records are still being studied to see how much time they cover.

Annual layers of snow and ice in glaciers provide a rich source of information about climate. These layers contain bubbles of atmospheric gases that can be sampled and studied directly and are the source of our most detailed information about changes in atmospheric composition. Glaciers in Greenland and Antarctica, for instance, record climatic events over the past 300,000 years.

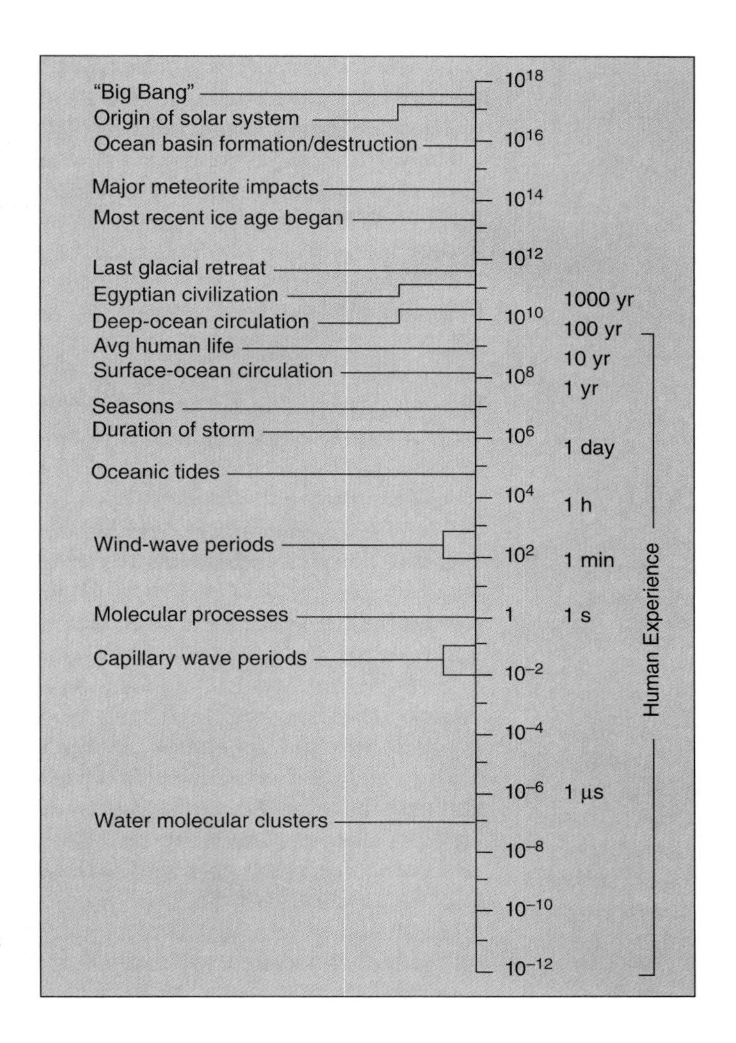

Figure P–3
Range of times that affect the Solar System, Earth, its ocean, and its biota. Note that the times are shown in a logarithmic scale. Each tick indicates a tenfold change in time from the adjacent tick.

Events occurring farther back in time are often recorded in sediment deposits. Cores of deposits from the ocean bottom provide records covering about 200 million years. Even older records can be obtained from deposits on land, which go back nearly 4 billion years. (We discuss the reasons for the preservation of longer records on land in Chapter 3.)

Space Scales

We must also work with an enormous range of space scales (Fig. P–4), the smallest of which involve atoms and molecules. These are the space scales involving chemical reactions that affect all aspects of the ocean and atmosphere and life on Earth. Here, too, we must rely on instruments, such as electron microscopes and their relatives, for observations and measurements.

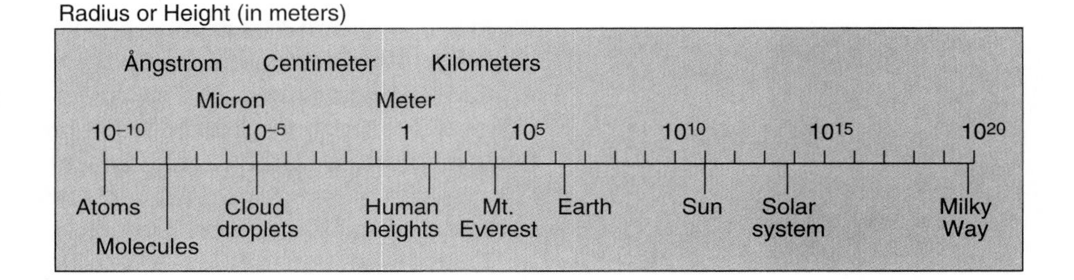

Figure P–4
Range of sizes and distances in the Solar system, Earth System, and universe. Note that the distances are shown in a logarithmic scale. Each tick indicates a tenfold change in size from the adjacent tick.

For very large space scales, our ability to make direct observations is again limited. We can directly experience items from a fraction of a millimeter (diameter of a human hair) to several kilometers—the horizon on a clear day or the height of a jet aircraft. Going beyond that to the outer limits of the Solar System or the Universe is truly beyond our ability to experience except in an abstract way. In fact, for great distances, we often speak not in terms of distance but in terms of the length of time it takes light to travel that distance.

Earth System Catastrophes

Major catastrophes have affected Earth many times in its past. Some of these left their marks in sudden extinctions of organisms, both on land and in the ocean. Large comets or asteroids (collectively called bolides) strike Earth every few ten to few hundred million years. Their immediate effect is to blast a crater up to 200 kilometers in diameter. These craters are later filled with water or covered by sediment deposits and eventually destroyed by processes we discuss in Chapter 3. However, they can still be recognized in many parts of Earth. Their effects can be seen in the sudden extinctions of organisms, such as the dinosaurs about 65 million years ago.

The other result of a bolide impact, a result that is less understood, is its effect on the atmosphere. The dust from the crater may cause global cooling for many months and sometimes even for a few years, and this cooling may be sufficient to cause the extinction of many plants and animals. The blast may also cause vast amounts of acidic compounds to form from atmospheric gases, with the presence of these compounds subsequently causing extinctions. The sudden release of energy in the atmosphere may also trigger global hurricane-scale storms.

In July 1994, 20 or more pieces of a disintegrating comet struck Jupiter. While the direct impact sites were not visible from Earth, the effects of the event on Jupiter's atmosphere left no doubt that such an event on Earth would have catastrophic effects.

Massive volcanic eruptions may also have substantial, even catastrophic effects. Such eruptions arise from processes that originate deep in Earth's interior. The eruptions of vast amounts of lava can change the composition of the atmosphere enough to cause dramatic changes in global climatic conditions.

Change, Stability, Instability, and Metastability

Change is a concept we use repeatedly throughout this book. We need to consider the factors that control whether a system changes easily, not so easily, or not at all. In other words, we have to understand some of the basic aspects of **system stability.**

A system is **stable** if it resists change and returns to its original state when disturbed. A simple stable system is a marble in the bottom of a round-bottomed cup, as shown in Figure P–5a. If the marble is moved away from the low point in the cup (we say that such a marble has been *disturbed*), it returns to its original position as soon as we stop holding it. If the walls of the cup are steep enough, we could have a system so stable that the only way to change it is to destroy the cup.

The other extreme is represented by inverting the round-bottom cup and carefully balancing the marble on the tip, as in Figure P–5b. Now the slightest disturbance is enough to disturb the marble. This is an example of an **unstable** system. Once disturbed, the marble cannot return to its original position.

The third concept we need to understand is **metastability.** Here a system may have several stable states. Imagine a marble on a hill that has several troughs on its side, as in Figure P–5c. The marble is stable any time it is in a trough, but it

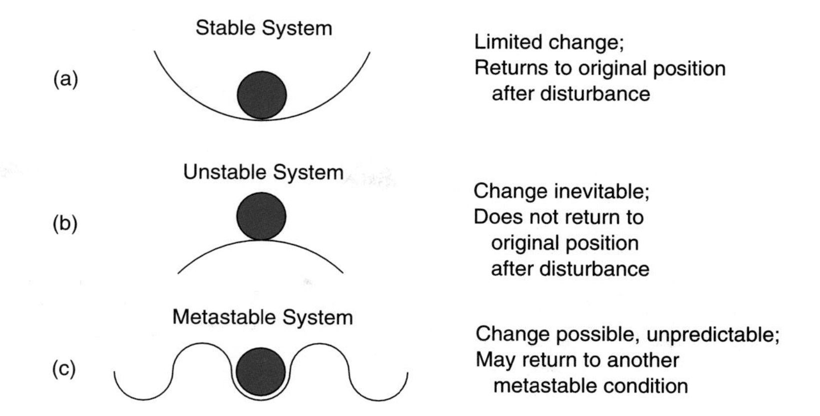

Figure P–5
Simple examples of (a) stable, (b) unstable, and (c) metastable systems.

is possible to disturb the marble enough to cause it to change from one metastable state to another. Also, if the marble is disturbed, it may not return to its original metastable state.

We shall see the importance of stability, instability, and metastability in many aspects of the ocean and related Earth subsystems. In discussing deep-ocean currents, for instance, we shall see how climatic conditions can form water masses that are unstable at the ocean surface and sink to find the level in the ocean where they are stable. Another example is the state of Earth's atmosphere. Under certain circumstances it may be stable, but it can readily change into another equally stable (or metastable) state.

The Scientific Method

Science dominates Western society's view of the world. Thus, it is worthwhile to briefly examine what we mean by the **scientific method.** Science has two premises: (1) that natural phenomena are controlled by simple relationships and (2) that these relationships can be discovered by careful observations or measurements. Observations are analyzed and then organized into statements—called **models**—of how such processes work. Such models allow predictions of future phenomena. Indeed, the ability to predict events or conditions, such as tides or weather, is a major benefit that science offers society. Models are used to describe systems and how they work. Models can be expressed in words *(conceptual models)* or mathematical formulas *(mathematical models).*

Predictive models begin with a **hypothesis**—an idea that must be tested. The hypothesis may come from analysis of data, or from hunches, or even from inspired guesses. For example, the English scientist Isaac Newton suggested that the gravitational attraction of the Sun and Moon acting on ocean waters caused the tides. According to legend, Newton's ideas about the force of gravity came to him when an apple fell on his head while he was sitting under an apple tree. Later, he showed how the gravitational attraction of the Sun and Moon causes ocean tides. Using this hypothesis, Newton could predict tides accurately, thus verifying his idea. This is probably the most familiar example of successful scientific predictions of ocean phenomena.

Generally, whenever a given set of phenomena can be modeled in more than one way, scientists prefer the simplest model that explains and predicts the phenomena. (One scientist calls this "the principle of least astonishment.") Thus, the simpler model of Earth orbiting the Sun replaced the earlier, more complicated model (Sun orbiting Earth), although ancient astronomers could predict planetary movements with both models.

Comparing model-based predictions with observations is one way of verifying hypotheses (Fig. P–6). A hypothesis confirmed by many tests is called a **theory.**

Figure P–6
A hypothesis is tested by making predictions using a model. If the prediction is satisfactory, the model may be used to provide commercial services, such as weather forecasts or tide tables. If the prediction is unsatisfactory, either the model must be changed through continued research or new data must be used in making the prediction. If more predictions are still unsatisfactory, the hypothesis must be modified or discarded.

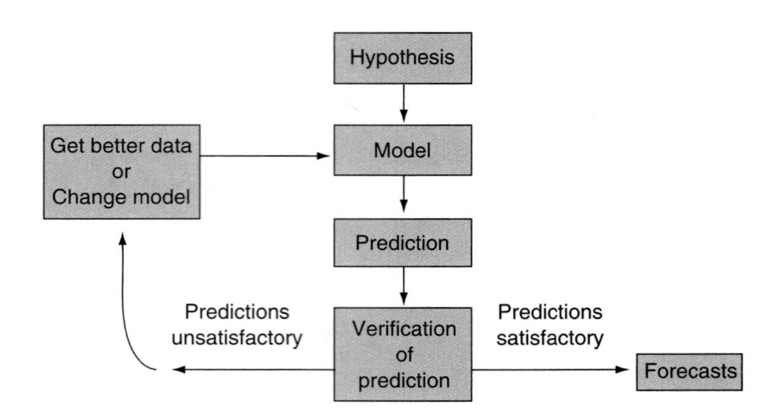

The utility of a theory is shown by its ability to answer old questions and to pose significant new problems for study.

Models and theories are revised whenever they fail to predict satisfactorily or to describe observed phenomena. If the predictions are nearly correct, the model may need only fine-tuning, but if the predictions are wrong, another model must be designed. If the new ideas are radically different from earlier ones, the process of designing the new model is called a *scientific revolution.* Such a major shift in thinking can change our view of the world.

A familiar example of a scientific revolution is Darwin's *theory of evolution,* which has dominated scientific thinking about relationships among species since its publication in 1857. A more recent example of a scientific revolution is the development of plate tectonic theory (discussed in Chapter 3).

The Advance of Technology

Technology—the application of knowledge gained through scientific studies and from practical experience—provides the instruments needed to observe the oceans and to analyze the data obtained from them. A striking example of technology in action is the improvement in our ability to observe oceanic phenomena using instruments carried on Earth-orbiting satellites. Availability of frequent ocean observations on a global scale has revolutionized ocean science. These instruments observe ocean conditions, such as wave heights or sea surface temperatures, over the entire global ocean within a few days. The vast amount of data obtained can be processed, stored, and manipulated only on the largest computers. The rate of advance in understanding the ocean depends heavily on the rate at which new instruments and measuring techniques become available.

Summary

Scientists now can treat Earth as a single system consisting of a series of interacting subsystems—atmosphere, cryosphere, hydrosphere, lithosphere, and biosphere. In dealing with the system and the various subsystems, one must be aware of the time and space scales involved. Enormous ranges of time and space are involved in dealing with the universe, Solar System, Earth, and its components.

Catastrophic events can markedly change the earth. For example about 65 million years ago, a large meteorite impacted the Earth, causing the extinction of the dinosaurs and subsequent rise of mammals and humans. Massive volcanic eruptions can also cause catastrophic changes.

Systems can be characterized as stable (return to original condition), unstable (do not return to original condition), or metastable (may return to original condition or to another metastable condition).

The scientific method is used to gather and interpret observations about the Earth System. It involves testing hypotheses (ideas on trial) by making predictions to see if the results are satisfactory. If they are satisfactory, a model can be used to make regular predictions, and a hypothesis is regarded as an accepted theory. If the predictions are unsatisfactory, the model must be changed or new data used to make new predictions. If the results are still unsatisfactory, the hypothesis is discarded and replaced by another one.

Technology is putting to practical use the knowledge gained from scientific studies. Technology contributes substantially to scientific progress.

Key Terms

system
Earth system
subsystem
feedback
positive feedback
negative feedback
system stability
stable
unstable
metastable
scientific method
models
hypothesis
theory
technology

Selected References

PETERSON, I., *Newton's Clock: Chaos in the Solar System.* New York: Freeman, 1993. A popular account of how Newton's theories replaced earlier ones about the Solar System, and how they are now being modified to account for chaos.

MACKENZIE, F. T. AND J. A. MACKENZIE, *Our Changing Planet.* Englewood Cliffs, NJ: Prentice Hall, 1995. Modern view of Earth system science and global environment change.

OBJECTIVES

Your objectives as you study this chapter are to understand:

- The ocean's role in human history
- How ocean science developed
- How ocean science contributes to human activities
- How advances in technology have changed the way we study the ocean

Instruments on earth-orbiting satellites measure global oceanic and atmospheric conditions several times a day. With such satellites, scientists can now obtain a global view unattainable using only ships. This satellite is about the size of a school bus and is powered by energy from the large, flat solar panels on each side. The control center and communications antenna are toward the front of the satellite; the remaining packages contain scientific instruments. (Courtesy NASA.)

1 The Origins of Oceanography

Oceanography—the scientific study of the ocean—is a modern activity, only about a century old. But humans have been sailing the ocean and harvesting its plants and animals for thousands of years. Long before recorded history, sailors and traders explored the ocean and its shores, looking for new lands to settle, new routes to take, or new products to buy and sell. The first maps useful for navigation were not available until 5,000 years later (Fig. 1–1).

Today, we study the ocean systematically in order to improve the techniques used to predict weather and climate, to protect fisheries and other resources, and to avoid disasters such as coastal flooding of densely populated areas.

Ancient Seafarers

One indication of the extent of ancient exploration and colonization is that humans settled all the continents except Antarctica. Reaching several of these regions required sea voyages, some quite long.

Refuse piles at ancient coastal villages are another indication of human use of the sea. These piles contain abundant remains of fish and shellfish, showing their importance in villagers' diets. Some also contain bones of open ocean animals, which suggests that ancient people had seaworthy boats. In fact, travel by river and through coastal waters was the primary mode of transportation for many peoples, because travel over land was slow, difficult, and often dangerous.

Figure 1–1
The earliest maps used for navigation (around 1275) were called *portolanos.* Drawn on animal skins, they indicated compass directions from one port to another as well as shoreline features. (Courtesy Library of Congress.)

Although we know that transportation by water was involved in early human migrations and settlements, little direct evidence remains of the boats of these earliest explorers, and we know virtually nothing of how they navigated. Materials commonly used in making primitive boats—woods, skins, reeds—do not survive long. The earliest evidence concerning boats comes from rock carvings in Egypt (Fig. 1–2) from about 2400 B.C.E. (before the common era). The earliest ship models come from Egyptian tombs. Viking rock carvings (Fig. 1–3) show the vessels of the earliest Norse sailors to be similar to those used by tenth-century Vikings in their raids and conquests. All this evidence suggests that both the Norsemen and the Egyptians were accomplished sailors.

Four kinds of boats appear in rock carvings or in ancient texts: dugouts made from logs hollowed and shaped by fire and simple tools, boats constructed of bundles of reeds lashed together, boats made of split sections of thin bark sewn together and stretched over a wooden frame (e.g., the birch bark canoe of Native Americans), and skin boats made of sewn animal hides stretched over a wood frame. Similar boats were still in use at the beginning of historical times. Boats like these—reconstructed from drawings and descriptions—have been used to successfully navigate the open ocean.

Early human migrations give us more circumstantial evidence of extensive seafaring. Around 10,000 years ago, people from Africa crossed the Strait of Gibraltar to colonize southern Europe.

Figure 1–2
Egyptians used sailing craft to navigate the Nile, as shown here in a tomb painting from around 3500 B.C.E. They used similar ships to sail along coasts. (Courtesy The Granger Collection.)

In the Mediterranean

The Phoenicians were superb sailors, navigators, and traders (Fig. 1–4). Their principal contribution to oceanography was celestial navigation: navigation using the stars as reference points. Phoenician fleets dominated trade in the Mediterranean and in the adjacent waters of the North Atlantic from about 1200 B.C.E. until about 200 B.C.E. They sailed as far as England to get tin for making bronze. In the beginning of the seventh century B.C.E., Phoenician ships circumnavigated Africa, nearly 2,000 years before the Portuguese would make the same voyage.

Most European sea trade in antiquity was limited to the Mediterranean, where the longest voyages were only a few days out of sight of land. Most ships put ashore at night, where crew and passengers slept.

In the Pacific

On the other side of the world, the Micronesians colonized many larger western Pacific islands between 2000 and 4000 B.C.E. Later, between 1000 and 800 B.C.E.,

Figure 1–3
Early Norwegian ships are shown in rock carvings. This memorial stone shows a typical Norse longship. (Courtesy Statens Historiska Museum, Stockholm.)

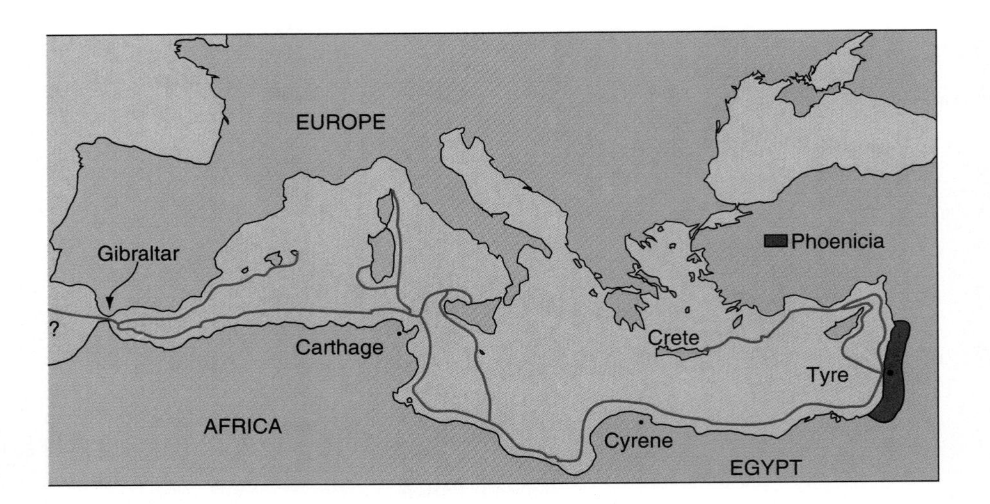

Figure 1–4
Phoenician trade routes included most of the Mediterranean Sea. These early mariners preferred sailing along coasts and made few open-water crossings.

Polynesians constructed elaborate double-hulled vessels (Fig. 1–5), which they used to explore and colonize much of the central Pacific, including Hawaii (Fig. 1–6). The largest of their ships had living quarters for people and domestic animals.

Much of our information about the seafaring traditions of these Pacific peoples comes from the accounts of the first European explorers to come into contact with them. One surviving example of their navigational skills is the stick charts (Fig. 1–7) used by the Micronesians. Shells mark locations of islands, and bamboo strips show wave patterns caused by the prevailing trade winds in these regions. Stars, cloud patterns, and winds were also used to navigate between islands.

Figure 1–5
Large sailing canoes were used by Polynesians for their trans-Pacific voyages between 1000 and 800 B.C.E. Such craft were still in use when the Europeans arrived in Tahiti in 1768, the first European contact with the Polynesians. (Courtesy Culver Pictures, Inc.)

Figure 1–6
Polynesian colonization of Pacific islands required long voyages in sailing canoes. (After F. Braudel, *The Structure of Everyday Life: Limits of the Possible.* New York: Harper and Row, 1981.)

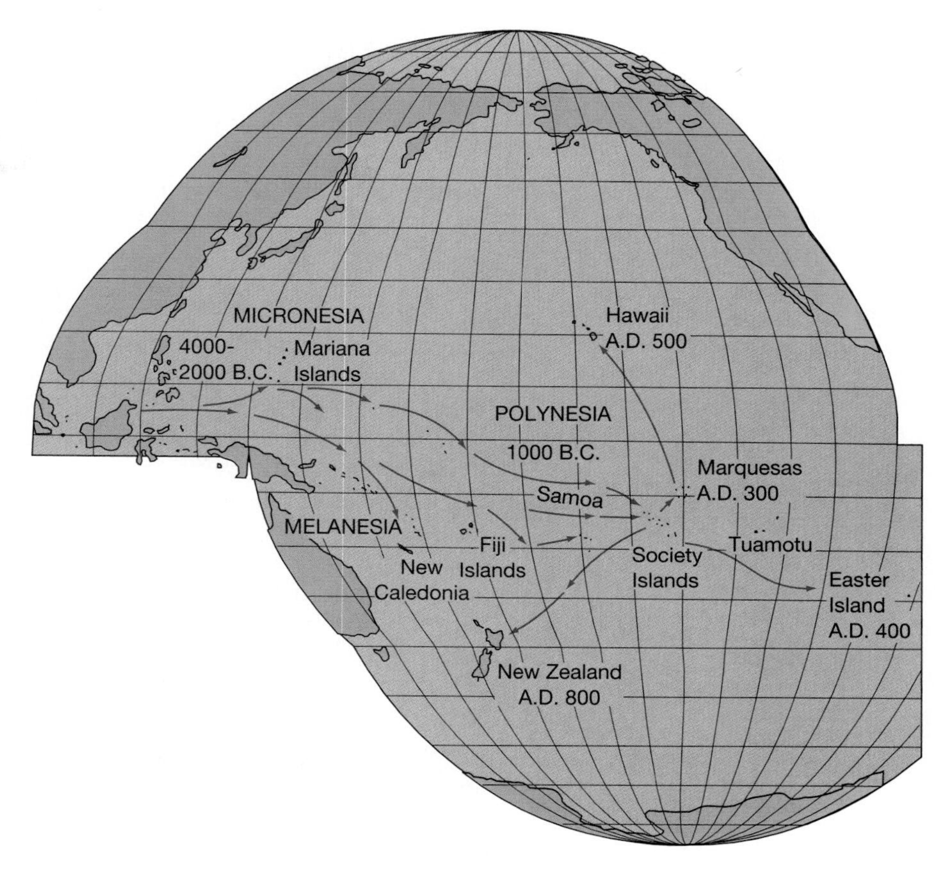

Figure 1–7
Micronesian navigators used stick charts to sail the Pacific. The shells represent islands, and the bamboo strips show prevailing wave patterns. Wave patterns around islands are shown by the bamboo strips curved around the shells at the lower right of the chart. (Courtesy Library of Congress.)

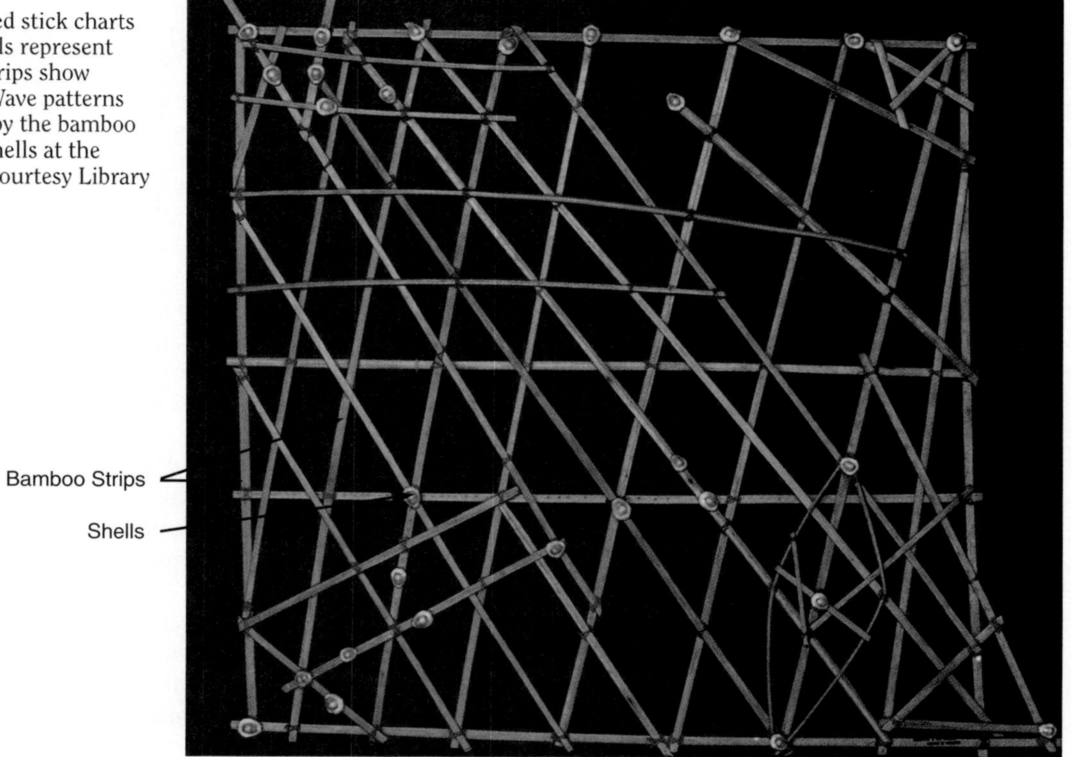

Ancient European Knowledge of the Ocean

Early seafarers must have known many ocean features and processes well, but we have few records of their knowledge. Stories of explorers and seafarers provide some clues, but little information was preserved, perhaps because it was so valuable that it was closely guarded and eventually lost.

Much of our insight into what the first seafaring Europeans knew comes from the writings of philosophers and theologians, writings based on observations reported by seafaring explorers. As early as 350 B.C.E., for instance, the Greek philosopher Aristotle, noting that the ocean neither dries up nor overflows, concluded that the amount of rainfall must equal evaporation over the earth.

In the seventh century, A.D., the English theologian and historian Bede knew that the Moon controls ocean tides. Tide levels at London Bridge were predicted routinely in the late twelfth or early thirteenth century.

Various simple navigational devices were used. Europeans began using magnetic compasses in the thirteenth century. The oldest surviving European chart for ship navigation dates from 1275 (Fig. 1–1). It provides compass directions and shows distances. To determine speed through the water, sailors threw a log overboard and counted how many knots on an attached rope ran out before all the sand had run through a timing glass. We still speak of nautical miles per hour as **knots.** (Tables for converting kilometers or statute miles into nautical miles appear in Appendix 1.)

To determine water depth as well as a ship's location, sailors threw into the water a lead weight coated with sticky wax and attached to a rope. The amount of line fed out before the weight touched bottom indicated water depth, and the type of material stuck to the wax could be compared with chart notations to locate the ship. Depths were measured in **fathoms** (6 feet or 1.8 meters), the distance between the left middle finger and the right middle finger of a large man standing with outstretched arms.

Medieval European Ocean Exploration

Leif Ericsson (980?–1001), a Scandinavian explorer and Viking leader, was the first European to reach North America (as far as we can tell from the evidence available today). Returning to Greenland from Norway in 1000, he lost his way and came ashore at what today is L'Anse-aux-Meadows in northern Newfoundland. The Vikings apparently lived there for several years, naming their colony Vineland. According to the ancient sagas, Ericsson or members of his family made four subsequent expeditions to Vineland. Because of continued conflicts with the natives, however, the Vikings finally abandoned Vineland and returned to Greenland.

For the next several centuries we have no record of further exploration by Europeans in North America. Their diminished interest in exploration and colonizing may have been caused by the substantial cooling that followed the Medieval Warm Epoch (1000–1300). Ultimately, their voyages had no effect on history or oceanography.

In 1410, however, western Europeans were stimulated by the publication of a Latin edition of a world map compiled in the second century by the Egyptian geographer Ptolemy. In this first scientific representation of the entire Earth, Ptolemy used **latitudes** (distances north and south of the equator) and **longitudes** (distances east and west of a prime meridian) much as modern maps do. His influence was so great that it took centuries before all his mistakes were eradicated from newer maps. Ptolemy's maps were copied uncritically until they were emended by the results of new explorations.

The great age of European ocean exploration began early in the fifteenth century, stimulated by the rediscovery of Greek and Arab geography that took place when Christian armies took control of southern Spain and thus had access for the first time to the great Islamic libraries. Along with Ptolemy's maps came two Greek ideas: (1) Earth was a sphere and (2) the ocean was navigable.

Chinese Ocean Exploration

The coastal regions around the northern Indian Ocean and the islands and coasts of southeastern Asia have a long history of trade and cultural relationships, and we are now learning more and more about this early history. Interest in the ocean in this part of the world occurred at various times and expressed itself in different ways in different cultures in the region.

One example of this interest is the brief period of large Chinese ocean voyages. Between 1405 and 1433, the government of the Ming dynasty organized seven voyages to the nearby Pacific and Indian Oceans (Fig. 1–8). These were the largest peacetime exploratory voyages ever undertaken, involving 37,000 men and 317 ships.

The Chinese ships were far bigger than any in western Europe at that time: the largest had nine masts, was 135 meters (444 feet) long, and had a beam (width) of 55 meters (180 feet). These ships had many modern features, such as transverse bulkheads dividing each ship into several watertight compartments. Thus, if a ship were damaged, the water could be confined to the damaged compartment. (Such construction is now standard.) Magnetic compasses and detailed navigation charts were also used.

The Ming dynasty expeditions were unlike any before or since in that they did not seek to conquer, collect treasure, make religious converts, or gather scientific information. Instead, their purpose was to extend and consolidate Chinese influence. China was then far more advanced technically than the rest of the world

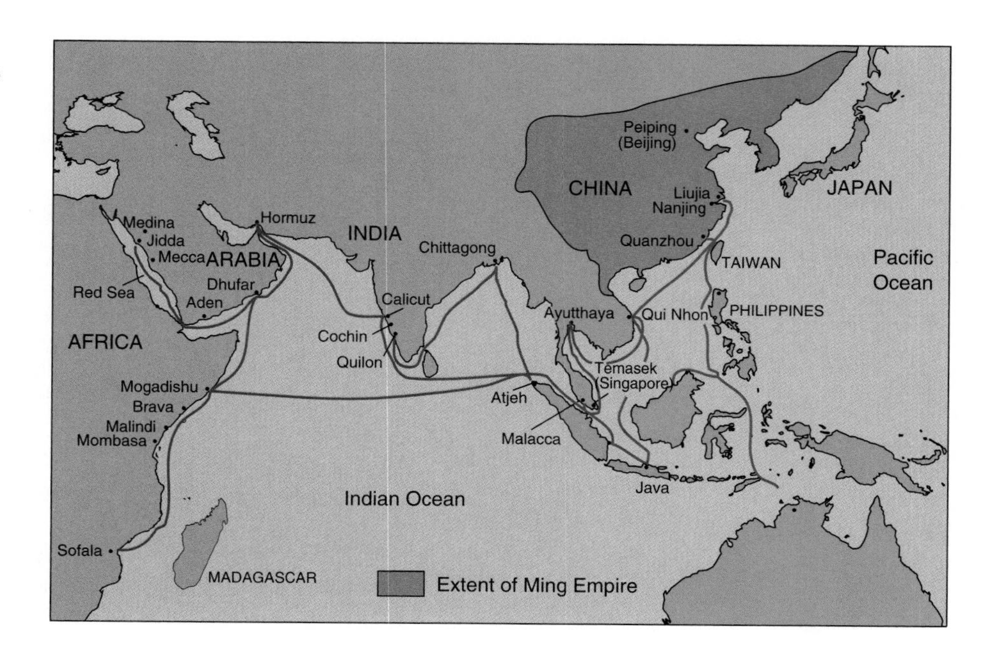

Figure 1–8
During the Ming dynasty, Chinese ships made expeditions to southeast Asia and the Indian Ocean to demonstrate the power and wealth of the Chinese. While these were perhaps the largest ocean expeditions ever, they made no lasting contributions to our knowledge of the ocean.

and was essentially self-sufficient for food and raw materials. In their own estimation, the Chinese had little or nothing to learn from the outside world. These expeditions simply displayed the splendor and power of the Ming dynasty.

Under the Chinese political system, client states brought gifts to acknowledge China as the most civilized country in the world. To prove their superior position, the Chinese were obligated to give back more than they received. Therefore, these states were a financial burden to the Chinese, with any expansion of Chinese influence simply increasing the drain on the national treasury. In the end, the expeditions brought back animals unknown to the Chinese to add to the imperial zoo, such as giraffes, but otherwise contributed little to China.

Bowing to internal politics and economic pressures, the emperor ended these voyages in 1433. As part of this increased isolation, the Chinese rebuilt the 1,600-kilometer-long (1,000-mile) Grand Canal so that boats could navigate it all year. This improved waterway system eliminated the need for coastal sailing to transport goods and grain around the country. Thus, as western Europe began expanding its horizons in the early 1500s, China turned inward. The Chinese government outlawed seafaring and began a period of isolation that lasted until the mid-nineteenth century. Because little new information was compiled during the great Chinese expeditions, they made no lasting contribution to our understanding of the ocean.

European Seafaring in the Fifteenth and Sixteenth Centuries

Although there was little contact between seafarers of the Indian Ocean and those of the western Pacific, these two areas of civilization were not completely isolated from one another. For more than 4000 years, they were connected by overland and sea routes (Fig. 1–9) that brought spices from India and Indonesia and silk from China to Europe. Roads were so poor, however, that moving goods a few tens of miles overland doubled their cost; transport by sea was far cheaper. The decline of the Mongol Empire in central Asia in the early fourteenth century and Constantinople's fall to Turkish armies in 1453 finally cut these ancient trade routes. After

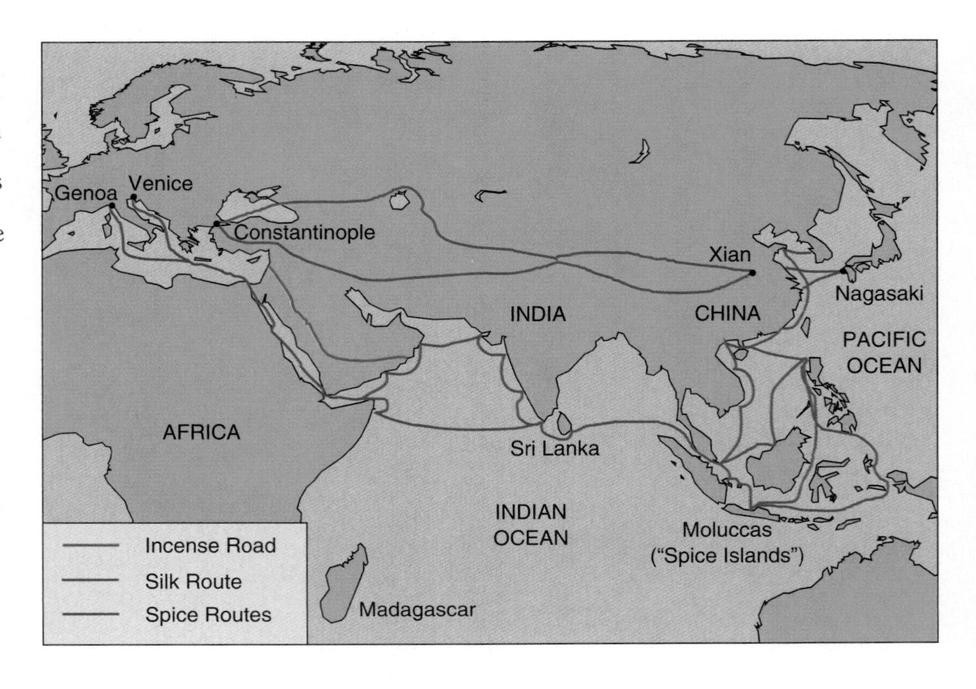

Figure 1–9
Europeans traded in the Far East for incense, silk, and spices, using a combination of sea routes and overland caravan routes across central Asia. Spices were traded primarily over the sea route across the northern Indian Ocean from India to southeastern Asia to the Moluccas ("Spice Islands"), now part of Indonesia.

Figure 1–10
Three-masted ships were used extensively by European explorers in the New World. Here the ship *Half Moon*, commanded by Henry Hudson, is shown exploring the Hudson River, which was named after him. (Courtesy Library of Congress.)

the fall of Constantinople, the eastern Mediterranean was dominated by Islamic peoples hostile to western European Christian countries.

Some of the impetus for ocean exploring came from the need of European traders to replace these caravan routes. Marco Polo (1254?–1324) was one of the few Europeans to make the entire 8,000-kilometer (5,000-mile) trip and to come back. His accounts of China, India, and southeast Asia were published after his return. His writings acquainted western Europeans with the riches of the region and increased interest in finding new ways to open trade routes.

Portugal led the early European exploration of the Atlantic, beginning in 1416 with the discovery of the Canary Islands off northwestern Africa. Discovery of the Azores, in the middle of the Atlantic, followed around 1430. Perhaps the most influential individual during this period was Portugal's Prince Henry the Navigator, who employed learned people to teach navigation to Portuguese sea captains. This educational program greatly stimulated Portuguese exploration.

In the voyages that followed the discovery of the Azores, Portuguese sailors explored the western coast of Africa, reaching the tip of South Africa (near present-day Cape Town) in 1488. Vasco da Gama sailed around Africa to reach India in May 1498. His pioneering voyage around Africa led to the establishment of profitable trade routes between Portugal and India and to the founding of the Portuguese empire.

Development of the three-masted ship (Fig. 1–10) was another important reason for European success in exploring the Atlantic and later the rest of the world ocean. These ships were large enough to carry men and supplies on long trading and exploring voyages. New sail designs were versatile enough to contend with variable winds and the stormy North Atlantic.

Newly developed cannons carried aboard these ships made them formidable weapon systems. This led to the control of the ocean by the navies of western Europe and to the conquest and colonization of coastal areas. Eventually, western European nations dominated both land and ocean for more than 400 years, from 1500 to about 1900. A description of the many explorers of this period is beyond the scope of this book, but some merit special mention.

Ferdinand Magellan was the first to sail around the world, crossing the Pacific in 1520–1521. Within a few years, regular ship service was available around the world.

Henry Hudson, among many others, explored North America as the major western European nations established colonies and divided up the New World.

Queen Elizabeth I of England encouraged British maritime development. During her reign, Sir Francis Drake sailed around the world, capturing Spanish ships and cargoes as he went.

Trade motivated much of this European exploration, with many navigators, such as Hudson, trying to find faster routes to the Pacific. They sought the elusive Northwest Passage, a route that would allow European ships to sail west but end up in the Far East, or more direct routes to China farther south.

Systematic Ocean Exploration

During these centuries of exploration, most advances in understanding the ocean came from solving practical problems. In turn, these improvements in navigation permitted the mapping of ocean shorelines. National offices for ocean mapping were established by France in 1770 and by England in 1795.

Despite the obvious benefits obtained from these early advances, there was no systematic exploration or study of the ocean until the mid-1800s. From that time until about 1950, the ocean was studied and charted by government-backed exploring expeditions or private companies. Universities had little involvement in ocean studies until after 1950.

In the early part of the nineteenth century, primitive measuring and sampling techniques caused serious misconceptions about deep-ocean conditions. It was believed, for instance, that the deepest ocean waters did not circulate but lay stagnant at the ocean bottom. Consequently, it was thought that no incoming oxygen would be available to support life there. Furthermore, an influential naturalist, Edward Forbes, working in the 1840s, collected animals in the Aegean Sea to compare his findings with those Aristotle had made nearly 2,500 years earlier. He noticed that fewer plant and animal species were caught as his nets went deeper. Forbes concluded that most deep-ocean waters must therefore be **azoic,** or devoid of animal life. This idea gained wide acceptance—so much so that occasional reports of animals taken from below 800 meters (2,600 feet) were ignored or discounted.

Forbes's azoic theory was not refuted until the 1860s, when the deep-ocean basins were surveyed during the laying of transoceanic telegraph cables. A cable in the Mediterranean Sea raised for repairs from a depth of 2,000 meters (6,600 feet) was encrusted with living animals (corals). This discovery came at a time when interest in deep-sea exploration had been aroused by Charles Darwin's theory about the origin of species through evolution. Scientists became interested in finding stable environments in which conditions had not changed over long time periods. Many scientists thought that the deep-ocean floor might be inhabited by ancient unknown species, the ancestors of modern organisms.

The Explorations of Captain Cook

Captain James Cook was a great eighteenth-century explorer and mapper of the Pacific Ocean. His three voyages, between 1768 and 1780, covered the entire Pacific basin, from the Bering Strait entrance to the Arctic Ocean on the north to the margins of Antarctica. Using the latest technology available—for example, the chronometer, an accurate ship clock—Cook determined the location of the many islands he found and accurately mapped the outlines of New Zealand and eastern Australia. Many of these islands and coastlines had been visited by earlier explorers, but their locations were not precisely known. Cook's explorations came during the time when England, France, Spain, and Holland were contending to expand their empires into the Pacific. Because he sailed under the flag of England, many of his "discoveries" were later made part of the British empire.

BOX 1–1
Christopher Columbus and the "Discovery" of America

We know that Columbus was not the first European to reach America. Nor did he understand what he had accomplished once he did reach America. He was convinced he had reached the Indies and, consequently, named the natives Indians. His voyages were a commercial success, however. His explorations and territorial claims made Spain rich and powerful for a century. It is an interesting insight into his times to consider how his voyages came about.

Columbus's first appeal for financing was made to King John II of Portugal in 1484, when Columbus was only 33. The Portuguese were interested in improving trade with India and southeastern Asia, together called the Indies. They had already supported 19 expeditions searching for a sea route to the Indies, but still had not reached the southern tip of Africa at the time Columbus made his proposal.

King John was unimpressed by Columbus but appointed a committee to examine his proposal. The members of this committee concluded (correctly) that his estimate of the distances involved was too low. Columbus estimated 2,400 miles to the Indies, while the committee estimated the trip at about 10,000 miles. By the time King John met with Columbus, Bartolomeu Dias had rounded Cape Good Hope, discovering a sea route around Africa to the Indies. Thus, the Portuguese had already found their route to the riches of the Indies.

Columbus then went to Spain, presenting his proposal to Ferdinand and Isabella, joint rulers of Aragon and Castile. Isabella was interested in the idea, but Spain was then fighting the Moors in Grenada, finally defeating them in January 1492.

Isabella set up a committee of theologians and lawyers to examine Columbus's ideas. They too concluded that he was mistaken about the distance, but had no doubt that the earth was round. A second committee, this one comprising astronomers, pilots, and philosophers, also recommended against the voyages because of the compensation Columbus demanded. Finally the keeper of King Ferdinand's privy purse (public monies allocated for the king's and queen's personal expenses) intervened, convincing Queen Isabella that Columbus's voyages would cost less than she spent in a week entertaining visitors. He argued that the possible returns to Spain were great and offered to finance the voyages himself. The queen eventually agreed to support his voyages, which cost about $5 million in today's terms. Contrary to legend, she did not have to pawn her jewels.

As a result of Columbus's discoveries, gold, silver, and other valuable resources flowed from the New World to Spain, strengthening her economy and providing the basis for centuries of world domination.

When Columbus died, he left a comfortable estate but not the huge riches he had hoped for. The Spanish court gave him the title "Admiral of the Ocean Sea," which is still in use in the Spanish Admiralty.

This was also a time of great interest in completing the exploration of Earth's surface. For instance, Cook's orders from the British Admiralty for each voyage specified that he explore the Southern Continent, what today we call Antarctica. This huge land mass had been postulated by the Greeks but never located. Here Cook was not successful, because his wooden sailing ships could not penetrate the icebergs and pack ice that surrounded the continent. The discovery of Antarctica was left to an American sealer, Nathanial Palmer, who in 1820 located and explored what we now know as the Palmer Peninsula, the northernmost part of the continent.

One of Cook's outstanding accomplishments was conquering scurvy, a disease caused by a deficiency in vitamin C. Before Cook, scurvy killed large numbers of sailors during long voyages exploring the Pacific. He experimented with foods—sauerkraut and spruce beer, to name just two—and finally settled on grasses and eventually oranges and limes. (The British navy's use of lime juice to prevent scurvy gave rise to the nickname "limey" for British sailors.) Whereas voyages during the time of Magellan might kill six out of seven men, Cook's crews usually lost only one or two. This statistic is especially remarkable considering the poor health of most sailors when they first joined the ship's crew.

Perhaps Cook's most famous "discovery" was the Hawaiian Islands in 1778. On his third voyage (1776–1780), he was looking for the Northwest Passage. On his way north through the Pacific, he came upon the Hawaiian Islands. After

Figure 1–11
Captain Cook's ships *Discovery* and *Resolution* arrive in Kealakekua Bay, Hawaii, on January 17, 1779. (Courtesy Bishop Museum.)

exploring much of the western coast of North America and passing through the Bering Strait into the Arctic Ocean, he returned to Hawaii. His ships put in at Kealakekua Bay on the Big Island of Hawaii (Fig. 1–11). At first, Cook was taken for a god by the Hawaiians, who consequently showed him great reverence. Tensions between Cook's crews and the natives grew, however, and on February 14, 1779, he was killed by the Hawaiians in a misunderstanding over a stolen boat. Kealakekua Bay is now a marine preserve, a favorite snorkeling spot for visitors; a monument marks the site of Cook's death (Fig. 1–12).

In addition to Hawaii, Cook also explored and mapped the many Pacific islands settled centuries earlier by the Polynesians. His maps of many of these islands were used until well into the twentieth century.

The Challenger Expedition

Figure 1–12
The Cook Monument in Hawaii marks the spot where Cook was killed.

By the 1860s, scientific interest in the deep ocean was growing. Answering new questions about the ocean required complex research projects, which even wealthy individuals could not afford. Therefore, the Royal Society of London began providing financial support. The British Admiralty also provided two ships for North Atlantic deep-sea studies during the late 1860s. As a result of these research efforts, old misconceptions about the ocean gradually disappeared. For instance, it was demonstrated that ocean waters move through deep-ocean basins.

Finally, the Royal Society was persuaded to sponsor the most ambitious ocean exploring expedition up to that time (Fig. 1–13). This voyage aboard HMS *Challenger* established a tradition of scientific ocean exploration that continues today. Between December 1872 and May 1876, *Challenger* traveled 109,000 kilometers (68,000 miles), collecting rocks and sediments from the ocean floor and taking soundings of water depths. All ocean basins except the Arctic were sampled.

The *Challenger* was formerly a sailing warship with an auxiliary steam engine, its guns replaced by scientific gear for the expedition (Fig. 1–14). The scientific party under Sir Wyville Thomson had been assigned to investigate "everything about the sea." They planned to study physical and biological conditions in every ocean, recording whatever factors or processes might influence the distribution of marine organisms. This task involved measuring the temperatures of

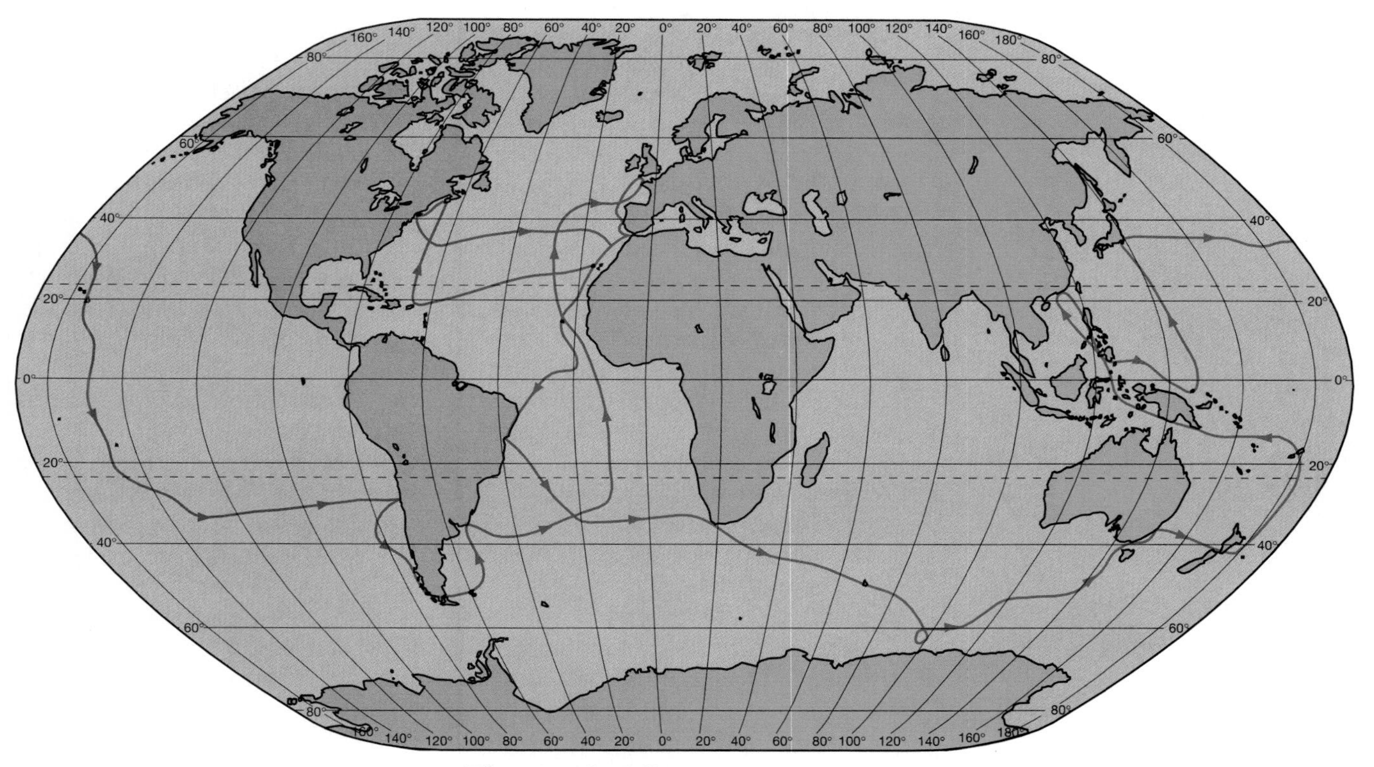

Figure 1–13
The track of HMS *Challenger* during the first comprehensive scientific study of the ocean, from 1872 to 1876. By comparison, today's satellites can make thousands of observations of the ocean in a single day.

Figure 1–14
The *Challenger* among floating sea ice and icebergs near Antarctica.

deep and surface waters, observing currents, and collecting samples to study sediments and identify new species.

Many misconceptions about the ocean were swept away by the *Challenger* results. The most famous, perhaps, was the debunking of *Bathybius,* a supposed primordial slime thought to represent ancient life on the ocean floor. Bathybius had been "discovered" by Thomas H. Huxley, a well-known biologist. In 1868 Huxley found a slimy substance in a jar containing a biological specimen several years old and concluded he had identified the substance from which primordial life originated in the deep ocean. He may have been overly anxious to provide new evidence for his friend Darwin's theory of evolution, which was then under attack. It was a fortuitous idea in that it stimulated interest in marine research. While at sea, however, a *Challenger* chemist revealed that the slime was precipitated by mixing alcohol with seawater in order to preserve samples, a discovery that was embarrassing to Huxley but did not discredit Darwin's theory.

After the expedition ended, a British government department, the *Challenger* Expedition Commission, was established to analyze and publish the results of the voyage. Data from the expedition filled 50 large volumes, which were a watershed in the development of oceanography and marked its beginnings as a science.

The commission was headed by Sir John Murray, a Canadian-born geographer and naturalist. After persuading the British government to annex uninhabited Christmas Island in the Indian Ocean, Murray made a fortune mining phosphate from the island. (This would now be considered an unacceptable conflict of interest.) He used his fortune to support many scientific activities. For example, he supported his own research laboratory and funded the Michael Sars expedition in the North Atlantic. After Murray's death, his estate funded a 1933–34 Anglo-Egyptian oceanographic expedition to study the Indian Ocean.

American Ocean Exploration

In the United States in the nineteenth century, government supported ocean research primarily to solve practical problems. The first U.S. ocean research vessel was the *Albatross,* built in 1882 for fisheries research. To protect its fisheries, the United States established a Fish Commission in 1871 and built a research center (now part of the National Marine Fisheries Service) on Cape Cod. This initial research center stimulated the later development of the Marine Biological Laboratory (1888) and the Woods Hole Oceanographic Institution (1930).

Wealthy individuals, as well as government agencies, sponsored research in the late nineteenth and early twentieth centuries. For instance, the American scientist Alexander Agassiz maintained a private laboratory in Newport, Rhode Island to study materials collected on his privately financed expeditions. He also outfitted the *Albatross* to dredge the deep-ocean bottom. On one of Agassiz's expeditions, materials were collected from water depths not surpassed until 1951.

Benjamin Franklin

Knowledge about the ocean was especially important in the American colonies. For instance, one of the earliest maps of the Gulf Stream—the strong current system along the U.S. Atlantic coast—was published by Benjamin Franklin (1706–1790) just before the American Revolution. Franklin had noted that the royal mail ships coming from England took much longer to arrive in the colonies than did American ships making the same trip. His cousin Timothy Folger, a Nantucket whaler, told Franklin about the Gulf Stream, which was well known among American whalers and presumably among other American ship captains as well (but evidently not to the British). From that information, Franklin published his map of the Gulf

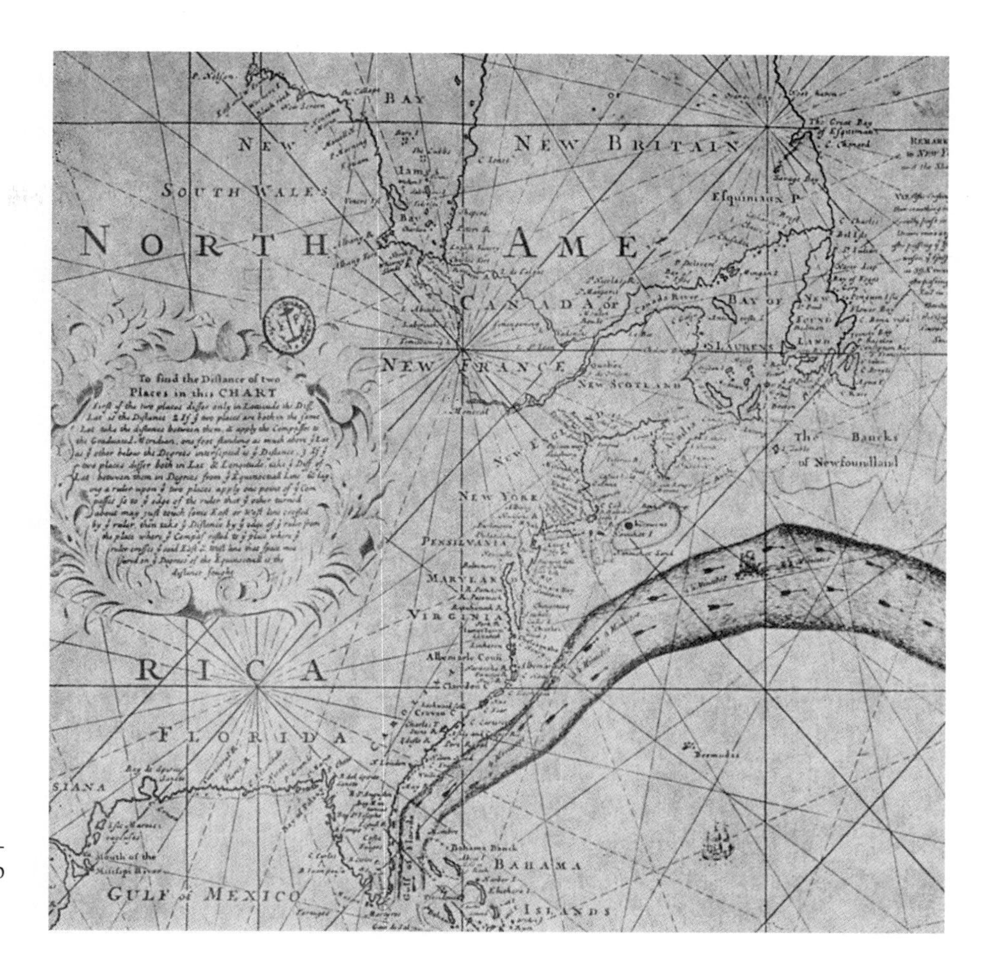

Figure 1–15
Benjamin Franklin's map of the Gulf Stream, published in 1770, was the earliest map of this major current. (Dr. Philip Richardson/Woods Hole Oceanographic Institution.)

Stream (Fig. 1–15). His instructions for locating the warm Gulf Stream waters were used by ship captains to sail with the strong currents on eastbound voyages and to avoid sailing against these currents on their westbound returns.

Figure 1–16
Matthew Fontaine Maury was the first superintendent of the U.S. Navy's Depot of Charts and Instruments. His *Physical Geography of the Sea* was the first oceanography book in English. He is called the "Father of Physical Oceanography." (Courtesy Library of Congress.)

Matthew Fontaine Maury

Major American concerns in the nineteenth century were ensuring the safety of passengers and goods aboard U.S. ships, maintaining adequate coastal defenses, and protecting fisheries. In 1830 the U.S. Navy created a Depot of Charts and Instruments to supervise the use of navigational instruments on government vessels. Matthew Fontaine Maury, a naval officer, was appointed superintendent in 1842.

Maury (Fig. 1–16) wanted to make ocean transportation safer and speedier. He subsequently organized the vast amount of data on winds, currents, and seasonal weather stored in ships' logbooks, mostly whaling ships. To gather more information, Maury furnished ships' captains with blank charts for recording weather and ocean conditions. Wind and current charts compiled in this way revolutionized navigation and cut weeks off transoceanic runs by clipper ships, the fastest ships of their time. In addition, Maury's sounding and bottom-sampling projects were used to map the Atlantic bottom.

In 1853 Maury organized the first international meteorological conference, which led to international cooperation in collecting weather information at sea. His popular and influential book, *The Physical Geography of the Sea* (1855), was the first major oceanographic work in English. He has been called the father of physical oceanography.

BOX 1–2
Underwater Archaeology

The seafloor holds a gigantic store of maritime history. The many thousands of shipwrecks are essentially time capsules, preserving the ship's equipment, passenger and crew belongings, and cargoes.

Invention of the self-contained underwater breathing apparatus (SCUBA) in 1943 was the starting point for modern shallow-water archaeology. More recently, the development of remotely operated vehicles (ROV) to locate wrecks, and of precise navigational systems to permit return to wreck sites, has opened up the ocean depths. Most of the work has been done in the Mediterranean, where millennia of shipborne commerce left thousands of wrecks on the bottom. More than 100 premedieval wrecks are located each year, most of them discovered by sport divers in shallow waters near ancient ports. Thousands of wrecks are known in waters less than 50 meters deep. Unfortunately, many wrecks are looted before they can be studied by archaeologists.

Many materials are recovered from wrecks. The hull's wooden timbers can be preserved if they are buried in an oxygen-free environment (below the sediment surface) where there are no boring organisms. For instance, a wooden warship, the *Mary Rose,* sank off Portsmouth, England, in 1545. Its hull, buried in the mud for 400 years, was salvaged in the 1980s and put on display.

Most recovered hulls are from commercial vessels because the weight of their cargoes caused the hulls to be buried in the sediments. Well-preserved wrecks of ancient warships are rare, because they had no cargoes and would float ashore, where they would be destroyed by waves and/or wood-boring insects.

The most common artifact found in the cargoes are pottery amphoras—disposable containers used for storing and transporting liquids, nuts, meat, and many other items. These vessels were cheaply produced and easily handled by one or two men. Their pointed bases permitted them to be stacked in interlocking layers. Often the lower hull of the wreck is preserved below the stacks of amphoras. The size, shape, and markings of the amphoras permit dating them and determining their source. Newly developed techniques permit analysis of amphora fragments to determine what they contained—wine or olive oil, for example.

Most wrecks discovered in the Mediterranean date from between the fourth century B.C.E. and the fifth century A.D., when the Roman Empire was flourishing. In those days, the Mediterranean was essentially a unified market free of piracy and warfare. Long-distance transport of food to feed Rome's million-strong population and of the large armies used to guard the empire's frontiers was common. Maintaining the political stability of the Roman Empire required continual handouts of grain, olive oil, and other staples to its citizens.

Most wrecks are found in the western Mediterranean, off southern France, Tuscany in Italy, Corsica, and southeastern Sicily. These are areas where seaborne commerce was important and where sport divers have been most active. Elsewhere, exploration has been sporadic or nonexistent.

Many wrecks have yielded rich finds of art. Bronze statues are especially well preserved. One first-century B.C.E. wreck produced a bronze "computer" made like a watch and functioning as a navigational calculator.

Some cargoes were quite ordinary. For example, the millstones highly prized in North African households and bakeries mostly came by ship from Italy. Apparently, millstones were carried south by ships that returned from North Africa loaded with grain for Rome. These stones make it possible to reconstruct ancient trade routes by combining evidence from shipwrecks, additional evidence from excavations on land, and a knowledge of the distributions of volcanic rocks around the Mediterranean.

Modern Global Ocean Studies

Military research dominated oceanography during World War II. Many advances were made in instrumentation, and our understanding of the ocean was greatly improved. For example, advances in predicting wave conditions were essential for the invasion of Europe and the many amphibious assaults on islands in the Pacific. Many ocean-basin features, such as the magnetic field, were mapped to improve capabilities to detect submarines. Research in ocean acoustics (the behavior of sound in the ocean) was also important in antisubmarine defense. These results were used for scientific purposes after World War II ended in 1945. Even more military surveillance capabilities were declassified and used for ocean studies after the Cold War ended in the early 1990s.

Modern ocean studies began in the 1950s. Government support for ocean research and education greatly increased. This expanded funding permitted universities, especially in the United States, to participate in ocean studies, a role that continues to the present day.

Modern ocean studies require many ships and scientists, frequently exceeding the resources of any single country. Thus, cooperative projects involving ships, equipment, and scientists from many countries became the norm. This trend was accelerated by the International Geophysical Year (1957–58), during which scientists from 67 nations studied Earth and various oceanic phenomena during a short period of intense scientific activity.

The International Decade of Ocean Exploration (IDOE) during the 1970s involved more than 50 countries from around the world. The intent was to learn more about the ocean as a potential source of food, fuel, and minerals, as land supplies of these commodities were being used up. In other words, IDOE projects were intended to solve practical problems. For instance, the ability to predict *El Niños* (periodic warming of surface waters in the eastern tropical Pacific) grew out of IDOE-supported projects.

Oceanic Field Experiments

Since the 1970s, expeditions to explore the ocean basins and their contents have been replaced by field experiments. Previously, samples were brought back and studied in laboratories on land; now, improved instruments and other new technology allow much more research to be conducted on ships at sea (Fig. 1–17). The ships both take samples and serve as an analyzing laboratory while the experiment is under way.

A typical field experiment begins with a thorough study of the processes of interest. The understanding gained from these studies is then expressed as a mathematical model. Such models are used to predict events because such predictions facilitate the subsequent observing of the events. Based on the results of tests using models, the field experiment is planned. Locations and timing for sampling or observations are picked to maximize the possibility of finding a given process active and being able to study it. Then the mix of ships, aircraft, satellites, and logistical support is planned and scheduled.

Figure 1–17
The North Atlantic Bloom Experiment used satellites for communications. Abundances and distributions of minute plants floating in the Atlantic were mapped by instruments on aircraft and satellites. Samples were collected and analyzed aboard ships. (US JGOFS/Institute of Oceanographic Sciences, U.K.)

During a field experiment, large areas of the ocean can be surveyed by aircraft and satellites and sampled by scientists on ships. These observations can be transmitted by communications satellite back to an operations center and the information used to adjust future sampling locations and timing. Results from such experiments are stored in large computers ashore for the use of the investigators taking part in the experiment and are later open to investigators around the world.

One large international field experiment in 1989–1991 was designed to investigate the timing and distribution of biological production in the North Atlantic (Fig. 1–17). The experiment involved six research ships from five nations and nearly 250 scientists. The study began before a rapid increase in the amount of microscopic floating plants in the surface-ocean waters, called the spring bloom, and followed it for seven months as it progressed northward.

A ten-year field experiment running from 1983 to 1993 studied interactions between ocean and atmosphere (Tropical Ocean Global Atmosphere—TOGA). Our understanding of the interactions between tropical ocean and atmosphere has permitted greatly improved long-range weather forecasts, which can be issued up to one year in advance. Using such forecasts to guide farmers in planting crops has saved billions of dollars in the southeastern United States and Florida alone.

Other long-term, international experiments describe the ocean of today so that the information obtained can be used as a baseline in future studies of changes. For example, the World Ocean Circulation Experiment—(WOCE) used an Earth-orbiting satellite called TOPEX/Poseidon (Fig. 1–18) to accurately measure the subtle contours of the ocean surface. From this, surface currents and wave heights can be mapped (Fig. 1–19). These results provided the first comprehensive maps of ocean surface features, analogous to the maps of atmospheric conditions that we see in newspapers and television weather forecasts. A companion study of the deep ocean made from research ships using floating buoys (Fig. 1–20) provided the first "snapshot" of the deep-ocean circulation ever made.

Figure 1–18
An Earth-orbiting satellite like this ERS-1, launched by the European Space Agency in the early 1990s, provides observations of the ocean surface at a level of detail impossible to achieve with ships. (Courtesy European Space Agency.)

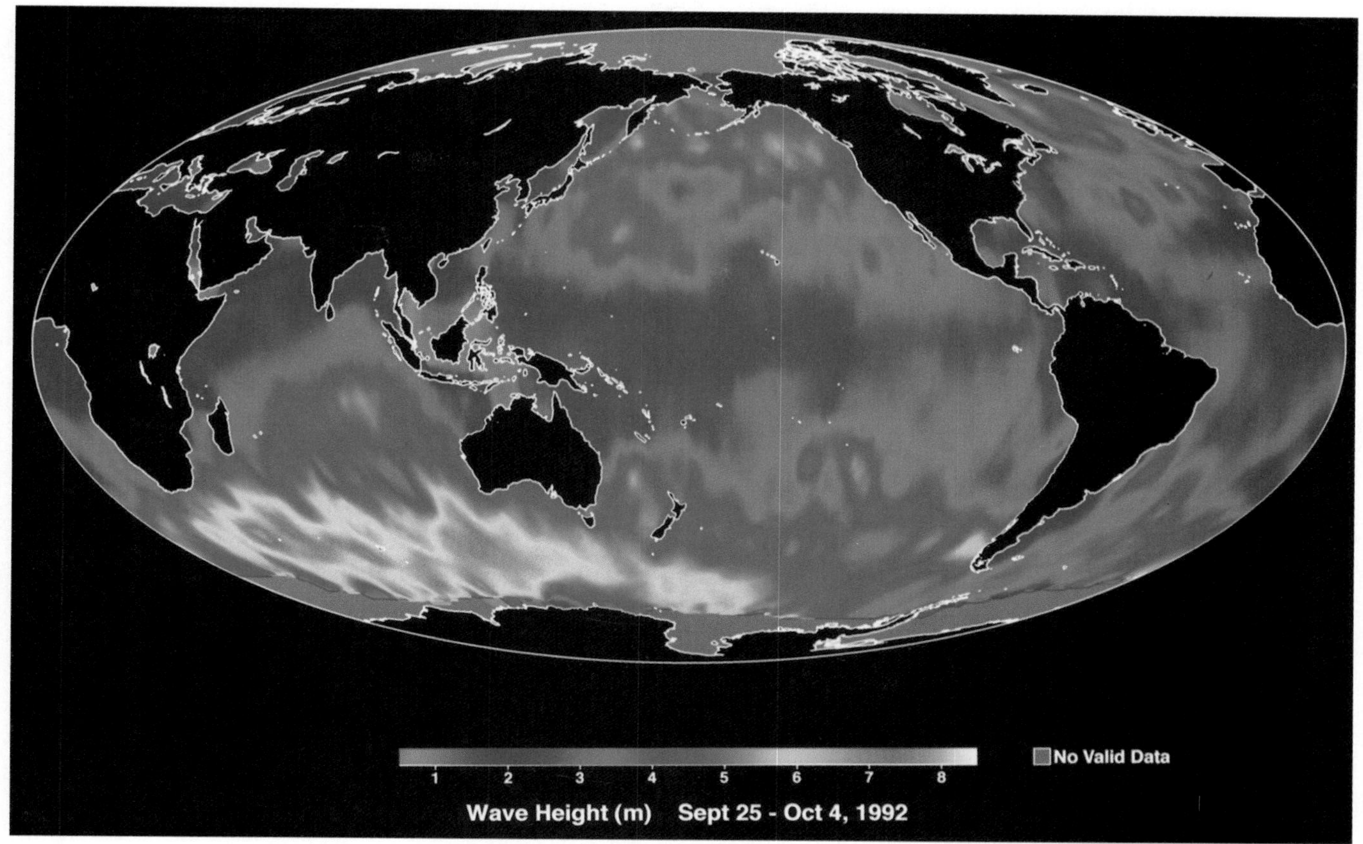

Figure 1–19
Wave heights measured globally by a radar altimeter on the Earth-orbiting U.S.–French Satellite TOPEX/Poseidon during a nine-day period in September–October 1992. (Courtesy California Institute of Technology.)

Figure 1–20
The research vessel *Thomas G. Thompson,* built in 1991, is operated by the University of Washington. This large ship is typical of modern oceanographic research vessels. The large frame on the stern is used to lower heavy instruments. The mast on the bow holds air samplers. (Courtesy University of Washington.)

Future Ocean Science

Ocean science continues to develop at an accelerating pace. Much of this acceleration results from availability of new instruments, especially remote-sensing ones, and from the release to scientists of large amounts of data previously collected for military and intelligence purposes. The increased amount of data, the rapid expansion of computing capacity to store and manipulate those data, and developments in communications have all changed ocean science. Now large databases can be easily accessed by scientists working anywhere in the world.

While ships and submersibles remain the workhorses of oceanographic research, they are being replaced by remote-sensing satellites (Fig. 1–18), which permit frequent and detailed observations and consequently more accurate forecasts and predictions. Only a few decades ago, it was necessary to average ocean observations made thousands of kilometers and many years apart. Indeed, it was necessary to assume that the ocean did not change; this assumption is called a steady-state approximation. We now know that the ocean changes and we can observe those changes as they occur.

As our understanding of ocean processes has improved, it has become increasingly apparent that the ocean plays a major role in controlling Earth's weather and climate. Much current oceanographic research is designed to study the role of the ocean in climate and how changes in Earth's atmosphere may affect the ocean. Eventually, surveys such as WOCE will be conducted on a regular schedule, just as the atmosphere all around the world is measured twice daily. The results of such studies will be used to predict changes in both the ocean and the atmosphere. A Global Ocean Observing System (GOOS) is being organized to provide such observations and to develop new predictive capabilities.

The Law of the Sea

After European exploration and colonization of the newly discovered continents in the sixteenth century, the "high seas" were free to all nations. The Dutch statesman, Hugo Grotius, formulated the legal doctrine of *mare liberum* (freedom of the high seas) in 1635. Until the 1970s all the world ocean outside a narrow territorial sea extending out three nautical miles (5.7 kilometers) from any shore was freely used by all nations.

By the 1970s this concept of an ocean belonging to everyone was challenged by nations interested in the immense quantities of oil and gas on continental margins and of metals on the deep-ocean floor. Uncertainties about the legal status of these resources inhibited the investment monies needed to exploit them.

In 1974, therefore, the United Nations convened a Conference on the **Law of the Sea** (UNCLOS). The nations of the world negotiated their many conflicting needs in order to develop a consensus, often ambiguous, to divide the ocean among them. The treaty, adopted in 1982, entered into effect in 1994 after 60 nations ratified it, markedly changed the legal status of the ocean margins. It extends national territorial seas to 22 kilometers (12 nautical miles) from shore. Within this zone, each coastal state has the same rights and responsibilities that apply on land.

With the extension of territorial seas to 22 kilometers, some major straits through which ships must pass going from one ocean to another now fall within the territories of coastal states (Fig. 1–21). Negotiations were necessary to ensure the rights of ships to pass through these areas. The status of some straits, such as the Northwest Passage through the Canadian Arctic, is still contested.

Another treaty provision was the recognition of **Exclusive Economic Zones** (EEZ) extending 370 kilometers (200 nautical miles) out to sea from the shore of each coastal nation. In its EEZ, a coastal state may regulate fisheries and resource exploration and exploitation. About one-third of the ocean falls in the Exclusive

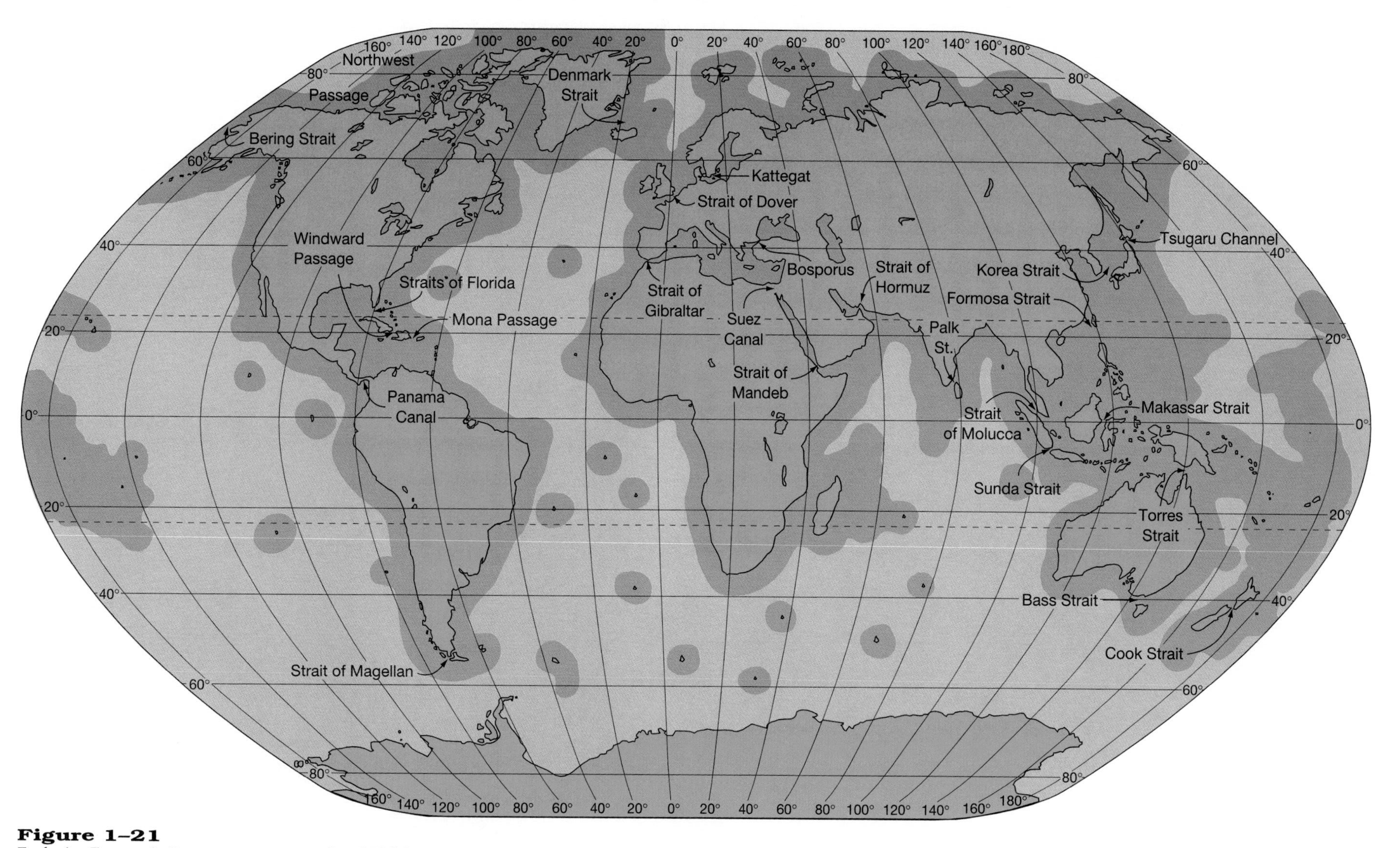

Figure 1–21
Exclusive Economic Zones are areas extending 370 kilometers out to sea from all shorelines of the world. Several large areas of the world ocean, such as the Gulf of Mexico and the Mediterranean and North Seas, are totally divided among several EEZs. Some strategic straits affected by the EEZs are labeled.

Polynesian Navigation—Using Winds, Currents and Stars

Ancient Polynesian navigators used their knowledge of the winds, waves, ocean conditions and stars to navigate across the Pacific. They sailed many thousands of miles from the Marquesas Islands south of the equator to Hawaii about one thousand years ago, bringing Polynesian settlers and their domestic animals. These navigators depended on the winds and currents to bring them to their desired destinations.

For many years, scientists thought that the Hawaiian Islands were discovered by accident when Polynesian boats were blown off course. In the last two decades, repeated voyages between Hawaii and the Marquesas, made in traditionally designed and constructed double-hulled Polynesian canoes, have demonstrated that the ancient Polynesians were, in fact, skilled sailors and navigators. These modern recreations of the trip take about three months using the traditional Polynesian navigation techniques. These techniques use the positions of the Sun, Moon and stars when they rise over the horizon, and the directions of the swells deflected by islands to determine their locations. The steady trade winds in the tropical Pacific cause very regular, predictable long waves, called swells, whose paths are bent by the islands in ways that the Polynesians recognized and recorded on their stick and shell charts. When they could not see the stars and Moon due to overcast skies, the navigators had to depend on the direction of the swells to steer their course. The navigator could only sleep a few hours each day and was the most important person on board.

The revival of traditional Polynesian navigation began in Hawaii. Now it has spread to other Polynesian islands.

Economic Zone of some coastal state. The North Sea, Mediterranean, Gulf of Mexico, and Caribbean are totally divided among coastal states. Enormous areas of ocean that were once free to all comers are now under national jurisdictions. The United States, for instance, now has responsibility for an ocean area approximately equal to its land area.

Summary

Humans crossed ocean areas to settle all the continents except Antarctica. They also fished and traveled extensively. Vikings explored and established colonies around the North Atlantic between 1000 and 1300. More than 1,000 years ago, Polynesians and Micronesians sailed thousands of kilometers across the Pacific to colonize Pacific islands from New Zealand to Hawaii. Chinese ocean voyages extended the influence of the Ming dynasty. None of these voyages contributed to modern ocean science.

European ocean exploration began in the fifteenth century with Portugal's exploration of the Atlantic. The Portuguese developed oceanic trade routes connecting Europe with India and Asia, to replace the older overland routes that were closed to European traders. European domination of the ocean lasted for more than 400 years.

In the late eighteenth century, Captain James Cook of the British Navy systematically explored the entire Pacific basin. He used the latest instruments available to accurately map islands and coastlines. Modern oceanography—systematic study of the ocean—began with the *Challenger* expedition in 1872–1876. During the nineteenth and early twentieth centuries, government agencies mapped coasts to locate or protect fisheries and to promote shipping. Ocean data were systematically acquired and shared internationally, led by the American M. F. Maury.

Ocean studies expanded greatly during the mid-twentieth century. New instruments were developed and international field experiments replaced the national, shipborne expeditions of the early part of the century. Future ocean studies will be increasingly concerned with global climate change and the ocean's role in that change.

The United Nations Conference on the Law of the Sea, ratified in 1994, changed the ocean's legal regime from one of freedom on the high seas to one where coastal-state control extends 370 kilometers (200 nautical miles) from shore. Consequently, exploitation of resources in these parts of the ocean has been radically changed. Materials produced from the deep-ocean bottom will also be affected. Many maritime boundaries and the status of important ocean areas remain contested.

Key Terms

oceanography
knot
fathom
latitude
longitude
azoic
Law of the Sea
Exclusive Economic Zone

Study Questions

1. Why didn't ancient sea captains contribute to ocean science, given their extensive experience with the ocean?
2. What did the Polynesians use to navigate in their voyages?
3. What did Prince Henry the Navigator contribute to oceanography?
4. What role did Captain Cook play in exploring the ocean?
5. Explain the importance of the *Challenger* expedition.
6. List the principal contributions to ocean science of M. F. Maury and Benjamin Franklin.
7. How did the United Nation's Conference on the Law of the Sea change the legal status of the ocean and the seafloor?
8. How did the dependence on ships restrict scientists' view of the ocean?
9. Why are computers necessary for modern oceanographic investigations?
10. Why did the Chinese Ming dynasty support large exploring voyages?
11. Why was knowledge of the Gulf Stream important to ships crossing the North Atlantic between America and Europe?
12. How did ocean science contribute to economic development in the fifteenth century? in the nineteenth century?
13. How does knowledge of the ocean contribute to economic development today?
14. How have observations from satellites influenced the modern view of the ocean?
15. *[critical thinking]* How is the concern about global changes in climate likely to stimulate ocean research and observations?

Selected References

Atlas of the Oceans. New York: Rand McNally, 1977. Good reference book.

BASS, G. F., *Archaeology under Water*. Harmondsworth, Middlesex, England: Penguin Books, 1970.

BOORSTIN, D. J., *The Discoverers*. New York: Random House, 1983. Includes history of early ocean exploration.

COUPER, A., ed., *The Times Atlas and Encyclopedia of the Sea*. New York: Harper & Row, 1990. Excellent reference book and atlas.

DEACON, M., *Scientists and the Sea, 1650–1900*. London: Academic Press, 1971. History of oceanography up to the twentieth century.

MAY, W. E., AND L. HOLDER, *A History of Marine Navigation*. New York: Norton, 1973.

MENARD, H. W., *Islands*. New York: Scientific American Books, 1986.

NATIONAL GEOGRAPHIC SOCIETY, *Into the Unknown: The Story of Exploration*. Washington, D.C., 1987.

STOMMEL, H. M., *Lost Islands: The Story of Islands that Have Vanished from Nautical Charts*. Vancouver, BC: University of British Columbia Press, 1984.

OBJECTIVES

Your objectives as you study this chapter are to understand:

- The origin of Earth and the solar system
- The principal features of Earth's surface and those of the ocean floor
- The principal differences between ocean basins and continents
- The distinctive features of the major oceans

"Earth rise" over the Moon. The presence of water in various forms gives Earth its distinct appearance, especially the blue of the oceans and the white clouds, which consist of water vapor. By contrast, the surface of the Moon seems featureless. *(Courtesy NASA.)*

2 Earth and Its Ocean

With the development of reliable rockets in the late 1940s, space-exploration programs became a reality. As a result of these programs, our understanding of the solar system has expanded enormously. Now we are able to recognize the uniqueness of Earth, its atmosphere, and its ocean. We can also compare the planets and learn why they differ from each other. In this chapter we learn about Earth and its neighbors so that we can better understand its ocean.

Formation of the Universe and Galaxies

The universe formed about 15 billion years ago (Fig. 2–1) in an enormous explosion today called the **"big bang."** Before this moment, all matter and energy in the universe were presumably compressed into a point. Then, at the instant of the big bang, the universe began expanding outward from the initial center point, and the expansion continues today. As the universe expanded, it formed a series of elements during the first few increments of time.

At first the universe was unimaginably hot, and even atoms or subatomic particles could not survive. As matter in the universe cooled, subatomic particles formed, and after about 1 million years simple hydrogen atoms could survive. After about 1 billion years, the first stars and galaxies formed. **Galaxies** (Fig. 2–2) are huge rotating aggregates of stars, gases, and dust held together by gravity. Each **star** is a massive sphere of hot gases held together by gravity. Stars form in **nebulae,** which are clouds of dust and

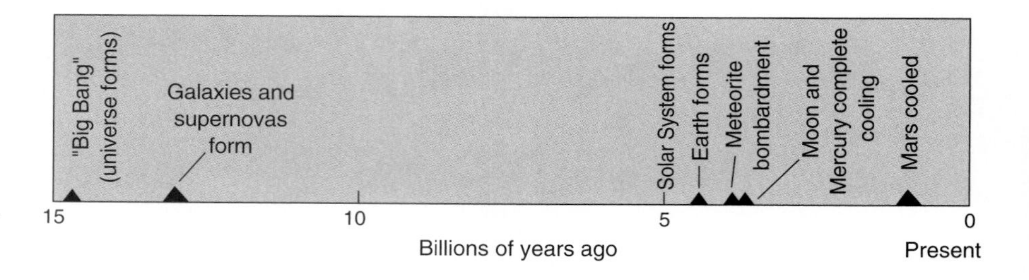

Figure 2–1
Time line for the development of the universe and the solar system.

gases. It is impossible to count the number of stars in a galaxy or the number of galaxies in the universe. Some astronomers estimate that the universe consists of perhaps 50 billion galaxies, each containing many billions of stars.

Star formation begins when parts of a nebula contract under the influence of gravity. As the star grows and compresses, the materials in its interior are heated. When the star's interior reaches a temperature of about 10 million degrees Celsius, nuclear reactions begin. Initially, hydrogen atoms combine to form helium, and energy is released in the process, which is called fusion.

The history of a star depends on its size. Stars such as our Sun follow a well-known evolutionary path. As fusion continues, the star reaches a stable size and then burns its fuel, hydrogen, at a steady rate. After many billions of years, the star has consumed all its hydrogen, converting it to elements as heavy as carbon and oxygen. As interior temperatures rise, the star expands and becomes a **red giant.** Eventually, all the various elements in the star have been converted to iron, and no more energy can be extracted from the fusion process. At this point, the red giant becomes unstable and begins to expand explosively, shedding its outer layers. Our Sun is expected to enter this red-giant phase in about 5 billion years.

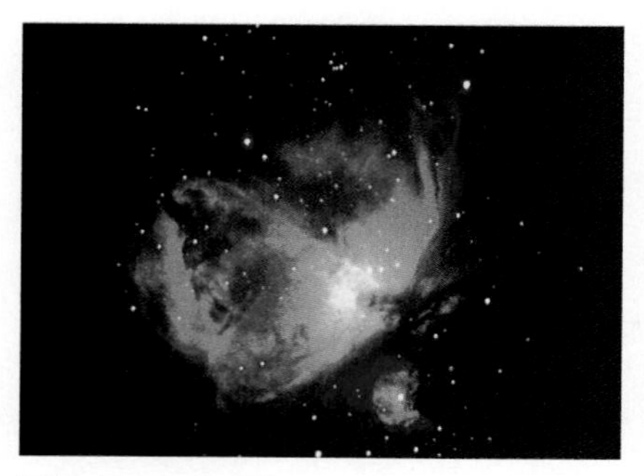

Figure 2–2
Stars and galaxies formed soon after the big bang. The Orion Nebula is a cloud of dust that shines because it has some very hot stars embedded within a cloud of hydrogen gas. It is typical of the materials from which stars and eventually planets formed. (Courtesy U.S. Naval Observatory.)

In the final stages, a dying star collapses in on itself because the diminished flow of energy from its interior is no longer able to keep the gases from contracting under the force of its enormous gravity. The rapid compression of the gases causes a catastrophic explosion called a **supernova.** Although the explosion lasts only a few seconds, it releases enormous amounts of energy. This is when heavier elements are formed. Virtually every element on Earth heavier than hydrogen has been through many supernovas. Material from supernovas is distributed throughout the universe.

Origin of Planets

The Sun, Earth, and the eight other planets that orbit the Sun—along with various moons—make up what is called the **solar system** (Fig. 2–3). The planets of the solar system formed about 4.6 billion years ago. They originated in a gigantic disk-shaped cloud of dust and gases orbiting the Sun. The dust grains came together, forming comets, asteroids, and planets that continue to orbit the Sun. The four planets nearest the Sun (Mercury, Venus, Earth, and Mars) consist largely of rocks and are called the terrestrial planets. The five farthest from the Sun are primarily gases and ice and are called the jovian planets (Table 2–1).

The rocks of the terrestrial planets melted soon after forming. The densest constituents, mostly metals, sank toward the center, while the less dense constituents remained at the surface. Early in their history (more than 4 billion years ago), the planets were bombarded by meteors that extensively cratered their surfaces. These meteors were the debris left over when the planets formed. Eventually, the debris was swept up and incorporated into the planets and the bombardment ceased. However, Earth and presumably the other planets as well are still occasionally struck by meteors. The craters formed during the meteor bombardment are still seen on the Moon (Fig. 2–4), Mercury, and Mars, but rarely on Earth. Earth's surface is constantly renewed by processes that eventually obliterate craters, as we discuss in Chapter 3. Only a few craters (Fig. 2–5) can be detected on Earth.

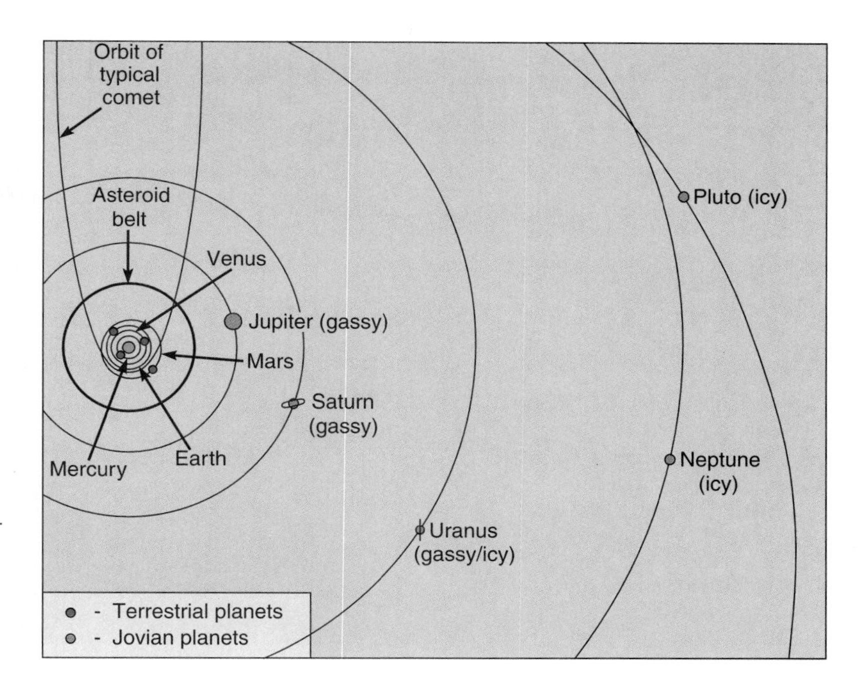

Figure 2–3
The four inner planets in the solar system, which are called the terrestrial planets, consist primarily of rock and metals. The five planets farther from the Sun, called the jovian planets, consist primarily of water and various gases, primarily hydrogen and helium.

Venus (Fig. 2–6) is hidden by clouds, but radar instruments on satellites observing the planet's surface (Fig. 2–7) have not found the extensively cratered surface so visible on the Moon. It appears that vast volcanic eruptions on Venus 300 to 500 million years ago may have obliterated any craters formed earlier.

Each planet has a different history, controlled primarily by its size. Mercury's and Earth's moons, because of their small sizes, cooled rapidly and therefore could not sustain prolonged volcanic activity. Thus, these small bodies retain their

TABLE 2–1
Planets in the Solar System

	Distance from Sun (in AU)*	*Diameter of Planet (in kilometers)*
Terrestrial planets (rocky)		
Mercury	0.4	4,878
Venus	0.7	12,104
Earth	1.0	12,756
Moon	1.0	3,476
Mars	1.5	6,796
Asteroids (rocky, metallic)	~2.8	Small
Jovian planets (gassy, icy)		
Jupiter	5.2	142,796
Saturn	9.5	120,660
Uranus	19.2	50,800
Neptune	30.0	48,600
Pluto	39.4	2,400–3,800

*One astronomical unit (AU) is the distance between Earth and the Sun, about 149,600,000 km.

Figure 2–4
The Moon's surface is scarred by craters formed by numerous meteorite impacts during its 4.5 billion years of existence. There is no water to erode the craters, and the lunar crust is not affected by mountain building. (Courtesy NASA.)

Figure 2–5
Earth has been struck by many meteors during its history, but the traces of craters are eventually destroyed by mountain-building processes, which we discuss in Chapter 3. The deeply eroded traces of the Manicouagan crater in northern Quebec are now occupied by a circular lake. (Courtesy NASA.)

Figure 2–6
Dense clouds of water vapor, sulfuric acid, and carbon dioxide obscure the surface of Venus. (Courtesy NASA.)

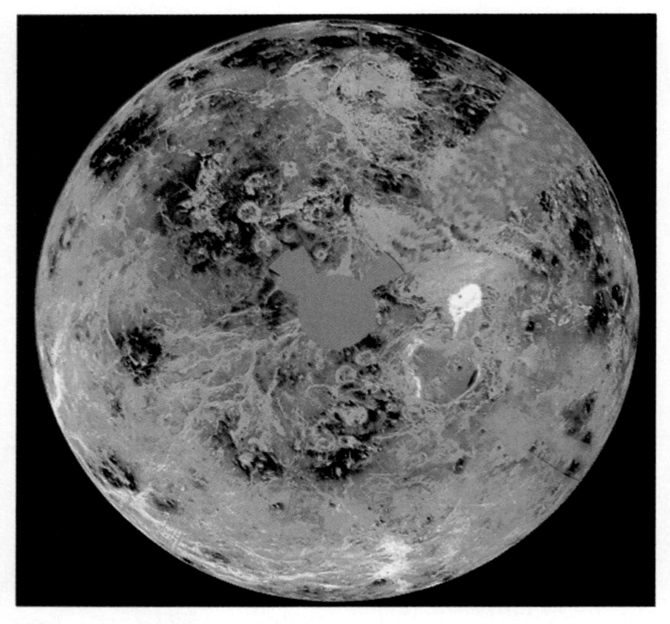

Figure 2–7
The surface of Venus, observed by radar on orbiting satellites, is covered by large plains of once-molten material formed by volcanic eruptions that developed within the last few hundred million years. Because there has been no extensive bombardment by meteors in this time period, few craters are visible. (Courtesy NASA.)

cratered surfaces. Larger planets, such as Earth and Venus, cool slowly, with the result that volcanic activity continues. Such activity continually renews and reshapes their surfaces, obliterating craters in the process.

Volcanic activity is a major source of Earth's ocean and atmosphere, because water and gases are released from molten rocks during eruptions. (Some water and gases are also derived from comets hitting the planet.) Earth exerts enough gravitational pull on the water and gases to keep them closely bound. Today, new supplies of water and gases are still released through volcanic eruptions, but the quantity is small compared with the amount already in the ocean and atmosphere.

Figure 2–8
Mars is too small to retain enough water to form an ocean. The river valleys (upper center) were formed by liquid water. Large volcanic craters are seen on the left side of this image. Meteorite impact craters are visible on the right side of the image. (Courtesy NASA/U.S. Geological Survey.)

Venus and Mars have histories radically different from Earth's. Because Venus is nearer the Sun than Earth, the Venusian ocean boiled away early in its development. This vaporization kept all the water vapor in the atmosphere, causing Venus to experience a runaway greenhouse effect. The water vapor and carbon dioxide in its atmosphere resulted in a dense atmosphere that traps the heat from the Sun. Today, the surface temperature on Venus is 470°C, hot enough to melt lead, and its atmospheric pressure is roughly 93 times greater than on Earth.

Mars (Fig. 2–8) apparently could not retain enough water vapor or carbon dioxide to form an ocean or a dense atmosphere. Instead, these gases were apparently lost to space after they were released in Martian eruptions, in part because Mars is so small that it could not exert sufficient gravitational pull on these gases. The valleys on Mars indicate that it did have liquid water at some time.

We still have much to learn about the history of all the planets, including Earth.

Origin of Life

The special feature of Earth is the presence of life. The origins of life on Earth are still hotly debated, but we know that simple one-celled organisms developed soon after Earth formed. As these early plants grew, they took carbon dioxide and water and used the energy of sunlight or energy-rich compounds to form organic matter and release free oxygen into the atmosphere. (We discuss these processes in Chapter 11.) If sunlight is involved, this process is called **photosynthesis;** if other energy-rich compounds are used, the process is called **chemosynthesis.** The result was that Earth's atmosphere changed from one dominated by carbon dioxide (like

Venus) to one dominated today by nitrogen and oxygen. This change occurred gradually:

2 billion years ago: atmospheric oxygen reached 1 percent of its present level.
700 million years ago: atmospheric oxygen reached 10 percent of its present level.
350 million years ago: atmospheric oxygen reached its present level.

The abundance of free oxygen in Earth's atmosphere led to the evolution of oxygen-breathing animals. These forms now dominate Earth's surface. Other forms that do not require oxygen dominate in oxygen-free (anoxic) waters and below Earth's surface. This reservoir of life under anoxic conditions may be extremely large but is poorly known.

Land and Ocean

From space, we see that Earth is nearly covered by ocean waters: 70.8 percent of its surface is covered by water (Fig. 2–9). The land is exposed in the remaining 29.2 percent, which is found mostly in the Northern Hemisphere (Fig. 2–10).

If we look at Earth from the South Pole, the ocean appears as three large basins extending northward from Antarctica, where they connect (Fig. 2–11). The continents separate the ocean basins. The boundary between the Atlantic and Indian Oceans runs from South Africa's Cape of Good Hope along longitude 20° E to Antarctica. The southern boundary between the Atlantic and Pacific is drawn between the Antarctic Peninsula and Cape Horn, South America. The Arctic Ocean is considered part of the Atlantic, and so the northern Atlantic–Pacific boundary is the *Bering Strait* between Alaska and Siberia. The boundary between the Pacific and the Indian Oceans runs through Indonesia and extends from Australia southward to Antarctica along longitude 150° E.

The Pacific Ocean dominates Earth's surface, holding 52 percent of its ocean waters (Fig. 2–12). The Atlantic and Indian Oceans account for the rest, except for the 2 percent of the ocean water that occurs as ice, primarily in the Antarctic and Greenland ice caps.

If we could drain the ocean basins and view Earth from space (Fig. 2–13), the most conspicuous features of its surface would be the enormous mountain ranges

Figure 2–9
Seen from space, Earth is a blue sphere (because of its ocean) with clouds (white areas) covering more than half its surface. Land is scarce in this Southern Hemisphere view. The ice-covered Antarctic continent is at the bottom center. (Courtesy NASA.)

Figure 2–10
The Northern Hemisphere, sometimes called the land hemisphere, is centered on western Europe and has roughly equal amounts of land and water. The Southern Hemisphere, where the ocean dominates, is centered on New Zealand; it is often called the "water hemisphere."

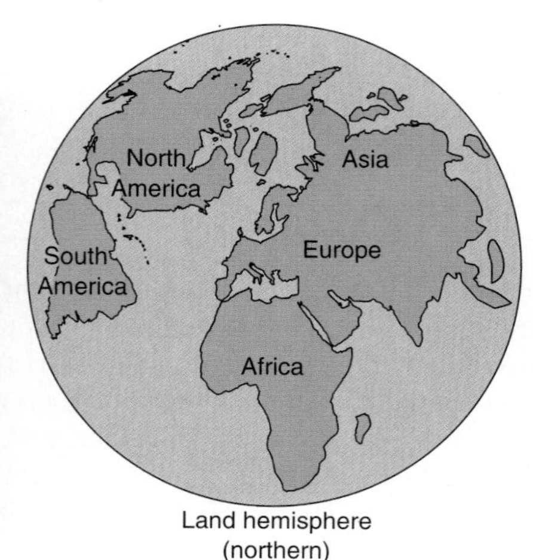

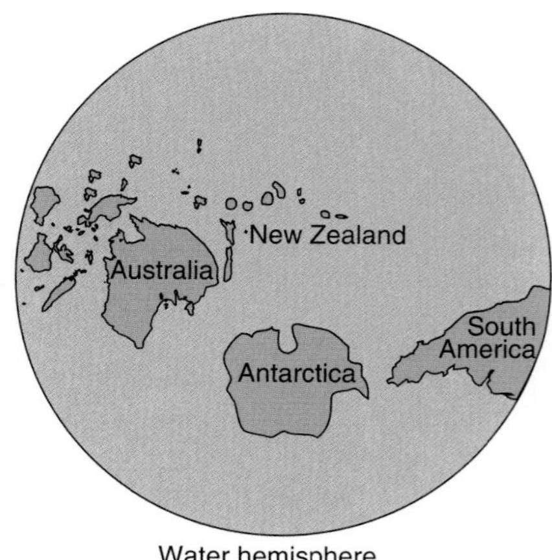

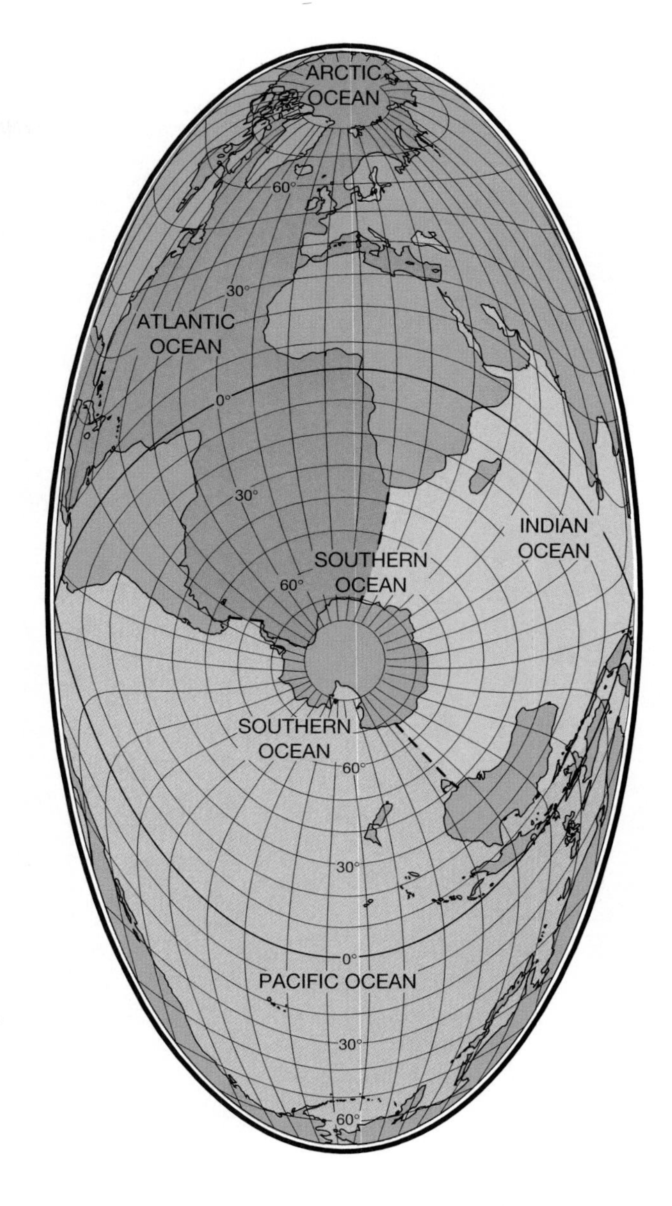

Figure 2–11
The world ocean is centered on Antarctica. For convenience, we delineate three major oceans that are connected around Antarctica. The Southern Ocean is a broad band of unrestricted ocean around Antarctica. The Arctic Ocean is considered to be part of the Atlantic.

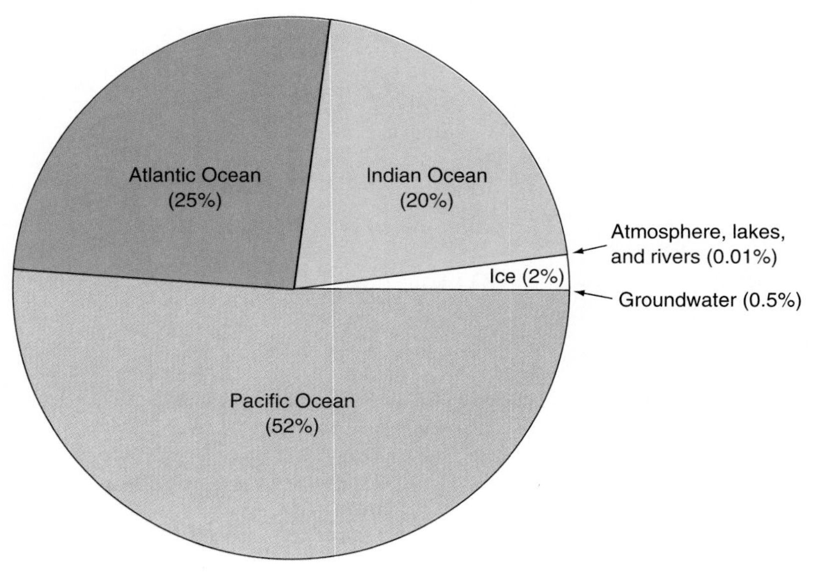

Figure 2–12
Distribution of water on Earth's surface. Only a minute fraction of the water occurs in lakes, rivers, and the atmosphere.

Figure 2–13
The topography of Earth's continents and ocean basins is shown by the various colors. The shallowest parts of the ocean basins (light blue) are near the continents and in the middle of the basins. The mid-ocean ridges are most conspicuous in the Atlantic, but comparable features occur in all the other basins. The deepest parts of the basins (dark blue) form

long, curved trenches. Fracture zones, oriented primarily east–west in the Atlantic and Pacific, offset the mid-ocean ridges. The high mountain ranges (red) on the continents form long belts that border the Pacific and cut across Asia. (Courtesy National Geophysical Data Center/NOAA.)

BOX 2–1
Exploring Ocean Basins

Most of Earth's surface remains unknown in detail. In fact, most of it is less well known than the surface of the Moon or Venus. This situation is likely to change dramatically during the next decades as a result of Earth-orbiting satellites and the declassification of once-secret data collected by the military during the Cold War.

Military operations need accurate maps of Earth's surface, including the ocean floor. Such information is required to program intercontinental missiles accurately, for instance. These data have been released for the Southern Ocean, an area where there were no military concerns during the era of the Cold War. When the full data set is available, it will be processed into accurate maps of ocean-bottom topography. Then costly and time-consuming ship-based surveys will be required only in critical areas where the satellite data were incomplete or inaccurate.

In addition, we are likely to see new Earth-orbiting satellites that monitor changing climate on Earth's surface, monitor pollution problems, and even make critical measurements of Earth's surface features.

in the middle of the Atlantic, in the eastern South Pacific, in the western Indian, and circling Antarctica. We call this set of globe-circling mountain ranges **mid-ocean ridges.**

The other principal feature of the ocean basin is the deepening of the ocean bottom (shown in Fig. 2–13 by the darker shades of blue) away from the mid-ocean ridges. Around the margins of the Pacific are the deepest parts of the ocean, the **trenches.** These are especially conspicuous near Asia and South America.

On land, the most conspicuous features are mountain ranges (shown in Fig. 2–13 as red browns). They form elongate north–south belts along the western margin of the Americas and an east–west belt across southern Asia. In short, the orientation of mountain ranges on land parallels the orientation of the mid-ocean ridges. There are two principal directions for each: north–south and east–west.

Heights and Depths on Earth's Surface

Earth's surface has two distinctly different levels (Fig. 2–14). The land stands about 840 meters above sea level, while the ocean bottom averages about 3,800 meters below sea level. Compared with human dimensions, the extreme heights and depths of Earth's surface are enormous. Mount Everest in the Himalayas of northern India stands 8,848 meters above sea level, while the deepest part of the Mariana Trench (in the North Pacific east of the Philippine Islands) is 11,035 meters below the sea surface. Even if we cut down the land and dumped the material into the ocean to make Earth a smooth sphere, it would still be covered by water 2,430 meters deep.

Compared with Earth's radius (6,370 kilometers), these heights and depressions are minuscule. In fact, relatively speaking, the heights of the mountains and depths of the trenches on Earth's surface are smaller than the imperfections on the best globe you can buy. Therefore, we can say that Earth is essentially a smooth sphere, with the ocean basins being very shallow depressions and the ocean a thin film of water interrupted here and there by continents.

Continental Margins

The principal boundary between any continent and ocean basin is the submerged area called the **continental margin,** a fundamental feature of the planet. A continental margin consists of a continental shelf, a continental slope, and a continental rise (Fig. 2–15). The **continental shelf** is the submerged upper part of a continental margin. From the shoreline, a shelf slopes gently toward the **shelf break** (the average

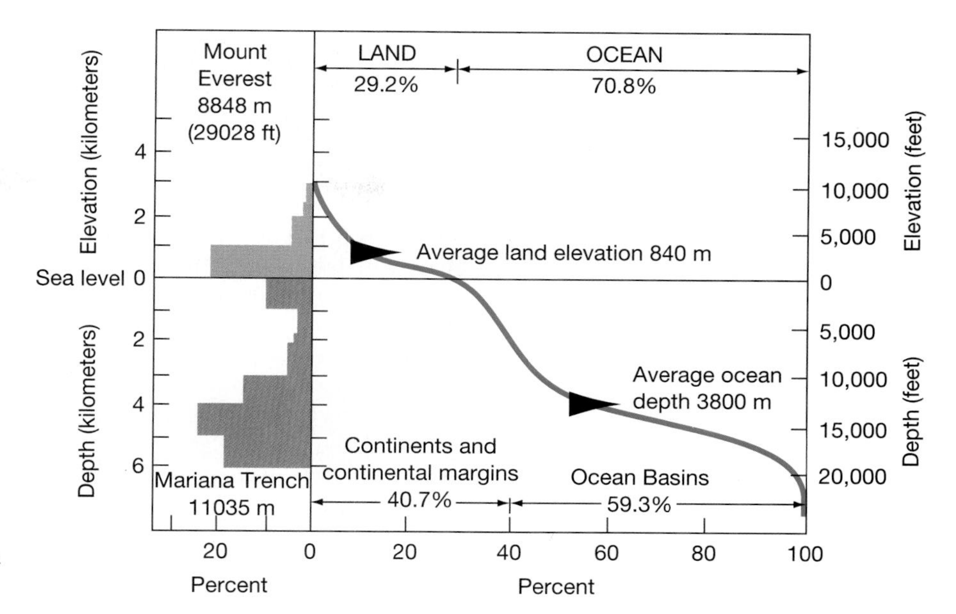

Figure 2–14
The fraction of Earth's surface in elevation or depth zones of 1 kilometer (left side). The right side shows the cumulative fraction of Earth's surface shallower than a given depth. Note that Earth has two dominant levels—one representing land and the other representing the ocean bottom.

depth of water above this boundary line is 130 meters). At the break, the slope of the bottom changes from gentle to much steeper, and this steeper area is the **continental slope,** which extends down to depths of 2 to 3 kilometers. At the base of the continental slope, the steepness disappears and the bottom begins to slope gently again; this second gently sloping area that begins at the base of the continental slope and extends seaward to connect with the deep-ocean floor is the **continental rise.** This area is usually divided into an upper and lower part, with the slope being different in the two.

Around much of the Pacific, where there is active mountain building, continental shelves are relatively narrow—only a few tens of kilometers wide. Much wider continental shelves occur where continental margins have not experienced mountain building for many millions of years. The shelf around Africa is unusually narrow because the continent stands higher than the other continents. (We discuss the reason in the next chapter.) A **coastal plain** is a former continental shelf that is now exposed above water.

Off Antarctica, the shelf break is about 500 meters deep because the weight of the ice on the land has depressed the continent and its margins. The shelf break

Figure 2–15
Schematic diagram of the North Atlantic margin of North America, showing the change in the character of the ocean bottom going from the land to the deep-ocean bottom. Note the vertical exaggeration.

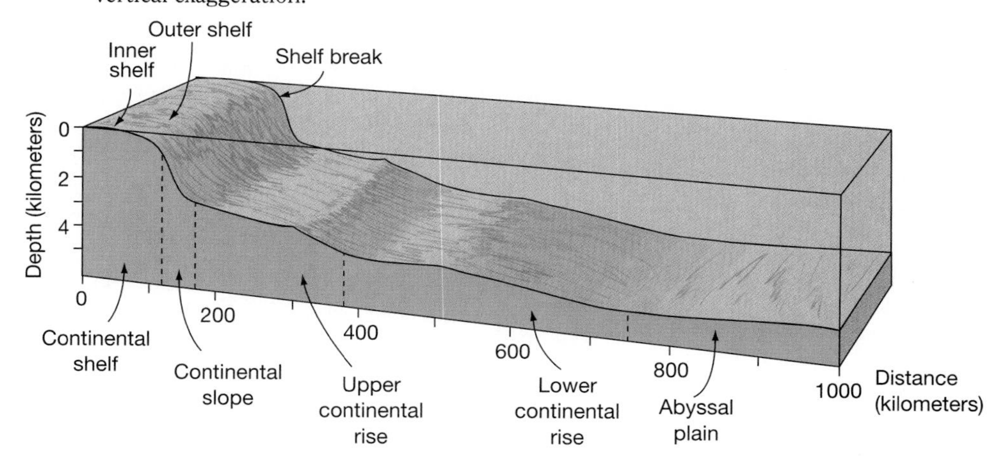

is also unusually deep around the Arctic Ocean because thick ice sheets covered the surrounding continents until only a few thousand years ago.

Continental slopes are the outer edges of continental blocks. On a waterless Earth, they would be its most conspicuous surface feature. Spectacular continental slopes occur where coastal mountains parallel a trench. Near the western coast of South America, for instance, the Andes reach elevations of around 7 kilometers, while the nearby Peru–Chile Trench is 8 kilometers deep. This is a total (vertical) relief of 15 kilometers within a horizontal distance of only a few hundred kilometers. Such steeply sloping, mountainous margins are common around the Pacific. Where mountains occur along the coast, as in southern California, continental margins are rugged, broken by submerged ridges and basins.

Submarine canyons sometimes cut across continental shelves and slopes. These canyons look much like river valleys on land. Some **submarine canyons** are as large as the Grand Canyon of the Colorado River. A few are associated with major rivers—for instance, Hudson Canyon or the Congo Canyon. In these cases, the upper parts of the canyons probably formed when sea level was much lower than its present level.

The World Beneath the Sea

The ocean floor is anything but uniform. Hidden by the ocean surface is a topography as spectacular and varied as any found on land. The ocean floor is covered with ridges of high mountains, vast plains, and deep trenches.

Mid-Ocean Ridges

We begin our study of deep-ocean basins with mid-ocean ridges, the most conspicuous features of the basins (Fig. 2–16). The combined length of all the ridges is 60,000 kilometers, and they cover 23 percent of Earth's surface.

The rugged *Mid-Atlantic Ridge* stands 1 to 3 kilometers above the deep-ocean floor. The most prominent feature of this ridge is its steep-sided central valley (Fig. 2–17), called a **rift valley.** It is 25 to 50 kilometers wide, 1 to 2 kilometers deep, and bordered by rugged mountains whose tallest peaks come to within 2 kilometers of the sea surface.

In the Pacific, the mid-ocean ridge is much less rugged than in the Atlantic, so much so that we call it not a ridge but rather the *East Pacific Rise*. Much of the East Pacific Rise has no rift valley. It intersects North America in the Gulf of California (Fig. 2–16).

In the Indian Ocean, the *Mid-Indian Ridge* (similar to the Mid-Atlantic Ridge) intersects Africa–Asia in the Red Sea. At its other end, the Mid-Indian Ridge joins the *Pacific–Antarctic Ridge,* which circles Antarctica (Fig. 2–16).

Small earthquakes occur frequently on the crests of mid-ocean ridges. The locations of these earthquakes coincide with the locations of the central rift valleys. Indeed, mapping of earthquake locations has been used to determine whether or not there are active ridges in poorly known ocean areas such as the Arctic Ocean.

The Deep-Ocean Floor

The **deep-ocean floor** (deeper than 4 kilometers) occupies about 30 percent of Earth's surface. Low **abyssal hills** (less than 1 kilometer above the surrounding ocean bottom) cover about 80 percent of the Pacific floor and about half of the Atlantic. These hills are also abundant in the Indian Ocean and are thus among the most common features on Earth's surface. They are typically about 200 meters high. Many appear to be extinct volcanoes.

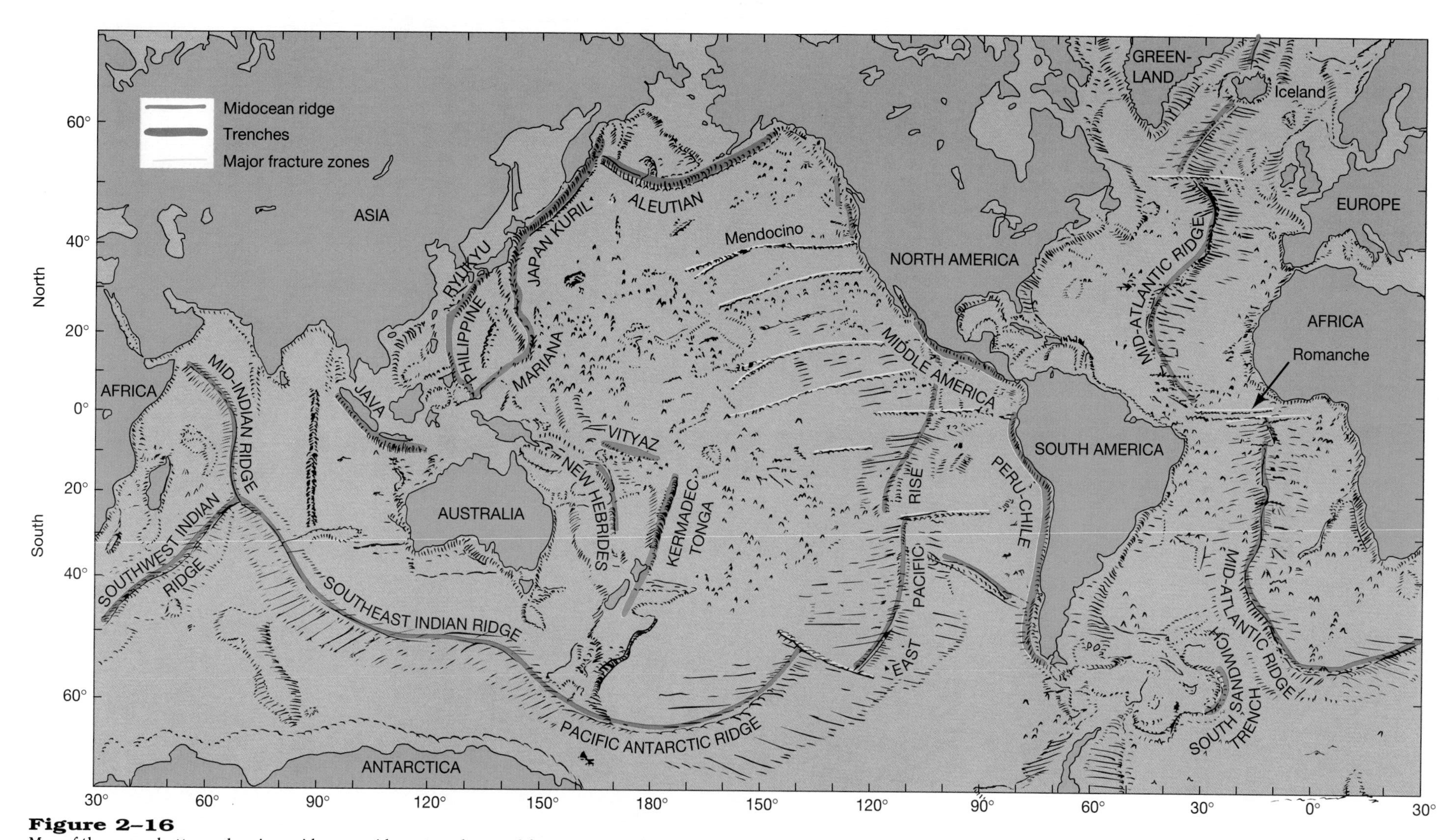

Figure 2–16
Map of the ocean bottom, showing mid-ocean ridges, trenches, and fracture zones. (Base map courtesy Hubbard Scientific.)

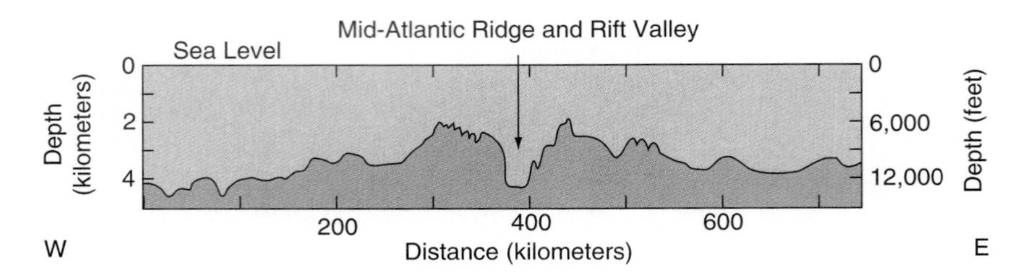

Figure 2–17
Profile of the Mid-Atlantic Ridge.

Immense areas of exceedingly flat ocean bottom, called **abyssal plains,** lie near the continents (Fig. 2–15). These are among the flattest parts of Earth's surface. They commonly occur at the seaward margins of the continental rises. Most abyssal plains appear to be covered with thick deposits of sediments that likely came from nearby lands.

Fracture Zones

Hundreds of **fracture zones** cut the ocean floor and offset mid-ocean ridges (Fig. 2–16). They are narrow elongated belts of rugged topography, 10 to 100 kilometers wide, that include steep ridges and valleys; they are roughly perpendicular to mid-ocean ridges and offset ridge segments. Some fracture zones are 4,000 kilometers long. They commonly parallel each other across large parts of ocean basins. Some of the deepest parts of the ocean floor occur in fracture zones. Seafloor depths usually change markedly across fracture zones. Fracture zones have few earthquakes and no volcanoes. They are quite different from trenches, which we discuss next.

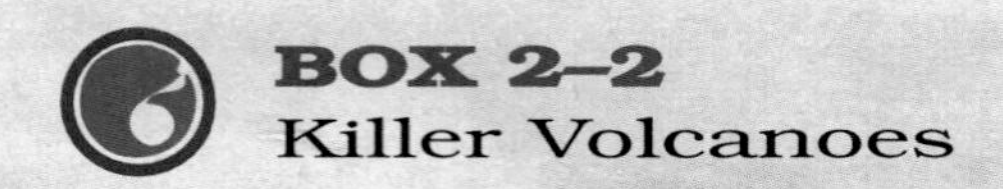

BOX 2–2
Killer Volcanoes

Each year several of Earth's 540 active volcanoes erupt, often forcing thousands to flee their homes to escape death. And usually a few to a few thousand persons are killed by the hot ashes from the eruptions or by tsunamis (large waves) generated by volcanic explosions. As populations increase, there are more people exposed to the risks of volcanic activity.

Most volcanoes occur in the Pacific *Rim of Fire,* an area associated with the many subduction zones around the Pacific. (A subduction zone is a place where two land masses collide and one sinks under the other.) A few volcanoes lie outside the Pacific Rim, such as those of the Hawaiian Islands. These mid-plate volcanoes do not have the violent eruptions that are so destructive.

Several of the worst killer volcanoes are in Indonesia, four of them on Java. In 1883, the long-dormant volcano Krakatoa erupted in a spectacular explosion that sent ash miles high into the stratosphere, causing brilliantly colored sunsets around the world for several years. The tsunamis generated by the violent explosions killed 96,000 persons.

It is now possible to monitor active volcanoes and in many instances to predict future eruptions so that people can evacuate the areas affected. For example, between 1980 and 1991, Mt. St. Helens in Washington erupted 22 times; 19 of these eruptions were predicted. Warnings also preceded eruptions of Alaska's Redoubt Volcano, which erupted in 1989. The 1991 eruption of Mt. Pinatubo in the Philippines was predicted far enough ahead that 15,000 persons could be evacuated from nearby Clark Air Force Base.

Scientists monitor earthquake activity near active volcanoes. From this earthquake information, they can detect movements of molten rock far below the surface and thus warn of impending eruptions. With the advent of new instruments that measure the heat given off by volcanoes, it has become possible to detect new activity from long-dormant volcanoes. This type of activity is often the most dangerous, because active volcanoes may remain dormant for centuries before erupting violently. Between eruptions, the volcano may become populated. Mt. Unzen in Japan was last active in 1792 before it erupted again in 1991.

Other sensors detect increases in the emissions of gases, such as sulfur dioxide, associated with volcanic eruptions. Even the shape of the volcano is monitored by lasers that can detect minute changes in the mountain's slopes caused by subsurface movements of molten rock.

Large fracture zones continue into the continents. The *Mendocino Fracture Zone*, for example, comes onto land at Cape Mendocino, California. From there, it continues as the San Andreas Fault, intersecting the East Pacific Rise in the Gulf of California.

One of the largest fracture zones is the *Romanche* (Fig. 2–16), which offsets the Mid-Atlantic Ridge by 950 kilometers near the equator. This fracture zone is part of a group of large faults (places where two pieces of crust slide past one another). The entire Romanche zone consists of deep valleys about 100 kilometers wide separated by ridges. The Romanche Fracture Zone contains the deepest part of the Atlantic, 7,960 meters. (Because of its great depth, it was thought for many years to be a trench.) This deep gap in the Mid-Atlantic Ridge permits bottom waters from the western Atlantic to flow into the deep basins of the eastern Atlantic.

Trenches

As we learned earlier, trenches are the deepest parts of the ocean floor, typically 3 to 4 kilometers deeper than the surrounding floor. They are relatively narrow, only a few tens of kilometers wide, but many thousands of kilometers long. Most occur in the Pacific, especially the western Pacific (Fig. 2–16). In fact, most of the Pacific is bordered by trenches. There are also trenches in the South Atlantic *(South Sandwich Trench)* and in the Indian Ocean *(Java Trench)*.

Trenches are associated with active volcanoes and earthquakes. Many trenches are near chains of volcanic islands (called **island arcs**). As we see in Chapter 3, trenches are important in Earth's mountain-building processes. Because of the large earthquakes associated with them, areas near trenches often suffer major catastrophes with substantial loss of human life.

Volcanoes

Volcanoes and volcanic islands are common in the ocean. They usually stand 1 kilometer or more above the surrounding ocean floor. Volcanic activity on land occurs primarily near the edges of ocean basins, especially around the Pacific.

Most volcanic eruptions occur—quietly and usually unnoticed—on mid-ocean ridges. There are two modes of oceanic eruptions. Some lavas move in tranquil flows in which the surface layer cools first. This surface cooling forms an insulated cover for the still-molten lava in the interior of the mass. This type of lava flow forms **pillow lavas** (Fig. 2–18), rounded masses of volcanic rock. The second type of oceanic eruption occurs in shallow waters and can be quite violent,

Figure 2–18
Toothpaste-like structures formed when molten lavas flowed out of fissures and were chilled by the cold oceanic waters. These formed during submarine volcanic eruptions along the Mid-Atlantic Ridge. These rocks are probably less than 10,000 years old. The arm of the submersible *Alvin*, visible in the foreground at the right, is picking up a rock. (Courtesy Woods Hole Oceanographic Institution.)

(a)

(b)

Figure 2–19
Kovachi, an active submarine volcano in the Solomon Islands, South Pacific. (a) When it erupted in October 1968, explosions ejected water and steam 90 meters into the air. Discolored surface waters extended 130 kilometers downstream. (b) An island formed by a 1961 eruption was eroded away, leaving a shallowly submerged bank. (Courtesy Smithsonian Center for Short-Lived Phenomena.)

with many explosions as the hot lavas explode on contact with much cooler seawaters (Fig. 2–19).

Curved chains of volcanic islands, which as we learned earlier are called island arcs, are associated with mountain building and trenches. For instance, the Aleutian Islands, southwest of Alaska, are a single chain of volcanoes having a trench on the seaward side, as shown in Figure 2–20.

The Indonesian–Philippine area has several trenches, including two large ones in which there are many active volcanoes (Fig. 2–21). Large, destructive eruptions occur in Indonesia, the most volcanic region on Earth. In 1815, Tambora

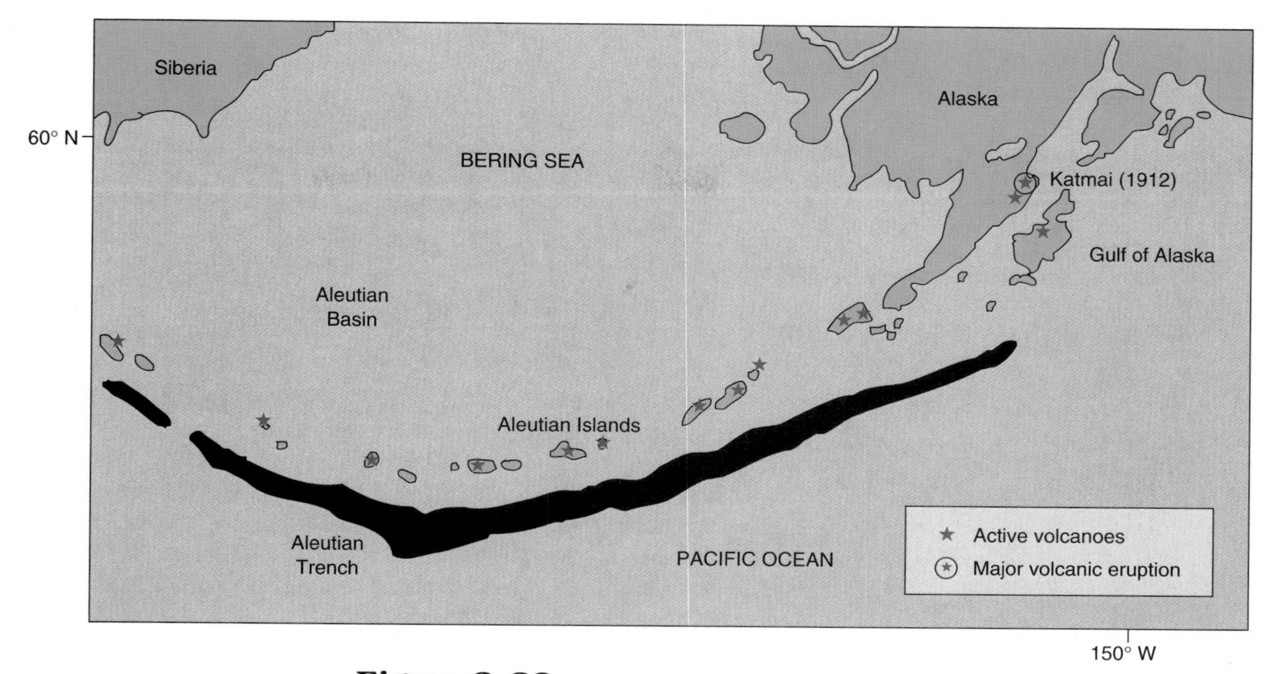

Figure 2–20
The Aleutian Islands, a simple volcanic island-arc system with a trench. Many of the volcanoes are active. Trench depths greater than 6 kilometers are shown in black.

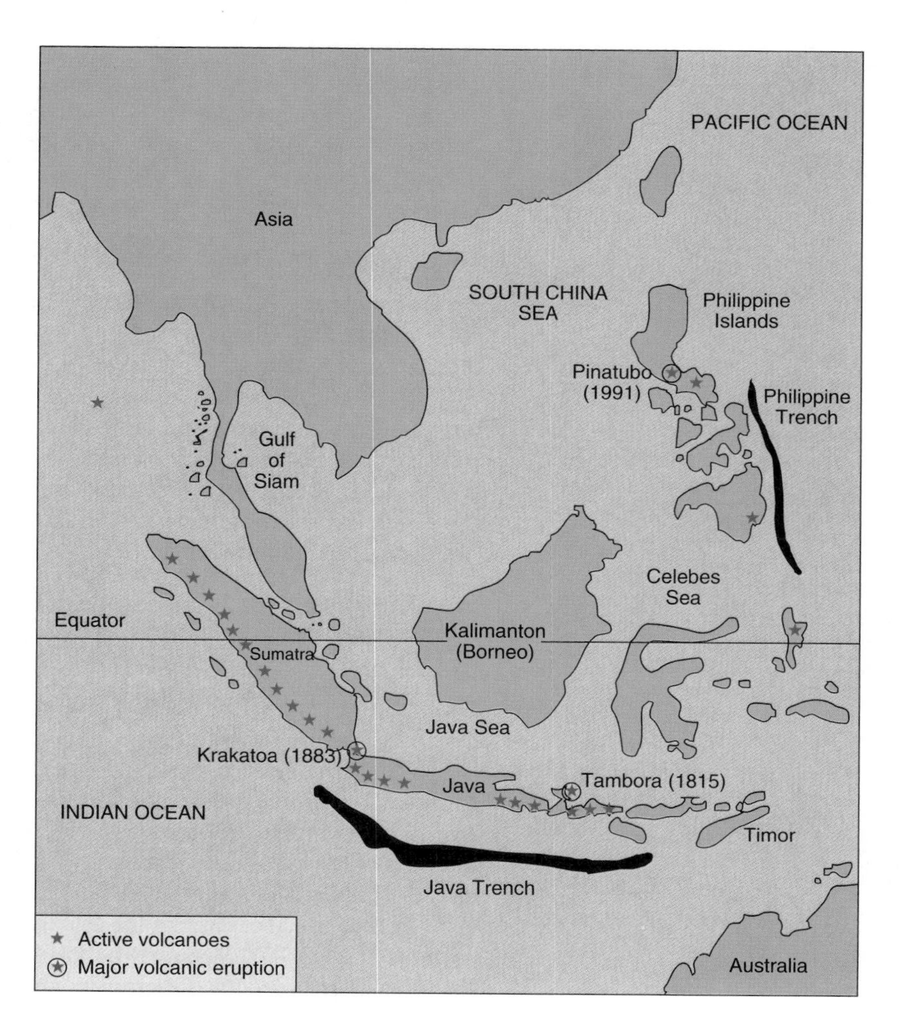

Figure 2–21
Indonesia, a complex island-arc system with several trenches lying between Australia and Asia, has the largest number of active volcanoes in the world. Trench depths greater than 6 kilometers are shown in black.

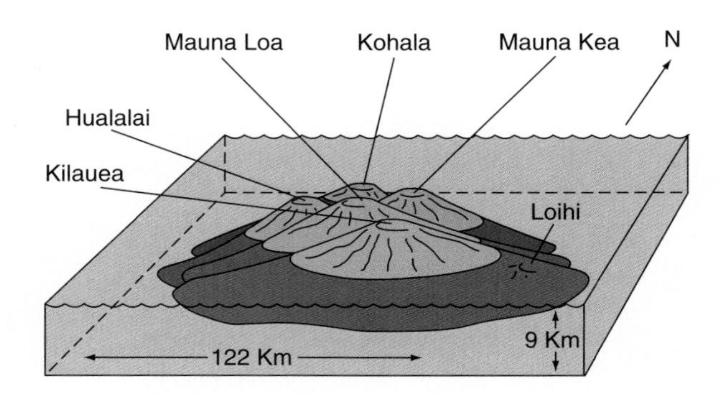

Figure 2–22
Schematic diagram showing the five shield volcanoes that coalesced to form the island of Hawaii. Loihi, a new volcano, is still submerged, but will eventually form a large crater that will build up above the ocean surface.

on the island of Sumbawa erupted, killing 92,000 persons. The large amount of ash this event injected into the atmosphere caused the summer of 1816 to be unusually cold throughout the Northern Hemisphere. In 1883, *Krakatoa*, a volcano located between Java and Sumatra, erupted, killing 36,000 and generating destructive sea waves. In 1991, Mt. Pinatubo in the Philippines erupted, injecting massive amounts of ash into the stratosphere. It was the largest volcanic eruption of the twentieth century and caused climatic cooling throughout the Northern Hemisphere that persisted for several years.

Active volcanoes also occur in the middle of ocean basins, far from mid-ocean ridges or island arcs. Some of these volcanoes—the largest on Earth—are called *shield volcanoes* (Fig. 2–22). They stand several kilometers higher than the deep-ocean floor and form high islands. Examples of such active volcanoes are on the island of Hawaii in the Pacific and Reunion Island in the Indian Ocean. While it is active, a shield volcano builds an enormous cone. After the volcano becomes dormant, the cone is eroded by wind, rain, and waves, often being eroded to a submarine bank within 10 million years. Once it is submerged, erosion is much slower, and subsurface volcanic cones can persist for tens of millions of years.

Oceanic Plateaus

Oceanic plateaus are isolated parts of the ocean floor that stand a kilometer or more higher than their surroundings. They occur in all ocean basins and constitute about 3 percent of the ocean floor (Fig. 2–23).

Because they contain rocks typical of continents, some plateaus appear to be fragments of continents and are called **micro-continents.** Others are associated with volcanic features, such as large volcanic islands. One of these is called the Walvis Ridge because it is elongated and is associated with the active volcano on the South Atlantic island of Tristan da Cunha. Another ridgelike plateau is the Ninetyeast Ridge in the Indian Ocean.

Pacific Ocean

The Pacific Ocean is the deepest and largest basin (Fig. 2–24), occupying more than one-third of Earth's surface. Mountain building dominates its margins.

Along the eastern margin, rugged mountains parallel the coastline. Here continental margins are narrow because the mountain slopes continue below sea level. The mountains on land block rivers from flowing into the ocean from the interiors of the continents, especially on the Pacific coast of the Americas. Therefore, the Pacific is less affected by land than either the Atlantic or the Indian.

Islands are abundant in the Pacific, especially in the southern and western portions (Fig. 2–25). Many of the islands are volcanoes, some still active. The Hawaiian Islands are good examples. They are the exposed tops of volcanoes in a

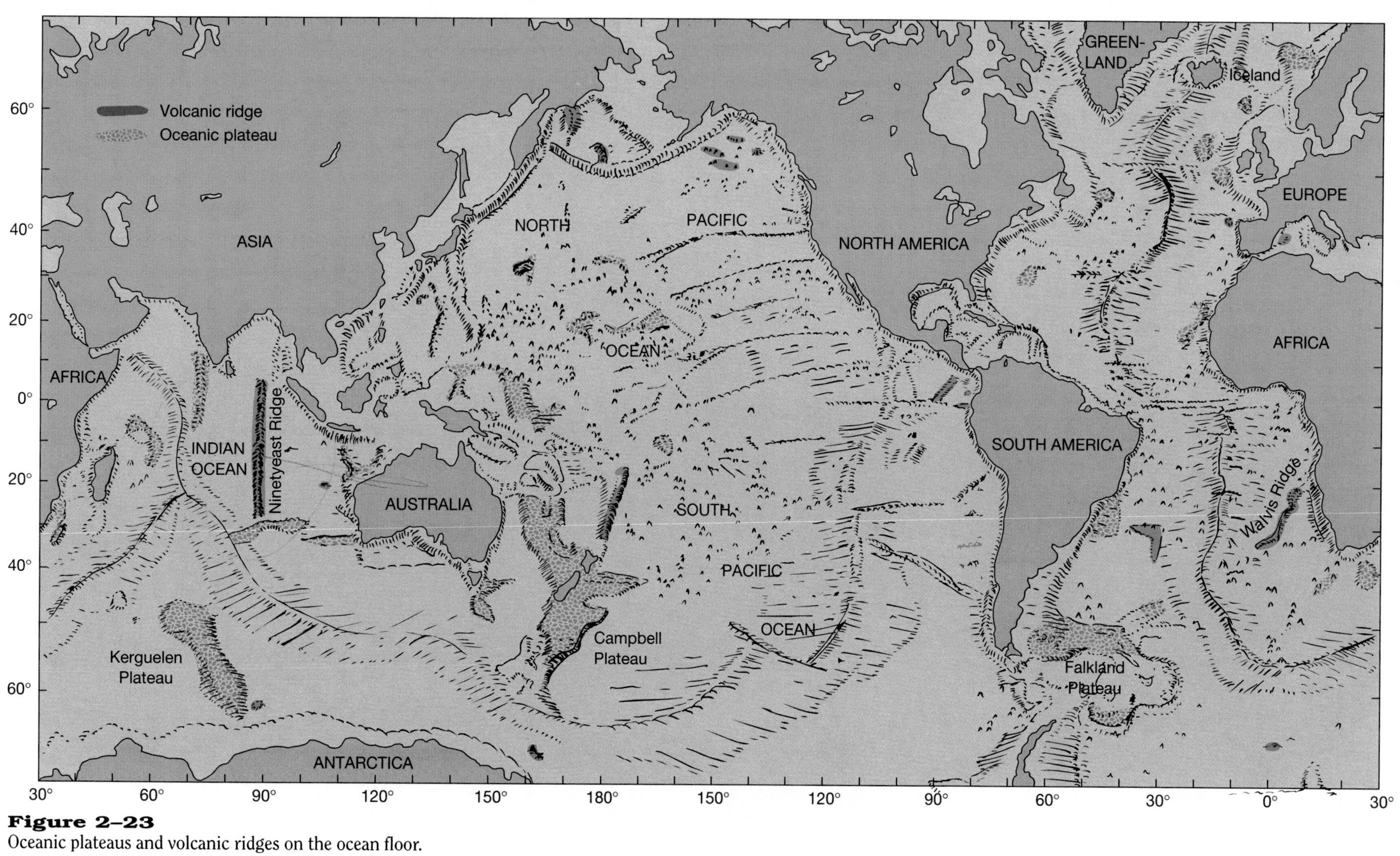

Figure 2–23
Oceanic plateaus and volcanic ridges on the ocean floor.

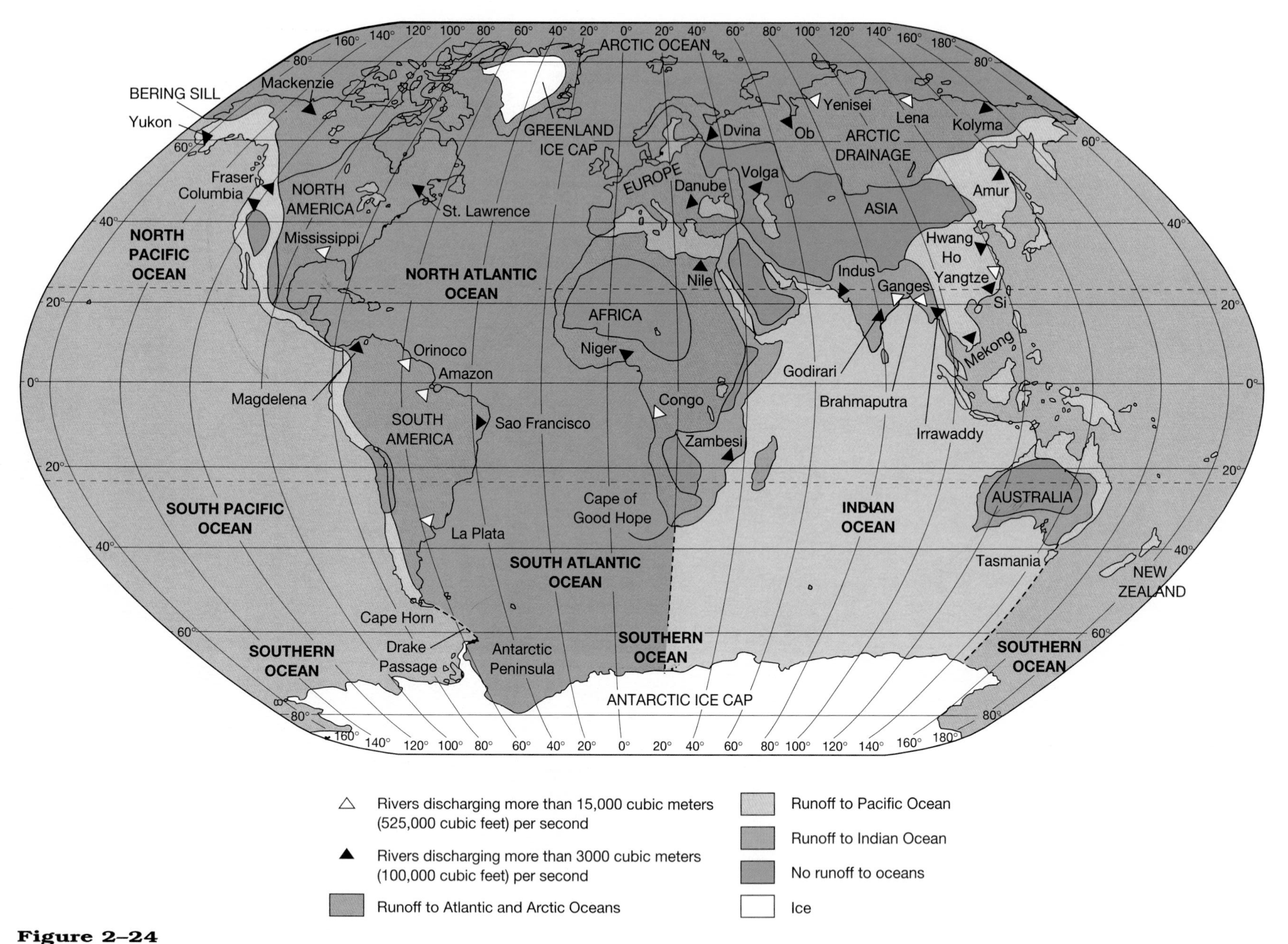

Figure 2–24
Boundaries between major ocean basins. Areas of the continents whose rivers drain into each of the ocean basins are shown, as well as the mouths of some major rivers. Note the large land area draining into the Atlantic and Arctic Oceans.

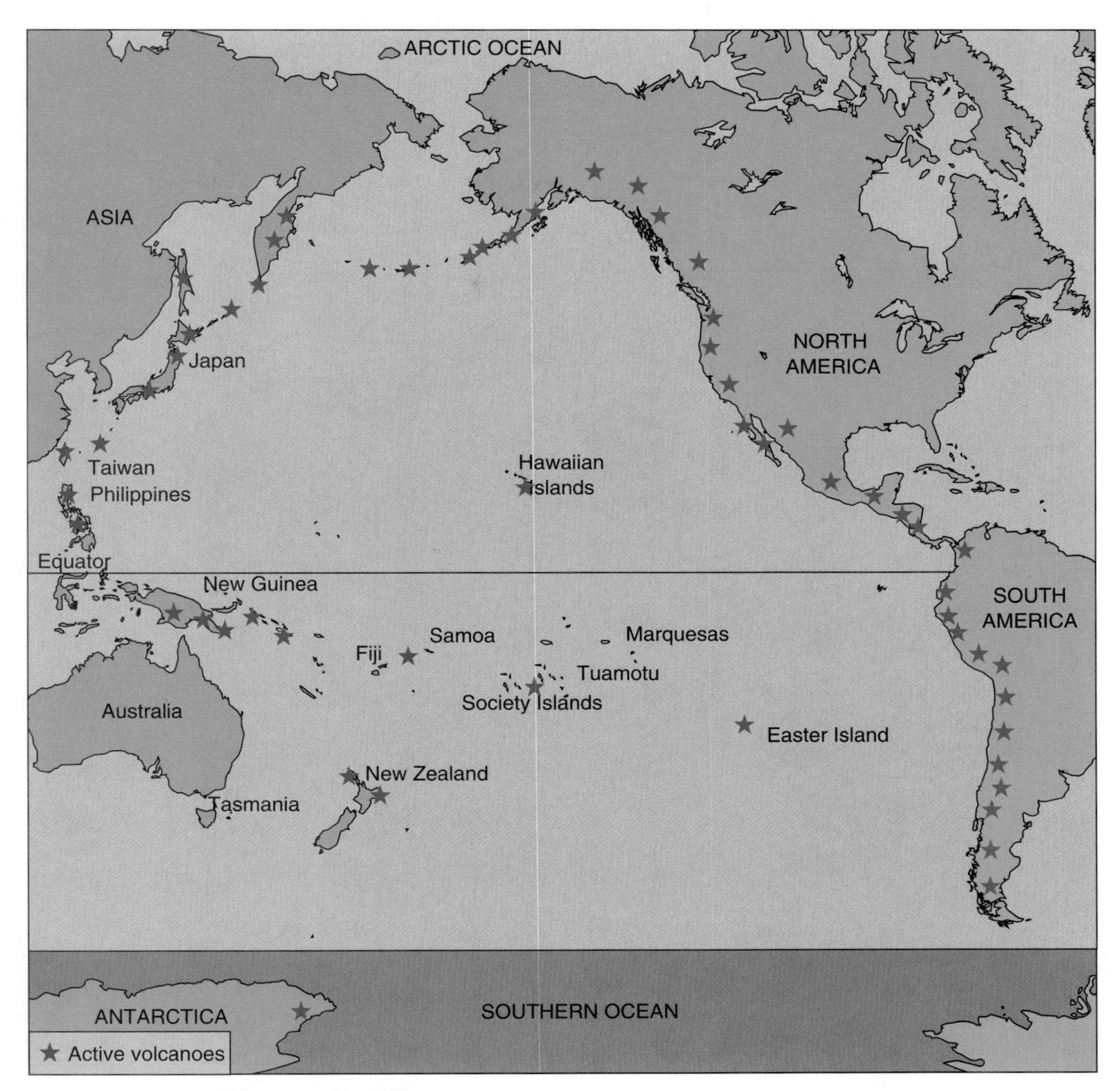

Figure 2–25
Major features of the Pacific Ocean.

3,000 kilometer-long chain which extends from Hawaii, with its active volcanoes, to Kure Island, a small sand island on top of a deeply submerged, eroded volcano more than 20 million years old. North of Kure, the chain changes direction and goes almost due north to the Aleutian Trench as a chain of submerged volcanoes called the Emperor Seamounts.

Island arcs are common in the western Pacific, extending from the Aleutians on the north to New Zealand on the south (Fig. 2–25).

Atlantic Ocean

The Atlantic is a relatively narrow basin connecting the Arctic and Antarctic Oceans (Fig. 2–26). It is also relatively shallow, averaging 3,310 meters deep. It has wide continental margins, the Mid-Atlantic Ridge, and several relatively shallow, semi-isolated seas such as the Gulf of Mexico and the Caribbean Sea. There are relatively few, mostly volcanic islands in the Atlantic (Fig. 2–26), except for Greenland, the world's largest island.

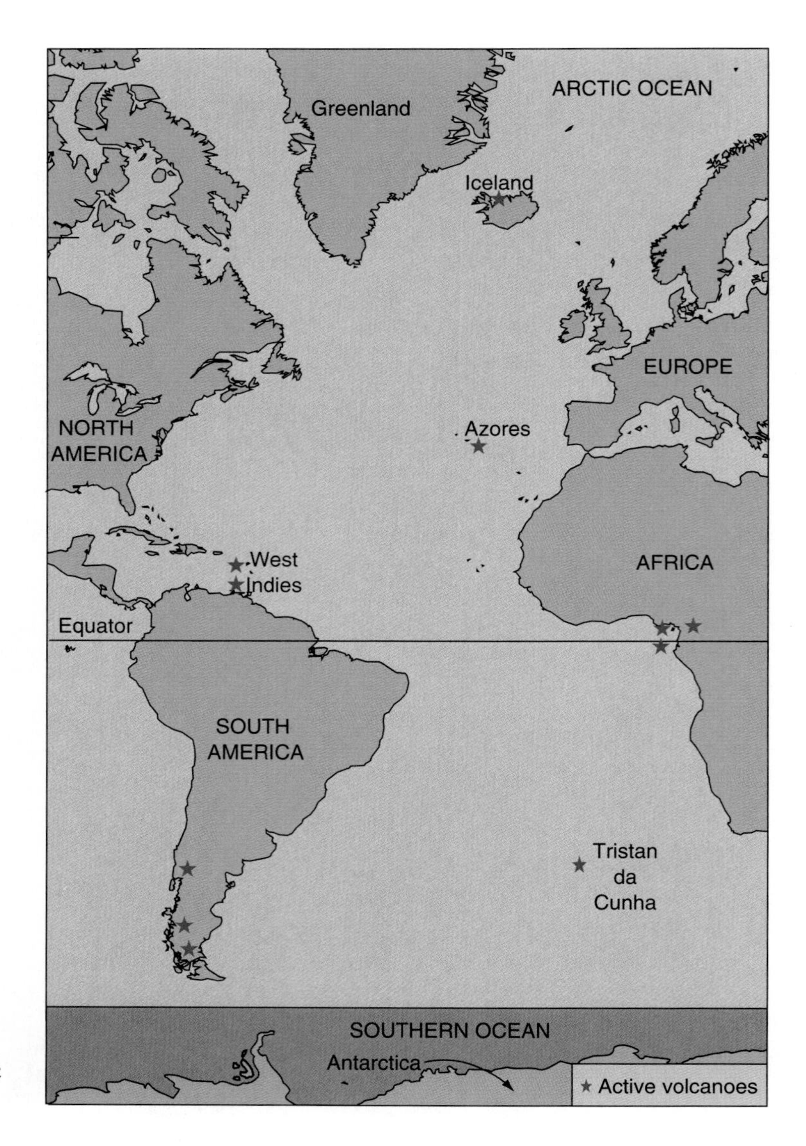

Figure 2–26
Major features of the Atlantic Ocean, a narrow basin connecting the polar Arctic and Southern Ocean areas.

The Atlantic Ocean receives large amounts of water and sediment from rivers (Fig. 2–24). The Amazon and Congo Rivers flow into the equatorial Atlantic. Together they discharge about one-quarter of the world's river flow to the ocean. Other large rivers flow into marginal seas in the Atlantic and into the Arctic Ocean.

Indian Ocean

The Indian Ocean lies primarily in the Southern Hemisphere (Fig. 2–27). It is the smallest of the three major ocean basins. Continental shelves around the Indian Ocean are relatively narrow, especially around Africa.

Three of the world's largest rivers (Ganges, Brahmaputra, and Indus) discharge into the northern Indian Ocean. This region has an abundance of both fresh water and sediment from the discharge of these rivers (Fig. 2–24). Thus, the northern Indian Ocean is the ocean most affected by nearby lands.

The northern Indian Ocean also has two major sources of warm saline water. The *Red Sea* and the *Arabian Gulf* receive little river discharge because they are surrounded by desert. Because of the intense evaporation, they produce warm saline subsurface waters that can be traced for hundreds of kilometers below the surface of the Indian Ocean.

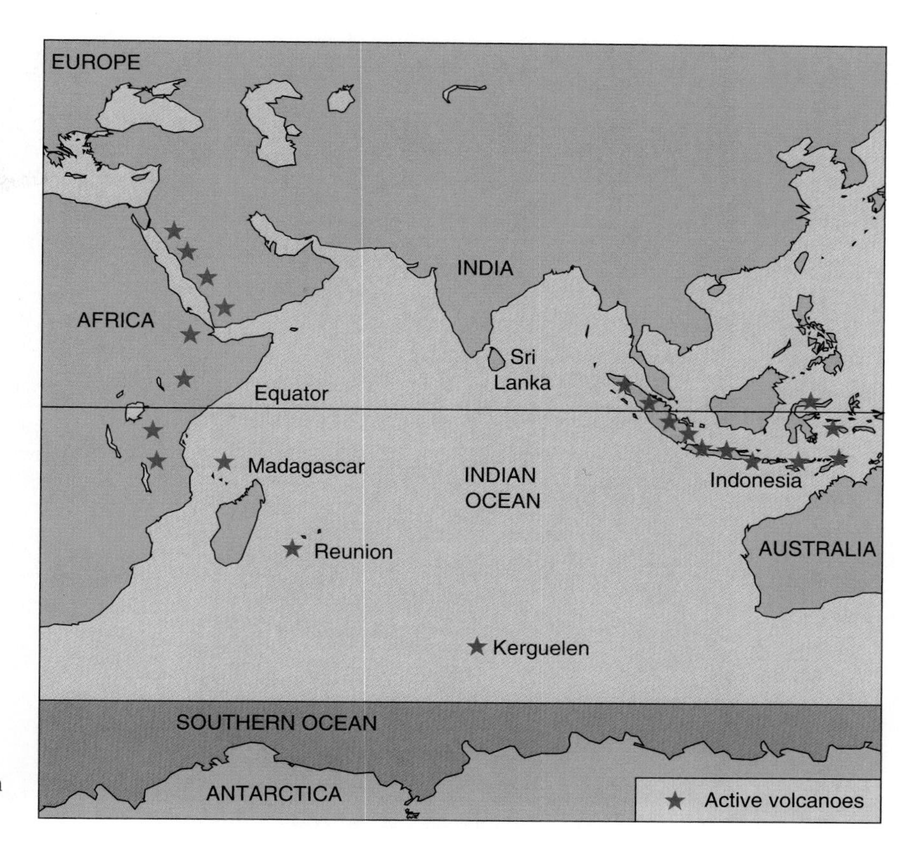

Figure 2–27
The Indian Ocean is primarily a Southern Hemisphere ocean.

There are few islands in the Indian Ocean (Fig. 2–27). Madagascar, the largest, was apparently once part of Africa. There are a few volcanic islands and several groups of low-lying limestone islands with extensive reefs.

Southern Ocean

As mentioned several times earlier, the three major ocean basins connect on their southern end (Fig. 2–11), and this region has special importance in ocean circulation and other Earth processes. It is often referred to as the Southern Ocean even though it does not occupy its own basin. Here, the absence of land permits strong winds and the ocean's largest currents to flow unimpeded. The only partial obstruction is the Drake Passage, the narrow space between Cape Horn (the southern tip of South America) and the Antarctic Peninsula.

The embayments on the margins of Antarctica, especially the Weddell Sea, are especially important to the rest of the ocean. Here the coldest and most dense waters form and flow northward into all three major ocean basins. (We learn more about this in Chapter 8.)

Thick ice sheets flow off Antarctica, forming ice shelves 100 to 200 meters thick. Large pieces of ice break off intermittently, forming icebergs that are then carried northward into the ocean basins. During winter the surface waters are chilled and sea ice forms. During summer most of this ice melts.

Marginal Ocean Basins

Marginal ocean basins are large ocean-bottom depressions that occur near continents, often in areas of active mountain building. These basins—typically more than 2 kilometers deep—are partially isolated from the nearby open ocean by submarine ridges or islands. The waters of the marginal seas that fill them are thus

Figure 2–28
The Red Sea, a marginal ocean basin, occupies a nearly rectangular depression between Asia and Africa. (Courtesy NASA.)

partially isolated from the nearby open ocean. As we see in later chapters, these marginal seas exhibit different characteristics, which vary substantially depending on local conditions. Several of these oceans are home to large human populations living around their shores and these receive large amounts of wastes. Because of their partial isolation from the rest of the ocean, they are especially vulnerable to suffering environmental degradation. We discuss this further in Chapter 14.

There are three types of marginal basins. Most common are basins formed by island arcs, such as the Bering Sea (Fig. 2–20). Here, submarine ridges between the volcanic islands separate the basins from the open ocean. These ridges can completely isolate the deeper waters in the basin. Another example of such a basin is the South China Sea, lying between Indonesia and southeast Asia (Fig. 2–21).

The second type is the long, narrow marginal sea surrounded by a continent. The *Red Sea* (Fig. 2–28) and the *Gulf of California* (Fig. 2–23) are examples of such basins. As we see in Chapter 3, these long, narrow ocean basins are formed by Earth's mountain-building processes.

A third type of marginal basin lies between two continents. The *Mediterranean Sea* lies between Europe and Africa, and the *Gulf of Mexico* and *Caribbean Sea* lie between North and South America (Fig. 2–23). As we shall see in Chapter 3, these basins have complex histories.

Arctic Ocean

As mentioned earlier, the Arctic Ocean (Fig. 2–29) is not a separate ocean but is usually considered to be a marginal sea, connected to the Atlantic. It is totally landlocked; its waters connect with the Atlantic only through the Fram Strait between Greenland and Norway (Fig. 2–29); and its several large rivers discharge large amounts of fresh water and sediment. Thus, the continental shelves which

Figure 2–29
The Arctic Ocean is nearly surrounded by the continents of the Northern Hemisphere.

make up almost half of the Arctic Ocean, as well as the deep-ocean floor in this area, have accumulated thick sediment deposits. Unlike any other ocean area, the Arctic is almost perpetually covered in ice—80 percent during summer and 90 percent during winter.

As we see in Chapter 14 when we discuss the coastal ocean, the Arctic is especially vulnerable to pollution from waste discharges. Russian rivers draining into the Arctic carry large waste loads from mismanaged oil fields and from nuclear waste dumps. In addition, the Arctic receives large amounts of airborne pollutants from all the industrialized regions of the Northern Hemisphere.

Meteorites, Earth, the Ocean, and Life

As scientists probe deeper into Earth's past to learn about its earliest history, the role of meteorites is becoming more apparent. These bodies colliding with Earth provided some of the mass of rock that now makes up the planet. They may also have provided water to the ocean, gases to the atmosphere, and even some forms of organic matter that formed heat-tolerant organisms. The importance of such events was underscored in July 1994, when pieces of a comet hit Jupiter. If this comet had struck Earth, it would have been an enormous catastrophe, killing many humans and possibly causing some extinctions.

Some of the earliest forms of life are now known to be able to tolerate temperatures that were previously thought to be lethal. Close relatives of these organisms are to be found living on the deep-ocean bottom near areas of recent volcanic eruptions. We will learn more about such organisms when we discuss oceanic life in later chapters. Determining the function and evolution of these organisms requires the latest and most sophisticated techniques of biochemistry.

Summary

The universe formed in the "big bang," about 15 billion years ago. The Sun formed from hot gases and the planets formed about 4.6 billion years ago. The inner planets, including Earth, consist primarily of metals and rocks; the outer planets are primarily ice and various gases. The history of each inner planet is determined primarily by its size. Small planets cool rapidly and are unable to retain an ocean of water or a thick atmosphere. Earth's history has been affected by the presence of life.

Earth's surface has two levels: The surface of the continents stands an average of 840 meters above sea level; the ocean bottom averages about 3,800 meters below sea level. Continents interrupt ocean basins, forming north–south barriers. Continental margins are the boundary between continents and ocean bottom.

The mid-ocean ridge is the world's largest mountain range. The Mid-Atlantic Ridge is a rugged mountain chain in the center of the basin. Similar ridges occur in all other ocean basins. Mid-ocean ridges are the most active volcanic regions on Earth; they have many shallow earthquakes. The deep-ocean floor is nearly flat or gently rolling. It is mostly covered by sediment deposits, especially near the continents. Fracture zones are elongate regions of mountainous topography. They connect offset segments of mid-ocean ridges. Trenches are the deepest parts of the ocean basins—maximum depths are about 11 kilometers. They occur primarily at the margins of the ocean basins, usually near continents.

The Pacific is the largest ocean basin, containing about half of Earth's free water. It is nearly surrounded by areas of active mountain building and, as a result, contains many active volcanoes and many earthquake belts. The Atlantic is a relatively narrow, S-shaped basin connecting the Arctic and Antarctic polar regions. The Indian Ocean is primarily a Southern Hemisphere ocean and is strongly affected by the lands around it. The Southern Ocean provides a connecting route for the currents of the other major oceans. It is especially important for the deep-ocean circulation throughout the other ocean basins.

Small, partially isolated marginal ocean basins occur on the edges of the major basins, especially in areas of active mountain building. The Arctic Ocean is nearly landlocked and partially isolated by submarine ridges from the rest of the Atlantic Ocean.

Key Terms

big bang
galaxies
star
nebulae
red giant
supernovas
solar system
photosynthesis
chemosynthesis
mid-ocean ridge, trench
continental margin
continental shelf
shelf break
continental slope
continental rise
coastal plain
submarine canyon
rift valley
deep-ocean floor
abyssal hill
abyssal plain
fracture zone
island arc
pillow lavas
micro-continent
marginal ocean basin

Study Questions

1 Describe the origin of the Universe, the Sun, and the solar system.
2. Compare and contrast the terrestrial planets, including Earth, with the outer (Jovian) planets.
3. Explain why a planet's size is an important factor in determining its history.
4. Describe the two levels of Earth's surface.
5. On an outline map, sketch the locations of earthquake belts, active volcanoes, and mid-ocean ridges.
6. Describe the characteristic features of trenches, mid-ocean ridges, and fracture zones.
7. List the principal similarities and differences of the three major ocean basins. Indicate where they connect and what separates them.
8. Describe the three types of marginal ocean basins. Where does each type occur?
9. What is the most conspicuous feature in the ocean basins?
10. Describe the three parts of a continental margin and explain what a shelf break is.
11. Where are the two largest areas of unusually high elevation on land?
12. Where is the continental shelf deepest? Why?
13. Discuss the relationships between trenches and active volcanoes.
14. Contrast the Indian Ocean basin with the other two ocean basins.
15. How are island arcs associated with trenches and volcanoes?
16. How is the Arctic Ocean connected to the rest of the world ocean?
17. Contrast the relationship between land and ocean basins in the Arctic and Antarctic regions.
18. *[critical thinking]* Discuss the principal reasons why Earth differs so much from its neighboring planets in the solar system.

Selected References

ALLÉGRE, C. J., and S. H. SCHNEIDER, "The Evolution of the Earth," *Scientific American, 271*(4): 66–75, 1995.

BALLARD, R. D., *Exploring Our Living Planet*. Washington, D.C.: National Geographic Society, 1983. Illustrated presentation of ocean features and their origins.

CONE, J., *Fire Under the Sea*. New York: William Morrow, 1991.

DIXON, R. T., *Dynamic Astronomy,* 6th ed. Englewood Cliffs, N.J.: Prentice Hall, 1992. Standard text on astronomy, discusses universe, solar system, Earth, and other planets.

FRIEDMAN, H., *Sun and Earth*. New York: Scientific American Library, 1985.

MORRISON, D., *Exploring Planetary Worlds.* New York: Scientific American Library, 1993. Covers knowledge of planets derived from space exploration.

PEEBLES, P. J., D. N. SCHRAMM, E. L. TURNERS, AND R.G. KRON, "The Evolution of the Universe," *Scientific American, 271*(4): 52–57, 1995.

WEINER, J., *Planet Earth*. Toronto: Bantam Books, 1986. Survey of ocean, Earth, atmospheric, and astronomic sciences.

OBJECTIVES

Your objectives as you study this chapter are to understand:

- Earth's principal features
- The principles of plate tectonic theory
- How Earth's evolution has affected ocean basins, seawater, and the atmosphere

A computer-generated view of global topography centered over the Atlantic Ocean. Clearly visible are the Mid-Atlantic Ridge, wide continental shelves on the western side of the Atlantic, narrow shelves on the west sides of Africa and South America, and the deep-ocean basins. *(Courtesy NASA.)*

3 Plate Tectonics

Earth is a dynamic planet because it is still cooling. Its surface—both continents and ocean basins—slowly but constantly changes. Instead of being one rigid piece covering Earth's interior, the surface of the planet consists of about 12 huge rigid pieces—each called a **plate**—that move around (very slowly, about the speed your hair grows) and collide with each other. New oceanic crust is formed as huge volcanoes erupt along mid-ocean ridges, and then the newly formed crust moves slowly away from its point of origin as part of a rigid plate. Over millions of years, newly formed oceanic crust cools, becoming denser as it ages, and moves toward the trenches. There, it sinks under an adjacent plate and is assimilated into the underlying material. In the process, continents are moved and ocean basins change their shape. Thus, the distribution of ocean basins and continents described in Chapter 2 is only a snapshot view of Earth's slowly changing surface.

Earth's Interior Structure

Earth consists of several concentric spheres (Fig. 3–1). At the center is the **core,** which is very dense because it contains metals, primarily iron and nickel. The **inner core** is solid iron, and movements of the liquid-iron **outer core** create Earth's magnetic field. The sphere outside the outer core, called the **mantle,** consists of dense rock and is mostly solid. Part of the upper mantle, called the **asthenosphere,** is partially molten and is therefore weak and can flow very slowly. Above the asthenosphere is a part of the upper

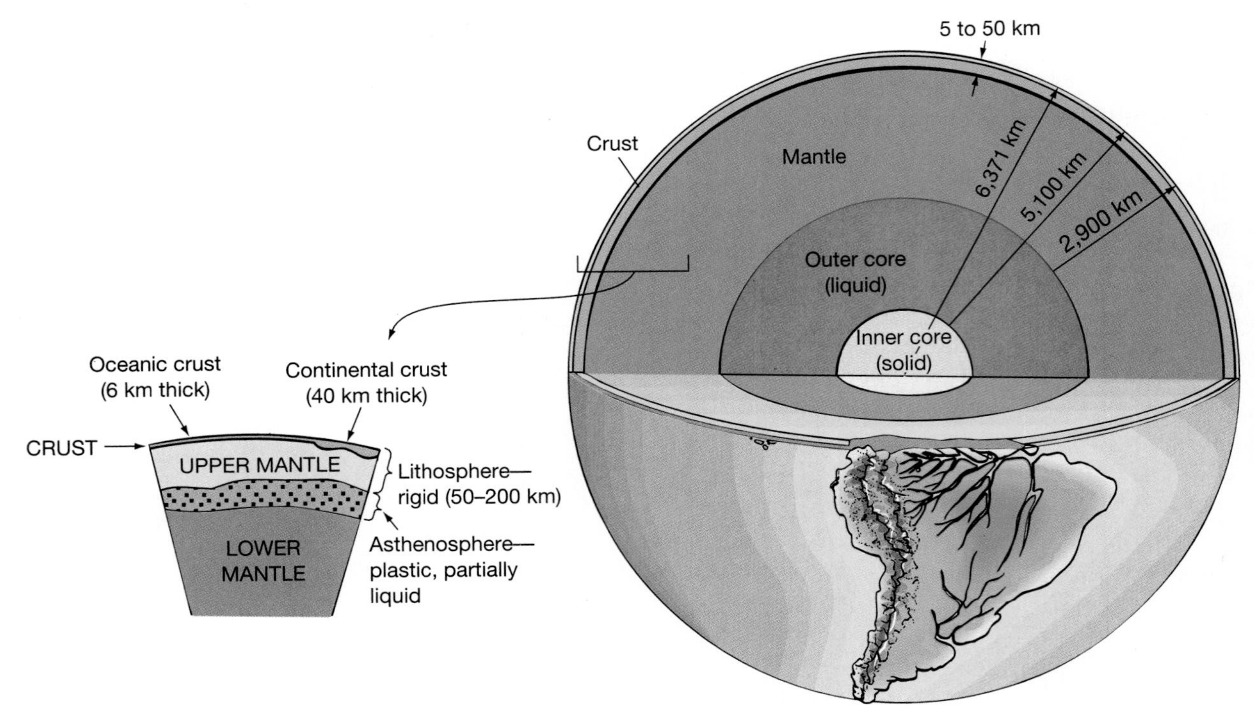

Figure 3–1
Earth consists of concentric layers.

mantle that is cooler and thus more rigid. The outermost sphere is the cold, rigid crust. The crust and rigid upper mantle are together called the **lithosphere,** and it is this layer, broken up into the plates mentioned previously, that floats on the asthenosphere.

Earth has two types of crust. Continents are made of **continental crust,** consisting of thick accumulations (30 to 40 kilometers) of many different kinds of rocks, generally rich in aluminum and silica. Ocean basins are made of thinner (5 to 7 kilometers thick) **oceanic crust,** consisting primarily of basaltic rocks, which are rich in magnesium and iron. The Mohorovičić discontinuity, or **Moho** for short, is the distinct structural change separating the base of the crust from the underlying mantle.

Isostasy

Lithospheric plates float in the asthenosphere in a balance called **isostasy** (*iso-*, Greek "the same"; stasis, Greek "standing"). The simplest way to visualize isostasy is to consider wood blocks of different thicknesses floating in water (Fig. 3–2). The thickest block stands higher above the water surface than the thinnest one. Thus,

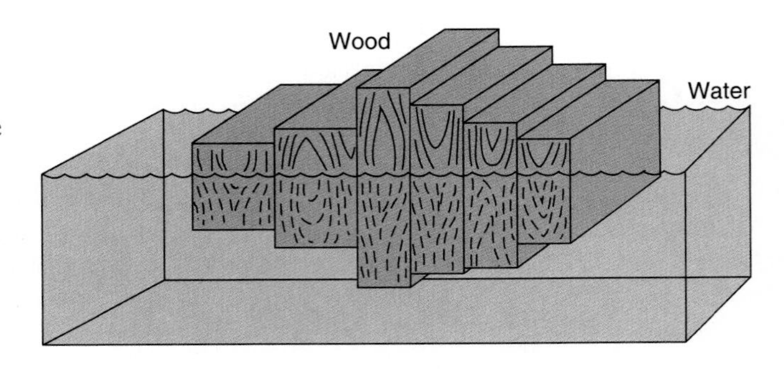

Figure 3–2
Isostasy is a condition of floating equilibrium represented here by wooden blocks floating in water. The blocks float because the density of the wood is lower than the density of the water. The tops of the thickest blocks stand highest above the water surface.

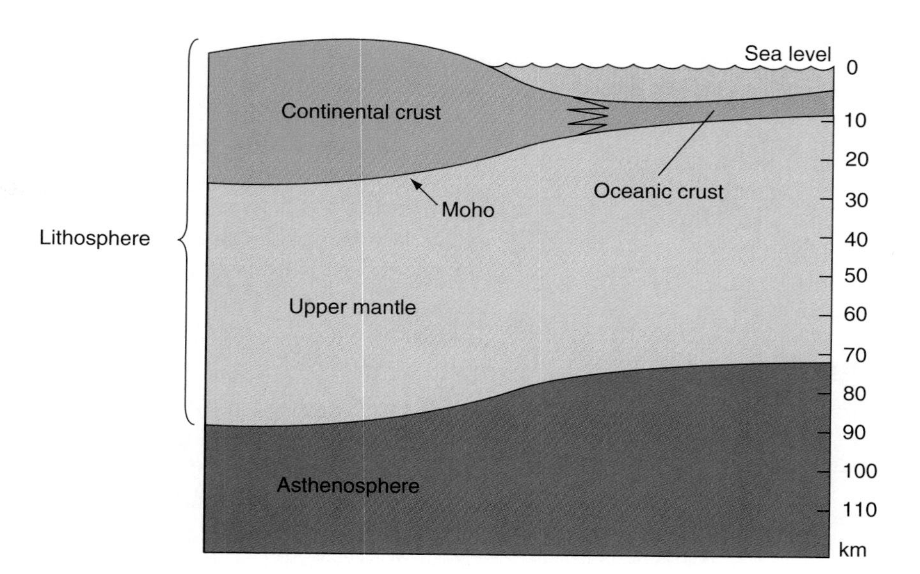

Figure 3–3
Schematic representation of the lithosphere (consisting of continental crust, oceanic crust, and upper mantle) floating on the asthenosphere, a partially molten layer in the mantle.

if oceanic and continental crust had the same density (as the wood blocks do), we would expect the top of the thicker continental crust to stand higher than the top of the thinner oceanic crust. There is, however, another factor. Continental rocks are less dense than the oceanic basalts, and so the continents stand even higher than expected above the ocean floor (Fig. 3–3).

In addition to floating horizontally in the asthenosphere, lithospheric plates also move vertically when loads, such as volcanoes or glaciers, are placed on them or removed by erosion or melting. When the Hawaiian Island volcanoes first formed, for instance, the Pacific plate was already about 80 million years old and fairly rigid. The weight of the volcanoes caused the plate to bend (Fig. 3–4). This bending created a moat around the base of the islands about 500 meters deeper

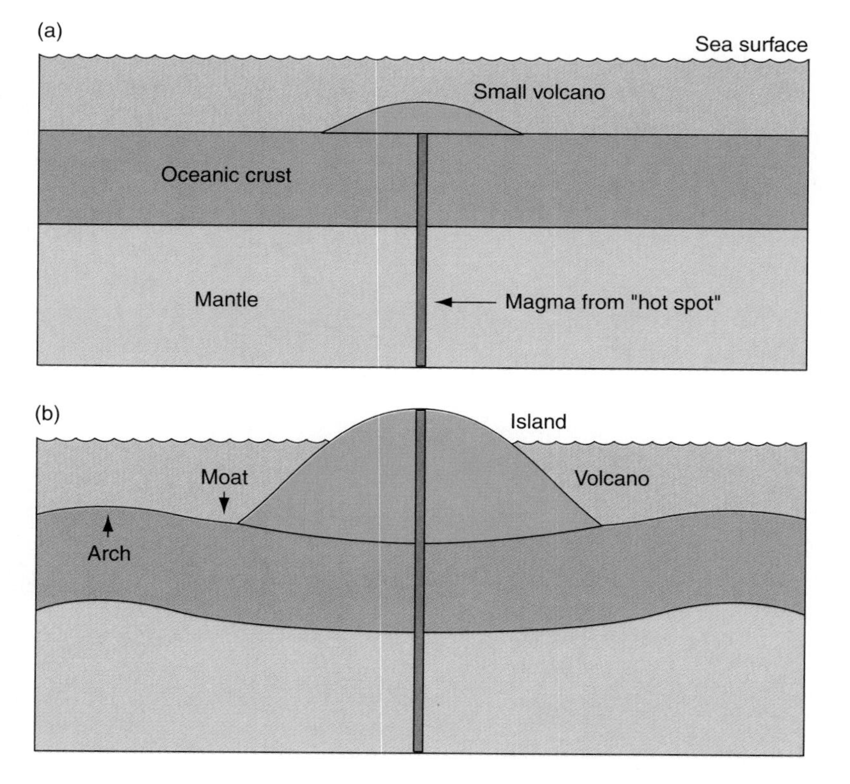

Figure 3–4
The weight of a small volcano (a) is supported without bending the rigid oceanic crust. (b) The weight of a large volcano causes the crust to bend, forming an uplifted arch and a depressed moat around the volcano.

than the general level of the ocean floor in that area. It also formed a broad arch now located about 250 to 300 kilometers from the island.

Large glaciers form during ice ages and then melt when the climate warms. Plates are depressed by the glaciers' weight, which accounts for the deeply submerged continental margins around Antarctica and the Arctic Ocean. After glaciers melt, the plates rise. Scandinavia is still rising as a result of the disappearance of its glaciers over the past 18,000 years. (We discuss the formation and melting of the continental ice sheets in Chapter 7.)

Plate Tectonics

As we have seen, the crust and the uppermost mantle make up the lithosphere, which is broken up into numerous smaller plates that float on the asthenosphere. The lithosphere consists of 12 large and many small plates that move because heat escaping from the core causes the upper-mantle rocks under the crust to move (Fig. 3–5). This concept is called **plate tectonics**—deformation of Earth by interactions among crustal or lithospheric plates.

Lithospheric plates have three kinds of boundaries (Fig. 3–5). Virtually all the deformation of Earth's crust occurs in narrow belts that are boundaries between plates. At mid-ocean ridges, plates move away from each other and the boundaries are **divergent** (also called *spreading centers*). At trenches, plates converge and the boundaries are **convergent.** At transform faults, which are places where two plates slide past each other, the boundaries are called **transform-fault boundaries.** They form the fracture zones that we encountered in Chapter 2.

Oceanic crust forms when molten rock (called **magma**) from the mantle either erupts onto Earth's surface and solidifies or fills fissures at a ridge crest and solidifies. New crust forms at the rate of about 20 cubic kilometers per year. Newly formed oceanic crust slowly moves away from the relatively shallow mid-ocean ridge. As it ages, oceanic crust cools, becomes denser and therefore floats lower

Figure 3–5
Schematic representation of the major processes of plate tectonics. Oceanic crust forms at mid-ocean ridges through volcanic eruptions of molten rock, derived from partial melting of mantle rocks. Rigid lithospheric plates move from mid-ocean ridges to trenches where they are subducted down into the mantle. There partial melting of the subducted plate supplies magmas to volcanoes that form either volcanic island arcs or contribute to building continental crust.

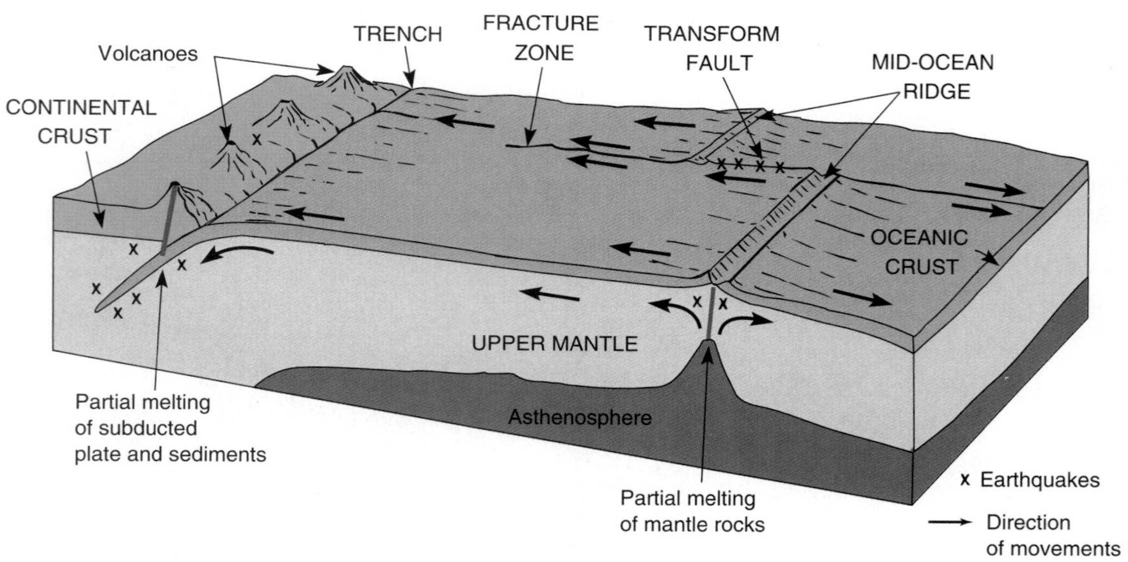

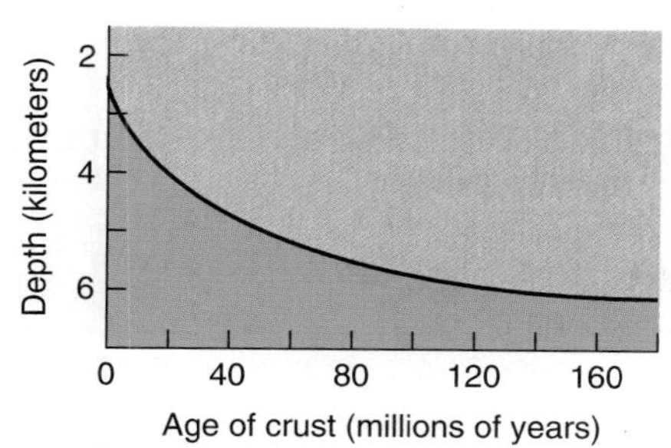

Figure 3–6
In the ocean basins, depth increases with the age of the crust as the crust becomes denser and sinks lower into the mantle. The shallowest ocean bottom is the most recently formed with still warm and buoyant crust at the mid-ocean ridges. The deepest, oldest crust occurs near land.

in the asthenosphere. The youngest crust lies about 2,500 meters below the sea surface, and the oldest is about 6,000 meters deep (Fig. 3–6). The lithosphere also thickens as it cools, increasing from a few kilometers in thickness when newly formed to about 150 kilometers in thickness for the oldest parts.

Lithosphere is resorbed into the mantle at the same rate as it forms. This occurs at trenches, where dense lithosphere sinks into the mantle and is assimilated, a process called **subduction.** In its simplest form, you can view the movement of a lithospheric plate as a gravity-driven conveyer belt, moving a plate from a shallow mid-ocean ridge, where oceanic crust forms, across a deep-ocean basin to an even deeper trench, where the plate material is incorporated into the mantle.

Continental crust forms at convergent margins. As plates are subducted into the mantle, they are heated in Earth's interior. The more easily melted materials in the rocks and sediment deposits melt and rise as magmas to erupt in volcanic island arcs and in volcanoes associated with trenches. These lavas are called andesites because of their common occurrence in the volcanoes of the Andes Mountains of South America. They are distinctly different from the lavas that erupt on mid-ocean ridges; they contain much more silica and aluminum than the iron-rich basaltic lavas of the mid-ocean ridges.

So far, we have treated plate movements as if they occurred on a flat surface, but they are actually on a spherical Earth. Plates on a sphere rotate about fixed poles (Fig. 3–7). One consequence of this rotation is that transform faults form perpendicular to the pole of rotation. The lowest rates of crustal formation occur near poles of rotation; the highest rates occur halfway between them.

Plates move from the mid-ocean ridges where they form to the trenches where they are resorbed into the mantle (Fig. 3–5). Offsets or discontinuities in the mid-ocean ridges occur at transform faults. These offsets are particularly conspicuous in the equatorial region on the Mid-Atlantic Ridge (see Fig. 2–13). In the active transform faults, where the plates are moving in opposite directions, earthquakes occur. In the inactive sections, on either side of the segments closest to the mid-ocean ridge, the plates are moving in the same direction, toward the trench,

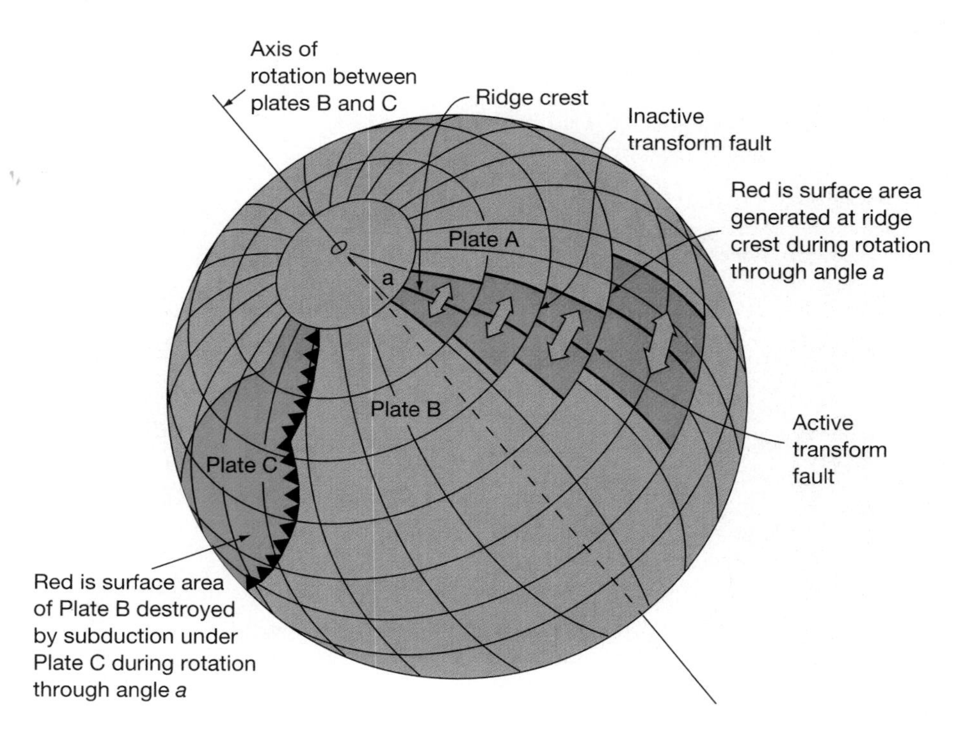

Figure 3–7
Representation of three rigid plates moving on the surface of a sphere. The relative rates of motion are least at the poles and greatest midway between them. Relative differences in motion are taken up along transform faults. As Plate B rotates through angle a (shown at top), new surface (rightmost red area) is added at the ridge crest and an equal amount of plate is destroyed by subduction in the leftmost red area. (After Kennett, *Marine Geology*. Englewood Cliffs, N.J.: Prentice Hall, 1982.)

BOX 3–1
Plate Tectonic Theory—History of a Scientific Revolution

Acceptance of plate tectonic theory is an example of a scientific revolution. This theory has dramatically changed how scientists view Earth and its history. In the process, the theory revolutionized the way scientists think about Earth processes, ranging from the origin of continents to the history of ocean circulation. It took earth scientists decades, however, to accept this simple but elegant idea.

Alfred Wegener, German meteorologist, balloonist, and polar explorer (Fig. B3–1–1), first advanced an idea he called *continental drift.* He had noticed many puzzling aspects of continents. For instance, the shorelines of South America and Africa match closely. Mountain ranges in Africa and South America cut off by shorelines match mountains on the other side of the ocean. Fossils in rocks on each side of the Atlantic as well as the characteristic features left by ancient glaciers also match. To account for all these findings, Wegener theorized that all continents were once part of a supercontinent, which he called Pangaea.

Figure B3–1–1
Alfred Wegener (1880–1930), photographed in Berlin just before he left for Greenland where he died. (UPI/Bettman.)

Wegener published his ideas in 1915, while recovering from wounds received during World War I. Continental drift attracted many early supporters in Australia, South Africa, and to some extent in Europe but was rejected in North America. The principal objections were that Wegener's evidence (fossils, paleoclimates) was not convincing. In particular, there were objections to the mechanisms he proposed for the movement of continents. Wegener proposed that continents were rigid bodies moving through the ocean basins. The driving forces for these movements were variations in the gravitational attraction of Earth's equatorial bulge and the westward drift due to the attractions of the Sun and Moon. For several decades there were heated arguments but no new evidence.

Until the 1960s, most North American geologists believed that continents and ocean basins were permanent features of Earth's surface. These ideas were increasingly challenged, however, by data gathered in the 1950s as ocean basins were mapped in detail and their geophysical properties measured.

Magnetic anomalies under the ocean basins had also been mapped in great detail because of the possibility of locating submarines through their disturbances of Earth's magnetic field. The striped patterns of the magnetic anomalies intrigued scientists for many years.

Based on his studies of these newly available data, *Harry H. Hess* (Fig. B3–1–2), a highly respected geologist, theorized in 1962 that Earth's outer shell moves. He speculated that these movements cause continents to break apart, move, and form new ocean basins. Hess further hypothesized that oceanic crust forms through volcanic activity at mid-ocean ridges and moves toward trenches, where it is destroyed (Fig. B3–1–3). Because of the strong feelings on the subject among his colleagues, he labeled his paper "geo-poetry." The American geophysicist *R. Dietz* called this idea seafloor spreading. The Canadian geologist *J. Tuzo Wilson* provided many of the basic concepts, including the role of transform faults.

Finally, *F. Vine* and *D. Matthews* of Cambridge University proposed that these patterns of magnetic anomalies

Figure B3–1–2
Professor Harry Hess of Princeton University postulated that oceanic crust is formed at mid-ocean ridges and moves toward trenches, where it is destroyed. (Courtesy Princeton University Geology Department.)

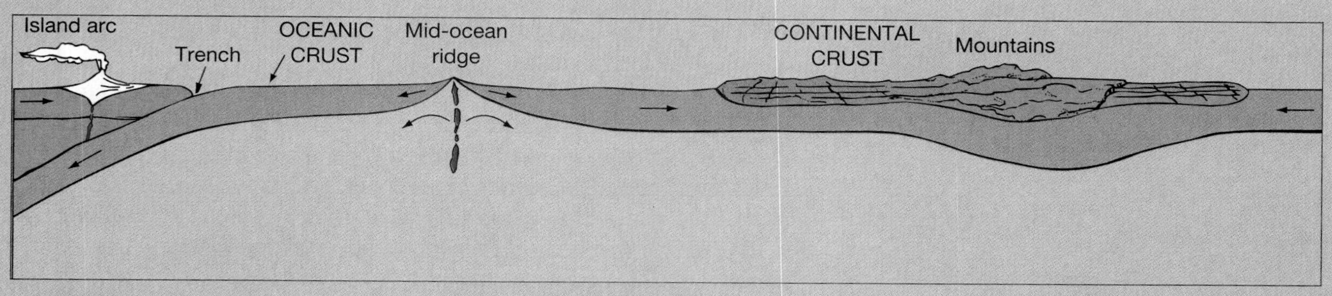

Figure B3–1–3
Major features of seafloor spreading postulated by Hess.

recorded ancient reversals in Earth's magnetic field. In their model, the newly formed crust acted like a gigantic tape recorder, indicating the orientation of the magnetic field at the time the crust formed. Thus magnetic anomaly patterns made it possible to date the age of ocean-floor formation. This technique is still widely used.

In 1965, the British geophysicist *Edward Bullard* showed that the continents could be fitted together along their 100- and 1000-meter depth contours, with the few areas of overlap corresponding to recently formed river deltas or coral reefs (Fig. B3–1–4). This match convinced many geologists that the continents could be fitted together in a reasonable way and, thus, that it was feasible that they once were all pieces of a single supercontinent, just as Wegener had proposed 50 years earlier.

Another development supporting the idea of seafloor spreading came from drilling the South Atlantic floor. Samples from beneath the sediment cover were progressively older moving away from the Mid-Atlantic Ridge. For most geologists, this was conclusive evidence.

Finally, other scientists put together the evidence from studies of earthquakes in subduction zones, along the transform faults between mid-ocean ridges, and along fracture zones. For instance, Wilson formulated many important aspects of the theory of plate tectonics. Since the late 1960s, its concepts have revolutionized thinking about how Earth works. This broad view was further assisted by the exploration of the solar system, which permitted comparative studies of the planets.

The precise positioning now available through the Global Positioning System makes it possible to detect plate movements as small as a few centimeters per year. Indeed, radar images from successive satellite orbits show details of land–surface movements following individual earthquakes. Such observations permit refinement of the theory. Original plate tectonic theory dealt primarily with the large-scale movements. These new observations permit studies of the more complicated movements of the continents and smaller blocks. Improved understanding will help scientists develop predictive capabilities and help alleviate the devastating effects of large earthquakes.

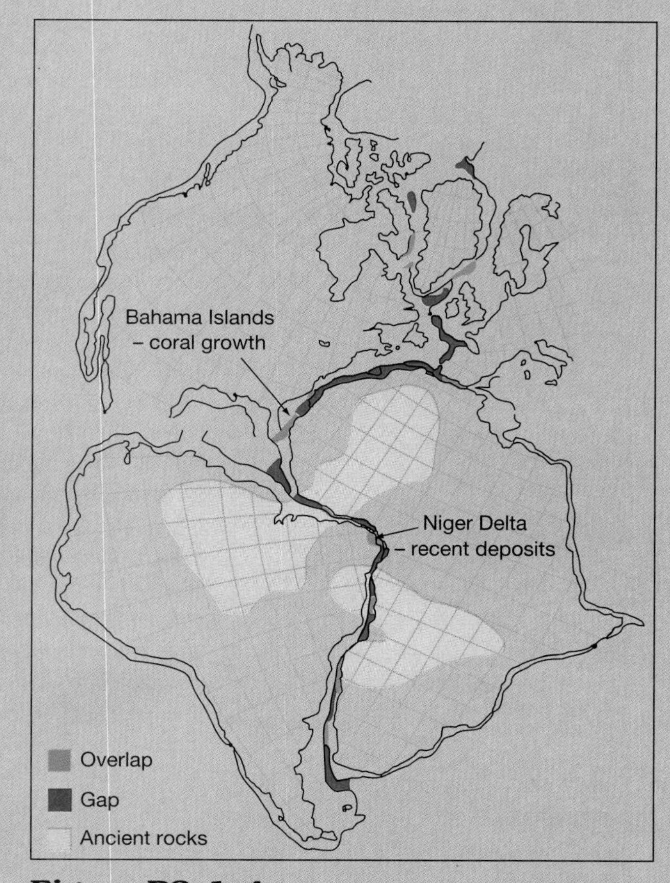

Figure B3–1–4
The continents on the two sides of the Atlantic fit along the 500-meter depth contour. Areas of overlap are sites of coral-reef growth or sediment deposition since the Atlantic basin formed. Areas of rocks older than 500 million years were split apart when the Atlantic opened.

and there are no earthquakes. The inactive sections of these faults are called fracture zones (Fig. 3–5). These fracture zones extend for thousands of kilometers across the deep ocean basins and are especially conspicuous on the East Pacific Rise, east of New Zealand (Fig. 2–13). These fracture zones are useful in reconstructing plate motions and how they may have changed in the past. In effect, they correspond to the small circles of rotation that we discussed in dealing with simplified plate movements on a sphere (Fig. 3–7).

Movements of plates on Earth's surface are interconnected. Changing the direction and/or rate of movement for one plate involves the others. Plate movements may change for many reasons, such as a continent-continent collision.

Wherever three plate boundaries come together, the intersection is called a **triple junction.** There are two principal types. Three intersecting subduction zones form an *unstable triple junction,* which moves as the plates are consumed in the subduction zones. Such an unstable triple junction occurs east of Japan (Fig. 3–8). Three intersecting ridges form a *stable triple junction,* which remains relatively fixed in location. Such stable triple junctions occur on the East Pacific Rise south of Easter Island. Another is in the South Atlantic on the Mid-Atlantic Ridge.

Hot Spots

Plumes of molten rock rise from deep within the mantle to erupt and cause **hot spots,** centers of prolonged volcanic activity (Fig. 3–8). Many hot spots occur on or near mid-ocean ridges; Iceland is an example. Others, such as Hawaii, occur in the middle of lithospheric plates, far from the edges where most active volcanoes are situated.

Locations of long-lived hot spots apparently remain fixed for up to 100 million years. As a plate moves across a hot spot, a chain of volcanoes is formed, and this chain is an excellent indicator of the direction of plate movement. Volcanoes on the youngest islands remain active while they are near the hot spot. After the crust has moved the volcano away from its source of magma, it becomes extinct. Eventually, extinct volcanoes are eroded by waves and rivers, first becoming small islands, then submerged banks, and finally seamounts.

The Hawaiian Islands formed over a hot spot (Fig. 3–9). Volcanoes on the island of Hawaii are still active because they are still near the hot spot. There is also submerged volcanic activity southeast of Hawaii, at a small crater called Loihii; a new island will eventually form there. The volcanoes northwest of Hawaii are extinct and deeply eroded. The oldest island is Kure, a low carbonate-sand island built on an eroded and deeply submerged volcano about 30 million years old. A chain of ancient submerged volcanoes (Emperor Seamounts) continues almost due north, marking a change in direction of plate movements about 40 million years ago. The oldest seamount in the chain is 65 million years old.

Magnetic Anomalies

Studies of Earth's magnetic field provided some of the most compelling evidence of plate movements. Patterns in these anomalies are now used to determine when parts of the oceanic crust formed. Let's see how.

Earth's magnetic field over the ocean exhibits long, irregular bands of deviations (either stronger or weaker) from the predicted magnetic field; these are called **magnetic anomalies** (Fig. 3–10). For many years these magnetic patterns were a great scientific mystery, but we now know that they are the result of reversals in the directions of Earth's magnetic field every hundred thousand years or so. Minerals in rocks record the orientation of Earth's magnetic field at the time the rocks cooled. During times of normal magnetic orientation, Earth's north and

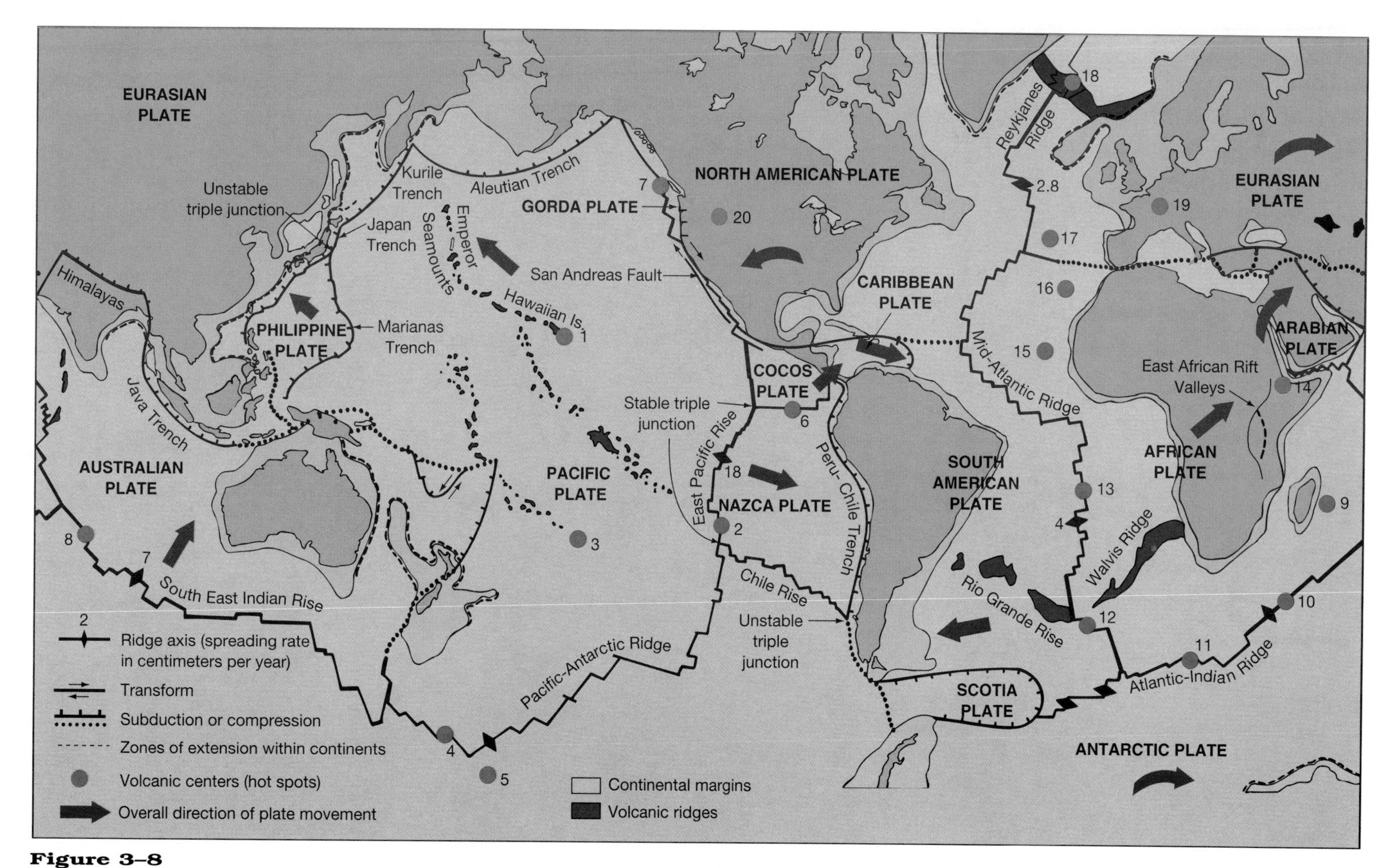

Figure 3–8
Major lithospheric plates and their boundaries, showing principal hot spots. 1. Hawaii, 2. Easter Island, 3. McDonald Seamount, 4. Balleny Island, 5. Mt. Erebus, 6. Galapagos Islands, 7. Cobb Seamount, 8. Amsterdam Island, 9. Reunion Island, 10. Prince Edward Island, 11. Bouvet Island, 12. Tristan da Cunha, 13. St. Helena, 14. Afar, 15. Cape Verde Islands, 16. Canary Islands, 17. Azores, 18. Iceland, 19. Eifel, 20. Yellowstone. [After D. L. Anderson, "The San Andreas Fault," *Scientific American,* 225(11), 1971.]

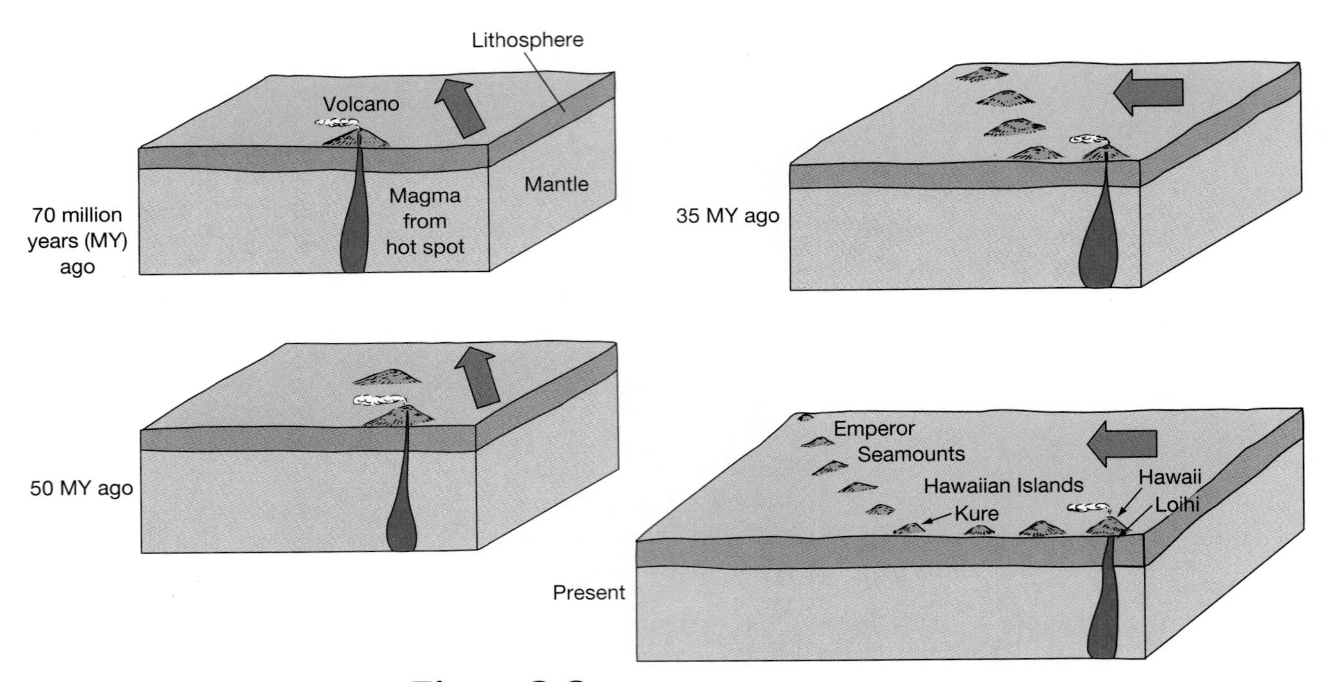

Figure 3–9
The Hawaiian Islands and the Emperor Seamounts formed as the Pacific plate moved over a hot spot. The lines of volcanoes record the change in the plate's direction of movement.

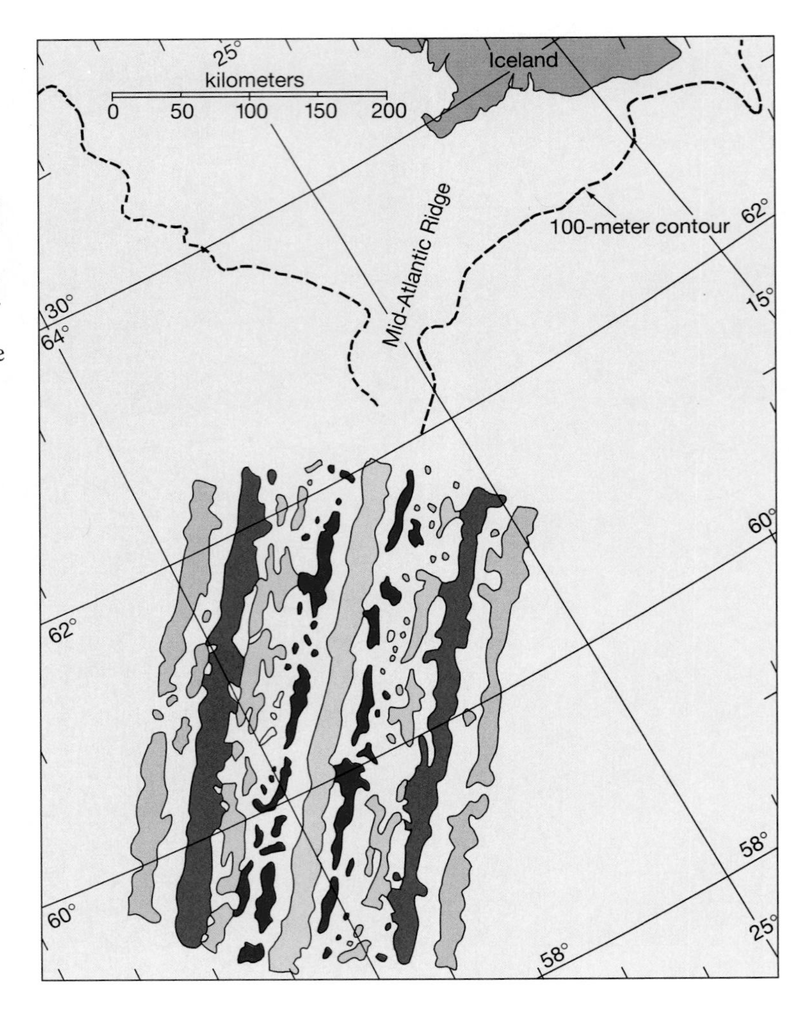

Figure 3–10
Magnetic anomalies on the Mid-Atlantic Ridge south of Iceland. These anomalies record the orientation of Earth's magnetic field at the time when the crust formed. The various darker-colored stripes indicate where the local magnetic field is slightly stronger than the average field of Earth. The rocks in these crustal segments formed when Earth's magnetic field had the same orientation as the present field. Light blue stripes indicate where the magnetic field is weaker. These crustal segments formed when the magnetic field was reversed from its present orientation.

south magnetic poles occupy their present positions. At times of reversed orientation, the pole locations are reversed. (No one knows what causes this reversal.)

The magnetic field is slightly weaker over crust formed when the poles were reversed than it is over rocks formed during times of normal orientation (the present orientation). Sensitive instruments, called *magnetometers,* towed behind aircraft or ships are used to map these variations. (These detailed maps of Earth's magnetic field over the oceans were collected as a tool to help locate ballistic-missile launching submarines during the Cold War.)

Studies of magnetic orientation of minerals in volcanic rocks of known ages provide a time scale for anomaly patterns. The relative lengths of time in each orientation provide a scale for determining ages of ocean-floor rocks.

Figure 3–11 shows crust forming at a mid-ocean ridge over a period of time during which Earth's magnetic field reversed several times. The crust records these reversals like a gigantic tape recorder. Because such anomaly patterns have been mapped over most of the ocean, we now know when the ocean floor formed (Fig. 3–12). As we would expect, the youngest crust occurs in a band centered on the mid-ocean ridges. The faster the spreading rate, the wider the band of young crust. The oldest oceanic crust (up to 190 million years old) occurs in the North Pacific near Asia and along the margins of the North and South Atlantic. About half the ocean bottom has formed in the last 80 million years.

Some parts of the oceanic crust show no magnetic anomalies. This crust formed during a time (80 to 120 million years ago) when Earth's magnetic field apparently did not reverse. Thus, at this time we have no way of determining the ages of these crustal segments. There is also evidence that deeply buried oceanic crust loses its magnetic record because the minerals are heated as a result of the thick sediment deposits overlying them. Since the oldest crust usually lies deeply buried near continents, this process may destroy the magnetic evidence of the oldest rocks in the ocean basins.

Hydrothermal Circulation

Much of the heat in newly formed, hot oceanic crust is removed by seawater flowing through the fractured rocks. Seawater enters through open fractures and permeable zones, penetrating several kilometers into recently solidified volcanic

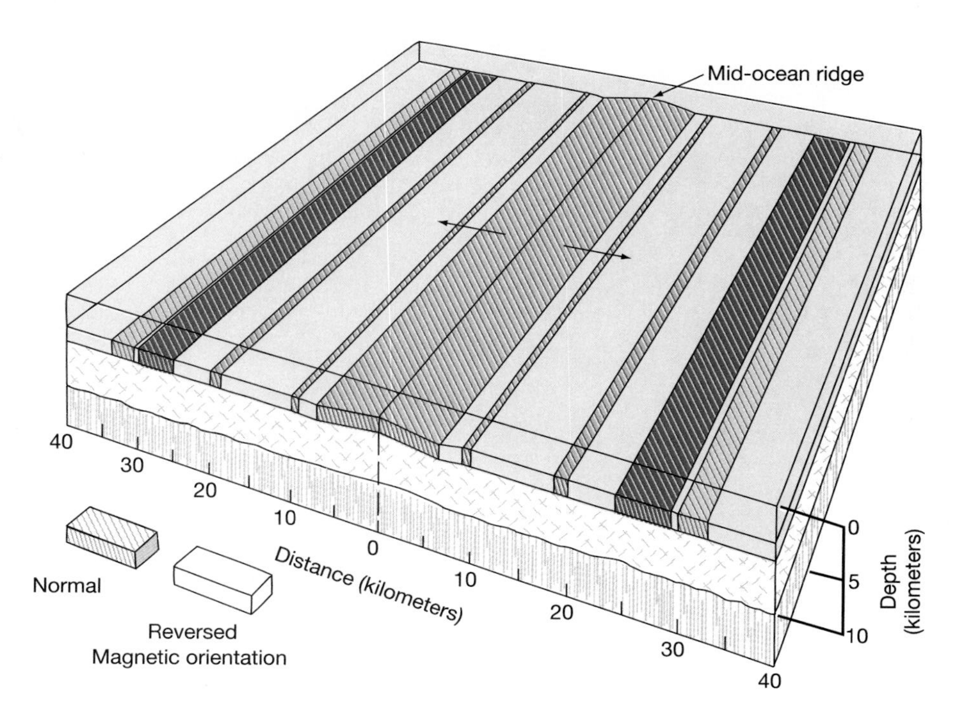

Figure 3–11
Striped patterns of magnetic anomalies result from minerals in oceanic crust recording the direction of Earth's magnetic field at the time of crustal formation.

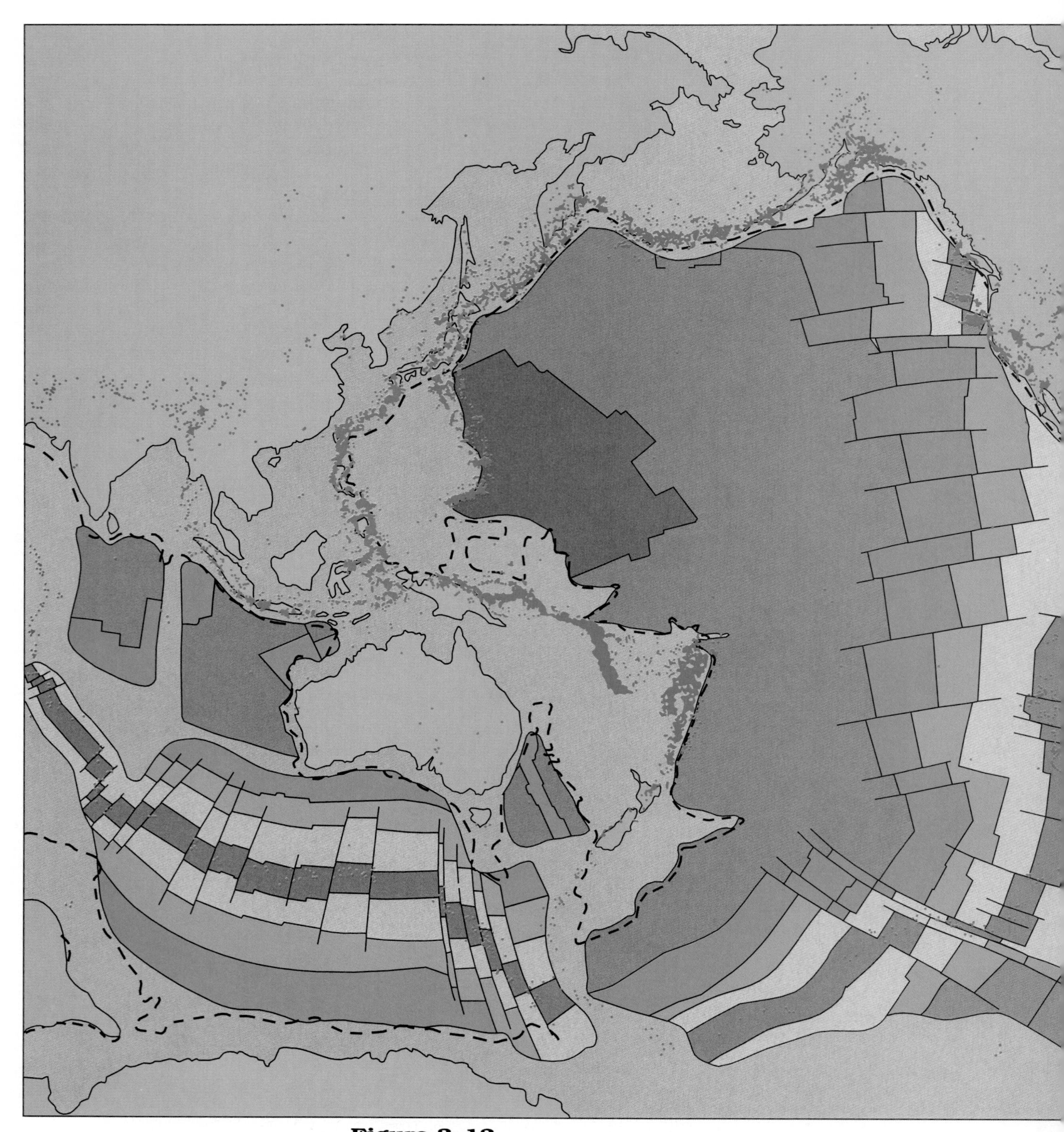

Figure 3–12
Ages of oceanic crust as determined from magnetic anomalies. Note the symmetrical patterns of crustal ages in the Atlantic Ocean. Large areas of relatively young crust have been subducted in the Pacific, east of the East Pacific Rise. Note that earthquakes occur primarily at subduction zones and along mid-ocean ridges.

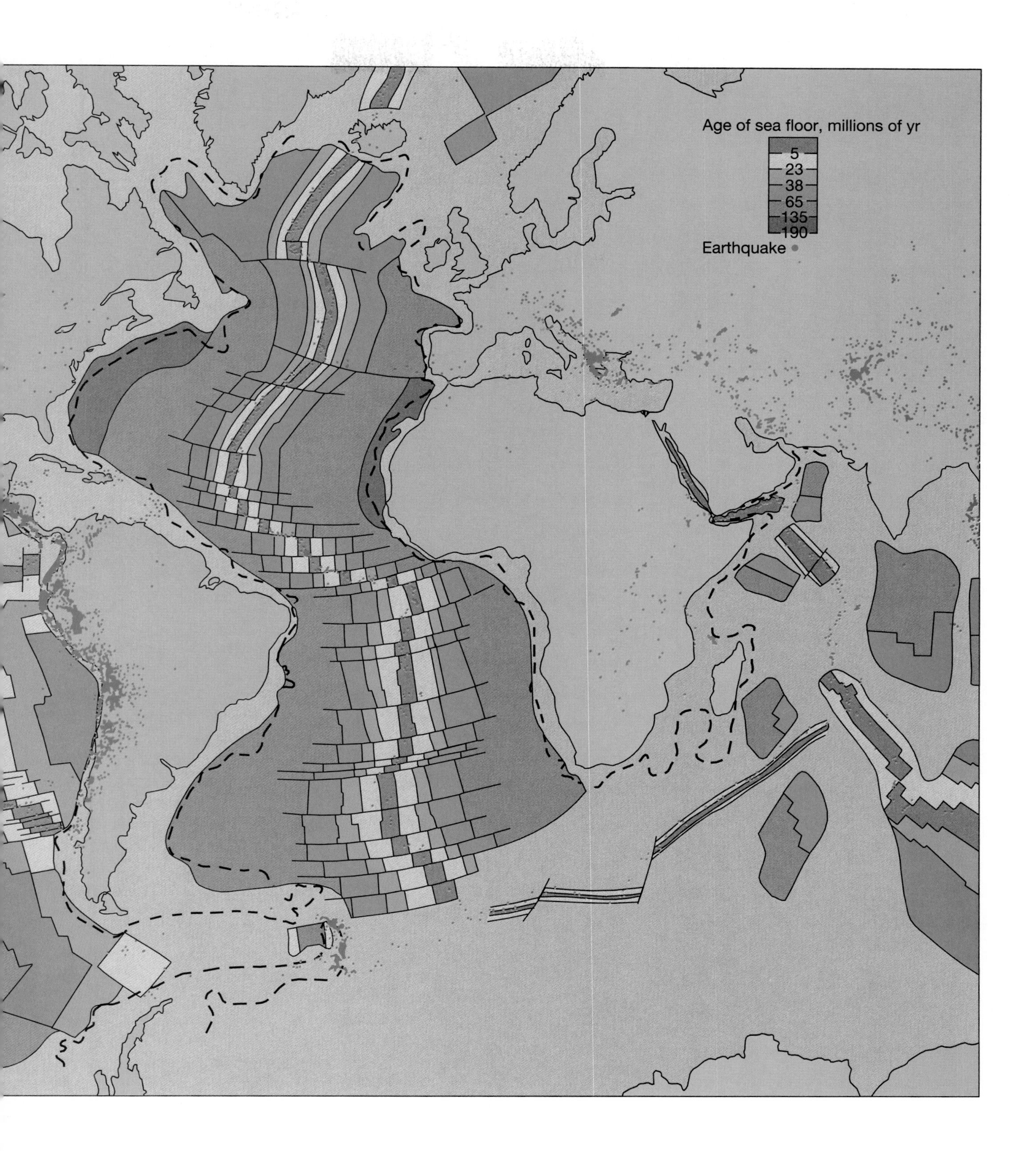
Age of sea floor, millions of yr
5
23
38
65
135
190
Earthquake

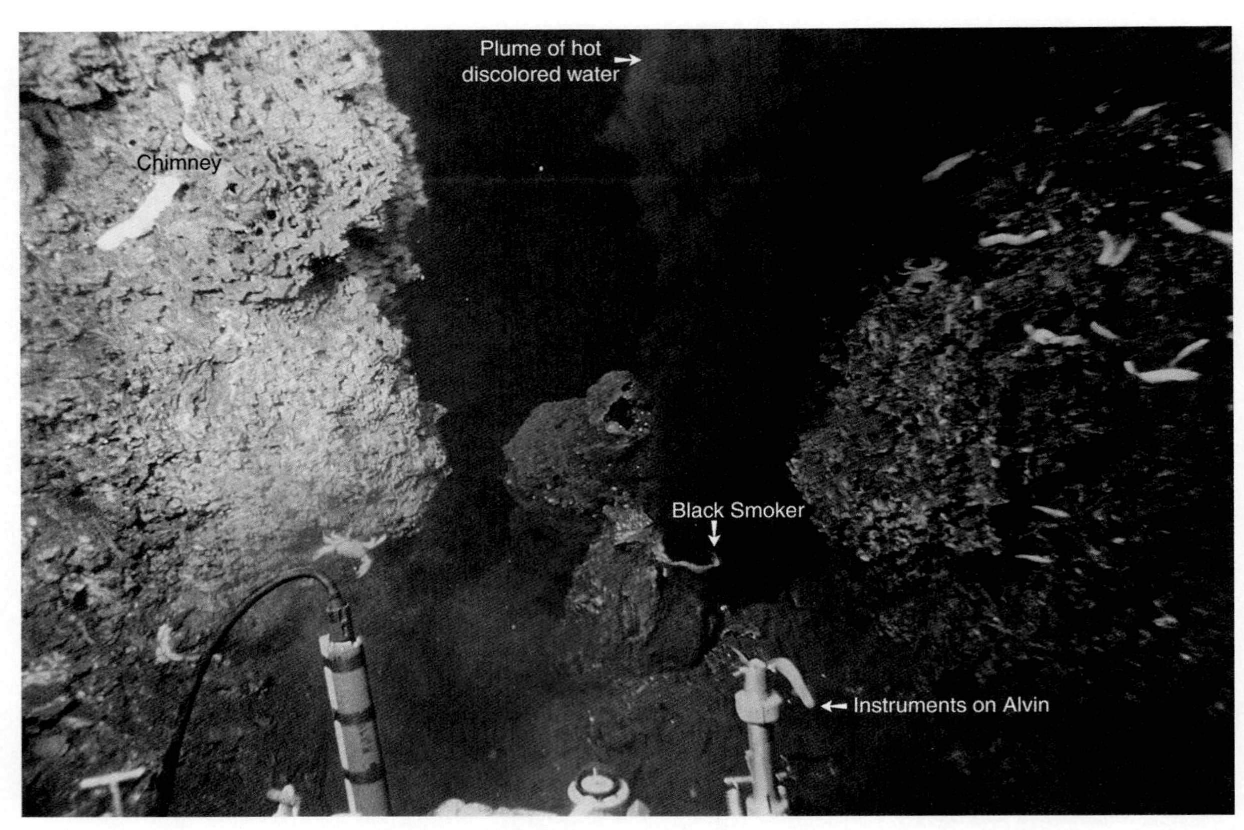

Figure 3–13
A black smoker on the East Pacific Rise. (Dudley Foster/Woods Hole Oceanographic Institution.)

rocks and picking up heat along the way. The hot water then flows out through crevices and vents on the ocean floor and quickly mixes with overlying water. The total flow of this **hydrothermal circulation** is immense, estimated at about 0.5 percent of the world's annual river flow. Another way to look at this circulation is to realize that all the water in the ocean circulates through newly formed crust every 5 to 10 million years. (Thus, the entire volume of the ocean has flowed through recently formed ocean floor many hundreds of times during Earth's history.)

These outflows of hot water through crevices and vents occur in areas of recent volcanic eruptions. Only limited parts of any mid-ocean ridge are active at any time. Volcanic eruptions at any location are separated by a few years to a few thousand years, depending on local spreading rates. Volcanic eruptions occur more frequently on rapidly spreading segments than on slowly spreading ones.

Three types of vents have been observed by scientists in submersibles. Most spectacular are *black smokers* (Fig. 3–13). They discharge superheated waters (300 to 400°C) at high flow rates, much like a fire hose. Because of their high temperatures, these waters are less dense than seawater; thus they rise, forming large plumes. These plumes are black because of chemical reactions that occur in the waters, forming sulfur-bearing minerals. Black smokers very rapidly form large, fragile, chimneylike mounds up to 10 meters high and made of porous silica (a glasslike substance), elemental sulfur, and sulfur-bearing minerals. The sulfur-bearing minerals color the mounds with yellows and blacks, like a Halloween decoration.

Cooler vents called *white smokers* are also common. These vents discharge waters between 25° and 250°C because the circulating fluids have mixed with cold ocean waters. Least spectacular is a third type of discharge where relatively cool (5° to 25°C) waters flow out through cracks and fissures in the ocean floor. These

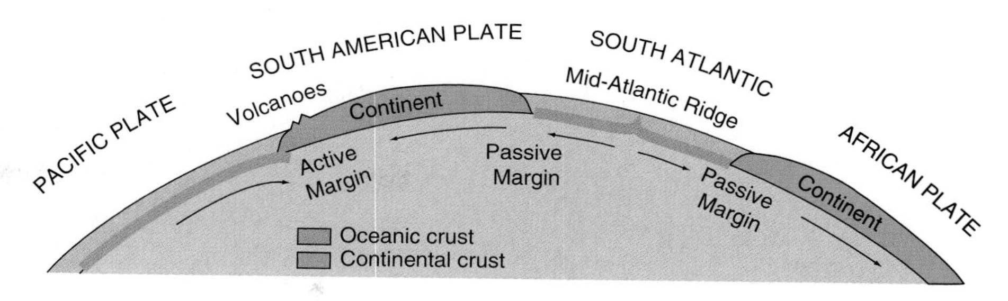

Figure 3–14
Continents move with the crustal plate around them, going away from the mid-ocean ridge toward trenches. Continental margins bordered by trenches are called active (or Pacific-type) margins. Such margins have many earthquakes and active volcanoes. The opposite continental margins are called passive (or Atlantic-type) margins; they have no earthquakes or volcanic activity. The plate boundary nearest the trench is also called the leading edge of the plate; the other margin is called a trailing edge.

waters are only a few degrees warmer than surrounding ocean waters. All three types of vents support abundant growths of bottom-dwelling organisms (discussed in Chapter 13).

Hydrothermal circulation continues for millions of years as the rocks cool. Eventually, fractures fill with mineral deposits and can no longer pass fluids. Furthermore, sediments accumulate on the ocean floor, covering areas where seawater flowed into the rocks. These processes eventually seal the rocks and stop the flows of hydrothermal fluids at that location.

Active Margins

On a plate that carries a continent, one of its continental margins moves away from a mid-ocean ridge while the opposite margin moves toward a subduction zone (Fig. 3–14). The continental margin moving toward a subduction zone is called either an **active margin** or a **Pacific-type margin** and is marked by active volcanoes, many earthquakes, and young mountains. Oceanic crust and overlying sediment are partially melted at an active margin as the plate is pulled down into the mantle; the remainder of the material sinks into the mantle. Island arcs are formed by the many volcanic eruptions. Through many millions of years, continents grow at active margins (Fig. 3–15) through volcanic eruptions and through accumulation of oceanic crust and other ocean-bottom structures welded onto the continental plate.

Earthquakes are common near Pacific-type margins. Shallow earthquakes are caused by the movements of the rocks sliding past each other along faults or fracture zones. The larger the rupture, the more energy released and the larger the earthquake. Deep earthquakes (100 to 660 kilometers below the surface) are characteristic of these margins. Deeper than around 700 kilometers, mantle rocks are too hot and, therefore, too plastic to fracture as they do in shallower earthquakes. The specific mechanisms involved in the deepest earthquakes remain a mystery. These earthquakes may be caused by changes in mineral structures under the high temperatures and pressures of these great depths.

In a subduction zone, the upper surface of the lower plate drags against the bottom surface of the upper plate, with the result that there is a great deal of friction at the interface. It is in such regions, at the top and bottom of subducting plates or on transform faults, that most earthquakes occur. The strains accumulate in the rocks as the plates move. Eventually the rocks fail. The resulting earthquakes release enormous amounts of energy as rocks break along narrow fault zones and then move for a few tens of seconds. These fault zones are marked by crushed rock fragments. After the stored energy is suddenly released, the rocks stop moving, but again accumulate more strain as plate movements continue. Earthquakes occur at irregular and unpredictable intervals.

At an active margin, the sinking plate and the materials on it melt as they sink into the underlying hot mantle. Some of the material on the plate melts and rises as magma to erupt on the volcanic islands that make up island arcs. In some circumstances, subduction can cause formation of new oceanic crust as a result of continued volcanic activity on small spreading centers.

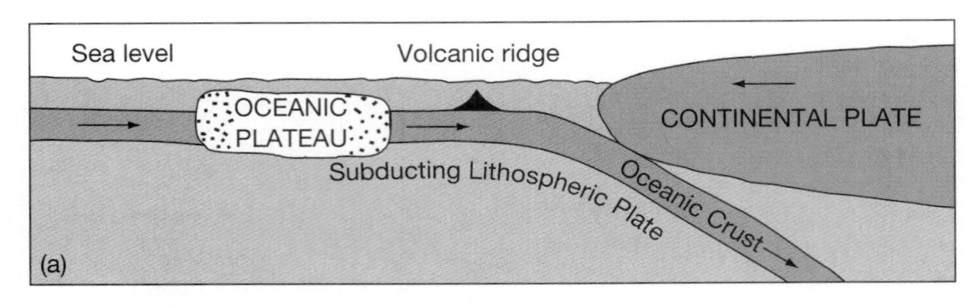

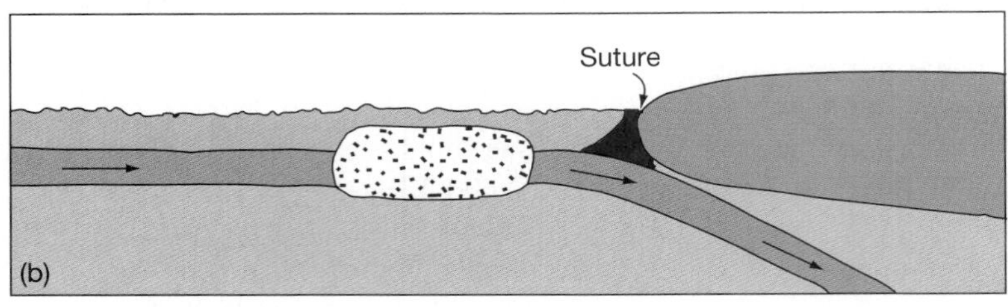

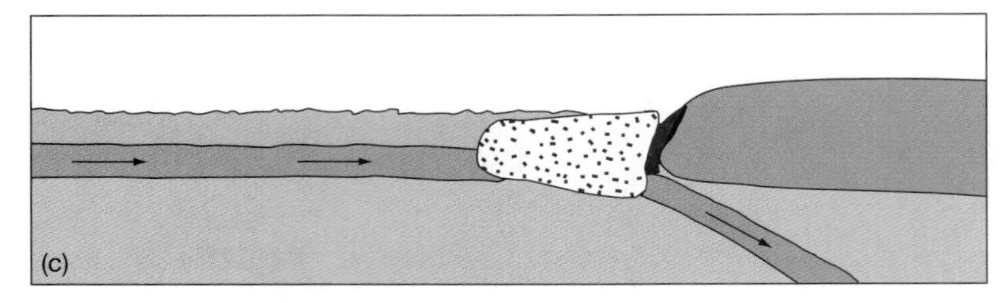

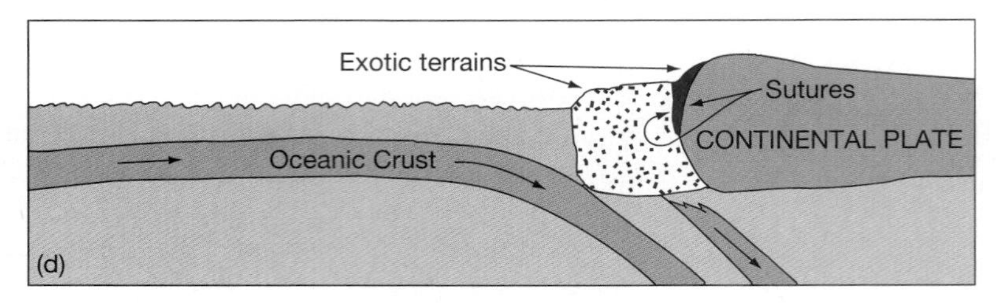

Figure 3–15
Continents grow at trenches or convergent margins by accreting oceanic plateaus, volcanoes, and other ocean-bottom structures. These are scraped off oceanic crust as it is subducted. These scraped structures are then welded to the continent on the overlying plate.

When the subducting plate is old and dense, it sinks into the mantle as a steeply dipping slab (Fig. 3–16). Such subduction occurs widely in the western Pacific basin, where the oldest oceanic crust occurs. In several of these subduction zones volcanic activity occurs in small spreading centers, called *back-arc spreading centers,* where new oceanic crust is formed (Fig. 3–16). The Mariana (Western Pacific) and Scotia (South Atlantic) island arcs both have back-arc spreading centers and actively forming small basins behind the island arcs.

Where the subducting crust is young and therefore still warm and relatively buoyant, the slab dips at a shallow angle (Fig. 3–17). This occurs along the eastern margin of the Pacific, near the East Pacific Rise, where the American plate is overriding recently formed crust of the Pacific plate. Volcanoes occur on land where young crust is subducting, and there are many earthquakes along the Pacific coast of North and South America for this reason.

The land near a subduction zone is elevated because it is underlain by the subducted plate. In addition to the volcanoes, young mountain ranges form from materials scraped off the subducting plate (Fig. 3–17). Both the chains of volcanoes and the young mountain ranges parallel the coastline.

Off Central America and southern Mexico, the Middle American Trench marks the location of the Cocos Plate subduction zone. Off the northwestern United States and southwestern Canada, the comparable subduction zone (where the Gorda Plate is subducted beneath the North American Plate) is buried by the large volumes of sediments derived from the land.

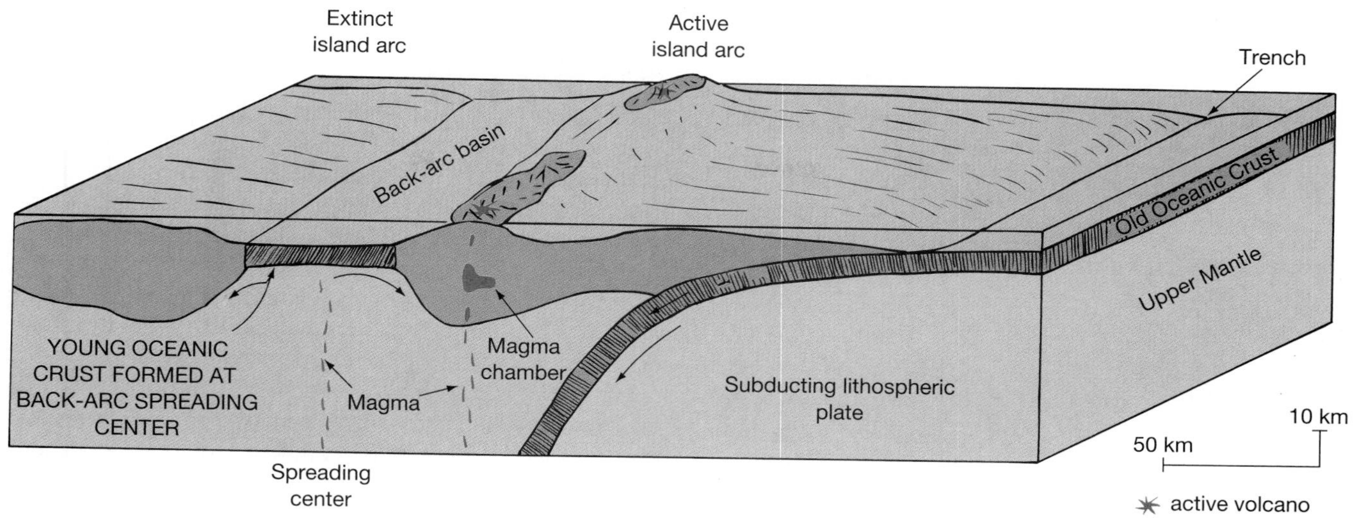

Figure 3–16
Subduction of old, dense oceanic crust exhibits steeply dipping (Mariana-type) subduction zones. Note that volcanic activity forms chains of volcanic islands, called island arcs. Volcanic activity also occurs on small spreading centers (called back-arc spreading centers), where young oceanic crust forms between the active island arc and older, extinct arcs or adjacent continents.

Small volcanoes on a subducting oceanic plate may be broken into smaller blocks and drawn down into the subduction. Larger volcanoes, volcanic ridges, and oceanic plateaus, however, are often too large for this to occur (Fig. 3–15). They are instead welded to the continent, forming an **exotic terrain** (so called because they came from somewhere else and were not originally formed in their present location). Many such blocks of former oceanic crust have been identified in North America from Alaska to Mexico. Other Pacific-type margins doubtlessly

Figure 3–17
Subduction of young, buoyant oceanic crust occurs on shallowly dipping subduction zones. Such subduction zones are typical on the Pacific coasts of North and South America. Note the thick wedge of deformed sediment deposits that is scraped off the subducted plate and accreted to the continent.

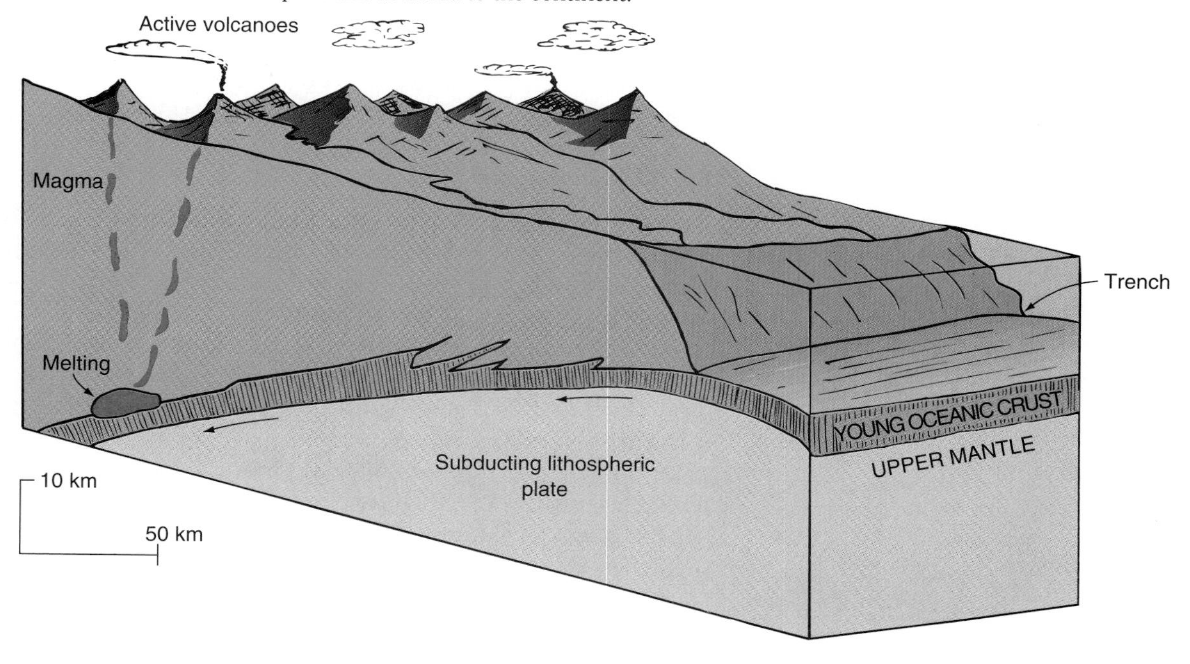

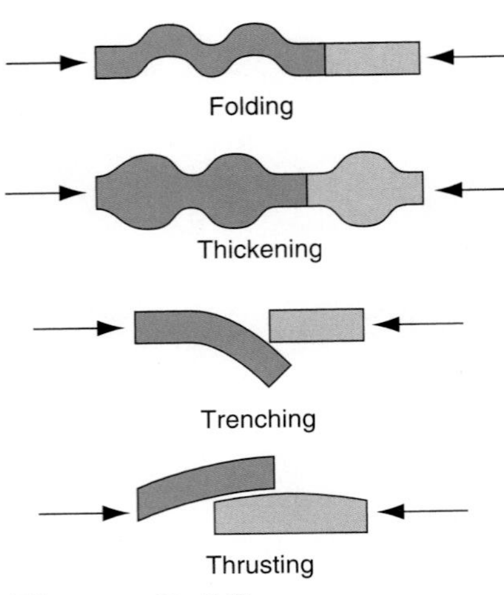

Figure 3–18
Mountain chains form as continents collide.

have similar blocks, but they have not yet been studied. Indeed, most continents consist of many such blocks, large and small, that have been welded together at ancient subduction zones.

Some pieces of ocean floor welded onto continents as exotic terrain contain deposits of sulfide minerals formed at spreading centers. These deposits have been mined for copper, lead, zinc, and silver since antiquity. Indeed, such deposits on the Arabian Peninsula provided King Solomon's wealth in biblical times. Similar deposits are still mined for copper on the island of Cyprus in the eastern Mediterranean.

When two continents collide (Fig. 3–18), the processes are different from what happens when oceanic plates collide or when an oceanic plate collides with a continental plate. Because in a continent–continent collision the rocks on both sides of the convergence zone are relatively light (low density), they are not easily subducted. Instead, the plates respond by folding and by thrusting slices of one plate on top of the other. This overriding and uplift is now happening as India collides with Asia, with the result that the Himalayas get taller every day. (More about this later in this chapter.)

The other possibility in a continent–continent collision is folding and thickening of the crust. This is happening in Turkey and nearby countries, where Africa and Asia are now colliding. The processes occurring at such margins are still poorly known.

Passive Margins

On a continent-bearing plate, the continental margin moving away from the mid-ocean spreading center is called either a **passive margin** or an **Atlantic-type margin,** and here there is no mountain building. They also occur in the interior of lithospheric plates and are usually formed when a continent splits early in a spreading cycle, which we discuss in a later section in this chapter. Thus, mountains near passive margins are usually very old, formed during previous spreading cycles. The Appalachian Mountains near the Atlantic margin of North America are one example. The rocks that folded during an earlier mountain-building cycle have been elevated and later eroded by other processes, not modern mountain building.

Thick sediment deposits and old oceanic crust are typical of passive margins (Fig. 3–19). The deposits usually cover the boundary between oceanic and continental crust. They also record the early history of ocean-basin formation and various changes in sea level.

Passive margins are important economically because oil and gas often accumulate in the thick sediments. Most of the world's largest oil and gas fields occur in such deposits, and most areas now being explored for oil and gas are on passive continental margins.

Mantle Convection

Lithospheric plates move because of **mantle convection,** which extends deep into Earth's interior (Fig. 3–20). The mantle convects, or overturns, because it is warmed at the core–mantle boundary; this warmed portion then rises to the surface, where it cools and sinks. The plates, because they float on the molten asthenosphere part of the mantle, move as a result of these convection currents. The energy causing mantle convection comes from the core (probably the inner core), which is still cooling. Mantle convection and associated plate movements are the primary means by which Earth cools.

Hot-spot volcanoes originate at unusually hot areas in the core–mantle boundary. There the mantle melts, forming plumes of magma that rise and penetrate the crust as volcanoes. These spots do not move with the mantle and, thus, remain fixed in location for tens of millions of years. (We have already seen how it is possible to reconstruct plate movements where chains of volcanoes formed as plates move over hot spots. See Fig. 3–9.)

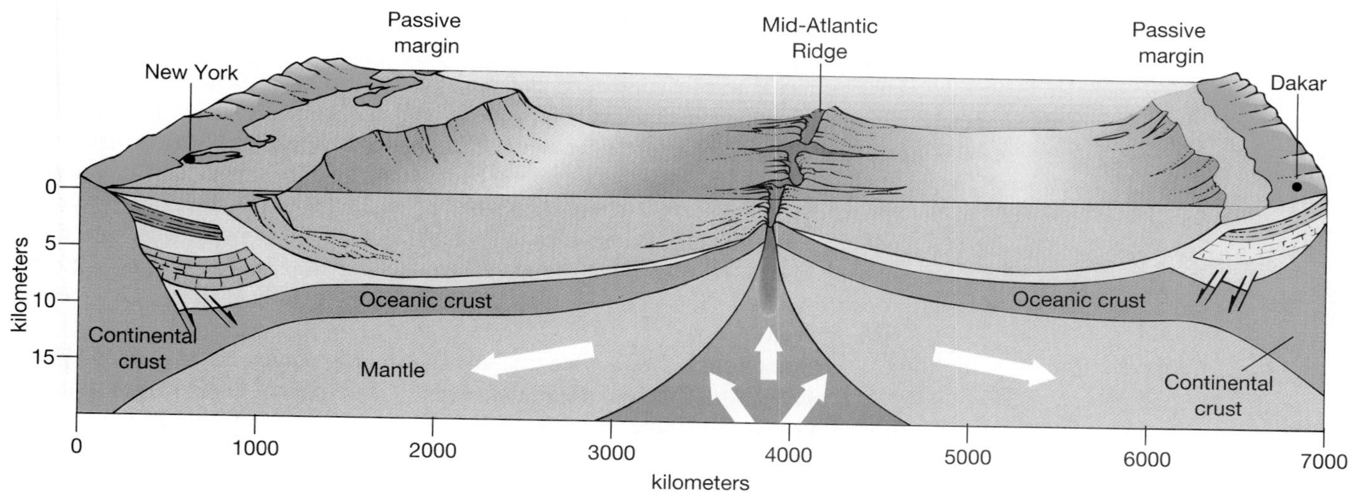

Figure 3–19
Passive or Atlantic-type margins form when a mid-ocean ridge breaks apart a continent and separates the continental fragments. Note that thick sediment deposits cover the transition between continental and oceanic crust.

Figure 3–20
Mantle convection causes the movements of lithospheric plates. The upper and lower mantles behave almost independently. Some of the subducting plates sink through the lower mantle and come to rest on the core–mantle boundary. Some of these materials return to the surface in the plumes of molten rock from the hot spots.

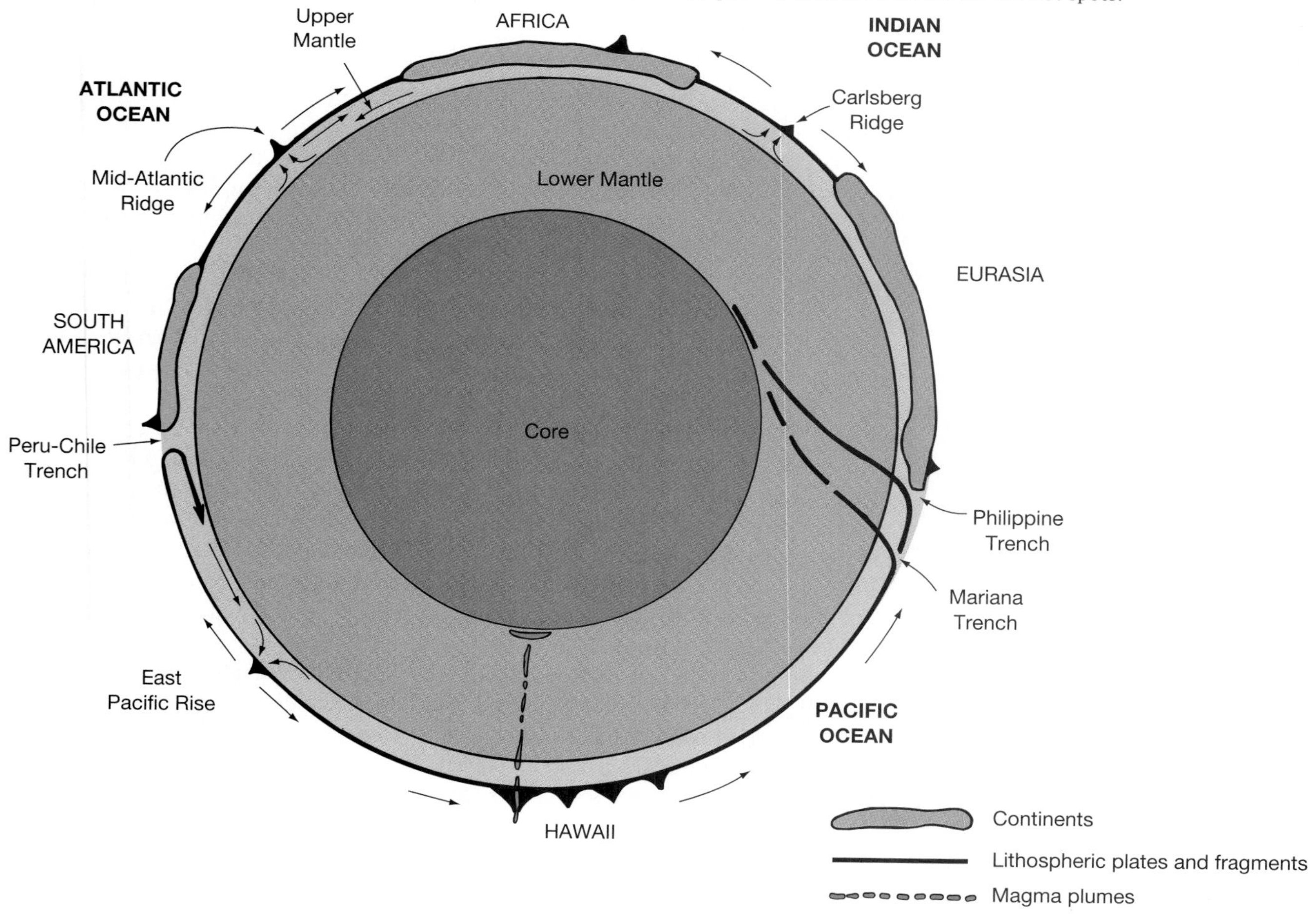

The mantle is cooled at the top by volcanic eruptions at mid-ocean ridges and hot spots, by seawater circulating through newly formed crust, and by heat conduction through the ocean bottom. The lithospheric plates thicken as the crust and upper mantle cool. When the cool, dense lithospheric plates sink at trenches, this too cools the mantle. Fragments of these subducted plates sink through the mantle; some eventually come to rest on the core–mantle boundary, which is much denser than areas higher up in the mantle. The plate fragments remain there for millions of years as they warm and their density decreases. Eventually some of the materials in these fragments may rise in the plumes of molten rock from the hot spots and begin the cycle again. (Details of these processes are still unclear.)

These plumes of rising magma and the sinking plates can be imaged by *seismic tomography.* This technique uses seismic waves from earthquakes that pass through Earth, the waves being detected by seismographs around the world. After analyzing how long it takes a wave to travel from the earthquake point of origin to a seismograph, computers map the regions of warm rock (having slower wave speeds) and cold rock (higher wave speeds). These maps show the rising and sinking portions of the mantle.

Movements in the mantle appear to be coupled to the source of Earth's magnetic field—the fluid outer core. At its deeper boundary, this sphere is heated by the hot inner core, and at its upper boundary it is cooled by dense lithospheric plate material that has sunk to the bottom of the mantle (Fig. 3–20). With increased ability to map the core–mantle boundary, it may someday be possible to determine what causes magnetic reversals.

Formation and Destruction of Ocean Basins

Ocean basins form through the breakup of continents (Fig. 3–21). This process begins when a continent remains stationary for several hundred million years. Thick continental crust impedes heat flow from Earth's interior. Eventually the excessively hot mantle under a static continent expands enough to uplift the overlying lithosphere. This is happening now in Africa, which has apparently remained in its present location for 100 to 300 million years. It now stands several hundred meters higher than the other continents, which have moved during this period.

Continued uplift stretches the overlying continental crust, as it is now doing in the East African rift valleys. The crust eventually breaks, forming narrow, fault-bounded valleys. In humid climates, these valleys fill with fresh water, forming deep lakes, such as Lakes Malawi and Tanganyika.

As valley floors subside (actually the crustal fragments on either side of the rift valley rise during the rifting process, leaving the valley floor at a lower level), they fill with sediment eroded from the steep valley walls and adjoining mountains. Eventually, as rifting continues, the valley floors subside to sea level and connections form with the existing ocean. At this point, freshwater lakes become long, narrow oceans as the continental fragments separate more and more. Modern examples of this stage of ocean-basin formation are the Red Sea and the Gulf of California.

The long, narrow seaway impedes ocean circulation and favors evaporation in the basin. High mountains on both sides cause winds to blow along the length of the basin. The uplifted basin sides divert rivers away from the basin, so that little or no fresh water flows into it. Thick salt deposits may form if the basin is partially cut off from the sea so that seawater cannot easily flow into it.

As spreading of the uplifted sides continues, the ocean basin widens. This is happening now in the Atlantic, which has been growing for approximately 190 million years.

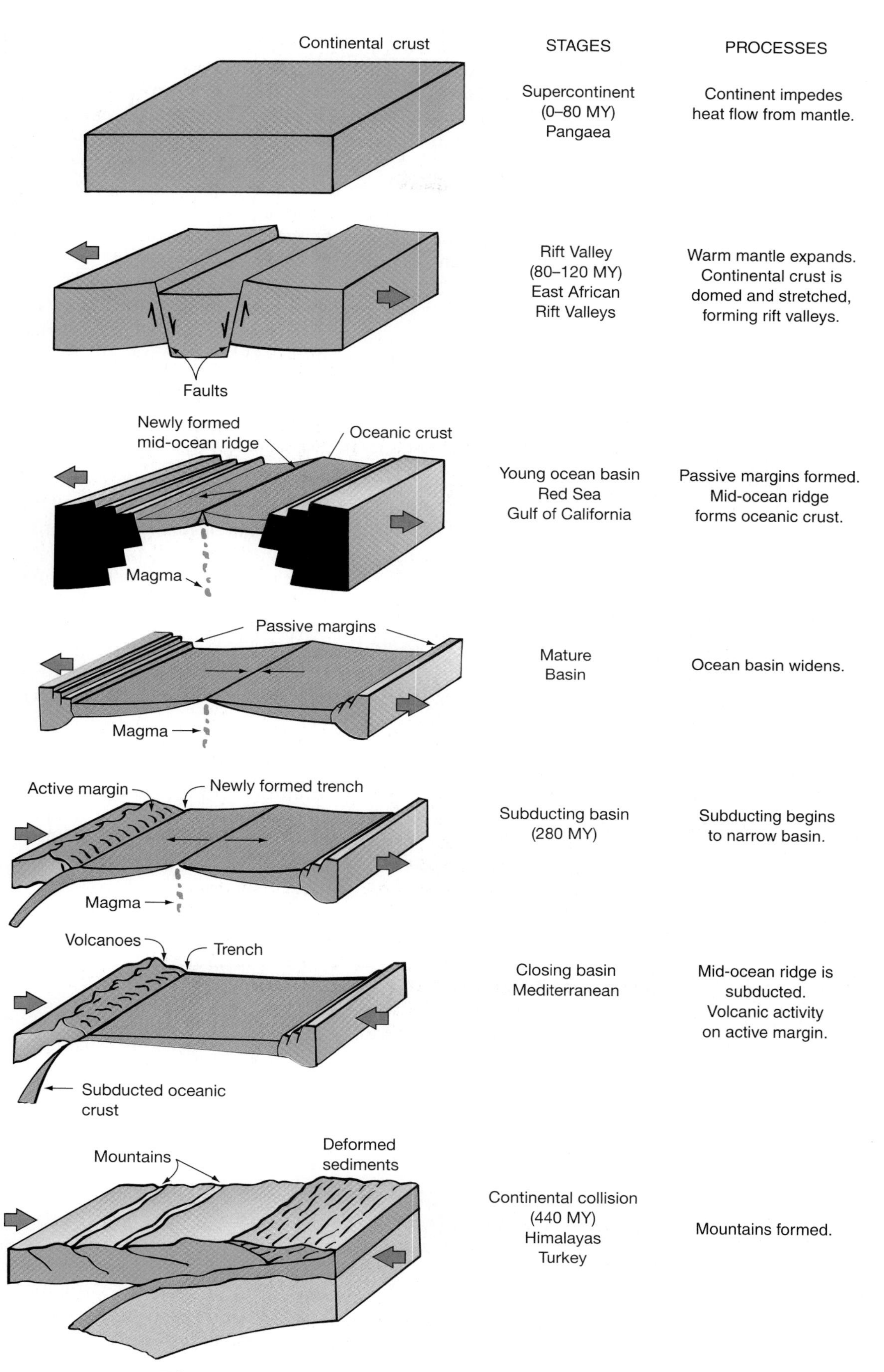

Figure 3–21
Schematic representation of the cycle of ocean basin formation and destruction.

Meanwhile the oceanic crust of the basin ages, cools, and grows denser. Eventually it becomes so dense that it sinks into the underlying mantle, starting a new cycle of subduction, as is now occurring in the West Indies. There Atlantic crust is subducting under the Caribbean plate. Similar processes are occurring in the South Atlantic in the South Sandwich Trench.

Through continued subduction, the ocean basin narrows, and eventually the two sides collide. In the process, the ocean disappears, and a mountain range forms where sediment deposits are squeezed between the two continental blocks. During the collision, sediments and large pieces of ocean floor are attached to the continental crust. In this way, continents grow by incorporating large pieces of ocean floor, oceanic plateaus, and fragments of former continents.

Each cycle takes about 400 million years. Thus, there have been about ten cycles in the roughly 4 billion years since Earth's surface solidified. Because we know most about the present cycle, we briefly discuss its history and its effects on each ocean basin.

The Present Spreading Cycle

The present cycle began about 200 million years ago. At that time there was a single land mass, or supercontinent, called *Pangaea,* after the Greek earth-goddess (Fig. 3–22). A single ocean, *Panthalassia,* the ancestor of the Pacific, covered the rest of Earth.

About 180 million years ago, Pangaea began to break up, initially forming a long east–west-trending basin called *Tethys* (Fig. 3–22a), partially separating Pangaea into two large pieces, called *Laurasia* and *Gondwanaland.* About 190 million years ago, the Mid-Atlantic Ridge developed forming narrow basins that eventually became the North and South Atlantic basins as they widened (Fig. 3–22b). This rifting of Pangaea eventually formed the three major ocean basins we have today (Fig. 3–22c).

The former Tethys area is now occupied by the Mediterranean and Black Seas. Land areas around the former Tethys area include Saudi Arabia, Kuwait, and other Middle Eastern countries, which contain about two-thirds of the world's oil. The basin was unusually well situated to form oil, and the reefs that grow around its margins provide excellent reservoirs. The Arctic basin was also positioned to form large amounts of petroleum and is now a promising area for future oil and gas exploration.

History of Ocean Basins

Each ocean basin on Earth today formed during the present spreading cycle by the processes just described. Let's now look at these basins to see how their individual histories have given them their present shapes and how these shapes have affected the rest of Earth.

Pacific Ocean

The Pacific is the oldest ocean, a remnant of earlier crustal spreading cycles. Its basin is nearly surrounded by subduction zones (Fig. 3–23); consequently, its size has decreased markedly since the present cycle of seafloor spreading began. Trenches are especially prominent in the North Pacific. The associated volcanic activity and earthquakes give the region its nickname of the "Rim of Fire." Another conspicuous feature are the several large fracture zones in the eastern North Pacific but also occurring in the South Pacific. These fracture zones show

Figure 3–22
Movements of lithospheric plates during the present spreading cycle. About 200 million years ago, the land formed one supercontinent called Pangaea (a). Early in the cycle, an east–west spreading center separated Laurasia on the north from Gondwanaland on the south. Remnants of that sea form the Mediterranean and Black Seas. About 180 million years ago, the north–south Mid-Atlantic Ridge formed, separating the Americas from Africa and Europe. The initially narrow basins eventually widened, forming the North and South Atlantic basins, seen here about 65 million years ago (b) and today (c). (After R. S. Dietz and J. C. Holden, "Reconstruction of Pangaea: Breakup and Dispersion of Continents, Permian to Present," *Journal of Geophysical Research,* 75:4939 [1970].)

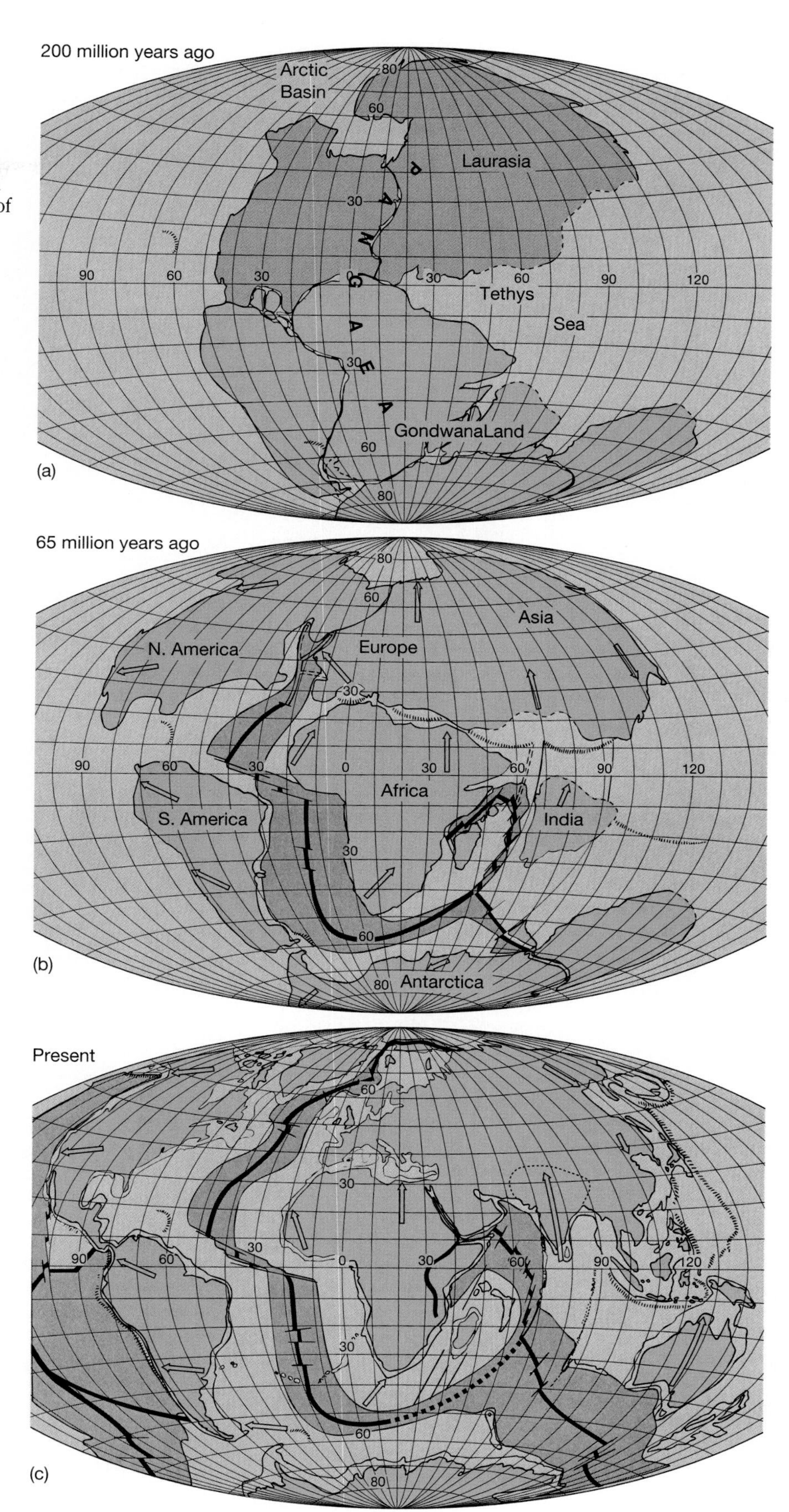

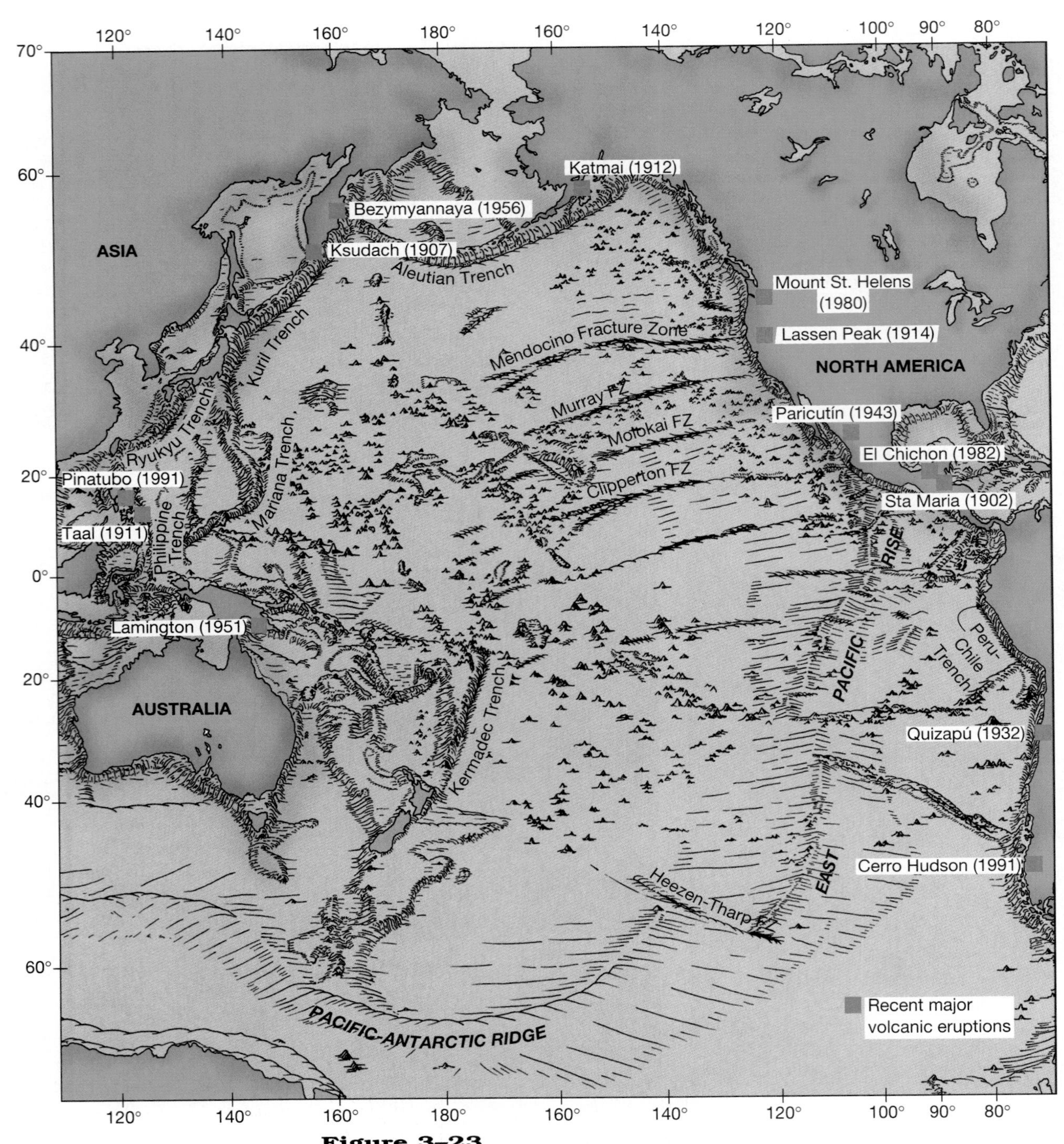

Figure 3–23
The Pacific Ocean basin is dominated by the East Pacific Rise, especially in the south; large fracture zones and mountainous topography dominate the western Pacific. Trenches occur around the Pacific margins except off North America.

the dominant direction of plate movement in the Pacific during earlier phases of the present spreading cycle.

The complicated ocean bottom between Australia and Asia is the result of the continuing collision between Asia and the northward moving Australia. The areas between many of the trenches form small microplates, which have had complicated histories, controlled by the movements of the larger plates around them.

The East Pacific Rise, a relatively young feature, is conspicuous in the South Pacific. The associated fairly simple and smooth bottom topography dominates the eastern portion of the South Pacific. The western portion of the basin is much more rugged. This mountainous and deep part of the basin is the oldest oceanic crust. The many western Pacific volcanoes formed during a time of extremely high levels of volcanic activity, about 60 million years ago.

The Hawaiian Islands and the Emperor Seamounts record two spreading directions for the Pacific plate as it moved over the Hawaiian hot spot (Fig. 3–9). An ancient north–south movement formed the Emperor Seamounts, and then the Hawaiian Islands formed during the present northwest–southeast movement. The change in spreading direction occurred about 40 million years ago. Many of the South Pacific Island chains also record these plates' movements and their changed directions. However, the many different islands obscure the relationship, shown in the Hawaiian Islands and the Emperor Seamounts.

Atlantic Ocean

The Atlantic is expanding east–west under the influence of the Mid-Atlantic Ridge (Fig. 3–24), which dominates the seafloor topography. Several large fracture zones offset the Mid-Atlantic Ridge. These deep valleys provide channels for waters flowing along the bottom to move through the Mid-Atlantic Ridge. This connection between the basins east and west of the Mid-Atlantic Ridge is especially important in the equatorial ocean where the Romanche Fracture Zone provides a path for deep waters to flow into the deep basins of the eastern side of the Atlantic.

Subduction occurs at two locations in the Atlantic—on the old crust east of the West Indies, a complex volcanic island arc, and in the South Sandwich Trench near Antarctica. Through time, more subduction is expected to begin in the oldest crust near the continental margins. When that occurs, the Atlantic will begin to close and eventually will be completely obliterated, as it was about 450 million years ago when the Appalachian Mountains in the eastern United States were formed.

Thick sediment deposits around the margins of this ocean basin have buried the early history of its formation. Thick salt deposits, which occur in many areas around the Atlantic, including the Gulf of Mexico and the eastern side of the Atlantic, also formed early in the basin's history.

About 100 million years ago, bottom waters in the Atlantic basin were stagnant several times. This stagnation of waters resulted in the bottom waters being depleted of their dissolved oxygen, which in turn meant that bottom-dwelling organisms could not survive to eat the organic matter that continually settles out of surface waters. (We discuss the processes involved in producing and destroying organic matter in Chapter 11.) With nothing eating the organic matter in the sediment, it accumulated as a carbon-rich deposit. These carbon-rich deposits were the source of much of the oil and gas now used as fossil fuels. (More about this in Chapter 5.)

The Atlantic now connects the two polar ocean areas. About 6 million years ago, the Iceland–Faeroe Ridge subsided (Fig. 3–24). This subsidence permitted cold, dense waters from the Norwegian Sea to flow south into the open Atlantic basin.

Indian Ocean

The Indian Ocean is the youngest of the three major ocean basins. It began to form with the breakup of Gondwanaland about 125 million years ago when Africa, Antarctica, India, and Australia separated. India moved north to collide with Asia, leaving conspicuous tracks of its movement on the sea floor.

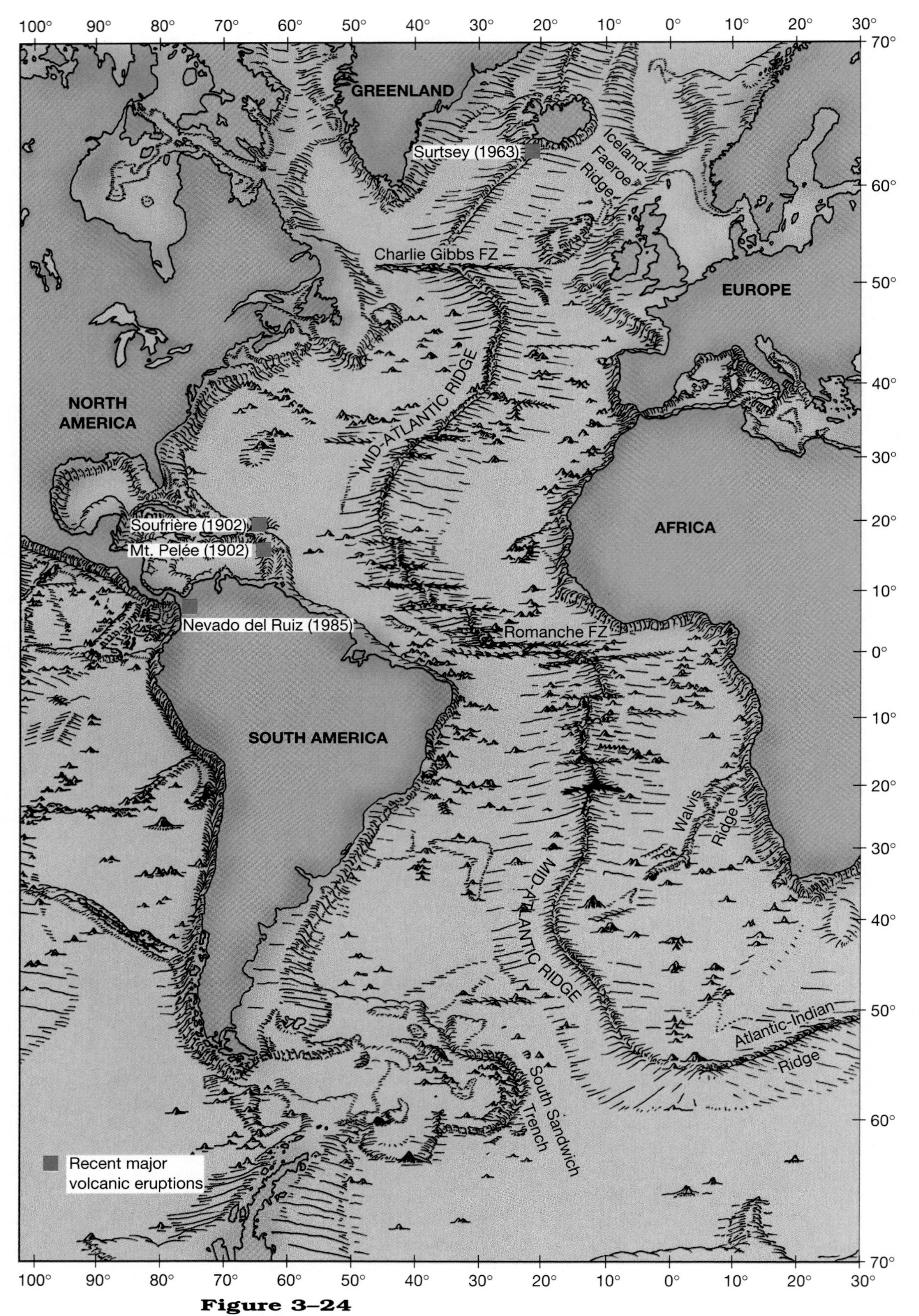

Figure 3–24
The Mid-Atlantic Ridge dominates the ocean bottom in the North and South Atlantic. Large fracture zones cut across the ridge. Subduction occurs in the area east of the West Indies and in the South Sandwich Trench north of Antarctica.

The basin is the shallowest of the major basins because of its young age. It reached its present shape and circulation about 15 million years ago, when Australia moved to its present location.

The Indian Ocean basin is also the most complex. It has three active mid-ocean ridges (Fig. 3–25). Two inactive volcanic ridges (the Chagos-Laccadive Plateau and the Ninetyeast Ridge) formed as India moved northward to collide with Asia. The north–south spreading direction associated with the Southwest and Southeast Indian ridges dominates. The Java Trench marks the subduction of the Indian plate under southeast Asia. The complex island arc of Indonesia is the result of the continued collision of Australia and Asia.

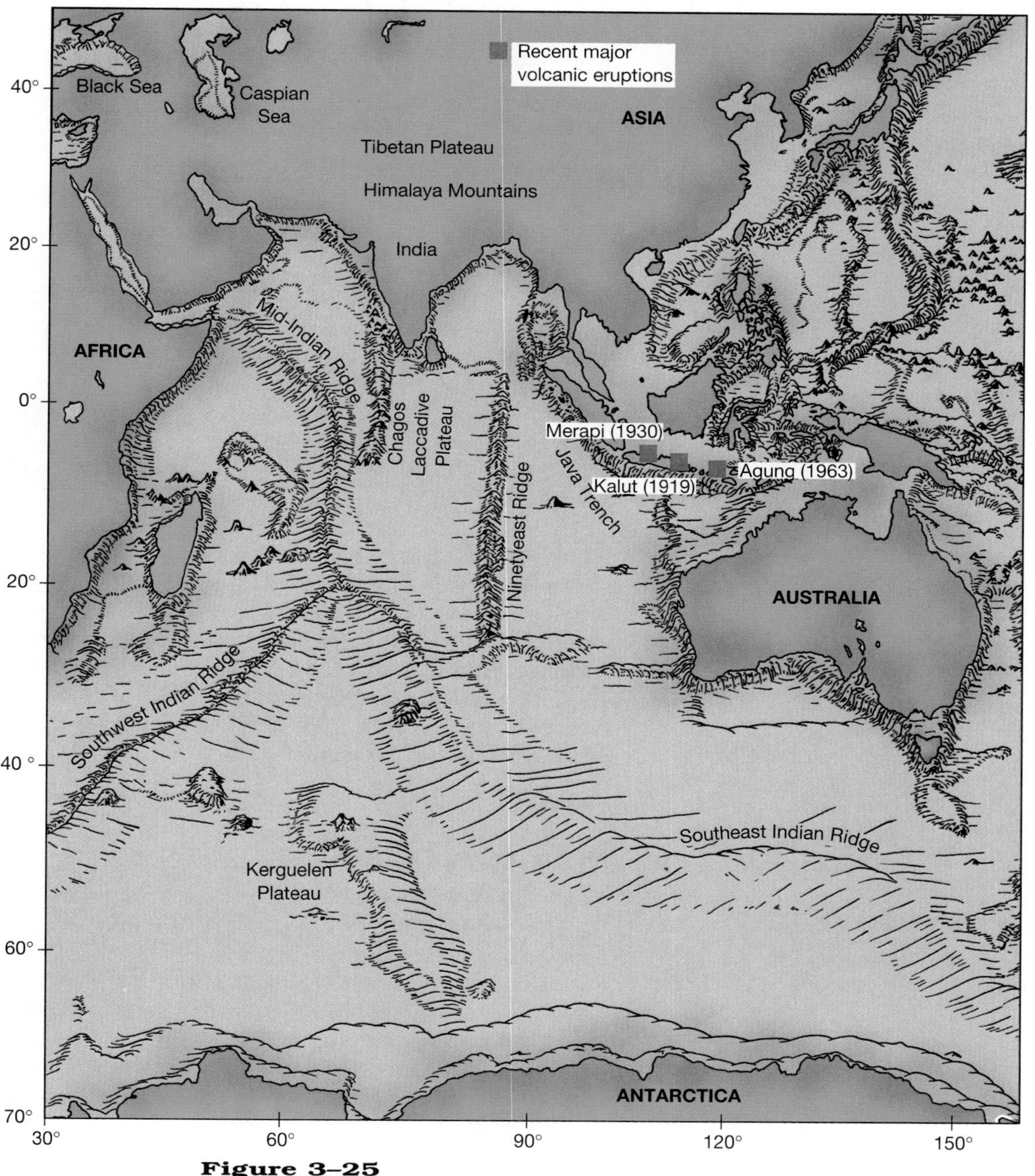

Figure 3–25
The Indian Ocean was formed by the northward movements of India and Australia. The Chagos Laccadive and Ninetyeast Ridges mark the traces of these movements. The mid-ocean ridges around Antarctica control the continued north–south spreading.

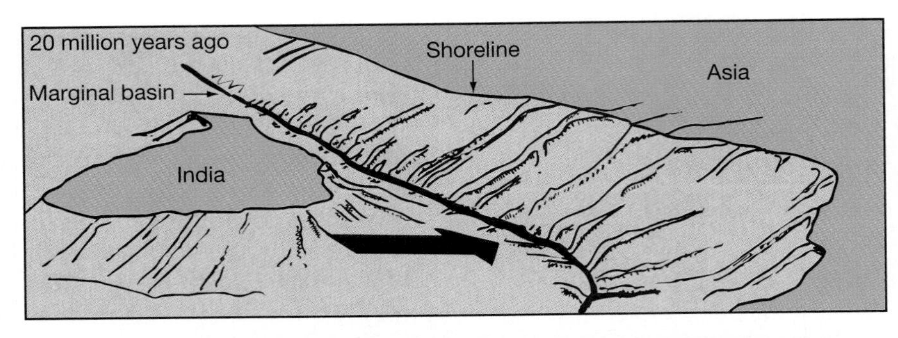

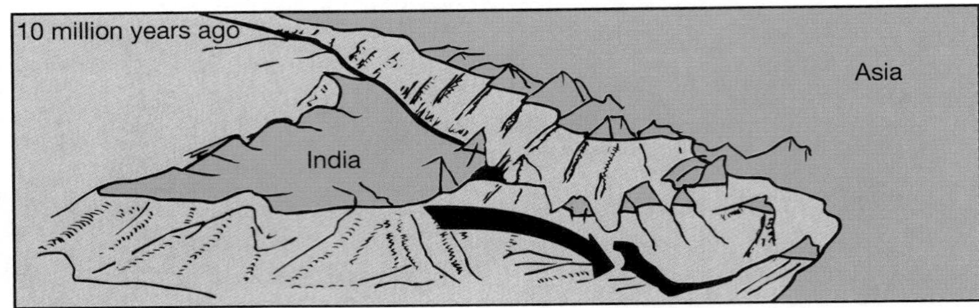

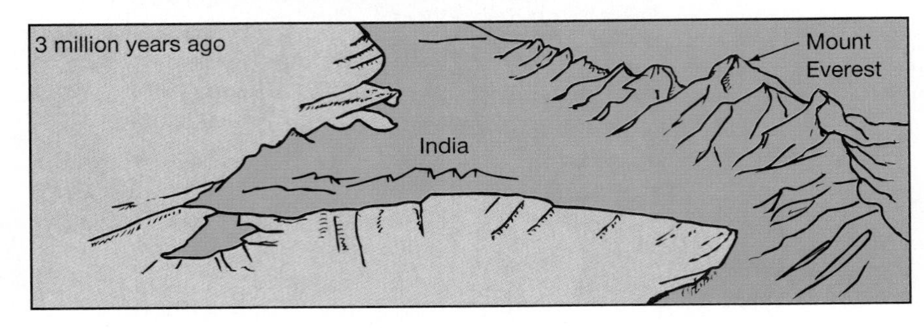

Figure 3–26
About 50 million years ago, the northern edge of the Indian subcontinent collided with Asia and began to underthrust it, forming the Himalaya Mountains.

The Himalaya Mountains are another result of the opening of the Indian Ocean (Fig. 3–26). They began to form about 50 million years ago when India, moving northward, underthrust Asia. This collision eventually formed continental crust about 70 kilometers thick. The former continental slope, now the edge of the Himalaya Plateau, rises sharply above the low-lying region to the south. The Tibetan Plateau, north of the Himalayas, is one of the highest elevations on Earth (Fig. 2–12). As we see in Chapter 7, formation of these high mountains and plateaus affected Earth's climate.

Southern Ocean

The Southern Ocean was formed by seafloor spreading at the mid-ocean ridges that now encircle Antarctica (Fig. 3–27). This basin began forming about 50 million years ago when Australia began moving northward, separating from Antarctica. New Zealand also moved to its present position at the same time. As the narrow seaway between the continental fragments widened, currents could flow between them. The opening of the Southern Ocean was accompanied by the northward movement of both Africa and South America.

Formation of the Southern Ocean over the past 50 million years caused Earth's climate to cool as Antarctica became more isolated from the rest of the land masses on Earth. When the Drake Passage between Antarctic and South America opened about 20 million years ago, the Southern Ocean assumed its

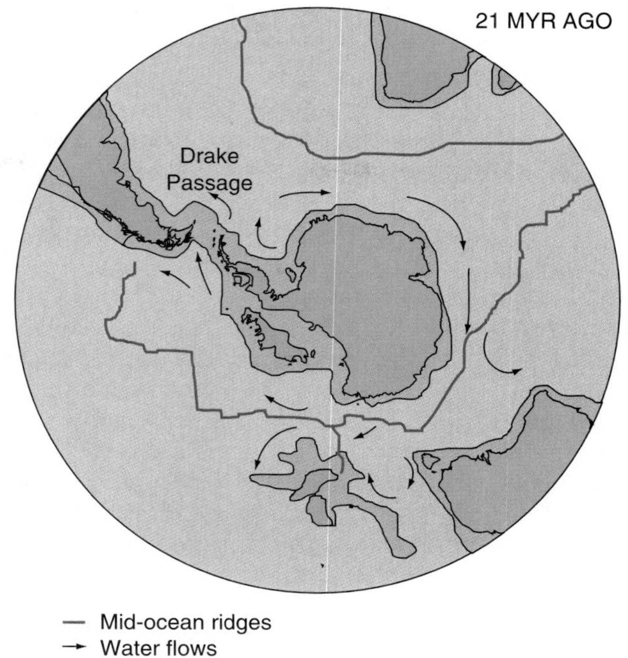

Figure 3–27
The Southern Ocean was formed by north–south plate movements controlled by the mid-ocean ridge that surrounds Antarctica. Opening of the Drake Passage permitted ocean currents to flow unimpeded around the continent.

present form. Beginning at this time, currents, including deep currents, could completely circle Antarctica for the first time. Also, the equatorial currents were blocked. This set of events marked Earth's change from a warm climate to the present Ice Age, and the Antarctic ice sheet formed about 14 million years ago. (We discuss this further in Chapter 7.)

Exploring the Deep-Ocean Floor

A legacy of the Cold War now permits scientists to listen for the sounds coming from the deep-ocean floor; this makes it possible to detect volcanic eruptions as they occur. Scientists can now use submersibles and cameras lowered from ships to explore the deep-ocean floor and study the processes involved in volcanic eruptions soon after they occur. Just a few years ago, most of these processes could only be guessed at, based on observations of similar processes on land.

Perhaps the biggest surprise to come from these studies is the previously unknown forms of life that these scientists discovered. As we shall see in several later chapters, these organisms greatly change our view of how life operates in the ocean. Furthermore, some of them have yielded valuable compounds that work extremely rapidly at high temperatures and pressures.

These studies have also shed light on the processes that form many important mineral deposits on land, especially copper deposits. The early stages of the ore-forming process can be observed directly. Better understanding of how such deposits form greatly improves geologists' ability to discover new deposits.

Summary

Earth is made of concentric spheres. The innermost sphere, the core, is rich in iron and nickel and very dense. The mantle consists of dense rock; the deepest part is rigid, but above that is the partially molten asthenosphere, which can flow very slowly. The outermost sphere, the crust, is Earth's rigid outer shell and includes both continental crust (thick, rich in aluminum and silica) and oceanic crust (thinner, mainly basalt rocks). The crust and underlying upper mantle constitute the lithosphere, Earth's rigid outer shell, which floats on the asthenosphere, a process called *isostasy.*

The lithosphere is broken into rigid plates that move more or less independently of each other. Plate movements form and destroy ocean basins, move continents, and add material to continental blocks. Plate boundaries are mid-ocean ridges (divergent boundaries), trenches (convergent boundaries), and transform faults (where plates slide past each other).

Oceanic crust forms at mid-ocean ridges and moves slowly away as it cools, becomes denser, and gradually sinks lower in the asthenosphere. Mid-ocean ridges are offset by transform faults, which parallel plate movements. Plates move toward trenches, where they are subducted and incorporated into the mantle.

Convection in the mantle supplies molten rock (magma) at mid-ocean ridges. The youngest parts of a plate are elevated at the mid-ocean ridges because of their high temperature and resulting low density. Old crust is cold, dense, and sinks at the trenches. Hot spots are persistent plumes of magma that rise at volcanic centers. Plates moving across hot spots form submarine ridges and chains of volcanic islands.

Earth's magnetic field reverses itself at irregular intervals. Rocks record the magnetic field orientation that existed when they formed at mid-ocean ridges. They retain this record, much like a gigantic tape recorder. Oceanic crust consists of alternating bands of rocks having different magnetic orientations, patterns that can be used to determine when a plate formed.

Seawater circulates through newly formed, hot crustal rocks and thereby removes heat. The hot water discharges in vents in the ocean floor, forming deposits of sulfide minerals.

Fracture zones are bands of irregular topography marking the extensions of the inactive portions of transform faults.

Active (Pacific-type) margins are areas of active mountain building at convergent plate boundaries. They are usually marked by trenches where there are many earthquakes and much volcanic activity. Many active margins, especially in the Pacific, have island arcs. Continental crust is formed at active margins by accreting oceanic plateaus.

Passive (Atlantic-type) margins of continents lie in the interior of plates and have no volcanic or earthquake activity. The boundaries between continental and oceanic crust at these old margins are covered by thick sediment deposits, which are often rich in oil and natural gas.

Ocean basin formation begins when a continent remains in one position for about 100 million years. Thick continental crust impedes heat flow, causing the underlying mantle to heat up and expand. This expansion stretches the continental crust and causes a long, narrow rift valley to form as the crust breaks in two. As spreading continues, the valley widens, forming a narrow ocean basin containing a mid-ocean ridge in the middle.

Newly formed crust is hot and therefore relatively buoyant. As it ages, it cools, becomes denser, and sinks into the mantle, initiating a subduction event that narrows the ocean basin. Finally, when the basin closes, mountains are built at the location of the former ocean basin.

The Pacific is the oldest basin—a remnant of the previous spreading cycle. The Atlantic and Indian Ocean basins are younger, both having formed in the present cycle. The Southern Ocean, the youngest major ocean feature, began forming about 50 million years ago. Its formation has profoundly affected Earth's climate.

Key Terms

plate
core
inner core
outer core
mantle
asthenosphere
lithosphere
continental crust
oceanic crust
Moho
isostasy
divergent boundaries
convergent boundaries
transform-fault boundaries
magma
subduction
triple junction
hot spot
magnetic anomaly
hydrothermal circulation
active (Pacific-type) margin
exotic terrain
passive (Atlantic-type) margin
mantle convection

Study Questions

1. Describe the layered internal structure of Earth.
2. Explain how reversals in Earth's magnetic field are used to determine the ages of specific parts of the ocean floor.
3. Explain the principal features of plate tectonics.
4. What makes lithospheric plates move?
5. Name the three types of plate boundaries and give an example of each.
6. Explain how a hot spot under an ocean basin forms an island arc.
7. Draw a diagram of an active (Pacific-type) continental margin, labeling the most significant features.
8. Draw a diagram of a passive (Atlantic-type) continental margin, labeling the most significant features.
9. Discuss the ways in which hydrothermal circulation affects oceanic crust.
10. Explain isostasy.
11. Describe the cycles of ocean basin formation and destruction.
12. Describe how lithospheric plates move on a spherical Earth.
13. Explain how hydrothermal circulation removes heat from newly formed crust.
14. Describe the characteristic features of the three types of hydrothermal vents.
15. Contrast the angle at which old oceanic crust is subducted with the angle at which young oceanic crust is subducted.
16. Discuss the causes of mantle convection and how it causes plate movements.
17. Contrast the history of the Atlantic and Pacific oceans.
18. *[critical thinking]* Discuss how the history of plate movements might be used to understand global changes in Earth's climate.

Selected References

ALLEGRE, C., *The Behavior of the Earth.* Cambridge, Mass.: Harvard University Press, 1988. Plate tectonics.

BALLARD, R. D., *Exploring Our Living Planet.* Washington, D.C.: National Geographic Society, 1983. Well illustrated.

DECKER, R., AND B. DECKER, *Volcanoes,* rev. ed. New York: W. H. Freeman, 1989. Introduction to volcanic phenomena.

GORE, R., "Living with California's Faults," *National Geographic,* (April 1995), pp. 2–35. Explanation of earthquakes based on plate tectonic theory.

KENNETT, JAMES P., *Marine Geology.* Englewood Cliffs, N.J.: Prentice Hall, 1982. Ocean basin geology.

SULLIVAN, W., *Continents in Motion. The New Earth Debate,* 2nd ed. New York: McGraw-Hill, 1991. Well-written treatment of plate tectonic theory and how it developed.

WEGENER, A., *The Origin of the Continents and Oceans* (trans. from the 3rd German ed. by J. G. A. Skerl). London: Methuen, 1924. Original statement of continental drift.

WOOD, R. M., *The Dark Side of the Earth: The Battle for the Earth Sciences,* 1800–1980. Winchester, Mass.: Allen & Unwin, 1985. Describes the scientific revolution brought about by plate tectonics.

WYLLIE, P. J., *The Way the Earth Works: An Introduction to the New Global Geology and Its Revolutionary Development.* New York: John Wiley, 1976. Introduction to plate tectonics.

WYSESSION, M., "The Inner Workings of the Earth," *American Scientist,* 83:134–147. Review of the principal concepts of plate tectonics and the contributions made by seismology.

OBJECTIVES

Your objectives as you study this chapter are to understand:

- How water's unique properties influence ocean processes
- What controls the composition of sea salts
- How sea salts influence physical and chemical processes in the ocean
- Conservative and nonconservative constituents of seawater
- The carbon cycle in seawater

The unique properties of water permit it to exist as a solid, liquid, or gas within the range of temperatures found on Earth.

4 Seawater

Water is the most common substance on Earth's surface. Most of it—98 percent, to be exact—fills the ocean basins as seawater, a mixture of about 96.5 percent water, 3.5 percent salt, and minute amounts of such other substances as dirt and dissolved organic matter. Originally, seawater and sea salts came mostly from rocks in Earth's interior, ejected during countless volcanic eruptions over the 4.5 billion years of Earth history. An unknown amount of these substances also came from comets that collided with Earth.

In this chapter, we study how water's unique properties influence how seawater behaves and how sea salts are added to and removed from seawater.

The Molecular Structure of Water

Water has many unique chemical properties (Table 4–1). Because these properties influence both the ocean and the marine organisms living in it, we begin our discussion of seawater by examining the reasons for water's uniqueness. We start with its molecular structure—in other words, with what holds it together. Here we are considering processes that occur on atomic scales in space and in tiny fractions of seconds in time.

As you know, molecules are made up of atoms, and every atom contains a nucleus that carries a positive electrical charge and a number of negatively charged electrons surrounding the nucleus, as Figure 4–1 shows. It is attractive forces between atoms that hold molecules together. These forces are called *chemical bonds,* and there are four kinds of them. The strongest,

TABLE 4–1
Some Properties of Water and Their Effect on Seawater and the Ocean*

Property	*Comparison with Other Substances*	*Importance in Ocean*
Heat capacity	Highest of all liquids except ammonia	Prevents extreme temperatures; heat transfer by currents is large
Latent heat	Highest except ammonia	Acts as thermostat; important in transferring heat and water to atmosphere
Surface tension	Highest of all liquids	Controls drop formation and such surface phenomena as capillary waves
Dissolving power	Dissolves more substances and in greater quantities than any other liquid	
Transparency	Relatively high	Absorption of infrared and ultraviolet radiant energy is large; absorption of visible-light radiant energy is small, making water "colorless"

*Modified after Sverdrup, H. V., M. W. Johnson, and R. F. Fleming, 1942, *The Oceans.* Englewood Cliffs, N.J.: Prentice-Hall.

covalent bonds, form when each atom in a molecule retains its electrons but shares them with adjacent atoms. When molecules having covalent bonds separate into their constituent atoms, each atom retains its own electrons, as Figure 4–2 shows.

Bonds called **ionic bonds** form when one atom gives up one (or sometimes more than one) of its electrons and this electron is captured and held by another atom. Both atoms are then called **ions.** The one that gave up electrons is a positive ion, the one that gained electrons is a negative ion, and ionic bonds result from mutual attractions between these opposite electrical charges. One example of ionic bonding is found in common table salt (NaCl), which is formed when sodium (Na) ions combine with chlorine (Cl) ions (Fig. 4–3). When the atoms in such a molecule separate, each will gain a positive charge (having lost an electron) or a negative charge (having gained an electron).

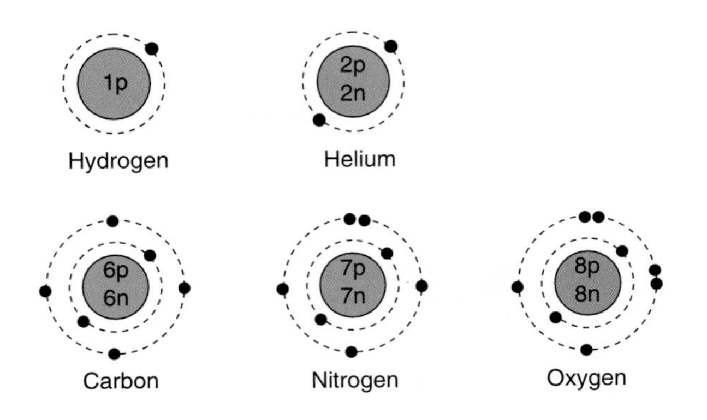

Figure 4–1
All atoms consist of protons (P), which have a positive charge, neutrons (N), having no charge, and electrons, which have negative charges. Protons and neutrons are located in the nucleus where the positive charge of the protons is balanced by an equal number of electrons, which occupy various levels or orbitals around the nucleus.

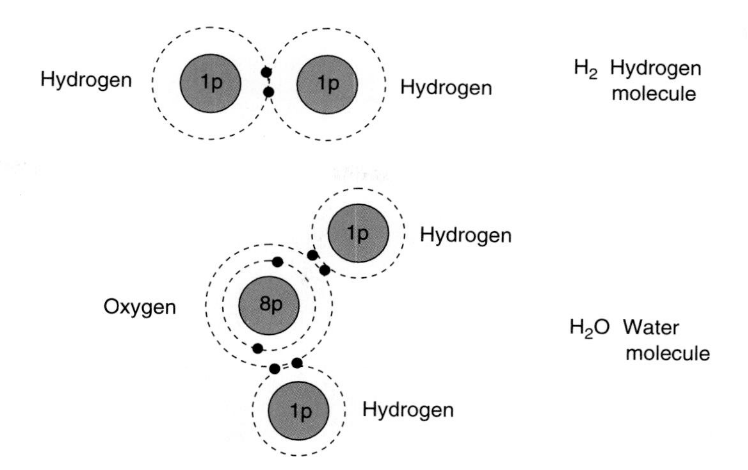

Figure 4–2
Covalent bonds form when atoms share electrons to form stable molecules.

Van der Waals forces, the third type of bond, are very short-range attractions between molecules. They are weaker than covalent and ionic bonds and, hence, are easily formed and easily broken.

The fourth type of force is a **hydrogen bond,** which is also a weak bond acting between two molecules, as Figure 4–4 shows. Hydrogen bonds form when the molecules involved have one side that is electrically positive and one side that is electrically negative. Such separation of charge exists in a molecule when the atoms involved are of very different sizes. In water, the oxygen atom is so much bigger than the hydrogens that the much larger positive charge in the oxygen nucleus pulls the shared electrons close to itself. This pulling of the electrons leaves the oxygen end of the molecule with a slight negative charge from the added electrons, and the other end with a slight positive charge because of the "missing" electrons. The net charge on the molecule is zero (the negative and positive charges are equal). However, on the small scale of molecular structures, each water molecule behaves as if it had a positive side and negative side. Such molecules are called **polar molecules,** and they form hydrogen bonds.

The two hydrogen atoms in one water molecule form hydrogen bonds with oxygen atoms in adjacent water molecules. These weak bonds are the reason for

Figure 4–3
Atoms form ions by gaining or losing electrons. When ions combine to form compounds, they share electrons, forming ionic bonds.

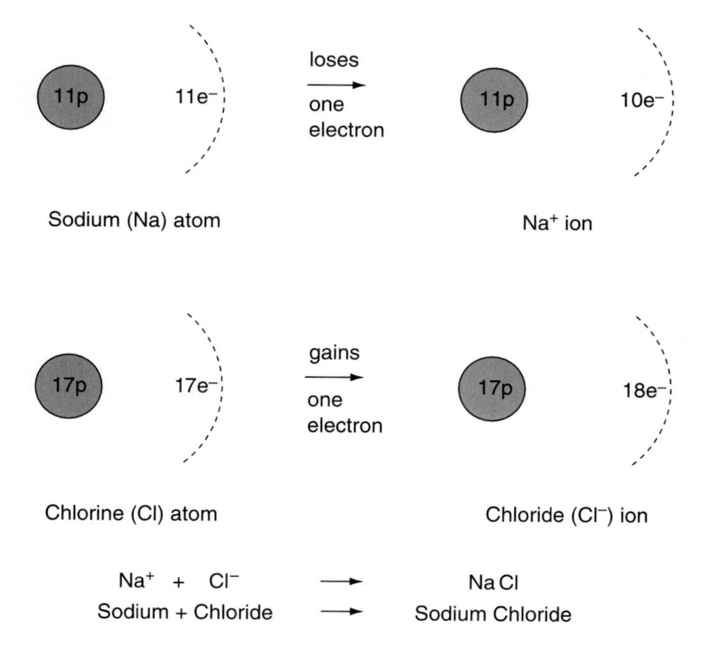

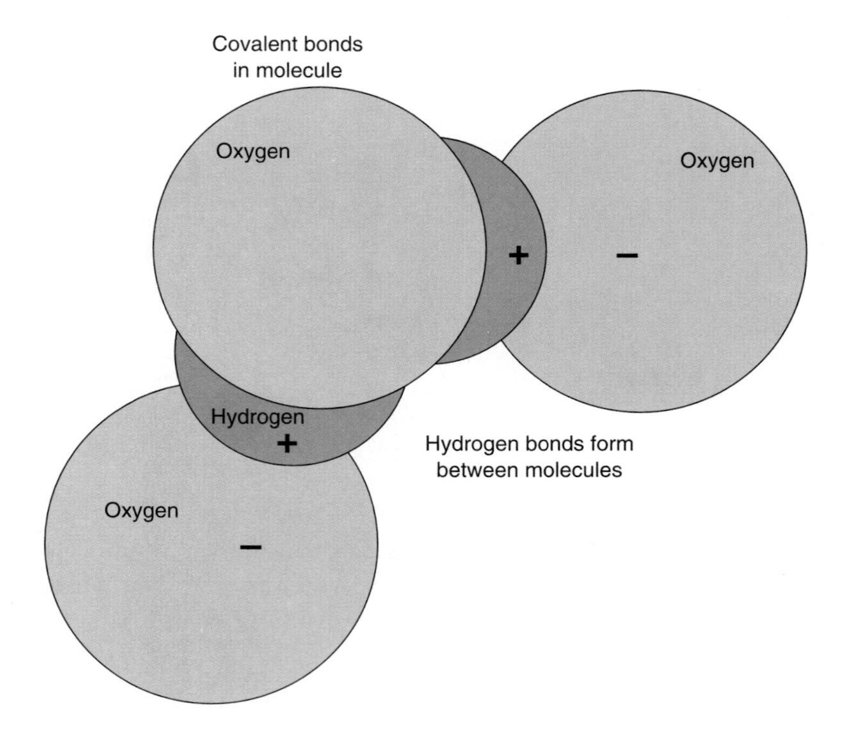

Figure 4–4
Schematic representation of a water molecule. The balls represent the space occupied by electrons associated with each atom.

water's ability to store large amounts of heat with small temperature changes. (Most of the properties of water listed in Table 4–1 arise from hydrogen bonding.)

To obtain an idea of the relative strength of these bonds, consider the amounts of energy (expressed in kilocalories, where 1 kilocalorie is the amount of heat required to raise the temperature of 1000 grams of water by 1°C) required to break all the bonds in 18 grams of water:

Covalent bonds:	100s of kilocalories
Ionic bonds:	10s of kilocalories
Hydrogen bonds:	4.5 kilocalories
Van der Waals bonds:	0.6 kilocalorie

Energy from the Sun's heating of the ocean surface is sufficient to break hydrogen and van der Waals bonds but not enough to break ionic and covalent bonds. In fact, the energy absorbed by evaporating water molecules (a process during which hydrogen bonds are broken) and released by condensing water molecules (a process during which hydrogen bonds are reformed) is one way heat is transferred back and forth between ocean and atmosphere. (We discuss this heat exchange further in Chapter 6.)

States of Matter

All matter on Earth exists in one of three states: gas, solid, or liquid. Each has distinctive properties arising from its molecular structure. The state in which a given sample of matter exists depends partially on pressure but mainly on temperature. If Earth were colder (maximum temperatures below 0°C), water could exist only as ice; if it were much warmer (minimum temperatures above 100°C), water could exist only as a gas. (We call water in the gaseous state "water vapor.") Thus, temperatures on Earth must remain within a relatively narrow range to sustain life as we know it.

Gases are the simplest state of matter to understand because gas molecules move independently of each other. Gases thus fill any closed container and have

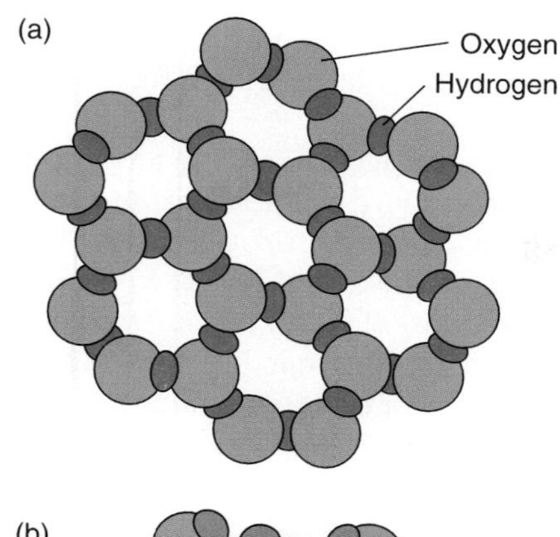

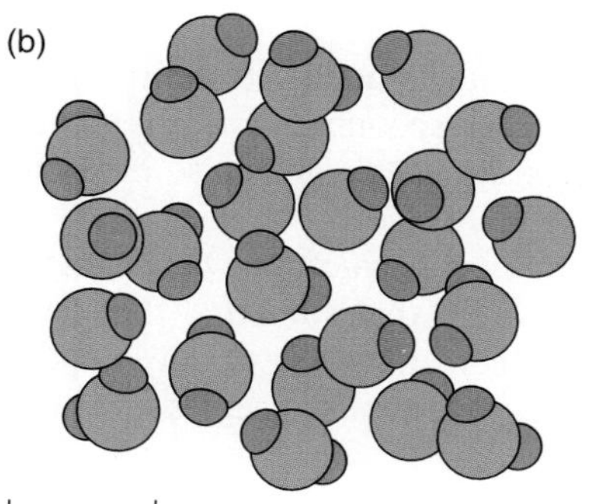

Figure 4–5
(a) The crystal structure of ice, showing the six-sided rings formed by 24 water molecules. (b) In the same volume of liquid water, there would be 27 water molecules. An angstrom (Å) is 10^{-8} centimeters.

neither a definite size nor a definite shape. Gas molecules striking the sides of a container exert pressure. If we add gas to a closed container or increase the temperature in a container holding a gas, this increases the rate at which molecules strike the sides, increasing the pressure. Alternatively, if we decrease the temperature of the gas, this makes molecules move more slowly and thereby reduces the rate at which they hit the sides, lowering the gas pressure.

Solids have definite size and shape. Furthermore, they break or bend when enough force is applied to them. Most solids have fixed internal structures (Fig. 4–5) in which atoms (or molecules) cannot readily move from their positions, although they can vibrate. Vibrations (and movements) of atoms and molecules increase as temperature rises.

Liquids are intermediate between solids and gases. They have a definite volume (size), but their molecules conform to the shape of any container into which a liquid is placed. In liquids, atoms and molecules are bound to each other more tightly than in a gas but not so tightly as in a solid (Fig. 4–5). These bonds are easily broken, allowing liquids to flow.

The Structure of Ice and Water

Hydrogen bonding between water molecules affects the structure of ice and that of liquid water, giving these two forms of water properties not found in most other liquids and solids. Water vapor, on the other hand, behaves like other gases because there is no bonding between molecules.

Water molecules in ice are held together by hydrogen bonds, as shown in Figure 4–6. The oxygen atoms form six-sided rings arranged in layers. The result is a fairly open network of atoms giving ice a lower density (0.92 gram per cubic centimeter) than that of water (1 gram per cubic centimeter). This is why ice floats in liquid water.

Despite this open structure, however, impurities such as sea salts do not easily fit into the holes in the ice structure. Thus, salts and gases dissolved in (liquid) seawater are excluded from the ice formed as water freezes. These sea salts and

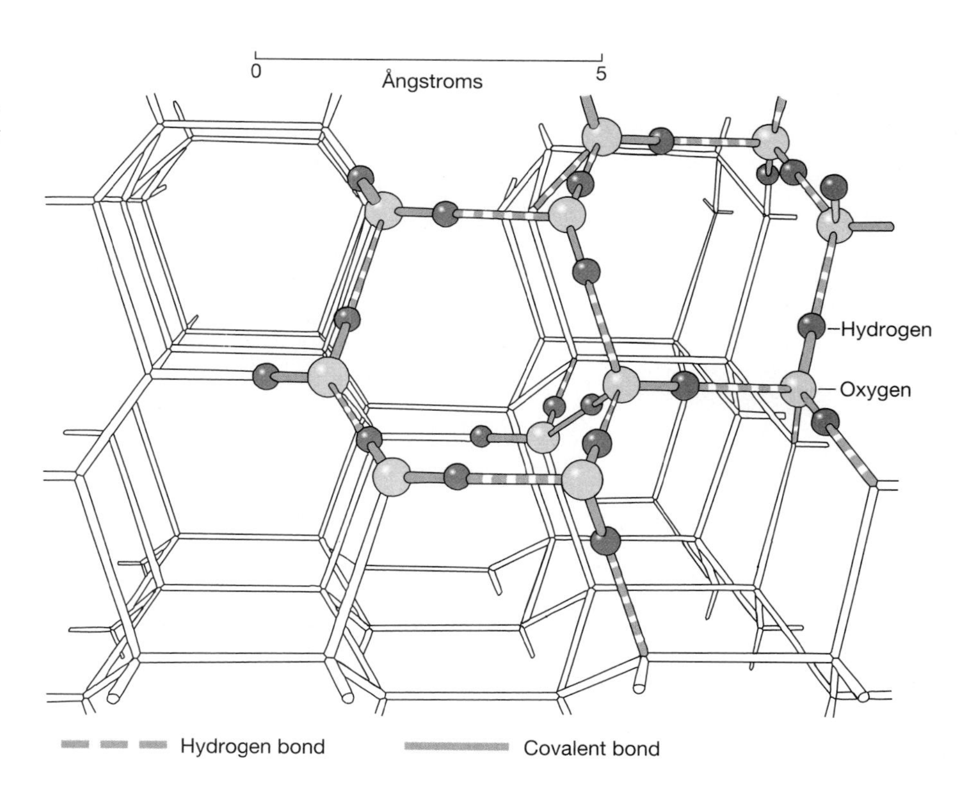

Figure 4–6
The ice lattice is shown using sticks to represent the bonds between water molecules and the covalent bonds of the water molecules.

dissolved gases remain in pockets of unfrozen very salty liquid (called brine) that migrate out of the ice. (We learn more about sea ice and its formation in Chapter 7.)

Liquid water consists of two types of molecular aggregates. The amount of each type present in a given sample of water is determined by the temperature and pressure of the water and the chemical composition of any dissolved salts.

One type of molecular aggregate consists of clusters of hydrogen-bonded water molecules, sometimes called the structured part of water. (We can visualize such clusters as resembling the crystal structure of ice, shown in Fig. 4–5a.) These clusters form and reform rapidly—10 to 100 times during one-millionth of a microsecond. Although the lifetime of any cluster is exceedingly short, each cluster persists long enough to influence water's physical behavior. At higher temperatures, clusters are smaller and have fewer hydrogen bonds.

Water's structured portions are less dense than its unstructured portions. When pressure exceeds about 1,000 atmospheres (corresponding to depths greater than 10,000 meters in the ocean), these clusters disappear. They also break up when water evaporates.

The other molecular aggregate found in water is unstructured, consisting of closely packed, "free" water molecules (see Fig. 4–5b) that surround the structured portions. These molecules move and rotate without restriction. In this unstructured portion of water, interactions between adjacent molecules are much weaker than in the structured portion.

Because of its unique structure, liquid water differs from the compounds that hydrogen atoms form with atoms similar to oxygen. Based on the behavior of these other compounds, water's melting point and boiling point are 90°C and 170°C higher, respectively, than expected. In other words, if water were a "normal" compound with no hydrogen bonding, it would occur only as a gas at Earth's surface temperatures and pressures.

How Temperature Affects Water

Temperature changes alter liquid water's internal structure and thus its properties. Much of the heat absorbed by water goes into changing its internal structure—breaking hydrogen bonds, for instance. Consequently, the temperature of water rises less than other substances after absorbing a given amount of heat. For this reason, we say water has a high **heat capacity.**

It is the water on Earth's surface that prevents wide variations in the temperature of that surface. As a comparison, surface temperatures on the Moon (no water) range from about +135°C at noon to about –155°C at night. The highest temperature recorded on Earth was 57°C at Death Valley, California, and the lowest was at the Russian research base (Vostok) on Antarctica, both locations far from the ocean's influence. Land areas near the ocean or Great Lakes have much narrower temperature ranges than those far from large water bodies.

Much of the solar energy that reaches Earth's surface goes into evaporating water and melting ice. In the ocean, water temperatures rarely exceed 30°C or go below –2°C, a much smaller temperature range than on land. Let us see what happens to water during these changes and how it affects Earth's heat balance.

First, we define a measure of heat, the **calorie**—the amount of heat required to raise the temperature of 1 gram of liquid water by 1 degree Celsius (1°C). This means that we can change the temperature of 1 gram of water by 50°C by supplying 50 calories of heat. Alternatively, we can change the temperature of 50 grams of water by 1°C with the same amount of heat.

To break bonds in ice and liquid water requires energy, and this energy is usually in the form of heat. Conversely, the bonds release energy when they reform. To illustrate this process, consider what happens when 1 gram of ice initially at –100°C is slowly heated. As we add heat to the ice, its temperature

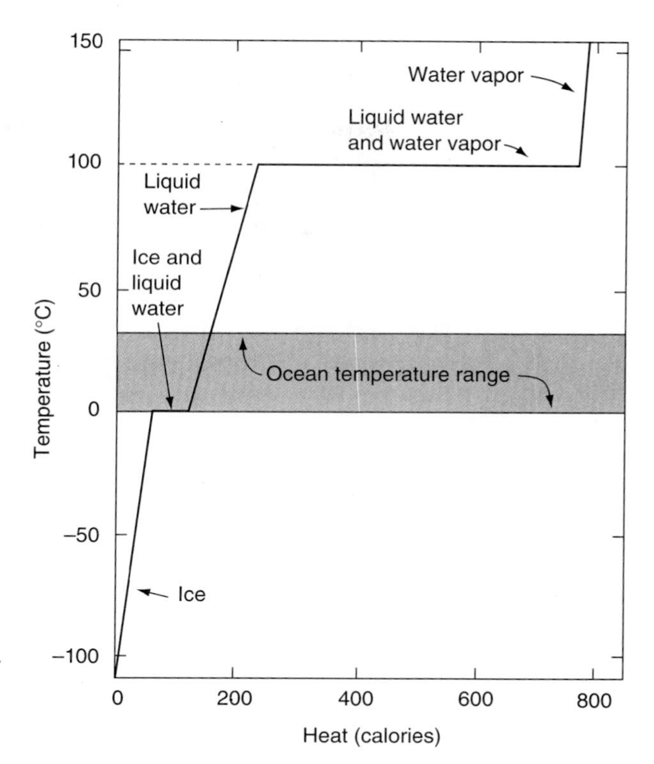

Figure 4–7
Temperature changes when heat is added to or removed from 1 gram of ice, liquid water, and water vapor. Note that the temperature does not change when ice and liquid water or liquid water and water vapor occur together.

increases about 2°C for each calorie of heat we add (Fig. 4–7). When the ice reaches its melting point (0°C), the temperature stops increasing and remains constant (indicated by the shorter horizontal part of the curve in Fig. 4–7) *even though we continue to add heat.* At this point, all the heat being added goes into breaking the hydrogen and van der Waals bonds. After we have added about 80 calories of heat, enough of these bonds have broken and the last bit of ice melts, leaving only liquid water.

As we continue to supply heat, the water temperature rises at a slower rate than in the ice portion of the curve, 1°C per calorie of added heat. This rate of temperature change remains constant between 0 and 100°C. At 100° (the boiling point of liquid water), the temperature remains constant as long as any liquid water is present. After we have added about 540 calories, all the hydrogen bonds in the liquid have been broken. Thus, each water molecule exists alone, in other words, as a gas. If we continue to heat the gas, the temperature rises again at 2°C per calorie of heat added.

If we started with water vapor and cooled it all the way down to ice at –100°C, we would reverse this process and gain heat each time there was a change in state.

Heat added to water is taken up in two ways. One is **sensible heat,** which is heat that we detect either through touch or with thermometers. Temperature changes caused by sensible heat results from the increased vibrations of molecules in solids or liquids or from increased molecular motions in the gaseous state.

The other form is called **latent heat,** which is the energy required to break bonds (*latent* comes from the Latin for "hidden"). As we just saw, at the melting and boiling points of water, there is no change in temperature as long as two states of matter coexist; the added (heat) energy goes into breaking bonds in the form that is "disappearing." We get back all the latent heat when the process is reversed. Condensing water vapor to liquid water releases 540 calories per gram at 100°C. Freezing water at 0°C releases 80 calories per gram.

The difference between the latent heats of melting and evaporation arises from the fact that only a small fraction of the hydrogen bonds are broken when ice melts but all of them are broken when water evaporates.

TABLE 4–2
Latent Heat of Evaporation for Water

Temperature (°C)	Latent Heat of Evaporation (Cal/Gram)
0	595
20	585
100	539

Although water freezes at 0°C and boils at 100°C, water molecules can leave a liquid surface and become gas molecules at temperatures other than 100°C and can be changed from liquid to solid at temperatures other than 0°C. The processes involved are similar to those just described, but the amount of heat released or taken up is different. The latent heat of evaporation changes as shown in Table 4–2.

Viscosity

Viscosity—defined either as internal friction or as the tendency of a fluid to resist flowing (two ways of saying the same thing)—is controlled by molecular structure. We can think of viscosity as the resistance of molecules to moving past each other.

We experience water's viscosity as the resistance, or drag, we feel when we move our hand through water. Because minute marine organisms are so much smaller than a human hand, water viscosity has a much more significant effect on their movements. (We learn more about this in Chapter 12.) On such a small scale, living in water is analogous to humans living in honey or molasses. Thus, tiny organisms must cope with swimming, sinking, and capturing food in a highly viscous fluid. Viscosity is not as important to processes involving humans and at large scales. This is an another example of where differences in time and space scales make it difficult for us to comprehend factors affecting minute organisms.

Water's viscosity decreases with increasing pressure because higher pressures favor the unstructured portions of water. With a greater amount of the less structured form of water present at higher pressures, a greater number of the water molecules, being free rather than trapped in a hydrogen-bond lattice, can move past each other more easily.

In near-surface waters, where viscosity is most important for floating organisms, it is affected by the amount of salt dissolved in the water and by water temperature. Water's viscosity increases with increased salt content because water molecules form hydration shells around each ion of salt (Fig. 4–8). Thus, when there are more salt ions in the water, there are more water molecules involved in

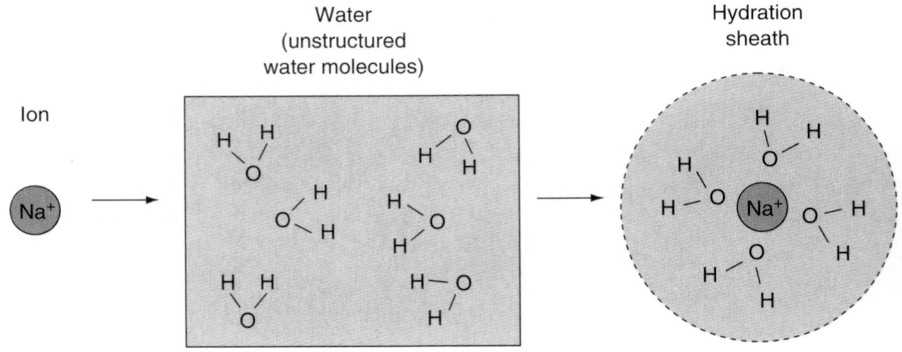

Figure 4–8
An ion placed in water causes water molecules to become associated with the ion, forming a hydration sheath.

hydration shells. Water in such shells does not move as readily as free water molecules, and so the viscosity of the water increases.

In near-surface waters, temperature effects are more important than the amount of salt present. Viscosity increases with lower temperatures because lower temperatures favor the structured portions of water.

Salinity

The amount of salts dissolved in a given volume of seawater varies from one part of the ocean to another, depending on whether fresh water has been added by precipitation or removed by evaporation. (We discuss this topic in Chapter 7.) To indicate how much salt is in seawater, oceanographers use **salinity** (abbreviated S), which is defined as the amount of salt (in grams) dissolved in 1000 grams (1 kilogram) of se water. Salinity is expressed in parts per thousand. In other words, seawater with a salinity of 10 contains 10 grams of salt in 1,000 grams of water—it is 1 percent sea salt.

When various materials dissolve in water, bonds between atoms and some molecules break because of interactions between crystal lattices and water molecules. Thus, salts dissolved in seawater form **ions,** which, as noted previously, are electrically charged and therefore can carry an electrical current. Salinity is determined by measuring seawater's **conductivity,** which is its ability to conduct an electrical current. The more salt in the water, the greater the conductivity. As we see in the next section, salinity is important because it affects the density of seawater. (In Chapter 8, we shall see how seawater density affects ocean currents.)

The six most abundant ions in seawater account for more than 99 percent by weight of all the salts found in the sea, as shown in Table 4–3. Note that the four most abundant constituents amount to more than 97 percent by weight.

Density and Stability

Density is the ratio of the mass of a substance to the volume of that substance and is usually expressed in grams per cubic centimeter. For instance, if a piece of metal has a mass of 500 grams and a volume of 200 cubic centimeters, its density is 500/200, or 2.5 grams per cubic centimeter.

Whenever materials of different densities are mixed together, they sort themselves out so that the densest is on the bottom and the least dense is at the

TABLE 4–3
Major Constituents of Sea Salts

Constituent	*Chemical Symbol*	*Percent by Weight*
Chloride	Cl^-	55.07
Sodium	Na^+	30.62
Sulfate	SO_4^-	7.72
Magnesium	Mg^{++}	3.68
Calcium	Ca^{++}	1.17
Potassium	K^+	1.10
TOTAL		99.36

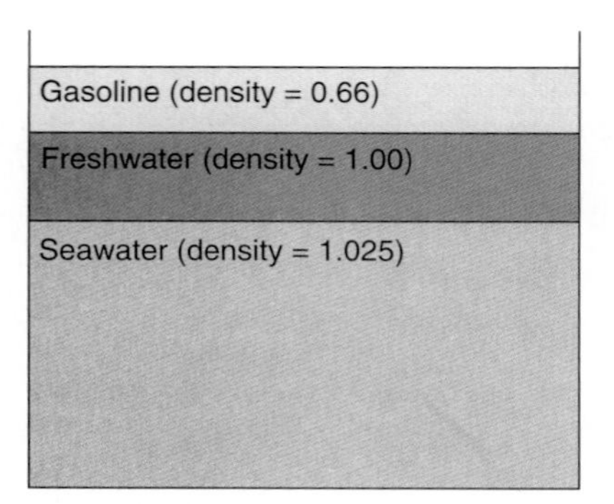

Figure 4–9
A stable, density-stratified system. Fresh water with a density between gasoline and seawater forms a layer between them.

Figure 4–10
Warm seawater is less dense than cold seawater. (Both have the same salinity, 25, shown by the dashed line in Fig. 4–11.)

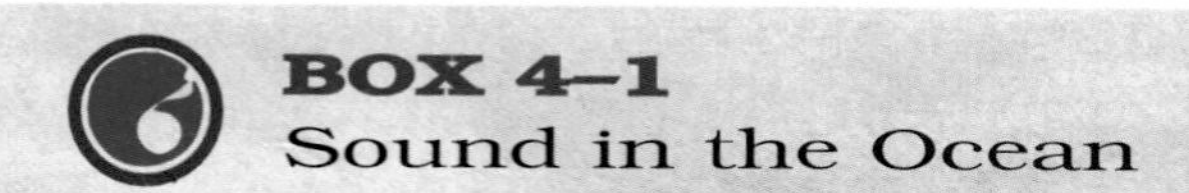

top, with the material of intermediate densities sandwiched between these two extremes. Such a system in which less dense liquids float on heavier ones is called a *stable density-stratified system.* An example is gasoline floating on fresh water that is floating on seawater (Fig. 4–9). (A stable system is one that returns to its original state after being disturbed. An unstable system, on being disturbed, returns to some state other than the original one.)

In an *unstable density-stratified system,* a parcel of water rises or sinks until it reaches a level of density comparable to its own—in other words, until the system stabilizes. Conversely, if its density is greater than its surroundings, it displaces less mass than its own and sinks.

Because the volume of any material changes with temperature, density changes too (remember, density is mass/volume). Colder seawater is denser than warmer seawater, for example (Fig. 4–10). Because the sun heats surface water, warm surface seawaters are less dense than cold, deep water (shown by the dashed line in Fig. 4–11). Whenever a parcel of seawater is placed in seawater, the mass either rises or sinks until eventually it reaches a level where the water below has a greater density and the water above has a lower density. At this point, there is no force acting on the mass and it remains at that level. A fluid parcel may spread out at the appropriate density level, forming a thin layer. Such behavior is common in the ocean. Indeed, such density differences drive the winds and deep-ocean currents, as we see in later chapters.

Increasing the salinity of seawater causes its density to increase (shown by the lines of constant density in Fig. 4–11), whereas increasing the temperature causes its density to decrease. (Remember that density is a ratio; it decreases because of either a decrease in the numerator or an increase in the denominator.) Warming water increases the vibrations of atoms and molecules. Because the more strongly vibrating atoms occupy more volume, the density of the water is decreased. (For a constant mass, density decreases as volume increases.) Increased pressure causes density to increase because water is slightly compressible. (For a constant mass, density increases as volume decreases.) However, we usually ignore water's compressibility in discussing ocean processes.

Ice, water vapor, and seawater, like most other materials, become less dense with increasing temperature. Pure water, on the other hand, is unusual. It has a density maximum at 3.98°C (Fig. 4–12). When liquid water warms from 0°C, the

BOX 4–1
Sound in the Ocean

Seawater is essentially transparent to sound. If seawater were as transparent to light as it is to sound, we could see the bottom of the ocean. Sound is used in sonar (sound navigation and ranging), the underwater equivalent of radar to detect submarines. Scientists use sound to determine ocean depths, to locate organisms living in the ocean, and to communicate. Animals use sound even more—to communicate, to locate food, and to detect predators. Sound in the ocean is as important as light is in the atmosphere.

Sound is a form of mechanical energy. It consists of the regular alternation of either pressure or stress in an elastic medium, in this case, water. The speed of sound in the ocean is about 1,480 meters per second. Changes in temperature have a marked effect on sound speed. Increasing temperature increases the speed of sound. Over most of the ocean, changes in temperature control sound speed. Sound speed also increases with depth as the water pressure increases. Salinity has little effect.

Sound speeds are thus relatively high in the surface zone because of warm waters and high at depth because of higher pressures. Consequently, there is a minimum in sound speed at a depth of around 1 kilometer in most ocean basins. Because sound waves bend toward regions of lower sound speed, this zone of reduced sound speed functions as a sound wave guide and is extremely efficient in transmitting sound. It was once proposed as an emergency signaling channel for airmen downed at sea.

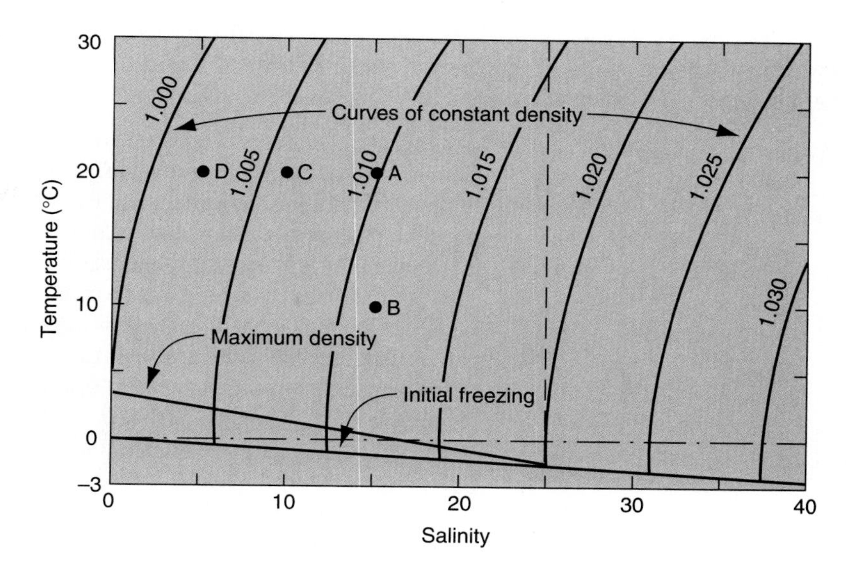

Figure 4–11
Seawater density (expressed here in grams per cubic centimeter) changes as temperature and salinity change. For example, the density of water that has salinity $S = 15$ is 1.010 g/cm^3 at 20°C (A), but 1.012 g/cm^3 at 10°C (B). The density of 20° water is 1.006 when $S = 10$ (C) but only 1.002 when $S = 5$ (D). (The dashed line illustrates the situation shown in Fig. 4–10.)

density increases slightly until it reaches a maximum at 3.98°C. Upon further warming, the density decreases.

The density maximum for pure water is especially important in lakes. In winter, surface waters cool until they reach 4°C. At that temperature, surface waters become denser than those waters below and, therefore, they sink. The process is called **overturning** or **convective sinking.** Overturning continues until the entire lake mixes and water temperatures are uniformly about 4°C. At that point, overturning stops, and bottom waters are protected against further cooling even though surface ice may form. Overturning therefore protects the deep lake waters from freezing. If water behaved like other liquids, lakes would freeze all the way to the bottom. Only shallow lakes and ponds normally freeze completely.

Dissolved salt eliminates water's density maximum. The temperatures of maximum density and of freezing both decrease as salinity increases (Fig. 4–11), but the temperature of maximum density drops more rapidly than the freezing temperature. (The two curves intersect at a salinity of 24.7.) Thus, seawater of average salinity (35) begins to freeze before it reaches the temperature of maximum density.

As previously discussed, temperature markedly influences the relative proportions of structured and unstructured portions of liquid water. Because these two constituents have different densities, water temperature strongly affects density.

The anomalous density maximum in liquid water is caused by the increase in the relative proportion of the structured portion of water below about 4°C. Water becomes less dense because the denser unstructured portion of the mixture is diminished. Addition of salt inhibits formation of the structured portion. This causes the temperature of maximum density to decrease sharply as salinity increases.

Like most other substances, ice becomes less dense with increasing temperature. At 0°C, it has a density of 0.917 gram per cubic centimeter. It is, therefore, about 8 percent less dense than liquid water at the same temperature and floats on water.

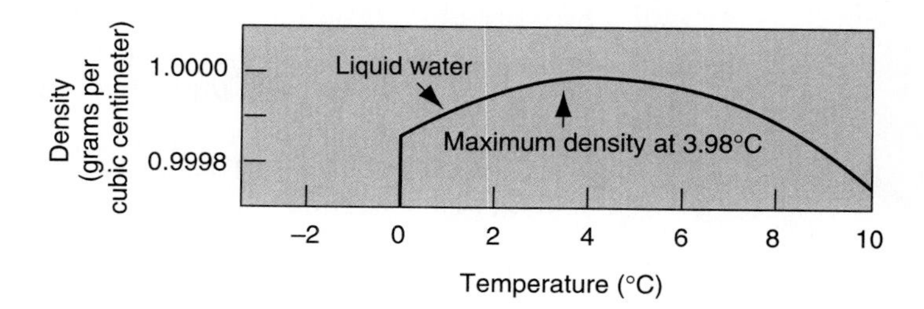

Figure 4–12
Effect of temperature on density of pure water.

Conservative and Nonconservative Properties of Seawater

The salinity of seawater changes whenever fresh water is added to the ocean in the form of rain or snow, river discharge, or thawing ice and whenever fresh water is removed through evaporation and freezing. These are all physical processes. They leave the proportions of the major ions relative to each other in seawater unchanged. This is an example of salinity as a **conservative property.**

Conservative properties are used to trace movements of waters after they leave the sea surface and move through the deep ocean. Below the ocean surface, both water temperature and salinity are conservative properties. That is, they change only through mixing with other water having different temperatures and salinities.

Salinity varies most near the air–sea interface, at current boundaries where water masses of different salinities mix, and in coastal ocean waters. Even when salinity changes are relatively small, they are useful in tracing water movements because salinity can be measured very accurately.

The relative abundance of many constituents in seawater is changed as they are consumed by growing organisms or released when those organisms die and decompose. Such constituents are **nonconservative.** Other nonconservative constituents participate in chemical reactions between seawater and sediment particles. As a general rule, the major constituents in seawater are conservative while most minor constituents are nonconservative.

Salinity Effects

Natural waters usually contain some dissolved salts. River waters contain, on the average, about 0.01 percent dissolved salts (salinity 0.1); average seawater contains about 3.5 percent various salts (salinity 35) (Fig. 4–13). Even rainwater contains small amounts of salts and gases that it picked up while falling through the atmosphere.

Dissolved salt affects the physical properties of water in several ways. We have already discussed the effect on temperature of maximum density and on freezing temperature. Other properties affected by salts are as follows:

Vapor pressure decreased by increased salinity.

Osmotic pressure increased by increased salinity (water molecules move through a semipermeable membrane from a less saline solution to a more saline solution; the opposing pressure necessary to stop this movement is called the osmotic pressure). This is important for marine organisms.

Figure 4–13
Major and minor constituents of seawater (by weight).

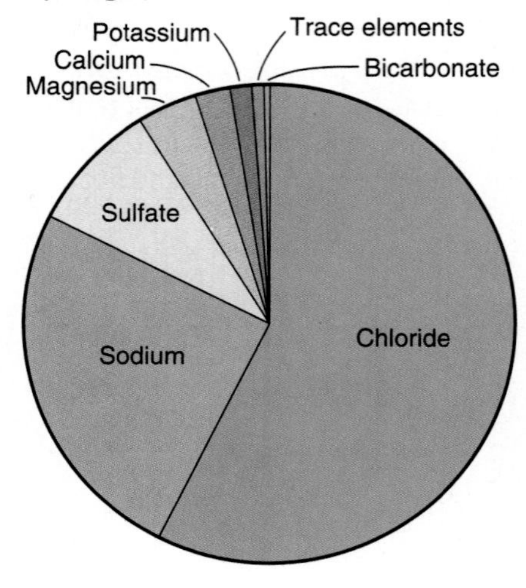

Consider the temperature of initial freezing. Pure water freezes at 0°C, and the temperature of the water–ice mixture remains fixed until only ice remains, as we learned earlier. In seawater the temperature of initial freezing is lowered by increased salinity, as we saw in Figure 4–11. Furthermore, salts excluded from the ice remain in the liquid, causing the brine to become still saltier. So the temperature of freezing for the remaining liquid is still lower. Then temperatures must drop before additional ice forms. The result of all this is that seawater has (a) a lower temperature of initial freezing than pure water and (b) no fixed freezing point.

Salts affect the internal structure of water. Some dissolved ions, such as sodium and potassium, shift the equilibrium toward water's unstructured phase. Other ions, such as magnesium, favor the structured portions. Changes in the relative proportions of these constituents affect such properties as *viscosity*.

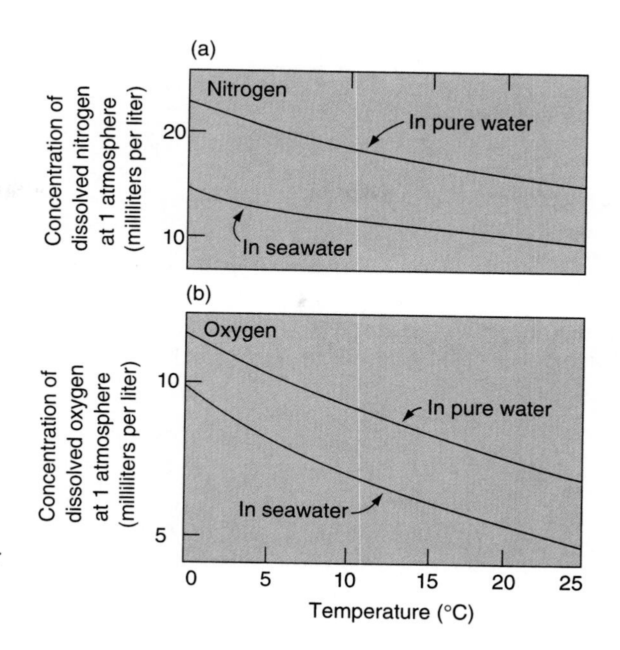

Figure 4–14
Solubility of (a) oxygen and (b) nitrogen in pure water and seawater. Note that both gases are more soluble in pure water than in seawater and that their solubilities decrease with rising temperatures.

Dissolved Gases

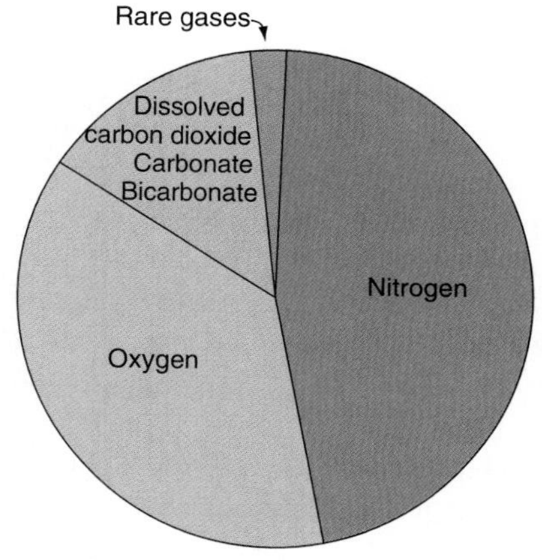

Figure 4–15
Relative abundances (by weight) of gases dissolved in surface seawater ($S = 30$, $T = 20°C$). Note the relatively high abundances of total carbonate and the rare gases.

Atmospheric gases dissolve in seawater at the air–water interface. Conversely, gases also pass through the interface in the opposite direction—back into the atmosphere. Water can dissolve only a limited amount of any substance at a given temperature and pressure. When that limiting value is reached, the amount of gas going into solution equals the amount going out. At that point, the water is *saturated* with the gas, which is then said to be present in *equilibrium concentration.*

Water temperature, salinity, and pressure affect saturation concentrations for a gas. In the normal range of oceanic salinity, temperature is dominant. Generally, such gases as nitrogen and oxygen or the inert gases helium and neon, which do not react chemically with water, become less soluble in seawater as temperature or salinity increases, as Figure 4–14 shows. Seawater at all depths is saturated with most atmospheric gases. Exceptions are dissolved oxygen and carbon dioxide, which are involved in life processes. Dissolved oxygen concentrations in the deep ocean are controlled primarily by biological processes. These are discussed in Chapter 12.

Nitrogen, the most abundant atmospheric gas, is also common in seawater (Fig. 4–15). It is not involved in most biological processes and thus remains near saturation throughout the ocean. Other rare atmospheric gases—argon, krypton, and xenon—behave like nitrogen and their concentrations vary little in seawater.

Acidity and Alkalinity

An **acid** is a compound that releases hydrogen ions (H^+) when dissolved in water. A strong acid is one that readily releases hydrogen ions. A weak acid does not release its hydrogen ions so easily. Another type of compound, called a **base,** releases hydroxyl ions (OH^-). Like acids, some bases are strong in that they readily release their hydroxyl ions; others are weak.

Both hydrogen and hydroxyl ions occur normally in seawater as water molecules break up (dissociate) and reform. The reversible reaction can be written as:

$$\underset{\text{Water}}{H_2O} \rightleftarrows \underset{\text{Acid}}{H^+} + \underset{\text{Base (hydroxyl)}}{OH^-}$$

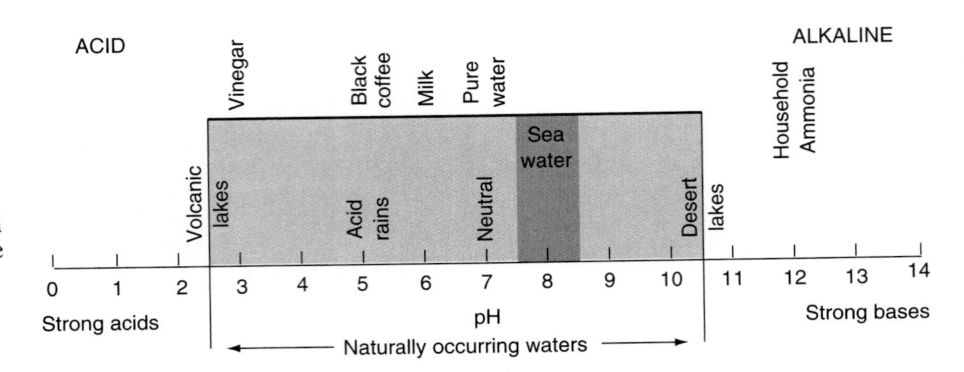

Figure 4–16
pH is a measure of the acidity or alkalinity of a solution. The pH of seawater has a narrow range because it is buffered by the carbonate–carbon dioxide system in the ocean and atmosphere. Freshwater systems are not so well buffered and, therefore, exhibit a wider range of pH values.

Water itself is neutral. That is, the concentrations of hydrogen and hydroxyl ions are equal. If we add an acid and increase the amount of hydrogen ions, the water becomes acidic. Conversely, if we add a compound that increases the amount of hydroxyl ions, the water becomes basic. (*Alkaline* is another word for "basic"; they mean exactly the same thing.)

The acidity or alkalinity of any solution is expressed using a **pH scale** (Fig. 4–16) based on the abundance of hydrogen ions. (The pH scale is a logarithmic scale, which means it is based on factors of ten. Thus, a change in one pH unit means a tenfold change in the concentrations of hydrogen and hydroxyl ions.) If the solution is neutral, it has a pH of 7. The most acidic solution has a pH of 0, and the most alkaline one has a pH of 14. Seawater is normally mildly alkaline, with a pH of about 8, because of the carbon dioxide dissolved in it.

The Carbon Cycle

Carbon has a complicated chemical behavior in the atmosphere, in the ocean, and in the biosphere. It is essential for life (as we see in Chapter 12) and is always present in abundance. Removal of carbon, as carbon dioxide, from the atmosphere by living organisms has given Earth its present atmospheric composition. Without this process, Earth would have a dense atmosphere like that of Venus, as we discuss in Chapter 6.

Carbon dioxide is highly soluble in seawater and occurs as a dissolved gas and in several other chemical forms (Fig. 4–17). Respiration by plants and animals also produces carbon dioxide throughout the ocean. In its simplest form, the carbon cycle is:

$CO_2 + H_2O$	$\rightleftarrows$	H_2CO_3	$\rightleftarrows$	$H^+ + HCO^{3-}$	$\rightleftarrows$	$2H^+ + CO^{3-}$
Carbon dioxide + water		Carbonic acid		Bicarbonate ions		Carbonate ions

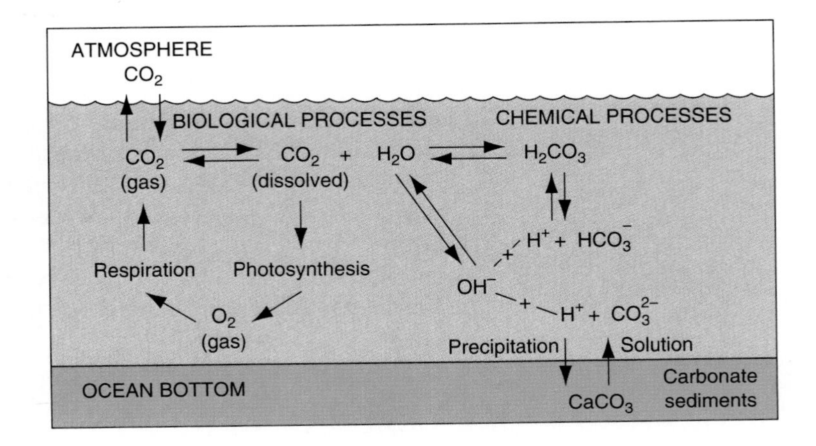

Figure 4–17
Carbonate–carbon dioxide cycle in the ocean.

When carbon dioxide dissolves, it forms a weak acid (H_2CO_3), which acts as a buffer to stabilize the pH of seawater between 7.5 and 8.5.

Carbonate and bicarbonate ions freely give up and accept hydrogen ions in seawater, thus buffering seawater against sharp changes in acidity. Any addition of acid (H^+) to the water creates more bicarbonate ion (HCO_3^-), which combines with the added H^+ to keep the acidity of the mixture unchanged. Because of this buffering action, respiration and decomposition processes producing carbon dioxide hardly affect the acidity of seawater, nor does removal of carbon dioxide during photosynthesis.

Because of its complex chemical behavior, carbon dioxide is involved in many processes in the ocean. Formation and destruction of the carbonate shells of ocean-dwelling plants and animals affect the abundance of carbonate in seawater. If carbonate is used in plant production, the chemical reactions cause bicarbonate to dissociate. Carbonate precipitation as calcium carbonate ($CaCO_3$) in plant and animal tissues causes a net loss of carbon dioxide.

The total amount of carbon dioxide (all forms) in seawater is about 60 times the amount of carbon dioxide in the atmosphere. Thus, the ocean is a major reservoir of carbon. In recent years, there has been increased interest in carbon because burning of fossil fuels (coal, oil, and gas) and deforestation have increased the carbon dioxide content of the atmosphere by approximately 0.2 percent per year. As we see in later chapters, this increased carbon dioxide content is predicted to cause Earth's surface temperatures to increase. Efforts to study how much and when the temperature will rise are discussed in Chapter 15.

Processes Controlling Sea Salt Composition

Some constituents dissolved in seawater come from gases (chlorine, sulfur dioxide) released to the atmosphere through volcanic eruptions. Most of these gases are quickly removed by rainfall and go into the ocean.

Volcanic eruptions on the seafloor supply calcium and other elements. When seawater reacts with hot volcanic rocks, these elements are extracted from the rocks and carried away by the circulating water. The importance of this process is still unknown.

The third major source of materials dissolved in seawater is decomposition (called **weathering**) of rocks exposed on the land. Rocks formed at high temperatures and pressures in the Earth's interior usually form in the absence of free oxygen. When they are exposed to atmospheric oxygen and rain, they slowly break down. (The effects of this process can be seen in the gradual obliteration of inscriptions on tombstones.) The insoluble mineral grains released first become part of the soils. They eventually enter the ocean as particles after being transported by winds, rivers, or ice. The soluble portions released during rock weathering are removed by running water and carried to the ocean by rivers. For instance, calcium and magnesium in the ocean come from the alteration of seafloor volcanic rocks and from decomposition of rocks on land.

Sea salt composition and probably seawater salinity have varied little during the past 1.5 billion years. Since large amounts of dissolved materials are supplied each year, some processes must be removing these constituents at roughly the same rate. Otherwise, salinity would rise and sea salt composition would change. The types and abundances of organisms living in the ocean would very likely have changed if salinity had been greatly altered, but there is no evidence in the fossil record of ancient marine organisms that this has occurred. (This is an example of the steady-state condition of the ocean. It is changing little, if at all, over many millions of years.)

Figure 4–18
Seawater flows into evaporating ponds on southern San Francisco Bay. Heat from the Sun evaporates the water, leaving behind brines that are then transferred to smaller ponds where salt crystallizes. The red color of brines is caused by the growth of one-celled algae. (James Hanley/Photo Researchers, Inc.)

The simplest process for removing dissolved constituents from water is the formation of **salt deposits** (Fig. 4–18). If an arm of the sea is partially isolated from the open ocean and located in an arid climate, as the water evaporates some of the salts that were originally dissolved in it precipitate, forming salt deposits as the brines continue to evaporate. Limestones and related rocks form first, followed by deposits of more soluble components—sodium, potassium, chloride, and sulfate. As we have already learned, such deposits are likely to be formed in newly created ocean basins. (Modern examples of such partially isolated basins are the Red Sea and the Gulf of California.)

Biological processes also remove dissolved constituents from seawater. Perhaps the most conspicuous example is the removal of calcium and carbon dioxide by shell-forming organisms, such as clams and oysters.

Finally, chemical reactions between cold seawater and hot, newly formed oceanic crust remove dissolved constituents. This process is especially important for removing the major constituents magnesium and sulfate.

Residence Times

A useful concept for characterizing substances in seawater is **residence time**—the time required to completely replace the amount of a given substance in the ocean. This concept works in two ways, using either the rate of addition or the rate of

TABLE 4–4
Estimates of Residence Times for Some Elements in Seawater

Element	*Millions of Years*	*Removal Process*
Sodium	68	Salt deposition
Chlorine	100	Salt deposition
Potassium	7	Chemical reactions with clays
Calcium	1	Shell formation by organisms
Lead	0.0004	Removal by absorption on sediment particles
Aluminum	0.0001	Absorption on clay particles

removal of elements incorporated in sediments depositing on the ocean bottom. Using the second option, we can define residence time:

$$\text{Residence time (in years)} = \frac{\text{Amount present}}{\text{Removal rate}}$$

An element's residence time in seawater is related to the chemical behavior of the element. Elements like sodium, which are little affected by chemical or biological processes, have residence times of millions of years. Elements used by organisms or readily incorporated in sediments, such as aluminum or iron, have much shorter residence times, ranging from a few hundred to a few thousand years. Table 4–4 gives some representative residence times.

We can also calculate a residence time for water. Each year, there is a net removal, due to evaporation (which does not fall back on the ocean surface as rain

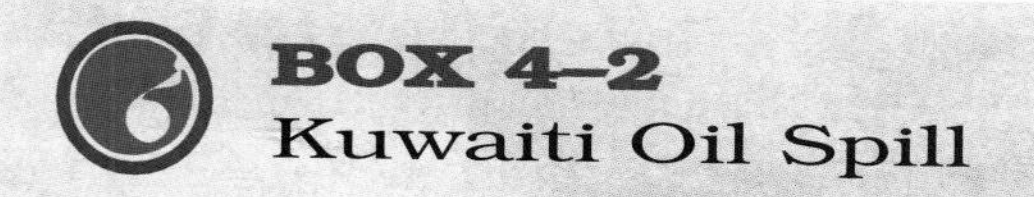

BOX 4–2
Kuwaiti Oil Spill

During the war between the United States and Iraq in 1991—the Persian Gulf War—enormous amounts of oil spilled into the Arabian Gulf after retreating Iraqi armies destroyed Kuwaiti oil-production and collection facilities and set fire to more than 500 wells.

Because of the volume of oil spilled and the dangers from mines left by both sides in the war, little effort was made to contain or remove the spilled oil. Thus, the Kuwaiti oil spill provides an opportunity to learn how the ocean and marine organisms respond to large amounts of spilled oil left untreated. Some critics of cleanup efforts during previous large spills, such as the 1967 *Amoco Cadiz* spill in Normandy or the 1989 *Exxon Valdez* spill in Prince William Sound, Alaska, contend that more damage was done to the ocean and to marine life by the cleanup effort than if the spilled oil had been left alone to decompose naturally.

Oil occurs naturally in the ocean. It is released through natural discharges from crevices in the ocean floor (called seeps) as well as from many small spills and the regular activities of ships and tankers. Mostly this oil decomposes through little known processes, apparently involving bacterial decomposition. Little oil is actually found in the ocean except along tanker routes and in coastal areas near ports where large amounts of oil are handled and some spilled routinely.

The countries in the Gulf region, with U.S. assistance, studied the results of the oil spill. The results of those studies help us to deal with such large oil spills, especially those in tropical waters.

right away), of a layer of water about 10 centimeters thick from the ocean surface. This water falls on the land and returns to the ocean through river discharge. Recall that the ocean has an average depth of about 4 kilometers. From these figures we find a residence time of 40,000 years for water.

The Ancient Mediterranean Salinity Crisis

Between 5 and 6 million years ago, the Mediterranean basin was partially cut off from the Atlantic. During that time, a worldwide drop in sea level of about 40 to 50 meters isolated the basin when the sea level fell below the depth of the ridge at Gibraltar that separates the Mediterranean from the North Atlantic. As a result, thick salt deposits accumulated in the basin. Such events of salt deposition have happened many times in the history of the ocean, but few have been so well studied.

The Mediterranean formed when Africa moved north and collided with Europe at Gibraltar. Today the Mediterranean is connected with the Atlantic by the narrow Strait of Gibraltar, but when sea level dropped 5 million years ago, this connecting waterway did not exist and, consequently, the Mediterranean basin was intermittently isolated. In any case, the volume of water flowing into the isolated basin was smaller than the volume evaporating, causing the water level in the Mediterranean to drop and salt deposits to form. The Nile and Rhone Rivers flowing into the basin cut deep gorges to reach the much lowered sea level in the basin. These gorges are now filled by thick deposits of sediments.

As water in the basin evaporated, the salinity increased. Eventually, salt concentrations reached levels where salt crystallized and accumulated on the bottom. If the present Mediterranean dried out, it would take about 1,000 years to evaporate the water, forming a salt deposit 70 meters thick. Such deposits in the basin are now 2 to 3 kilometers thick. Thus, the volume of salt there is equivalent to filling the basin with seawater and then drying it out 30 to 40 times.

It is likely, however, that the basin never totally dried out, although isolated areas may have been saline lakes or even deserts for short periods. When sea level rose again about 5 million years ago, the reestablished connection with the Atlantic permitted salinities to drop, so that salt deposits no longer formed.

Removal of so much salt, by having it locked up in salt deposits when the sea level of the Mediterranean was lower, resulted in a reduction of seawater salinity worldwide by about 2. This reduction caused a salinity crisis that affected the evolution and distribution of animals around the Mediterranean.

Similar events must have occurred many times in other bodies of water, too, as indicated by the large salt deposits found now in the Gulf of Mexico and around the margins of the Atlantic.

Resources from Seawater

Salt (sodium chloride) is the most common material taken from seawater. Seawater is pumped from the coastal ocean into large evaporating basins located in many dry coastal regions and then solar energy is used to evaporate the water (Fig. 4–18). The evaporation is carefully controlled so that only one type of salt—only sodium chloride, say, or only potassium sulfate—precipitates out at a time. Salts recovered from seawater must be treated further to remove magnesium sulfate (a strong laxative, Epsom salt) and calcium carbonate. Sea salt is used by the

Figure 4–19
Single-stage condensation apparatus (still) to recover fresh water from seawater.

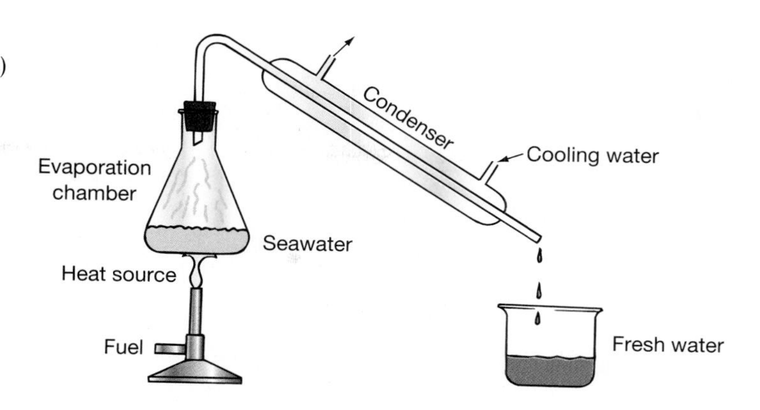

chemical industry, as is magnesium, a lightweight metal. Bromine extracted from seawater is used as a component of antiknock compounds in gasoline.

Despite the variety and large amount of valuable materials dissolved in ocean water, their low concentrations make it prohibitively expensive to extract most of them. Gold is an example. Its concentration in seawater is extremely low, although there are about 5 million tons of gold in the entire world ocean. The cost of energy for pumping seawater, added to the cost of chemical treatment, far exceeds the value of the gold that could be recovered. Another problem in many areas of the world is the difficulty of obtaining undiluted, uncontaminated seawater.

Growth of cities in arid regions makes water from the ocean an increasingly important resource. One simple way to produce fresh water from seawater is in a greenhouse. Large, shallow pans of seawater are placed in the greenhouse. During the day, when the sun is shining through the glass, the water in the pans evaporates into the air inside the greenhouse. At night, this water condenses back to the liquid state when the air cools. This evaporation/condensation cycle creates a humid, nearly tropical environment where plants can be grown. Such artificially made oases are useful for small-scale production of high-value products like fresh vegetables and fruits.

Fresh water is easily extracted from seawater with a simple still (Fig. 4–19). Seawater is fed into an evaporation chamber and heated. The salt-free water vapor that forms rises up through the top of the chamber into a condenser, which is one tube inside another tube, with cold water circulating in the space between the two. There the water vapor condenses, and the fresh water that forms is collected. The salt remaining behind and most of the original seawater (typically 90 percent) are discharged as a hot brine.

Recovery of fresh water from a single-stage evaporation process like that in Figure 4–19 is impracticable, however. Too much heat is lost in the cooling water, and high energy costs restrict this process to small-scale or emergency uses.

One way to increase the efficiency is to use the seawater for cooling before it is fed into the evaporation chamber. In this way, energy consumption is reduced. It is also possible to reduce the boiling point of the seawater by carrying out the process under a slight vacuum, a technique known as flash evaporation.

Another way of desalinating seawater is to freeze it in order to separate out the salts. A coolant is added to the seawater, and as this coolant evaporates, it removes heat from the water, causing it to freeze. The ice crystals are removed, washed free of brine, and then melted to obtain fresh water. The coolant is recycled, and the remaining brine is about twice as salty as it was at the beginning of the process. Such a process is particularly well suited to northern areas because here the ocean water is already cold, a condition that reduces the overall costs of the operation.

Chlorine: An Environmental Hazard?

Synthetic organic chemicals containing chlorine and various metals have been released to the environment in massive quantities since the 1950s. Large amounts of these substances end up in the ocean. Some of them—especially DDT, PCBs, and dioxins—interfere with the reproductive systems of fishes, birds, and mammals, including humans. These effects were first noted when DDT interfered with calcification of birds' eggs. The thin-shelled eggs broke easily and populations of bald eagles, ospreys, and other birds were greatly reduced. The banning of DDT in North America has helped these bird populations recover during the last twenty years.

Many other chlorinated hydrocarbons mimic hormones and interfere with the hormones of fishes, birds, wildlife, and humans. For instance, the increased incidence of breast and testicular cancers in humans may be due to exposure to chlorinated hydrocarbons. Exposure to these chemicals may decrease sperm counts, diminishing fertility. Such effects have been demonstrated in mink that are fed fish from the Great Lakes. Many toxic chemicals are known to be concentrated in fatty tissues as they are passed from one trophic level to another in a food web. By the time they reach the top predator (such as the mink, or humans in many cases) they are present in sufficient concentrations to cause serious metabolic problems. The severity of this widespread problem is just beginning to be appreciated.

Summary

Seawater consists of 96.5 percent water, 3.5 percent sea salts, some dissolved organic matter, a few solid particles, and some dissolved gases.

Water's molecular structure causes its unique physical and chemical behavior. Strong convalent bonds hold the molecule together. Weak bonds between water molecules account for its high heat capacity. Water can occur as a gas, liquid, or solid (the three states of matter), depending on its temperature. In water vapor (gas), molecules move freely and expand to fill any volume. In ice, water molecules are bonded together, forming a lattice that is less dense than liquid water. Voids in the ice lattice are too small to accommodate gases or salts in the structure, so they are excluded as water freezes. Liquid water contains both structured and unstructured portions; changing temperature, pressure, or salt content alters the relative abundances of the two. Increased pressure favors the denser, unstructured phase.

Heat is absorbed by a substance as the bonds in that substance are broken and released when bonds form. Because of the large amount of bonding in liquid water, it has a high heat capacity, which means that water can absorb large amounts of heat with only a small temperature increase. The size of the structured clusters of water molecules decreases as water temperature rises.

Viscosity—internal resistance to flow—is sensitive to the molecular structure of water. Water's viscosity increases with higher salinities, lower temperatures, and lower pressures, all three conditions favoring the more structured component.

Sinking (or floating) is determined by relative density (mass per volume); less dense objects float in more dense ones. Seawater density is controlled by temperature, salt content, and pressure. Water density decreases with increasing temperature because the water volume expands. Fresh water has a density maximum at 3.98°C. Seawater begins to freeze before it reaches its density maximum.

Six constituents account for 99 percent of sea salts: chloride, sodium, sulfate, magnesium, calcium, and potassium. Salinity, the total amount of dissolved salts, expressed in parts per thousand, is used to indicate the amount of salt dissolved in seawater.

The major constituents in seawater occur in constant proportions; these are called conservative substances. Substances occurring in low concentrations that vary as a result of chemical or biological processes are called nonconservative substances.

The lower the temperature and salinity of seawater, the greater the amount of gas that can be dissolved. Surface seawater exchanges gas with the atmosphere. When the amount taken up equals the amount lost, seawater is saturated with the gas. Nitrogen is near saturation levels throughout the ocean. Oxygen is taken up from the atmosphere at the surface and is also released by plants; it is near saturation in surface waters and depleted in deep-ocean waters.

Seawater's acidity–alkalinity balance is controlled by the carbon dioxide–carbonate cycle in the ocean.

Carbon dioxide has a complicated chemical behavior in seawater. It occurs as a dissolved gas and combines with water in the form of bicarbonate and carbonate.

Sea salt composition is controlled primarily by chemical reactions between seawater and hot, newly formed crust. Some constituents are removed by biological processes; others, by evaporation. Sodium and chloride have the longest residence times.

Sea salts deposited when the Mediterranean was isolated caused salinities to decrease worldwide.

Fresh water and various salts are the primary resources recovered from seawater.

Key Terms

covalent bonds
ionic bonds
ions
van der Waals forces
hydrogen bonds
polar molecule
heat capacity
calorie
sensible heat
latent heat
viscosity
salinity
ions
conductivity
density
overturning (convective sinking)
conservative property
nonconservative constituents
pH scale
weathering
salt deposits
residence time

Study Questions

1. Draw a diagram of a water molecule. Explain why it is polar.
2. Describe and contrast the molecular structures of ice, liquid water, and water vapor. Discuss the significance of the different structures on the amount of energy taken up or given off as water changes from one state to another.
3. List the six major constituents of seawater and give the percentage of each.
4. Define residence time. Discuss its significance and give examples explaining elements having long residence times and short residence times.
5. Explain the carbon dioxide–carbonate cycle in seawater without living organisms or carbonate solids.
6. Describe why fresh water in the bottom of large lakes does not get colder than 4°C.
7. Discuss the difference between the conservative and nonconservative constituents of seawater.
8. Discuss the processes controlling the composition of sea salts.
9. Why is the concentration of carbon dioxide dissolved in seawater greater than in the atmosphere?
10. What resources are extracted from seawater?
11. How do changes in the viscosity of seawater affect marine organisms?
12. Describe how changing the salinity of seawater affects its physical properties.
13. Compare the range of acidity–alkalinity of ocean waters with that of freshwater lakes.
14. Discuss how weathering and evaporite formation affect the composition of seawater.
15. Why is the composition of the atmosphere different from that of the gases dissolved in seawater?
16. Explain how the isolation of the Mediterranean basin and resulting salt deposition affected the ocean worldwide.
17. *[critical thinking]* How do organisms affect the chemical composition of sea salts?

Selected References

BERNER, E. A., AND R. A. BERNER, *The Global Water Cycles.* Englewood Cliffs, N.J.: Prentice Hall, 1987.

BROECKER, W. S., *How to Build a Habitable Planet.* Palisades, N.Y.: Eldigo Press, 1987. Discussion of the processes forming Earth and the ocean.

DEMING, H. G., *Water: The Foundation of Opportunity.* New York: Oxford University Press, 1975. General discussion of water and its properties.

HSU, K. S., *The Mediterranean Was a Desert: A Voyage of the Glomar Challenger.* Princeton, N.J.: Princeton University Press, 1983. Lively account of the discovery of the Mediterranean salinity crisis.

OBJECTIVES

Your objectives as you study this chapter are to understand:

- How sediment deposits form and how they relate to other ocean processes
- Methods of deciphering ocean history recorded in sediment deposits
- Resources in sediments

5 Sediments

The ocean drilling ship JOIDES Resolution *collects samples of sediments and rocks from all the ocean basins to study their history.*
(Courtesy Ocean Drilling Program.)

Ocean basins receive rock fragments of all sizes that have been either washed or blown off the land. These particles mix with the shells and bones of marine organisms that have sunk to the ocean bottom, and the accumulated mixture forms what we call **sediment deposits.** In addition to these two sources of sediments deposits, there is occasionally a third source. In some locations where few other materials accumulate, chemical reactions in seawater produce particles that coat any hard surfaces on the bottom.

The fossils in these sediment deposits are important to us because they record Earth's history over hundreds of millions of years, of the present ocean basins, and the history of life in them. Comparable sedimentary rocks preserved on land record similar changes, including changes in Earth's climate over billions of years.

Sediment deposits from the continents cover about one quarter of the ocean floor, with accumulations thickest in marginal ocean basins. Although ocean margins account for only 2 percent of total ocean-floor surface area, the deposits on these margins contain about one-sixth of all oceanic sediment and are important sources of oil and gas.

Sources of Sediment Particles

We begin our study of ocean sediments with an examination of their sources and of the ocean processes that affect their distribution and abundance.

Terrigenous Sediments

The most common constituents in sediment deposits are **terrigenous** (land-derived) particles. They include rock fragments and mineral grains formed during the chemical and physical weathering of rocks on land and from volcanic eruptions.

Although we think of rocks as immutable, an inspection of old monuments or cemetery headstones demonstrates how quickly rocks can be broken down by weathering (Fig. 5–1). The reasons for this ready decomposition are easily understood. Most rocks form deep inside Earth, at high temperatures and pressures under conditions in which there is no oxygen and little free water. When the rocks are later exposed at Earth's surface, the minerals in them are unstable at surface temperatures and pressures and in the presence of air and water; their decomposition is called *weathering*.

As the rocks weather chemically, their water-soluble constituents dissolve and move with river waters to the ocean. The resulting particles are carried in suspension by turbulent waters or dragged along the bottom of stream beds by swift currents. In physical weathering, rocks are broken into small particles by such physical processes as freezing and the abrasive action of glaciers. Rivers, glaciers, and winds transport these particles. Some minerals, such as quartz, for example, resist chemical alteration and enter the ocean essentially unchanged. Others, such as the weathered remains of mica minerals, interact with seawater to form clay minerals, which are common in deep-sea muds. Any deep-ocean deposit containing 70 percent or more by volume terrigenous particles is called a **deep-sea mud.**

Terrigenous deposits dominate continental margins and much of the deep-ocean bottom. In the Atlantic, iron-stained particles from several deserts form the

Figure 5–1
Weathering of a marble headstone in Madison Cemetery, Connecticut has nearly obliterated the inscription. This illustrates how rapidly marble weathers in a humid climate. (Courtesy S. Judson and S. M. Richardson, *Earth: An Introduction to Geologic Change,* Englewood Cliffs, N.J.: Prentice-Hall, 1995.)

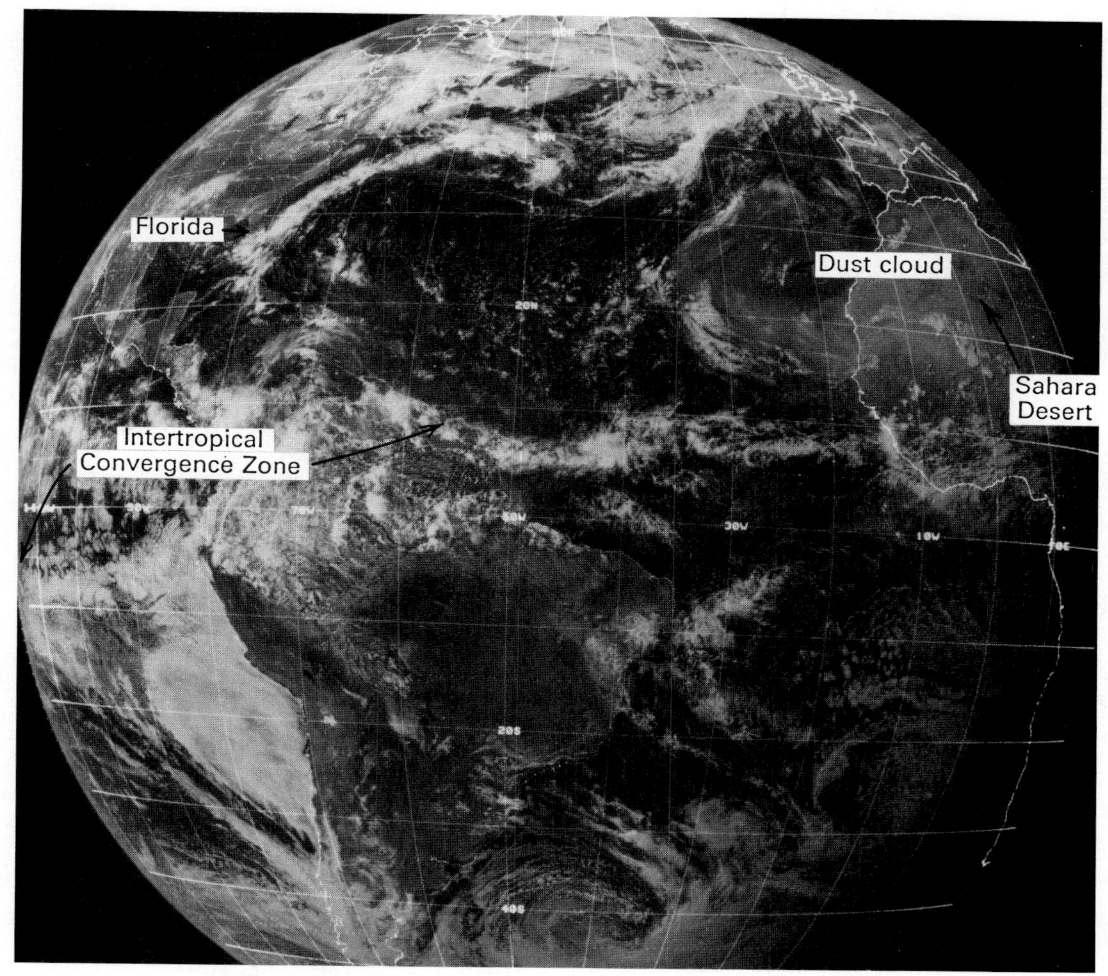

Figure 5–2
Storms carry dust particles from the land and deposit them far out into the oceans. (Courtesy NASA.)

red clays that are common on the deep-ocean floor. These particles are carried from the deserts in windborne dust clouds that can be seen by satellites (Fig. 5–2).

Each year, rivers carry about 20 billion tons of sediment from land to the ocean (Fig. 5–3). It comes primarily from mountainous areas in active continental margins; half of all this sediment comes from Pacific islands alone, with another third coming from southeastern Asia. In Asia, four rivers supply about one-quarter of all the sediment reaching the world's ocean from the continents. The many small rivers draining mountainous island arcs also carry large sediment loads. In addition to these natural sources, human activities, especially farming and deforestation, have greatly increased erosion and sediment discharges.

Little riverborne sediment enters the major ocean basins at present, a time of high sea level, for the following reason. Rivers carrying small sediment loads deposit most of the sediment in their estuaries and, therefore, little of it escapes to nearby ocean basins (Fig. 5–4a). Rivers carrying large sediment loads have already filled their estuaries and, therefore, most of the load reaches the mouths of the rivers, where deltas form as the sediments accumulate in the shallow coastal ocean. In the past, when the sea level stood lower because the continents were covered by glaciers (Fig. 5–4b), rivers entered the ocean near the edge of the continental shelf. Then sediments moved down submarine canyons onto the ocean

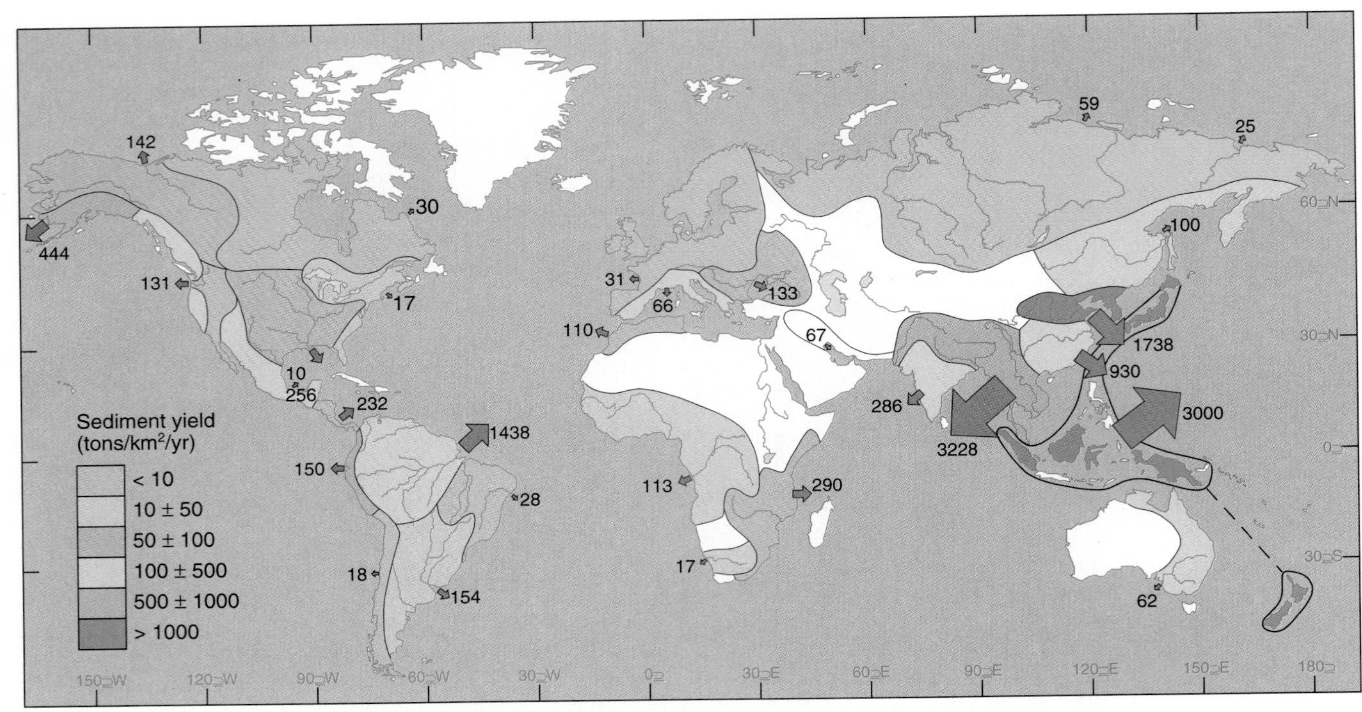

Figure 5–3
Large rivers carry about 20 billion tons of sediment from their drainage basins into the oceans. Small rivers draining areas of active mountain building carry especially large sediment loads. [After J. D. Milliman and R. H. Meade, "Worldwide Delivery of River Sediment to the Oceans," *Journal of Geology,* 91, (1983) p. 16. (Used with permission of University of Chicago Press.)]

floor. At times of low sea level, therefore, the amount of sediment entering the deep-ocean basins was much greater than it is today.

Around Antarctica, sediments transported by glaciers dominate so that deposits there consist largely of rock fragments released by melting ice. Even large boulders can be rafted out and then released as the ice melts. These sediments and rocks form **glacial–marine deposits.** Such deposits, which are also common in the Arctic basin and around its margins, show that the great ice sheets covering Iceland and North America until about 10,000 years ago released large pulses of icebergs, with the release perhaps triggered by sharply rising sea levels. This sudden calving of icebergs from the edges of the glaciers increased the amount of sediment they transported to the ocean.

Biogenous Sediments

If a deposit contains more than 30 percent by volume biogenous particles (which means particles from once-living plants and animals), it is called a **biogenous sediment.** The scientists on the *Challenger* expedition called such sediments *oozes* because of their fine-grained nature; that name is sometimes still used. Biogenous sediments are dominated by the shells of one-celled plants and animals, mixed with the shells and skeletal fragments of larger marine animals (Fig. 5–5). These deposits cover more than half the deep-ocean bottom (Fig. 5–6). Three processes interact to cause this distribution: production of shells and skeletal fragments by organisms living in surface waters, dilution by other kinds of particles, and destruction of shells, mostly by dissolution while they sink through the water or after they have landed on the sea floor. (Remember that biogenous sediments contain more than 30 percent shells by volume.)

Figure 5–4
Rivers transport sediments through estuaries onto the coastal ocean. (a) During times of high sea level, sediments are first deposited in estuaries and on deltas at river mouths. Little or no sediment crosses wide continental shelves to reach the deep-ocean floor. (b) During times of low sea level when coast lines lie near the edge of narrow continental shelves, sediments can easily move out onto the deep-ocean floor by moving across the shelves and down submarine canyons.

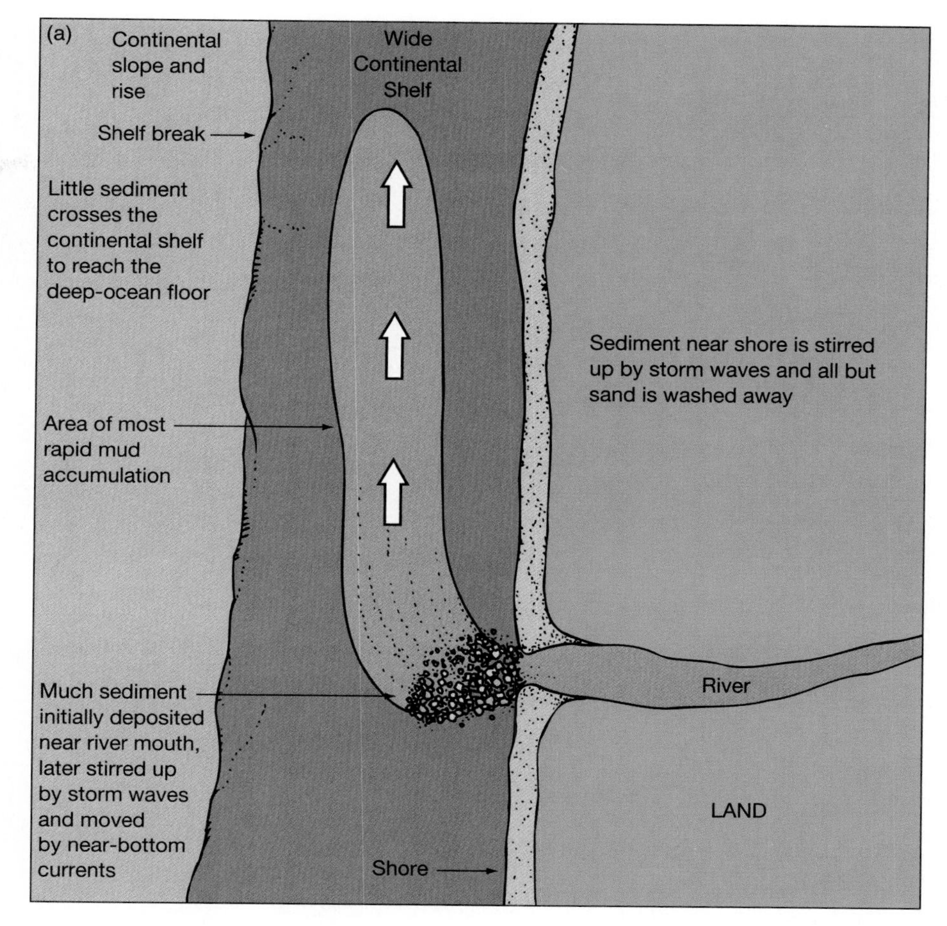

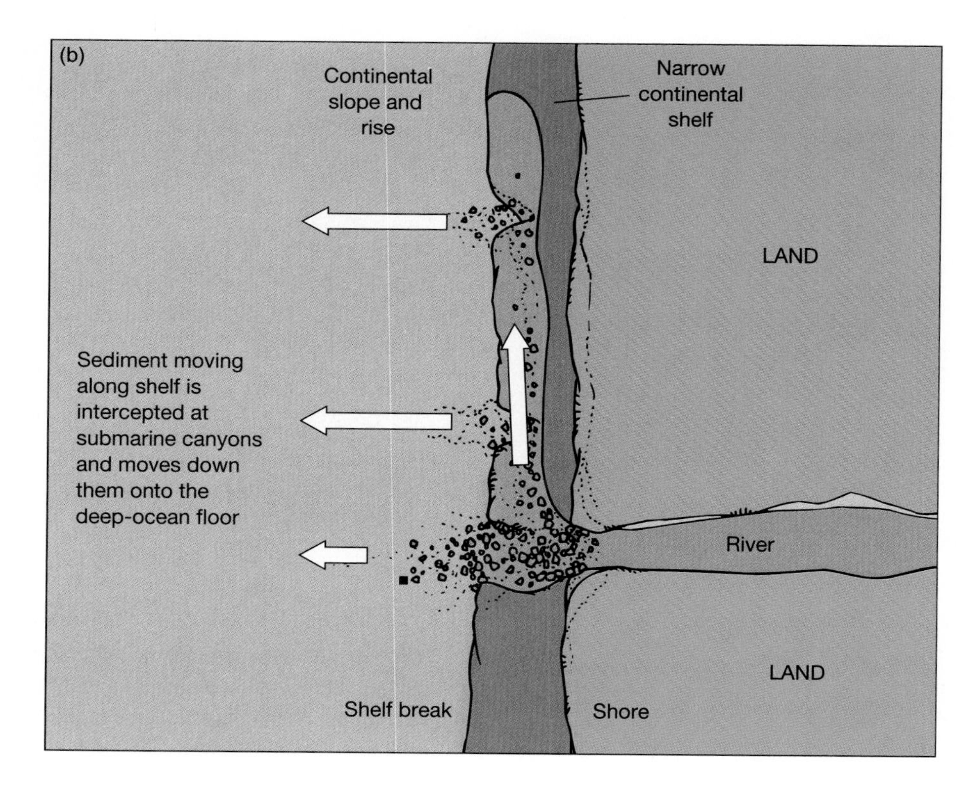

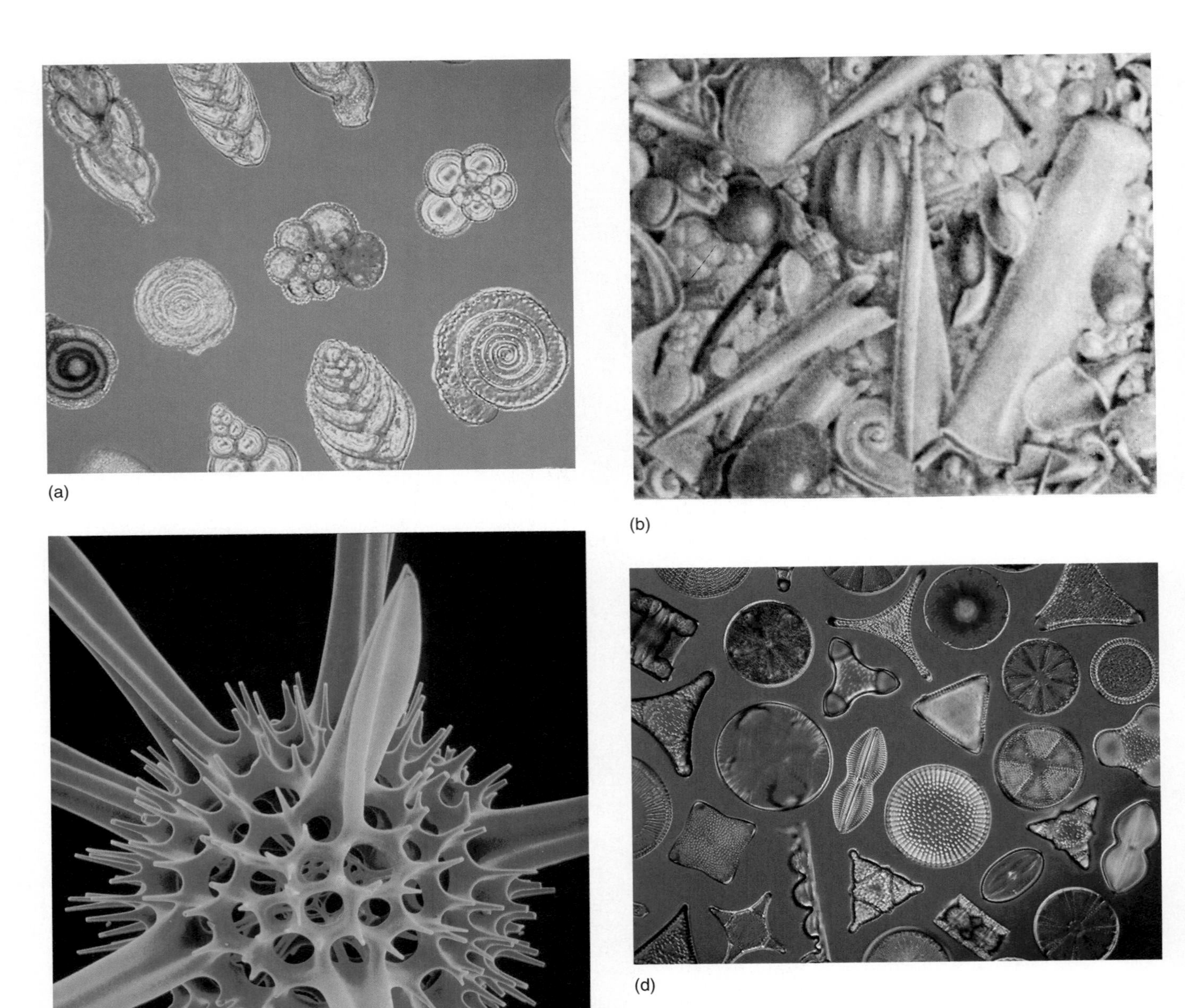

(a)

(b)

(c)

(d)

Figure 5–5
Some of the animal remains that make up biogenous sediments. (a) Calcareous foraminiferan; (b) Pteropod shells in deep-ocean sediment; (c) The skeleton of a siliceous radiolarian (latticelike structures); and (d) Siliceous diatom skeletons. These particles are about 100 micrometers in diameter, about the thickness of a human hair. [(a) Astrid & Hans Frieder-Michler/Science Photo Library/Photo Researchers, Inc.; (b) "Report on the Scientific Results of the Voyage of the H.M.S. *Challenger,* Vol. 1, Deep Sea Deposits"/New York Public Library; (c) and (d) Manfred Kage/Peter Arnold, Inc.]

Calcareous deposits, the most common type of biogenous sediment, consist of the shells of one-celled animals that form calcium carbonate shells, the principal constituent of limestones and chalk (Fig. 5–5a). These deposits are especially common on the shallower parts of the deep-ocean floor. (The reasons these deposits are found at shallower depths are explained later in the chapter.)

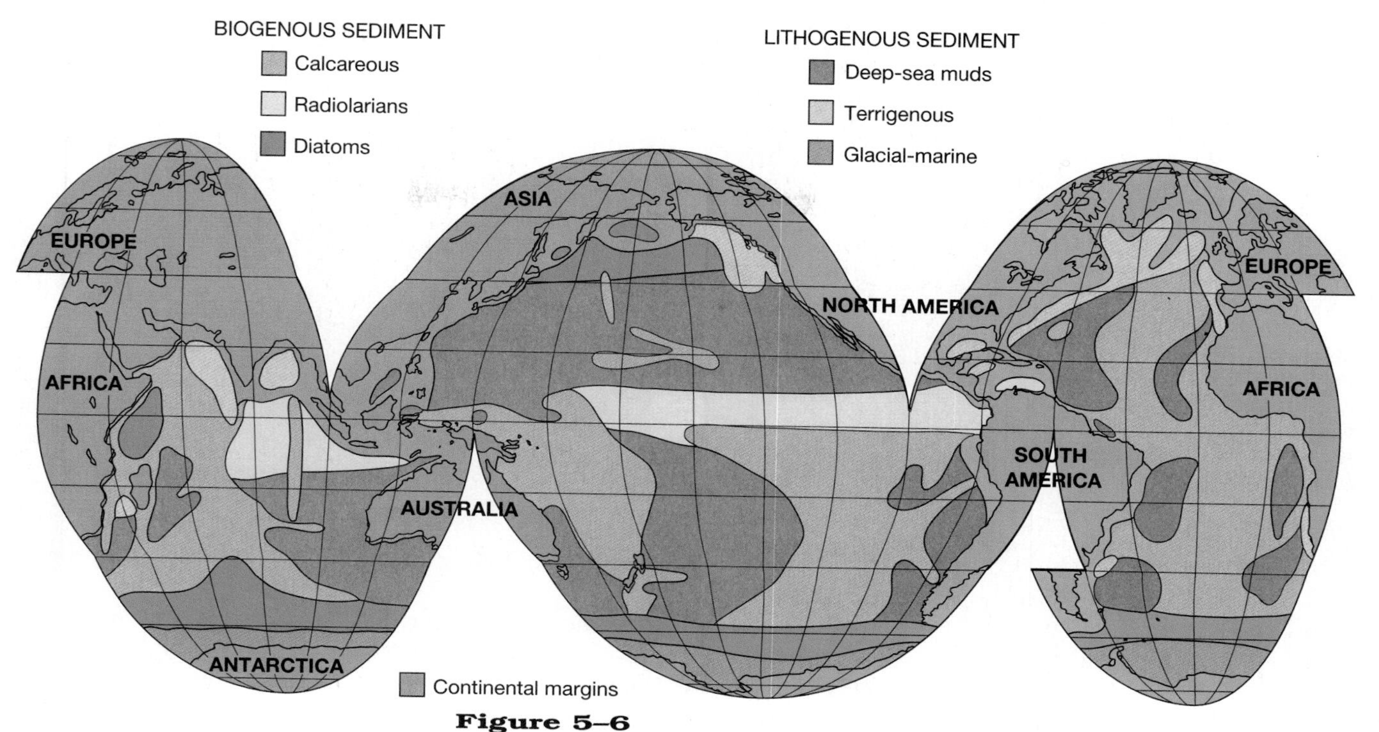

Figure 5–6
Distribution of deep-sea sediment deposits. (Those on continental shelves and slopes are not shown.)

Siliceous sediments are formed by accumulations of glasslike shells of microscopic, one-celled algae (diatoms) and animals (radiolarians) that grow in highly productive surface waters. (We discuss the causes of this high productivity in Chapter 13.) In the equatorial Pacific and Indian Oceans, bottom deposits are dominated by the radiolarians (Fig. 5–5c). In the subpolar North Pacific and around Antarctica and under productive coastal areas, glasslike shells formed by diatoms are the dominant constituents (Fig. 5–5d).

Biogenous particles are diluted by mixing with materials from other sources, primarily terrigenous materials. This is particularly common in the North Atlantic.

Destruction of Biogenous Particles

Siliceous shells as well as the bones and teeth of these animals, which are made of phosphate materials, dissolve everywhere in the ocean. Thus, deposits of such particles occur only below biologically highly productive waters where the shells, bones, and teeth are deposited so rapidly that they are not completely dissolved before they reach the bottom and are covered by more material from above. In areas that are not biologically productive, most of the shells dissolve and only the most robust ones survive to form marine deposits. Some highly resistant phosphatic fragments, such as fish teeth or whale's earbones, resist dissolution and persist for long times on the deep-ocean floor. Fish bones are easily destroyed and, therefore, are rarely found in sediment deposits.

Effects of high biological productivity in the upper ocean can be seen on the deep-ocean bottom and on continental shelves. On the deep-ocean bottom, bands of diatomaceous muds in high latitudes and radiolarian muds in equatorial regions directly reflect the high biological productivity of surface waters in these regions. On continental shelves, the abundance of recent carbonate sediment in

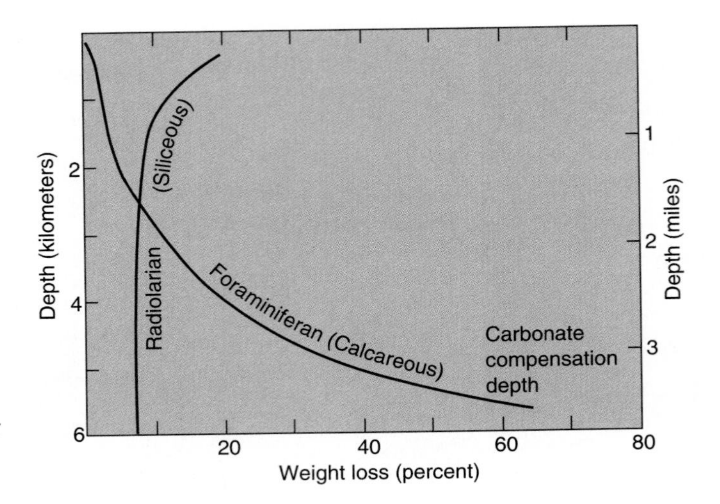

Figure 5–7
Percent weight loss of siliceous and calcareous particles due to dissolution at various depths in the central Pacific after the particles have been in the water for four months.

tropical waters is again a result of locally abundant growth of carbonate-shelled organisms.

Some one-celled organisms form their shells from calcium carbonate. The warm surface waters in which these organisms live are supersaturated with calcium carbonate and, therefore, the carbonate shells cannot dissolve. Below the surface zone, however, the water is no longer saturated with calcium carbonate and so calcium carbonate minerals dissolve. This dissolution occurs most rapidly at the low temperatures and high pressures in the deep ocean. The **carbonate compensation depth** (depth at which carbonate particles dissolve), typically occurs at depths of 4 to 5 kilometers (Fig. 5–7). At this depth, rates of weight loss for calcareous shells begin to increase markedly. Below the depth at which seawater is saturated by carbonate, carbonate particles totally dissolve and do not form calcareous sediments except below highly productive waters.

The shells of some organisms, such as pteropods (pelagic or swimming snails, shown in Fig. 5–5b and discussed more fully in Chapter 12) consist of an especially soluble carbonate mineral. Such shells dissolve rapidly as they sink through the ocean and, consequently, accumulate only on shallow volcanic peaks, primarily in the Atlantic (Fig. 5–8).

Hydrogenous Sediments

Chemical reactions in seawater involving iron and manganese form tiny particles that stick to hard surfaces, forming either a complete coating or patches of nod-

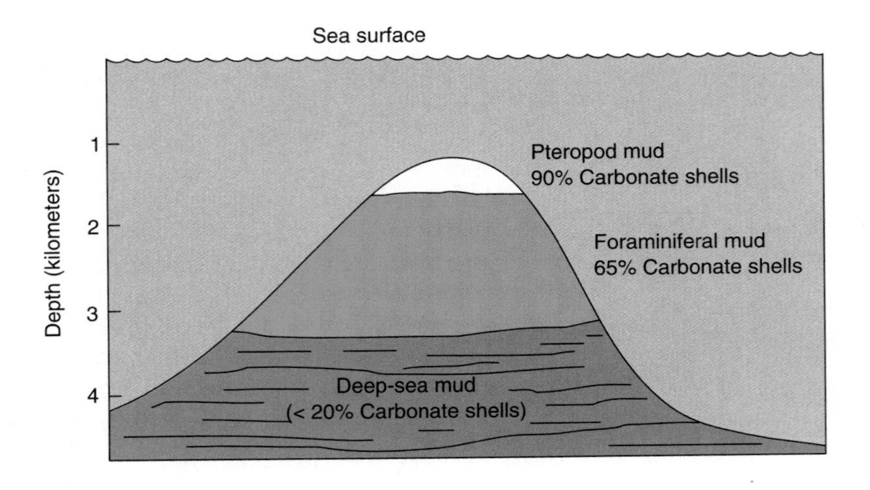

Figure 5–8
Carbonate deposits are common on shallow portions of seamounts and ridges. These deposits contain easily dissolved components, such as pteropod shells.

Figure 5–9
Fist-sized manganese nodules. (Tom McHugh/Photo Researchers, Inc.)

ules; these are called **hydrogenous sediments.** In ocean areas receiving little other sediment, manganese nodules as large as tennis balls are often conspicuous on the ocean bottom (Fig. 5–9). In the central Pacific, metal-rich coatings on the surfaces of extinct volcanoes are another type of hydrogenous deposit.

Hydrogenous nodules form around objects that are very resistant to being dissolved in water, such as volcanic rock fragments and resistant skeletal fragments. Over millions of years, iron-manganese coatings collect on such objects, forming layer upon layer of nodules so that a cross section through a nodule reveals roughly concentric rings (Fig. 5–10).

Slow-growing manganese nodules are buried by sediments that accumulate more rapidly. Thus, nodules are rare in the Atlantic because sediments accumulate rapidly there, but in the central Pacific, where sediments deposit slowly, nodules cover an estimated 20 to 50 percent of the ocean bottom. Tunneling and churning

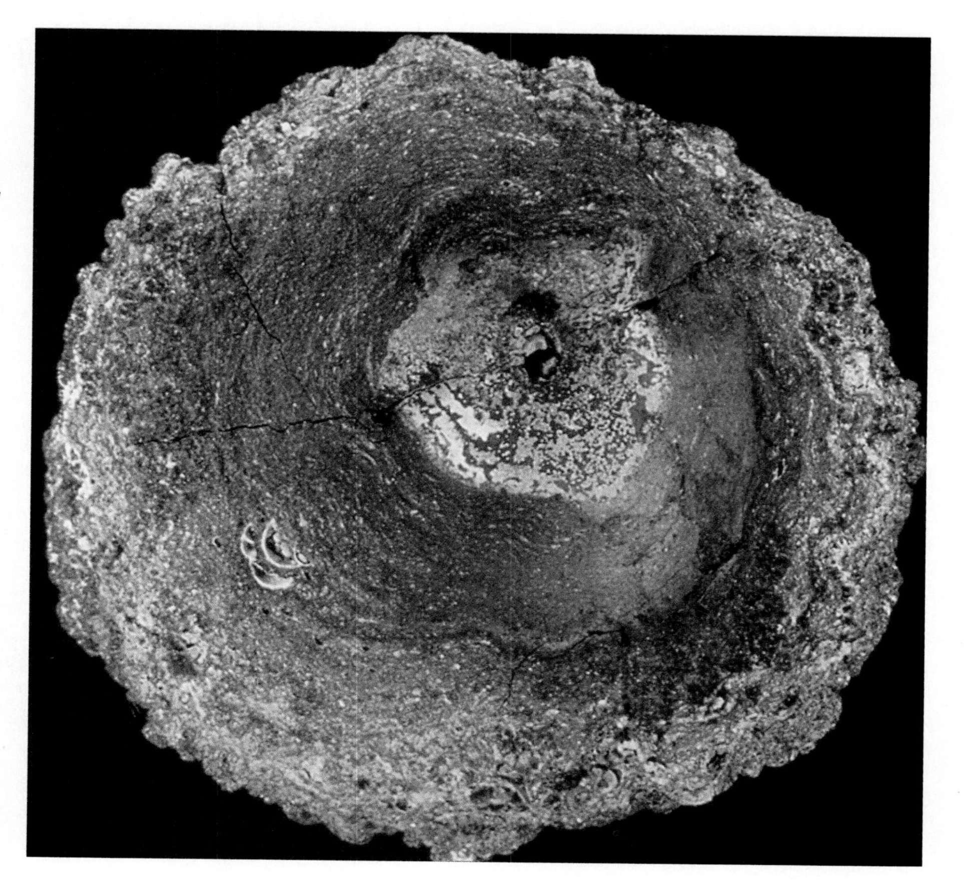

Figure 5–10
Growth rings in a cross section of a manganese nodule. The white layers form when organisms encrust the outer surfaces of the slowly growing nodule. Each surface is gradually covered by later deposits, thus forming growth rings. (J&L Weber/Peter Arnold, Inc.)

BOX 5–1
Oil and Gas from the Sea Floor

Oil and natural gas—the most valuable resources taken from the ocean floor—are the remains of organic matter buried in sediments eons ago and later altered by heat and pressure. In areas where biological productivity was exceptionally high in the upper ocean and bottom-water circulation was sluggish, dissolved oxygen was used up by decomposing marine plants. Growth of plant-eating organisms was thereby inhibited, permitting organic matter to accumulate in the sediment deposits. By contrast, today's ocean has few (if any) areas favorable for oil and gas formation.

Heat from Earth's interior transforms buried organic matter to oil at temperatures of 100° to 150°C and at depths of around 5 kilometers. At higher temperatures or with very prolonged heating, all the organic matter may be transformed to natural gas or even destroyed; both increased temperature and increased pressure speed up the reactions.

Through time, sediments compact under the weight of overlying deposits. Fluids (water, oil, and gas) are expelled and move from the sediments (such as delta deposits) into nearby porous rocks, where they accumulate (Fig. B5–1–1). Porous reefs and permeable buried beaches are especially favorable sites for the accumulation of oil. Eventually the movement of these fluids is stopped by impervious sediments through which the fluids cannot readily move. Thus, three factors are necessary for oil and gas deposits to form:

1. Deposition of sediments rich in organic matter (source beds).
2. Heating (combined with high pressures) of source beds to form oil or gas from the organic matter.
3. Permeable, porous reservoir rocks to hold oil and gas where it can be extracted.

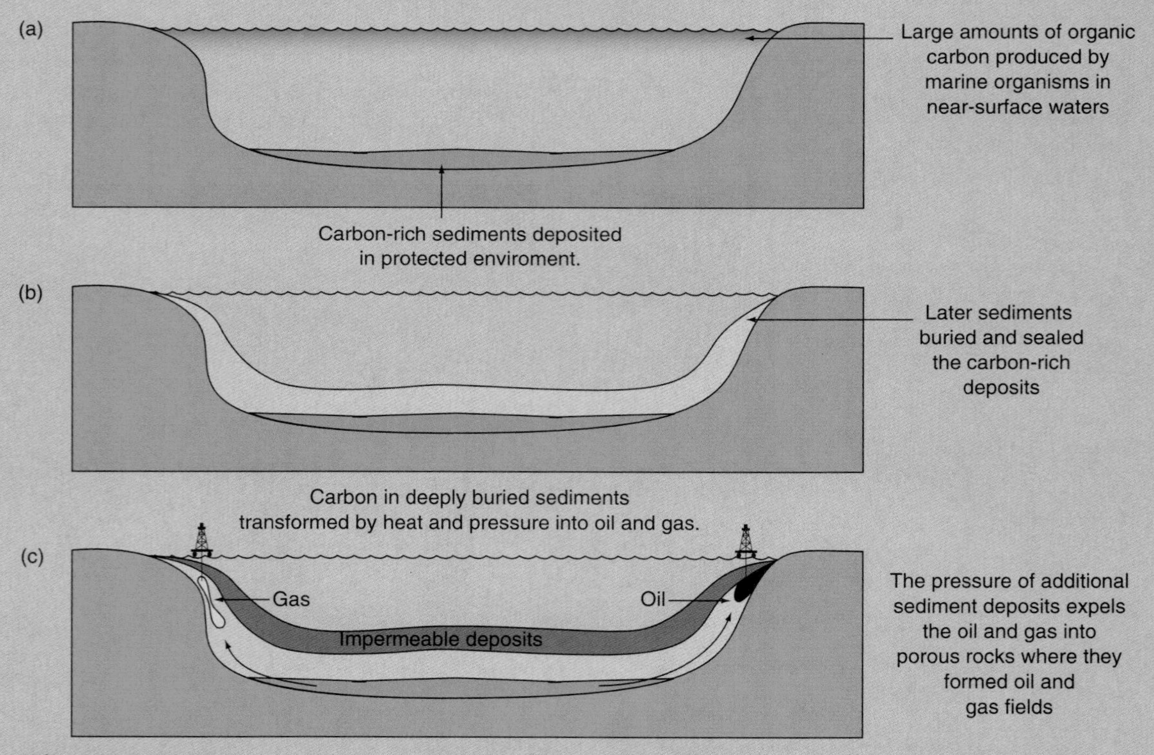

Figure B5–1–1
Oil and gas forms in marine sediments in protected and usually isolated ocean basins. (a) Initially carbon-rich materials, formed in the surface waters, accumulate in deep waters where they cannot be consumed by other marine organisms. (b) Later accumulations of sediments seal the carbon-rich materials. The high temperatures and pressures transform the organic matter into oil and gas. (c) Accumulations of more sediment compress the original deposits, expelling the oil and gas which migrate into permeable rocks, usually sands. These accumulations of oil and gas can then be drilled and produced.

Estimates of undiscovered petroleum resources are shown in Figure B5–1–2. Geologists believe that about 70 percent of the undiscovered oil and gas lies under continental shelves and shallow marginal ocean basins. The continental shelf beneath the North Sea is a major oil- and gas-producing area, for instance, and oil and gas from the continental shelves around western Indonesia and southeastern Asia are now coming into production. Other areas, such as the North Atlantic shelf off Newfoundland, are expected to begin producing soon.

About 24 percent of the as-yet undiscovered oil and gas is expected to come from the deeper parts of continental slopes. These pools are likely to be exploited in the future as new production techniques are developed. Relatively little petroleum is likely to come from the deep-ocean floor, where little organic matter is now preserved. Also, costs of drilling there would be extremely expensive, well beyond the limits of present technologies.

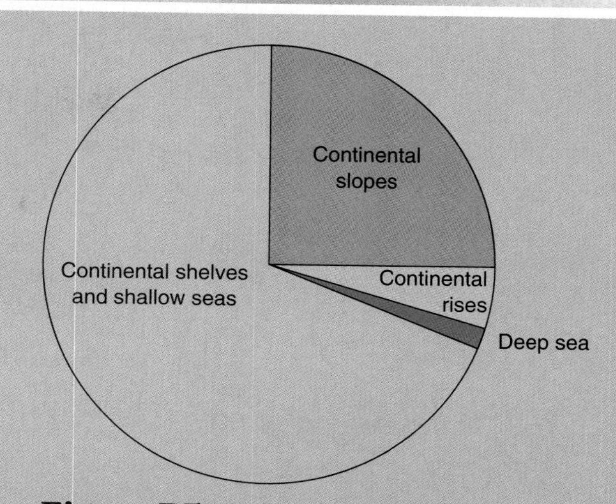

Figure B5–1–2
Relative amounts of undiscovered petroleum resources in the ocean. (After NAS/NRC, 1975.)

by sediment-feeding organisms help keep nodules at the surface so that they continue to grow.

Cosmogenous Sediments

Meteorite fragments from space accumulate in oceanic sediments and in ancient glacial ice in Antarctica, and these accumulations are called **cosmogenous sediments.** An estimated 30,000 metric tons per year of such particles fall into the ocean, derived mostly from meteorites and comets that burn up in Earth's atmosphere.

Cosmic particles were first recognized in marine sediments by Canadian-born John Murray when he studied deep-ocean sediments collected by the *Challenger* expedition. The cosmogenous particles are about 200 to 300 micrometers in diameter (Fig. 5–11). Most are magnetic and consist primarily of iron or iron-rich minerals. Others consist of silicate minerals, which make the grains harder to recognize and differentiate from other types of terrigenous grains. Many have distinctive surface features that formed as they melted while passing through the atmosphere (Fig. 5–12).

Figure 5–11
Spherical cosmogenous particles from deep-ocean deposits. (*Challenger Reports.*)

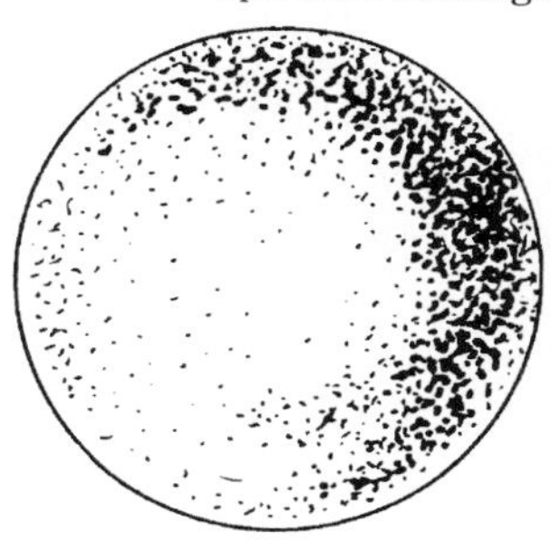
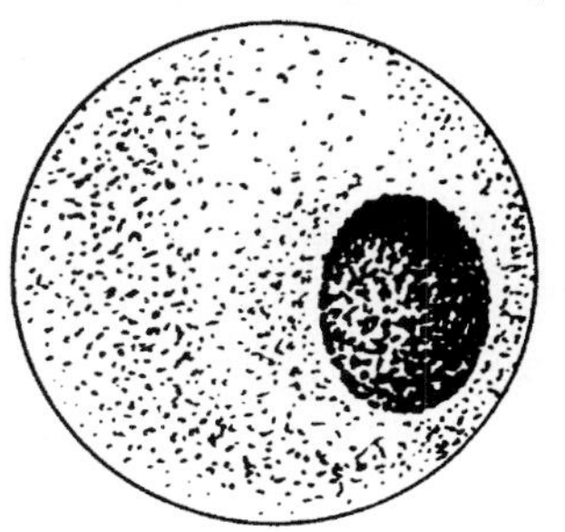
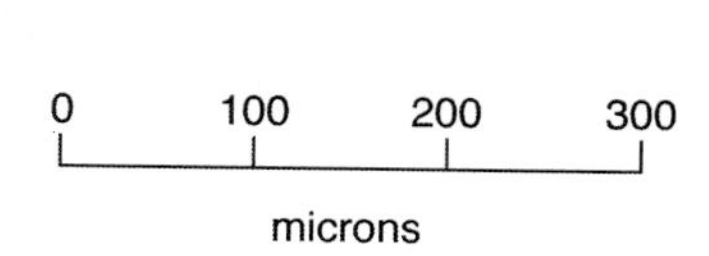

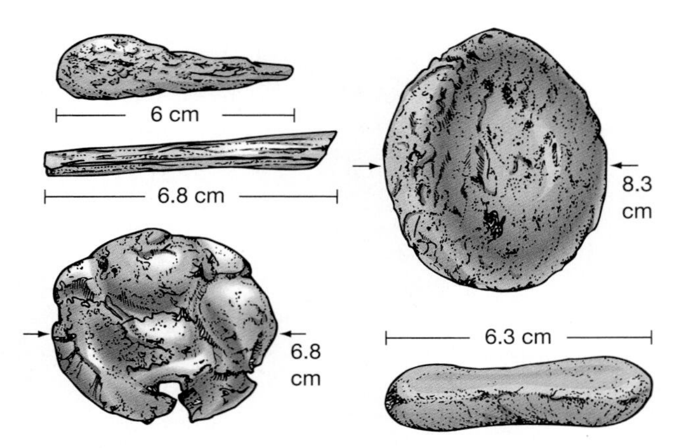

Figure 5–12
Some cosmic materials melt as they pass through the atmosphere, forming distinctive teardrop shapes. Some flatten when they hit the water, forming tektites.

Violent collisions between meteorites or asteroids and Earth's surface cause fragments of Earth's crust to melt and spray outward from the impact crater. This crustal material remelts as it falls back through the atmosphere and forms glassy particles called **tektites.** Tektites, up to about 1 millimeter in size, form a very small component of some ocean sediment layers, but they provide compelling evidence of past meteor impacts.

Particle Transport

Particles are transported in the ocean by physical processes, which are controlled primarily by grain size and current speed. Consequently, sediment particles are also classified according to size, with different sources producing particles of different sizes, as indicated in Figure 5–13.

Figure 5–13
Particle size distributions in (a) different size classifications, in (b) biogenous sediment. (c) Relative amounts of the various particle types found in a typical sediment deposit. The solid portions of the lines indicate the dominant size ranges; the dashed portions indicate extremely large or small-sized fractions.

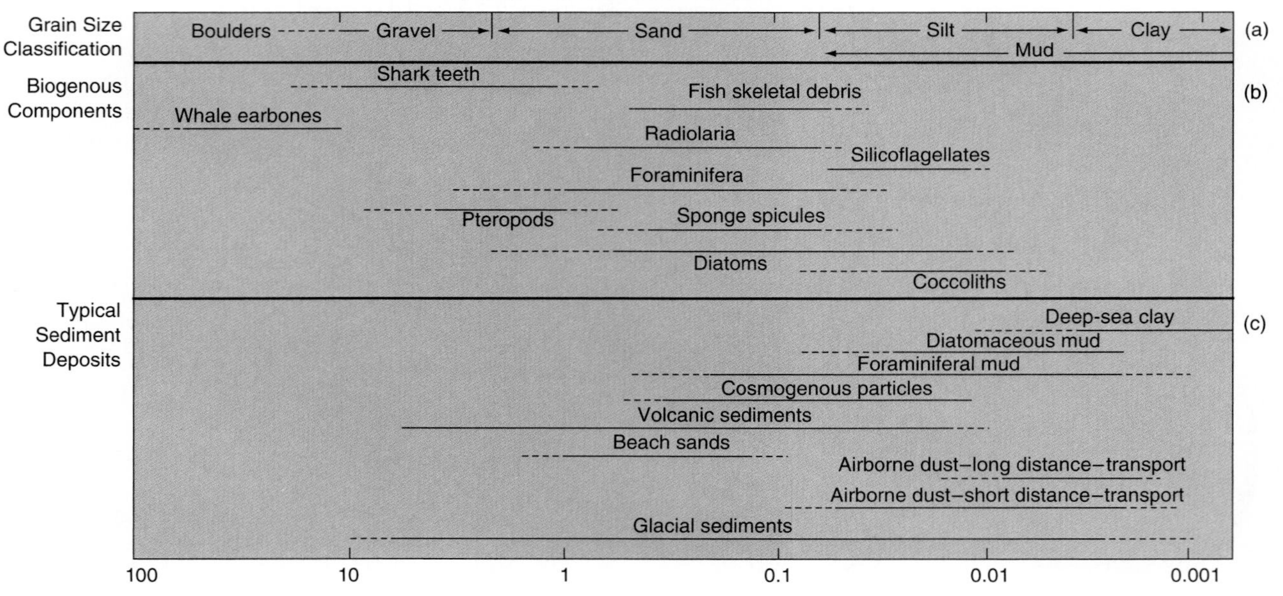

TABLE 5–1
Settling Speeds for Various Particles

Particle Type	*Particle Diameter (μm)*	*Settling Speed (cm/s)*	*Time Needed to Settle 4 km*
Sand	100	2.5	1.8 days
Silt	10	0.025	185 days
Clay	1	0.00025	50 years

Sinking of individual particles through seawater is easily visualized; a particle's size determines its settling speed. Large particles sink more rapidly than small ones, as shown in Table 5–1. Thus, large grains settle near the spot where they enter the ocean. Small grains, on the other hand, are easily transported great distances while they slowly settle to the deep-ocean floor. Very small particles are dispersed by currents throughout the ocean during the decades that they may take to sink to the deep-ocean bottom. Thus, settling times are increased by water movements and by turbulence.

Not all particles that enter the ocean settle to the bottom, because some are removed by organisms that filter ocean water to obtain their food (discussed in Chapters 11, 12, and 13). The inedible particles are eventually compacted into pellets and excreted. These **fecal pellets** sink to the bottom within a few days. As they sink, they may be eaten by other organisms and then reexcreted. They may also be partially decomposed by bacteria. In nearshore waters, oysters, mussels, clams, and many other animals remove particles from suspension and bind them into pellets that accumulate at the bottom of the nearshore.

Turbidity Currents

Turbidity currents are dense mixtures of water and sediment particles. Because they are much denser than normal seawater, these currents move like avalanches along the ocean-floor slope, transporting large amounts of sediment particles (Fig. 5–14) onto the deep-ocean floor.

Earthquakes can cause turbidity currents. In November 1929, the Grand Banks earthquake, near Newfoundland, triggered a turbidity current that hurtled down the continental slope at speeds up to 100 kilometers per hour, breaking submarine telegraph cables in its path. Sediments deposited by this turbidity current covered a large part of the nearby North Atlantic ocean bottom.

Turbidity currents are also triggered by sudden large discharges of riverborne sediment during floods.

As a turbidity current moves, it picks up particles of all sizes, eroding the bottom as it passes. The largest of these erosion particles settle out first. Later, as its flow slows, the current can transport only finer particles, which eventually settle out on top of the earlier-deposited coarser-grained materials. The result is a characteristic deposit—called a **turbidite**—with the coarsest particles at the bottom and finest at the top in a layered arrangement called **graded bedding** (Fig. 5–14b). Turbidity currents can also carry shallow-water organisms and plant fragments onto the deep-ocean floor. Plant materials—some still green—have been recovered from the ocean bottom by crews repairing submarine cables broken by turbidity currents.

Turbidity currents occur most commonly along narrow continental shelves and are least common off wide shelves. In the Atlantic and northern Indian Oceans, turbidity-current deposits near the basin margins have buried the older

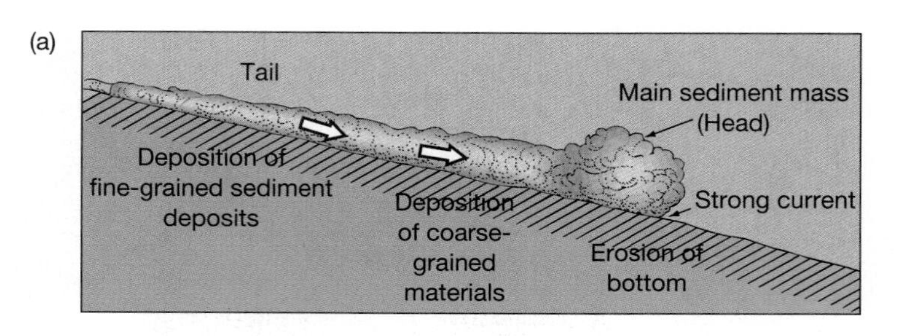

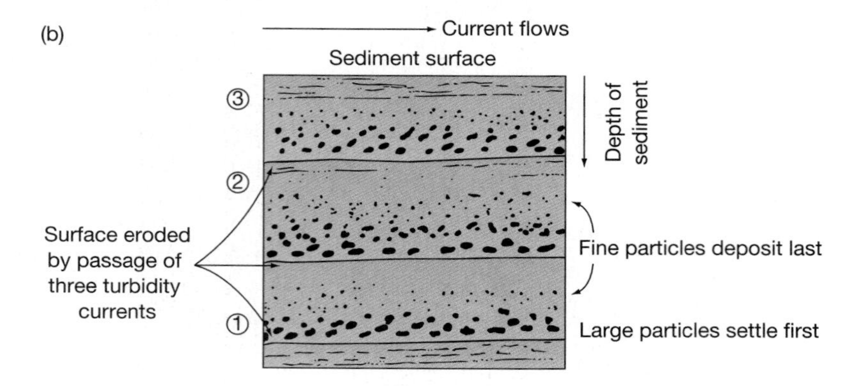

Figure 5–14
(a) Schematic representation of a turbidity current moving down a slope. (b) Cross section of deposits formed by the passage of three successive turbidity currents. A turbidity current forms graded bed 1 by dropping first its largest particles and then smaller and smaller ones. Some time later, a second turbidity current comes along and deposits graded bed 2 in the same fashion. Still later, a third turbidity current forms graded bed 3.

seafloor topography. As a result of this coverage, turbidity currents can now flow far out onto the deep-ocean floor because all bottom features that might impede their movements have been buried. In the Pacific, on the other hand, trenches and ridges (formed by continued mountain building) around the basin margins trap turbidity currents before they can reach the deep-ocean floor.

Besides carrying sediment and burying ocean-bottom topography, turbidity currents also erode submarine canyons that indent the continental margins. In many areas, turbidity currents remove sediment deposits from canyons, keeping them open. The currents break submarine cables that cross submarine canyons off the mouths of large sediment-transporting rivers, such as the Congo in Africa and the Ganges–Brahmaputra in Asia. Cables near such rivers break as often as once a year.

Atmospheric Transport and Volcanic Eruptions

Winds carry about 500 million metric tons of dust to the open ocean (Fig. 5–2) each year, primarily from deserts (Fig. 5–15) and high mountains. Windborne particles from the large deserts in Asia and North Africa dominate deep-ocean deposits in remote open-ocean areas in the mid-latitudes of the Northern Hemisphere. Windblown dust is less important in the Southern Hemisphere, where there is little land and few deserts.

Volcanic eruptions (Fig. 5–16) also contribute immense quantities of ash to the atmosphere, and this ash can be windborne for long distances before settling out in the ocean. Volcanic ash particles smaller than 20 micrometers in diameter are carried thousands of kilometers by winds. Ash particles smaller than about 10 micrometers move with the winds around the world after explosive eruptions inject them into the stratosphere.

Numerous ash layers in deep-ocean sediment cores attest to many large volcanic eruptions throughout Earth's history. Only a few have been dated and their sources identified, however. Enormous volcanic eruptions in southeastern Asia have left numerous easily recognizable ash deposits on the nearby deep-ocean floor. In 1815, for instance, the Indonesian volcano *Tambora* erupted, injecting an estimated 80 cubic kilometers of ash into the atmosphere. These ash clouds disrupted weather around the world, causing 1816 to be the "year with no summer"

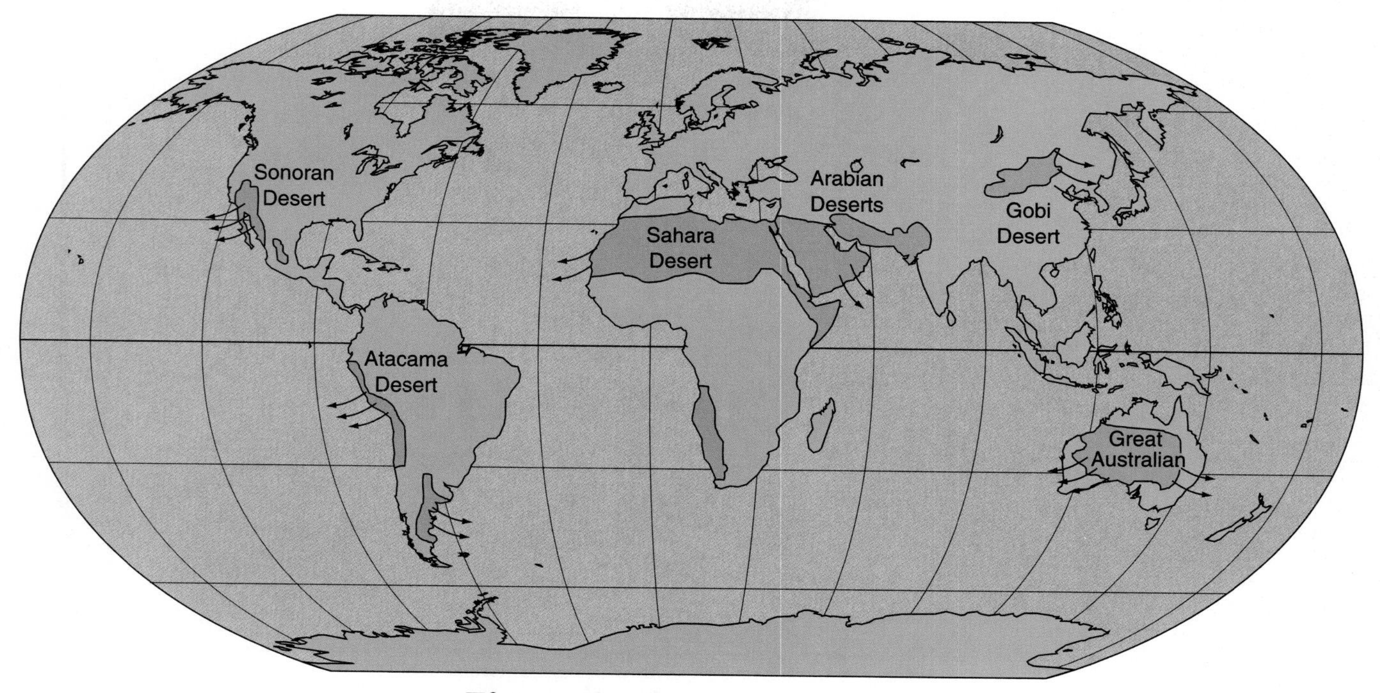

Figure 5–15
Winds from the major arid regions of the world carry dust clouds that deposit sediments in the ocean.

in New England where snow fell in July. The eruption left a halo of volcanic ash deposits that is still recognizable in deposits on the deep-sea floor surrounding the volcano. On August 26, 1883, *Krakatoa,* a volcano erupting between Java and Sumatra in Indonesia, put about 16 cubic kilometers of ash into the atmosphere, causing brilliantly colored sunsets for several years. Ash deposits cover 4 million square kilometers of ocean bottom around the volcano. The June 1992 eruption of the long-dormant Philippine volcano *Mt. Pinatubo* was comparable to Krakatoa, and oceanographers have no doubt that Mt. Pinatubo's ash deposits will be found on the nearby ocean floor. (See Table 5–2.)

Generalized distributions of deep-ocean sediments are shown in Figure 5–17.

Figure 5–16
A cloud of ash from the Kliuchevskoi volcano on Russia's Kamchatka Peninsula is carried by winds out over the North Pacific Ocean. (Courtesy NASA.)

TABLE 5–2
Major Volcanic Eruptions and Ash Injections to the Atmosphere

Volcano (Country)	*Date*	*Volume Ejected (km³)*	*Effect*
Toba (Indonesia)	70,000 B.P.	1,000	May have triggered glaciation
Tambora (Indonesia)	1815	100	Caused "year without summer"
Santorini (Greece)	1470 B.C.E.	10	Source of Atlantis legend
Krakatoa (Indonesia)	1883	10	
Laki (Iceland)	1783	10	Studied by B. Franklin
Mt. Pinatubo (Philippines)	1992	10	Caused worldwide cooling
Mt. St. Helens (U.S.)	1980	1	Regional destruction

Sources: *The New York Times* and H. Stommel and E. Stommel. *Volcano Weather.* Newport: Seven Seas Press, 1983. 177 pp.
B.P. = before present.
B.C.E. = before the common era.

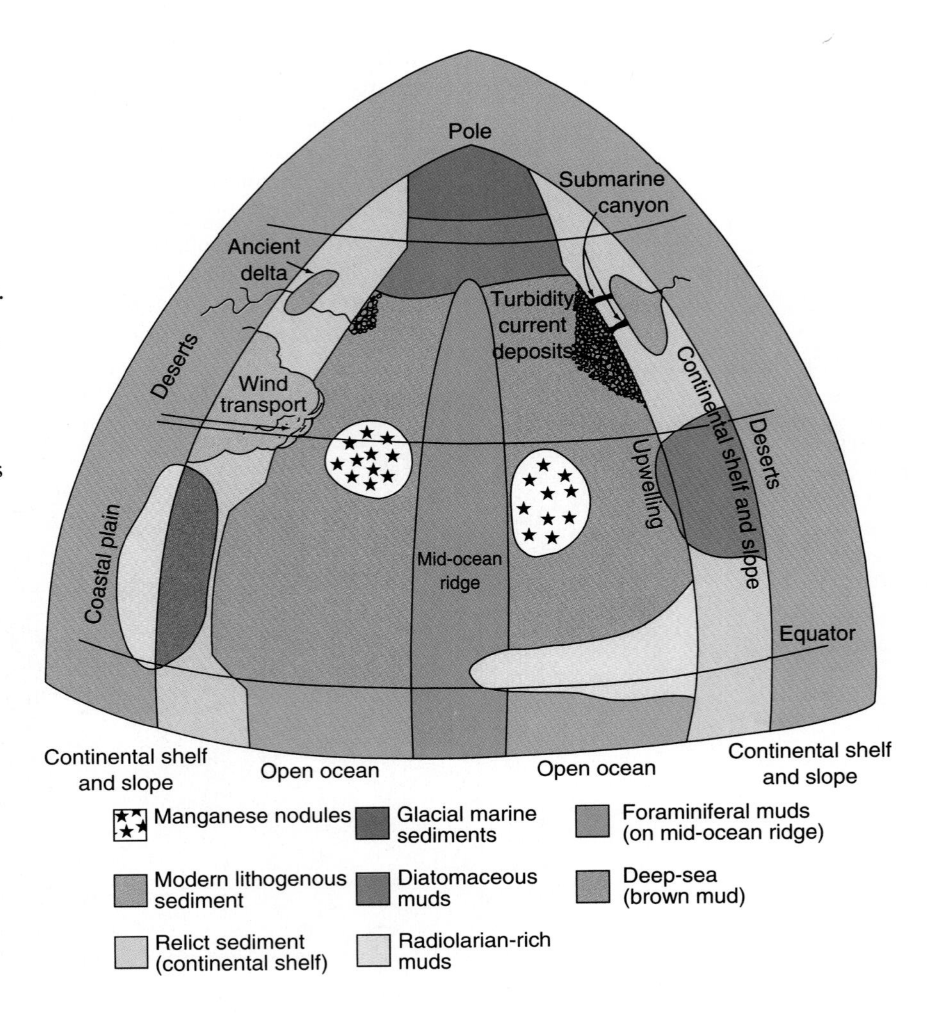

Figure 5–17
Schematic representation of the general distribution of sediments in the deep and coastal oceans. Sediment deposits are thickest near land where they are transported by rivers onto continental shelves and then by turbidity currents down submarine canyons on the deep-ocean floor. Sediments are thinnest on the recently formed oceanic crust of mid-ocean ridges. Winds transport sediment into remote ocean areas far from land. Ocean bottoms receiving little sediment from other sources are often covered by manganese nodules. Biogenous sediments are most abundant in areas of upwelling near coasts and under the equator. (We discuss the processes involved in Chapter 12.)

Sea Level and Sediment Deposits

Sea level has changed many times in Earth's history. During the present Ice Age, for instance, when continental ice sheets reached their maximum extent about 18,000 years ago, sea level stood about 130 meters lower than now because the water frozen in the ice sheets came from the ocean. When Earth's climate was warmer, between the various glacial periods, the polar ice caps were greatly reduced by melting, and sea level stood many meters higher than its present level. These as well as many other comparable sea-level changes are recorded by the types of plants and animals (fossils) preserved in sediment deposits worldwide.

Volcanic activity on the mid-ocean ridges also affects sea level. During periods of high submarine volcanic activity, mid-ocean ridges expand because the large amounts of recently formed oceanic crust are shallower (Fig. 3–7). This effectively reduces the volume of the basin. Having the same amount of water in a smaller basin, of course, means the water level is higher. Such times of high sea level are recorded in the sediment deposits that formed on the continents when large continental areas were flooded by shallow seas (Fig. 5–18). These ancient high sea stands were about 150 meters above present sea level.

Figure 5–18
Sea levels are low when rates of crustal formation are low (a). Higher sea levels occur during times when rates of crustal formation are high (b) because large areas of recently formed crust are relatively shallow and decrease the volume of the basin. Sea levels return to lower level when rates of crustal formation decrease. (c) These changes in sea level are recorded in sediment deposits on continental margins. Terrestrial deposits form during times of low sea levels and are covered by marine deposits when sea level is high. Deep-ocean sediment deposits are thickest near the continents and thinnest over the recently formed crust at mid-ocean ridges.

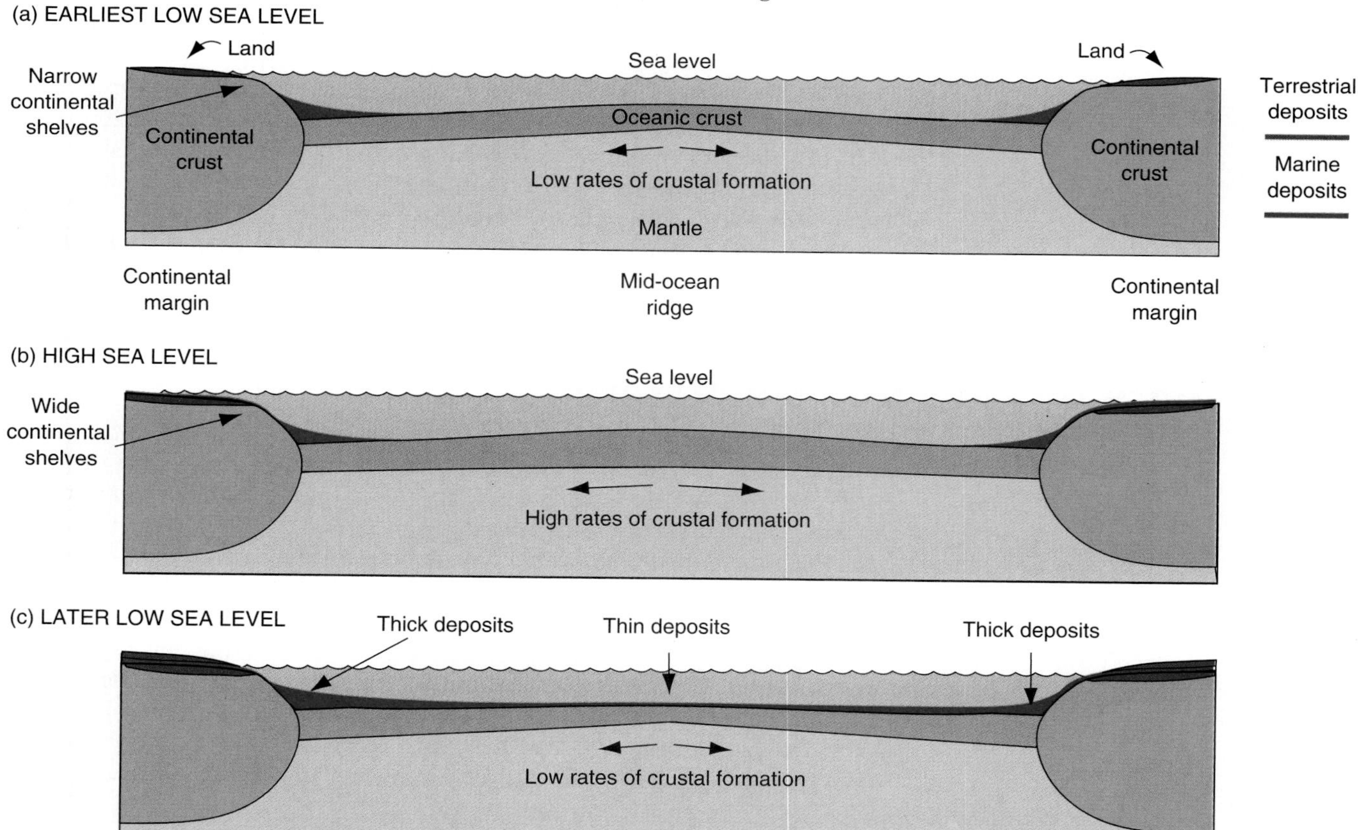

TABLE 5–3
Typical Sediment Accumulation Rates

Location	*Average Accumulation Rate (cm/1,000 years)*
Deep ocean	
Deep-sea muds	0.1
Marginal ocean basins	10–100
Continental margins	
Continental shelf	30
Continental slope	20
Estuary, fjord	400
Delta (Fraser River, Canada)	700,000

Deep-Sea Sediments

Much of the deep-ocean floor is covered by **pelagic sediments**—deposits that accumulate very slowly, particle by particle. Typical accumulation rates are between 0.1 and 1 centimeter per thousand years (Table 5–3). These deposits blanket the bottom, preserving the original outlines much like snow on a hilly countryside.

Since deep-ocean sediments accumulate slowly, particles may spend years sinking through the water before they are finally buried. As a result, there is ample time for any coatings on the particles to react chemically with seawater. For instance, dissolved oxygen in deep-ocean waters reacts with iron on the particles to form a coating of rust (iron oxide), and exposure to oxygenated waters and consumption by benthic organisms destroy any organic matter attached to the particles.

The thinnest sediment deposits generally occur on recently formed crust along mid-ocean ridges (Fig. 5–18). Because strong currents on the bottom sweep ridge crests, sediment deposits there can accumulate only in protected pockets. Sediment deposits thicken away from the ridges and are usually thickest over the oldest crust near the edges of the ocean basins. And, as we have noted, sediment deposits are thickest on continental rises and in marginal ocean basins near the mouths of major sediment-transporting rivers.

Sediments on Continental Margins

Particle sources and transport processes control sediment distributions on continental margins. In general, most river-transported particles that reach the ocean today remain on the continental margins. In some places, turbidity currents may carry the sediments into the deep ocean.

Deposits formed under conditions no longer existing are called **relict sediments** (Fig. 5–19). These sediments have distinctive features, such as remains of organisms that no longer live at the location of a given relict sediment. For example, oyster shells occur on the U.S. continental shelf, deposited during times of lower sea levels. Other features of relict sediments include iron stains or coatings on grains that would not form under present marine conditions. About 70 percent of the world's continental shelves have such relict sediments. Many of these are submerged ancient landscapes not yet covered by later deposits. Again, this is a

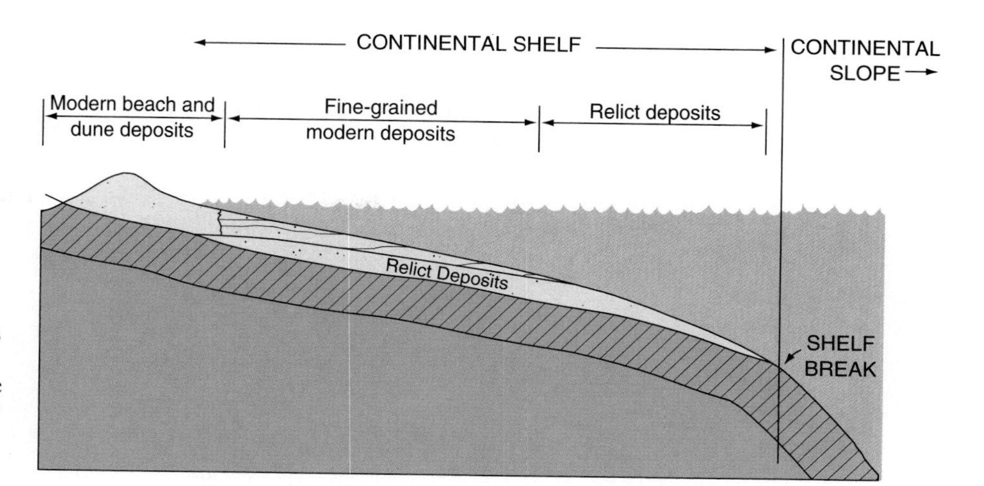

Figure 5–19
A mid-latitude continental shelf often has modern deposits on the shallow ocean floor that are thickest near the shore. The outer edge of the shelf is often covered by relict sediments deposited under conditions no longer prevailing in the area.

consequence of the recent rise of sea level and the fact that little sediment now escapes estuaries and deltas.

Where sediments from today's rivers do get as far as the coastal ocean, relict sediments and ancient coastal features are buried by modern deposits. Also, waves rework relict materials as shorelines move across the continental shelf as sea level rises and falls.

As already mentioned earlier in the chapter, most riverborne sediment particles remain in estuaries. For example, on the U.S. Atlantic coast, rivers transport about 20 million metric tons of sediment each year to their estuaries. But virtually no sediment now reaches the outer portions of wide continental shelves.

Wave-generated longshore currents (discussed in Chapter 9) move sand deposited in nearshore waters, especially near river mouths. The association of sand beaches and river mouths is obvious on many coasts. For example, the large area of beaches and dunes on the Washington–Oregon coast is near the Columbia River mouth.

Most sediments deposited on continental shelves accumulate rapidly and, consequently, there is usually not enough time for the particles to react chemically with seawater. Terrigenous coastal-ocean sediments, therefore, retain many characteristics acquired during weathering. These rapidly accumulating deposits also contain organic matter, causing them to be dark-colored—gray, greenish, or sometimes brownish.

On the continental margins in polar regions, glacial-marine sediments are common. Sea ice normally contains sediment only if it has gone aground and incorporated sedimentary materials that freeze to the bottom of the ice. This happens often in the Arctic Ocean. When such sea ice melts, its sediment load—usually poorly sorted—sinks to the bottom.

Beaches

Beaches are the sediment deposits we know best because they are used so extensively for recreation. We are most familiar with beaches composed of sand grains (diameters between 0.062 and 2 millimeters). Most beaches are dominated by highly stable mineral grains such as quartz, a form of silica. In Hawaii, black sand beaches result from the erosion of lava. Where silicate rocks are rare or absent, sand-sized grains of other stable minerals form beaches. In the tropics, broken and ground carbonate shells and skeletons of marine organisms dominate the beaches. Such beaches—in Bermuda, for example—are often white or slightly pink.

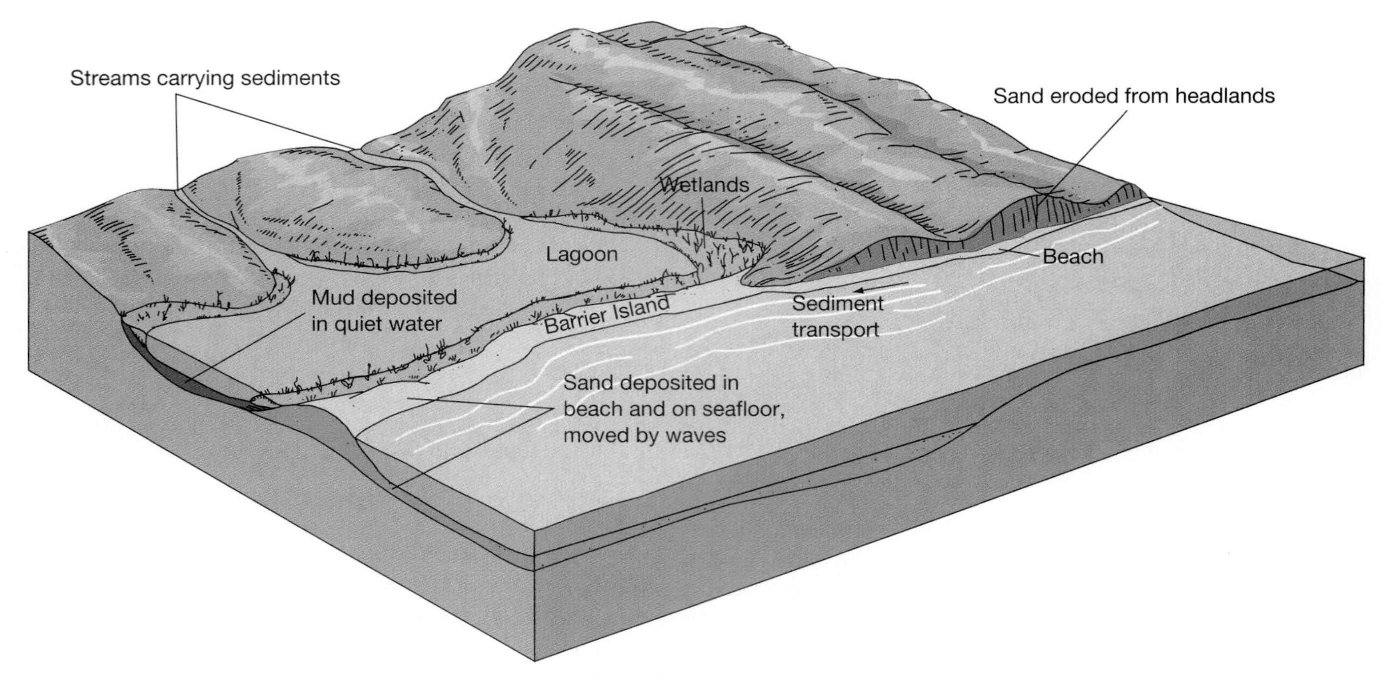

Figure 5–20
A barrier island isolates a lagoon from the adjacent coastal ocean. Muds accumulate in the lagoon while sands eroded from the headlands are moved along the beach by waves and longshore currents.

Sand beaches and sandy barrier islands border the U.S. Atlantic coast (Fig. 5–20) from New York to the tip of Florida. On the Gulf Coast, barrier islands and lagoons dominate. On mountainous or recently glaciated coasts, beaches are usually small and restricted to low-lying areas between rocky headlands.

A **beach** is actually a sediment deposit in motion. Movements of sand grains in the surf may be obvious, but the rest of the beach appears quite stable. During major storms, however, large segments of beaches are moved. Higher parts, protected by plants, may move only during exceptionally powerful storms.

Not all beaches are sandy. Where wave and current action is especially vigorous, sand may be washed away faster than it is being brought in, leaving behind gravel or boulders. Where wave action and currents are strong but not strong enough to remove all the sand, beaches consist of mixtures of gravel and sand.

Beaches are accumulations of locally abundant materials not immediately removed by waves, currents, or winds. Beach sands and gravels can be derived from erosion of glacial deposits, originally containing unsorted gravels, sands, silts, and clays. Only the gravels and sands remain on the beaches. Silts and clays are usually washed out of beach areas by even weak waves or tidal currents. Fine-grained sediments tend to accumulate in areas where there is little wave action or weak tidal currents as, for example, on continental shelves at depths below about 30 meters and in lagoons, bays, and tidal marshes.

Beaches typically form near sediment sources—at the base of cliffs or near river mouths (Fig. 5–20). In addition to these landward sources of beach sand, there is another source: Sediment is moved onto beaches by waves and currents, replacing materials either moved out into deeper water or transported along the coast.

On the North Atlantic coast of the United States, virtually no riverborne sediment now escapes the estuaries to enter the ocean, and so beaches along this coast are formed primarily by materials eroded from nearby cliffs (Fig. 5–20) or brought onshore from offshore sand deposits. Beaches also form where sand transport is interrupted by an obstruction, such as a headland or a breakwater.

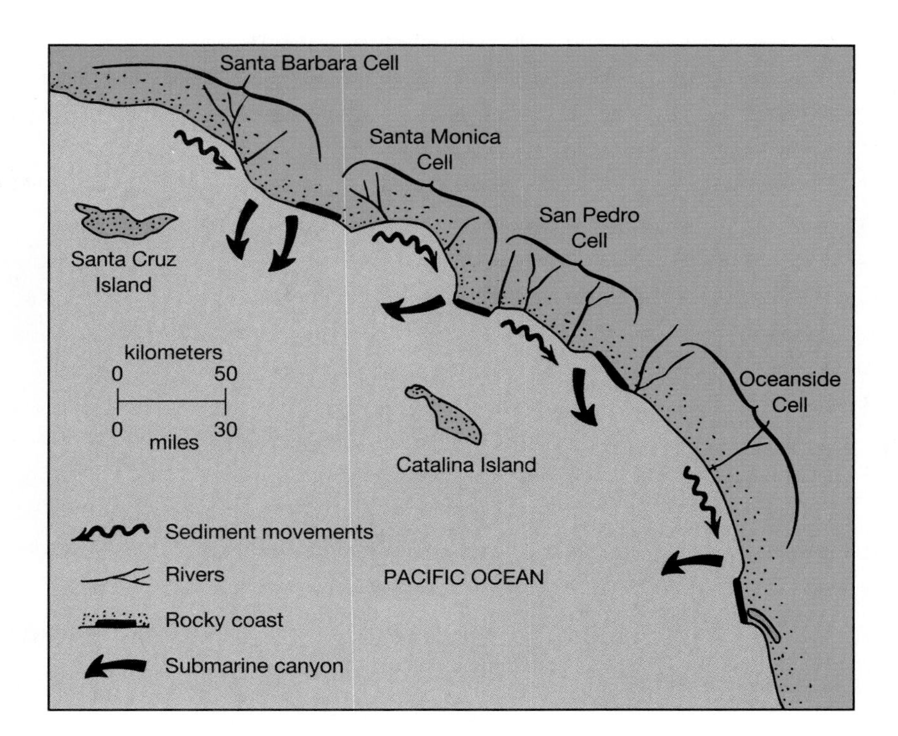

Figure 5–21
The southern California coast has four littoral cells. Beach sands move from the sources (primarily rivers) first to the submarine canyons and from there to offshore, out of reach of the beach processes. [After D. L. Inman and J. D. Frautschy, 1966. Littoral processes and the development of the shoreline. Proc. Coast, Eng Specialty Conf. *ASCE* (Santa Barbara, CA.) p. 511.]

Sediment budgets are useful in determining major sources and losses of sand along beaches. Major sand sources are usually either rivers or eroding sea cliffs. Major losses occur through longshore transport out of a particular region, offshore transport (down submarine canyons), and losses due to wind transport to form sand dunes or to deposit sand in marshes behind the beach.

In southern California the coast is divided into four cells (Fig. 5–21). Each coastal cell consists of a river (or rivers) providing sand, a current along the shore, and a submarine canyon that comes in close to the shore to trap the sand flow, carrying it offshore. Each cell is separated from adjoining cells by stretches of rocky coast devoid of large beaches because the sand has moved down a submarine canyon into water too deep to be moved by waves.

Inlets and lagoons behind barrier islands (Fig. 5–20) also trap sediment moving along a coast. Sand is stirred up and put in suspension by waves and then moved into the inlets or lagoons by tidal currents. When these currents slacken or when sediment encounters the dense vegetation of the tidal flats, it settles out. Without resuspension by waves, tidal currents cannot move this sediment back out of the inlet.

Deltas

Deltas are another important type of very large sediment deposits. They are often heavily populated by humans because their soils are highly productive and easily farmed. Indeed, most of the world's civilizations originated on deltas. Among them were the Egyptians on the Nile delta (Fig. 5–22), the Chinese on the Yangtze delta, and the Middle Eastern civilizations on the delta of the Tigris-Euphrates and Ganges Rivers (Fig. 5–23).

The first step in delta formation is the filling of the lower reaches of the river with sediment deposits. Rivers that have large embayments where they enter the sea and moderate sediment loads are still filling their estuaries (Fig. 5–20). Until

Figure 5–22
The Nile delta was formed by sediments transported by the Nile River and deposited when it flowed into the ocean. The delta is marked by the dark color of the vegetation growing on the fertile soils. (Courtesy NASA.)

the lower reaches of the river are completely filled, little or no sediment reaches the river mouth, and no delta forms.

As a delta forms, the river builds channels across it called **distributaries,** through which water flows on its way to the ocean (Fig. 5–23). Often a series of these channels extends across the delta as long, radiating, and branching fingers. An active distributary continually builds its mouth farther seaward until the distance to the sea is so great that the river flow can no longer maintain that channel. At this time the river shifts course, often during a flood, and the flow cuts a new channel through a different set of distributaries to reach the ocean and the whole process begins again.

This forming of distributaries has happened many times in the building of the present Mississippi delta, which has many abandoned distributaries, each with its own subdelta. This process forms a complex lobate delta (Fig. 5–24). Viewed in three dimensions, such a delta consists of complexly interleaved sheets of sediments. Only the modern subdeltas of the Mississippi (Fig. 5–25) and Atchafalaya Rivers are growing at present.

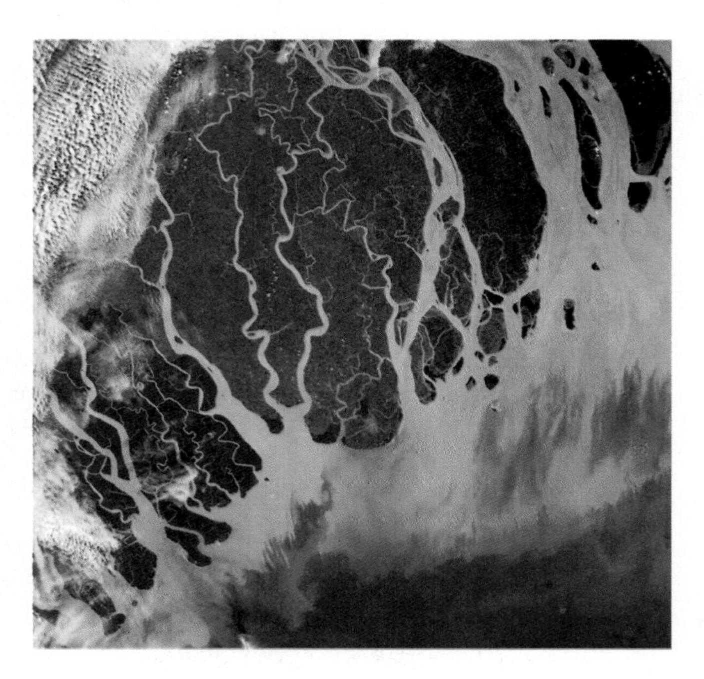

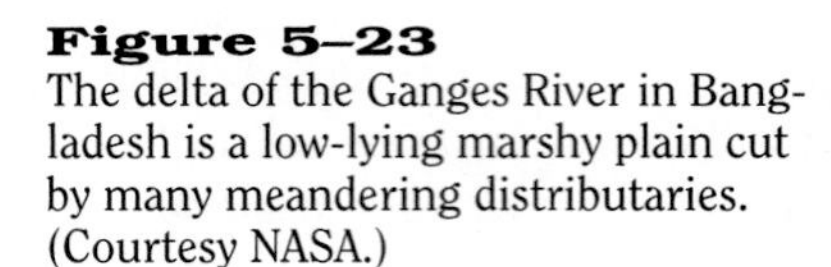

Figure 5–23
The delta of the Ganges River in Bangladesh is a low-lying marshy plain cut by many meandering distributaries. (Courtesy NASA.)

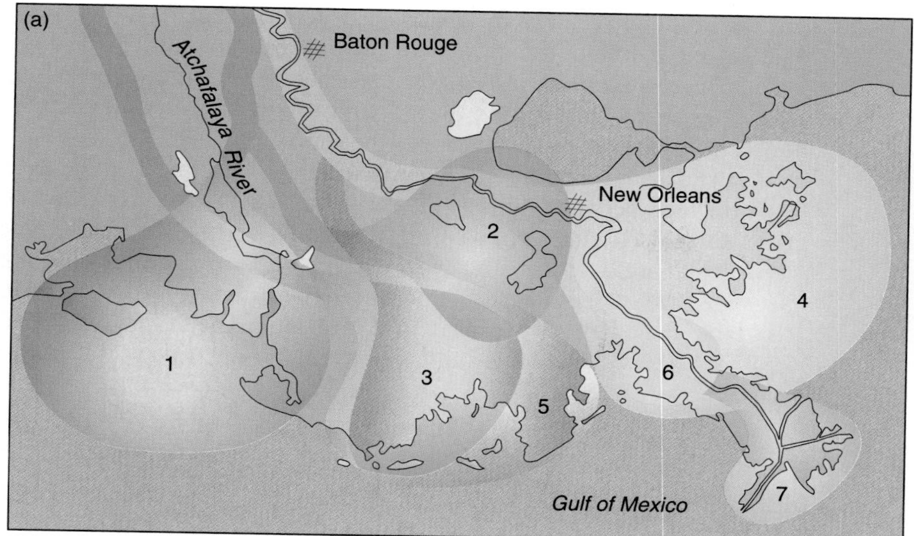

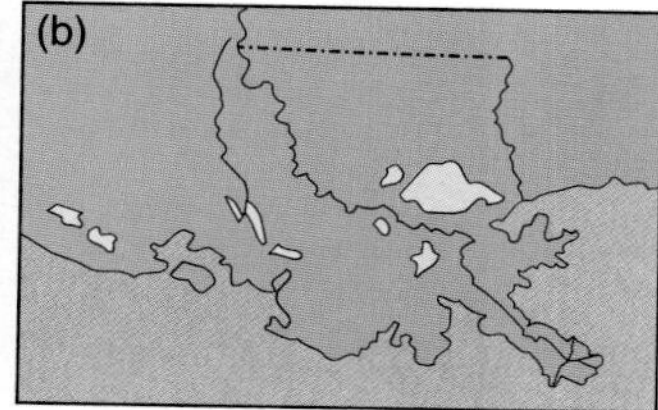

Figure 5–24
The modern Mississippi delta plain has grown over the past 7,000 years as six delta complexes were first active and later abandoned. (Adapted from C. R. Kolb and J. R. Van Lopik, "Depositional Environments of the Mississippi River Deltaic Plain," in M. L. Shirley, ed. *Deltas in Their Geologic Framework,* Houston: Houston Geological Society, 1966, p. 22. Adapted by permission.)

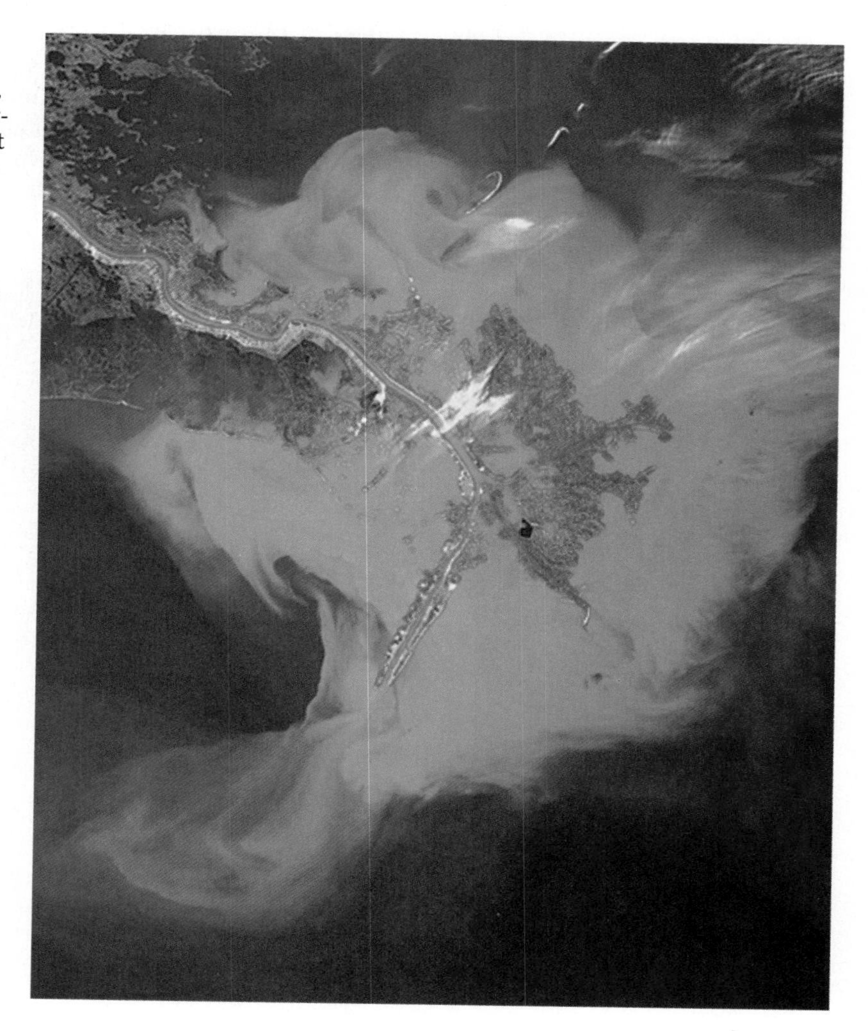

Figure 5–25
The modern Mississippi River subdelta grew between 1874 and 1940. Since then, its growth has slowed because soil conservation practices have reduced the amount of sediment carried by the river. Note the birds-foot shape resulting from each distributary building seaward with only occasional flooding between distributaries. This unusual shape is caused by the limited tidal range and wave action in the Gulf of Mexico and by the levees built to protect low-lying areas from flooding. Because of reductions in sediment supplies after successful soil-erosion control practices in the drainage basin, areas between the distributary mouths are now eroding rapidly. (Courtesy NASA.)

Distributary abandonment may occur suddenly during a flood or gradually as one channel becomes too shallow to carry a large amount of water and another gradually receives more of the flow and becomes enlarged. The Yellow River of China has changed its path to the sea many times, often resulting in catastrophic loss of life due to flooding of the low-lying, densely populated lands beside the river.

Delta deposits compact as they accumulate, expelling some of the water that deposited between the grains in the deeper layers. As this compaction occurs, the delta surface subsides unless it is constantly resupplied with new sediment deposits. Where sediment supplies are lacking or have been cut off, the delta surface subsides and becomes a shallowly submerged part of the continental shelf. This type of submergence is currently occurring in the older subdeltas of the Mississippi. We shall study this submergence more in Chapter 14 when we discuss coastal ocean problems.

Ocean History Recorded in Sediment Deposits

As mentioned at the beginning of the chapter, sediment deposits and the fossils in them record the history of ocean basins and of Earth's climate. Scientists use such deposits to study ancient ocean conditions much as archaeologists use broken pottery and refuse piles to study ancient humans. Sediments can be sampled by various techniques, such as drilling through rocks or by lowering a weighted pipe into the ocean floor to core soft sediments, much like coring an apple. Various layers in such cores are then sampled for detailed study (Fig. 5–26).

Samples are analyzed by many techniques. Relative ages of layers are determined by studying the fossils they contain. The idea is that the presence (or absence) of certain fossils can tell us when the layer was deposited; this information gives relative ages. That is, we can determine which layer is oldest, which is

Figure 5–26
Scientists on board the drill ship *JOIDES Resolution* sample a sediment core from the deep-ocean bottom to study its composition and ocean history. (Courtesy Ocean Drilling Program.)

BOX 5–2
Earth Catastrophes

One of the most intriguing questions of Earth's history is what causes the sudden extinction of large numbers of organisms, which has happened many times. Indeed, geologists use sudden changes in the fossil record to mark major periods in Earth's history.

The most recent major extinction event was that of dinosaurs and many other life forms about 65 million years ago, for which there are two explanations. The more dramatic is that Earth was struck by a mountain-sized asteroid (or possibly a comet). The amount of ash injected into the atmosphere by the impact caused dramatic changes in climate that led to the disappearance of the dinosaurs and many other organisms.

There are many lines of evidence supporting a meteor impact. One persuasive bit of evidence is the unusually high concentration of the rare element iridium found in a thin black layer of sediment that lies over much of the older ocean bottom. Iridium is rare in crustal rocks but common in asteroids and comets. No other obvious source of iridium has been identified. Large numbers of unusually deformed quartz grains have also been found in deposits around the world. These too are consistent with a meteor impact.

The most convincing evidence for a meteor impact—the crater formed by the impact—is still missing, however. A small crater of the appropriate age has been found in Iowa buried beneath much later glacial deposits, but it is too small to have been the principal crater. A large craterlike structure has been identified in the Caribbean, north of Yucatán, Mexico. Research on its age and structure indicates it is a promising candidate for the site of a major meteorite impact. If the meteor origin of this crater is confirmed, it will be the second largest crater known in the solar system. Further evidence from the many deposits around the Caribbean and Gulf of Mexico indicates that a huge wave, possibly formed by the meteorite impact, struck islands and coastal areas around the Gulf of Mexico and the Caribbean.

The second explanation for the extinction of the dinosaurs is based on the fact that, at several times during Earth's history, there have been sudden (within a million years) and very large outpourings of molten rock that have left huge areas covered by volcanic rocks. Some of these rocks in northern India correspond to the time of the dinosaurs' disappearance. Such large eruptions may cause the climate to cool enough to eliminate many groups of plants and animals.

Further investigations of the Yucatán structure and studies of the massive volcanic eruptions will be necessary to answer this question. The resulting information will help scientists to understand even larger mass extinctions earlier in Earth's history.

youngest, and their approximate ages. We can also correlate with other deposits that contain the same sequences of fossils. We cannot, however, determine absolute ages. In other words, we cannot, using fossils alone, say that a particular deposit is so many million years old.

Radioactive substances in particles are used to determine absolute ages. Each **radionuclide**—any radioactive isotope of an element—decays at a fixed and known rate. The time elapsed since that particle formed is indicated by the amount of the decay product in it compared with the amount of the original radionuclide present. The type of radionuclide used in dating depends on the expected age of the sediment. For very young deposits (less than 30,000 years), radioactive carbon-14 is used. This isotope is formed in the upper atmosphere as cosmic rays constantly bombard Earth. All living organisms contain carbon-14 in their tissues. While animals or plants are alive, they constantly exchange carbon with the atmosphere and their carbon-14 levels remain constant. When an organism dies, however, it stops taking up carbon-14. Instead, the carbon-14 in its tissues at the moment of death begins to decrease at a steady rate through radioactive decay. One-half is gone in 5,600 years (its **half-life**). After 11,200 years, only one-quarter of the original amount remains. Thus, by comparing the amount of carbon-14 per unit of carbon in a fossil with comparable living organisms, it is possible to determine when the organism that produced that fossil died. Assuming that the organism was incorporated into the deposit when the deposit formed, we can determine the age of the deposit.

Dating older minerals requires radionuclides having half-lives longer than that of carbon-14. Most rock-forming minerals contain some potassium, including the radioactive potassium-40, which has a half-life of 1.3 billion years. This isotope is useful for deposits or rocks that are hundreds of millions of years old. Uranium and thorium are also used for dating deposits of a few million years.

Evidence other than isotope counts is also used to date parts of the ocean floor. We discussed (in Chapter 3) how magnetic reversals can be used to determine when a part of the ocean floor formed. A comparable record of magnetic reversals is contained in sediment cores, permitting them to be dated as well.

Sediments also record changes in ocean currents. Perhaps the most striking case is changes in the carbonate compensation depth (CCD). At present, the CCD lies at an average depth of around 4 kilometers (although it varies both from one place to another within one ocean basin and from one basin to another).

Sediment deposits recovered by drilling show that the CCD has varied through time by as much as 2 kilometers. This fluctuation has been explained as the result of differences in deep subsurface currents, changing the chemical composition of the water making up the currents. It has also been speculated that the depth changes result from differences in sea level. When sea level is high, carbonate sediments are retained on continental shelves, thus reducing the rate at which this material is supplied to the deep ocean. As a consequence, the CCD rises because there is less carbonate to dissolve. Conversely, when sea level drops, more carbonate sediment is transported to the deep-ocean basins and the CCD is lowered.

Records of Climatic Changes

Sediment deposits record changes in Earth's climate over the past 200 million years. This sediment record tells us that during most of its history, Earth's climate was warm and humid, very similar to the climate in the present-day tropics.

One of the most useful records of past climate changes is the minute changes in the isotopic composition of the oxygen contained in shells of foraminiferans (single-celled organisms that have calcium carbonate shells). By determining the differences between the isotopic composition of the shells and of the water in which they grew, it is possible to determine the water temperature within a fraction of a degree.

Foraminifera living in near-surface waters record Earth's surface temperatures, and those living on the ocean bottom record near-bottom temperatures. There is some uncertainty about the composition of the waters, which leads to uncertainty of a few degrees. Thus, only large temperature changes can be measured.

Beginning about 50 million years ago, Earth's climate began to cool. The fossil record of foraminiferan shells tells us that near-bottom waters went from being relatively warm to having temperatures near the freezing point of seawater. Bottom waters in polar regions apparently cooled about 10°C, while mid-latitude surface waters cooled only about 5°C. There was some change in surface-water temperatures in equatorial regions, but how much is still disputed.

Resources from Sediments

Sand, gravel, and shell are mined from the shallow ocean bottom. Large quantities of such materials are used to pave roads, construct buildings, and fill low-lying areas. Sand is used to restore beaches damaged by storms and rising sea level.

Sediment Records of the Ice Ages

Deep-ocean sediments contain the records of Earth's history during the last phase of the Ice Age. Thin layers of pebbles recovered in cores from the bottom of the North Atlantic record several times during the last stage of the Ice Age when enormous groups of icebergs were released from glaciers on Canada and Iceland. As these glaciers melted, they dropped their sediment loads to ocean bottom where they were discovered. These iceberg events may have been the result of dramatic climatic changes, which caused the glaciers to surge and form unusually large numbers of icebergs. Furthermore, the melting of these glaciers may have contributed vast amounts of freshwater to North Atlantic, possibly triggering other changes in ocean circulation.

Scientists will compare the results of the studies of the sediment cores with the results obtained from analysis of ice cores obtained from glaciers on Greenland and Antarctica. From all these records, the story of the dramatic climatic changes in the last Ice Age—which may have played a role in triggering human evolution—can be reconstructed.

Heavy minerals—gold, tin, chromium, and titanium—are also recovered from the shallow ocean bottom. These minerals are much denser than sand and, therefore, are concentrated and left behind when less dense sand grains are removed by river flows or by waves. Channels formed by rivers that cut across continental shelves when sea level was lower sometimes contain heavy minerals—for example, tin near Indonesia and Malaysia. Other heavy-mineral deposits formed on beaches when waves and longshore currents separated heavier particles from the lighter sand grains.

Phosphorite nodules, which contain the element phosphorus (used in fertilizers), occur on continental shelves and therefore are easily dredged. The phosphorus content of the nodules tends to be relatively low, but billions of tons are available. These nodules are not exploited now because there are huge phosphorus deposits on land that are cheaper to mine and easier to transport.

Manganese nodules may some day be mined for their copper, nickel, and cobalt. The most interesting region for nodule production is in the central Pacific, south of the Hawaiian Islands. At present, however, the high cost of mining materials from the deep-ocean floor and low metal prices have kept manganese nodules from being exploited.

Muds rich in zinc and copper occur on the bottom of the Red Sea, and deposits containing high concentrations of sulfides of zinc, copper, and iron have been found in samples taken from active hydrothermal vents in the Atlantic and Pacific Oceans. Ancient marine deposits are mined for copper on the island of Cyprus in the Mediterranean, but none have been commercially mined from the ocean bottom.

Summary

Terrigenous sediments (derived from rocks on land) and biogenous sediments (from skeletons of plants and animals) cover most of the ocean bottom. Sediment deposits are thin or absent near the crest of mid-ocean ridges, on newly formed crust, and wherever strong currents prevent sediment accumulation.

Biogenous sediment deposits cover more than half the ocean floor. They accumulate most rapidly in areas of high biological productivity, especially near the equator. Destruction of soluble or fragile remains leaves only sturdy forms at great depth. Destruction of shells is especially important for carbonate deposits, which occur primarily in shallow ocean areas. At the greatest depths in the oceans, most of the biogenous particles have been dissolved, leaving only terrigenous materials.

Terrigenous sediment deposits are thickest near continents. They are also carried by winds and deposited in open-ocean areas. Rivers transport about 20 billion metric tons of sediment each year to the ocean. Much of the riverborne sediment is deposited in estuaries and on continental shelves near river mouths. Most continental shelves are covered by deposits that formed when these areas were exposed before being submerged by the rising sea level.

Turbidity currents—dense, sediment-laden waters—transport sediments out onto the deep-ocean floor. They form characteristic deposits, with the coarsest materials on the bottom and fine-grained materials on top.

Hydrogenous sediments (precipitated from seawater) occur in areas where there are no other major sediment sources. Manganese nodules, the most frequently formed hydrogenous sediment, are common in areas of slow sediment accumulation, such as the North Pacific.

Cosmogenous sediments (derived from space) also occur in marine sediments and in Antarctic ice. They are thought to be the remains of meteorites that survived passage through the atmosphere.

Ocean sediments contain a history of ocean-basin development, of changes in sea level and climate, and of life in the ocean. Ages of sediment layers are determined from the presence or absence of fossils and from the amounts of radioactive constituents. Magnetic reversals recorded in cores also provide useful indications of sediment age.

Key Terms

sediment deposits
terrigenous sediments
deep-sea mud
glacial–marine deposits
biogenous sediments
carbonate compensation depth
hydrogenous sediments
cosmogenous sediments
tektites
fecal pellets
turbidity currents
turbidite
graded bedding
pelagic sediments
relict sediments
beach
distributary
radionuclide
half-life

Study Questions

1. What techniques are used to determine the ages of sediment deposits?
2. Describe the three major categories (by composition) of particles in oceanic sediment deposits. Where in the deep-ocean basins is each sediment type most likely to predominate? What is the primary source of each?
3. Describe where turbidity currents form, how they transport sediment, and the type of deposits they form. Where are turbidity currents most common?
4. Explain why calcareous biogenous deposits are rare or absent on the deepest parts of the ocean floor.
5. Define relict sediments and explain why they are common on many continental shelves.
6. Draw a cross section through an ocean basin showing the distribution of various types of sediments in relation to continental margins and mid-ocean ridges.
7. Describe how oil and gas deposits form.
8. List some minerals recovered from the ocean bottom.
9. What are the most significant resources recovered from the ocean bottom?
10. Where is atmospheric transport of sediment to the ocean most important? Where is it least important?
11. Where is most of the undiscovered oil under the ocean floor expected to be found?
12. Discuss how ocean-bottom topography affects the types of sediments deposited on the ocean bottom.
13. Discuss the relationship between biological productivity in the surface ocean and the type of sediment deposited on the bottom.
14. How can grain size be used to locate the source of sediments?
15. Discuss the relationship between sediment sources and large beaches.
16. Why are land areas affected by subduction zones prolific sediment producers?
17. Why are trenches often filled with sediment deposits?
18. Why are oil and gas fields common beneath deltas?
19. Why are delicate pteropod shells found only at shallow deposits?
20. Where in the ocean are hydrogenous deposits most likely to occur?
21. Why are airborne particles most common in deposits in the centers of major ocean basins?
22. What types of sediment deposits are most common on the floor of the Arctic Ocean?

23. What types of sediment are most common around Antarctica?

24. How can fossil shells in marine sediment be used to determine relative ages of various sediment layers?

25. *[critical thinking]* Why did increased production of oil and gas from the seafloor lead to more frequent maritime-boundary disputes and eventually to the Law of the Sea Convention and Treaty?

Selected References

DAVIS, Jr., R. A., *The Evolving Coast.* New York: Scientific American Library, 1994. Sedimentary processes in the coastal ocean.

SIEVER, R., *Sand.* New York: Scientific American Library, 1988. Earth science from the perspective of a sand grain.

OBJECTIVES

Your objectives as you study this chapter are to understand:

- How ocean and atmosphere interact
- How Earth's rotation affects winds
- Daily and seasonal weather patterns
- The differences between weather and climate and the processes that affect them

Hurricane Andrew approaches the Mississippi Delta from the Gulf of Mexico. Note the distinct eye of the storm.
(Courtesy NOAA and Louisiana State University Earth Scanning Laboratory.)

6 The Atmosphere

Together, the ocean and atmosphere make up Earth's fluid outer layer. We now examine the atmosphere, its interactions with the ocean, and the processes that control the winds.

Atmospheric Components

Earth's atmosphere—what we call "air"—is a mix of transparent, odorless gases, predominantly nitrogen (78 percent) and oxygen (21 percent). Also found in the atmosphere, but in variable amounts, are gaseous water vapor, ozone, and carbon dioxide as well as solid dust particles (Fig. 6–1).

Atmospheric water vapor is especially important because, as we see later, it affects the way the atmosphere absorbs heat from the Sun and because it is involved in heat transport. Evaporating water vapor from the ocean surface transfers latent heat energy from the ocean to the atmosphere. Conversely, condensing water vapor releases the latent heat energy to the atmosphere. Let's look at this process in more detail.

When a water surface is heated, water vapor escapes into the overlying atmosphere. At a given temperature, a certain amount of water vapor is in equilibrium with the water surface (Fig. 6–2). At this equilibrium point, as much water leaves the surface as returns through condensation. When there is less than the equilibrium amount of vapor in the overlying dry atmosphere, the liquid water evaporates. When there is more than the equilibrium amount, the vapor condenses, forming either rain or fog. As the temperature

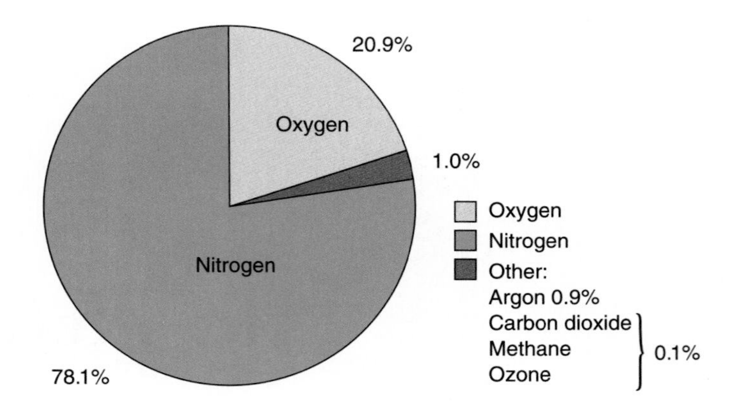

Figure 6–1
Relative abundance (by volume) of gases in Earth's atmosphere.

of the system rises, the equilibrium amount of water vapor the atmosphere can hold increases. For instance, an increase of temperature from 0 to 10°C doubles the amount of water vapor in the atmosphere at equilibrium.

Evaporation removes energy from the water. The higher the temperature, the more energy removed, as we can see by looking at two processes: warming water from 20°C to 30°C and vaporizing 100°C water. Warming 1 gram of water from 20°C to 30°C adds 10 calories of heat energy to the water. Changing liquid water at 20°C to water vapor at 20°C requires, as you should recall from Chapter 4, about 585 calories. Thus much more energy is required to evaporate water than to warm it. Removal of so much energy from the ocean surface by evaporation is the primary reason that surface water is rarely warmer than 30°C.

Adding water vapor to the atmosphere makes the atmosphere less dense and causes vertical air movements (which we call **convection**). We can see why this is

(a) Equilibrium

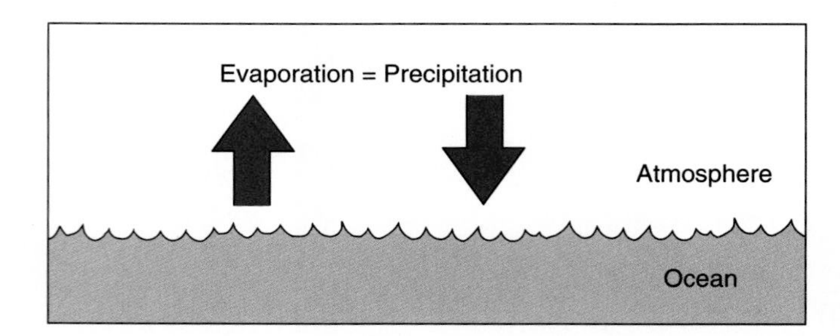

(b) Dry atmosphere

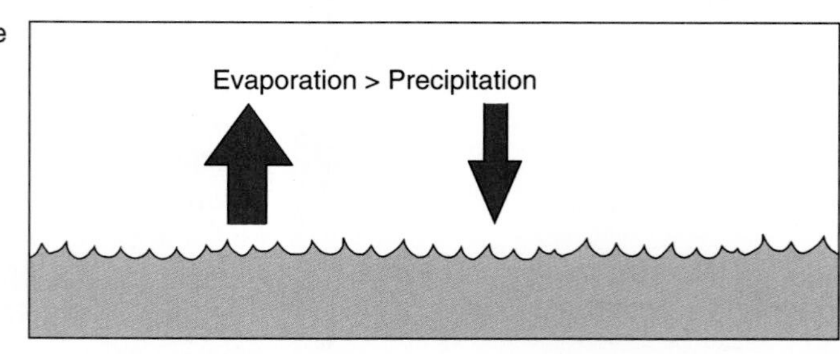

(c) Wet atmosphere

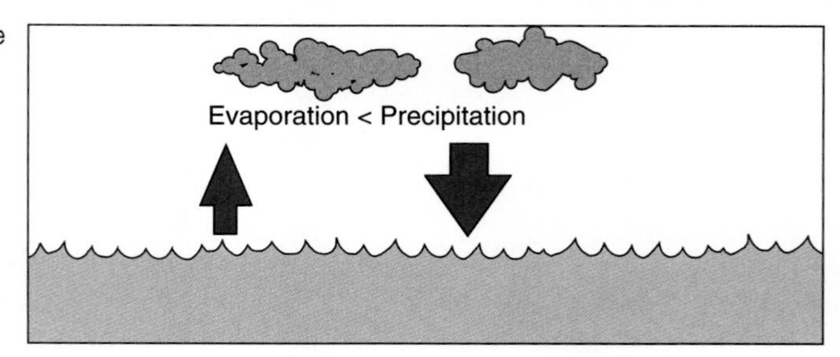

Figure 6–2
In an atmosphere in equilibrium with a water surface (a), equal amounts of water are evaporated and precipitated. In a dry atmosphere (b), more water evaporates than precipitates. Finally, in a wet atmosphere with clouds or fog (c), more water precipitates than evaporates.

true by comparing the molecular mass of water with that of nitrogen, the main component of air. The molecular mass of a water molecule (H_2O) is 18 (oxygen = 16, hydrogen = 1), while that of one molecule of nitrogen gas (N_2) is 28 (nitrogen = 14). Thus adding water vapor lowers the density of the air.

Ozone—a molecule made up of three oxygen atoms—is an important but variable constituent of the atmosphere. Near Earth's surface, ozone occurs as a pollutant in smog. It also occurs naturally in the upper levels of the atmosphere, where it absorbs damaging radiation from the Sun.

Carbon dioxide, another variable constituent, is an important absorber of radiation in the atmosphere. Its increased concentration in the atmosphere caused by the burning of fossil fuels (coal, oil, and gas) is a major contributor to the apparent warming of Earth's surface, a phenomenon called the greenhouse effect.

Finally, *dust* (which is a solid rather than a gas) occurs in the atmosphere, primarily over land. It is not a major factor in atmospheric circulation.

Atmospheric Structure

Earth's atmosphere is density-stratified, with the densest air nearest the ground. Since gases are compressible, the higher pressure near the ground compresses the gases, making them denser than those higher in the atmosphere. Atmospheric pressure decreases by half for every six-kilometer increase in altitude.

The layer of air closest to the ground (and therefore the densest layer of air) is called the **troposphere** (Fig. 6–3a). It extends from sea level to an altitude of about 11 kilometers, and its topmost portion is called the **tropopause.** Beginning at the top of the tropopause and extending upward to about 45 kilometers is the second-densest layer, the **stratosphere.** Within the stratosphere, is the **ozone layer,** so-called because of its relatively high concentration of ozone gas. Above the stratosphere lies the **mesosphere.**

Figure 6–3
The structure of the atmosphere (a) and its temperature (b) both change with increasing altitude.

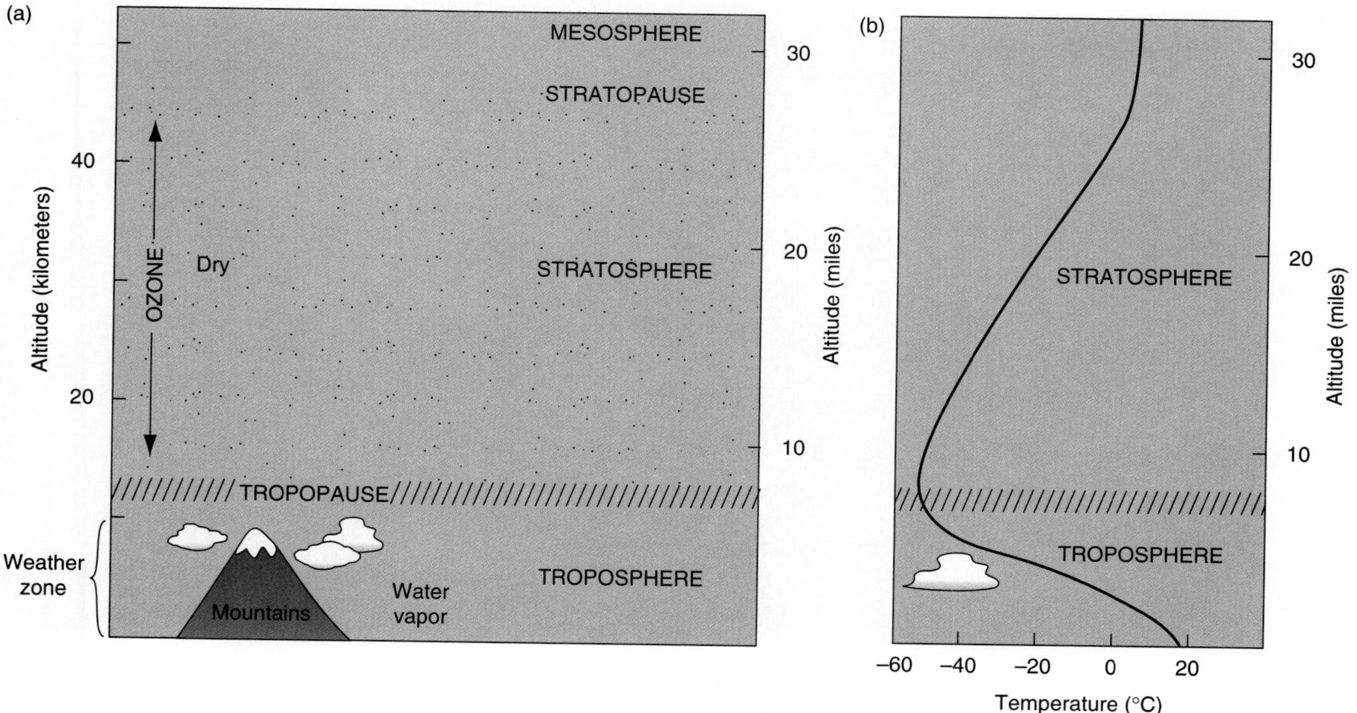

There is little vertical movement in the stratosphere, unlike in the troposphere, where most of our weather occurs. Thunderstorms are familiar examples of vertical motions in the troposphere. Substances injected into the stratosphere by volcanic eruptions persist for very long times because there are few ways to remove them and the removal processes operate slowly.

In the troposphere, air temperatures decrease with increasing elevation as Figure 6–3b shows. In the stratosphere, temperatures rise with elevation due to interactions with particles and radiation from the Sun—the solar wind.

Depletion of the Ozone Layer

Temperatures continue to rise in the ozone layer because ozone molecules absorb ultraviolet radiation from the sun. In other words, the ozone layer shields Earth from ultraviolet radiation. In the last 20 years or so, reduction in the thickness of the ozone layer observed in the atmosphere over the Antarctic and Arctic has caused concern, because a thinner layer of ozone removes less ultraviolet radiation.

Humans exposed to excessive ultraviolet light may develop cataracts and skin cancers. Other organisms can be damaged, too, including those living in the near-surface ocean waters.

The protective ozone shield over the Antarctic has been found to be greatly thinned during the southern spring when ozone levels have been observed to be as much as 40 percent below normal. Year by year, what scientists call the "ozone hole" has become larger (Fig. 6–4).

The cause of the ozone hole is still under study, but the major points are generally agreed on. The culprits are synthetic chemicals called chlorofluorocarbons (CFCs). (One brand name is Freon.) They are nearly chemically inert and are used in many applications—as coolants in refrigerators and air conditioners, as propellants in aerosol sprays, and as foaming agents and cleaners in the electronics industry. When released, they either dissolve in seawater or go into the

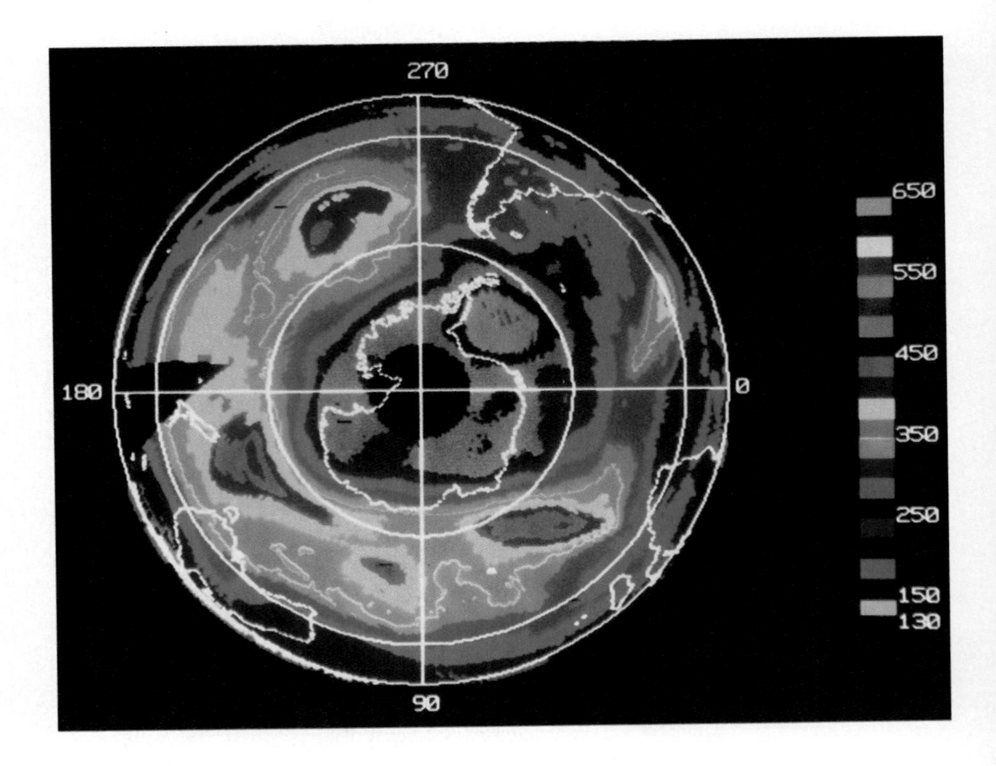

Figure 6–4
The September 1986 minimum in the Antarctic stratospheric ozone layer is shown by the gray areas in the image. (Courtesy NASA.)

troposphere, where they remain for many decades. In the ocean and in the troposphere, the CFCs pose no problem. Some escape into the stratosphere, however, where they absorb ultraviolet light and form harmful chlorine atoms as a result. The chlorine acts as a catalyst (that is, it is not affected by the chemical reaction) that destroys the ozone molecule by causing it to form ordinary oxygen. A single chlorine atom can destroy 100,000 ozone molecules before it is destroyed or returns to the troposphere.

Polar stratospheric clouds are involved in the process. In the extreme cold of the Antarctic, the clouds remove nitrogen compounds from the stratosphere. These compounds would normally inactivate chlorine, thereby reducing ozone depletion. When the nitrogen compounds are removed, the chlorine is left to destroy ozone molecules.

While the results of the ozone-destroying processes are most conspicuous in the Antarctic, similar conditions are now known to exist in the Arctic stratosphere. In the more densely populated northern latitudes, an Arctic ozone hole would expose more humans to ultraviolet radiation, but would have less impact on the ocean (which covers much less of Earth's surface in the north). There is also the possibility that the process might affect the stratosphere in other areas of the globe, in which case the risk to the ocean and to people over larger areas would increase.

Heat Budgets

One way to study Earth processes is to construct *budgets,* which show where energy or materials come from (*sources*) and where they go (*sinks*). We start with the atmosphere's heat budget. Directly beneath the Sun (Fig. 6–5), the top of the

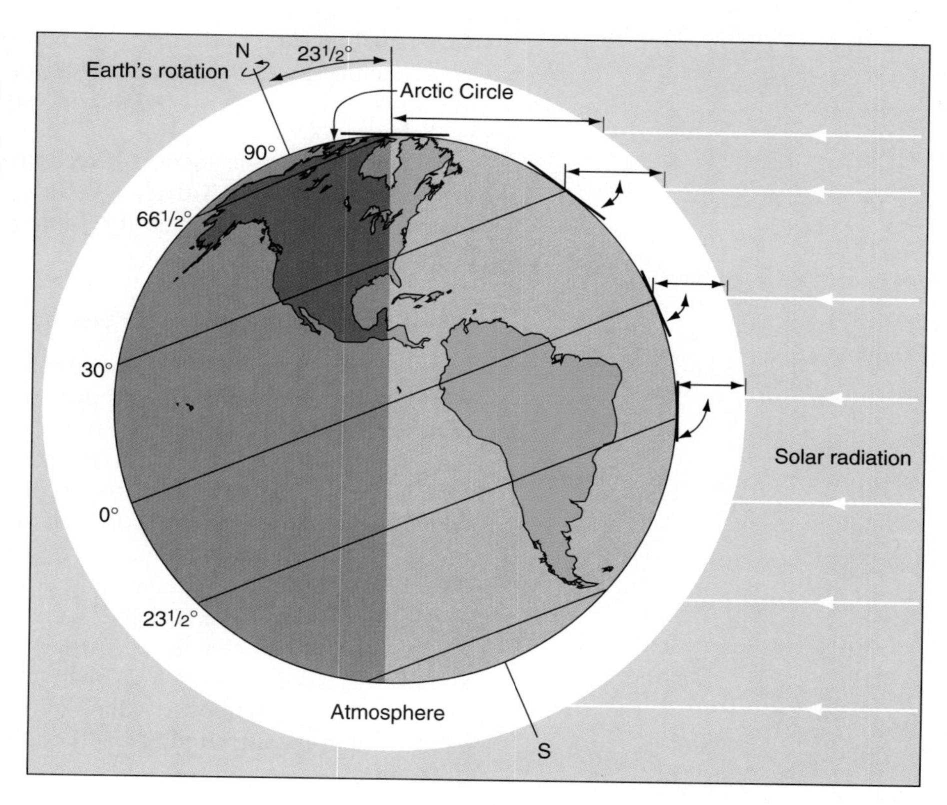

Figure 6–5
Incoming solar radiation per unit area of Earth's surface is greatest when the Sun's rays strike perpendicular to the surface and least when the rays are parallel to the surface at the Arctic Circle. Energy is also lost when the incoming radiation must penetrate more of the atmosphere (Lutgens and Tarbuck, 1995).

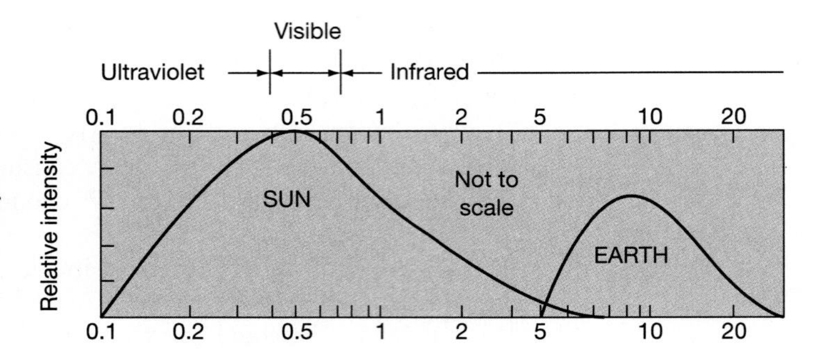

Figure 6–6
Emission spectrum of the Sun and Earth. Note that Earth's surface is much cooler than the surface of the Sun, and so the Earth radiates infrared energy, whereas the Sun radiates primarily visible-light energy.

atmosphere receives about 2 calories per square centimeter per minute. This incoming solar energy is distributed over Earth's surface as the planet rotates. On the average, solar radiation strikes any given point at the top of the atmosphere at the rate of about 0.5 calorie per square centimeter per minute.

Despite the solar heating, average surface temperatures over the past few centuries (where we have records) have been nearly constant. This indicates that Earth radiates back to space as much energy as it receives from the Sun.

The Sun's surface is extremely hot (approximately 6,000°C) and radiates a wide spectrum of wavelengths, primarily in the visible part of the spectrum. Absorbing this energy warms Earth's atmosphere and ocean. Earth radiates heat to space primarily in the long-wavelength infrared part of the spectrum (Fig. 6–6) because its surface temperature (18°C) is much cooler than the Sun's. Much of the long-wavelength infrared energy from Earth is radiated out into space by cloud tops. Clouds reflect back to space about one-fourth of the solar radiation striking the top of the atmosphere (Fig. 6–7).

On an average day, about half of the Sun's radiation striking the top of the atmosphere reaches Earth's surface (Fig. 6–8). About 5 percent of Earth's heat loss is due to long-wave radiation from the surface. Clouds absorb most of this long-wave radiation from Earth's surface. Without clouds and the atmosphere, Earth's surface temperatures might drop to as low as –20°C, the average temperature of the cloud tops. Instead, ocean surface temperatures average about 17.5°C and the land about 14°C.

The lower troposphere is heated primarily by latent heat released by condensing water vapor. Relatively little heat is transferred directly by sensible heat, in which a volume of warm air flows into a region and an equal volume of cold air flows out.

The Greenhouse Effect

Earth's atmosphere is nearly transparent to visible light. Thus sunlight easily penetrates the air and reaches the ground surface. Atmospheric gases do absorb some of the radiation coming from the Sun, however, and they also absorb some of the long-wavelength infrared energy radiated back to space from Earth's warm surface. As we saw earlier in the chapter, Earth's surface is in equilibrium with the radiation it receives from the Sun, radiating back an equivalent amount of energy each year so that surface temperatures stay essentially within a narrow range.

Gases released by the burning of fossil fuels (coal, oil, and gas) and from agriculture and manufacturing are altering the composition of the atmosphere. This compositional change, in turn, is changing how the atmosphere absorbs and transmits solar energy. This phenomenon is called the **greenhouse effect** (an analogy with a greenhouse, which allows light to enter, but traps heat and is warmer than its surroundings, although the processes involved are somewhat different).

Carbon dioxide (CO_2) is the most important greenhouse gas. It readily transmits the shorter wavelengths of incoming solar radiation while absorbing much of

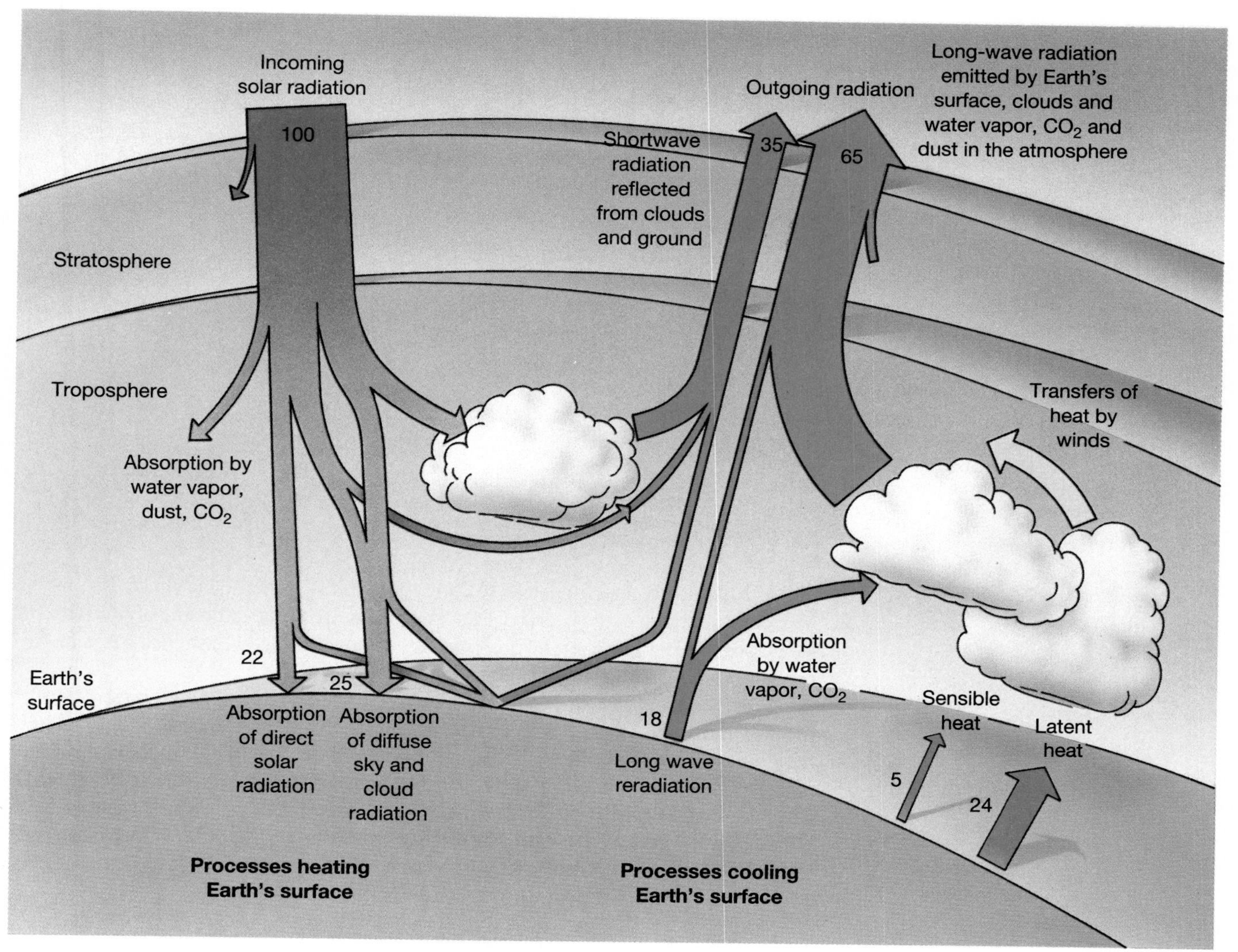

Figure 6–7
A schematic representation of Earth's average annual heat budget, showing how light (short-wave radiation) from the Sun is transformed into heat and transported in the atmosphere. On an average day, about 35 percent of the incoming solar radiation is reflected by clouds and Earth's surface. About 65 percent is radiated (emitted) back to space from clouds, Earth's surface, and various components in the atmosphere. On average, incoming solar radiation equals the amount reflected plus the amount radiated back to space. Thus, the heat budget is balanced.

the long-wavelength infrared radiation emitted by Earth. This dual transmission/absorption capability of CO_2 keeps Earth's surface warmer than it would be if there were less carbon dioxide in the atmosphere.

Since the 1750s, when the Industrial Revolution began, the burning of coal and oil has resulted in increased carbon dioxide concentrations in the atmosphere (Fig. 6–9a). Within the next century, atmospheric carbon dioxide levels will likely be double preindustrial levels. Unless effective controls on the burning of fossil fuels are negotiated between nations and enforced, carbon dioxide levels in the atmosphere will continue to rise.

About half the carbon dioxide released during fossil-fuel combustion dissolves in the ocean or goes into vegetation; the rest remains in the atmosphere. This atmospheric increase has already caused Earth's surface to warm by about 0.6°C since 1850. Further warming—up to 1°C by 2025 and 3°C by 2100—will likely occur if emissions of carbon dioxide are not controlled.

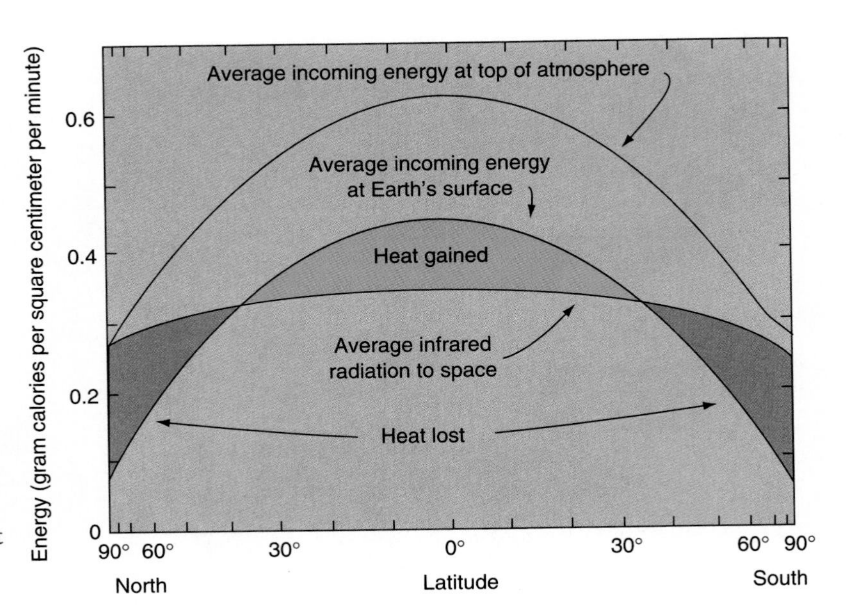

Figure 6–8
Primary areas of heat loss and heat gain on Earth's surface. On the average, about four times more solar energy reaches the surface at the equator than at the poles. The radiation of heat back to space is nearly the same over the entire Earth. Net heat loss occurs primarily in polar and subpolar regions, between 50° and 90°.

Other atmospheric gases also contribute to the greenhouse effect (Fig. 6–9b). Methane, nitrous oxides, and CFCs are released by agricultural and industrial processes. Molecule for molecule, they are as effective as greenhouse gases as carbon dioxide, but are much less abundant.

The effects of warming Earth's surface are not known. One concern is that the warming will cause polar ice caps to melt, raising sea level. Flooding of low-lying coastal lands could ensue. Warming would also be likely to change rainfall patterns, seriously affecting agriculture, especially in the interiors of continents. Grain-growing regions of North America and Asia might shift northward as local climates warmed and growing seasons lengthened. On the other hand, deserts in the interior of continents might spread to cover much larger areas.

Figure 6–9
(a) Atmospheric carbon dioxide concentrations have steadily increased since the beginning of the Industrial Revolution. The solid circles are data on CO_2 content of the air bubbles trapped in the Antarctic ice since 1750; the open circles are data collected since 1958 at Mauna Loa, Hawaii. (b) The relative contribution of the major greenhouse gases to global warming. While their chemical effects may be the same on a molecular basis, their abundances dictate their significance in the greenhouse effect.

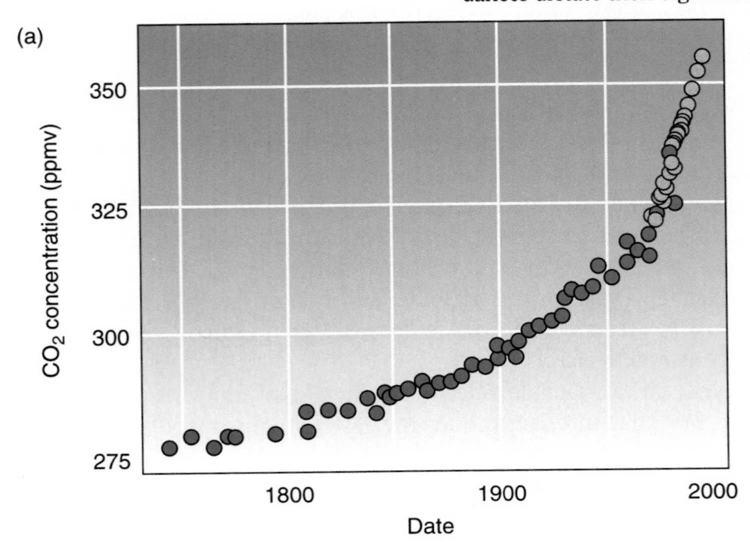

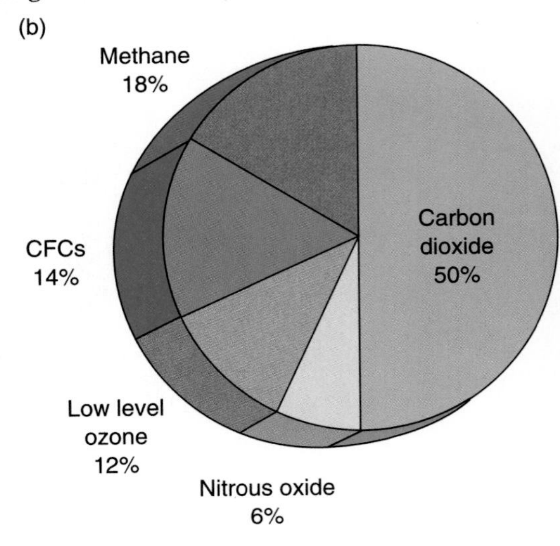

As we have already seen, the cooling of the ocean surface by convection limits sea surface temperatures to about 30°C, but warm tropical ocean waters could spread farther north and south. This spreading of warm waters might increase the number and strength of hurricanes or *El Niños* (Chapter 7) and may cause increased year-to-year variability in weather. How this global warming will affect the cycles of droughts and other climatic phenomena is more difficult to predict. In general we can say that, through our continuing dependence on fossil fuels, we are engaging in a planetary experiment for which the outcome is uncertain.

Winds on a Water-Covered, Nonrotating Earth

The winds are driven by the unequal heating of Earth's surface; the surface heats unevenly because land and water absorb heat from the sun at different rates. On a nonrotating Earth completely covered by water, wind patterns would be quite simple (Fig. 6–10). The atmosphere would be warmed near the equator and would rise. Because the rising air would cool, water vapor would condense and fall as rain. The air, now drier, would flow toward the poles, where it would cool further. Near the poles, the cold, dry air would sink to flow along the surface to the tropics, where the process would repeat. On such an Earth, there would be a simple two-celled circulation. In the Northern Hemisphere, surface winds would blow from north to south; this would be reversed in the Southern Hemisphere. Such a simple circulation pattern does indeed occur near the equator.

Winds on a Rotating Earth

Earth rotates and this rotation affects atmospheric circulation. We begin by considering the Coriolis effect, which arises from Earth's rotation. Later we consider the more complicated weather patterns caused by the presence of land and mountains.

Coriolis Effect

When we throw a snowball, we expect it to travel in a straight line and hit our target because on this small scale we experience Earth as a nonrotating system. But Earth does rotate and if our snowball could travel far enough, we would see its

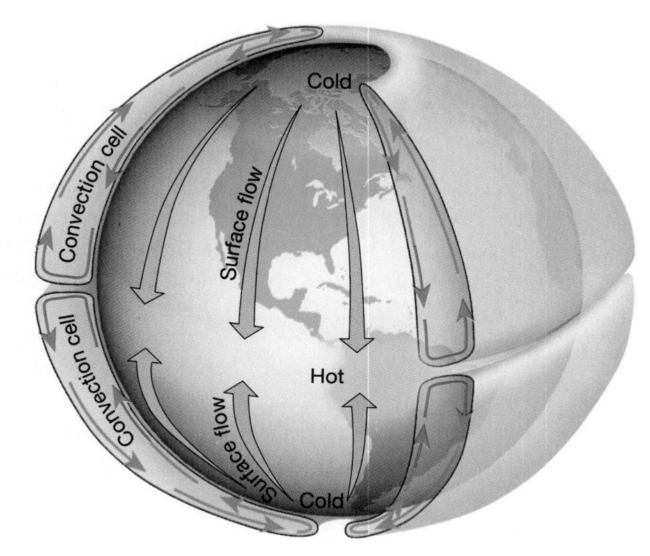

Figure 6–10
On a nonrotating Earth, unequal heating of Earth's surface causes warm air near the equator to rise. The air then flows north or south, cooling as it moves towards the poles. Cold air sinks near the poles and flows along Earth's surface, warming as it goes. This circulation forms a convection cell in the Northern and Southern Hemispheres (Lutgens and Tarbuck, 1995).

BOX 6–1
The Gaia Hypothesis

We know for certain that Earth's atmosphere is very unlike the atmospheres of its neighbors but why this is so is still an unanswered question. One theory explaining Earth's unique atmosphere is the **Gaia hypothesis** (named after a Greek goddess of the earth and pronounced *GUY-yah* with a hard *g*), advanced in 1979 by the British biochemist J. E. Lovelock. His main idea is that physical and chemical conditions on Earth's surface are controlled by biological processes, and that these same biological processes control the composition of atmospheric gases.

Lovelock hypothesized that these biological processes are self-regulating. In addition to regulating the composition of atmospheric gases, these processes have effectively regulated Earth's surface temperatures; they have remained nearly constant—between 0°and 20°C—since life first appeared, about 3.8 billion years ago. The constancy of the planet's temperature over this great span of time is difficult to explain otherwise, because the Sun is now approximately 30 percent brighter than when Earth formed. Without some form of temperature regulation, one would expect either the newly formed Earth to have been ice-covered or the present Earth to be extremely hot, similar to Venus. Neither has happened, as shown by the records left in sediments and by fossils.

Lovelock argues that if there were no life on Earth, the atmosphere would consist primarily of carbon dioxide, just as is true for Venus and Mars. Also, without life, surface temperatures on Earth would be far too high for liquid water to exist.

As we saw in Chapter 4, life in the ocean affects sea-salt composition. Lovelock estimates that the ocean on a lifeless Earth would have a salinity of 130. The excess salt would primarily consist of nitrogen compounds now in the atmosphere.

Atmospheric Composition and Surface Temperatures on Venus, Mars, and Earth (with and without life)

			Earth	
	Venus	*Mars*	*Lifeless*	*With Life*
Carbon dioxide (%)	98	95	98	0.03
Nitrogen (%)	1.9	2.7	1.9	79
Oxygen (%)	Trace	0.13	Trace	21
Argon (%)	0.1	2	0.1	1
Surface temperature (°C)	477	–53	ca. –20	18

path appear to bend. Because of the planet's rotation, winds or currents that are freely moving (not bound to Earth's surface) appear to be deflected to the right in the Northern Hemisphere and to the left in the Southern Hemisphere.

This effect is called the **Coriolis effect,** after the French mathematician Gaspard Gustave de Coriolis (1792–1843). Coriolis explained that these apparent deflections are a consequence of observing a rotating system while assuming that it is not rotating. Under these circumstances, freely moving objects appear to move in circles at constant speed unless acted on by an outside force. Thus, Earth's rotation causes the curved paths of rockets and artillery shells as seen by an observer on the ground. An observer on the Moon viewing the situation would not have to consider the Coriolis effect and would see that rockets travel in a straight line.

To understand the Coriolis effect, refer to Figure 6–11. If we stand on a stationary turntable and throw a snowball, it moves in a straight line and hits the target. If the turntable is moving counterclockwise, however, (this is the direction in

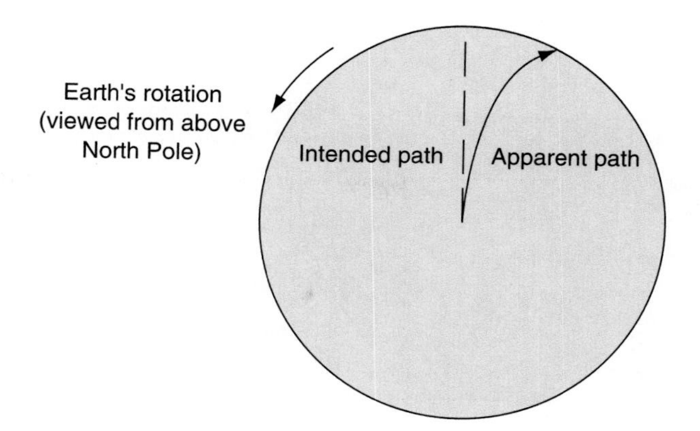

Figure 6–11
The Coriolis effect is caused by Earth's rotation. To an observer standing on a rotating turntable, the path of a freely moving object is apparently deflected to the right as the turntable rotates under the path of the ball.

which Earth rotates when viewed from above the North Pole), the snowball seems to deflect to the right. Actually the snowball traveled a straight line, but the target moved as the snowball was in flight. From this example it is easy to understand why the apparent deflection is reversed (to the left) in the Southern Hemisphere. Viewed from above the South Pole, Earth appears to rotate clockwise.

At the equator, there is no Coriolis effect for an object moving east or west because there is no effect of Earth's rotation there. In fact, little effect of rotation is felt within 5° on either side of the equator (Fig. 6–12). Apparent deflection increases with the speed of the object.

As we have already seen (Fig. 6–10), unequal heating of Earth's surface causes warm air to rise at the equator and to sink in polar regions after it has cooled. As we learned, on a simple, water-covered nonrotating planet, this would cause a very simple circulation pattern for the winds. But Earth's rotation and the Coriolis effect complicate this pattern of winds. Let's see how this happens by following the path of a bit of air.

If a ball is moving on a sloping surface, as in Figure 6–12a, it is the pull of gravity that is making the ball roll down the slope. As the ball moves, it is apparently deflected to the right in the northern hemisphere. Eventually the ball moves along a path where the downward pull of gravity matches the apparent deflection due to the Coriolis effect. At this point, the ball is moving essentially parallel to the slope (Fig. 6–12a).

In the atmosphere and in the ocean, this situation occurs where winds or currents are deflected by the Coriolis effect and by gravity. Such winds (or currents) are called **geostrophic** ("Earth turned") **winds** or currents. (We will discuss geostrophic currents in Chapter 8.)

Particles moving freely over the surface of a rotating Earth, with no outside influences, follow circular paths, as shown in Figure 6–12b. As we shall see in later sections, other forces are usually acting on the system as well.

Wind Patterns

After the unequal heating of Earth's surface by incoming solar radiation sets the atmosphere in motion, the Coriolis effect causes a complex six-celled atmospheric circulation (seen on the right side of Fig. 6–13). Let us see how this complex wind pattern is formed.

When the atmosphere is warmed in equatorial regions, air rises and begins to move toward the poles, as it did in the simple nonrotating situation (Fig. 6–10). The Coriolis effect deflects the rising air to the right in the Northern Hemisphere and to the left in the Southern Hemisphere.

Along the equator, a band of clouds and high rainfall marks areas of persistently rising air. This band, called the **intertropical convergence zone** (abbreviated

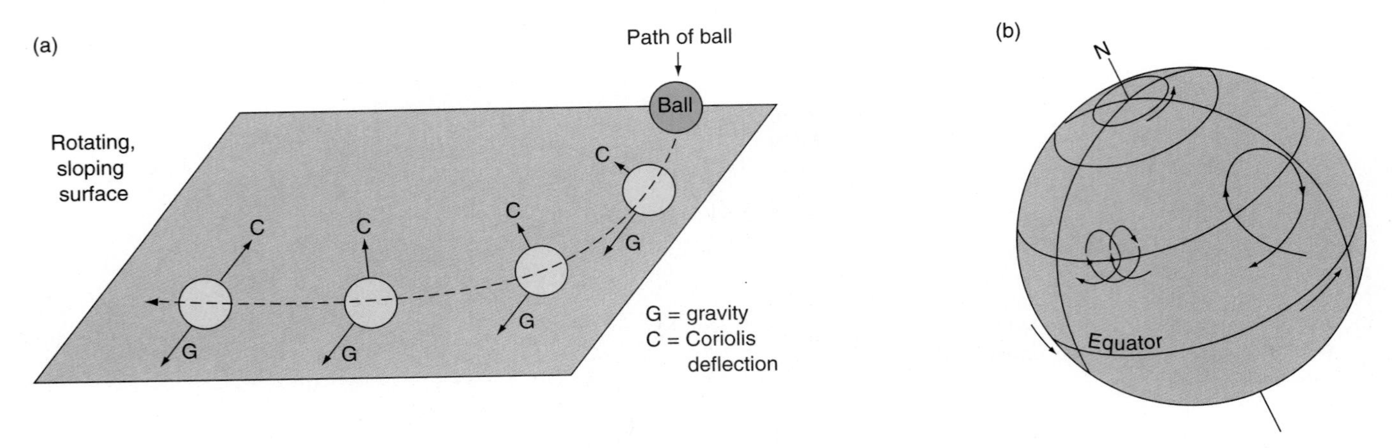

Figure 6–12
The Coriolis effect deflects the paths of objects moving on rotating bodies. (a) The path of a ball rolling down a sloping, rotating surface is deflected. (b) Objects moving freely over a rotating sphere, such as the Earth, apparently follow complicated circular paths except when moving east or west near the equator.

Figure 6–13
Global wind patterns form three major convection cells in each hemisphere, due to the influence of the Coriolis effect.

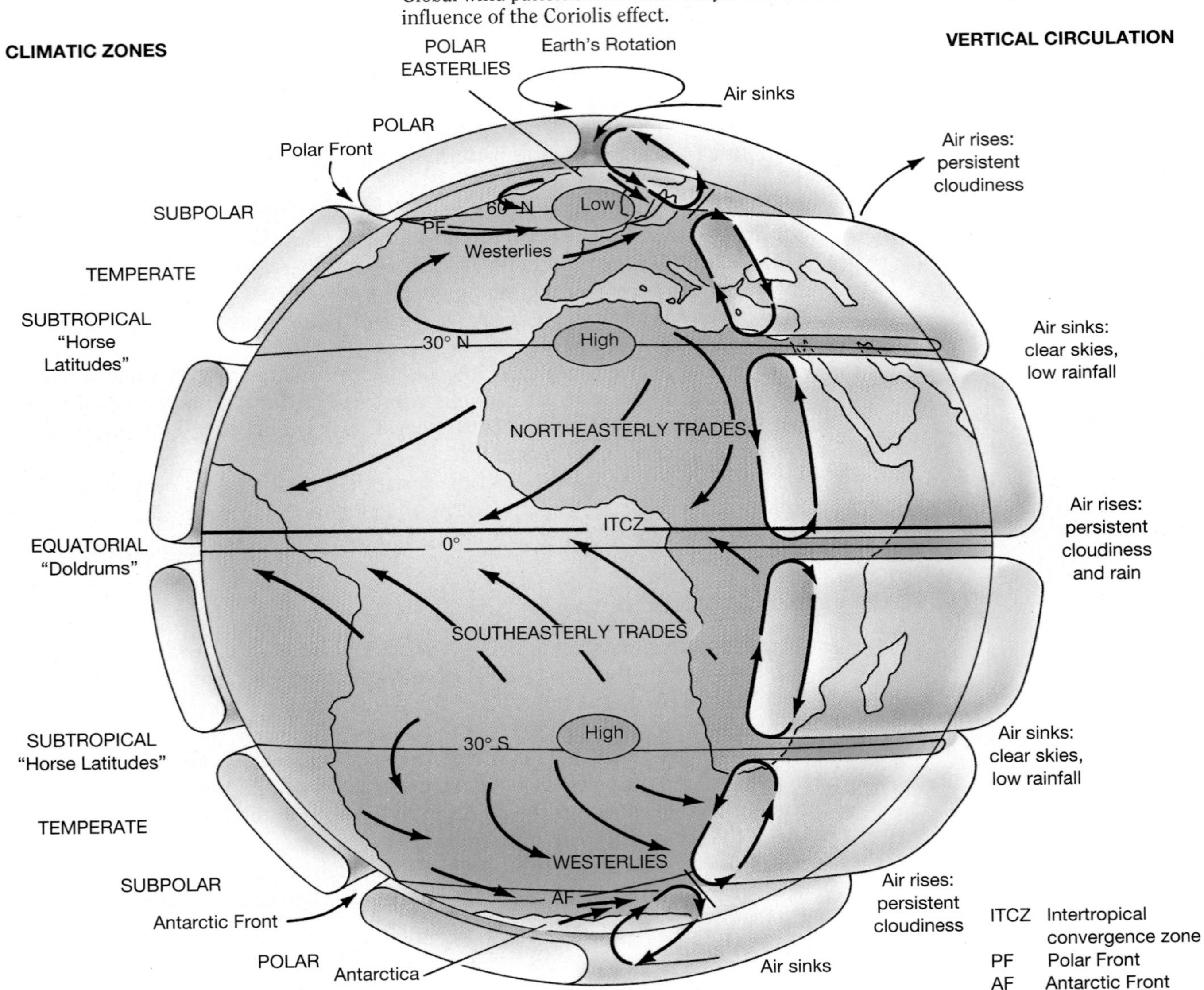

ITCZ), is usually visible in satellite photos of the equatorial ocean and moves seasonally (Fig. 6–14). Winds cool at high altitudes then sink at around 30° north and south, forming a belt of high pressure. After the air in these zones sinks, it spreads along Earth's surface, causing the prevailing surface winds shown in Figure 6–15. Part of this air moves toward the equator as the **trade winds.** As this air moves along the surface, it is warmed, picking up water vapor. Along the equator, it rises to continue the process.

The rest of the air flows generally toward the poles. The *westerlies* are a band of surface winds blowing from the west. (Note that winds are named for the direction from which they are blowing.)

As mentioned earlier, there are six cells in the global air-circulation pattern. The two low-latitude cells include the trade winds. The two mid-latitude cells include the westerlies, and the two polar cells include the polar easterlies.

At about 50°north and south, the westerlies meet colder, denser air that is coming from the polar regions and moving toward the equator. This convergence zone is the *polar front* (called the *Antarctic front* in the Southern Hemisphere), a persistent boundary between the polar and warmer air masses. The latitudinal band affected by the polar front experiences highly variable weather, namely a succession of relatively warm, moist subtropical air masses and cold, dry polar air masses. Such weather is typical of much of North America and Europe.

The polar front is made up of a succession of large waves that appear on weather maps as curved warm or cold fronts (shown schematically in Fig. 6–13). Narrow bands of strong winds, called **jet streams,** occur at elevations of around

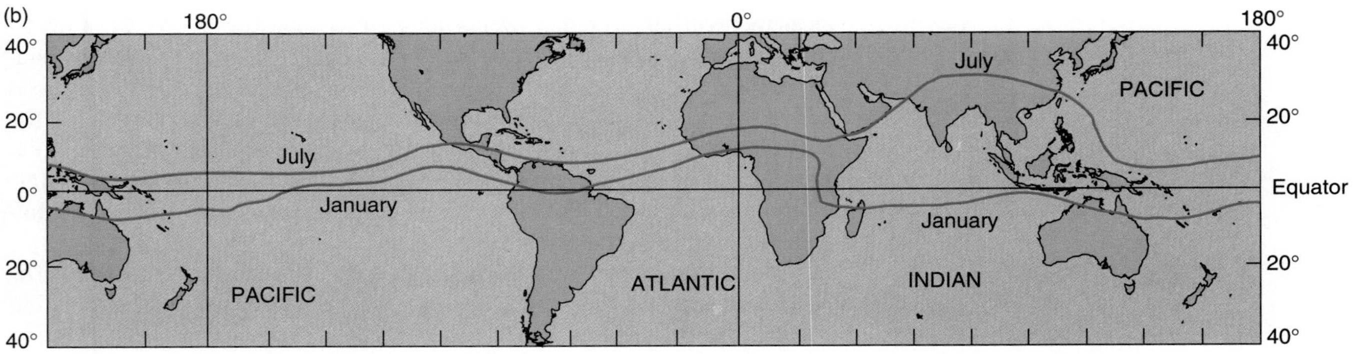

Figure 6–14
(a) The Intertropical Convergence Zone (ITCZ) appears as a prominent band of clouds near the equator. (European Space Agency/Science Photo Library/ Photo Researchers, Inc.) (b) The position of the ITCZ shifts seasonally. It reaches its most northerly position in July and is in its most southerly location in January.

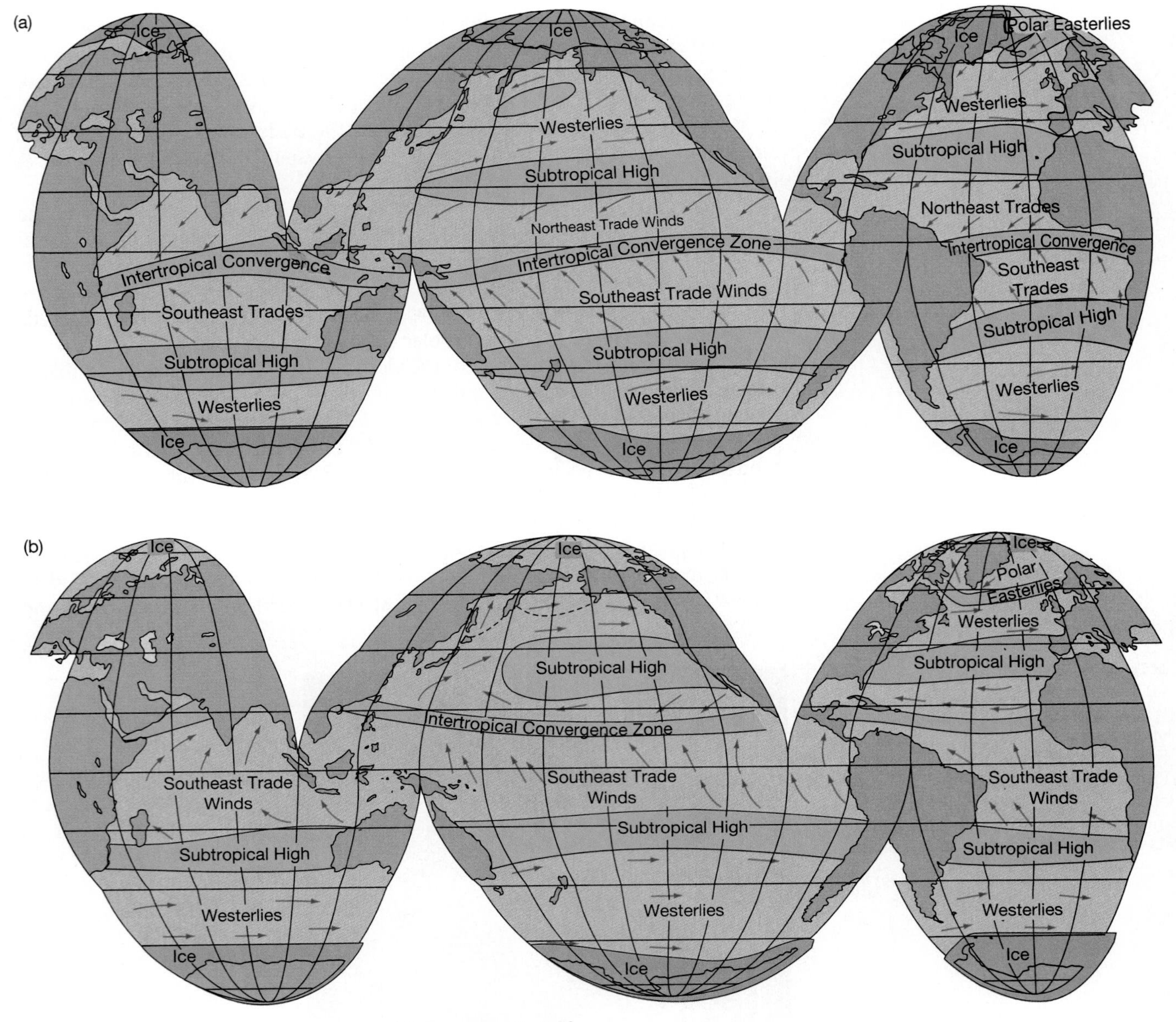

Figure 6–15
(a) Generalized surface winds over the ocean in February. (b) Generalized surface winds over the ocean in August.

10 kilometers. They are associated with the polar front—comparable jet streams are found on the Antarctic front.

Climate (weather averaged over a long time, typically 30 years) is determined by the long-term average locations of rising or sinking air masses. In mid-latitudes, where air masses commonly sink over the ocean, the climate is characterized by low rainfall, high evaporation, light variable winds (called the **horse latitudes**) and high atmospheric pressure. Near the equator, where the air is rising because it is warm and contains a lot of moisture, the climate is characterized by high rainfall, much cloudiness, light and variable winds (called the **doldrums**) and low atmospheric pressure.

Because of Earth's rotation, the winds in each cell follow spiral paths as the warm air rises, cools, and then sinks. In addition, there is an east–west circulation (not affected by the Coriolis effect) along the equator, called the *Walker cell,* named after Gilbert Walker (1868–1958), a British meteorologist who studied the weather systems of India.

Weather Systems

We have discussed atmospheric motions on a planetary scale, but most of our experience with weather involves smaller atmospheric motions. Winds move across Earth's surface in a series of turbulent systems. These seemingly chaotic motions involve swirling motions called **eddies.** We are familiar with the small-scale turbulence visible in a column of cigarette smoke. The large-scale atmospheric motions may be as large as enormous hurricanes that occupy the entire Gulf of Mexico.

Eddies are caused by differences in atmospheric pressure, temperature, and humidity. Each eddy obeys physical laws that also control movements of ocean waters. Predicting these movements is the basis of weather forecasts. The unpredictable aspects of atmospheric motions limit weather forecasts to only a week to ten days. Beyond that, these chaotic motions cannot be predicted.

Where upper atmospheric winds diverge, or flow away from a location, they cause areas of low pressure, called **lows** or low-pressure systems (Fig. 6–16). Where upper-level winds converge, they cause areas of high pressure, called **highs** or high-pressure systems. Air moves from high-pressure areas toward low-pressure areas. By mapping distributions of highs and lows and the atmospheric pressures, meteorologists are able to predict movements of storm systems and to predict local weather.

Air masses form over both land and ocean. Staying in one spot for several days or weeks, they acquire the characteristics of their source regions. In winter, cold, dry air masses form over the interior of Canada and Siberia, for example; when these air masses move out of these regions, they are initially cold and dry but are later warmed as they move over warmer land or ocean. Likewise, the tropical maritime air masses that form over equatorial waters are distinguishable by warm air and high humidity; when these masses move into other regions, they are modified, and their temperatures and humidities are altered. Tropical maritime air masses lose moisture by precipitation and cooling, especially when they move over the land.

When two air masses collide, the boundary between them is called a **front** and is marked by changes in temperature and humidity. A cold front is the leading

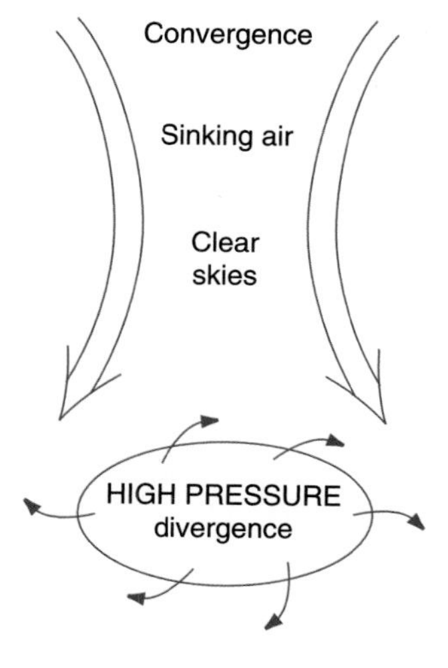

Figure 6–16
Areas of low and high pressure cause surface winds. These areas are caused by divergence and convergence of winds in the upper atmosphere. Such areas are zones of rising and sinking air, respectively.

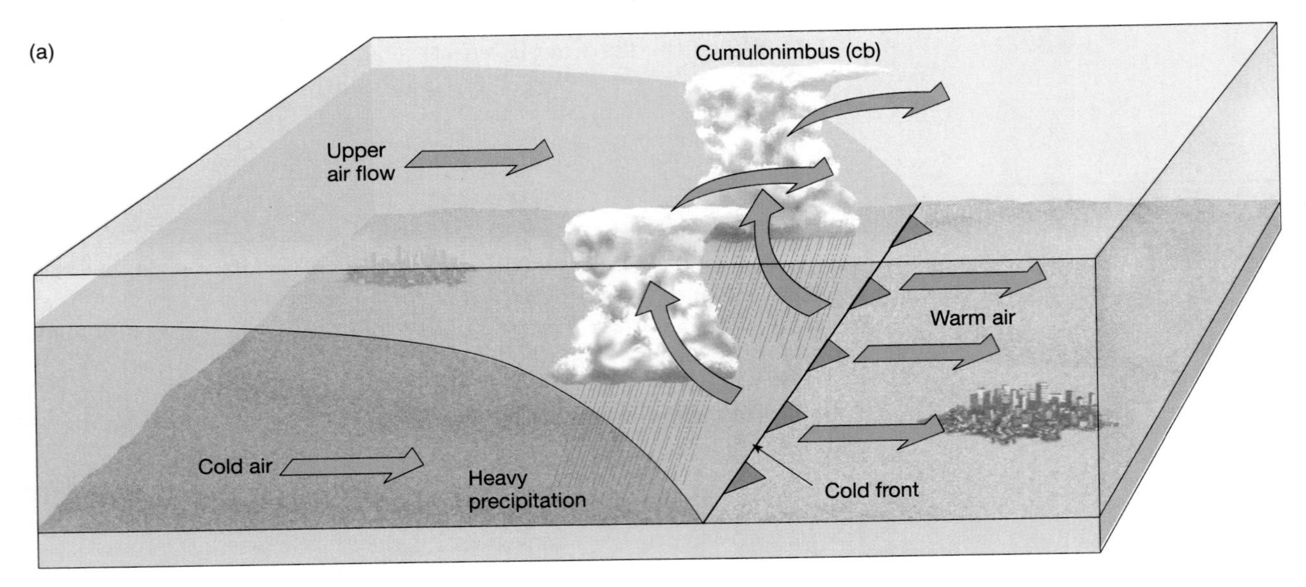

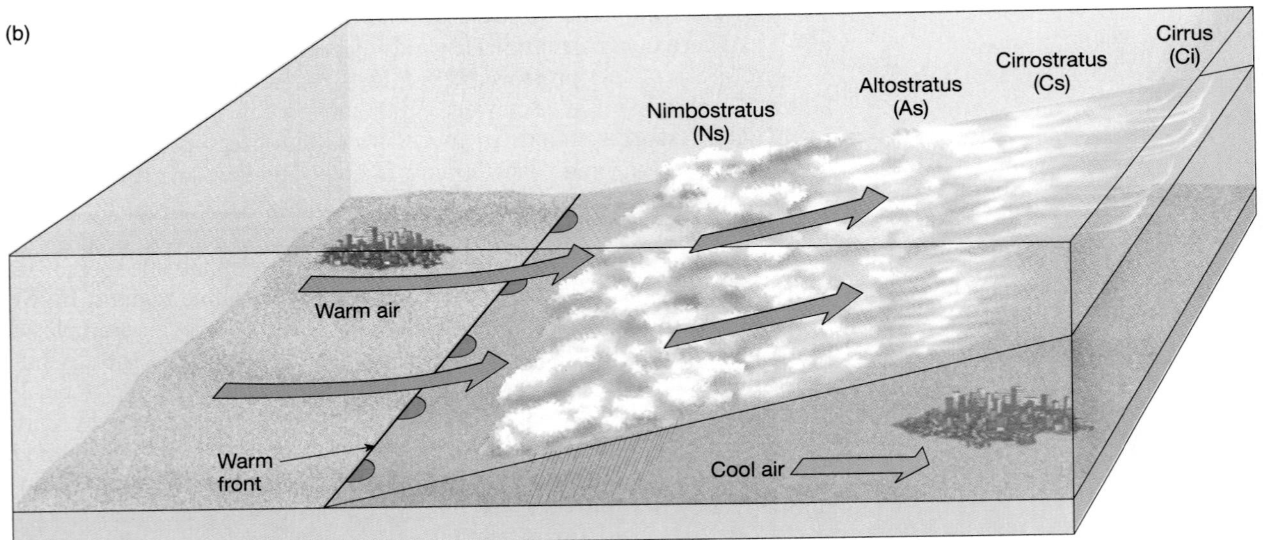

Figure 6–17
Air masses having different characteristics are separated by fronts. (a) A cold front causes warm air to rise and, when the air is humid, may result in the formation of thunderstorms. (b) An advancing warm front rises over a cold air mass (Lutgens and Tarbuck, 1995).

edge of a cold-air mass advancing on a warmer air mass (Fig. 6–17a). Being denser than warm air, cold air pushes under it. The rising warm air cools and often loses some of the water vapor it was carrying; in other words, an advancing cold front usually means rain (or snow if the temperature is low enough). A warm front is the leading edge of an advancing mass of warm air, displacing colder air (Fig. 6–17b). Interactions between such air masses cause much of the weather that we experience in the mid-latitudes.

Ocean Influence on Weather Patterns

Large masses of unusually warm or cold surface ocean waters influence weather on land in two ways. First, the presence of large masses of warm water on the western side of the equatorial Pacific Ocean causes *El Niños* and the Southern Oscillation, which we discuss in the next chapter. Second, masses of either warm or cold water steer weather patterns in the mid-latitudes. Such warm or cold water masses occur in the North Pacific, where they have been extensively studied. These water masses are 1,000 to 2,000 kilometers across and 200 to 300 meters deep. They persist for many years as they move across the ocean basin. The size of these water masses and the energy they contain permit them to steer weather systems over the continents.

Large cold-water masses in the North Pacific cause shifts in the prevailing westerly winds blowing across eastern North America. Cold, dry air from Canada displaces warm, moist air from the Gulf of Mexico and the tropical Atlantic. As a result, winter temperatures in the southeastern United States are much colder than they would be in the absence of the North Pacific cold water. Shifts in the relative positions of such water masses can bring about the opposite effect.

It is thought that prolonged droughts in Africa may be caused by unusually warm surface waters in the equatorial Atlantic. If this hypothesis is proven, it will allow meteorologists and oceanographers to predict persistent weather patterns such as multiyear droughts. Such predictions may help governments alleviate the suffering caused by famines like those that occurred during the African droughts of the 1980s.

Extratropical Cyclones

Two types of powerful storms affect the ocean: extratropical cyclones in the mid-latitudes and hurricanes in the tropics. Cyclones (of which hurricanes are one type) are intense storms around low pressure centers.

Extratropical cyclones form at polar fronts in the mid-latitudes of both hemispheres, where cold, dense air moving toward the equator meets warm, humid air moving poleward. Their formation is shown in Figure 6–18. The winds forming an extratropical cyclone blow around a low-pressure area along a stationary front (Fig. 6–18a). A wave develops on the front (b) and begins to rotate (c). Eventually, part of the front is cut off (d), forming an intense storm (e). At this stage the low-pressure cell intensifies greatly as its atmospheric pressure drops and winds grow stronger. The intensified low-pressure cell with its associated winds (Fig. 6–19) moves eastward.

Such cyclones have strong winds with speeds greater than 120 kilometers per hour (75 miles per hour), comparable to the wind speeds in hurricanes. Extratropical cyclones are common in winter when contrasts in air temperature across the polar front are largest, often causing widespread damage in coastal areas. These storms are much larger and move more slowly than most hurricanes, causing extensive erosion of beaches and flooding of low-lying areas.

Hurricanes

Hurricanes (which are called *typhoons* in some parts of the world) are intense tropical cyclones—the most damaging storms on Earth. They occur in local summer and autumn (June through November in the Northern Hemisphere, peaking in September) and they form in all tropical oceans except the South Atlantic and the eastern South Pacific (Fig. 6–20). Hurricanes form when winds converge and concentrate the rotation that comes from their participation in Earth's rotation. This convergence and concentration can cause low-pressure disturbances in the large-scale winds, such as thunderstorms, to intensify, causing local winds to converge more strongly. As warm, moist air sweeps in, it rises in a helical (twisting) pattern. As the hurricane develops, wind speeds build to 300 kilometers per hour. The fastest winds at the sea surface form a band surrounding the hurricane's eye—a relatively cloud-free area of low atmospheric pressure in the center of the storm, typically about 25 kilometers across (Fig. 6–21). A typical hurricane lasts about nine days before dissipating.

Most hurricanes form between latitudes 15° and 20° in each hemisphere (Fig. 6–20). None form within 5° of the equator, since the Coriolis effect is weak

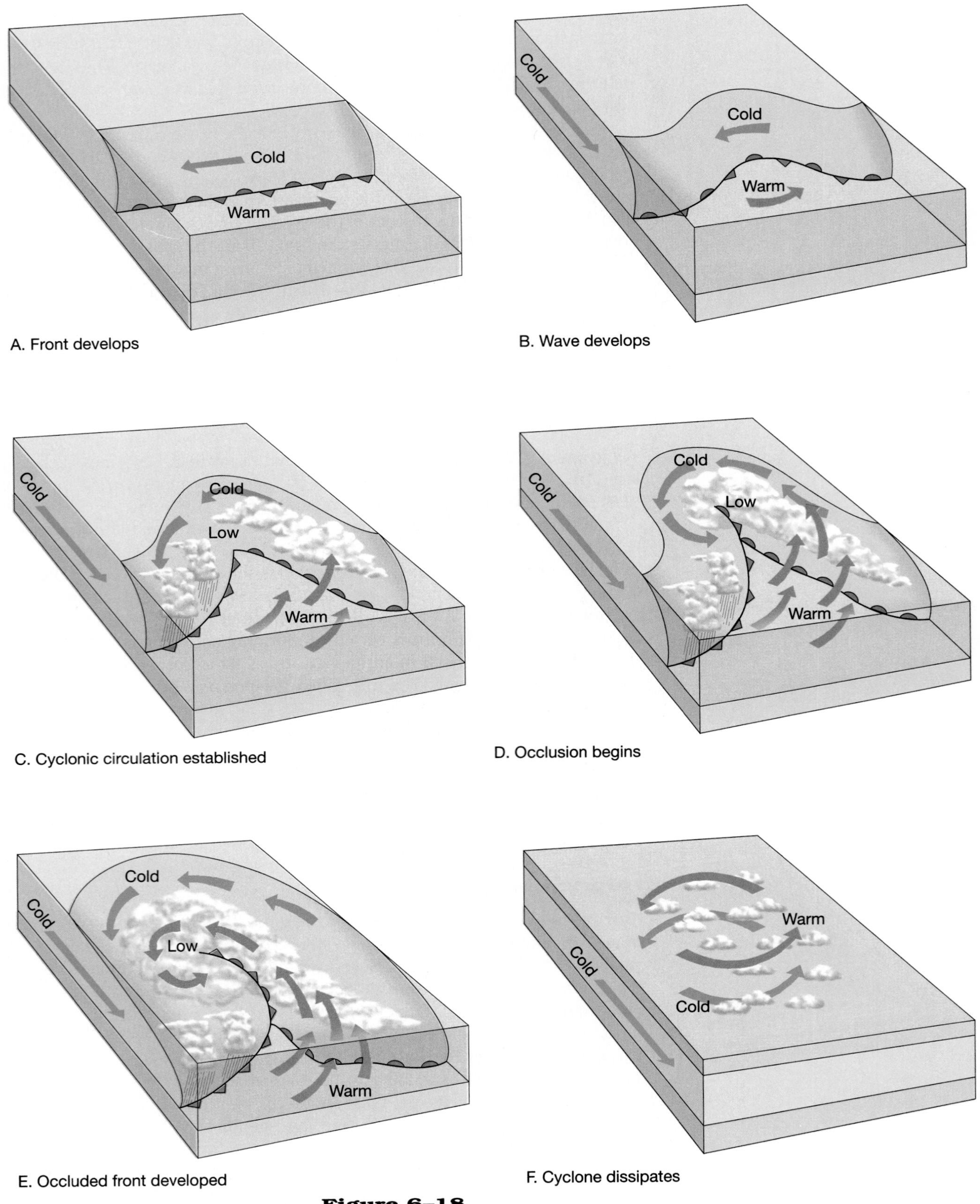

Figure 6–18
An extratropical cyclone in the mid-latitudes develops from a wave on a stationary front into an intense storm before finally dissipating (Lutgens and Tarbuck, 1995).

Figure 6–19
A large extratropical cyclone. Note the spiral structure which resembles that of a hurricane. (Courtesy NASA.)

Figure 6–20
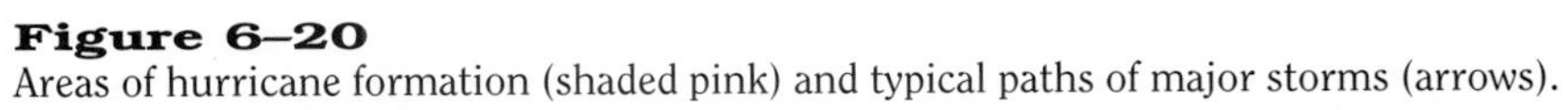
Areas of hurricane formation (shaded pink) and typical paths of major storms (arrows).

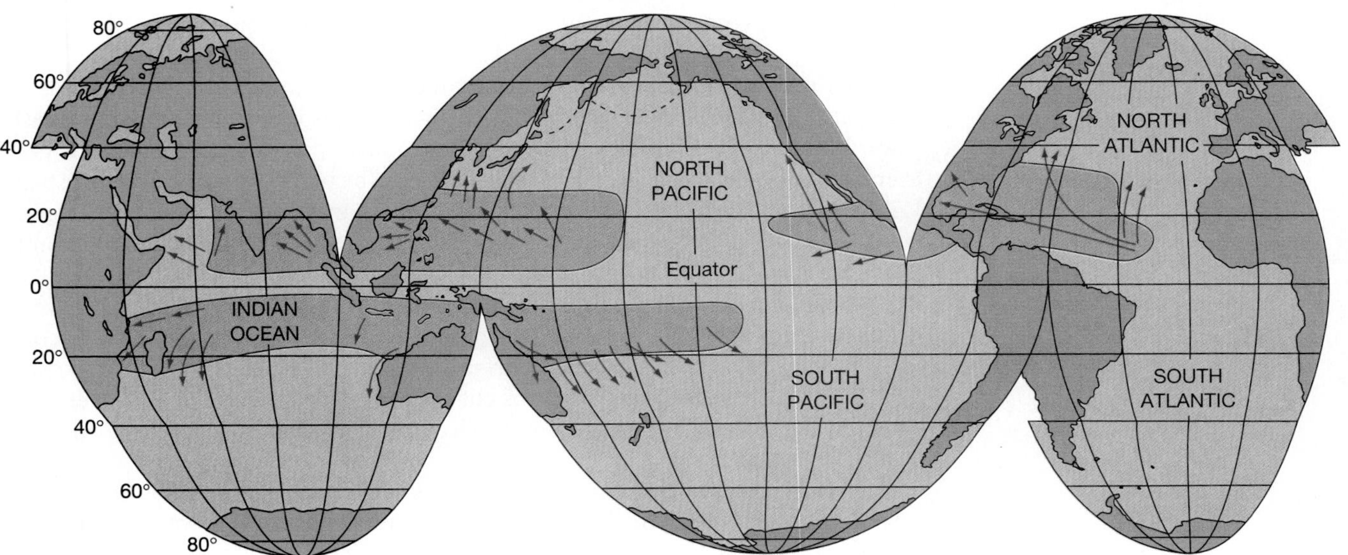

Figure 6–21
A hurricane near Hawaii observed by a satellite. The dashed line indicates the track of the hurricane. Note the eye of the hurricane and the circular winds around it. (Courtesy NASA.)

there. Hurricanes form where sea surface temperatures exceed 28°C, usually in the western parts of ocean basins. About 40 hurricanes occur each year. A hurricane's power comes from condensing water vapor in the rising air currents in the eye.

Northern-hemisphere hurricanes move westward in the trade winds at around 10 to 30 kilometers per hour; Southern-hemisphere hurricanes move eastward (Fig. 6–20). Eventually each hurricane moves poleward, carried by the general atmospheric circulation and usually steered by high-level winds. At this stage, its speed and direction often become erratic. Some hurricanes follow looping paths; others become stationary; still others move rapidly, around 1,000 kilometers per day.

When a hurricane moves over either land or relatively cold surface water, it rapidly loses strength. Winds blowing across the land experience more friction than winds blowing over the sea surface. So storms lose energy once they reach land, and their water vapor supply is cut off. As a result, the storm dissipates most rapidly over land. Cold waters also deprive storms of their energy source and they quickly dissipate.

Hurricanes release enormous amounts of energy, typically 2,000 billion kilowatt-hours each day, which equals the total annual electrical power consumption in the United States. An average hurricane precipitates 10 to 20 billion tons of water every day, often causing disastrous flooding.

Hurricanes cause millions of dollars in damage each year through the effects of winds, associated waves, and elevated sea levels. In 1992, Hurricane Andrew, the most damaging storm in U.S. history, caused about $20 billion in insurance losses in southern Florida and the Gulf coast states.

Many aspects of hurricanes still remain unexplained, such as their origins or distributions, but there has been some success in studying them. For instance, it is now possible to predict paths of hurricanes, which has decreased hurricane death tolls substantially. For instance, Hurricane Gilbert in September 1988 was the strongest hurricane in this century to strike the Caribbean and Mexico. Yet observations by satellites and aircraft permitted detailed predictions that saved the lives of many people who evacuated low-lying areas before the hurricane struck.

Hurricanes may also spawn tornadoes, which also cause much damage. Due to increasing populations and buildup of coastal areas, property damage from hurricanes continues to increase.

Seasonal Wind Patterns

Earth's equator is tilted 23.5° to the plane of Earth's orbit around the Sun. Thus the point directly beneath the Sun (where the incoming solar radiation is greatest) changes seasonally (Fig. 6–22). At the summer solstice (June 21), the Sun is directly overhead at 23.5°N. At this time, the Northern Hemisphere receives much more solar energy per unit area than does the Southern Hemisphere. At the winter solstice (December 21), the situation is reversed.

As described in the section "Winds on a Water-Covered Nonrotating Earth" the presence of continents breaks up the surface of the world ocean and alters wind patterns. These alterations are especially pronounced in the Northern Hemisphere because of the large amount of land there. As the summer begins, land warms faster than the ocean. As the air over the land warms, a low-pressure area forms (Fig. 6–23a). Because the ocean surface temperature is lower than the land surface temperature, air over the ocean is cooler than air over land and forms high-pressure areas. This is quite different from the simple north–south bands predicted by the model of a water-covered Earth shown in Figure 6–10.

In winter, the situation reverses (Fig. 6–23b). Air over land cools markedly, forming large high-pressure areas. Ocean surface waters retain their warmth so the air overlying them warms, forming low-pressure areas. Averaged over a year, air flows from high-pressure areas to low-pressure areas to form the general patterns shown in Figure 6–15.

In the Southern Hemisphere, there is relatively little land to disturb large-scale wind patterns. Therefore, there is little seasonal change in wind patterns there.

Monsoons

Large-scale seasonal changes resulting from differential heating of land and ocean are called **monsoons.** These seasonal changes are most obvious in the northern Indian Ocean. In summer in Southeast Asia and India, the warm land causes air to rise over the continent (Fig. 6–24). These vertical air motions draw cooler, moisture-laden winds from the ocean to replace the rising air. As this onshore flow (meaning flowing from the water to the land) rises over the continent, it produces heavy monsoon rains. Summer monsoon rains are essential to rice crops in India and Southeast Asia.

In winter, the winds reverse. Cold air from over the continent is drawn seaward by warm air rising over the ocean. This produces cool, dry weather on land, known as the winter monsoon. For centuries, sailors of the northern Indian Ocean depended on these seasonal wind reversals to carry them back and forth on their trading expeditions between India and Africa. (This was part of the Silk Route between Europe, China, and India, discussed in Chapter 1.)

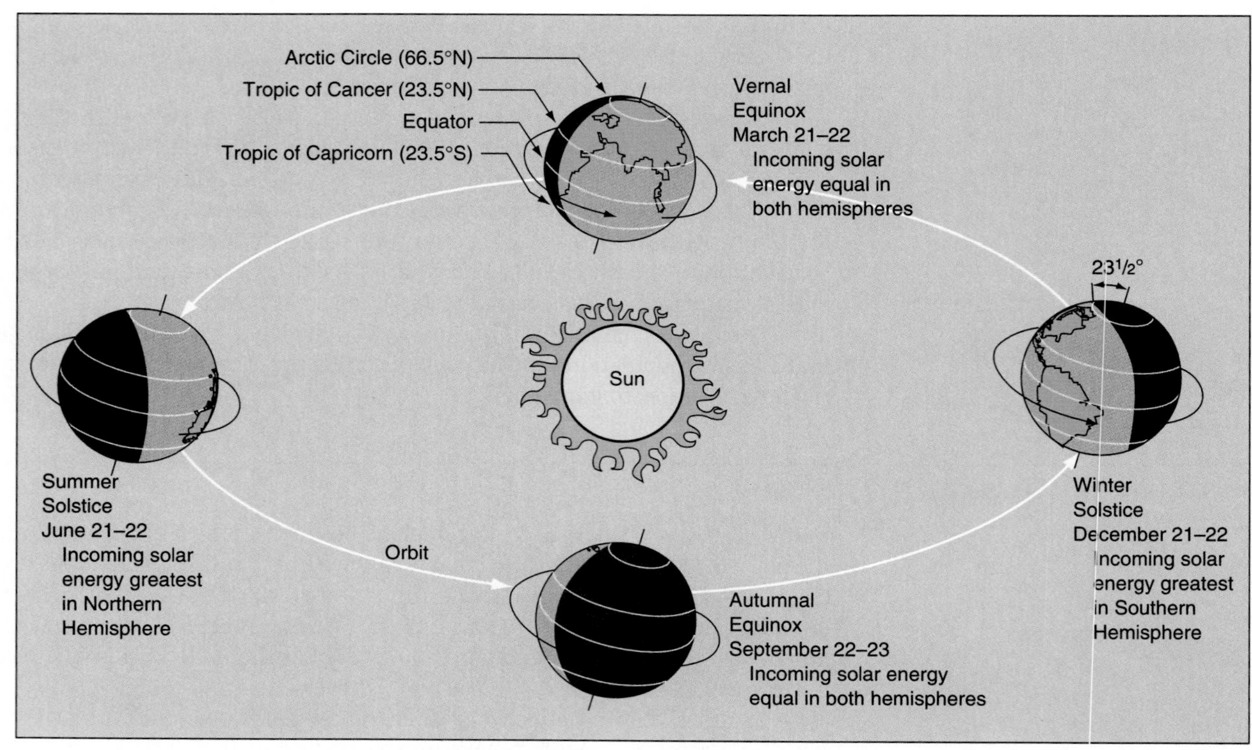

Figure 6–22
The tilt of the Earth's axis of rotation with respect to the plane of its orbit about the Sun causes the amount of solar energy coming to the Northern and Southern Hemispheres to change seasonally. During the summer solstice, the Northern Hemisphere receives more solar energy that the Southern; during the winter solstice the situation reverses. The two hemispheres receive equal amounts of solar energy during the equinoxes. (Lutgens and Tarbuck, 1995.)

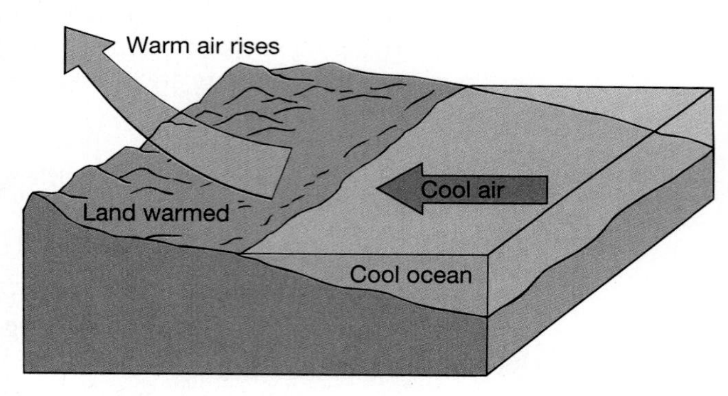

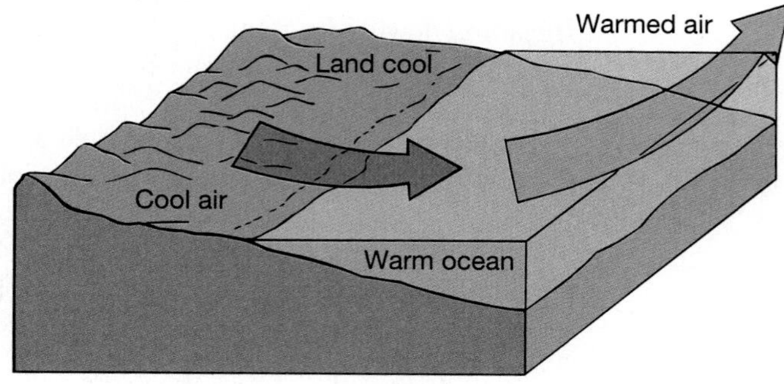

Figure 6–23
(a) Sea breezes occur during the day, when air warmed by the land rises and is replaced by cool air from the ocean. A sea breeze flows from sea to land. (b) Land breezes occur at night, after the land is cooled below the ocean surface temperatures. Warm air then rises over the ocean and is replaced by cool air from the land. A land breeze flows from land to sea.

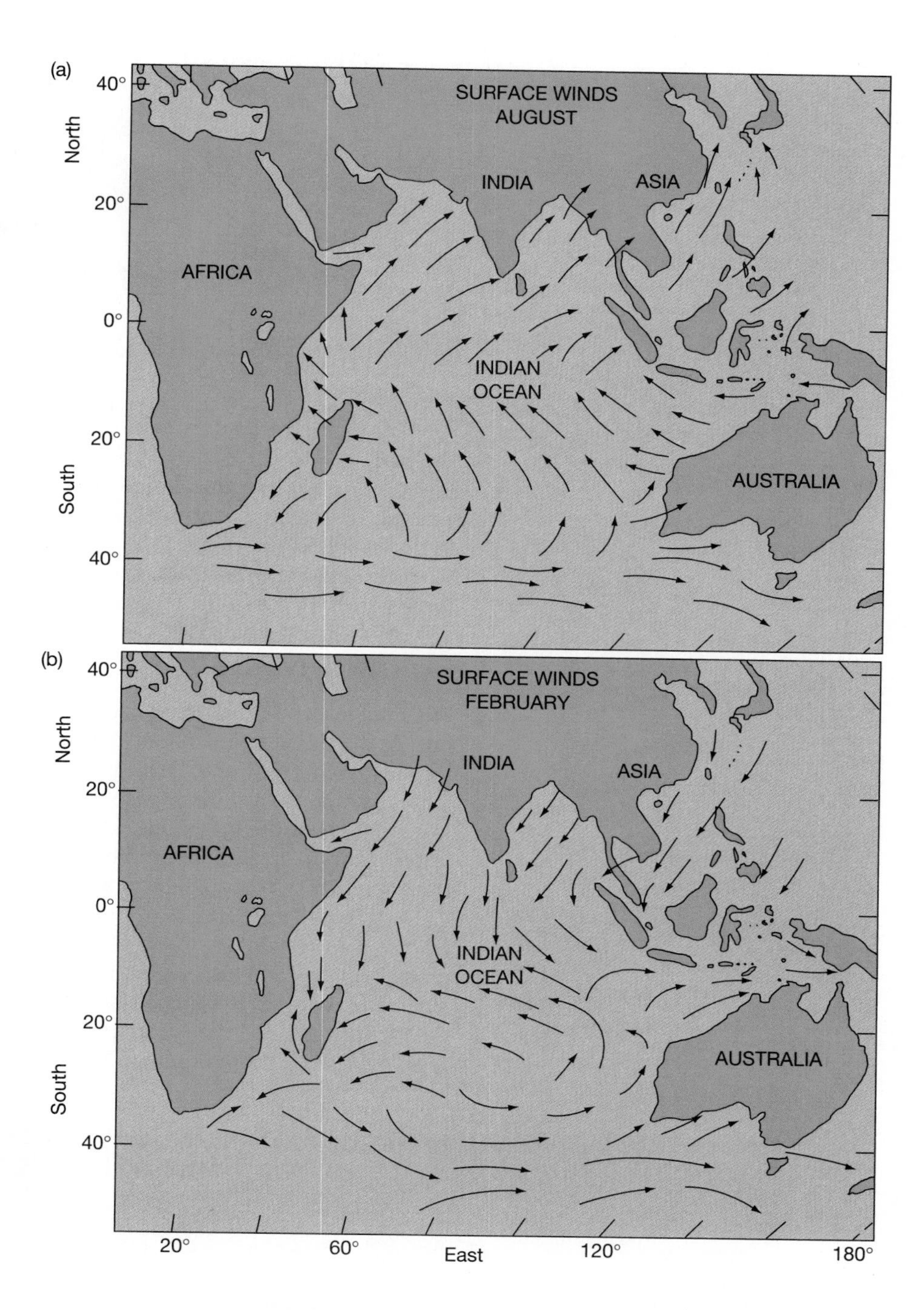

Figure 6–24
(a) In summer, surface winds flow from the Indian Ocean onto India and southeastern Asia, bringing the summer monsoon. (b) In winter, surface-wind flow is from land to sea, and the result is the cool, dry weather known as the winter monsoon.

Winds drive surface ocean currents. Thus, these wind reversals cause major shifts in currents in the northern Indian Ocean, as we shall see in Chapter 7. Monsoon-like wind patterns are also common in other parts of the world. For instance, much of the summer rainfall over the United States mid-continent comes from monsoon-like atmospheric circulation with water-laden winds coming from the Gulf of Mexico and from the Pacific.

Wind Patterns over Land

Differential heating of land and ocean surface also affects coasts daily, but on a smaller scale. During the day, the land warms more quickly than the ocean. Thus during the afternoon, air over the land rises, pulling in cooler air from the ocean. This is known as a **sea breeze.**

Figure 6–25
Moisture-laden air from the ocean rises over a mountainous island, causing clouds and rain on the windward side. As dry air flows down the mountain, it warms, causing the leeward side to be sunny, warm, and dry.

At night, the land cools more rapidly than the sea. Now the wind blows from the land toward the ocean. This is a **land breeze,** which is strongest in the late night and early morning hours. The Florida peninsula in summer provides an example of this. Frequently there is a north–south line of vigorous thunderstorms in the middle of the state.

Around Southeast Asia, fishing boats make use of these daily wind reversals to take them to sea in the morning and bring them home in the afternoon.

Island Effects

The presence of islands affects oceanic winds, especially if the islands are mountainous. As they blow across mountains, winds are forced upward (Fig. 6–25). These rising air currents cool, and the lower temperatures cause water vapor to condense and fall as rain on the windward side of mountains. The winds blow across the mountains and down the other side. In the process, the air warms and can hold more water vapor. In areas of descending winds, there is little or no rain. Sides of mountains opposite the prevailing winds, called the *leeward side,* are often quite dry.

An excellent example of mountains affecting precipitation is the Hawaiian Islands. On the windward side of all the mountains in this group of islands, there is frequent rain and lush vegetation. On the side away from the winds (the leeward side), the weather is sunny, and there is little rain and sparse vegetation (Fig. 6–26). This is called the *island effect.* Because of such effects, islands and coastal areas often have more rain than the surrounding ocean.

Figure 6–26
Two locations on the Hawaiian island of Kauai illustrate the island effect. (a) The wettest spot on Earth has rainfall 365 days a year. (b) Waimea Canyon on the leeward side of the island has a much drier climate and desert-like vegetation.

Hurricane Predictions

Hurricanes can now be predicted many months in advance. The number and intensity of storms in a season can be predicted for a region, based on ocean and atmospheric conditions, though their exact locations and paths are still unpredictable.

Most of the hurricanes that hit the Gulf Coast region and the Atlantic coast of the United States form off northwestern Africa. The meteorological conditions in this region of Africa are used in predictions of the North American hurricane season. During wets years in western Africa, more hurricanes form; during drought years, fewer form.

A prime factor in determining the number and intensity of hurricanes is whether or not a large pool of unusually warm water exists in the western equatorial Pacific; as we see in later chapters, this condition is called an *El Niño.* These large pools of warmer-than-usual waters in the western equatorial Pacific cause strong lower-level westerly winds in the tropics. Such strong winds inhibit hurricane formation by tearing storm systems apart before they can develop into hurricanes. Thus, *El Niño* years have fewer hurricanes than normal in the Atlantic Ocean.

Equatorial stratospheric winds also influence hurricane formation. These winds that circle Earth reverse their direction every 12 to 16 months, blowing from the east and then from the west. When these winds blow from the west, hurricane activity in the Atlantic is roughly as intense as when they blow from the east.

Still another factor used in the prediction of hurricanes is the presence of unusually warm surface ocean waters nearby. For example, warmer-than-normal surface waters between West Africa and the Caribbean increase the probability of hurricanes striking North America.

In addition to the improvements in making seasonal hurricane forecasts, it is also possible to improve the accuracy of forecasts for individual storms. Better observations, more powerful computers, and more elaborate models of atmospheric processes all help make the forecasts more accurate.

Summary

The atmosphere consists of a mixture of gases, primarily nitrogen and oxygen, with traces of ozone, carbon dioxide, dust, and water vapor. The atmosphere is density-stratified—densest near the surface. The layer nearest the ground is called the troposphere, the next highest layer is the stratosphere, and above that is the mesosphere. The boundary between troposphere and stratosphere is called the tropopause. There are strong vertical motions of the air in the troposphere; most weather occurs here. The troposphere is warmed at Earth's surface and cooled at the top. There is little vertical motion in the stratosphere. Materials injected there remain for a long time.

The atmospheric heat budget describes sources and sinks of energy. Earth's heat comes from the Sun, averaging 0.5 calorie per square centimeter per minute. This amount varies seasonally because of Earth's inclined axis of rotation. Most of the heat is received in the equatorial zone and least, near the poles, where heat is lost. Water's latent heat of evaporation is the main driving force for heat-transport processes.

On a nonrotating Earth completely covered by water, the atmosphere would have a simple two-celled circulation pattern. Warm air would rise at the equator and sink at the poles. On a rotating Earth, winds appear to be deflected by the Earth's rotation—to the right in the Northern Hemisphere, to the left in the Southern. This is called the Coriolis effect. The effect is greatest near the poles and absent along the equator. Because of the Coriolis effect, Earth has a complicated six-celled circulation pattern—three cells in each hemisphere. The simple circulation pattern persists near the equator, with warm air rising there and sinking 30° north and south. A polar front separates warm and cold air at each pole.

Patterns of prevailing winds shift seasonally and are deflected by the continents. This deflection is greater in the Northern Hemisphere than in the Southern Hemisphere because there is more land. The intertropical convergence zone shifts northward in northern summer and southward in the northern winter.

Extratropical cyclones form at mid-latitude polar fronts when cold air moving toward the equator meets warm air moving poleward. These storms move generally eastward, relatively slowly. Wind speeds in these storms may reach hurricane strength.

Hurricanes are intense, fast-moving cyclones. They form over water warmer than about 28°C in all tropical oceans except the eastern South Pacific and the South Atlantic. Storm tracks are controlled by upper-air winds. The storms are powered by the condensation and release of heat from water vapor. Hurricanes lose power over land or cold waters.

Monsoons are seasonal wind shifts caused by warm air rising over continents in summer. The rising air is replaced by cool, moisture-laden air from the ocean, which produces heavy rains on land. In winter, the circulation is reversed, causing cool, dry weather on land. Comparable wind shifts occur on a daily basis in coastal areas. Onshore winds are called sea breezes; offshore winds are called land breezes.

Near mountainous islands and continental coasts, winds are forced to rise as they blow over the mountains. This causes local precipitation on the windward side of the island and dry weather on the opposite side. Comparable effects occur over mountains on land.

Key Terms

convection
troposphere
tropopause
stratosphere
ozone layer
mesosphere
greenhouse effect
Gaia hypothesis
Coriolis effect
geostrophic winds
intertropical convergence zone
trade winds
jet streams
climate
horse latitudes
doldrums
eddies
lows
highs
front
extratropical cyclones
hurricanes
monsoons
sea breeze
land breeze

Study Questions

1. Sketch a vertical cross section showing the major layers and boundary zones of the atmosphere.
2. Describe how budgets are used to study atmospheric processes.
3. Discuss the atmospheric heat budget.
4. What causes seasonal changes in atmospheric temperature?
5. Describe the greenhouse effect and what causes it.
6. Describe what atmospheric circulation would be on a nonrotating Earth completely covered by water.
7. Sketch the pattern of prevailing winds on Earth.
8. Describe how the Coriolis effect deflects winds.
9. Describe the causes of the monsoon winds, land breezes, and sea breezes.
10. Where do hurricanes form? What controls their paths?
11. How does the ocean influence weather on land?
12. Describe the Gaia hypothesis.
13. How do satellites permit more accurate predictions of hurricanes?
14. Discuss the causes of the ozone hole over Antarctica.
15. *[critical thinking]* How might global warming affect the distribution and frequency of hurricanes?

Selected References

EAGLEMAN, J. R., *Meteorology: The Atmosphere in Action,* 2d ed. Belmont, CA.: Wadsworth, 1985.

Firor, J., *The Changing Atmosphere.* New Haven, CT: Yale University Press, 1990. Human activities and their impacts on Earth's atmosphere and climate.

INGERSOLL, A. P., "The Atmosphere," *Scientific American,* September 1983.

LUTGENS F. K. AND E. J. TARBUCK, *The Atmosphere: An Introduction to Meteorology.* Englewood Cliffs, N.J.: Prentice Hall, 1995.

LOVELOCK, J. E., *Gaia: A New Look at Life on Earth.* Oxford: Oxford University Press, 1979.

MUSK, L. F., *Weather Systems.* Cambridge: Cambridge University Press, 1988.

SIMPSON, R. H. AND H. RIEHL, *The Hurricane and Its Impact.* Baton Rouge: Louisiana State University Press, 1981.

WEBSTER, P. J., "Monsoons," *Scientific American,* August 1981.

WELLS, N., *The Atmosphere and Ocean: A Physical Introduction.* London: Taylor & Francis, 1986.

7 Ocean and Climate

OBJECTIVES

Your objectives as you study this chapter are to understand:

- The ocean's role in transporting heat
- The relationship between Earth's heat and water budgets
- How ocean structure affects oceanic processes
- How and where water masses form and move
- How sea ice forms and its effect on Earth's climate

The ocean plays an important role in controlling Earth's climate. Cold, dense water masses which fill the ocean interior form in the high latitudes.

In this chapter, we begin by considering an ocean without currents. In other words, we examine the effects of ocean processes averaged over many years and over thousands of kilometers—essentially a static ocean. We also examine how the ocean affects climate.

Two approaches are used to study long-term behaviors of ocean waters; we use both in this chapter. First is the budget approach, wherein heat and water budgets are constructed to show major sources and sinks. Regarding the water budget, it is important to remember that (except over hundreds of millions of years) the amount of Earth's surface water is fixed. For this reason, we do not need to consider new sources of water when constructing a budget; all we must account for is the redistribution of water on Earth.

The second approach determines connections between surface and subsurface waters. This procedure involves studying distributions of water temperature and salinity.

Warming and Cooling the Ocean

Solar radiation striking Earth's surface lights the surface waters, warming them and the lower atmosphere as we saw in Chapter 6. Part of this radiation is in the visible part of the light spectrum. After passing into the surface of the ocean, most of this light energy is converted to heat energy,

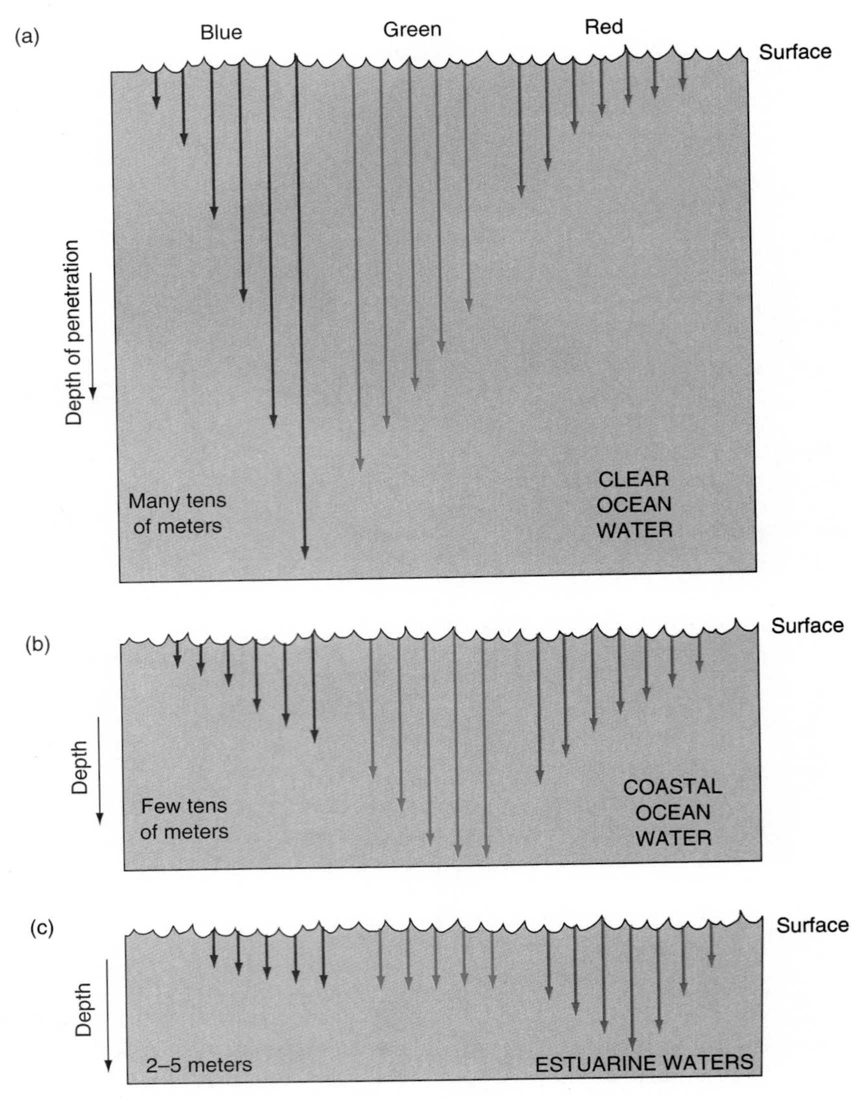

Figure 7–1
Penetration of different colors of light below the surface in (a) open ocean, (b) coastal ocean, and (c) estuarine waters.

with the heat either raising the water temperature or evaporating some of the water.

Energy from the Sun is filtered first as it passes through the atmosphere and again in surface ocean waters. Within the first 10 centimeters below the ocean surface, virtually all the infrared portion of the incoming sunlight is absorbed and changed to heat (Fig. 7–1). Of the remaining radiation, about 60 percent is absorbed within the first meter of seawater below the surface and about 80 percent is absorbed in the first 10 meters. Only about 1 percent remains at 140 meters in the clearest subtropical ocean waters. In short, the ocean is nearly opaque to light but, as we saw in Chapter 4, nearly transparent to sound.

In coastal waters, abundant marine organisms, suspended sediment particles, and dissolved organic substances absorb much of the light in shallow depths. In the coastal waters near Cape Cod, Massachusetts, for instance, only 1 percent of the surface light commonly penetrates to 16 meters. In such waters, the color of the light that penetrates to the greatest depths shifts from the bluish light typical of clear oceanic waters to longer wavelengths, as shown in Figure 7–1. In turbid coastal waters, the light that penetrates the farthest is in the yellow range, because all other colors are absorbed closer to the surface. In highly turbid waters, all light is absorbed within a few centimeters of the water surface.

Far from the coast, open-ocean waters often have a deep luminous blue color quite unlike the greenish or brownish colors common to coastal waters. The deep blue color indicates an absence of particles. In these areas, the color of the water results from the scattering of light rays by water molecules or particles in the water. Such scattering is also responsible for the blue color of the sky seen through a clean atmosphere.

The amount of light reflected from the ocean is controlled by the state of the sea surface and the angle at which the Sun's rays strike it. Waves on the sea surface generally increase the amount of light reflected by as much as 50 percent. When the Sun is directly overhead, only about 2 percent of incoming radiation is reflected; the remainder enters the water. When the Sun is near the horizon, nearly all incoming radiation is reflected when the sea surface is calm.

Sea-Surface Temperatures

In an ocean without currents, bands of equal surface temperatures would run east–west. Water temperatures would be highest along the equator (because there Earth is most warmed by solar radiation) and become cooler toward the poles (Fig. 7–2). There are, however, complications due to currents transporting cold waters toward the equator and warm waters toward the poles. Here we deal only with general temperature distributions in a hypothetical ocean without currents, putting off until Chapter 8 our coverage of how currents affect temperature distribution.

Ocean temperatures change with the seasons, as a result of variations in incoming solar radiation (Fig. 7–3). In the equatorial ocean, water and air temperatures change little seasonally, because the amount of incoming solar radiation does not vary much during the year. In the high latitudes, seasonal changes in surface-water temperatures are also small, but here the lack of temperature variation is due to the year-round presence of ice. Seasonal differences in ocean surface temperatures are greatest in the mid-latitudes. (By contrast, on land the largest temperature differences occur in the high latitudes, where ice covering the ocean surface prevents ocean water from moderating air temperatures as elsewhere.)

Sea-Surface Salinities

Sea-surface salinity distributions are quite different from temperature distributions. Highest salinities occur in the centers of the ocean basins, where there is no dilution by river discharge (Fig. 7–4). Lowest salinities occur in high latitudes, where the ocean receives fresh water from melting ice, and near continents, where fresh water comes from river discharges.

Evaporation of water from the sea surface is greatest in subtropical areas where there are clear skies, little rain, and strong prevailing winds. The highest surface-water salinities therefore occur in the subtropics in the centers of ocean basins, where there is no dilution by river discharges. The sides of ocean basins, where dry winds blow off the continents, are also areas of high evaporation; here, however, river outflows often dilute the surface waters, lowering the salinity and thus obscuring effects of high evaporation rates. The relatively low salinities of waters in the high northern latitudes, shown in Figure 7–5, are due, in part, to low rates of evaporation there.

Salinity differences among the major ocean basins are caused by regional variations in evaporation and precipitation. The relatively high salinities in the

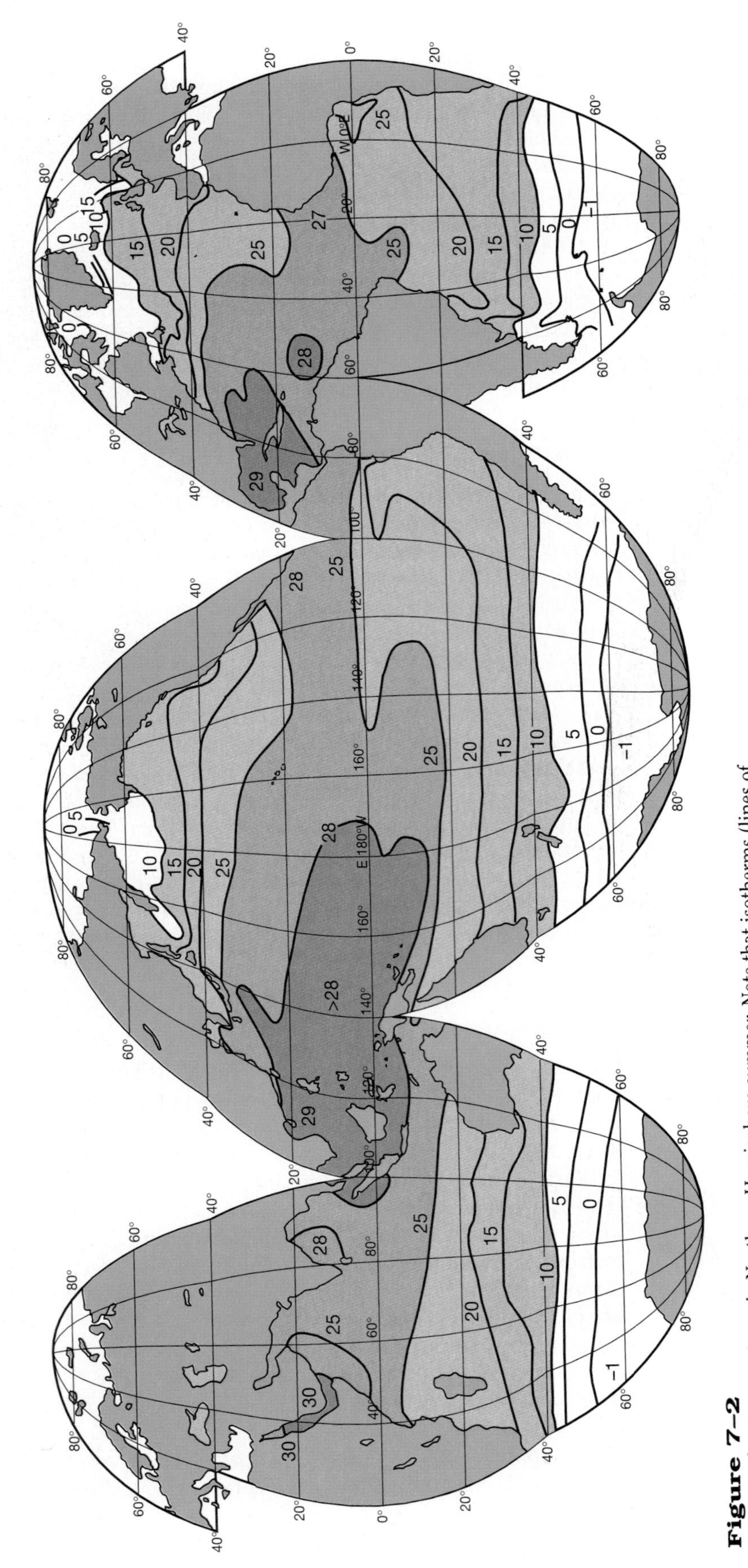

Figure 7–2
Ocean-surface temperatures in Northern Hemisphere summer. Note that isotherms (lines of equal temperature) generally parallel the equator. (After H. U. Sverdrup and others: *The Oceans.* Englewood Cliffs, NJ: Prentice Hall, 1942.)

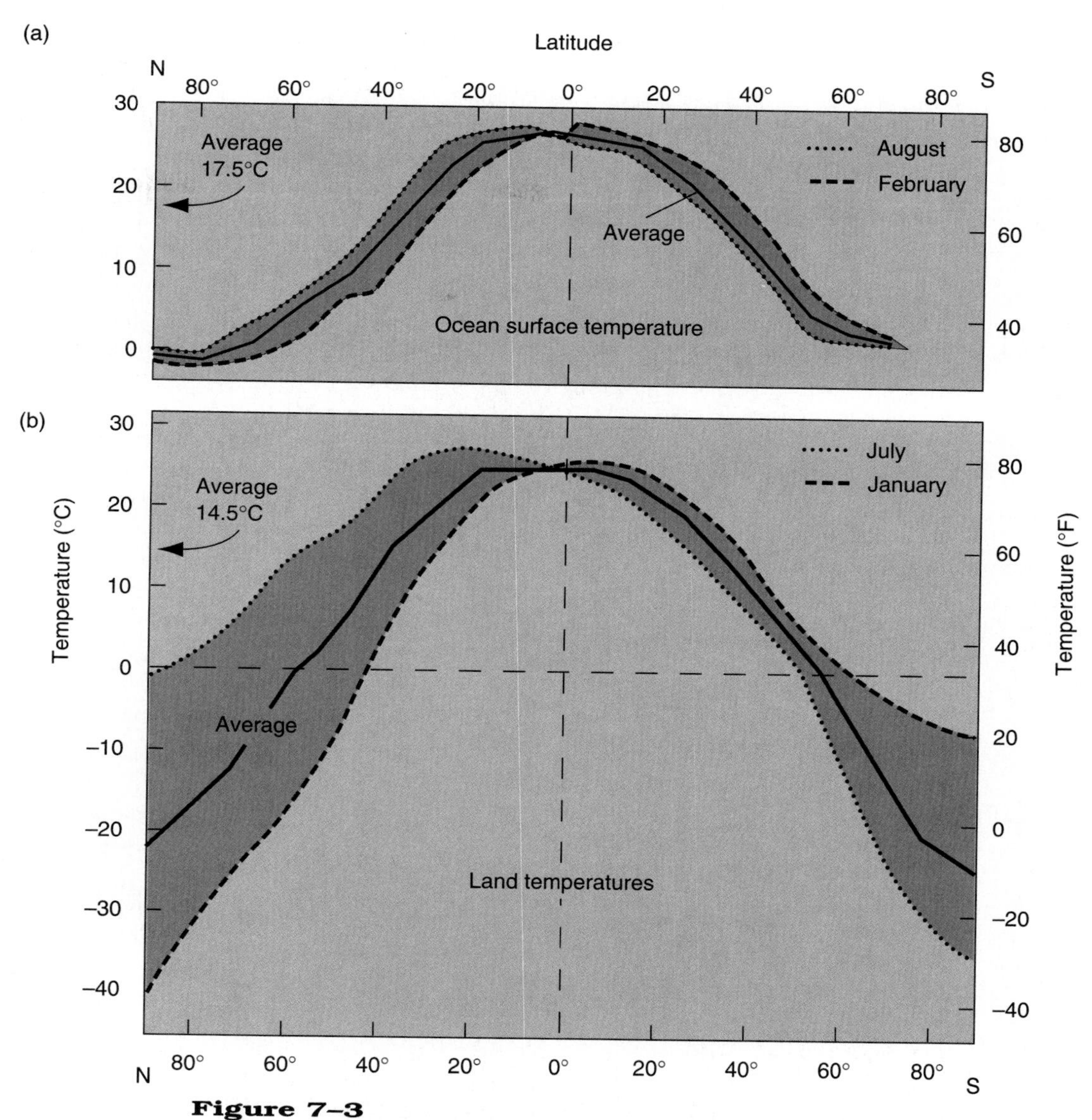

Figure 7–3
(a) Average temperature ranges for the ocean surface. The greatest seasonal changes in temperature occur in the mid-latitudes, around 40°N and 40°S. (After G. Wüst and others, 1954.) (b) Average temperature ranges on land. The most extreme ranges occur in the high latitudes and polar regions, in contrast to the oceans.

Atlantic, for instance, result from interactions between prevailing winds and the mountain ranges of North and South America. Winds blowing across the western mountain ranges lose water vapor on the Pacific side of the mountains (Fig. 7–6). Thus, winds blowing toward the Atlantic are relatively dry, with the result that there is little precipitation over the Atlantic. In the equatorial region, winds carrying water vapor from the Atlantic blow across the low-lying Central America region to carry water vapor into the Pacific, causing evaporation rates to be high over Atlantic water. The combination of little precipitation and high evaporation leads to high salinity. Finally, highly saline waters from the Mediterranean flow into the North Atlantic through the Strait of Gibraltar.

The highest salinities occur in the nearly landlocked Red Sea and Arabian Gulf (Fig. 7–4). There the loss of water due to evaporation is especially large, and there are no river outflows to supply freshwater.

Precipitation is highest just north of the equator, and therefore salinity is relatively low there. This region lies between the northern and southern trade

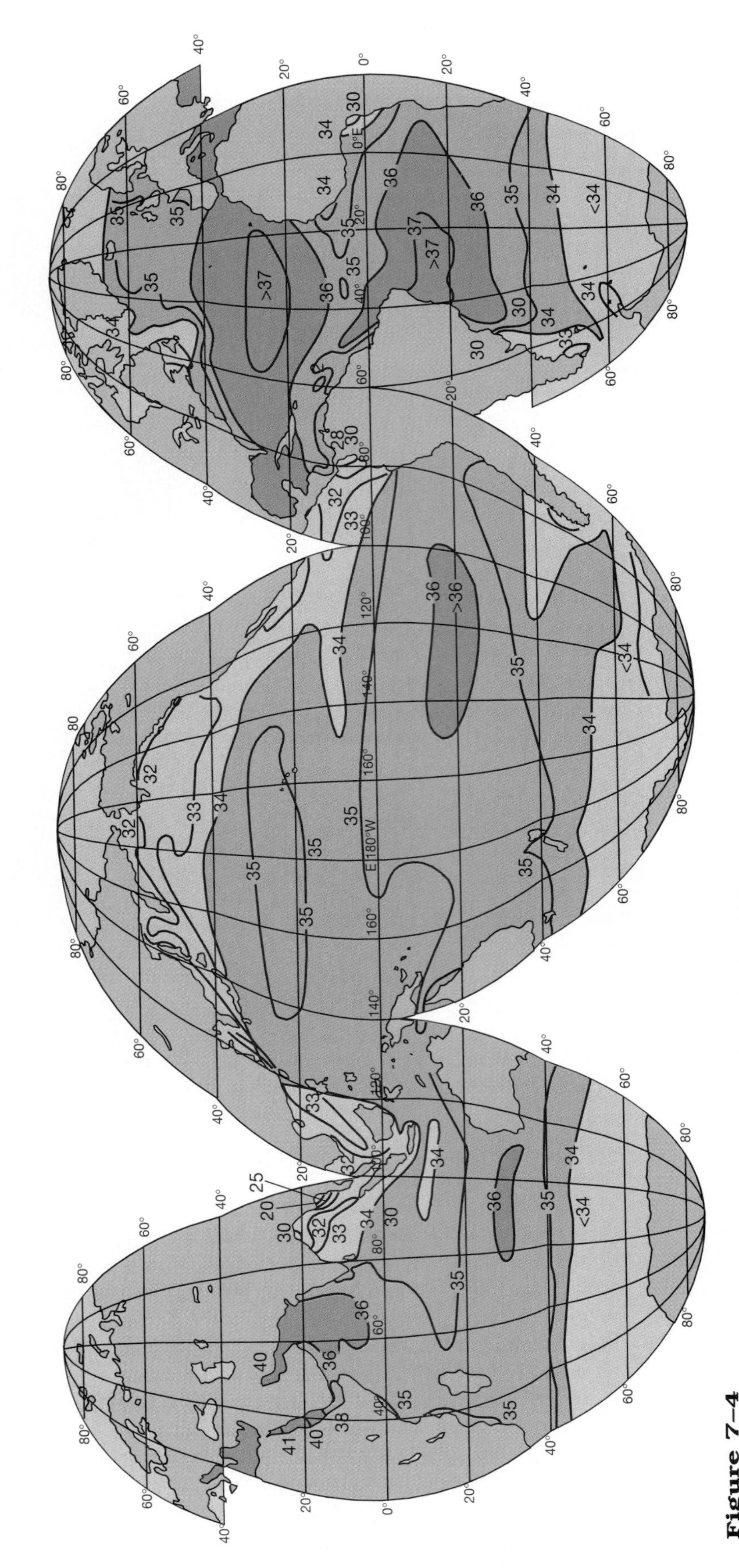

Figure 7–4
Salinity of surface waters in northern summer. (After Sverdrup and others, 1942.)

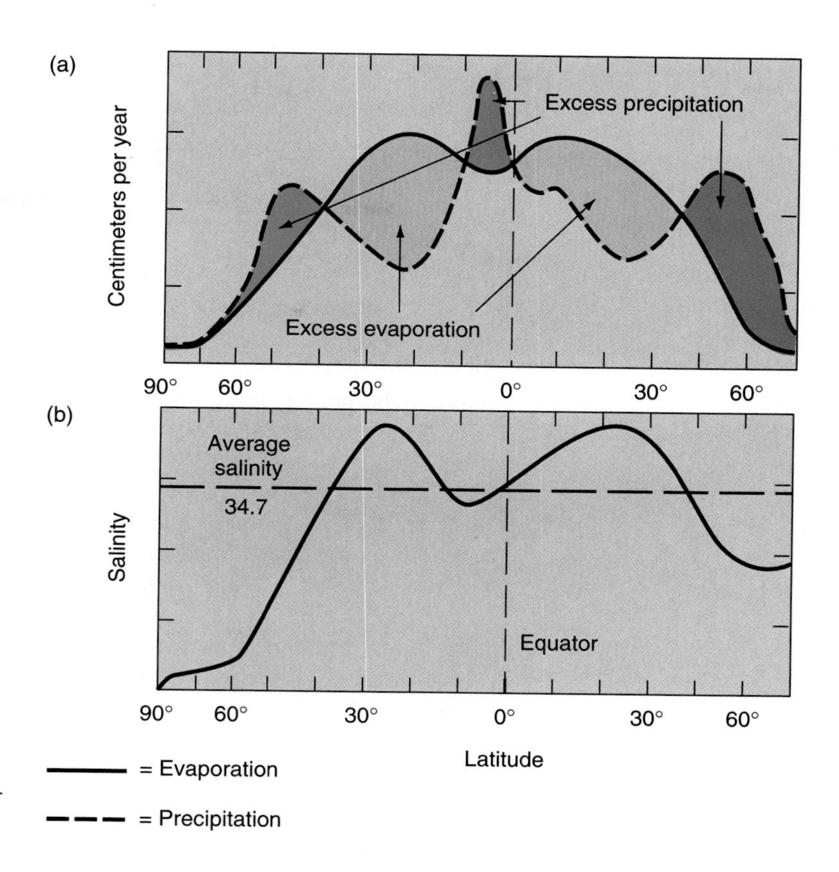

Figure 7–5
Relationships between oceanic evaporation/precipitation and surface water salinities. (After Wüst and others, 1954.)

wind belts. Heavy rainfall, resulting from the condensation of warm, moist air rising in the equatorial atmospheric circulation, dilutes equatorial surface waters. In the marginal seas of Southeast Asia, low salinities are due to not only to high precipitation, but also to large river discharges there, especially during monsoons.

In summary, highest salinities occur in areas of excess evaporation, either in subtropical central regions of the ocean basins or in the landlocked seas of arid regions. Lowest salinities occur where precipitation exceeds evaporation, primarily coastal or equatorial regions (Fig. 7–5).

Figure 7–6
Schematic representation of prevailing westerly winds blowing off the Pacific Ocean and across North America. When the winds rise to cross the mountains, they are cooled, causing heavy precipitation on the windward side of the mountain ranges. Blowing across the continent, the winds are warmed and dried. Then they cause unusually high rates of evaporation over the North Atlantic.

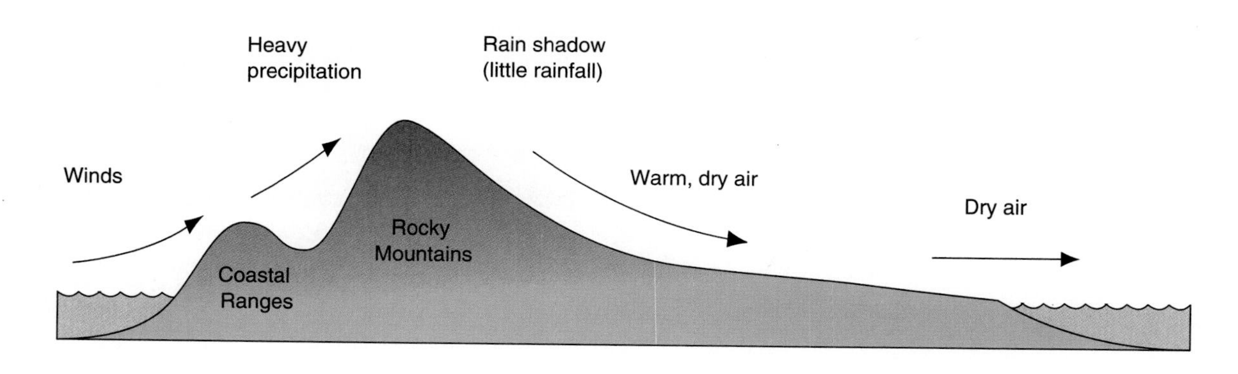

Water and Heat Budgets

The distributions of water and heat are intimately related, as we saw in Chapter 6. Evaporation from the ocean surface removes heat from the upper ocean. This latent heat is later released to the atmosphere when water vapor condenses and falls back to Earth as rain or snow. We know that the amount of water on Earth's surface has remained constant for nearly a billion years. So we can make a budget to keep track of where the water goes, how it moves, and where it is stored.

The ocean holds most of the water on Earth's surface. The small quantities of water in rivers, lakes, and the atmosphere are in transit back to the ocean, returning within a few days to a few years. Groundwater also returns to the sea, but it takes much longer, decades to centuries.

One convenient way to visualize Earth's water budget (Fig. 7–7) is to consider the thickness of the layer of water evaporated each year from the ocean surface. About 97 centimeters of water is evaporated each year. About 88 centimeters, 91 percent of what evaporated, returns to the ocean as rain (or snow if it is cold enough). The remainder falls on land (along with rain from local evaporation) and eventually makes its way through rivers to return to the sea.

Another way to visualize Earth's water budget uses the volume of water rather than the thickness of the evaporated layer. The volume of water involved in evaporation and precipitation is expressed in thousands of cubic kilometers, as in Figure 7–8.

A water budget for a region is written

$$\text{Evaporation} = \text{precipitation} + \text{runoff} + \text{current transport}$$

If the region under consideration is large (the North Atlantic, for example), the amount of river runoff is small compared with the amount of seawater. In addition, the currents in most regions do not transport much fresh water. Thus, we can ignore both terms, and the water budget for a large ocean region simplifies to

$$\text{Evaporation} = \text{precipitation}$$

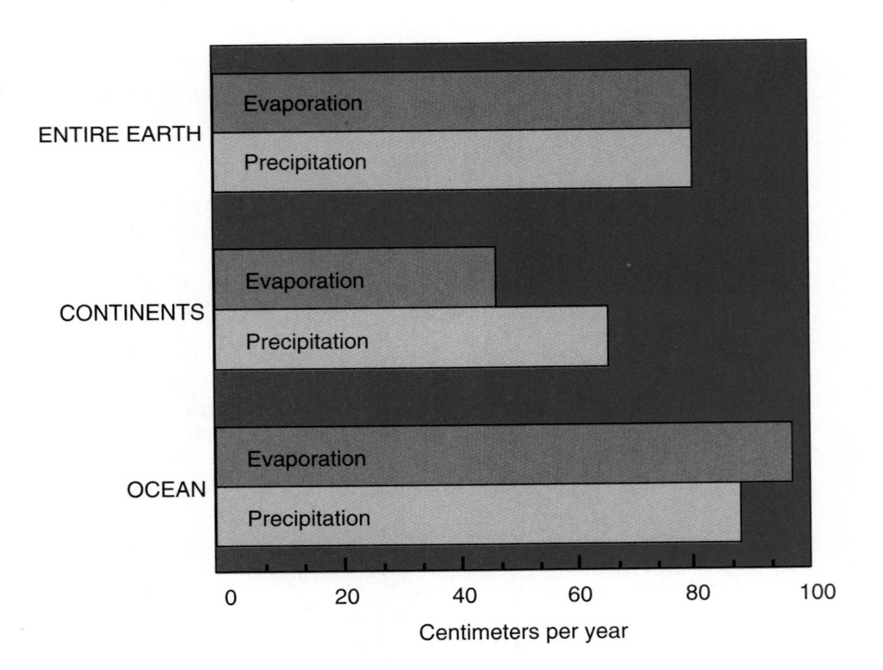

Figure 7–7
Earth's water budget in its simplest form. (After A. Defant: *Physical Oceanography,* vol. 1. Pergamon Press, 1961, p. 235.)

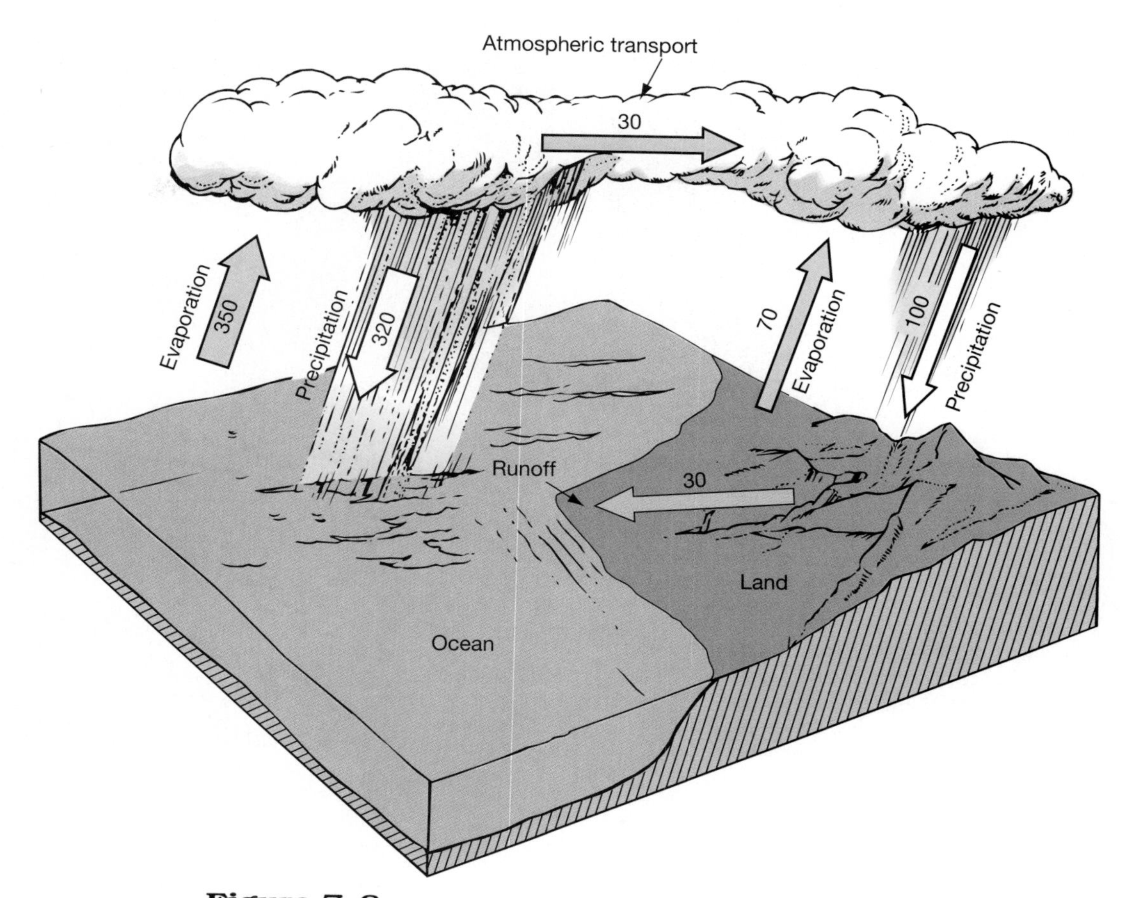

Figure 7–8
Schematic representation of Earth's water budget, showing amounts of liquid water involved in each process, in thousands of cubic kilometers.

We can also make a heat budget, because we know that each year Earth loses as much heat as it gains. Heat budgets are slightly more complicated than water budgets, however, because of interactions between ocean and atmosphere. In its simplest form, the heat budget for a region is

Heat gained = heat lost

There are more terms that must be considered in the full heat budget, which is written

Heat from solar radiation = latent heat lost during evaporation
+ heat lost in longwave radiation
+ heat lost to overlying atmosphere
+ heat lost via current transport

If the region covered by the heat budget is large (say an entire ocean basin), the amount of heat gained or lost by current transport (Fig. 7–9) can be ignored. Thus, the heat budget says that the ocean loses heat through evaporation, long-wave radiation back to space, and direct heating of the atmosphere.

Regional salinity differences in surface ocean waters are manifestations of the movements of heat and water in the ocean and atmosphere over the entire planet. Areas of high surface salinity supply water vapor to the atmosphere, and this vapor

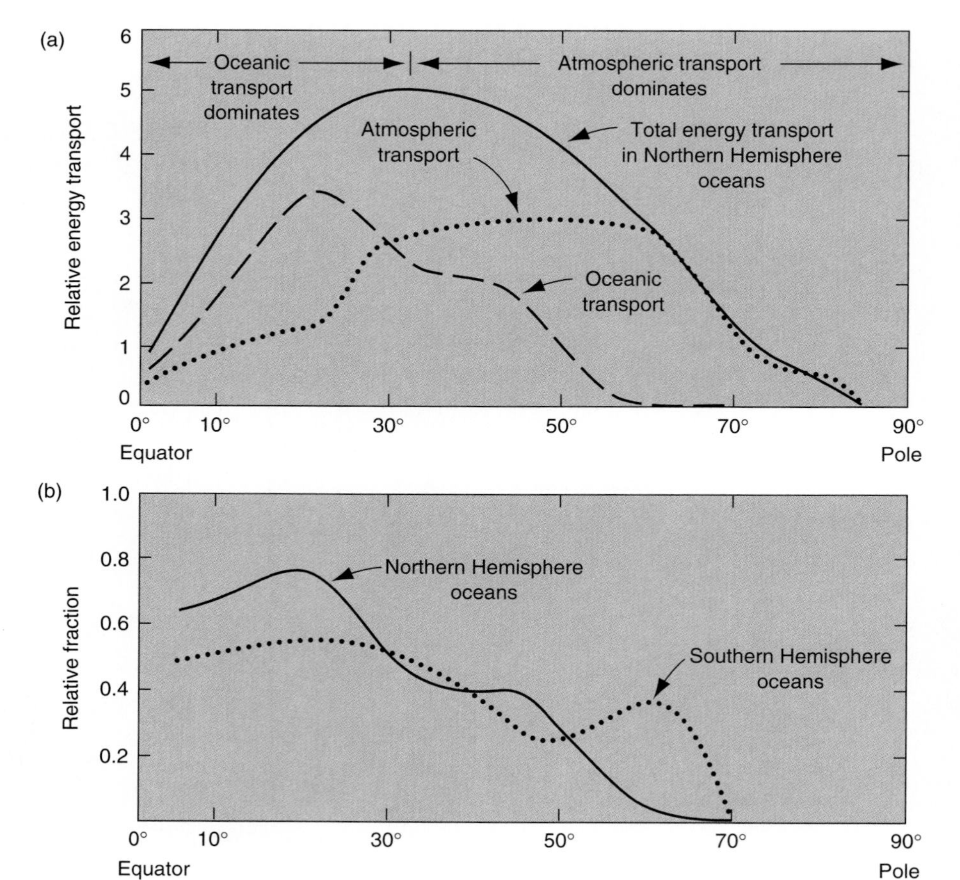

Figure 7–9
Near the equator (between about 30°N and 30°S), ocean currents carry more than half the total energy transported poleward. Nearer the poles, atmospheric transport exceeds oceanic transport of heat. [After T. H. Von der Haar and A. H. Oort: New estimate of annual poleward energy transport by Northern Hemispheric oceans. *Journal of Physical Oceangraphy.* 3:169–72 (1973).]

is transported by winds to equatorial or subpolar regions. There the vapor condenses, forming either rain or snow and releasing heat to the atmosphere. Thus, the poleward transport of heat and water is reflected in low surface-water salinities in high latitudes. Precipitation near the equator also depresses surface salinities there—this feature is especially noticeable in the Atlantic (Fig. 7–4).

In summary, heat is transported by the ocean and atmosphere. The ocean dominates heat transport in the low latitudes, and the atmosphere dominates heat transport in the mid- and high latitudes.

Oceanic Depth Zones

Now we consider the ocean's vertical structure. There are three principal depth zones: surface, pycnocline, and deep (Fig. 7–10).

The ocean surface zone and the overlying atmosphere are intimately linked. For instance, water temperatures and salinities in the surface zone change seasonally because of variations in precipitation, evaporation, cooling, and heating. The surface zone contains the warmest and least dense waters in the ocean (Fig. 7–11). Average surface-water temperature is 17.5°C (Fig. 7–12).

The **surface zone** is 100 to 500 meters thick and contains about 2 percent of the ocean volume. Near-surface waters are well mixed by winds, waves, and cooling or heating of the surface. For this reason, the surface zone is also called the **mixed layer,** because the waters there move vertically very easily. These vertical motions are mainly wind-driven, as we shall see in Chapter 8.

The **pycnocline** is where water density changes markedly with depth (Fig. 7–11). The top of this zone corresponds approximately to the 10°C contour

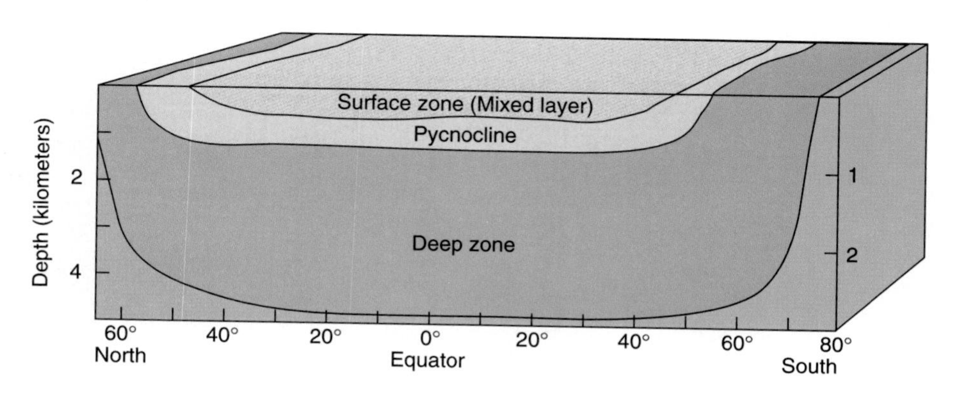

Figure 7–10
Schematic representation of the ocean's vertical structure. The pycnocline isolates the deep ocean from contact with the atmosphere in the equatorial and mid-latitude areas. Only near the North and South Poles do deep ocean waters come into contact with the atmosphere.

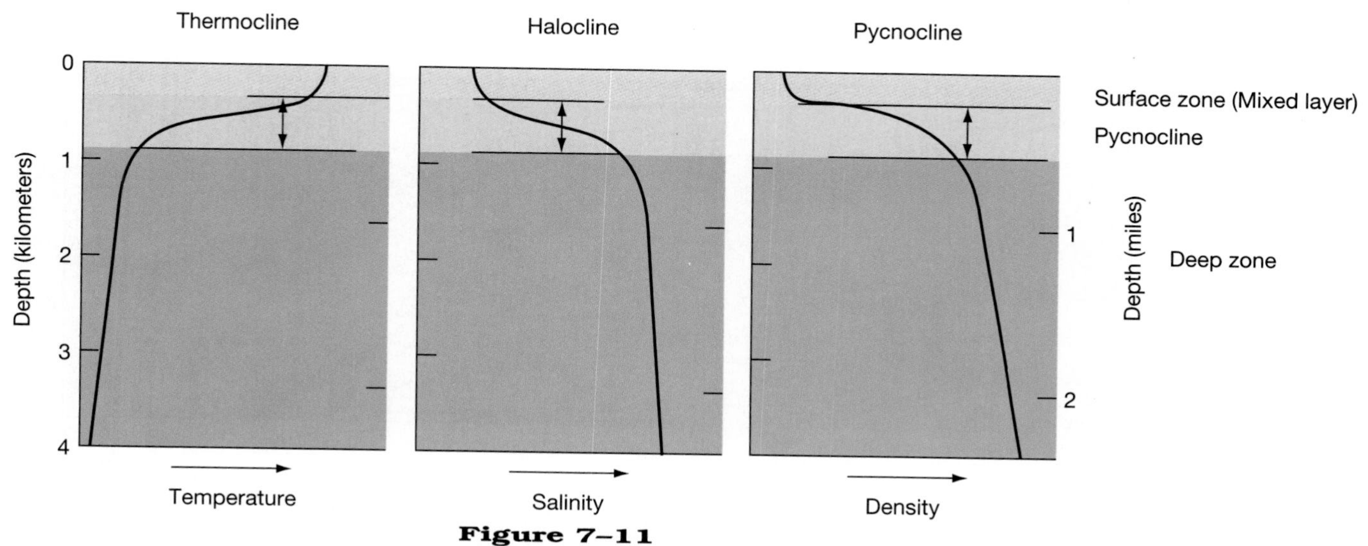

Figure 7–11
Marked variation in temperature with depth defines the thermocline. Marked variation in salinity with depth defines the halocline. Marked variation in density with depth defines the pycnocline. This can be caused by changes in either temperature or salinity.

Figure 7–12
The temperature and salinity of 99 percent of the water in the world ocean fall within the area enclosed by the 99 percent contour. The temperature and salinity of 75 percent of the water fall within the area enclosed by the 75 percent contour.

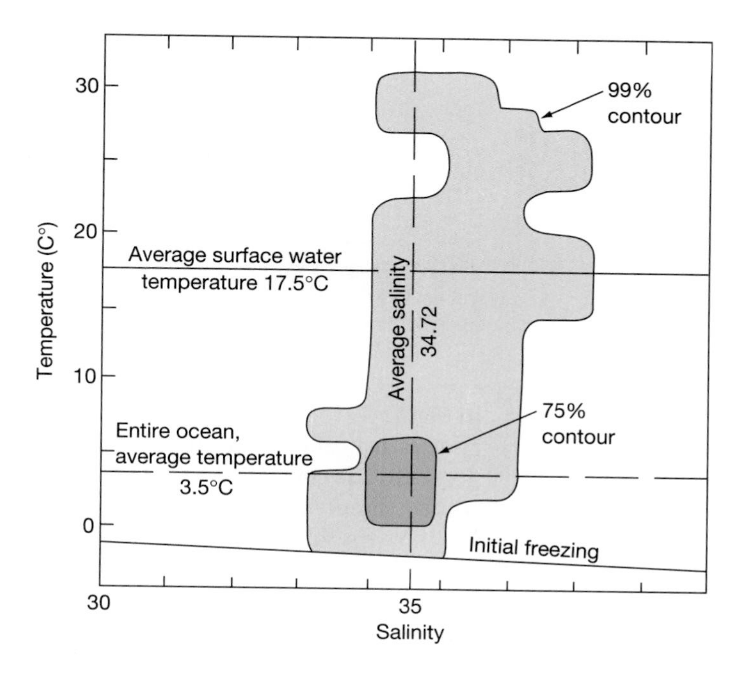

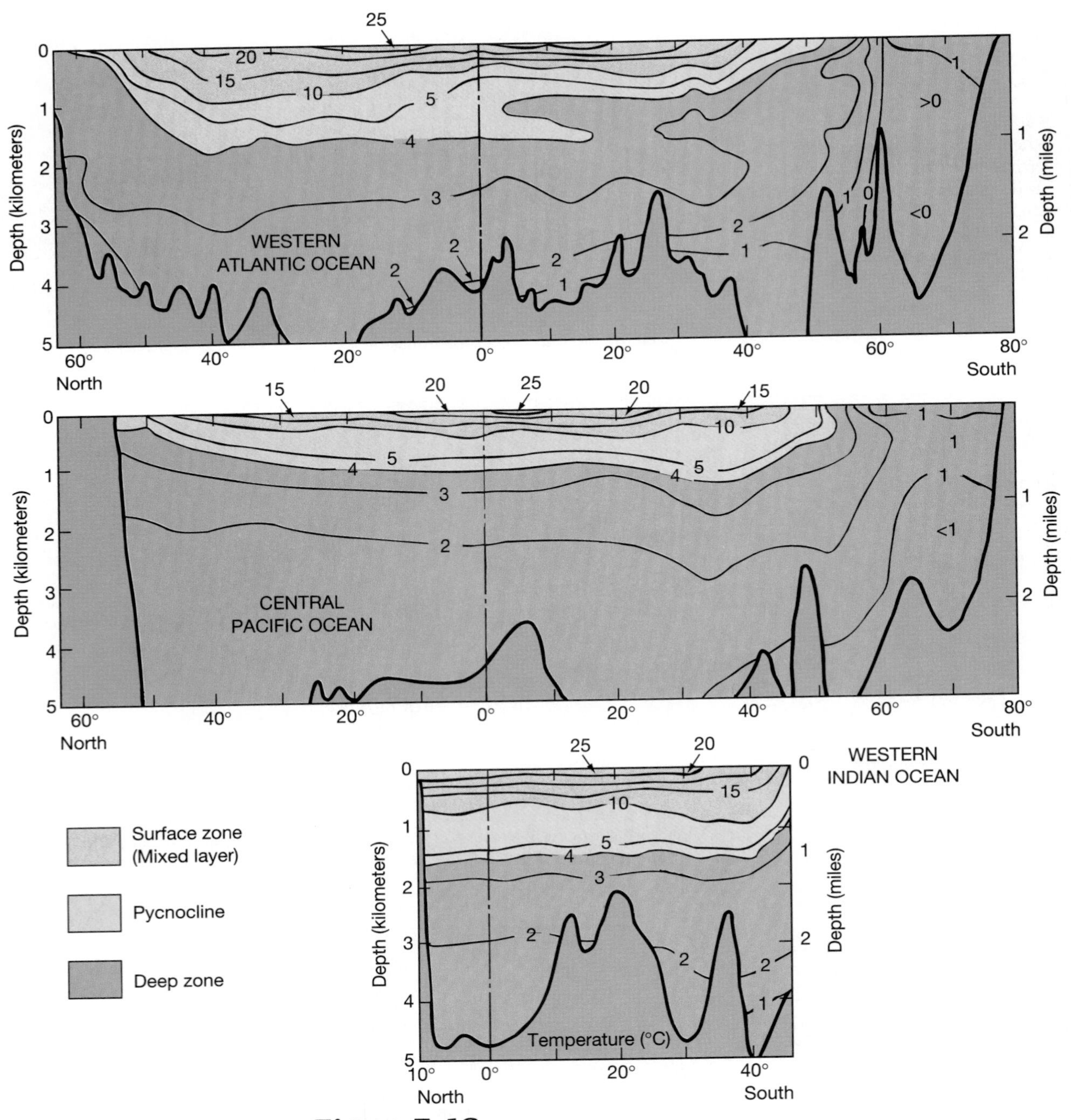

Figure 7–13
Vertical distribution of temperature in the three ocean basins. Vertical exaggeration is approximately one thousand times. (After Dietrich and others, 1980.)

in Figure 7–13, and its bottom can be taken as the 4°C contour. Thus, the pycnocline is typically about one-half to one kilometer thick. The exact depth of the pycnocline is controlled by those factors which influence the density of seawater, namely temperature and salinity.

Where the seawater density is controlled primarily by changes in temperature, the pycnocline coincides with a zone of marked temperature change, called a **thermocline** (Fig. 7–11). The zone where seawater density is controlled by marked changes in salinity is referred to as a **halocline.** Because temperature changes are more important in the open ocean, where salinity changes little, the depth of the open-ocean pycnocline is controlled by a thermocline. In coastal ocean areas, where salinity changes dominate and temperature changes are less important, a halocline controls the depth of the pycnocline.

The pycnocline is a zone of great stability. The vertical movements of waters in the surface zone and seasonal changes in their temperature or salinity do not penetrate the pycnocline. Waters in and below the pycnocline move primarily horizontally along density surfaces. Except in the high latitudes, there is little or no vertical movement below the pycnocline.

Below the pycnocline is the **deep zone,** which contains about 80 percent of the ocean's volume. Except in the high latitudes (Fig. 7–10), the deep zone is separated from the atmosphere. This isolation of the deep zone prevents interactions with the atmosphere and warming of the deep ocean water by solar heating. Thus, the deep zone retains its low water temperature—3.5°C—characteristic of the surface waters in the polar regions. Since the temperature and salinity of deep-ocean waters are unaffected by surface processes (Fig. 7–11), temperature and salinity are conservative properties.

Vertical distributions of temperature (Fig. 7–13) and salinity (Fig. 7–14) are similar in all three ocean basins. They show the deep zone exposed in the high latitudes, except in the North Pacific, where the pycnocline extends up to the northern continental margin. The cross sections also show some marked differences from our simplified ocean in Figure 7–10, especially in the North Atlantic. These differences are caused by deep-ocean currents, which we discuss in the next chapter.

In brief, the open ocean has a three-layer structure: surface zone, pycnocline, and deep zone. The surface zone responds quickly to changes in the overlying atmosphere. The pycnocline inhibits exchanges between atmosphere and deep zone. The deep zone is exposed to the atmosphere only in the high latitudes, which causes its waters to be cold. Vertical distributions of temperature and salinity are similar in all three ocean basins. In the North Atlantic, the surface layer and pycnocline are thicker than in the Pacific and Indian oceans.

Temperature–Salinity Relationships

Temperatures and salinities of surface waters differ markedly from those in the deep ocean. Yet, as Figure 7–11 shows, temperature and salinity values always fall within limits determined by various physical processes. For example, we discussed in Chapter 4 how ice affects seawater temperatures. Indeed, we found that the temperature of the coldest waters is controlled by the temperature of initial freezing of seawater. The temperature of the warmest waters in the ocean is controlled by the evaporation of water, which removes so much heat from the surface that sea-surface temperatures rarely exceed 30°C (as we discussed in Chapter 6) and then only in very restricted areas.

Salinity also shows a relatively limited range, between 33 and 37 (Fig. 7–12). Waters with higher or lower salinities usually occur in small quantities and in isolated basins. Thus, they have little effect on the ocean as a whole.

Below the surface zone, temperatures and salinities tend to vary together in characteristic ways. These variations reflect the conditions in the region where the waters in a particular layer formed. Since both temperature and salinity are conservative properties in the deep ocean, these characteristic relationships can be used to identify water masses (having fixed temperatures and salinities). The curve of the temperatures and salinities, called a *T–S curve* (Fig. 7–15), can be used to identify any given water mass. For instance, the relatively high salinities of waters from the Mediterranean can be used to identify them as they flow through the Strait of Gibraltar (Fig. 7–16) and throughout the North Atlantic (Fig. 7–17). (The T–S curve for Mediterranean waters is shown in Figure 7–15 in the middle of the right-hand side.)

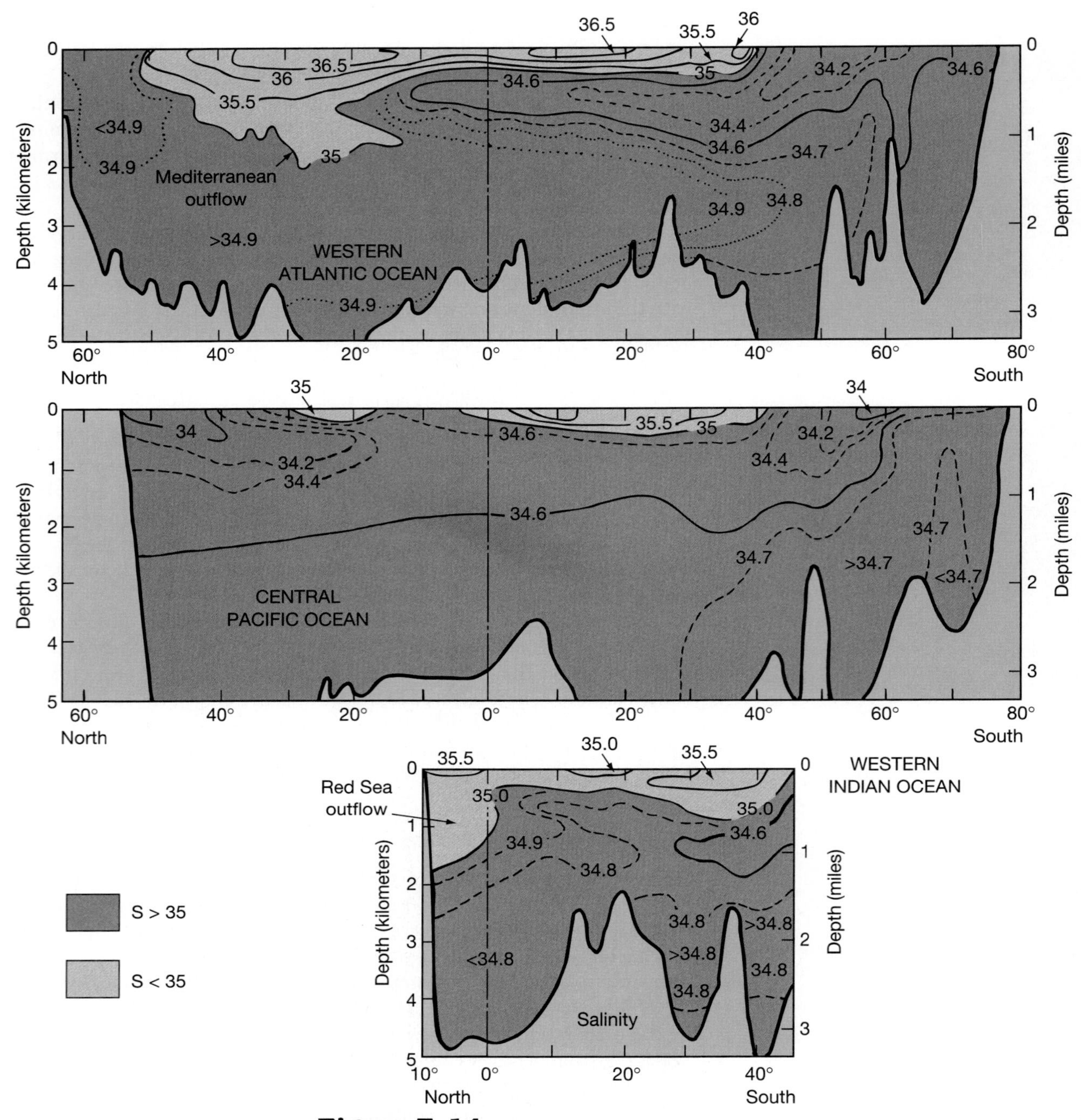

Figure 7–14
Vertical distribution of salinity in the three ocean basins. Vertical exaggeration is approximately one thousand times. (After Dietrich and others, 1980.)

T–S curves can also be used to determine how water masses mix. To see how this works, consider the two water masses represented in Figure 7–18a. As these water masses mix, the newly formed mass will have a temperature and salinity intermediate between those of the two original water masses. Adding a third water mass (Fig. 7–18b) complicates the situation, but the same general principles hold. We use this technique in the next chapter to study movements of subsurface water masses.

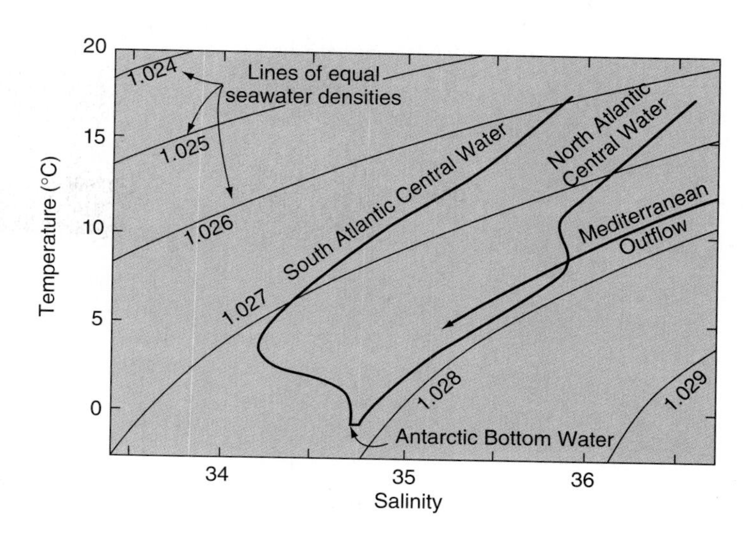

Figure 7–15
Temperature–salinity curves for major water masses in the Atlantic Ocean. The South Atlantic and North Atlantic central water masses form in the subtropics (which is why the temperatures and salinities of these water masses are relatively high). The Mediterranean waters are recognizable by their high salinities. The other water masses form in the high latitudes and so have relatively low temperatures. (After Sverdrup and others, 1942.)

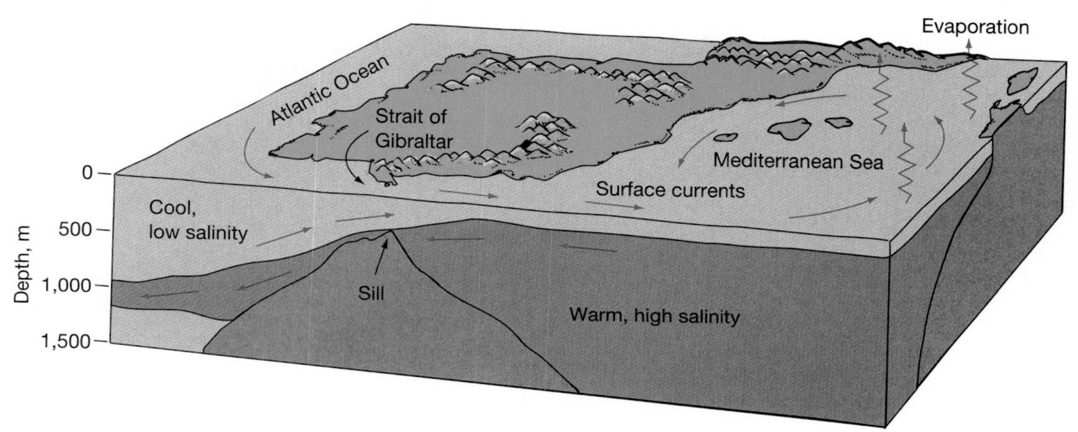

Figure 7–16
Waters from the Mediterranean Sea flow through the Strait of Gibraltar and into the North Atlantic Ocean to form a thin layer about 1 kilometer below the surface.

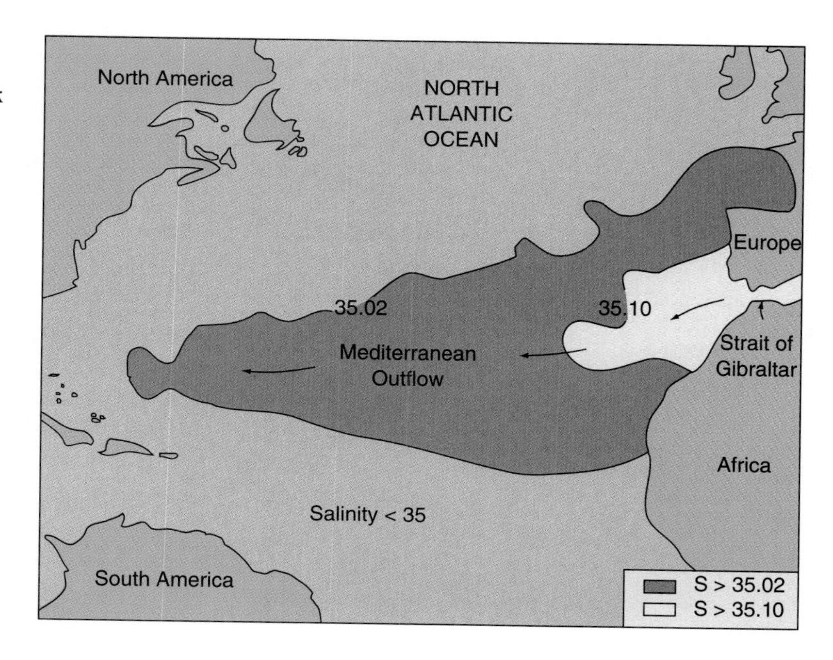

Figure 7–17
Waters from the Mediterranean spread out as a layer a few hundred meters thick that can be recognized across the North Atlantic by their relatively high salinity (greater than 35). (After L. V. Worthington and W. R. Wright. North Atlantic Atlas. *Woods Hole Oceanographic Institution Atlas Series,* vol. 4, 1970.)

(a) Mixing of two homogeneous water bodies

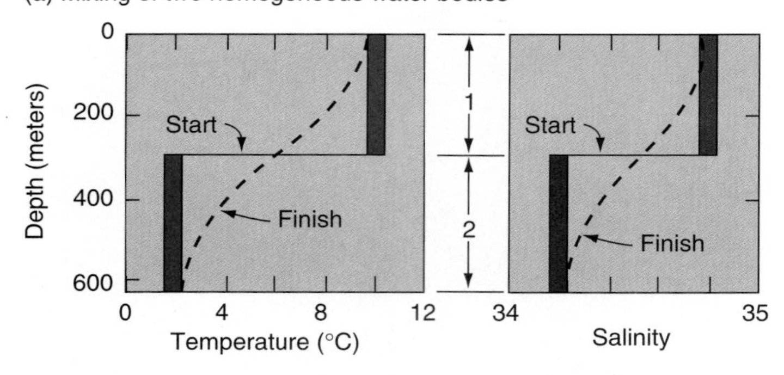

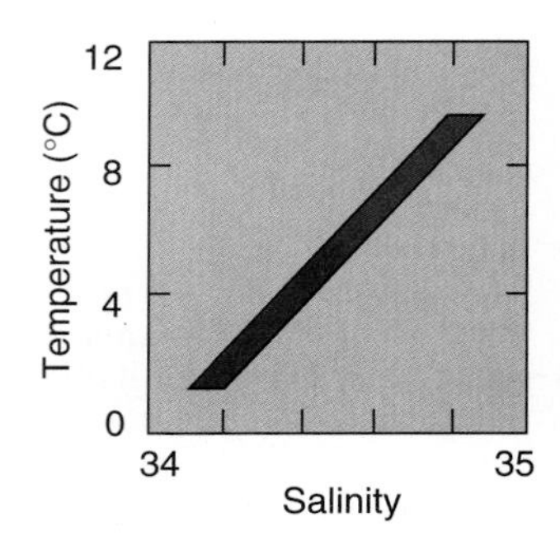

(b) Mixing of three homogeneous water bodies

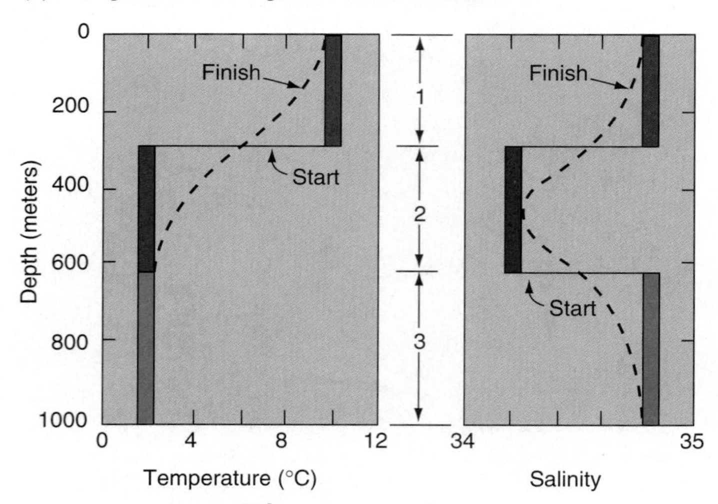

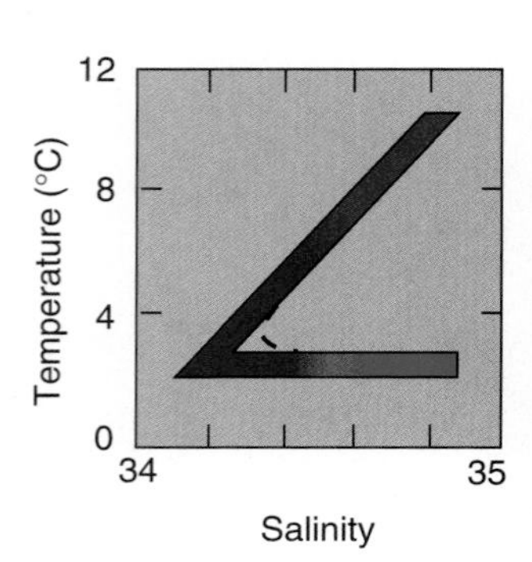

Figure 7–18
Temperature–salinity relationships in mixing water masses. (a) Water mass 1 (depth 0 to 300 meters; T = 10°C, S = 34.8) mixes with water mass 2 (depth 300 to 600 meters; T = 2°C; S = 34.2). Initially, the boundary between the two water masses is sharp and represented by marked changes in temperature and salinity with depth. After mixing, the boundary is more diffuse, and the changes in temperature and salinity occur over a depth range of nearly 600 meters. On the T–S diagram, which is the graph on the extreme right, mixing of the two water masses is shown by the straight line connecting the T–S points characteristic of the two original masses. (b) Mixing of three water masses. Water mass 3 (depth 600 to 1,500 meters; T = 2°C; S = 34.8) mixes with water mass 2 but not with water mass 1. The sharp boundaries between water masses are gradually obliterated. On the T–S diagram on the extreme right, the mixing is shown by two straight lines, indicating that water masses 1 and 2 mix and that water masses 2 and 3 mix, but not 1 and 3.

Sea Ice

Sea ice forms where seawater cools below its initial freezing point. Each winter, sea ice covers the entire Arctic Ocean and completely surrounds Antarctica. It also forms in bays and along the coast of Alaska, the western coast of Canada, and the Atlantic coast of North America as far south as Virginia. In spring, much of the ice melts, but large areas in the polar oceans remain ice-covered throughout the year. The annual expansion of ice-covered ocean areas and the retreat of the ice in local summer affect climate worldwide. In this section, we discuss how and why sea ice forms.

When seawater chills to its temperature of initial freezing, clouds of tiny, needlelike ice particles form, making the water surface slightly turbid. The ocean surface dulls and no longer reflects the sky. As ice particles grow, they form hexag-

Figure 7–19
Newly formed ice makes the sea surface look greasy.

onal spicules 1 to 2 centimeters long. The needles and spicules of newly formed ice are called *frazil ice.* When the surface is stirred by winds and waves, the ice forms a soupy-looking layer known as *grease ice* (Fig. 7–19).

As sea ice continues to form, ice crystals eventually form a blanket over the water surface. When the surface is calm, an elastic layer of ice forms. It is only a few centimeters thick and easily moved by winds.

Since salt is excluded from ice, the remaining water becomes more saline and its freezing point is lowered. Some brine pockets remain trapped in the ice. Salinities of newly formed ice are typically 7 to 14, but this value depends on temperature. The more slowly the ice forms, the easier it is for brines to escape, resulting in lower ice salinities. Conversely, at very low temperatures, ice forms rapidly, and the salty brines cannot easily escape. This quick freeze results in higher ice salinities. Salinity of sea ice is always lower than that of the surrounding waters. As sea ice ages, the brines are expelled. Thus, multiyear ice typically has salinities around 0 at the top and around 4 at the bottom.

As freezing continues, rejected salts mix with underlying waters, making them more saline and therefore denser. Very cold water from the ice also chills the underlying waters. Eventually, convection occurs underneath the ice as the dense waters sink. Less dense water from below rises in finger-shaped parcels to replace the sinking brines. Such processes occur in both the Greenland and Labrador seas as well as near Antarctica, especially in the Weddell Sea. (As we see later, these processes form the coldest and densest water masses in the ocean.)

Newly formed ice is broken by waves. As the pieces smash into one another, they form *pancake ice,* so called because individual pieces resemble pancakes. As pancakes coalesce, their outlines can still be seen in the young sea ice (Fig. 7–20).

Waves and especially winds break ice sheets into large pieces called *floes.* Floes constantly move and shift, freeze together and break loose, buckle up, or flatten out as the ice moves.

Snow accumulates on top of, and freezes to, the ice surface. Thus, sea ice grows from both top and bottom. First-year ice is flat and usually snow-covered. During a single winter, new sea ice can reach a thickness of two meters. Where sea ice never completely melts, multiyear ice continues to grow, and older ice has a rough, hilly surface. In the central Arctic, multiyear ice reaches thicknesses of three to four meters. Ice melts during the summer, down to about two meters in the central Arctic. The fresh water released by melting ice forms a thin layer of low-salinity surface water.

Figure 7–20
Pancake ice near Antarctica.

Currents and winds move large pieces of sea ice together, forming mounds called **hummocks** or **pressure ridges** (Fig. 7–21) that are the ice pack's most conspicuous features. At these pressure ridges, the ice is deformed and thickened, up to 20 meters thick. These ridges can extend many tens of meters below the ice and are hazards to submarines moving under the ice.

When floes move apart, they expose open waters in area called **leads.** Leads range from a few centimeters to many hundreds of meters wide and can extend for many kilometers. Ships moving through sea ice utilize leads where possible to avoid having to break ice. Mammals stay near the leads and near holes in the ice. This permits them to catch fish and other food in the underlying waters (Fig. 7–22).

In shallow coastal waters, sea ice forms readily during cold winters and is blown by the winds (Fig. 7–23). Ice that is attached to the shore and therefore doesn't move is called *fast ice.* **Pack ice** forms at sea and moves with currents and winds. An area called the *marginal ice zone* occurs seaward of the fast ice. It is a mixture of open water, some first-year ice, and floes of multiyear ice, typically 3 to 5 meters thick. The main polar pack consists of floes, some tens of kilometers across. Pack ice covers the central Arctic (Fig. 7–24) and surrounds Antarctica (Fig. 7–25).

Figure 7–21
Pressure ridges in first-year ice near Antarctica.

Figure 7–22
Killer whales trapped in an open area in the Antarctic sea ice.

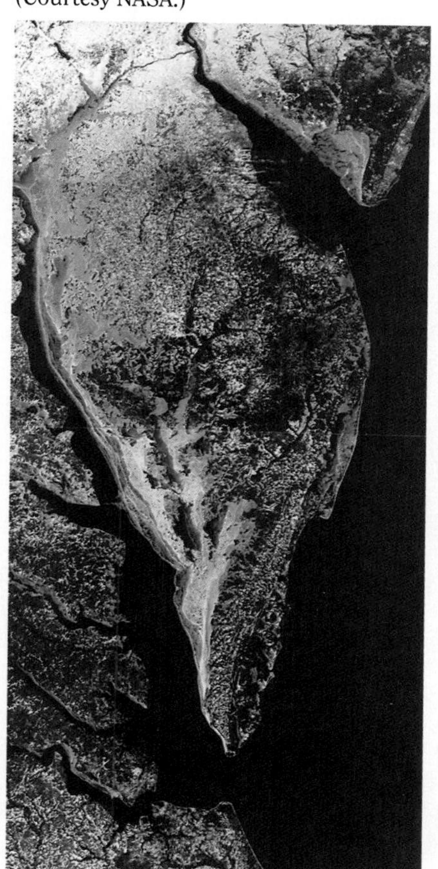

Figure 7–23
Sea ice (light blue) on Chesapeake and Delaware bays was pushed to the eastern shore by strong westerly winds during an exceptionally cold February 1977. (Courtesy NASA.)

Strong currents and winds near Antarctica usually keep the pack ice close to the continent (south of 55°S) except in the coldest winters (Fig. 7–25). During the southern summer (December through February), much of the Antarctic pack ice melts, except in large embayments.

To sum up, sea ice is a major feature of the ocean. Its freezing in winter and melting in summer dominate surface waters in the polar oceans. Sea ice also influences the deep-ocean circulation. The coldest and densest water masses form in the polar oceans. Freezing sea ice expels salt, and this excess salt in the remaining liquid water increases the density of water masses, which is especially important near Antarctica.

Polynyas

Satellites have shown that areas of open water much larger than leads, called **polynyas,** occur within the ice pack. Polynyas can be as large as the state of Colorado and persist over several years. There are two kinds: coastal and open-ocean. *Coastal polynyas* occur along the coast of Antarctica, formed by winds blowing the ice away from shore (Fig. 7–26). These were known before satellite images were available. Satellite images show them to be more common and larger (50 to 100 kilometers across) than was previously thought. Coastal polynyas are important because of the large amount of sea ice that forms in them during the winter. Most of the ice in the ice pack is thought to form in coastal polynyas. They are also called *latent-heat polynyas* because latent heat is released to the atmosphere as ice forms in the polynyas.

Open-ocean polynyas form in the midst of the ice pack (Fig. 7–25). They often re-form in roughly the same location over several years. Then in other years, they do not appear at all. The forces controlling their formation are more complicated than those controlling the formation of coastal polynyas.

Convection cells are vertical water movements in which warm subsurface waters come to the surface, cool, and sink toward the bottom. The cells may be many kilometers across. The heat they bring to the surface melts the sea ice and keeps the surface ice-free—in other words, these cells are what form open-ocean polynyas. There are probably many individual cells involved in forming and maintaining a single polynya.

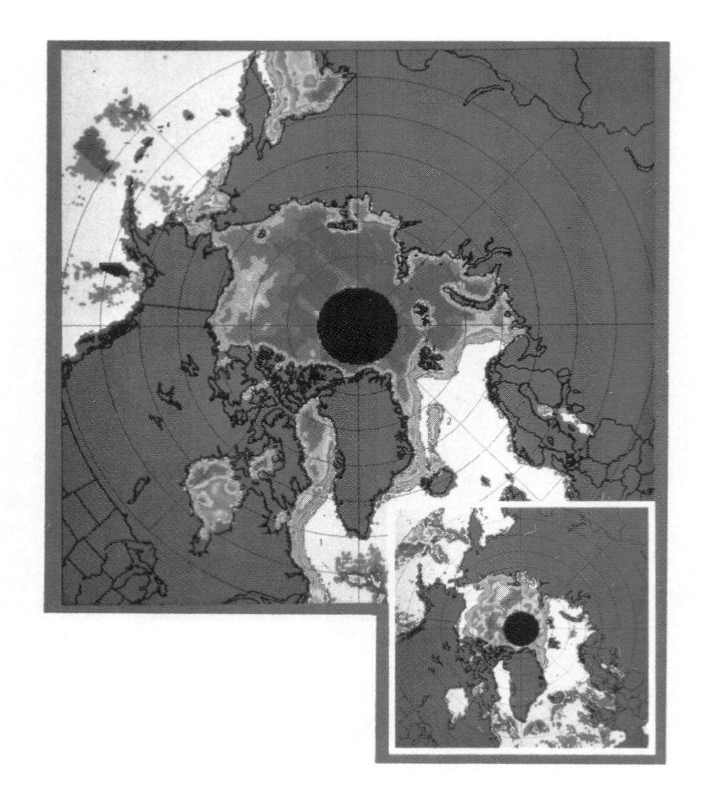

Figure 7–24
Sea ice covers the Arctic Ocean during winter (larger map) but is reduced to the central part of the basin during summer (lower inset). (Courtesy NASA.)

Open-ocean polynyas are called *sensible-heat polynyas* because the heat is directly associated with changes in temperature; that is, it can be sensed. Heat from these polynyas is transferred to the atmosphere. Since the waters involved also come to the surface, they can exchange gases, such as carbon dioxide.

The polynya shown in Fig. 7–25 was exceptionally large (350 by 1,000 kilometers) and persisted from 1974 through 1976. Its effects were detectable for several years in elevated water salinities and lower temperatures. The location of

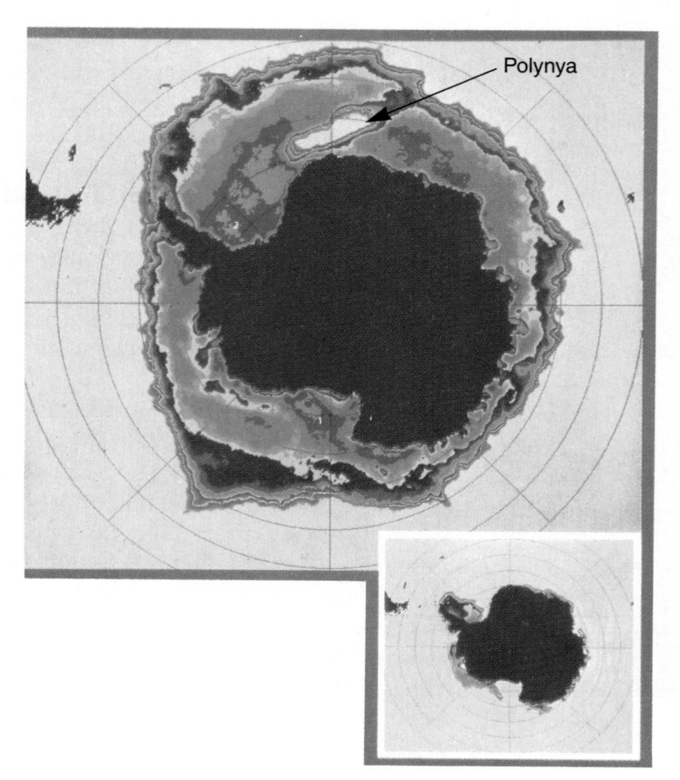

Figure 7–25
Sea ice occurs around Antarctica, expanding greatly during winter. Maximum ice coverage is shown in the upper map; minimum coverage during the Southern Hemisphere summer is shown in the small inset map. A large coastal polynya is shown in the upper map where the light blue area (representing open water) is surrounded by the sea ice, shown by reds and purples. (Courtesy NASA.)

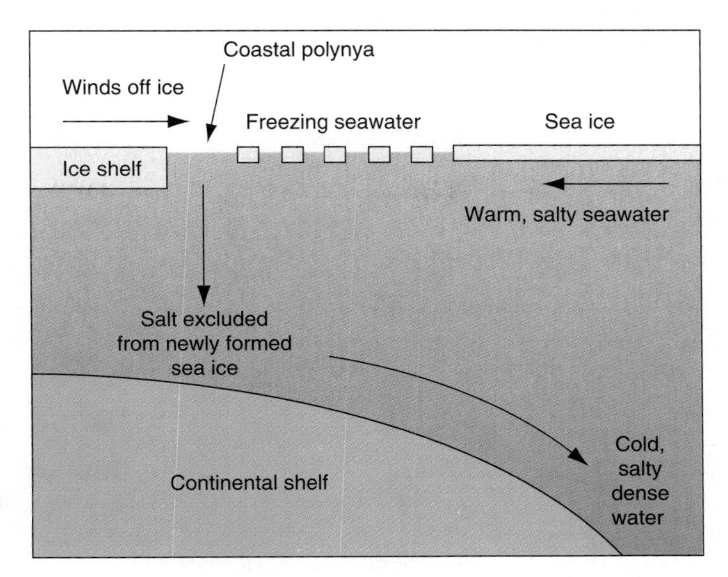

Figure 7–26
Schematic representation of the processes forming coastal polynyas and dense water masses around Antarctica.

open-ocean polynyas is apparently controlled, at least in part, by ocean-bottom topography, and this persistent polynya was associated with an Antarctic submarine ridge.

Arctic coastal polynyas are important to native peoples because the open waters permit them to hunt seals and whales, which are important sources of food.

Icebergs

Masses of freshwater ice in the ocean, called **icebergs,** break off the glaciers that cover Greenland and Antarctica. The process of iceberg formation is called *calving*. Icebergs from the glaciers of western Greenland (Fig. 7–27) are irregularly shaped because they come from mountain glaciers along the coastal fjords. In the Antarctic, the icebergs (Fig. 7–28) come from the ice sheets that flow from the continent out over the coastal ocean waters. These floating *ice shelves* are flat-topped, and the flat-topped icebergs they form are called *tabular icebergs* (Fig. 7–28). Antarctic icebergs are usually much larger than those from Greenland;

Figure 7–27
Icebergs in the North Atlantic come from mountain glaciers in western Greenland which discharge into Baffin Bay. They are then carried southward by currents into the North Atlantic, frequently entering the busy shipping lanes. (William W. Bacon III/Photo Researchers, Inc.)

Figure 7–28
Tabular icebergs are typical around Antarctica. They break off floating ice shelves and are carried by currents around Antarctica.

some are as big as the state of Rhode Island. Some scientists speculate that more of these gigantic icebergs will break loose if ocean waters warm as a result of climatic changes.

Icebergs typically last about four years before they melt, and older ones are rare. They can be moved long distances by currents. However, few move out of the polar regions. Around Antarctica, currents keep icebergs relatively close to the continent. Some move northward past the tip of South America to about 40°S in the Atlantic but rarely reach even 50°S in the Pacific. In 1894 an iceberg reached 26°S, only about 350 kilometers from the Tropic of Capricorn, the edge of the subtropical ocean.

In the Atlantic, currents move icebergs southward along the North American coast. Large ones can travel as far as 2,500 kilometers, reaching the Grand Banks off Newfoundland. The *Titanic* sank in April 1912 after striking a large iceberg near the Grand Banks. The accident took more than 1,517 lives.

As a result of the sinking of the *Titanic,* the International Ice Patrol was formed. It now tracks icebergs and issues regular bulletins during iceberg season. These predictions have greatly reduced the number of collisions with icebergs—only one ship has been lost since the formation of the patrol. This loss occurred during World War II when patrol activities were curtailed and radio communications with ships were restricted.

Oceanic Climate Regions

Open-ocean climatic zones extend primarily east–west. These regions, indicated in Figure 7–29, have characteristic ranges in surface-water properties. Temperatures and salinities of surface waters in these zones are controlled primarily by incoming solar radiation and by the relative amounts of evaporation and precipitation.

The Arctic Ocean and the band around Antarctica are the **polar ocean areas.** Sea ice occurs most of the year, keeping surface temperatures at or near freezing despite long days of strong sunshine during local summer. In winter, there is little sunlight. Relatively low surface salinities and an ice cover inhibit mixing in the polar oceans.

Subpolar ocean regions are affected by seasonal formation of sea ice. In winter, chilling of surface waters and freezing of sea ice lead to the formation of dense water

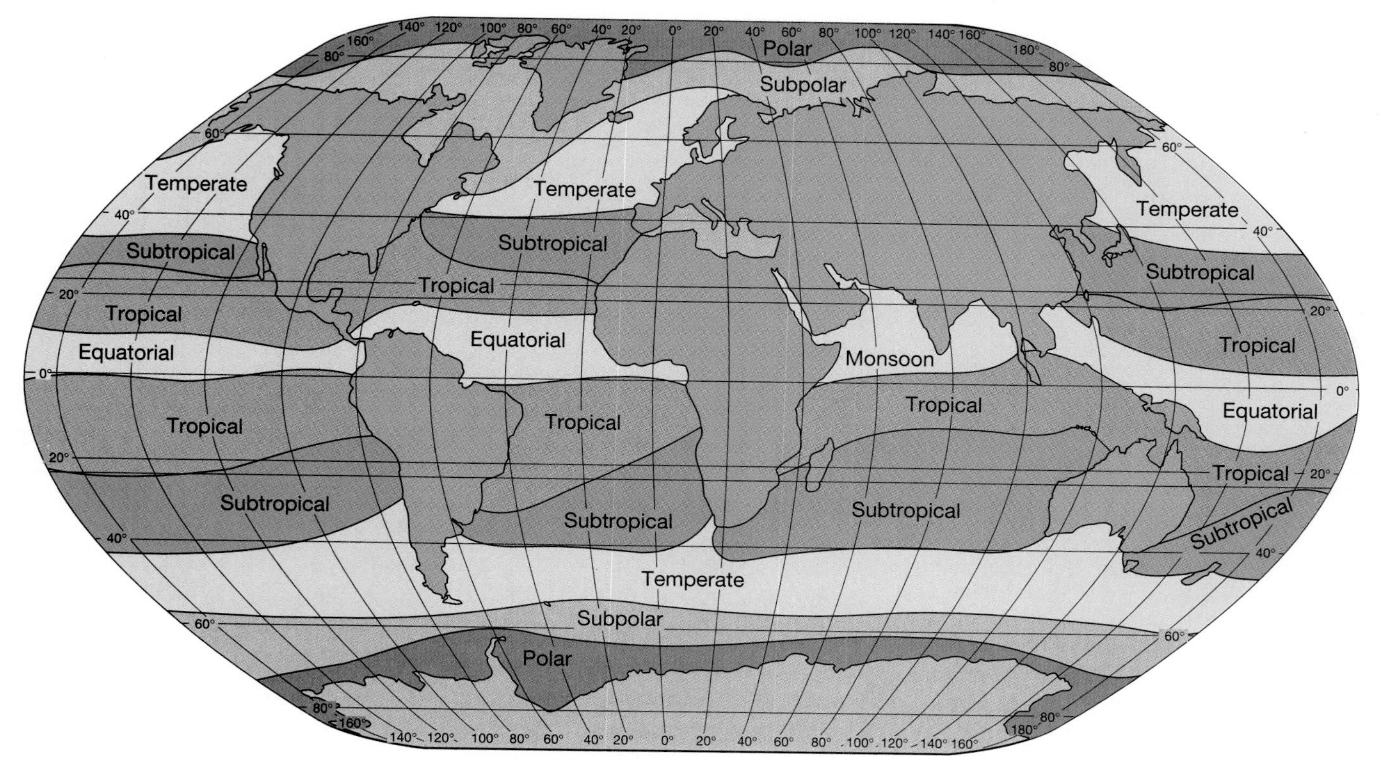

Figure 7–29
Climatic zones over the ocean. [After D. V. Bogdanov: Map of the natural zones of the ocean. *Deep Sea Research* 10:520–23 (1963).]

masses that flow into the deep-ocean basins. Sea ice disappears during summer as solar heating melts the ice and raises surface-water temperatures to about 5°C.

The **temperate region** corresponds to the band of strong westerly winds. It has many storms, especially during winter.

Subtropical regions coincide with the nearly stationary mid-latitude high-pressure cells. Winds are weak here, and so are surface currents. Because clear skies, dry air, and abundant sunshine cause extensive evaporation, surface salinities are generally high. This zone was called the *horse latitudes* by sailors. Tradition has it that sailing ships carrying horses often had to throw them overboard when they ran out of animal food after their ships were becalmed in subtropical waters.

Tropical regions (also called trade-wind regions) have persistent winds blowing from the northeast in the Northern Hemisphere and from the southeast in the Southern Hemisphere. These winds cause the equatorial currents that we discuss in Chapter 8. Tropical ocean waters come from areas of high evaporation rates in the subtropical regions. They are therefore more saline than average seawater. Near the equator, precipitation increases, causing lower surface salinities there.

In **equatorial regions,** surface waters remain warm throughout the year. Annual temperature variations are small. Warm, moist air generally rises near the equator, causing heavy precipitation and therefore relatively low surface salinities. In the equatorial regions of the Atlantic and much of the Pacific, winds tend to be weak. Sailors named this region the *doldrums* because sailing ships were often becalmed there.

Land Climates

The ocean moderates climate on land. Coastal land areas typically have **maritime climates,** where annual temperature ranges are much smaller than in inland

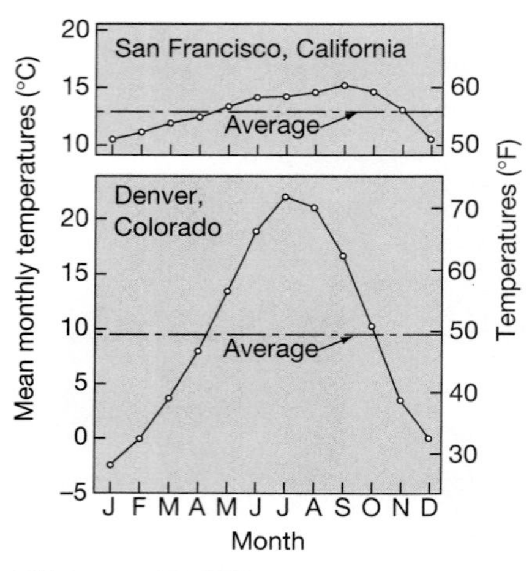

Figure 7–30
Yearly variations in temperature at two locations. San Francisco has a maritime climate with cool summers and relatively warm winters. Denver has a continental climate with cold winters and warm summers.

areas, as shown in Figure 7–30. Far from the ocean, **continental climates** prevail, with characteristically large temperature ranges.

The ocean stores large amounts of heat because of water's large heat capacity. This permits substantial energy storage without an accompanying large temperature increase. Furthermore, the ocean (or any other large water body) stores heat in a layer several meters thick, due to mixing of surface waters by winds and waves. This transfer of heat to below the surface retards the loss of heat at night by keeping the water surface relatively cool. In contrast, rocks and soils have low heat capacities. Thus, land surfaces become quite hot during the day, and the heat is not readily transferred to rocks beneath the surface. Most of this heat is lost at night by radiation from the surface. Such a situation can be observed in deserts, where intense daytime heat is followed by chilly nights.

Mountain ranges can block winds off the ocean, thereby affecting climate. For instance, the north–south ranges of the western Americas (Fig. 7–6) partially block winds coming from the Pacific Ocean. Along these mountains, the climate is coastal on the western (seaward) side of the ranges and continental on the eastern side. In contrast to the high rainfall on the ocean side of the mountain ranges, the interior is often desert. (The process is the same as in the island effect.) In Europe, the highest mountains run east–west and thus do not block winds coming from the Atlantic. There, maritime climates extend far inland.

Ancient Climates

Over the past 100 million years, Earth has had two distinctly different climates. During most of that time, Earth has had a warm, humid climate with little temperature difference between equator and poles (Fig. 7–31). For example, during Cretaceous time (about 100 million years ago—see Appendix 7), crocodiles and palm trees lived within 1,000 kilometers of the North Pole.

Earth's second climate is glacial, much like the climate of today and of the past three to five million years, beginning perhaps 20 million years ago. During glacial times, there are marked differences between polar and equatorial temperatures.

It is still not clear what causes such climatic shifts. Increased rates of volcanic activity may have raised carbon dioxide levels in the atmosphere. (Carbon dioxide is one of the principle greenhouse gases.) Furthermore, the high levels of volcanic activity associated with greatly increased rates of seafloor spreading

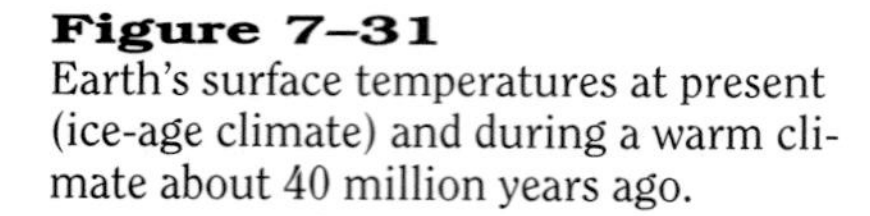

Figure 7–31
Earth's surface temperatures at present (ice-age climate) and during a warm climate about 40 million years ago.

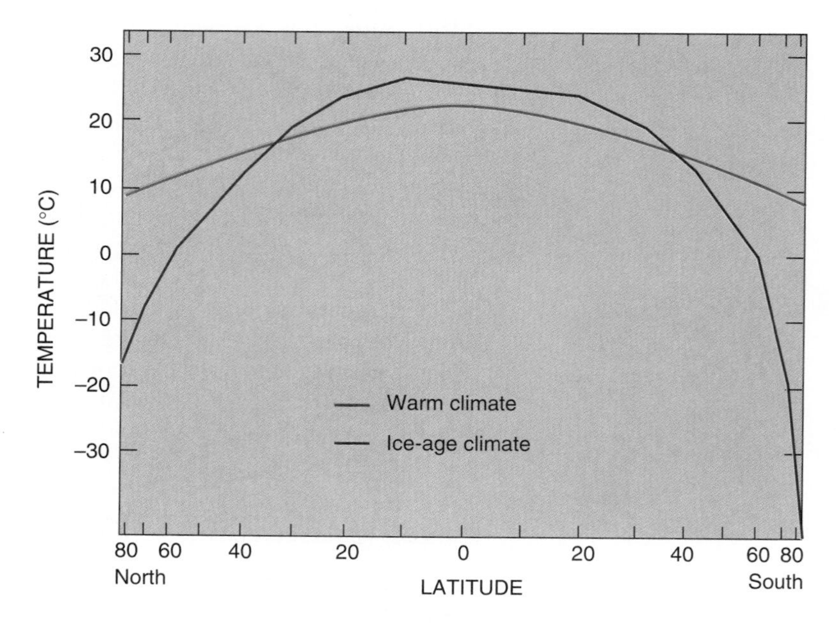

caused the ocean basins to be shallower, flooding the continents. With water formerly confined to the ocean basins spread out over a much larger surface area, evaporation rates increased and the result was a moister atmosphere. Thus, warm climates occur during times of high sea level.

Large amounts of water vapor in the atmosphere could shift heat-transport mechanisms between subtropical and polar regions. During times of warmer climate, heat transport by latent heat would be far more important than at present. Heat taken up during evaporation would be released in high latitudes through precipitation. Cooler, low-salinity waters would then return toward the equator in surface currents. Thus, heat transport through the atmosphere would be more important than the present ocean circulation, where surface currents move warm waters poleward and subsurface currents return dense, cold water from polar regions toward the equator.

Since our knowledge of ocean processes comes from studies made over the last few decades, during a time of cold climate, it is difficult for us to comprehend how the ocean worked during times of warm climates. We can only base our theories on the evidence of past climatic states found in the ocean sediments (as discussed in Chapter 5).

Ice-Age Climates

The most marked recent climatic change was the last glacial stage of our present Ice Age, a stage that ended about 10,000 years ago. The record of that glacial stage has been studied both to determine what causes climatic shifts and to improve our ability to predict them.

Fossils in marine sediments were used to reconstruct distributions of ocean surface salinity and temperature 18,000 years ago, when the most recent glacial advance was at its maximum. Using samples from carefully dated layers in sediment cores, scientists analyzed the chemical composition, abundance, and distribution of the fossils of microscopic organisms. Knowing the conditions under which the organisms that formed these fossils now grow, scientists were able to determine the salinity and temperatures of the ocean surface waters at the time the sediments were deposited.

The ocean surface during the last glacial stage was quite different from what it is today (Fig. 7–32). Surface waters were cooler by about 2.3°C on the average, and the land was about 6.5°C cooler. But the waters in the centers of the major ocean basins were little changed. The Gulf Stream flowed directly eastward from the Carolinas to Spain. Cold polar waters advanced southward; thus temperature changes between subtropical and subpolar waters were much more pronounced than at present. There was much more upwelling along the equator, causing the equatorial ocean to be much colder than it is today. Temperature changes were especially noticeable in the North Atlantic, which was markedly cooler (Fig. 7–33).

Changes in Earth's orbit around the Sun seem to be a major factor controlling the waxing and waning of glacial stages. These orbital changes (called Milankovitch cycles) affect Earth's climate in periods of 22,000, 40,000, and 100,000 years. Because of these changes, the volume of ice on Earth reaches a maximum every 100,000 years. At present, Earth's surface is warmer than it has been for 98 percent of the past half million years. Future increases in levels of carbon dioxide (and other gases) in the atmosphere may cause still warmer climates and higher sea levels.

Climatic Effects of Mountains

During the past 50 million years, large areas of Asia and North America have been uplifted to form the Himalayan Mountains, the Tibetan Plateau, and the Rocky Mountains. These high mountainous areas influence planetary winds, strengthen-

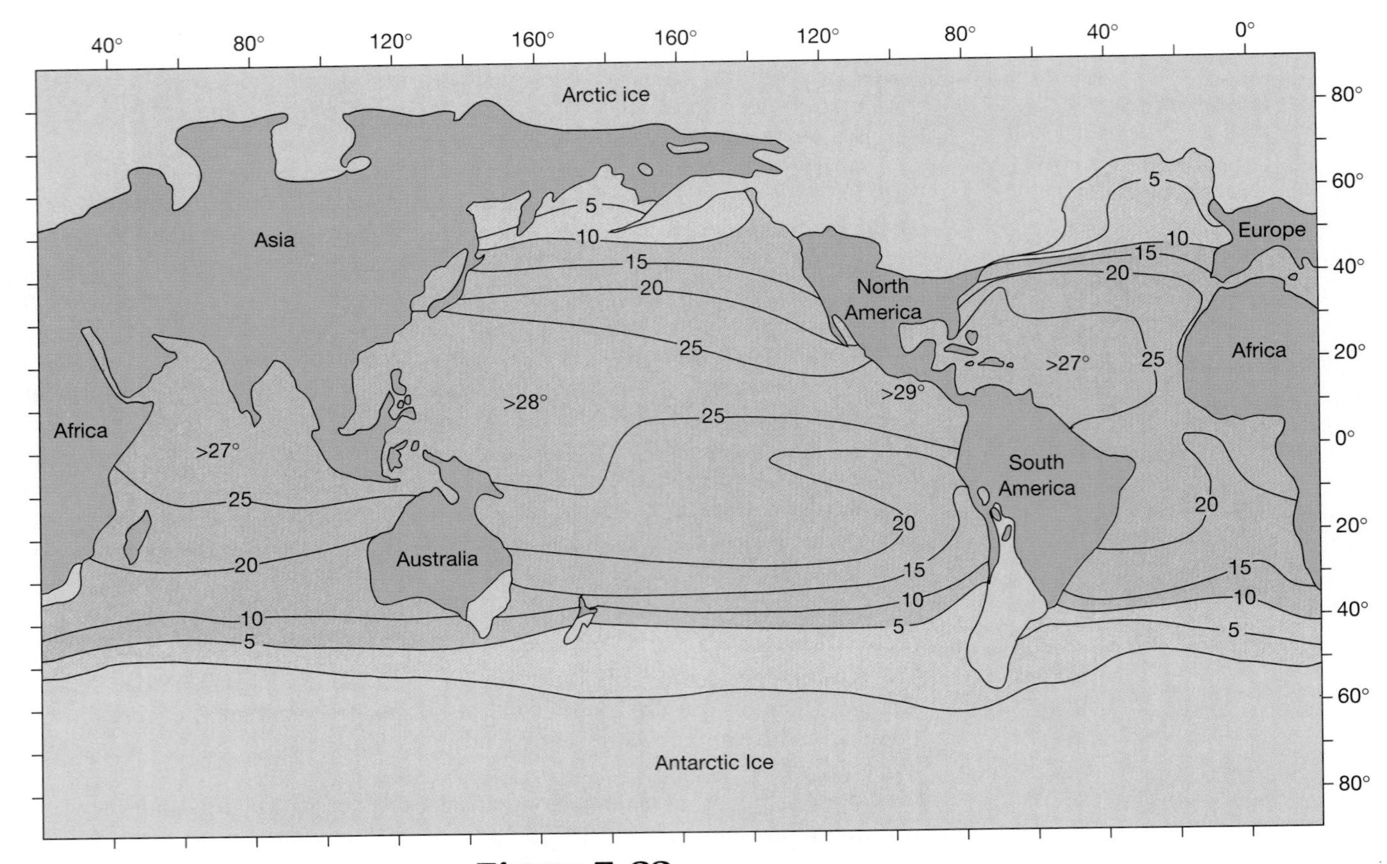

Figure 7–32
Sea surface temperatures (°C) and the extent of ice during an average August 18,000 years ago, when the continental ice reached its maximum expansion. Continental outlines show sea level 85 meters below its present level. [After CLIMAP: The surface of the ice-age Earth. *Science* 191:1131 44 (1976).]

ing the monsoons of southeastern Asia, and they may have initiated the present Ice Age in the Northern Hemisphere.

The Himalayan Mountains began to form about 40 to 50 million years ago when India collided with Asia. That event also began elevating the Tibetan Plateau, a vast area nearly the size of the western third of the United States. Its average height is higher than the highest mountains in the lower 48 states of the United States. This large area reached its present elevation within the past 15 million years.

The Rocky Mountains also began to be uplifted during the past 55 million years, with most of that uplift occurring during the past 10 million years. Before the uplift, the Rocky Mountain region was relatively low and probably had little influence on planetary wind patterns.

To investigate the effects of these uplifts, numerical models of the atmospheric circulation were run on the largest, fastest computers available—so-called supercomputers. These models incorporate the physical processes affecting the atmosphere. By studying wind patterns with and without mountains in the models, it is possible to determine what effect their appearance had on Earth's climate.

These mountain masses are high enough and large enough to deflect prevailing wind patterns on a planetary scale. For example, the Tibetan Plateau is a major heat source in summer and a major area of heat loss in winter. Both of these effects greatly strengthen the monsoons.

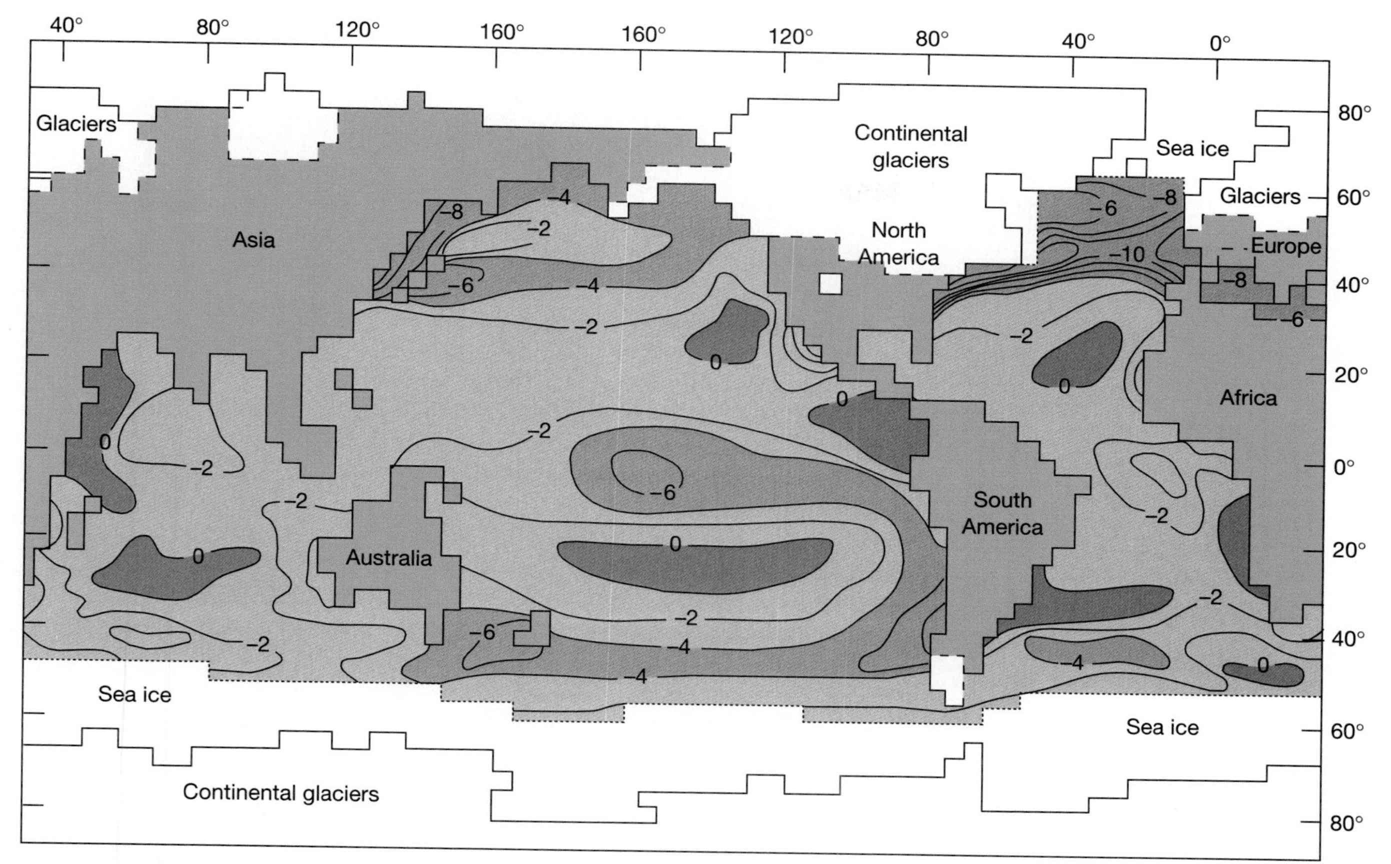

Figure 7–33
Differences between sea surface temperatures in an average August 18,000 years ago and an average August today. Each contour line represents a temperature difference (in degrees Celsius), with the minus sign indicating that today's sea surface temperatures are warmer than temperatures 18,000 years ago. Dotted lines show the extent of sea ice. Dashed lines indicate the limits of glaciers on land. [After CLIMAP: The surface of the ice-age Earth. *Science* 191:1311–44 (1976).]

The shifts in planetary winds could have also brought on the Ice Ages of the past several million years. Shifts in the average path of storms could have left large areas of land much colder in winter, leading to the formation of continental ice sheets. Further, the shifts in wind patterns could have increased the winter formation of very cold water masses off Greenland, a condition that would strengthen the present deep-ocean current patterns.

El Niños

Ocean currents, winds, and weather patterns are closely linked, especially in equatorial regions. These relationships influence weather over most of Earth. The best understood case of this linkage between oceanic conditions and weather patterns involves recurrent changes in the western Pacific Ocean, its overlying atmosphere, and the weather in North and South America.

Waters warmer than 28° C normally cover one-third to one-half of the tropical oceans, primarily in the equatorial eastern Indian and the western Pacific oceans (Fig. 7–2). Cooler surface waters normally dominate the Pacific coast of South America and the equatorial ocean. Every three to seven years, however, this

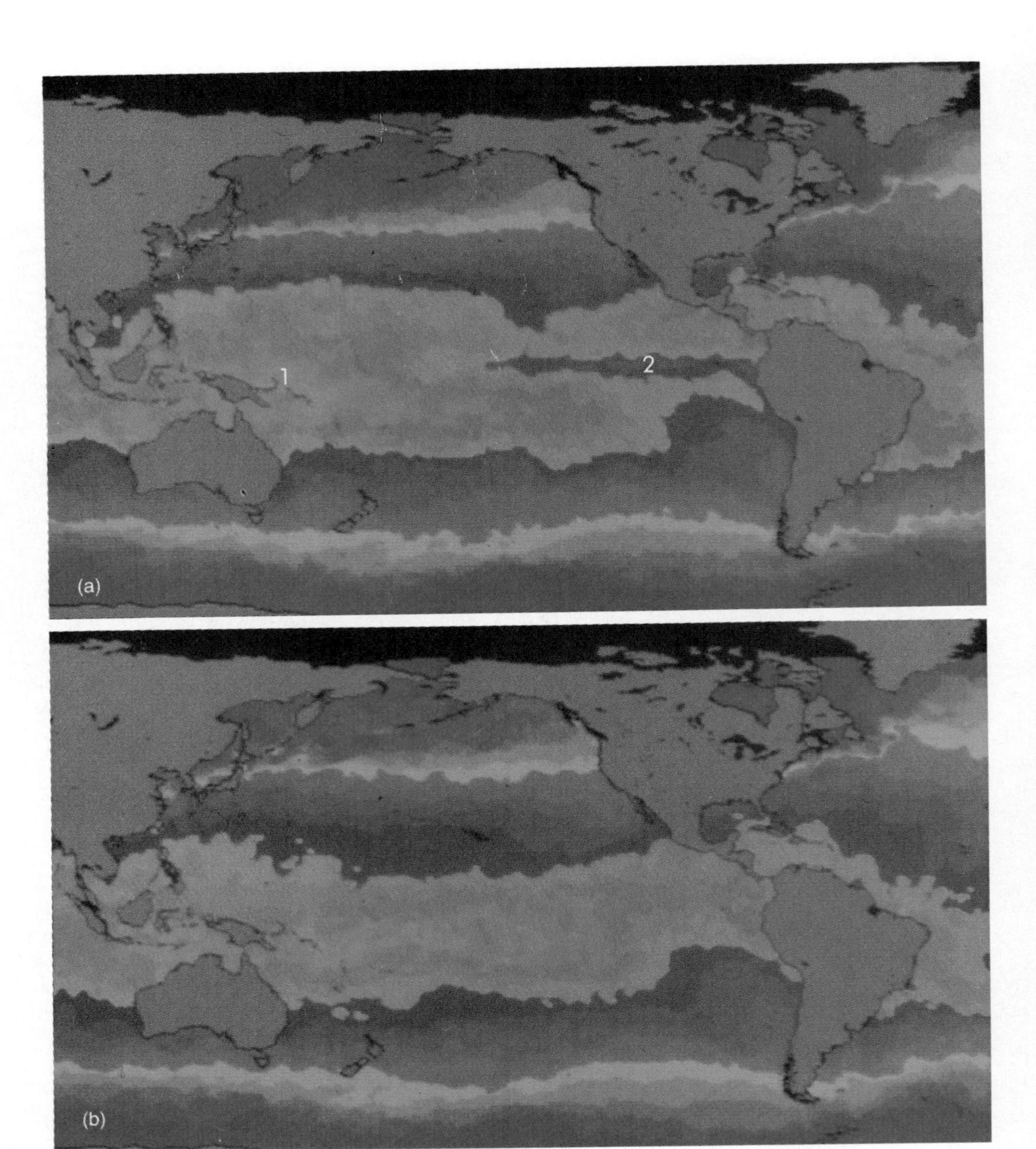

Figure 7–34
The 1982–1983 *El Niño* was the strongest *El Niño* event of the century. It altered sea-surface temperatures from their normal patterns over large areas of the Pacific. (a) In the normal distribution of sea-surface temperatures, colder waters, shown in green and labeled (2), occur along the equator in the eastern tropical Pacific and warmer (red) surface waters (1) occur in the western tropical Pacific, north of Australia. (b) During the *El Niño,* warm surface waters occurred throughout the tropical Pacific, replacing the normally cold tongue of eastern Pacific surface waters along the coast of South America.

pattern of oceanic and atmospheric conditions changes dramatically. Warm waters replace the normally cold surface waters along the equator and the western coast of South America. Wind patterns and rainfall distributions shift, and the effects are felt globally (Fig. 7–34). Since this change usually occurs around Christmas, local fishermen called it ***El Niño***—Spanish for "the child."

Before satellite observations were available, the first indication of such changes came when unusually warm surface waters replaced the cold surface

BOX 7–1
El Niños and Mt. Pinatubo

A strong *El Niño* occurred in 1991–1992, modifying weather and affecting marine life around the world. The event was not as strong as the 1982–1983 event—the strongest in this century. Still, the 1991–1992 event changed distributions of rainfall and marine life along the Pacific coast of North America.

It was difficult to recognize the effects of this *El Niño* because of two other processes occurring at the same time. First was the continued warming of Earth's atmosphere due to the steadily increasing concentrations of carbon dioxide released by our burning of fossil fuels. The 1980s were the warmest years in this century, and the 1990s will likely continue this trend. So far, the warming effect is estimated to be about 0.5°C since the late 1800s.

The second competing event was the powerful June 1991 eruption of Mt. Pinatubo in the Philippines. The cooling that resulted because volcanic ash shielded Earth's surface from the sun was estimated at approximately 0.5°C (Fig. 7B–1–1). Sorting out the individual effects of these three processes was difficult for those studying climatic change.

Figure 7B–1–1
The eruption of Mt. Pinatubo injected a massive cloud of volcanic ash and other particles into the stratosphere encircling the globe in a period of eight weeks. The cloud reduced the incoming solar radiation, resulting in a slight cooling that persisted for several years. The eruption of Mt. Pinatubo was the largest volcanic eruption of the twentieth century. (Courtesy James J. Mori/U.S. Geological Survey.)

waters along the Peruvian coast. Now the processes involved can be monitored as they occur. Using satellite observations and sea-level measurements on central Pacific islands, scientists can now monitor the buildup of warm waters and observe the typhoons that initiate *El Niños.* This permits prediction of an *El Niño* up to 9 months before it affects the coast of South America.

To understand the processes causing *El Niños,* we begin with the currents, which normally move warm surface waters westward along the equator, where they accumulate in the western Pacific north of Indonesia. Apparently, ocean currents cannot transport and disperse these warm waters effectively. Thus, they accumulate in equatorial regions on the western side of the Pacific basin. The accumulation of a large pool of warm water in the western Pacific north of Australia around Indonesia somehow causes the prevailing trade winds to reverse. (This process is not well understood.) Such wind reversals last for a few days to a few weeks and usually occur in November through April. These reversals are sometimes accompanied by twin "super typhoons"—unusually strong typhoons that occur simultaneously north and south of the equator.

The presence of this large pool of warm water in the equatorial western Pacific affects the atmosphere over the entire South Pacific. When it expands, the atmospheric high-pressure zone usually located off South America weakens, as does the low-pressure zone located over Indonesia. In short, the atmospheric circulation is changed across the South Pacific. These atmospheric changes are called the Southern Oscillation. The entire set of processes involving *El Niños* and the Southern Oscillation are called "ENSO events."

The twin typhoons cause pulses of warm water to move eastward along the equator at speeds of a few tens of kilometers per day. (These movements are propagated by a special kind of wave that can exist only along the equator.) These pulses moving eastward along the equator dissipate the warm waters accumulated in the western Pacific. They encounter the South American coast about nine months after their formation. These warm waters off Chile and Peru (where surface water temperatures are normally cool) are what the fishermen call *El Niño.* They are caused by the upward flow of cold sub-surface water along the coast (this process, known as upwelling, is discussed in Chapter 8).

After an *El Niño* (which normally lasts only one to two years), the large pool of warm water in the western equatorial Pacific is depleted. Normal current patterns are re-established in the equatorial ocean, and the atmosphere resumes its normal circulation. The upward flow of cold subsurface waters resumes along the equator and off South America. Warm waters again begin to accumulate in the western Pacific. Then the process begins again, leading to another *El Niño.* (Recently, the frequency of *El Niño*s may have increased. Some scientists cite this as evidence of global warming.)

During an *El Niño,* the changes in weather patterns are felt worldwide, especially along the equator. The results can be locally catastrophic. Areas of heavy rainfall shift from Indonesia eastward to the normally dry islands of the central equatorial Pacific. Sea level rises because the warmer (less dense) surface layer is thicker. Higher sea level combined with heavy rains floods low-lying islands. There are fewer hurricanes in the Atlantic during *El Niños;* therefore, those areas which depend on hurricanes to supply summer rains experience droughts.

There was an especially severe *El Niño* in 1982–1983, the strongest in the twentieth century. Droughts and fires in Australia caused billions of dollars in damages through massive losses of crops and livestock. Heavy rains and storms caused flooding and destroyed highways and bridges in normally arid coastal regions of South America. Another strong *El Niño* in 1991 lasted until 1993, unusually long for such an event. It recurred in the winter of 1994–1995, again bringing severe storms to the California coast.

El Niños also disrupt biological processes. Thick layers of warm water at the surface disrupt the processes that cause the abundant growth of microscopic marine organisms, affecting local fish populations. This reduces fish catches with catastrophic economic results. For example, in 1972 a strong *El Niño,* combined with overfishing, destroyed Peru's anchovy fishery, then the world's largest.

Effects on other organisms can be equally dramatic. Sea birds living on equatorial islands provide one of the earliest and clearest indications that an *El Niño is* occurring. With the loss of their fish prey, adult birds must abandon their nests and chicks and go far to find new food supplies to survive. In addition, seabird nests on the ground are flooded and destroyed by the combination of higher sea level and heavy rains. This results in the death of many adult birds and virtually all the chicks hatched that year. Recovery of seabird populations and other organisms takes years, especially after an intense *El Niño.*

After an *El Niño,* surface currents usually return to their former condition, but they can "overshoot," resulting in unusually cold waters along the Peruvian coast and in the eastern equatorial Pacific. This happened in 1988, following the

Monitoring Global Ocean Changes

The ability of seawater to transmit sound for thousands of kilometers is being used to detect signs of global ocean warming. Sound speed in seawater varies with water temperature—the higher the temperature, the higher the sound speed. Repeated measurements of the travel time between transmitter and receiver provides a measure of the temperatures of the water through which the sound pulses have passed. Monitoring over long periods permits the detection of changes in the temperatures of the subsurface waters. The large heat capacity of the ocean averages out the seasonal and daily fluctuations that plague similar measurements in the atmosphere.

One of the most ambitious schemes uses powerful sound transmitters—underwater loudspeakers—to transmit sound pulses that can be detected across entire ocean basins. The feasibility of the approach was demonstrated by an experiment conducted in January 1990. A sound source in the southern Indian Ocean, near Heard Island, was detected by sensitive listening devices in all three oceans. A followup experiment will use a similar sound source in deep waters off the central California coast, with listening stations on various Pacific islands. This experiment will monitor deep-ocean temperatures over a large portion of the ocean basin.

The powerful sound sources have raised concerns about their possible effects on marine mammals, which also can use sound for locating food and mates and for communicating. The dispute was resolved by locating the sound sources in deep water on the side of an exinct volcano off the California coast. In addition, more research will be done to determine the sensitivity of whales and other mammals to loud sounds.

Still another means of monitoring the ocean to detect possible climatic changes is to listen for sounds of increased cracking of sea ice in polar ocean regions. The Artic Ocean seems especially attractive for such studies. Listening arrays built by the U.S. Navy to detect and track Soviet submarines might be used for such purposes.

relatively mild 1986–87 *El Niño.* The unusually cold water along the equator caused the intertropical convergence zone to shift northward. This in turn displaced the jet stream, disrupting weather patterns in the United States and Canada. This set of circumstances contributed to an unusually severe drought in the summer of 1988.

Summary

Solar radiation is the source of energy for heating the ocean. Infrared radiation is absorbed in near-surface waters. Turbidity increases light absorption and scattering. Pure ocean water is most transparent to blue-green light.

Sea surface waters are warmest near the equator and coldest near the poles. Bands of equal sea surface temperatures are oriented east–west. Seasonal changes in water temperatures are greatest in the mid-latitudes.

Sea-surface salinities are highest in the centers of the ocean basins. Lowest salinities occur near the coasts, near river mouths, and in tropical areas, where there is much precipitation. Salinities are controlled by the balance between evaporation and precipitation. Evaporation is highest in subtropical regions; precipitation is highest in equatorial regions and in high latitudes. Poleward heat transport is reflected in low surface-water salinities in high latitudes.

The ocean is divided into three zones: surface zone, pycnocline, and deep zone. The pycnocline is sometimes called the thermocline (in which case we are interested in how temperature changes with depth) and sometimes called the halocline (in which case we are interested in salinity changes with depth). The pycnocline is a zone of marked change in density, caused by changes in either temperature, salinity, or both.

Temperature–salinity relationships are used to identify water masses. They are also useful in studying mixing between different water masses. Finally, temperature and salinity relationships can be used to study how waters move below the surface zone.

Sea ice forms in chilled surface waters. The first stage is freezing of individual crystals, which later form a plastic layer. Eventually, pancakes form and freeze together, forming floes, which are moved by winds and currents. Icebergs are freshwater ice broken off glaciers.

Climatic regions on Earth are generally oriented east–west. Nearness to the ocean moderates climates on land. Ocean surface waters affect land climates over large areas.

Earth has experienced two climatic regimes over the past 600 million years. One is a glacial climate similar to the climate of today. The other is a warm climate with little variation over Earth. During the last glacial advance, ice on both land and sea extended much closer to the equator than at present. There was little change in temperatures at the equator. The greatest temperature changes occurred in the mid- and high latitudes. Times of warm climate generally correspond to times of high sea level; glacial periods correspond to low sea levels.

Ocean waters affect weather over large areas by steering storm tracks. Pools of unusually warm or cool water in the North Pacific affect weather over large areas of North America. *El Niños*—occurrences of warm surface waters off the Pacific coast of South America—are caused by shifts in winds and movements of warm equatorial waters. They affect weather patterns over large areas, causing droughts over North America and Africa.

Key Terms

surface zone
mixed layer
pycnocline
thermocline
halocline
deep zone
hummocks
pressure ridges
leads
pack ice
polynya
iceberg
polar ocean areas
subpolar ocean regions
temperate region
subtropcial regions
tropical regions
equatorial regions
maritime climates
continental climates
El Niño

Study Questions

1. Describe how the color of light changes with depth below the ocean surface. What causes these changes?
2. What factors control seawater density?
3. Why are seasonal temperature differences much greater over land than over ocean?
4. Draw a diagram showing the layered structure of the ocean.
5. Sketch the general distributions of surface-water temperatures in the world ocean.
6. Draw a diagram showing the generalized distribution of surface salinities.
7. Explain why most of the bottom water of the ocean forms in the North Atlantic and near Antarctica.
8. Describe how sea ice forms.
9. Describe how temperature–salinity diagrams are used to identify water masses and trace their movements through the ocean.
10. Where do icebergs come from? Where are they most abundant?
11. Describe the different types of polynyas and discuss how they form.
12. Why are satellite observations so important in studying sea ice?
13. Contrast today's ocean conditions with those prevalent during the most recent glacial stage.
14. In what ways may the uplifting of large mountain ranges have affected global weather patterns?
15. Describe the causes of *El Niños.*
16. How do *El Niños* affect weather?
17. *[critical thinking]* How might global warming affect the frequency of *El Niños?*

Selected References

BROECKER, W. S. AND G. H. DENTON, "What Drives Glacial Cycles?" *Scientific American,* January 1990.

BURROUGHS, W. J., *Watching the World's Weather.* Cambridge, England: Cambridge University Press, 1991.

MUSK, L. F., *Weather Systems.* Cambridge, England: Cambridge University Press, 1988.

OFFICER, C., AND J. PAGE, *Tales of the Earth*. New York: Oxford University Press, 1993. Well-written account of events affecting Earth and its ocean.

PERRY, A. H. AND J. M. WALKER, *The Ocean-Atmosphere System.* London: Longman, 1977. Discussion of air-sea interactions.

PICKARD, G. L., *Descriptive Physical Oceanography, 3d ed.* Elmsford, N.Y: Pergamon, 1979. General discussion of ocean features.

RAMAGE, C. S., *"El Niño," Scientific American,* June 1986.

RUDDIMAN, W. F., AND J. E. KUTZBACH, "Plateau Uplift and Climatic Change," *Scientific American,* March 1991.

STOMMEL, H., AND E. STOMMEL, "The Year Without Summer," *Scientific American,* June 1979.

HUDSONS
BAY
TERRA DE LABRADOR
NEW BRITAIN
NEW
SOUTH WALES
NORTH AMERICA
CANADA or
NEW FRANCE
NEW SCOTLAND
BAY OF
S. LAURENS
NEW FOUND LAND
The Bancks of Newfoundland
To find the Distance of two
Places in this CHART
NEW YORK
PENSILVANIA
MARYLAND
VIRGINIA
FLORIDA
Bermudas
GULF of MEXICO
Mouth of the Missisipi River
BAHAMA
ISLANDS
Bay of Campeche
Iucatan
CUBA
HISPANIOLA
HONDURAS
COSTA RICA
CARIBBE ISLANDS
Barbados
TERRA FIRMA

8 Currents

OBJECTIVES

Your objectives as you study this chapter are to understand:

- The ocean's surface and subsurface currents
- How winds cause surface currents
- The processes driving subsurface currents
- The major features of ocean circulation
- How ocean currents have changed through time

In 1769 the postmaster of the American colonies, Benjamin Franklin, used reports from the captains of mail ships crossing the Atlantic to draw the first map of the current which came to be known as the Gulf Stream. Today, Earth-orbiting satellites can observe surface ocean currents globally on a weekly basis.
(Dr. Philip Richardson/Woods Hole Oceanographic Institution.)

Currents—large-scale water movements—occur everywhere in the ocean, both on the surface and far below it. There are two entirely different ocean current systems. We are most familiar with surface currents, which are driven primarily by winds. Subsurface currents, driven by chilled waters sinking in the polar and subpolar oceans, are less familiar. These deep-water masses spread out to flow through all the oceans and eventually return to the surface to be warmed by incoming heat from the Sun.

Both surface and subsurface currents are caused by the unequal heating of Earth's surface. Ocean currents of both types transport heat from the tropics toward the poles, partially equalizing surface temperatures over Earth. They also play an important role in controlling climate in the long term.

Until recently, our only knowledge of ocean currents came from surface-current records compiled by nineteenth-century sailing ships, records that mapped only large average currents. Over the past few decades, that limitation has been removed by developments in technology. Today, our knowledge of ocean currents is increasing because of sophisticated current meters and global observations made by remote-sensing instruments on Earth-orbiting satellites. Although our knowledge is growing rapidly, we still know little about short-term variations in currents, aside from Indian Ocean monsoon currents.

Open-Ocean Surface Currents

We classify the major surface ocean currents as being either open-ocean currents or boundary currents. This section discusses open-ocean currents; boundary currents are treated in the following section.

The surface patterns of **open-ocean currents** are similar in the three major ocean basins (Fig. 8–1) and closely resemble surface-wind patterns (Fig. 8–2). The equatorial ocean is dominated by westerly flowing waters in the *north* and *south equatorial currents*, driven primarily by the trade winds. Between them is the narrow, eastwardly flowing *equatorial countercurrent*, which occurs in a region of light and variable winds called the doldrums. Across the bottom of the Southern Hemisphere is the eastward-flowing *Antarctic circumpolar current,* and across the top of the Northern Hemisphere are the eastward-flowing *North Pacific* and *North Atlantic currents.* In each hemisphere, the combination of westward equatorial current and eastward high-latitude current forms a current **gyre**—a nearly closed current system (Fig. 8–2). The east–west-elongated gyres are centered in the subtropical regions north and south of the equator.

In the subpolar and polar regions of all ocean basins, there are also smaller current gyres (Fig. 8–2). The direction of circulation in these high-latitude gyres is opposite to that of the subtropical gyres, which means the subpolar gyres are counterclockwise in the Northern Hemisphere and clockwise in the Southern Hemisphere. Because of the positions of the continents, subpolar gyres are especially well developed in the Northern Hemisphere. They occur in the Southern Hemisphere as well, primarily near Antarctica. Because there are no land barriers there, currents flow uninterrupted around Antarctica in the Antarctic Circumpolar Current. As discussed in Chapter 2, this is the primary connection between the three major ocean basins.

Ekman Spiral

Surface winds and surface ocean currents are intimately related (Fig. 8–2), but how winds drive currents is not so obvious. As we shall see, Earth's rotation plays a complicating role, just as it did with the winds.

The process begins when winds blow across water and drag on the surface. This surface drag sets into motion a thin layer of water, a few centimeters thick, which in turn drags on the thin layer beneath, setting it in motion. This process

BOX 8–1
Predicting Currents and Ocean Conditions

Modern computers now permit us to predict future currents. Scientists begin with a computer model of the present state of the ocean. They enter today's wind patterns into the model to find out how the ocean responds to both winds and heating from the Sun. Then the computer model predicts how the ocean will respond to changed conditions at some time in the future.

At present, these predictions are made only for the open ocean, where the systems are simpler because we do not have to be concerned with shoreline complexities. Also, the currents are often predicted only for the surface zone so that the model can ignore complexities of the sea floor. However, British scientists were able to model the currents at several levels around Antarctica, including the bottom circulation. U.S. Navy forecasters predict near-surface thermal conditions, which control propagation of the sound used by sonar systems to locate submarines.

The next step is for scientists to develop the same predictive capability for the coastal ocean. More powerful computers will be able to handle the complexities of the shorelines and the shallow continental shelf. Wind charts detailed enough to show coastal complexities will also be required. Within a few years, it should be possible to predict coastal currents for recreational fishing and various other activities.

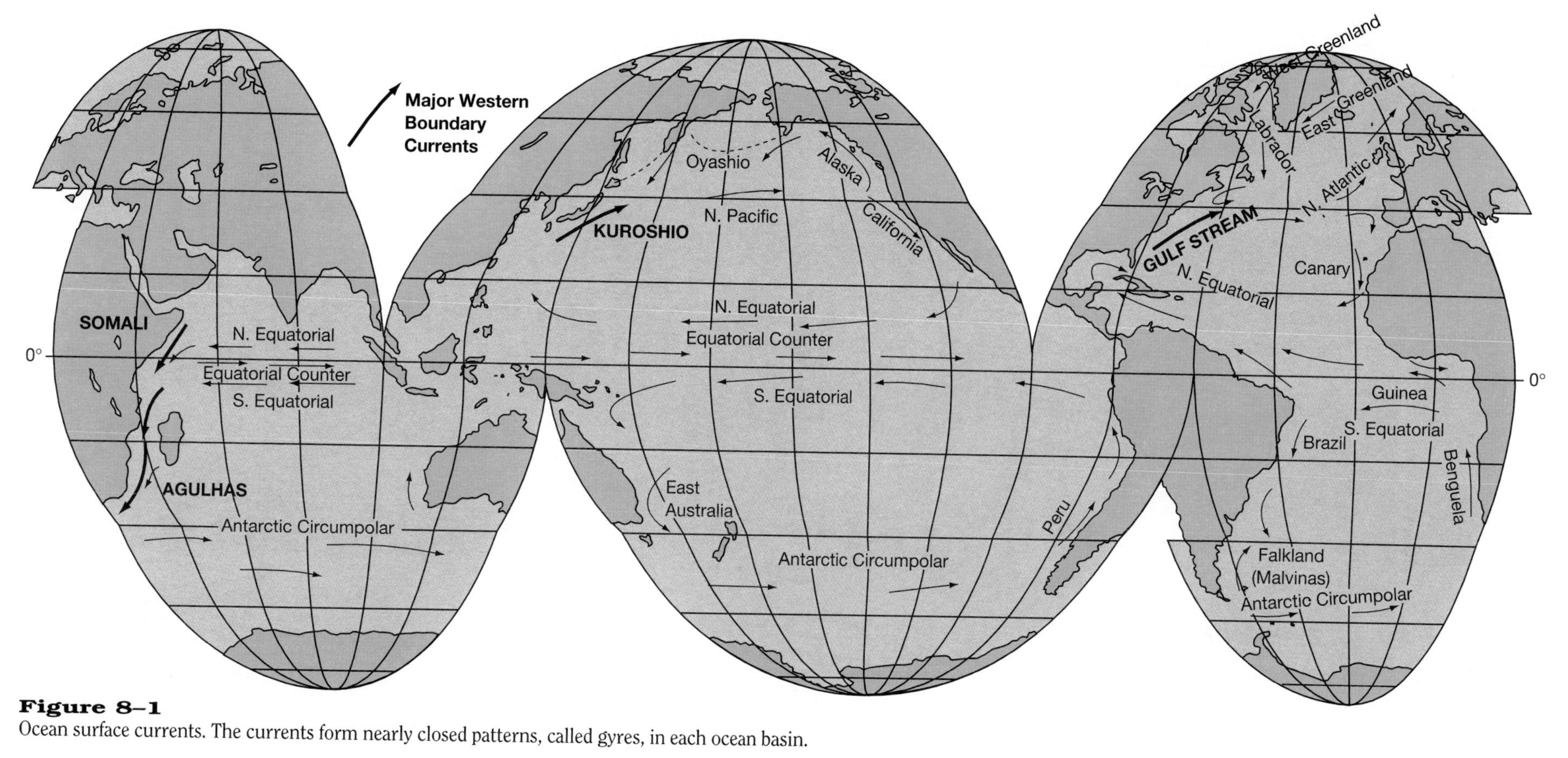

Figure 8–1
Ocean surface currents. The currents form nearly closed patterns, called gyres, in each ocean basin.

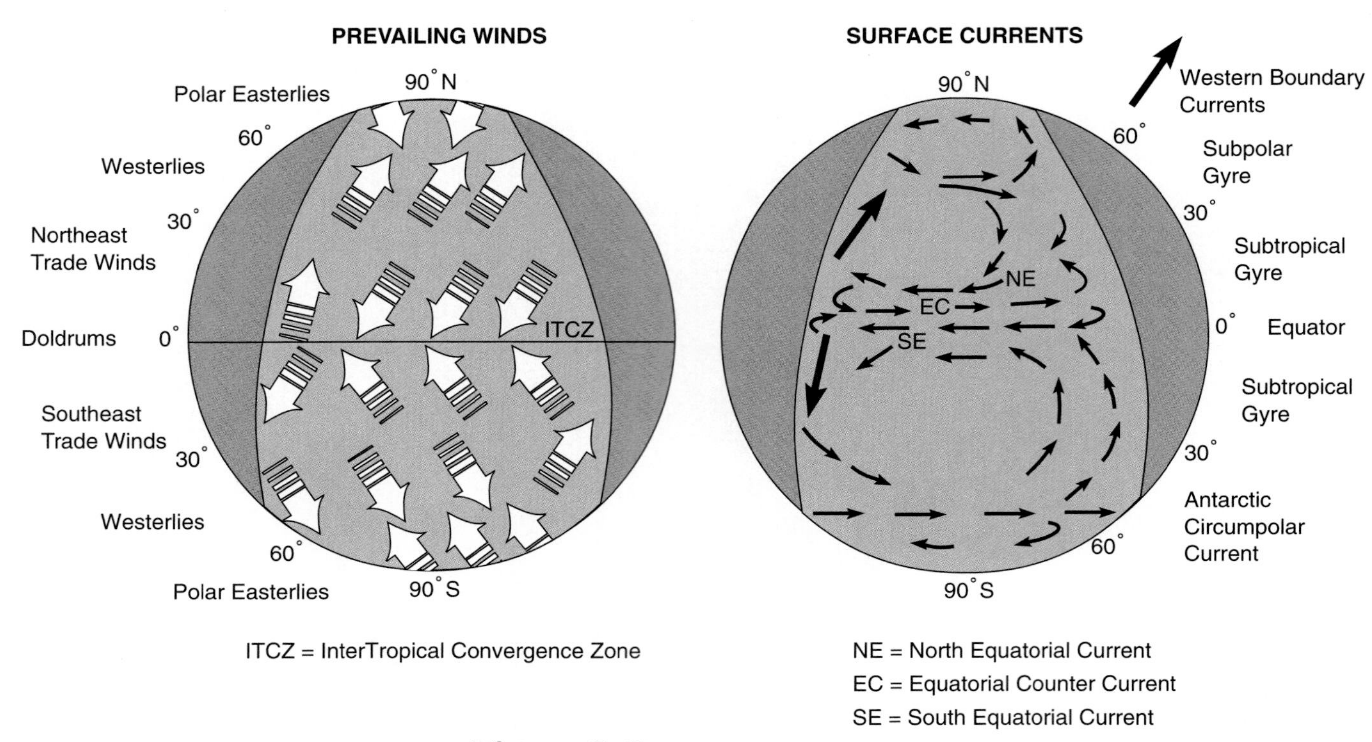

Figure 8–2
Prevailing surface winds and generalized surface ocean currents. Note the current gyres.

continues downward, transferring momentum (product of mass and velocity) to successively deeper layers. Such transfers of momentum from one layer to the next are inefficient, however, and therefore energy is lost in the process. As a result, current speed decreases with increasing depth. Surface currents move at about 2 percent of the speed of the wind that caused them.

In an infinite ocean on a nonrotating Earth, water would always move in the same direction as the wind that set it in motion. Because Earth rotates, however, surface waters are deflected to the right of the wind (viewed with your back to the wind) in the Northern Hemisphere. This is another manifestation of the Coriolis effect. This rightward movement of surface water masses was first noted during the *Fram* Expedition (1893–1896) by Fridtjof Nansen, a Norwegian oceanographer and Arctic explorer, while studying the drift of Arctic ice. He found that the ice moved 20° to 40° to the right of the wind.

To explain this shift, we assume a simple, uniform ocean with no boundaries. In such an ocean, the motion of each deeper layer is deflected to the right of the one above. These movements can be represented by arrows *(vectors)* whose orientation shows current direction and whose length indicates current speed. The change in current direction and speed with increasing depth forms a spiral when viewed from above. This is called the **Ekman spiral** after the Swedish physicist V. W. Ekman, who first explained the phenomenon. Figure 8–3 shows an Ekman spiral for the Northern Hemisphere. A spiral for the Southern Hemisphere has the opposite sense of deflection, but current speeds still decrease with increasing depth.

Water-column stability controls the depths to which wind effects penetrate, with a strong pycnocline inhibiting transfer of momentum from the surface to waters below the pycnocline. The limit for wind effects is usually taken to be the depth at which the direction of the subsurface current is exactly opposite the direction of the surface current. A current driven by a strong wind may be as deep as 100 meters below the surface.

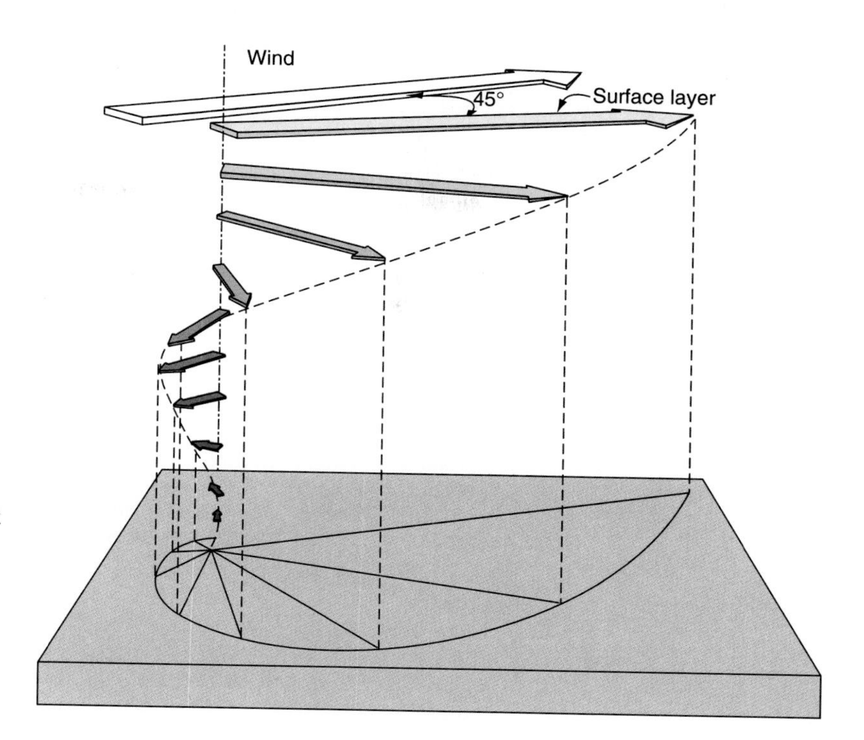

Figure 8–3
An Ekman spiral in a wind-driven current in deep water in the Northern Hemisphere. Current speeds decrease with increased depth, and the water in a layer moves more to the right as one goes deeper. The speed of surface currents is about 2 percent of the speed of the wind that set them in motion.

In shallow waters, wind-generated currents are deflected less than predicted by Ekman's model. The ocean is obviously not uniform, which is one of the assumptions in the Ekman spiral; we have already mentioned the effects of the pycnocline. The wind-driven water movements do not penetrate the pycnocline. Furthermore, winds do not always blow long enough (one to two days of steady winds) to produce a fully developed Ekman spiral. Under these conditions, the deflection is less than is predicted by the theoretical case shown in Figure 8–3. The net movement of waters affected by the Ekman spiral is 90° to the right of the wind in the Northern Hemisphere (again, viewing with the wind to our back). This is called *Ekman transport* (Fig. 8–4).

Geostrophic Currents

Prevailing winds move surface waters toward subtropical regions both from tropical areas and from high-latitude areas. This is an example of Ekman transport. As Figure 8–2 shows, these mid-latitude areas—the subtropical gyres—are zones of **convergence** where surface waters accumulate. Two things happen in a convergence (Fig. 8–5): water piles up, forming slopes (as on a hill), and the surface layer thickens. This process of winds moving waters toward or away from a gyre center is sometimes called *Ekman pumping*.

Figure 8–4
Relationship between wind, surface current, and net water movement (also called Ekman transport) in the Northern Hemisphere. Current speeds are approximately 2 percent of the wind speed.

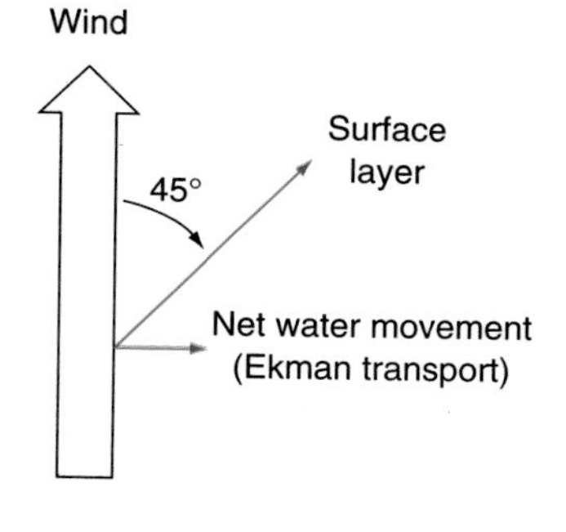

In the opposite case, a **divergence,** winds blow surface waters away from an area, as illustrated in Figure 8–6. In a divergence, subsurface waters move upward to replace waters that have moved away from the region, causing the surface layer to be thinner than normal. A prominent divergence occurs around Antarctica. Divergence also occurs in the subpolar gyres and at the equator (Fig. 8–7).

As a result of convergences and divergences, the ocean surface has a subtle topography—low hills in areas of convergences and shallow valleys at divergences. (This topography is exaggerated in the figures in this chapter for the purposes of explanation.) The difference in height between the mid-latitude convergences on the western side of the ocean basins and the divergence surrounding Antarctica is about 2 meters. This difference occurs over a distance of about half Earth's circumference. Thus these are very gentle slopes compared with those on land and

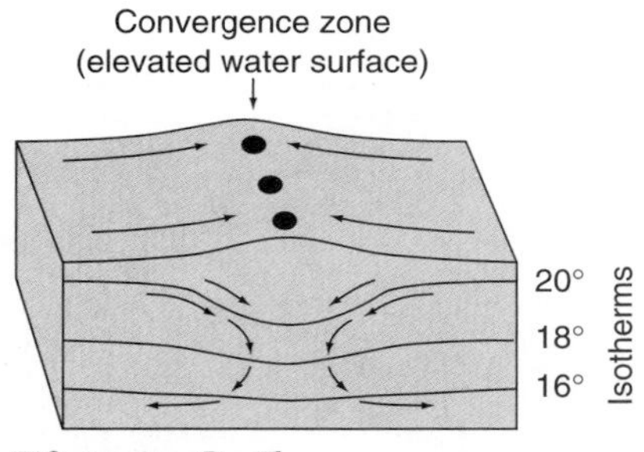

Figure 8–5
In a convergence zone, a hill of water forms and the surface layer thickens because of the accumulation of surface waters. The arrows indicate the direction of water movement. *In this figure and others in this chapter, the sea surface slopes have been greatly exaggerated in order to emphasize the water movements.*

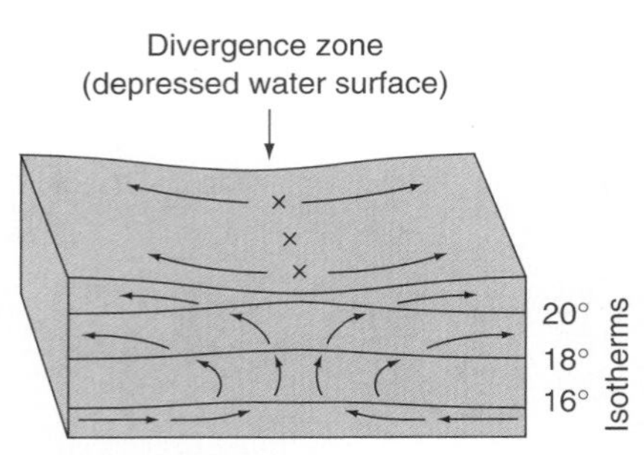

Figure 8–6
In a divergence zone in the open ocean, the surface layer is thinned because water moves away from an area.

were impossible to measure directly until satellite observations became available. Now *altimeters*, precision radar on satellites, can measure differences in surface elevations of only a few centimeters across an entire ocean basin. Furthermore, such observations can be made over the entire ocean in about ten days.

These slopes of the ocean surface, which are controlled by density distributions, respond very slowly to wind shifts. Consequently, ocean currents are equally slow to change. Ocean topography responds to winds over many days, thus acting like a flywheel that stores pulses of energy from winds and storms.

Water responds to the slopes of the ocean surface as it would on land—by running downhill. Consider the case of a hill in the Northern Hemisphere. Let us follow a water parcel (Fig. 8–8) to see how Earth's rotation changes its path. The easiest way to visualize this process is to imagine a water parcel moving initially down slope. The Coriolis effect deflects it to the right. This deflection continues until the water follows a path that allows it to flow downhill just enough to keep moving. Most of its motion parallels the side of the hill.

If our hill showed contour lines of equal elevation, as Figure 8–8 does, the path of the water parcel would nearly parallel them, with any deviation from this

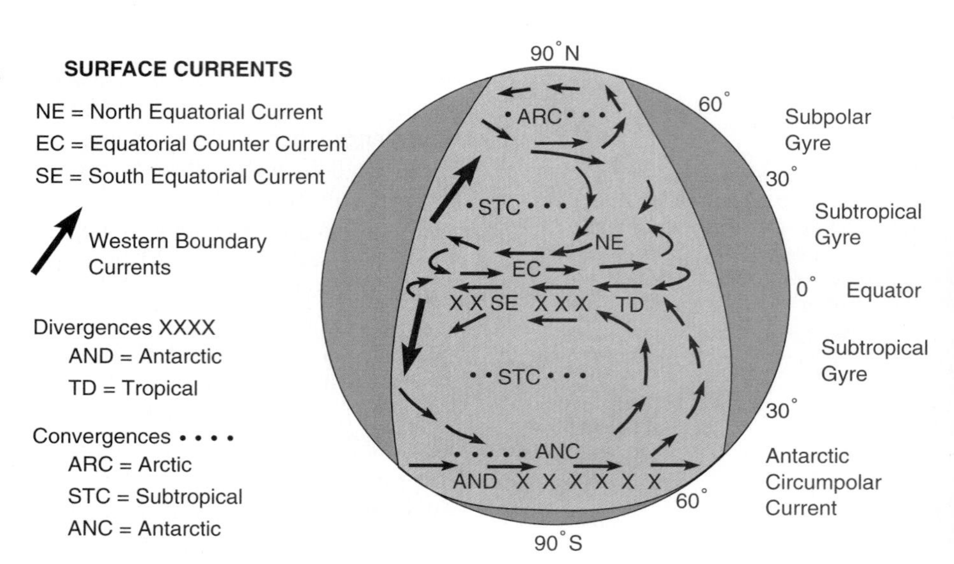

Figure 8–7
Schematic representation of relationships among surface currents, gyres, and zones of convergence and divergence.

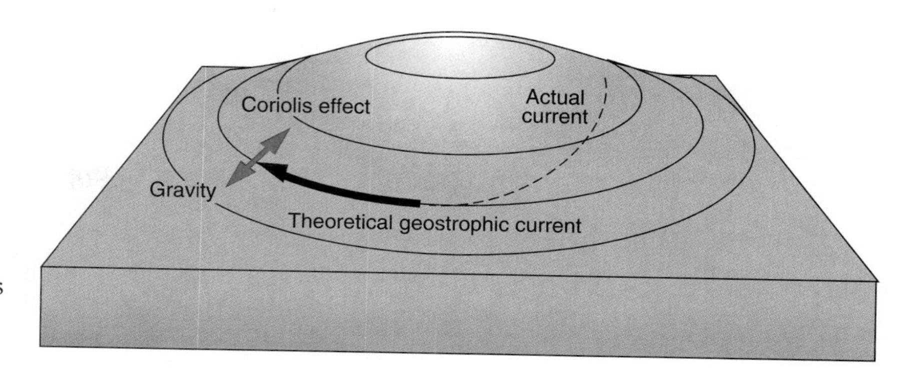

Figure 8–8
Water flow and balance of forces (red arrows) in a geostrophic current in the Northern Hemisphere. The surface waters flow gradually down the gently sloping sea surface.

parallel path being due to friction in the water. In other words, on a frictionless ocean, water movements would exactly parallel the side of the hill. This balance—between the downslope component of the gravitational force and the Coriolis effect that deflects it—gives rise to **geostrophic currents.** (These are analogous to the geostrophic winds discussed in Chapter 6.) Such currents are strongest on the steepest slopes and weakest on gentle slopes. (Remember that sea surface slopes are extremely slight, and so forces acting along sea surfaces are essentially horizontal.)

Where a strong current passes near land, it is possible to measure such steep sea surface slopes directly. Thus we know that the *Florida Current* (part of the *Gulf Stream* system between Cuba and the Bahamas Banks) has slopes of about 20 centimeters over about 200 kilometers. At this location, current speeds are 150 centimeters per second (about 3 miles per hour). This rapid current has a much steeper slope than other currents, where waters move only a few kilometers per day.

Boundary Currents

Boundary currents flow close to and parallel to the continental margins, usually north–south (Fig. 8–9), and dominate surface-current patterns on the world ocean. The two principal types of boundary currents—western and eastern—are compared in Table 8–1. (These currents are named according to which side of an ocean basin they appear on.) **Western boundary currents** are the strongest surface currents. They are especially conspicuous in the Northern Hemisphere—for instance, the Gulf Stream in the Atlantic and the *Kuroshio* in the Pacific. Because the Gulf Stream has been studied most, we focus on it.

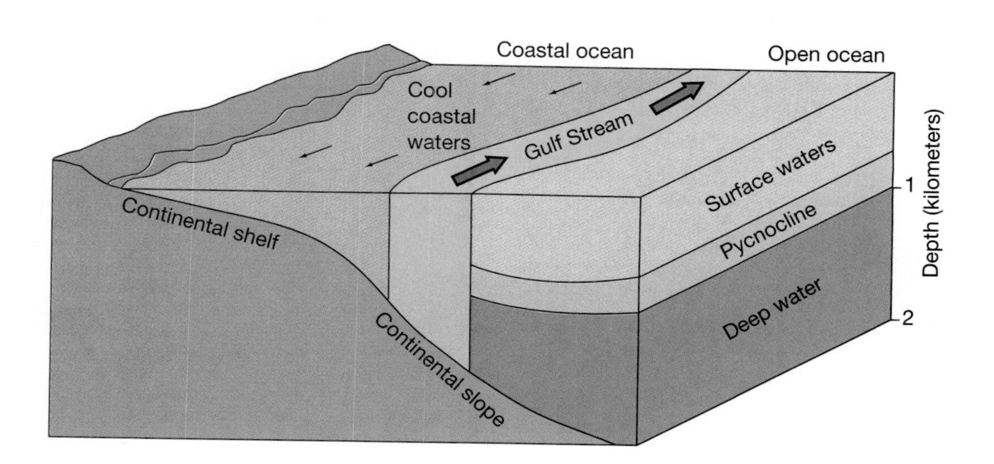

Figure 8–9
The Gulf Stream, a western boundary current, is so deep that it is unable to come up onto the continental shelf.

TABLE 8–1
Northern Hemisphere Boundary Currents

Type (example)	*General Features*	*Speed*	*Transport*
Eastern boundary (California, Canary)	Broad (about 1,000 km), shallow (about 500 m)	Slow (10–100 km/day)	Small (10–15 million m^3/s)
Western boundary (Gulf Stream, *Kuroshio*)	Narrow (less than 100 km), deep (2 km)	Swift (100-1,000 km/day)	Large (more than 50 million m^3/s)

The Gulf Stream separates open-ocean waters from coastal waters (Fig. 8–10). The salinity and temperature of coastal waters change with the seasons. In contrast, Gulf Stream waters just offshore are warm (20°C or higher) and have relatively high salinities (around 36) year round. There is some convergence of Gulf Stream surface waters toward the center of the North Atlantic subtropical gyre. This area, known as the *Sargasso Sea,* is bounded by currents.

Western boundary currents extend to depths of around 1 kilometer, too deep to come up onto a continental shelf; thus they flow along the continental margin (Fig. 8–9). The Florida Current, part of the Gulf Stream system, is relatively fast, with speeds of 100 to 300 centimeters per second. Speeds are highest at the surface in a relatively narrow band, 50 to 75 kilometers wide.

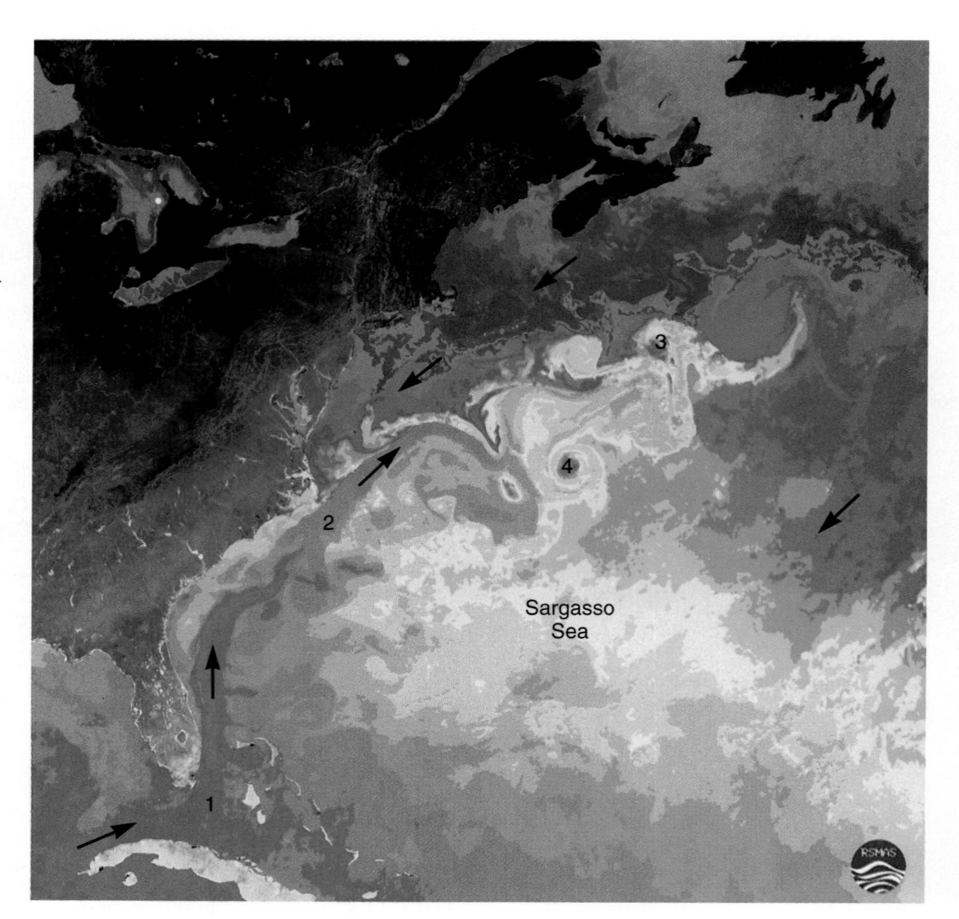

Figure 8–10
A satellite image made with infrared (heat-sensitive) film allows us to see differences in sea surface temperatures. The warm waters (reds and oranges) are the Florida Current (1) and the Gulf Stream (2). The cooler waters (blues and purples) are flowing southward along the Atlantic coast of North America in the Labrador Current. Warm-core rings (3) (yellow surrounded by green) occur north of the Gulf Stream while cold-core rings (4) (green surrounded by orange or yellow) are seen south of the Gulf Stream. The Florida Current and the Gulf Stream separate the cool coastal waters from the warmer, open-ocean waters. Arrows indicate current directions. (Courtesy Rosenstiel School of Marine and Atmospheric Science, University of Miami, and Goddard Space Flight Center, NASA.)

BOX 8–2
The Bermuda Triangle and Mysterious Sinkings

The Bermuda Triangle off the Atlantic coast of North America is the locale of many mysterious disappearances of ships and aircraft. Many supernatural explanations have been advanced to explain them. A comparable area off Japan has a similar reputation—the Dragon's Triangle, also known as the Sea of the Devil. Here, legends of sea monsters developed to explain the disappearances.

The Bermuda Triangle first came to public attention on December 5, 1945, when five U.S. Navy Avenger torpedo planes disappeared under mysterious circumstances. A search plane trying to find them also vanished. No wreckage was ever found.

Since then there have been many mysterious disappearances in the Bermuda Triangle, usually small craft operated by their owners. The Dragon's Triangle is similar, except that most of the disappearances there have involved large commercial or naval ships, aircraft, even Soviet submarines. What can we say about the ocean conditions that might explain these phenomena?

First, the two areas are heavily traveled by ships and aircraft. Even a small percentage of unexplained disappearances would amount to many such reports over a few years. Many of the disappearances happened before modern communications were available, at night, or during bad weather.

Most important, the two areas border two of the strongest currents in the ocean—the Bermuda Triangle lying seaward of the Gulf Stream and the Dragon's Triangle lying near the *Kuroshio*. This proximity to strong currents undoubtedly plays a major role in the mysteries.

When cool, dry air from the continents meets the warm tropical waters in the two currents, the result is the quick formation of strong, often violent storms. With modern communications, small craft caught in such storms can report problems as they occur. Search-and-rescue ships and aircraft are often successful in rescuing the survivors. The Bermuda Triangle is frequently hit by hurricanes coming up from the Caribbean, while the Dragon's Triangle experiences similar storms, called typhoons.

Another possible factor is unusually large waves—called rogue waves—formed when large waves encounter strong currents. Such waves occur off South Africa—another area where ships mysteriously disappear—where large waves encounter the strong Agulhas Current, a current similar to the Gulf Stream and the *Kuroshio*. Off Japan, there are reports of what are described as triangular waves, where waves advance on a ship from three directions and overwhelm it.

There are other possible explanations for ship disappearances off Japan. Naval mines laid during World War II claimed many ships, and some of these mines may still be active.

There are also many active submarine volcanoes near Japan. Gases discharged by an erupting volcano can lower the water density so much that a ship simply sinks. This happened to one Japanese research ship that was lost with all hands, and it nearly happened to a U.S. research ship sampling gases discharged by an erupting submarine volcano.

Another possible culprit in the Dragon's Triangle disappearances is tsunamis, resulting from the frequent large earthquakes in the western Pacific.

Separating the Gulf Stream from adjacent waters that move more slowly are oceanic fronts, analogous to the weather fronts we learned about in Chapter 6. Oceanic fronts are narrow zones marked by changes in surface-water characteristics, such as color, temperature, and salinity. At the northern edge of the Gulf Stream, water colors change from the green typical of coastal waters to the intense cobalt blue of subtropical waters. Figure 8–10 shows another example: changes in temperature displayed in a false color satellite image.

Eastern boundary currents are much weaker than their western counterparts. In eastern currents the waters move much more slowly, and the fronts are more diffuse and more complicated (Fig. 8–11). Eastern boundary currents are relatively shallow and therefore can readily flow over continental shelves.

Boundary currents are more changeable than open-ocean currents, mainly because the many irregularities in continental coastlines cause the direction of flow of boundary currents to change frequently.

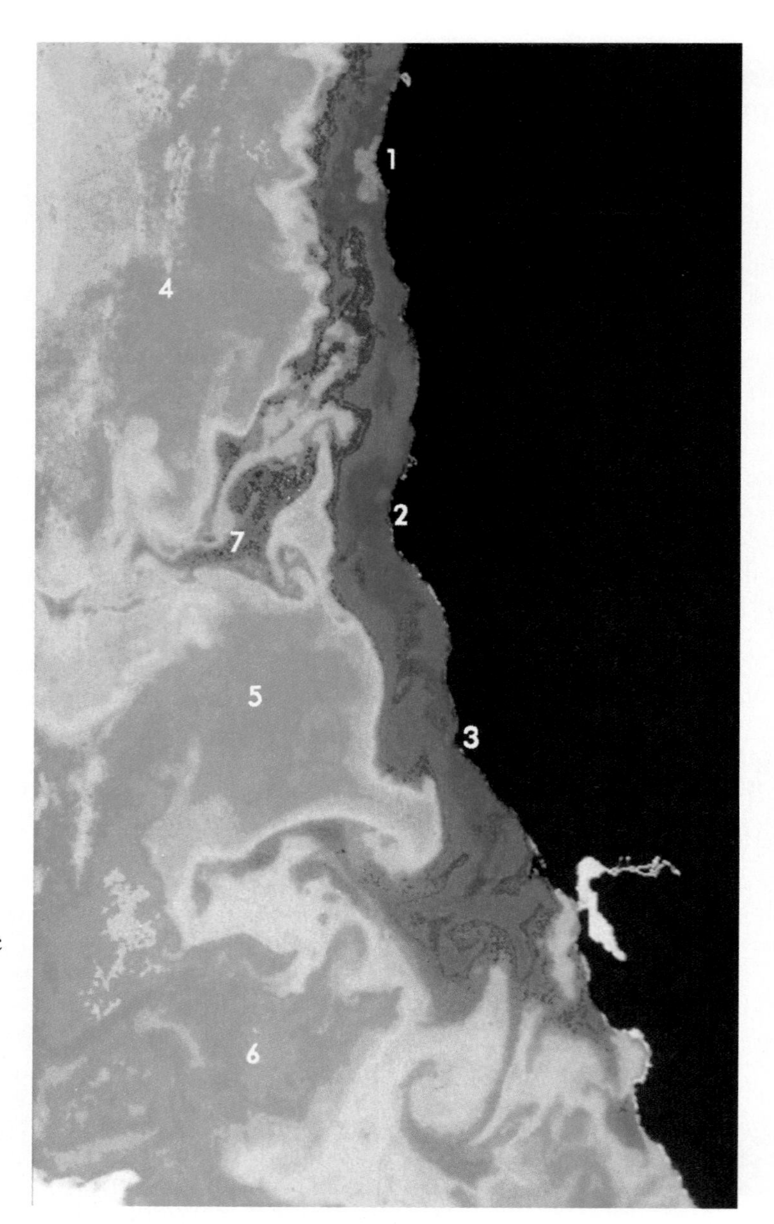

Figure 8–11
Surface-water temperatures on the Pacific coast of North America are dominated by upwelling, especially at capes along the shore (1, 2, 3); the cold upwelled waters are shown as blues and purples. The warmer waters of the offshore California Current are shown as oranges and reds. Several large meanders (4, 5, 6) are visible, as well as long filaments of cool upwelled water extending out from the coast (7). (Courtesy NASA.)

Western Intensification of Boundary Currents

Several processes acting together intensify western boundary currents. First is Earth's rotation, which displaces gyres toward the west. This displacement compresses the gyres against the continents. Consequently, sea-surface slopes are steeper on the western than on the eastern side of an ocean basin (Fig. 8–12).

Prevailing wind patterns over the open ocean are the second factor (Fig. 8–2). Trade winds, as we learned in Chapter 6, blow generally westward along the equator. This piles up the surface waters on the western sides of equatorial oceans. Put another way, the pycnocline is deeper on the western than on the eastern side of gyres. The strong westerly winds in the higher latitudes, coupled with the Coriolis effect, force surface waters in the mid-latitudes to flow toward the subtropical gyre. This flow toward the center of the gyre occurs over a broad zone on the eastern side of ocean basins and is not confined to eastern boundary currents

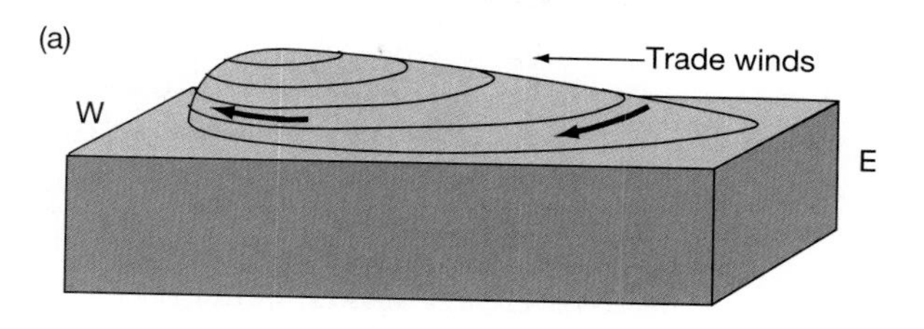

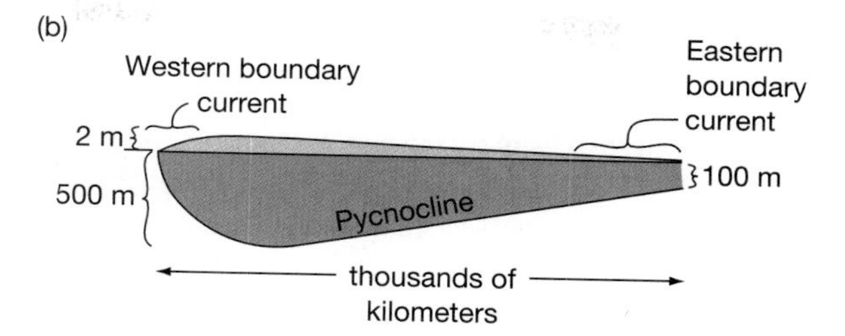

Figure 8–12
(a) In a northern hemisphere current gyre, water piles up on the western side because of Earth's rotation; the sea surface has a steep slope there. (b) Strong currents are associated with relatively steep slopes and weaker currents with gentler slopes.

(shown schematically in Fig. 8–2). It contributes to the accumulation of water in the gyre, making the slopes of the sea surface steeper on the sides of the gyre.

The last factor involves apparent changes in the state of rotation of objects transported north or south along Earth's surface. To illustrate this effect, imagine a bicycle wheel at the equator, sitting motionless on frictionless bearings (Fig. 8–13). When we carefully transport the wheel (by its support base) to the North Pole (being careful not to start the wheel rotating) and view it from above, we find that it now appears to rotate clockwise once every 24 hours. Actually the wheel remains motionless as Earth rotates counterclockwise beneath it. At intermediate latitudes the wheel appears to rotate more slowly than once a day.

The same thing happens to parcels of air or water moving north or south. Their rotational state appears to change because they move from an area rotating around Earth's axis at one speed to an area rotating at another speed. On the western

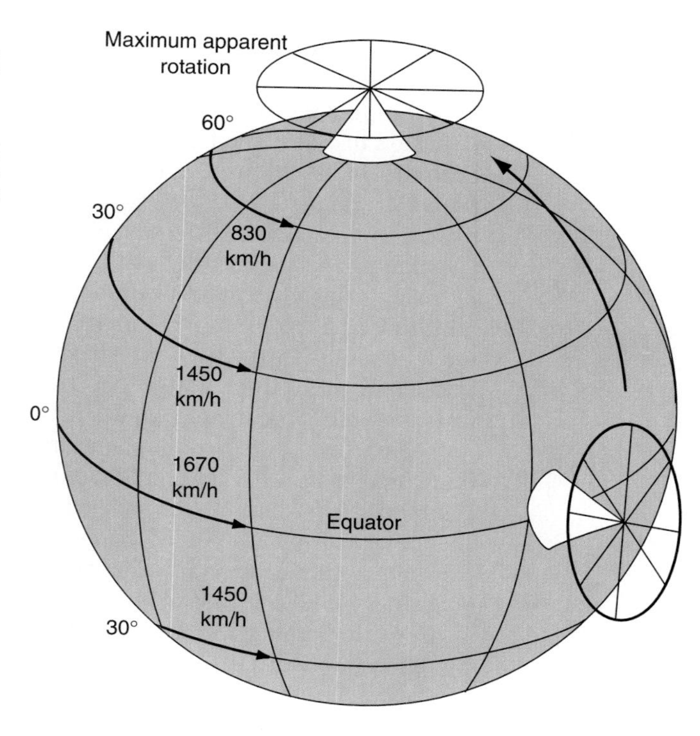

Figure 8–13
At the equator, a motionless bicycle wheel on a frictionless stand has no apparent rotational motion. At the North Pole, this same wheel appears to rotate clockwise as viewed from above. In actuality, the wheel is still motionless and the ground beneath it is rotating counterclockwise. (After K. Stowe: *Ocean Science*, 2d ed. New York: John Wiley, 1983.)

side of ocean basins, the currents and the apparent spin due to Earth's rotation are in the same direction. This gives western boundary currents a higher-than-average speed. On the eastern side of a basin, the waters are moving toward the equator, so current speeds are lower than expected.

Meanders and Rings

Satellite observations of the ocean surface reveal ocean processes as they happen. One example is the development of current meanders and rings in the Gulf Stream and the North Atlantic Current. A current **meander** is a bend in a current's flow path, and a current **ring** is a large, whirl-like eddy formed by a cut off meander in a strong current, such as a western boundary current.

As the red–yellow interface off the mid-Atlantic states in the satellite image in Figure 8–10 shows, a pronounced temperature change separates the cooler coastal waters from the warmer Gulf Stream waters. Moving eastward on this map, we see that the Gulf Stream forms the northern boundary of the Sargasso Sea, where the waters are slightly cooler than in the Gulf Stream.

Gulf Stream meanders occur frequently off the U.S. coast, especially after a storm (Fig. 8–14). The meanders move slowly northeastward with the Gulf Stream at speeds of 8 to 25 centimeters per second (7 to 22 kilometers per day). If a meander becomes too large, it forms a ring and detaches itself from the Gulf Stream. Such rings move with the waters, flowing slowly (5 to 10 km per day) southwestward on either side of the Gulf Stream.

Rings, which are mainly associated with western boundary currents, are 100 to 300 kilometers across and are bounded by swift currents (80 kilometers per day). Rings and their associated currents extend to depths of 2 kilometers; consequently they cannot intrude onto the shelf. Currents induced by the passage of such rings can influence currents on the shelf.

Rings form on both sides of the Gulf Stream. Those on the northern side have a warm-water core surrounded by colder slope water (Fig. 8–14a). Cold-core rings form on the south side of the Gulf Stream and inject cooler water into the Sargasso Sea. As cold-core rings move southwestward, their surface waters warm up. And as the surface-water temperatures reach those of the surrounding waters, the rings become more and more difficult to detect.

A cold-core ring was tracked for seven months as it moved from south of Cape Cod to near Cape Hatteras, North Carolina, where it was resorbed into the Gulf Stream. When the ring formed, its waters were nearly 10°C cooler than surrounding waters. These surface waters off Cape Cod were coastal waters rather than open-ocean waters. The ring had a complex path, remaining nearly stationary for a month, briefly merging with the Gulf Stream (Fig. 8–14b), then forming a meander for nearly a month before it again separated and resumed its southwesterly drift. Finally after nearly 7 months, the ring disappeared when it was resorbed into the Gulf Stream off Cape Hatteras (Fig. 8–14c). Similar rings have been tracked for up to three years. These rings were studied extensively because they interfered with acoustic techniques used to track submarines.

Rings transport heat, momentum, dissolved constituents, and weakly swimming organisms. They apparently do not transport water, however, because of the complex way in which they form and because their direction of movement is opposite the direction of the Gulf Stream.

Small current rings are called **eddies;** they are common in the western portions of the Atlantic (Fig. 8–15) and Pacific oceans. Such eddies extend from the sea surface to the ocean bottom, are 200 to 400 kilometers across, and take several months to pass a location. Currents associated with eddies are weaker than those in rings, a few tens of centimeters per second. Still, currents in eddies are a hundred times more energetic than average deep-ocean currents. Eddies are abundant near western boundary currents. They are the equivalent of atmospheric storms in the deep ocean.

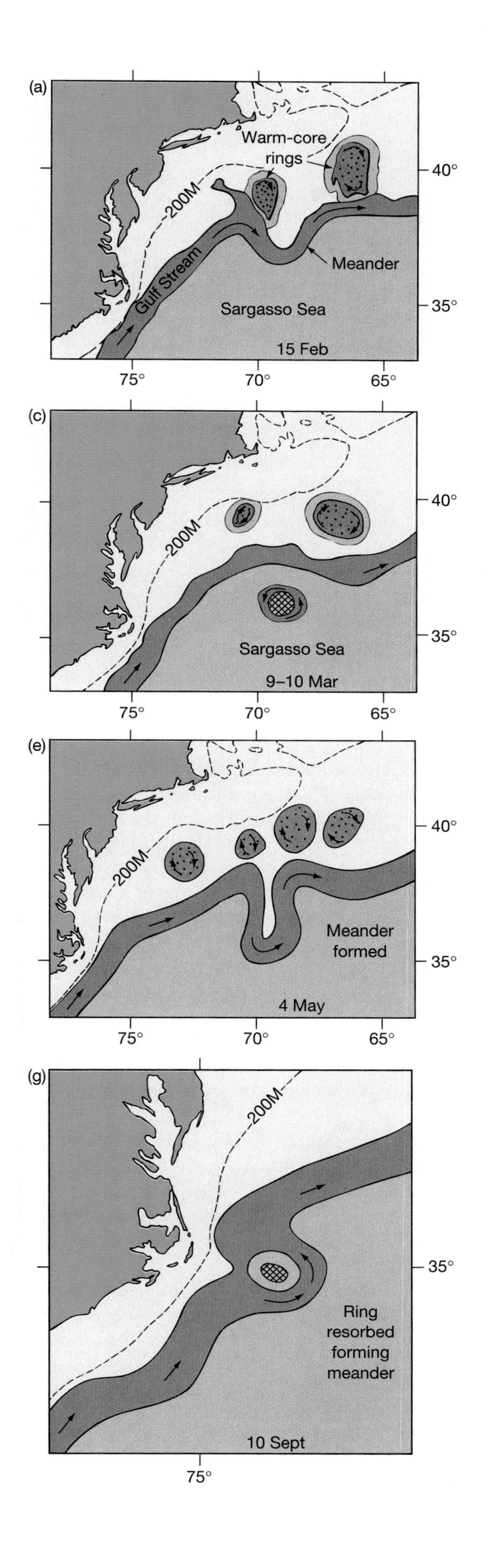

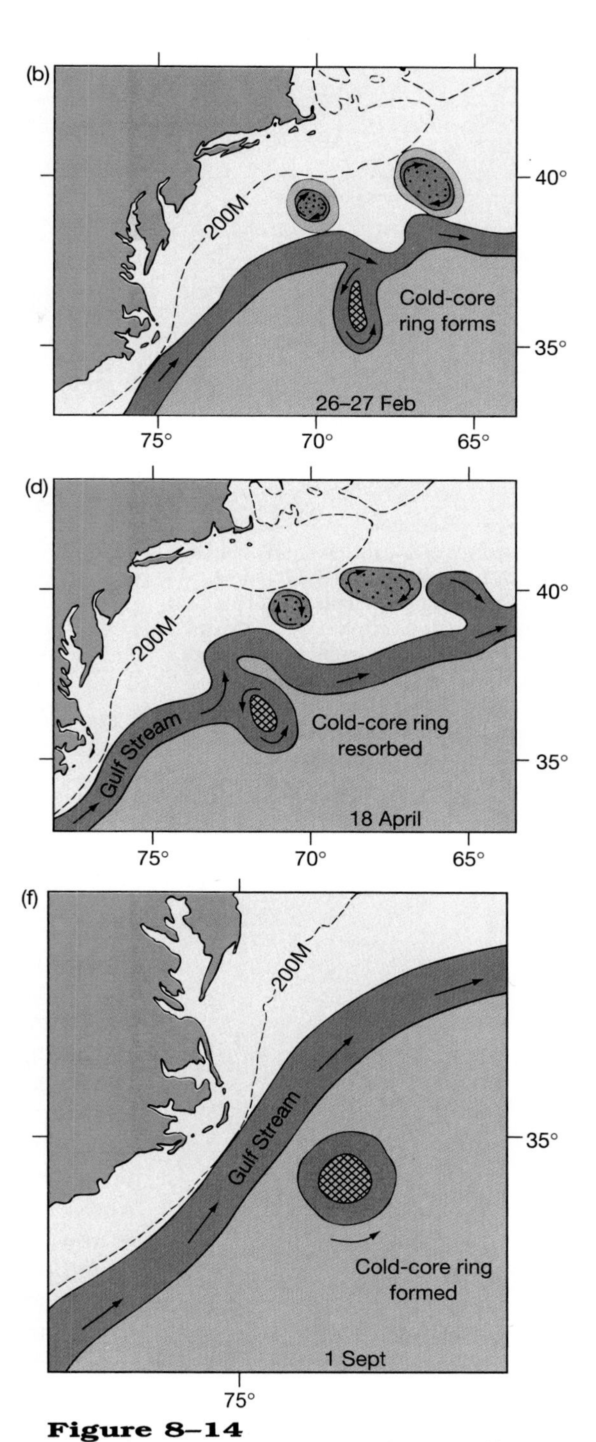

Figure 8–14
A Gulf Stream meander (a) evolved into a cold-core ring (b) and moved southwestward (c) through the Sargasso Sea. Subsequently it was resorbed (d), again forming a meander (e) before the ring reappeared (f) and was finally resorbed by the Gulf Stream (g). [After P. L. Richardson: "Gulf Stream Trajectories," *Journal of Physical Oceanography, 10* (90), 1980.]

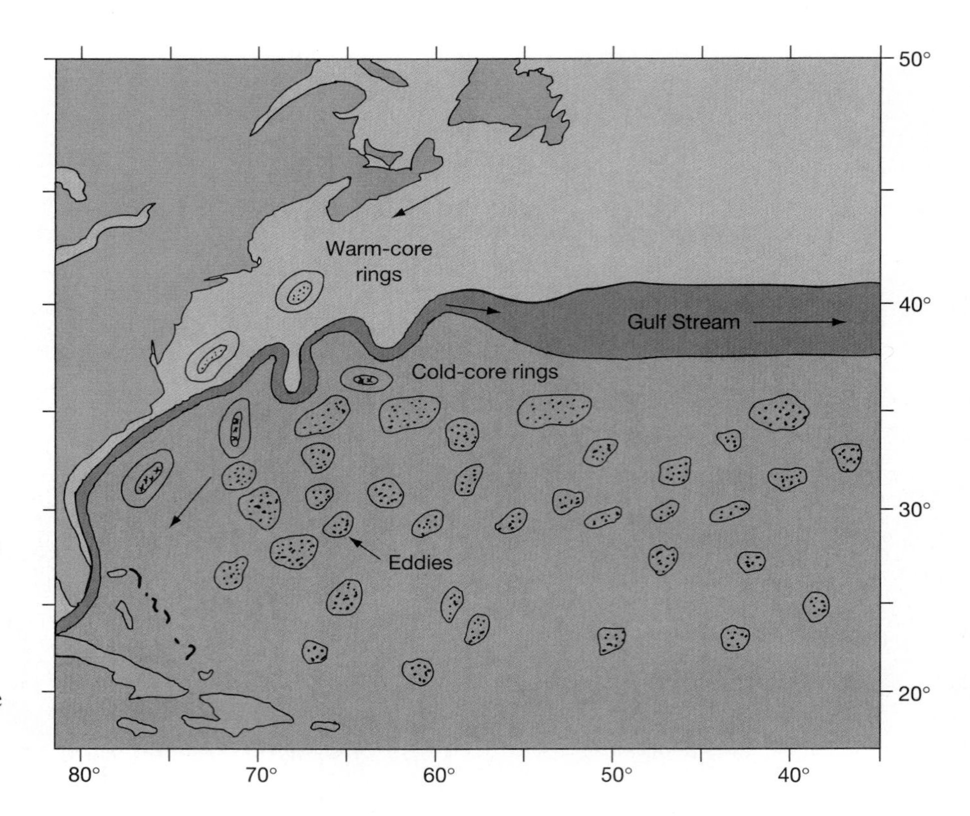

Figure 8–15
Relationships of warm-core and cold-core rings to the Gulf Stream and to eddies. Rings have strong currents and are closely associated with the Gulf Stream. Eddies have weaker currents and are not obviously associated with the Gulf Stream; they cover 15 to 30 percent of the Sargasso Sea, southeast of the Gulf Stream. (Courtesy National Science Foundation.)

Upwelling and Downwelling

Winds cause vertical water movements in addition to driving the horizontal surface currents. To understand vertical water movements, we must again refer to the Ekman spiral (Fig. 8–2), which causes net movement of the surface layer—the combined movement of the many thin layers shown in Figure 8–3—to be 90° to the right of the wind in the Northern Hemisphere (Fig. 8–3). Near a coast, winds blowing parallel to the shoreline cause a layer of surface waters several meters thick to move either away from or toward the coast. In the first case, when surface waters move away from the coast, as in Figure 8–16a, subsurface water flows upward to fill the space vacated by the seaward-moving surface water. This exposes subsurface waters, which then move upward toward the surface and flow seaward. This is called **upwelling.** Where surface waters move toward the coast, as in Figure 8–16b, they cause the surface layer near the land to thicken and depress the pycnocline, a process called **downwelling.** In both cases, Coriolis effects and the sloping sea surface create geostrophic surface currents parallel to the coast, but in different directions.

Upwelling is especially conspicuous on the eastern side of ocean basins, where surface layers are relatively thin. Much upwelling occurs near coasts, usually in areas a few tens of kilometers across, often located near capes or other irregularities in the coastline (Fig. 8–16c). The cells of upwelled water often have plumes of cold waters extending offshore for tens of kilometers (Fig. 8–11). The plumes move with coastal currents.

Subsurface waters upwelling to the surface from below the pycnocline are rich in nutrients and can therefore support abundant plant growth, which in turn can support large populations of marine animals. For this reason, upwelling regions provide about half the world's fish catch.

The equator is another major upwelling area (Fig. 8–17). Consider the trade winds near the equator (see Fig. 6–13 for the location of the winds). North of the

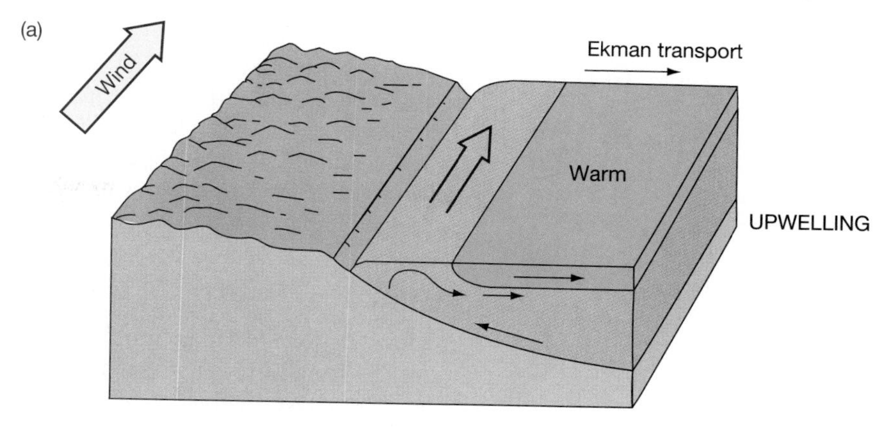

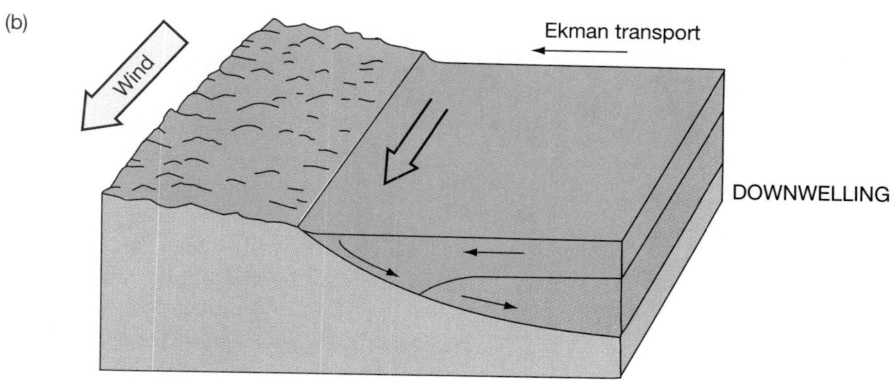

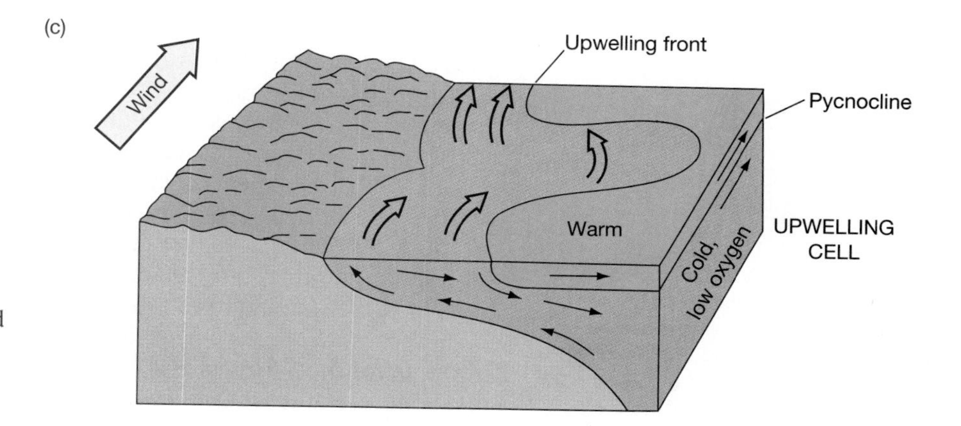

Figure 8–16
(a) Upwelling and (b) downwelling caused by winds blowing along a shoreline. Upwelled waters come from below the pycnocline.

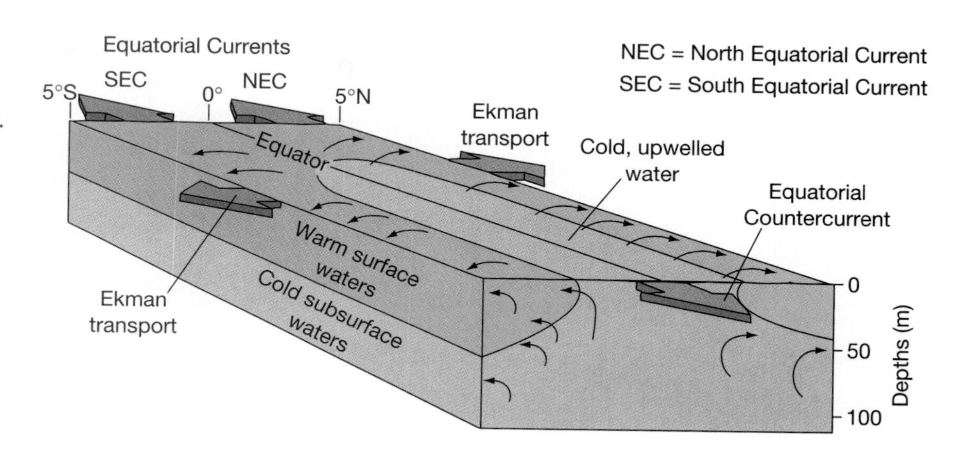

Figure 8–17
Ekman transport of surface waters away from the equator causes a band of cold, upwelled water to occur along the equator.

equator, the Ekman transport is to the right of the winds in the northeast trades (in other words, the current is to the northwest), so surface waters move away from the equator. South of the equator, the Ekman transport associated with the southeast trades is to the left of the winds (toward the southwest) and again away from the equator. Thus the equator is a divergence zone, where subsurface waters are brought to the surface. In Chapter 11 we shall see how this upwelling causes increased biological productivity along the equator.

Langmuir Cells

In addition to the Ekman spiral, winds cause small-scale near-surface currents known as **Langmuir cells** after their discoverer, Irving Langmuir. In Langmuir cells, water moves with screw-like motions, in alternately right- and left-handed helical vortices (Fig. 8–18a). The long axes of these cells generally parallel the wind direction. Langmuir cells tend to be regularly spaced and are often arranged in staggered parallel rows. Because the rotation direction alternates in rows, alternate convergences and divergences form at the surface. Floating debris and foam collect in the convergences (Fig. 8–18b). Between the lines of convergence are lines of divergence, where water moves upward and along the surface of each cell toward the next convergence.

Langmuir cells form when wind speeds exceed a few kilometers per hour. The higher the wind speed, the more vigorous the circulation. Wherever evaporation and cooling of surface waters have increased water density, convection takes place, and in such situations Langmuir cells can form at relatively low wind speeds. Increased stability resulting from surface warming or lowered salinity tends to inhibit cell formation, and so in these stable situations, stronger winds are required for Langmuir cell formation.

Streams of water move downward under the convergences. In large lakes, where Langmuir circulation has been extensively studied, downwelling streams have been observed to extend 7 meters below the surface. Their speeds have been measured at about 4 centimeters per second. In the adjacent divergences, upwelling waters moved at speeds of about 1.5 centimeters per second.

The vertical dimension of Langmuir cells depends on wind speed and on the vertical density structure of the upper layers. If the mixed layer is deep, there is relatively little obstacle to how deep a cell can extend. If the mixed layer is shallow, however, the downwelling waters may not be able to penetrate the pycnocline, and as a result the Langmuir cell cannot extend very far into the water.

Vertical water movements in Langmuir cells may partially control the depth of the pycnocline. This is also an important mechanism for transporting heat, momentum, and substances from the surface to layers a few meters deep.

Figure 8–18
(a) Divergence and convergence zones created by Langmuir circulation set up by a strong, steady wind. (b) Foam lines caused by the presence of Langmuir cells are seen near an oil drilling platform in the Gulf of Mexico. (Lowell Georgia/Photo Researchers, Inc.)

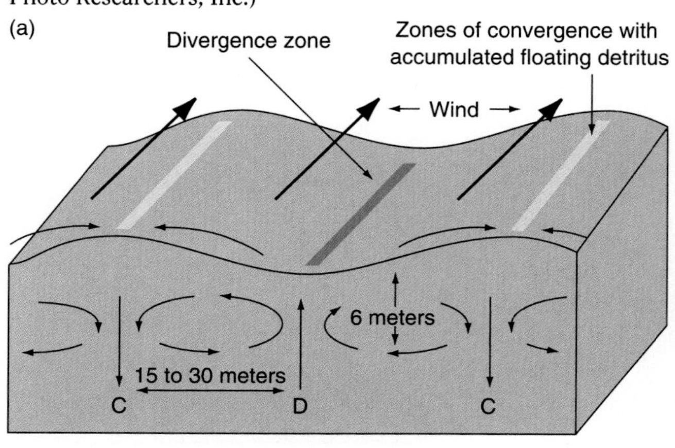

Langmuir cells are an example of short-term responses (occurring over minutes to hours) of the ocean surface to winds blowing across it. Wind waves—considered in the next chapter—are another short-term ocean response. In contrast, Ekman transport is an example of a response to winds occurring over hours to a few days.

Thermohaline Circulation

Below the pycnocline, currents are slow and only their general patterns of movement, directions, and rates are known. The patterns are mapped from density distributions in the deep waters. (The same indirect technique is used for subsurface currents as for surface currents.)

Over most of the ocean, subsurface flows differ markedly from those at the surface. East–west currents dominate the surface ocean. North–south flows dominate the deep ocean circulation. Unlike the surface currents that are driven by the winds, the deep ocean currents are driven by intense cooling of waters in the polar regions. These waters gradually warm as they slowly move through the ocean basins and eventually return to the surface many hundreds to a thousand years later. Return flows are poorly known, but appear to occur primarily in upwelling areas. Let's examine the processes involved.

Vertical water movements occur primarily in high latitudes, where dense water masses form and sink. These vertical movements control the temperatures and salinities of deep waters throughout the ocean. They also drive currents along the bottom and in the mid-depths. These density-driven currents are the **thermohaline circulation** (Fig. 8–19), so called because temperature and salinity control the differences in seawater density that causes the currents.

A sinking water mass reaching the appropriate density level spreads out horizontally, forming a thin layer. It takes less energy to move a parcel of water along a surface of constant density than to move it vertically through an area of changing density. Therefore most subsurface flow is horizontal, because of the lower energy required to carry out the movement.

Thermohaline circulation transports cold waters from polar regions toward lower latitudes. The deep waters in these currents slowly return to the surface (after hundreds of years). Some move up through the pycnocline throughout the ocean. Others come to the surface through upwelling along the equator and in coastal regions.

The deep currents are especially strong along the western sides of the Atlantic. While these subsurface currents are slower than the surface western boundary currents they are far faster than most other subsurface water movements.

Bottom-Water Formation

The densest water masses form in only a few locations in the polar oceans. These polar areas are partially isolated from the rest of the world ocean and subjected to

Figure 8–19
Thermohaline circulation in the deep Atlantic Ocean.

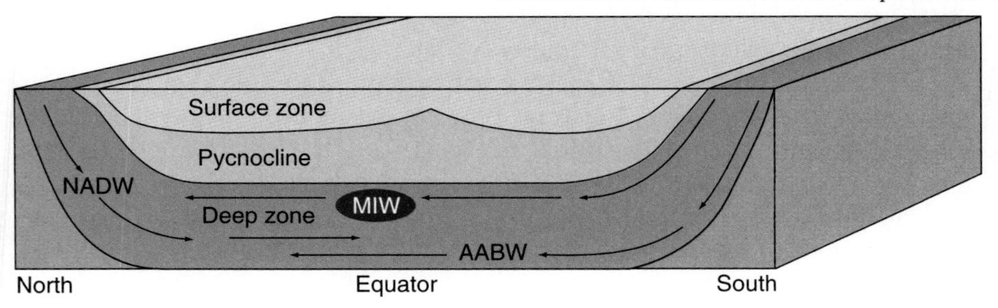

AABW = Antarctic Bottom Water
NADW = North Atlantic Deep Water
MIW = Mediterranean Intermediate Water

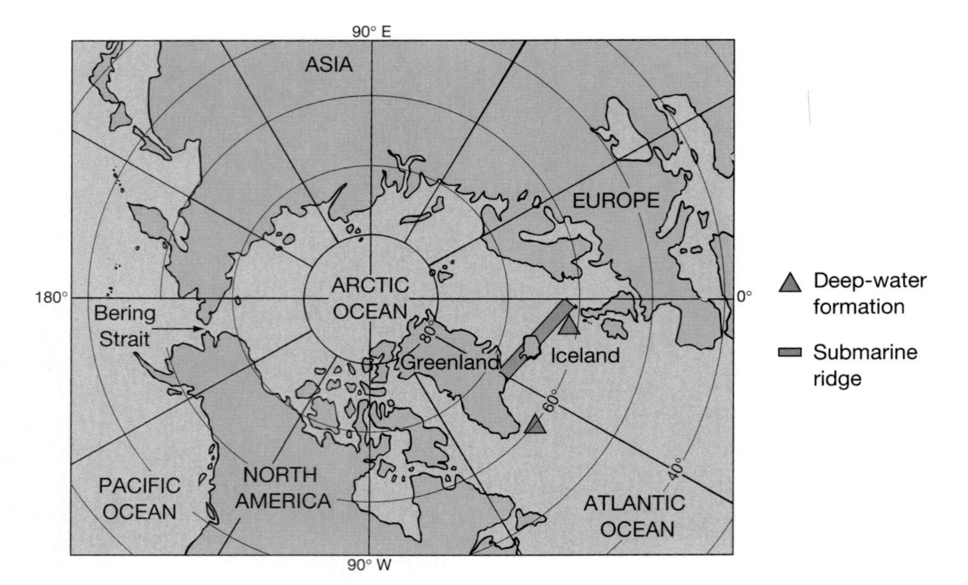

Figure 8–20
Deep-water masses form near Greenland and Norway in winter, when salty surface waters are intensely chilled, thereby increasing their density. The submarine ridge between Greenland, Iceland, and the Faeroe Islands (north of Great Britain) prevents dense waters from the Arctic from flowing into the North Atlantic.

intense cooling of high-salinity waters or the freezing of sea ice. When conditions are right, large volumes of bottom waters apparently are formed from seawaters having high salinities, more than 35.

Because of the outflow of warm saline waters from the Mediterranean, the waters of the North Atlantic are the saltiest of the major ocean basins. Salty surface water (S>35) is carried into high latitudes by the Gulf Stream. Near Greenland, it is intensely cooled so that its density increases. Evaporation at the water surface also intensifies this process. When this water, originally from the Mediterranean, becomes dense enough it sinks and flows south as a mass along the bottom of the North Atlantic basin, especially along the western side. Such high density waters apparently form intermittently in very cold winters, and they flow south in the deep North Atlantic.

Submarine ridges between Greenland and Scotland (Fig. 8–20)—the region forming the entrance to the Arctic basin—prevent bottom waters formed in the Arctic Ocean from entering the main part of the Atlantic basin. On the other side of North America, the Bering Sill effectively isolates the Arctic from the Pacific Ocean (Fig. 8–21); so deep, cold water masses enter the North Pacific only from Antarctica (Fig. 8–22).

Large quantities of *Antarctic Bottom Water* form in the Weddell Sea, a partially isolated embayment in Antarctica (Fig. 8–22). There, surface waters are chilled to –1.9°C. At this temperature, with a salinity of 34.62, the water sinks to the bottom where it forms the densest water mass in the open ocean. In the process of sinking, this water mixes with other waters and is warmed to 0.9°C.

After circulating around Antarctica and mixing with other water masses, cold, dense Antarctic Bottom Water moves northward into the deeper parts of all three major ocean basins, as far as 45°N in the North Atlantic and as far as the Aleutian Islands (50°N) in the Pacific. In both basins, these waters have moved more than 10,000 kilometers. Antarctic Bottom Water, being denser, flows under the southward-moving North Atlantic Deep Water.

Deep-water movements through ocean basins are affected by ocean-bottom topography. The Romanche Fracture Zone, for example, provides a path for Antarctic Bottom Water to flow into the deep basins of the eastern South Atlantic; direct entry of this water mass into the eastern side of the South Atlantic is blocked, however, by the Mid-Atlantic and Walvis ridges on the sea floor. The various ridges which obstruct flows along the bottom can be seen in Figures 8–22, 2–16, and 2–23.

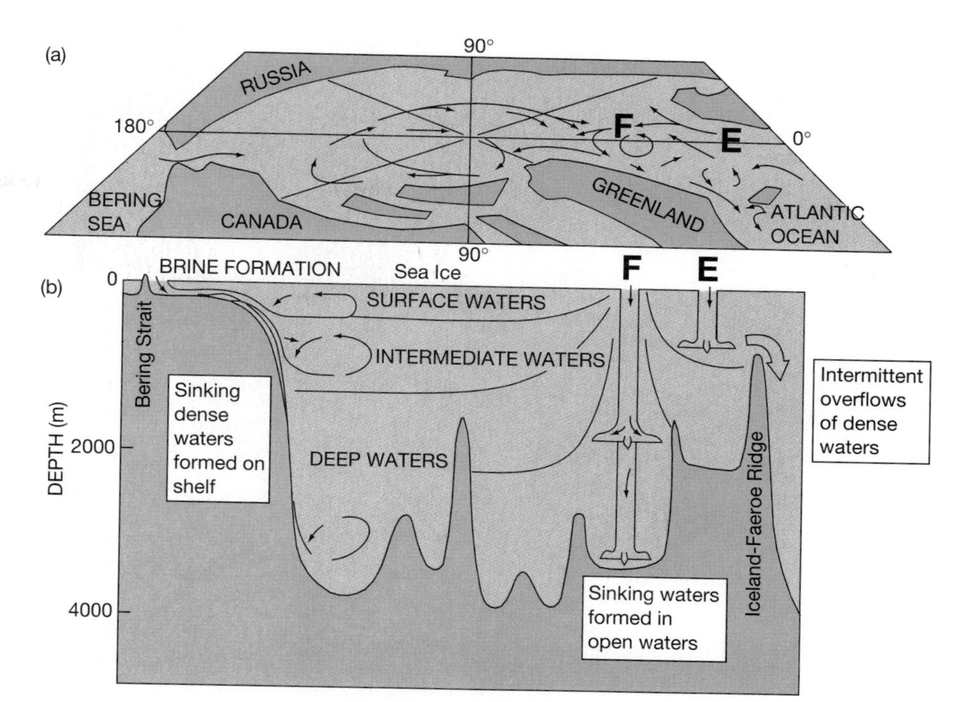

Figure 8–21
(a) Schematic surface circulation and (b) deep ocean structure in the Arctic basin and adjacent seas. Relatively warm waters carried northward (E and F) are cooled and sink to spread at intermediate depths. Some dense waters flow over the Iceland–Faeroe Ridge into the North Atlantic. [After A. Aagaard, J. P. H. Swift, E. C. Carmack, 1985. "Thermohaline circulation of the Arctic Mediterranean Seas," *Journal of Geophysical Research, 90*(C5), 4833–4846.]

Near-bottom currents move much more slowly than surface currents. Speeds of 1 to 2 centimeters per second are typical—except along the western basin margins, where speeds of 10 centimeters per second occur. This is another manifestation of the strong boundary currents along the western side of ocean basins.

Tracking Deep-Water Movements

In addition to temperature (Fig. 8–22) and salinity, various other properties and substances (Fig. 8–23), called **tracers,** are used to map movements of deep waters. Changes in water temperatures (Fig. 8–22) are used to trace the general directions of bottom water movements but do not indicate the amount of time involved. Other tracers, such as changes in concentrations of dissolved oxygen give only a general sense of the speeds of water movements. The presence and amount of radioactive substances coming from atmospheric testing of nuclear weapons are also used to map water movements and to determine current speeds.

Tritium, a radioactive hydrogen isotope, is one such tracer. It is especially useful because we can detect minute quantities of it in seawater. Produced in enormous quantities during above-ground hydrogen bomb tests in the 1960s, tritium immediately reacted with oxygen to form radioactive water, which then fell as rain or snow and eventually moved into the ocean. A small amount of tritium occurs naturally in the ocean because of cosmic rays bombarding the upper atmosphere, but bomb-produced tritium is now far more abundant in ocean surface waters and is used to trace subsurface water movements.

Tritium distributions in the deep ocean illustrate how oceanic processes transport substances entering the ocean through its surface. Tritium from nuclear bomb tests was incorporated in river waters. When these waters flowed into the ocean, they initially stayed in the surface zone (Fig. 8–23). In the high latitudes, newly formed water masses incorporated some radioactive water when they sank. In the western Atlantic, such water masses formed, sank to the bottom, and flowed southward as the North Atlantic Deep Water. The high tritium values around 50°N (Fig. 8–23) came from outflows of Arctic Ocean surface waters, which received especially large amounts of tritium from Russian bomb tests.

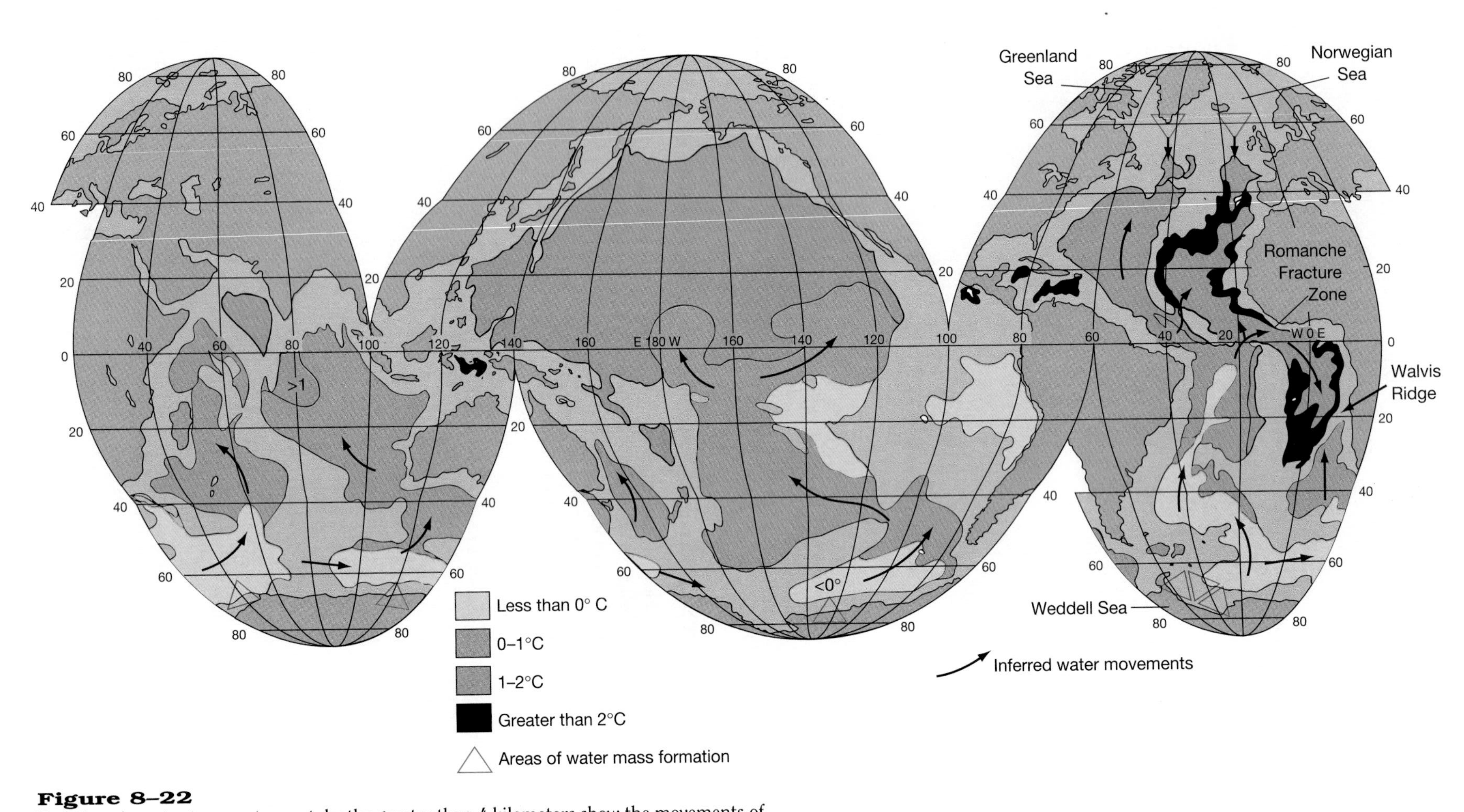

Figure 8–22
Variations in water temperatures at depths greater than 4 kilometers show the movements of bottom waters. [After G. Wüst: Die Stratosphere. Deutsche Atlantische Exped. *Meteor*, 1925–1927. *Wiss.–Erg.* 6(1):288 (1935).]

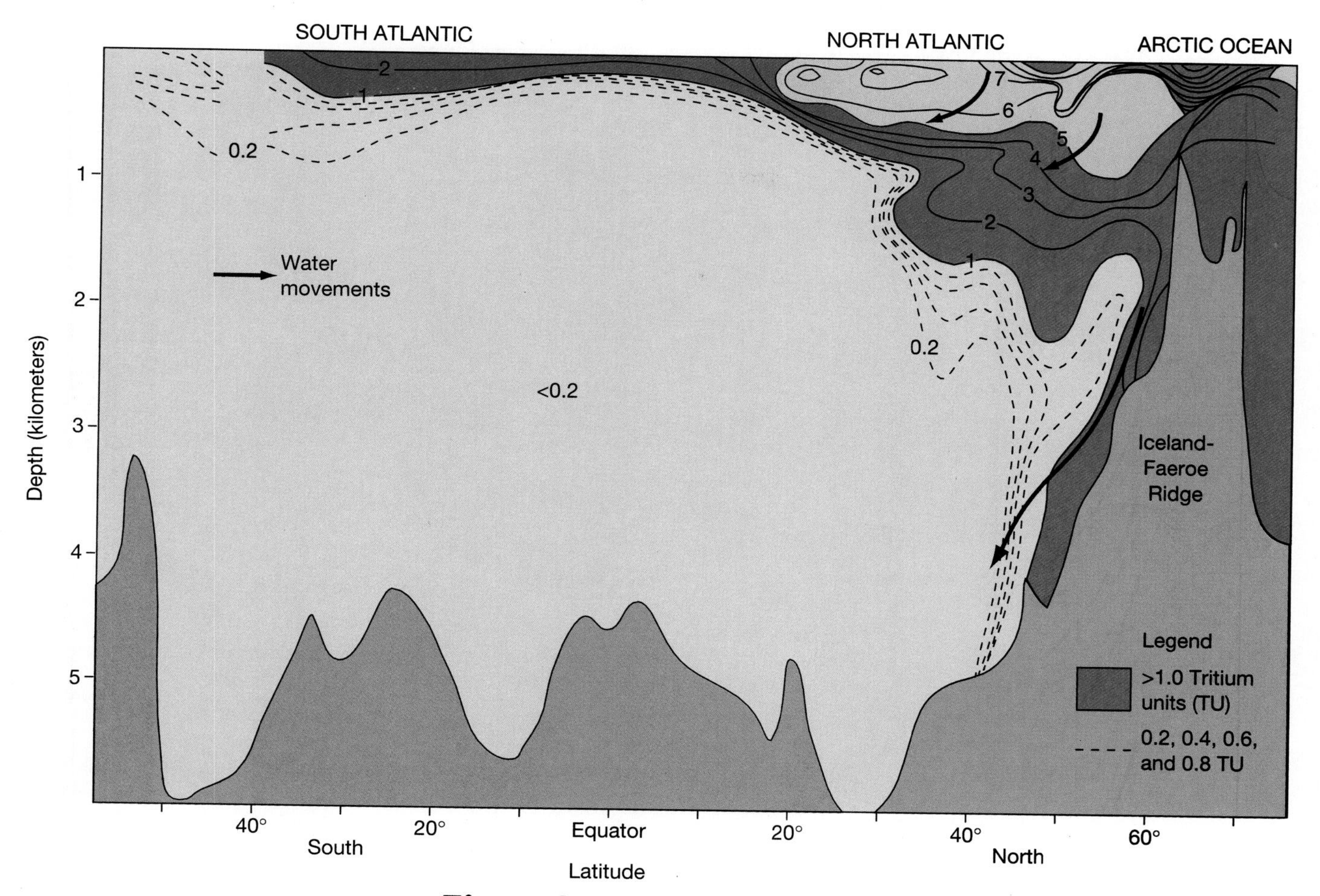

Figure 8–23
Tritium distributions in the western Atlantic, 1972–1973. Note the high concentrations in surface waters around 60°N and the penetration of tritium into the intermediate and bottom waters of the North Atlantic. No tritium had then penetrated into the depths of the South Atlantic. (Data from G. Ostlund, Rosenstiel School of Marine and Atmospheric Sciences, University of Miami.)

Tracer studies also provide information about the ocean's role in taking up the carbon dioxide released by the burning of fossil fuels. About half the carbon dioxide released to the atmosphere since the 1850s has apparently gone into the ocean. The remainder either has stayed in the atmosphere or is stored in vegetation on land.

Carbon dioxide injected into deep waters does not return to the atmosphere for hundreds of years. Thus, the deep ocean provides long-term storage, thereby reducing the effects of carbon dioxide in the atmosphere. Any carbon dioxide that remains in the surface waters is able to exchange freely with the atmosphere, however, and therefore provides no long-term storage. Thus, it is important to know where carbon dioxide is stored in order to predict the long-term effects of fossil-fuel burning.

The Great Global Conveyer

The oceanic thermohaline circulation has been portrayed as a **global conveyer** (Fig. 8–24) by W. S. Broecker and his colleagues at Columbia University. His model explains the major features of thermohaline circulation we have discussed. The basic feature is that the deep ocean circulation is driven by the formation of highly

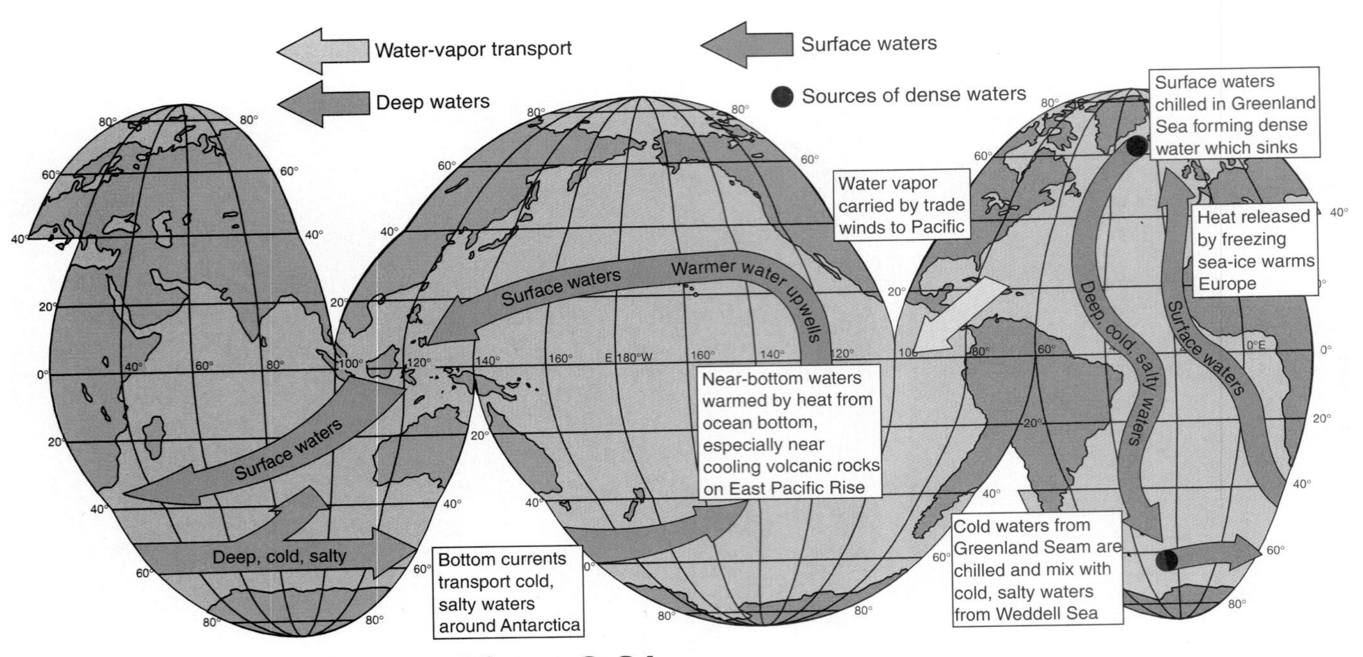

Figure 8–24
The global conveyer system shows how deep-water formation, deep currents, and surface currents act together to warm northern Europe's climate.

saline deep waters in the North Atlantic. Evaporation at the surface removes fresh water from the ocean, leaving more saline water behind. The resulting water vapor is eventually transported by winds away from the Atlantic Ocean and its drainage basins into the Pacific Ocean.

The conveyer begins in the North Atlantic near Iceland and Greenland. There, surface waters are chilled by cold, dry winds from the Canadian Arctic. Seawater is also evaporated, leaving behind the salt. Both processes make the surface waters denser, so they sink. These waters form the North Atlantic Deep Waters (NADW), which then flow southward through the Atlantic basin. The heat released by the freezing of sea ice helps to keep northern Europe warmer than would seem normal for its high-latitude position.

Once they reach Antarctica, the North Atlantic waters flow eastward around the continent, carried by the deep currents there, and get mixed in the process (1 part NADW, 2 parts Antarctic water). By the time it is halfway around the continent, the NADW has lost its identity. From there, the deep waters flow northward into the Indian and Pacific oceans.

The deep waters gradually warm and mix with other waters. Some return to the surface in the massive upwelling around Antarctica. The surface waters return to the Atlantic, but the details of the paths are still unclear. Some water returns by flowing through the islands of Indonesia. From there it flows in the Agulhas Current around Cape Good Hope at the southern tip of Africa to return to the Atlantic. Eventually, the waters reach the area off Greenland, where they begin the cycle again.

Interest in the conveyer-belt hypothesis comes from the fact that conveyer-type circulation may exist at some times and not at others. In other words, the ocean's thermohaline circulation can exist in two states. When the conveyer is operating, as it is today, a vigorous thermohaline circulation in the Atlantic warms northern Europe. When the conveyer is not operating, there is a marked difference in salinity between Atlantic surface and subsurface waters, just as we find today in the North Pacific. Under nonconveyer conditions, no deep waters would form in the North Atlantic, and the North Atlantic and the North Pacific would be more similar than at present. Furthermore, the climate of northern Europe would be

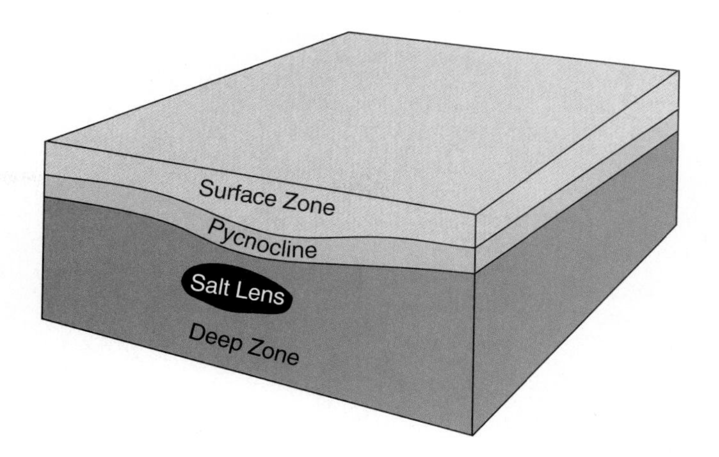

Figure 8–25
A salt lens is a relatively thin water mass in the deep ocean, having a distinctive temperature and salinity. Currents around these lenses apparently isolate them from the surrounding waters.

much colder than it is at present. There is even speculation that past prolonged cold periods in northern Europe, such as the Little Ice Age (1650–1850), may have been caused by weakening or cessation of the conveyer circulation.

Salt Lenses

Oceanographers usually think of subsurface water movements in terms of thin water layers moving between adjacent water layers—like cards inserted into a stack of other cards. We now know that subsurface waters can also move in distinct units called **salt lenses** (Fig. 8–25). Differences in temperature, salinity, and nutrient concentrations show that water parcels retain the properties characteristic of the region in which they formed. Strong currents (up to 25 centimeters per second) surround the lenses, containing them by inhibiting their mixing with adjacent waters. Thus, subsurface waters move in current-bound lenses, much like the rings and eddies found in surface waters, but apparently smaller.

Lenses of relatively warm saline waters have been found in the North Atlantic. They form from the waters flowing out of the Mediterranean and persist for several years, moving with the subsurface currents. This suggests that our view of subsurface water movements must be revised, just as our ideas about surface currents were altered by the discovery of rings and eddies.

Instruments on Earth-orbiting satellites can map only the ocean surface. Thus we must use other techniques to detect lenses. One technique is *acoustic tomography,* in which sound pulses traveling through subsurface waters are timed. Waters of different temperature and salinity can be detected by changes in speed of sound pulses passing through them.

Ancient Current Patterns

Surface currents are driven by prevailing winds, and their patterns are modified by each ocean basin's shape. As we have seen, prevailing planetary winds are primarily controlled by the heating of Earth's surface in the low latitudes and its cooling near the poles (Chapter 6). Thus, prevailing wind patterns change little over time, despite the changing positions of continents and ocean basins. This relatively unchanging nature of winds allows us to reconstruct ancient surface currents with some confidence, using our knowledge of the shapes and positions of ocean basins and continents.

Ancient Surface Currents

In narrow basins, winds and wind-driven currents tend to parallel the sides. In the newly-formed Atlantic, for instance, high mountains bordering the narrow basin

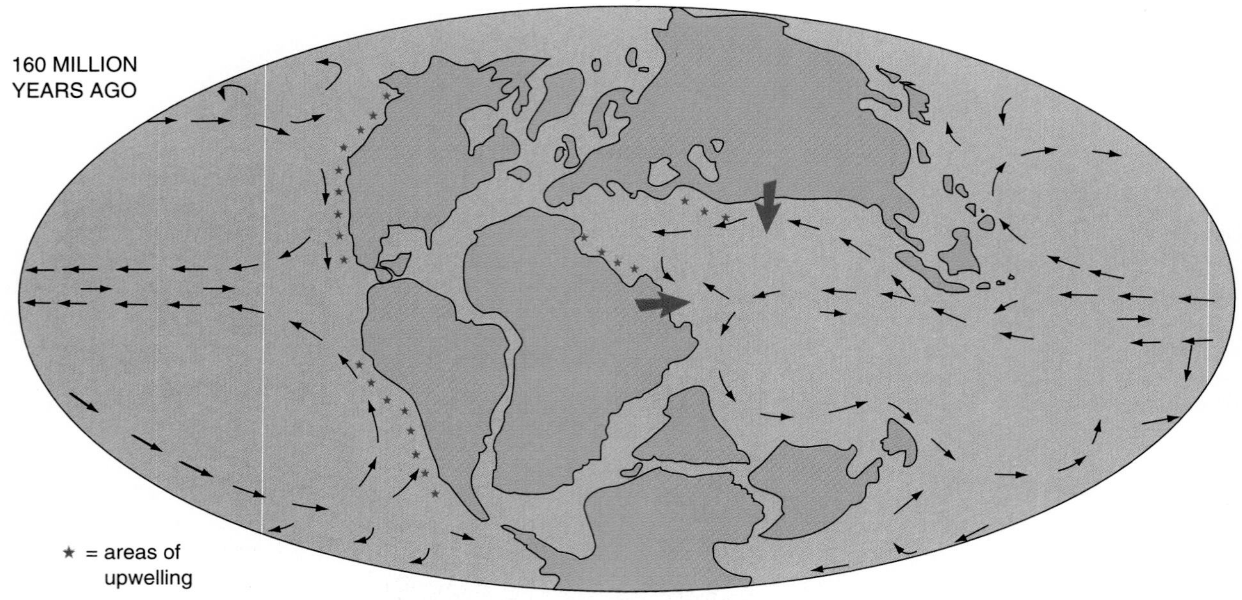

Figure 8–26
Inferred surface currents about 160 million years ago, just after the North Atlantic opened up into a long, narrow basin. The red arrows indicate probable places where warm, saline waters discharged from shallow coastal evaporating basins. [After B. U. Haq: "Paleo-oceanography: a synoptic overview of 200 million years of ocean history." In *Marine Geology and Oceanography of Arabian Sea and Coastal Pakistan,* B. U. Haq and J. D. Milliman (eds.). New York: Van Nostrand Reinhold, 1984.]

restricted prevailing winds to a north–south movement. In such a basin, surface currents would also have paralleled the basin sides (Fig. 8–26) because of the effects of the high valley walls funneling the winds down the narrow basin.

As the basin widened, winds blowing along its north–south axis weakened, and more complicated current systems developed. These would involve the Ekman spiral discussed earlier in this chapter, for example. Eventually, the current patterns resembled those now found in the Atlantic.

Long after the Americas separated from Eurasia and Africa, the connection between North and South America remained submerged (Fig. 8–27). Once the Tethys Seaway opened between Africa and Eurasia, therefore, there existed a globe-circling equatorial current system. This current pattern continued until the ends of the Mediterranean Sea closed as Africa and Eurasia came together about 30 million years ago. Only a few million years ago, the land connecting North and South America emerged and further disrupted these equatorial currents.

A major development in the evolution of the present pattern of surface currents was the formation of the Circum-Antarctic Current (Fig. 8–28). The location of this current system was determined by spreading of the mid-ocean ridge surrounding Antarctica. About 40 million years ago, Australia separated from Antarctica, permitting a very-high-latitude connection between the Pacific and Indian oceans. Then about 30 million years ago, the Drake Passage between South America and Antarctica deepened, permitting the Circum-Antarctic Current to flow around Antarctica. This strong current system isolated Antarctica, probably initiating the formation of the present Antarctic ice cap, marking the onset of the present glacial climate.

Ancient Subsurface Currents

Because locations of bottom-water formation are controlled by details of basin shape and climate much more than surface circulation is, we have great difficulty reconstructing ancient subsurface currents and bottom-water formation sites.

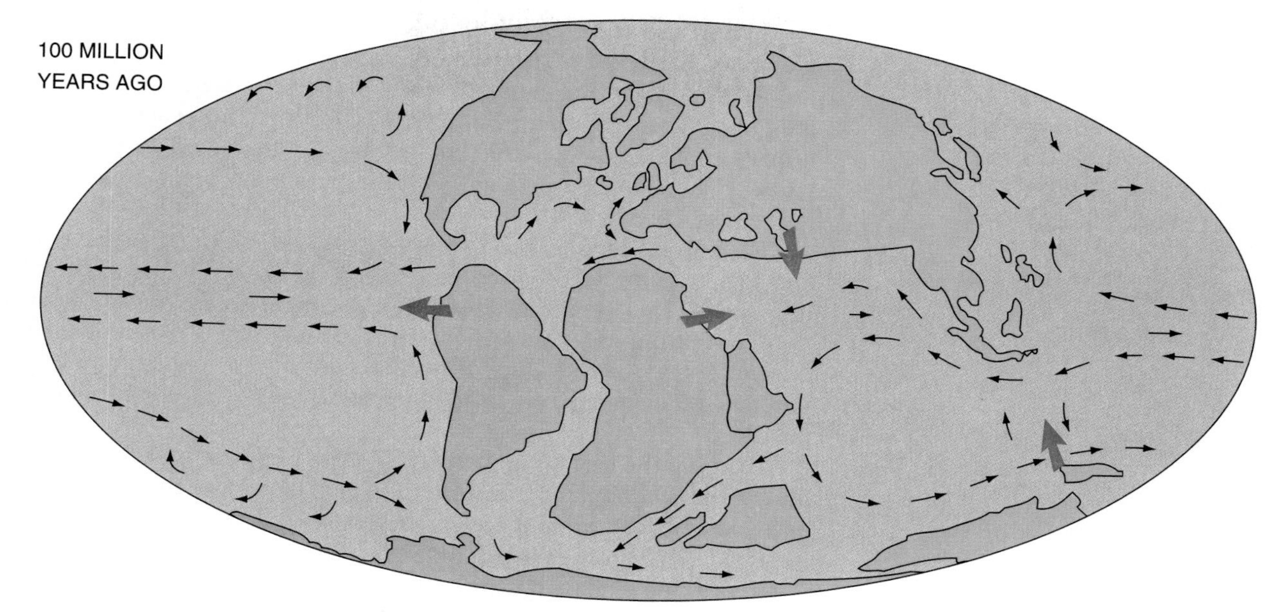

Figure 8–27
Inferred surface currents about 100 million years ago, after the opening of the South Atlantic. The opening of the Strait of Gibraltar and the submergence of Central America permitted currents to flow around the globe along the equator. Warm saline bottom waters still formed in the midlatitudes. [After B. U. Haq: "Paleooceanography: a synoptic overview of 200 million years of ocean history." In *Marine Geology and Oceanography of Arabian Sea and Coastal Pakistan,* B. U. Haq and J. D. Milliman (eds.). New York: Van Nostrand Reinhold, 1984.]

Figure 8–28
Inferred surface currents about 30 million years ago. Deepening of the Drake Passage between South America and Antarctica and submergence of the Tasman Ridge south of Australia permit the Circum-Antarctic Current to flow around Antarctica. Cold bottom waters (red arrows) formed around Antarctica and flowed north into all the ocean basins. [After B. U. Haq: "Paleooceanography: a synoptic overview of 200 million years of ocean history." In *Marine Geology and Oceanography of Arabian Sea and Coastal Pakistan,* B. U. Haq and J. D. Milliman (eds.). New York: Van Nostrand Reinhold, 1984.]

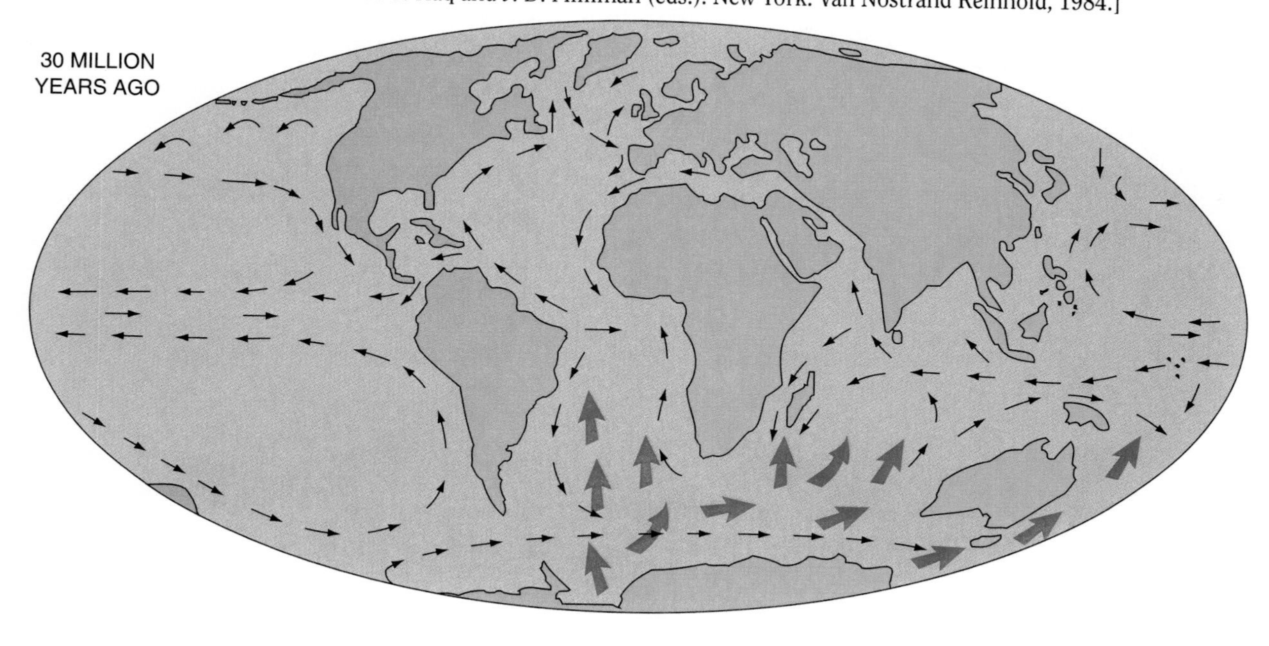

When the present cycle of mid-ocean ridge spreading began (200 million years ago), Earth had a warmer, more equable climate than at present. While tropical temperatures were similar to today's, the poles were much warmer than at present, and, as a result, little or no sea ice formed. Thus, conditions did not favor formation of cold, dense water masses because there was no chilling of surface waters and no increase in salinity resulting from salts being excluded from newly formed sea ice.

Any dense bottom waters formed during these times probably were warm and highly saline, originating in evaporating basins in the arid mid-latitudes (Fig. 8–29). The present Mediterranean and Red seas are probably similar to the areas of bottom-water formation found on the planet 200 million years ago. Ancient bottom waters came from the sides of the mid-latitude basins. These waters were warm and contained much less dissolved oxygen than bottom waters now do.

During the past 50 million years, Earth's climate has cooled. At the beginning of the cooling trend, ocean bottom waters (probably the coldest waters in the open ocean) were around 13°C and have cooled to their present temperature (around 0°C). After the Circum-Antarctic Current isolated the shallow seas around Antarctica, the climate cooled enough to permit formation of sea ice. Antarctic Bottom Water could then form in the shallow, partially isolated embayments around Antarctica such as the Weddell Sea. From there it could flow northward into all the deep-ocean basins (Fig. 8–28).

Submergence of the Iceland–Faeroe Ridge between Greenland and Scotland, which took place about 40 million years ago, permitted subsurface Arctic waters to flow south into the North Atlantic. Thus, for the past 40 million years, any dense waters formed in this region during cold winters could flow southward in the Atlantic as North Atlantic Deep Water. This marked the development of the three-layered ocean in which the coldest waters are at the bottom (Fig. 8–30).

Figure 8–29
Sources of bottom water at present (glacial climate) (a) and during times of warm climate (b) when warm saline bottom waters formed in the mid-latitudes.

(a) Present

Deep water masses form in polar regions

N

Chilling high-salinity waters

Cold (3°) Intermediate salinity (34.7)

High dissolved oxygen
Strong subsurface
Western boundary currents

Chilling surface waters

Sea ice formation

Equator

Antarctica

S

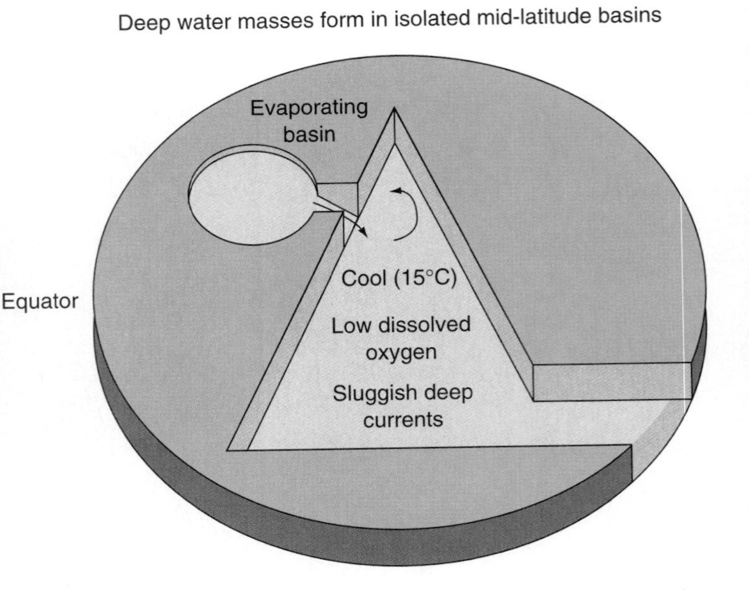

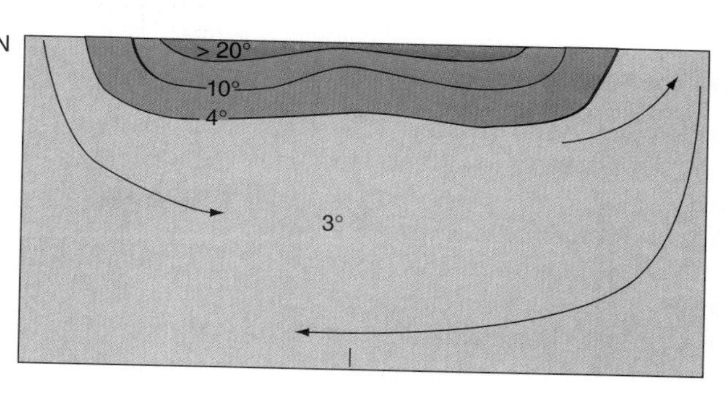

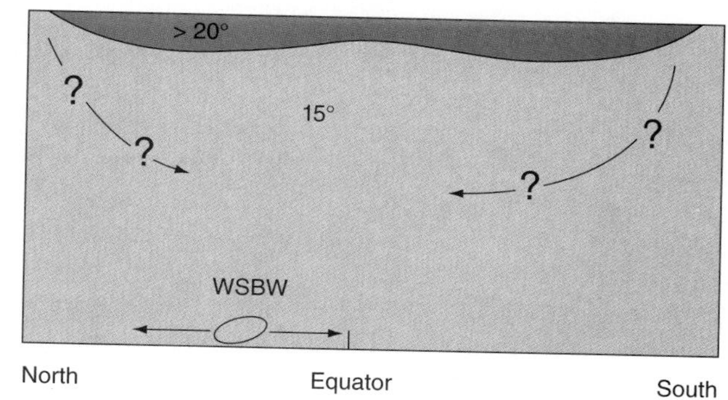

Figure 8–30
Comparison of present subsurface current patterns (a) with those involving warm saline water masses (b).

Ocean Currents from Space

Observers on the space shuttles and high-flying aircraft can often see ocean features that are impossible for scientists on ships to study. Some of these, such as tides, are so large that they cannot be seen well enough at the surface to work out the processes involved. Others are so transitory that they can rarely be observed while they are active. Indeed, our understanding of currents is still dominated by the averages of observations made from ships over many years and often separated by many thousands of kilometers. We often still think of currents as "rivers in the sea," the way the Gulf Stream was first described by Benjamin Franklin. Space-based observations are providing us with a new view of oceanic processes such as tides.

The availability of observations from space is substantially changing our view of currents and their effects. A striking current pattern was observed in the equatorial Pacific Ocean, where a sharp line on the ocean surface has been observed to stretch nearly 7,000 kilometers, from the Galapagos Islands near South America to an area south of Hawaii. This line, only a few hundred meters to a few kilometers wide, is formed by the abundant growth of tiny floating plants where the cool waters of the north meet and mix with the warm waters of the equatorial ocean. The extreme abundance of plants growing in this narrow zone is due to the presence of nutrient substances in the cooler waters. (We further discuss the processes involved in Chapter 11.) The result is a surface layer so rich with life that it has been likened to a soup.

This feature has actually been observed by ships passing through the region. Not knowing what it was, sailors reported that there were shallow reefs with extremely abundant life. When other ships tried to locate these features—which are transient—they could not find any areas of shallow water. Thus, for years charts of these regions carried notations that some ships had encountered shallow areas when, in fact, they had simply passed through one of these transient current features. New observations from spacecraft show these features in their entirety, so we can begin to understand such transient current features.

Summary

Currents—large-scale horizontal water movements—occur everywhere in the ocean. They are driven primarily by winds and, like winds, result from Earth's being heated near the equator and cooled near the poles.

Open-ocean surface-current patterns are similar in all oceans. All form gyres, each gyre being a nearly closed set of currents. Continents deflect east–west currents. Variable currents occur wherever prevailing winds shift seasonally, as in the northern Indian Ocean.

Winds blowing across a water surface set the upper water layer in motion. When times and distances are short, waters move in the same direction as the wind. When wind-driven water movements continue over longer times and distances, currents are affected by Earth's rotation. In such water movements, as depth increases there is a steady decrease in current speed and a change in direction of flow—a current pattern called an Ekman spiral. The reduction in speed results from loss in momentum as energy is transferred from a layer to the one below it.

Convergences of surface waters resulting from prevailing winds cause water to accumulate in certain ocean areas, forming slight elevations of the ocean surface. Water responds to this topography by flowing downhill, but these movements are deflected by the Coriolis effect. Eventually the water flows around the hills, so that Coriolis deflection is balanced by the pull of gravity. Because these elevations are so small, forces acting on currents are essentially horizontal. Currents in which downslope (gravity) flows are balanced by the Coriolis effect are called geostrophic currents.

Where surface currents flow away from a region, the water surface is depressed. Such areas are called divergences. Upwelling is often associated with divergences.

Western boundary currents are the strongest currents in the ocean. They separate coastal ocean waters from the open ocean. Most western boundary currents are so deep that they must flow along the continental margins. Eastern boundary currents are much weaker and shallower. Western boundary currents are intensified by several factors. Earth's rotation displaces gyres to the west. Trade winds and the westerlies contribute to the accumulation of water in the subtropical gyre. The apparent spin of western boundary currents to Earth's rotation is in the same direction as the current flow.

Meanders and rings are especially conspicuous in western boundary currents. These isolated bodies of water are enclosed by strong currents. Rings move with the currents, and some are resorbed back into boundary currents. Other smaller, current-bounded, ring-like structures called eddies occur widely throughout the western halves of ocean basins.

Upwelling and downwelling are vertical water movements caused by winds. The net water movement in the surface layer of an Ekman spiral is 90° to the right of the wind in the Northern Hemisphere and 90° to the left in the Southern Hemisphere. When surface waters are blown away from a coastline, subsurface waters flowing upward replace them; this process is called upwelling. When winds blow surface waters toward the coast, they accumulate there, forming a thicker-than-usual surface layer. This process is called downwelling.

Langmuir cells—organized sets of horizontal corkscrew-like water motions in the surface layer—are caused by winds blowing across the ocean surface.

Thermohaline circulation is caused by differences in water density, differences that drive deep ocean circulation. Bottom waters form when surface waters are chilled and freeze near Antarctica and in the North Atlantic near Greenland. Salt released during evaporation or sea-ice formation increases water density. The dense waters sink to the bottom and flow toward the equator. The return flow occurs primarily in upwelling zones. This system of water movement has been called the great global conveyer, and has been linked to sudden climatic changes, primarily in Western Europe.

Changes in dissolved oxygen concentrations, salinities, water temperatures, and introduced anthropogenic substances are used to trace bottom-water movements.

Some deep-water masses are bounded by strong currents that inhibit mixing with surrounding waters. These salt lenses persist for thousands of kilometers as they move below the surface.

Ancient surface currents were affected by the changing positions of the continents. The equatorial currents, once a globe-encircling system, were later blocked by the emergence of Central America and the closing of the Mediterranean. Ancient subsurface currents involved warm saline bottom waters during times when Earth's climate was warmer.

Key Terms

currents
open-ocean current
gyre
Ekman spiral
convergence
divergence
geostrophic current
boundary current
western boundary current
eastern boundary current
meander
ring
eddy
upwelling
downwelling
Langmuir cells
thermohaline circulation
tracer
global conveyer
salt lenses

Study Questions

1. Draw a diagram showing the principal open-ocean surface currents.
2. What causes the principal open-ocean surface currents?
3. Draw an Ekman spiral. Why do the currents change direction and slow with increasing depth?
4. Contrast eastern and western boundary currents.
5. Explain Ekman transport and its role in upwelling.
6. How are warm-core rings formed by the Gulf Stream?
7. Describe geostrophic currents.
8. What causes intensification of western boundary currents?
9. Describe and contrast upwelling and downwelling.
10. Diagram the formation, movement, and fate of warm-core rings associated with the Gulf Stream.
11. Describe thermohaline circulation.
12. Describe Langmuir circulation. What causes it?
13. Describe salt lenses.
14. Why don't the dense waters formed in the Arctic Ocean flow into the North Atlantic or North Pacific?
15. How does ocean-floor topography affect near-bottom currents?
16. Contrast deep-ocean circulation of today with that of 100 million years ago.
17. Discuss the importance of the formation of the Circum-Antarctic Current. What controlled its formation?
18. *[critical thinking]* Explain why cessation of the global conveyer-belt circulation would be especially important for the climate of Northern Europe.

Selected References

HUYCHE, P., "The Storm Down Below," *Discover,* November 1990.

MCLEISH, W. H., "Painting a Portrait of the Gulf Stream," *Smithsonian 19*(12):42 (1989). Popular account of the causes and effects of the Gulf Stream.

PICKARD, G. L., AND W. J. EMERY, *Descriptive Physical Oceanography,* 5th ed. Pergamon Press, 1990.

WOODS HOLE OCEANOGRAPHIC INSTITUTION, "Physical Oceanography," *Oceanus 33*(2) (1992).

OBJECTIVES

Your objectives as you study this chapter are to understand:

- The basic features of waves and how they form
- The differences between shallow-water waves and deep-water waves
- Wave effects in shallow water and on beaches

Waves, from tiny ripples to giant tidal waves, are complex but universal features of the ocean surface and interior.

9 Waves

Waves—disturbances of water surfaces—can be seen virtually everywhere, on any water surface. Despite the ubiquitousness of waves in water, however, oceanographers still have much to learn about them. Ancient sailors knew that waves are generated by winds, but it was not until the nineteenth century that the first mathematical descriptions of waves were developed.

Water patterns on the ocean surface are complex, continuously changing, and never repeating. Ocean waves come in all sizes and shapes, ranging from tiny ripples formed by light breezes through enormous storm waves, tens of meters high, to the tides (which are also waves, as we see in Chapter 10). Some waves—called rogue waves—are so large that they can sink the largest ships.

Because of their complexity and ever-changing patterns, ocean waves do not lend themselves to accurate description or complete explanation in simple terms. Nevertheless, oceanographers commonly work with simplified explanations and descriptions that help them understand wave phenomena. Most advances in understanding waves have come through the use of appropriate simplifications. Ever more powerful computers and satellite data have also helped.

In this chapter we study the features of waves, how they form, and some of the ways they affect the ocean. The chapter discusses the features of ideal waves, deep-water and shallow-water waves, the processes causing waves, waves in shallow water and on beaches, and energy from waves.

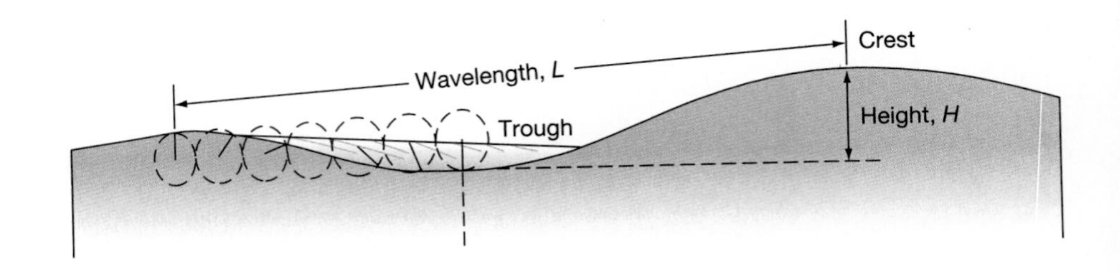

Figure 9–1
Features of a simple wave.

Simple Waves

We can make simple waves in water by steadily bobbing the end of a pencil in a basin of water or a still pond surface. The waves we produce move away from the disturbance that caused them (in this case, the moving pencil), and so we call them **progressive waves.** A sequence of waves moving together in this way is called a **wave train.** Each wave consists of a crest (the highest point of the wave) and a trough (the lowest point of the wave) (Fig. 9–1). The vertical distance between any crest and the succeeding trough is the **wave height *(H)*.** The horizontal distance between successive crests or successive troughs is the **wavelength *(L)*.**

The time it takes for successive crests or troughs (i.e., one wavelength) to pass a fixed point is the **wave period *(T)*.** We can express the same information by counting the number of waves that pass a fixed point in a given length of time. This is the **frequency** (1/***T***), usually expressed in terms of *events per second*. For individual progressive waves, the **wave speed *(C)*** (in meters per second) can be calculated by dividing the wavelength by the wave period in seconds, or $C = L/T$.

Where wave height is low relative to its length, crests and troughs tend to be rounded. Such waves may be approximated mathematically by a **sine wave** (a smooth, regular oscillation, shown in Figure 9–2. As wave height increases, sea waves normally have more sharply pointed crests. (We shall see later how simple waves can be combined to make complicated wave patterns.)

So far we have considered only movements of the water surface—crests and troughs moving together in a wave train. But what happens to the water as waves pass? How is the motion of the water related to the motion of the waves? We can study this relationship by using water-filled tanks with small floats on the surface, or dyed water or oil droplets at various depths below the surface.

When small waves move through deep water, individual bits of water move in vertical, circular orbits that are nearly closed (Figs. 9–3 and 9–4). Any given small parcel of water moves forward as a crest passes, then downward, and finally backward as a trough passes. The orbits are retraced as each subsequent wave passes. After each wave has passed, the water parcel is found nearly in its original position. There is some slight net movement of the water in the direction of wave travel, however, because the water moves forward slightly faster as the crest passes than it moves backward under the trough. This results in a slight forward displacement of the water in the direction of wave motion and perpendicular to the wave crests, as shown in Figure 9–3b. This net water movement is very slight. Indeed, if there were large forward water movements caused by the passage of waves, ships could not withstand the forces exerted by them.

Figure 9–2
Relatively small waves (and large ocean swell) can be described as simple sine waves.

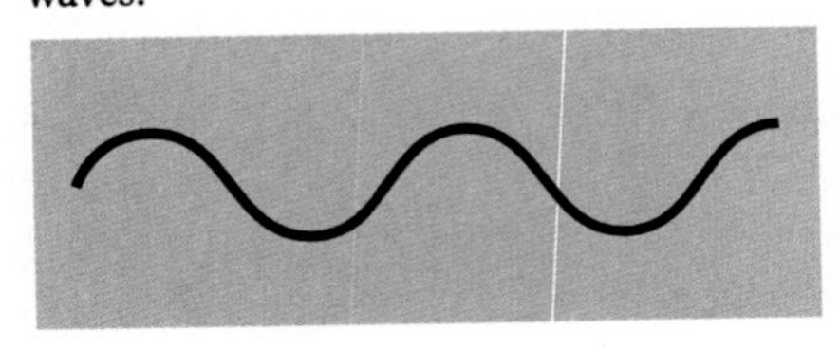

You have probably experienced wave motion. When you float beyond the breakers, you experience only a gentle rocking back and forth as waves pass under you because there is little net water movement. However, you try to stand in very shallow water where waves are breaking, the pounding by breakers convincingly demonstrates large-scale water movements. The water in breakers moves with the waves. Even large ships are damaged when hit by large breaking waves. Consequently, ships are routed to avoid areas of high waves.

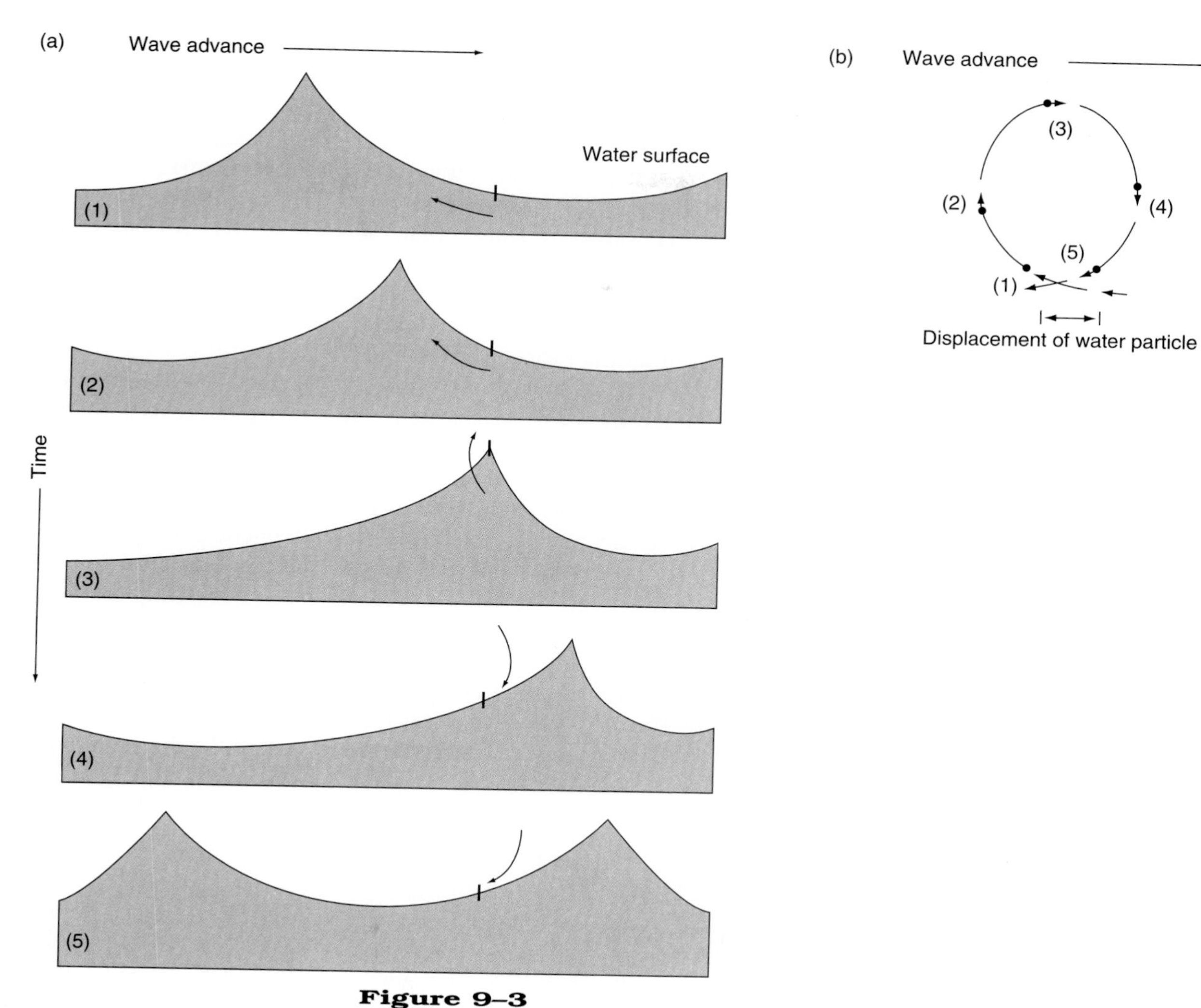

Figure 9–3
(a) Orbital motion and displacement of water particles as a wave passes. (b) Passage of the wave causes only a small horizontal displacement of water particles.

Figure 9–4
Movement of water particles caused by the passage of deep-water waves. The particles tend to move in circular orbits, with orbital diameters becoming smaller with depth. Little orbital motion occurs at depths greater than half the wave length.

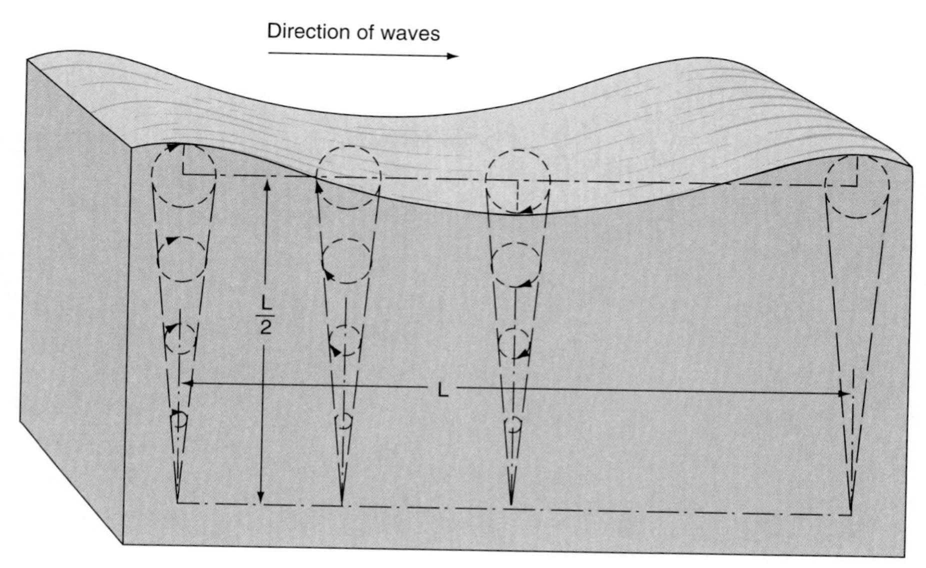

In deep water—water depths greater than half the wavelength ($L/2$)—water parcels move in nearly stationary circular orbits. Such waves, unaffected by the bottom, are called **deep-water waves.** Waves in water depths of less than half a wavelength are called **shallow-water waves.** Movements of water parcels in shallow-water waves are substantially affected by the presence of the bottom. (More about shallow-water waves later.)

The diameter of the orbits of water parcels at the surface approximately equals the wave height; this diameter decreases to one-half the wave height at a depth of $L/9$ and is nearly zero at a depth of $L/2$ (Fig. 9–4). Water at depths greater than $L/2$ is moved very little by passing waves. Thus, a submarine is undisturbed by waves when submerged to depths greater than $L/2$. When the water is much shallower (less than $L/20$) waves are greatly affected by the bottom.

An observed wave profile of a sea is shown in Figure 9–5a. This group of waves has been analyzed to determine which frequencies of waves are occurring. At any given time, many different factors (winds, distant storms, nearby land masses, etc.) influence the patterns of waves observed at the sea surface. The various components (each a simple sine wave) shown in Figure 9–5b combine to cause the complex wave patterns normally seen on the sea surface. Combining these simple waves results in the observed wave profile (a).

Waves usually occur as wave trains—a system of waves of many wavelengths. Because wave speed is related to wavelength by the equation $C = L/T$, as noted earlier, each wave in a train moves at a speed corresponding to its own wavelength. This separates waves by wavelength, a process called **dispersion.**

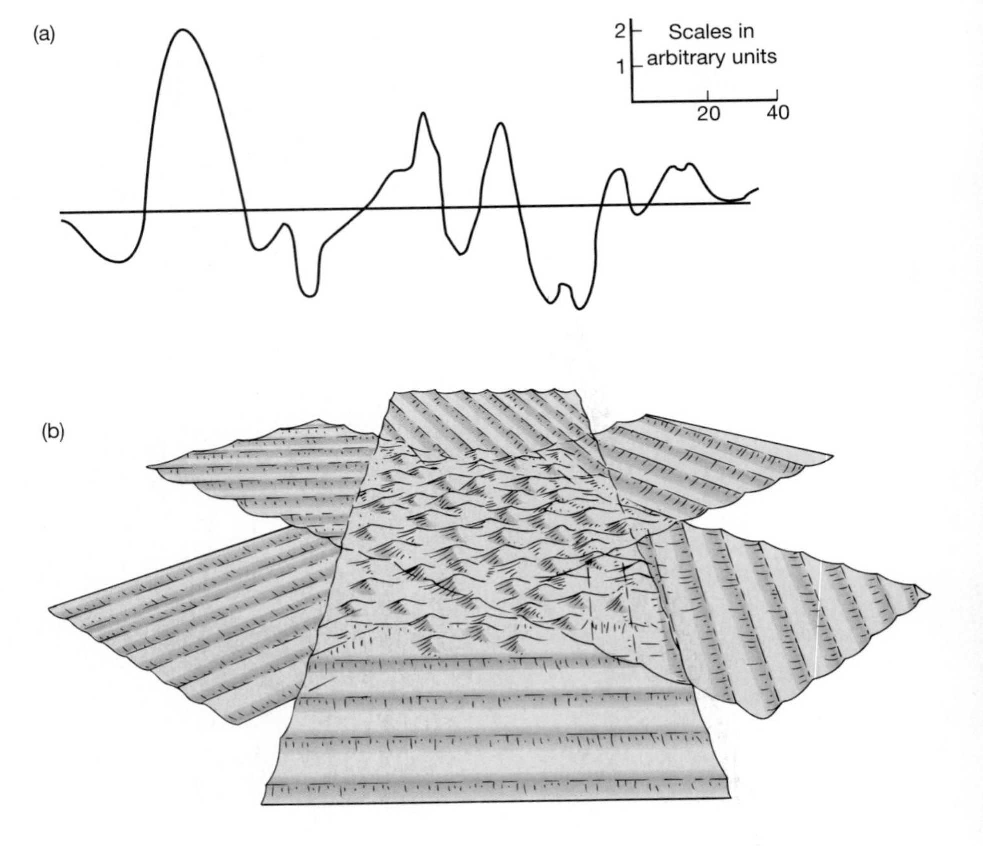

Figure 9–5
(a) An observed profile of a complex wave in a sea. (b) The complex wave of part (a) can be considered as consisting of many different simple waves superimposed on each other.

Forces Causing Ocean-Surface Waves

Wave formation involves two types of forces: those that initially disturb the water (called wave-forming forces) and those that act to restore the equilibrium (still-water) condition (called **restoring forces**) (Fig. 9–6). Let us look at some wave-forming forces: earthquakes or explosions, winds, and the gravitational attraction exerted on the ocean by the Sun and the Moon.

We are all familiar with the waves formed when someone tosses a pebble into a pond. If the water surface is initially still, we observe a group of waves changing continuously as they move outward in a circle away from the point at which the pebble entered the water—another example of wave dispersion. The ocean counterpart of a tossed pebble in a pond is a sudden impulse, such as a volcanic explosion or submarine earthquake. Such impulses cause some of the longest waves in the ocean. If the disturbance affects only a small area, the waves move outward from that area in a circle, much as the waves moved away from our pebble.

Winds cause most ocean waves (note in Fig. 9–6 the large amount of energy associated with wind waves). Because winds are highly variable, wind waves vary greatly, too. (More about wind waves in the next section.)

The third disturbing force—the attraction of the Sun and Moon on ocean water—cause the longest waves of all, the tides, which we discuss in Chapter 10. Because these forces act continuously on ocean water, the tides are not free to move independently the way the waves created by, say, an earthquake or a meteor impact do.

Figure 9–6
Schematic representation of the relative amounts of energy in waves that have different periods and originated from different wave-forming forces. Note that most of the energy in waves comes from winds. (After B. Kinsman, *Wind Waves: Their Generation and Propagation on the Ocean Surface*. Englewood Cliffs, NJ: Prentice Hall, 1965.)

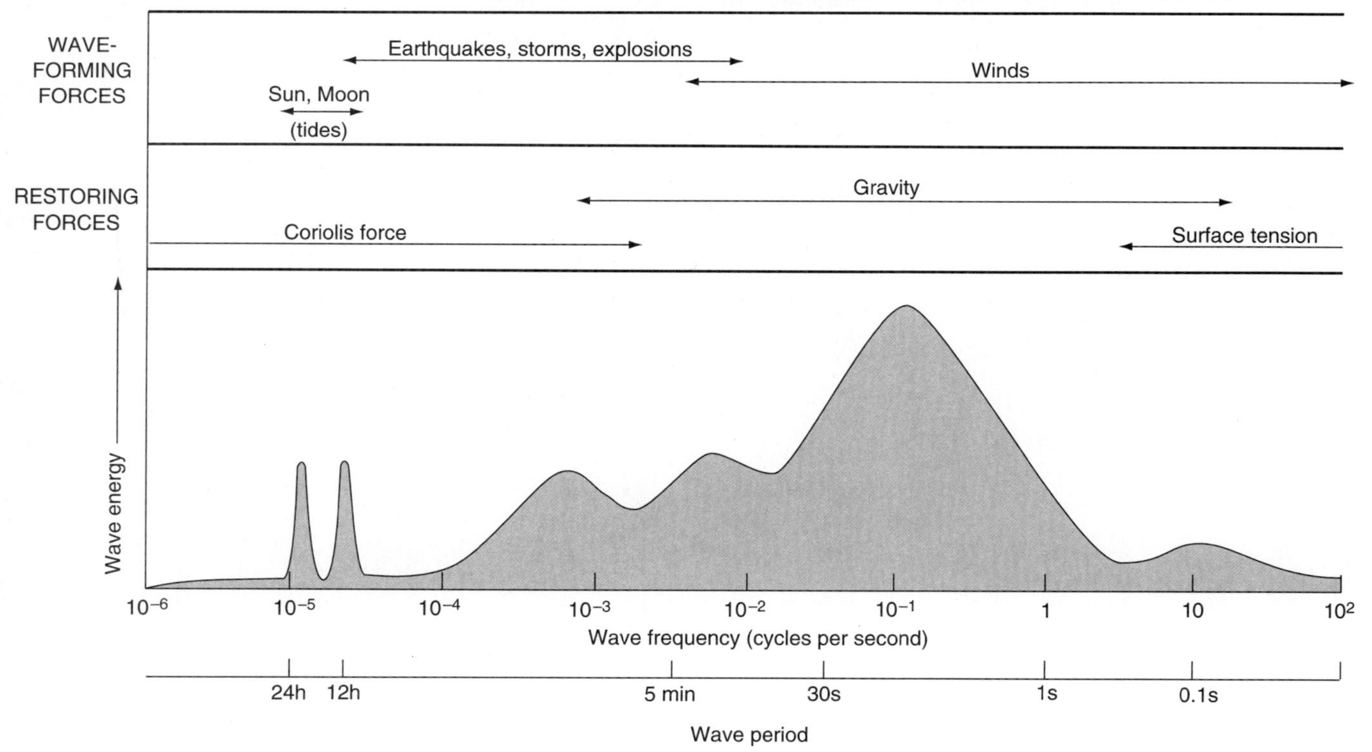

Where the disturbing force is continuously applied, the resulting waves are known as **forced waves**—in contrast to **free waves,** which move independently of the disturbance that caused them. Thus the tides are forced waves, and impulse waves are free waves. Wind waves have characteristics of both free and forced waves.

The smallest waves (wavelength less than 1.7 centimeters, period less than 0.1 second) are called **capillary waves** because they are affected by water surface tension, the same force that causes water to rise in a capillary tube. These waves are formed by light winds blowing across a water surface—the familiar "cat's paws" on the sea surface as a breeze begins to blow lightly after a period of calm.

BOX 9–1
Energy from Waves

Waves are a potential energy source. To give you an idea of the amount of power involved, waves in a 10-second, 2-meter swell approaching a coastline dissipate approximately 400 kilowatts across each 10 meters of wave front. Thus, the energy potentially available on some coasts is enormous. The western coast of Great Britain, to name just one example, is a prime candidate for generating power from waves.

Many designs exist to extract energy from waves. Most involve floating devices that are moved by the waves. Because seawater is so corrosive to metal parts, though, and because the environment along wave-swept coasts is so severe, devices having no moving parts in the water are most desirable.

The simplest scheme for getting energy from waves, shown in Figure B9–1–1, uses a partially enclosed chamber in which the rise and fall of the sea surface pumps air into and out of a chamber through a pipe containing a turbine connected to an electric generator. Such devices would be built as part of a breakwater system to protect a harbor or coastline from wave attack.

Such systems are expensive to build and vulnerable to extreme wave conditions, which can quickly destroy them. Because of the high construction costs and problems with maintenance, these energy-capturing devices are used primarily in remote locations where other energy sources are extremely expensive, such as offshore buoys to power horns and whistles. Japan and Norway lead in their development.

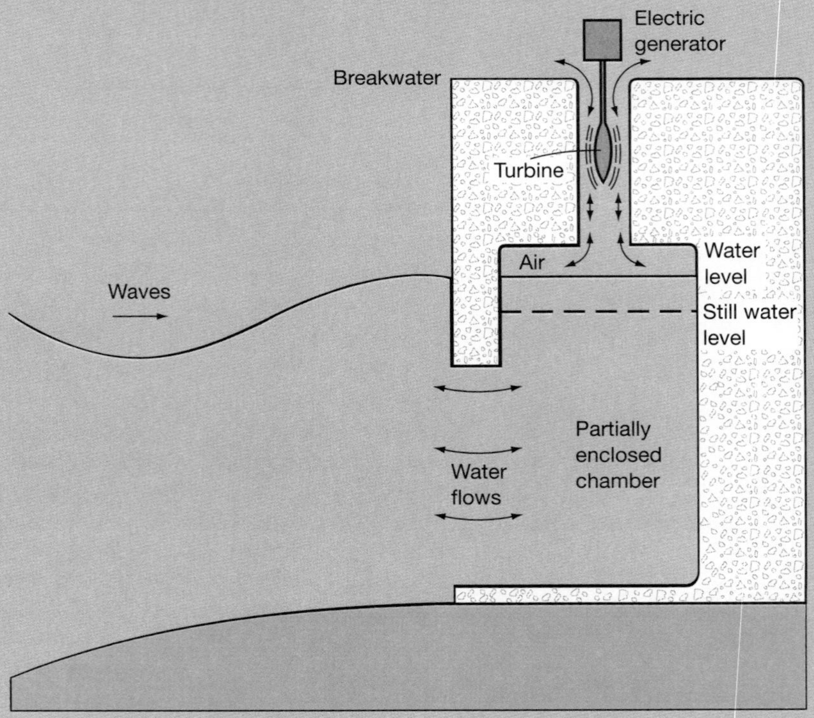

Figure B9–1–1
Simple scheme for generating electrical power from the energy contained in waves.

Depending on the size of the wave, different restoring forces are involved. For capillary waves, the dominant restoring force is surface tension. The water surface tends to act like a drum head, smoothing out the waves. For waves having periods between 1 second and about 5 minutes, gravity is the dominant restoring force. This range includes most of the waves we see. Such waves are known as **gravity waves.** Longer waves are restored by the Coriolis effect. For this reason, predictions of tsunami movements or storm surges must include the Coriolis effect.

The Energy in Waves

Waves transmit energy gained from the disturbances that formed them. This energy is in two forms. The first is **potential energy,** which is energy that depends on the position of the water above or below the still-water level. The other form of wave energy, known as **kinetic energy,** is energy the water has because it is moving. As a wave moves through the water, there is a continuous transformation of potential energy to kinetic energy and vice versa.

The total energy in a wave is proportional to the square of its height. The higher the wave, the greater the potential energy. In addition, the orbits of the movement of water particles are larger in a higher wave and thus, the kinetic energy is also greater.

An enormous amount of energy is contained in each wave. A wave 2 meters high, for example, has energy equivalent to 1,200 calories per square meter of ocean surface. A 4-meter-high wave has 4,800 calories per square meter. Nearly all this energy is dissipated as heat when waves strike a coastline. (The heat is so well mixed into the nearshore waters that the warming is imperceptible.) Incoming waves have so much energy that a sensitive seismograph (a device that detects earthquakes) will record surf hitting distant beaches as faint tremors.

Sea and Swell

Ocean waves are manifestations of energy moving across a water surface. Now we see how energy in winds forms wind waves.

Wind-wave formation is easily observed. Even a gentle breeze forms ripples (also called capillary waves) in arcs, often on top of earlier-formed waves. Ripples play an important role in wind-wave formation by providing the surface roughness necessary for the wind to pull or push the water. In short, they provide grip for the wind.

Ripples are short-lived; when the wind dies, they disappear almost immediately. If the wind continues to blow, however, ripples grow into short, choppy waves. These newly formed waves, called **seas,** are locally generated waves. They continue to grow as long as they continue to receive from the wind more energy than they lose through such processes as wave-breaking. The water gains energy as a result of the wind's pushing and dragging effects. The amount of energy gained depends on such factors as sea roughness, wave shapes, and relative speed of both wind and waves. Choppy, newly formed seas provide a much better grip for the wind than smooth-crested waves.

The largest wind waves commonly encountered are formed by large storms, because wave size depends on the amount of energy supplied by winds (Fig. 9–7). The factors operating here are wind speed, length of time the wind blows in a constant direction, and **fetch**—the distance over which the wind blows in a constant direction. As shown in Figure 9–8, for a constant wind speed, the wave height increases rapidly in the first few hours and within the first few hundred kilometers of fetch. As the duration and fetch increase, the growth of the wave height is much slower.

Usually some older waves are present on the ocean when new ones begin to form. Either the older waves are destroyed by the storm, or newly formed waves

Figure 9–7
A chaotic sea surface is caused by combined waves of all sizes. (Rex Ziak/Tony Stone Images.)

are generated on top of the old ones. The waves interact continuously. Wave crests coincide, forming momentarily new and higher waves. Seconds later, the wave crests may no longer coincide but instead cancel each other; then the wave crests disappear.

When winds change directions, waves produced under previous wind systems are destroyed or greatly modified. Thus, the major difference between ocean areas is the maximum fetch over which winds can act. In the North Atlantic, for example, the maximum effective fetch is about 1,000 kilometers. With a 1,000-kilometer fetch, a wind blowing about 70 kilometers per hour can produce waves about 11 meters high. With an unlimited fetch, the same wind could produce waves about 15 meters high. A longer fetch and higher wind speeds are needed to produce gigantic waves.

The largest waves are formed by strong, steady winds blowing for long times in the same direction over large bodies of water. Such waves occur most frequently at stormy latitudes, where storms tend to come in groups traveling in the same direction, with only short periods of time separating them. Thus, the waves of one

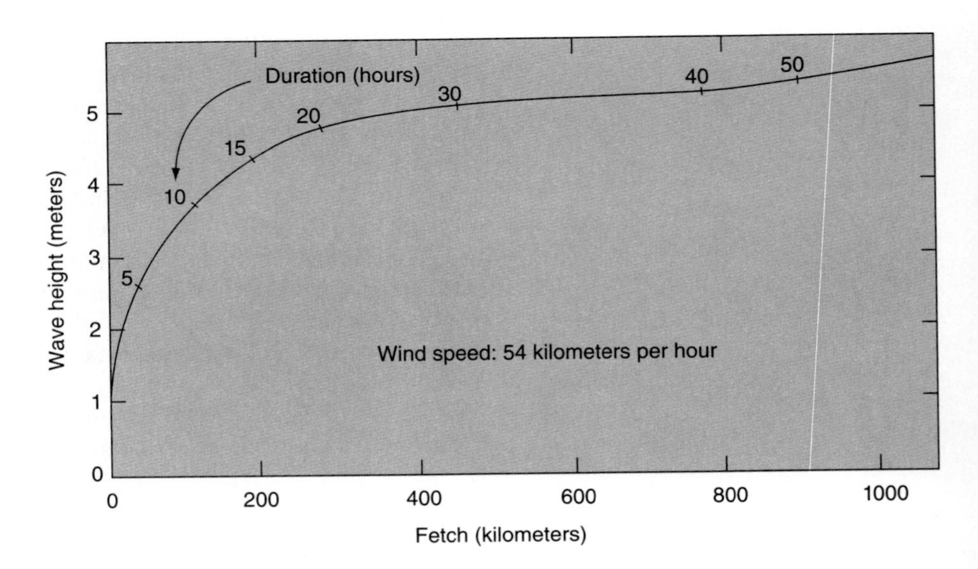

Figure 9–8
Growth of wave height under a constant-direction, constant-speed wind, blowing for different lengths of time and over different fetches. (After H. U. Sverdrup, N. W. Johnson, and R. H. Fleming, *The Ocean*. Englewood Cliffs, NJ: Prentice Hall, 1942.)

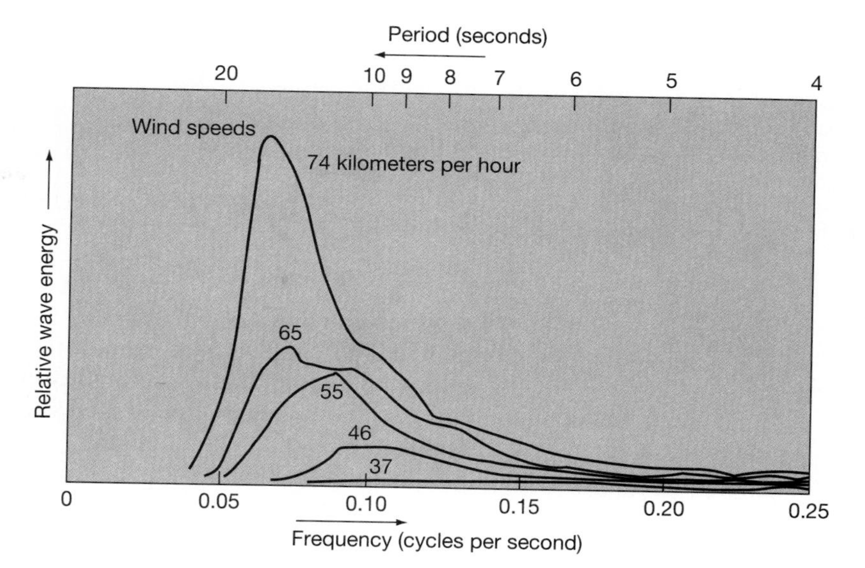

Figure 9–9
In a fully developed sea, most of the wave energy occurs in a relatively restricted range of wave periods. Note that changes in wind speeds cause marked changes in wind energy and wave period. (After G. Neumann and W. J. Pierson, *Principles of Physical Oceanography*. Englewood Cliffs, NJ: Prentice Hall, 1966.)

storm often have no chance to decay or travel out of the area before the next storm arrives to add still more energy to the waves, causing them to grow still larger. (It is interesting that, despite their reputation for ferocity, many typhoons and hurricanes do not form exceptionally large waves because their winds, although very strong, move in a circular pattern around the eye of the storm. They do not blow long enough from one direction to form overly large waves. In addition, such storms move very rapidly.

In contrast, the common North Atlantic winter storms (extratropical cyclones, locally called "nor'easters") do generate large waves which can cause serious beach erosion. This is because these storms move much more slowly; their winds, while weaker than hurricanes, may persist for days in a single location.

As winds continue to blow, the waves they generate continue to grow until they reach a maximum size, determined by the point at which the energy supplied by the wind is equal to the energy lost by breaking waves, called **whitecaps.** When this condition is reached, we refer to it as a **fully developed sea.** Initially, wave heights increase markedly as the winds continue to blow, but after about ten hours, wave heights do not increase as much, and after about thirty hours, there is little increase in wave height regardless of fetch length (Fig. 9–8).

Waves of many different sizes and periods are present in a fully developed sea, but waves having a relatively limited range of periods dominate for a steady wind with a fixed speed (Fig. 9–9). Knowing wind speed, wind duration, and fetch, therefore, we can predict the size of waves generated by a storm. Such predictions are complicated, however, because winds almost never blow at a constant speed; winds are just as likely to be gusty at sea as on land and just as likely to change direction.

Initially, waves in a sea are steep and sharp-crested, often reaching the theoretical **limit of stability** ($H/L = 1/7$), as Figure 9–10 shows. Then, either the waves break or their crests are blown off by the wind. As waves continue to develop, their speed approaches, then equals, and finally exceeds the wind speed. As this happens, wave steepness decreases as wavelength increases. As waves travel out of the generating area, or if the wind dies, the sharp-crested, mountainous, and unpredictable sea is gradually transformed into smoother, long-crested, longer-period waves called **swell.** These waves can travel long distances because they lose energy only gradually if they do not encounter shallow bottoms, new winds, or waves.

Let us examine some of the factors that cause waves to change from sea to swell. One factor is that waves spread as a result of variations in the direction of the winds that formed them. Unless destroyed or influenced in some way by ocean boundaries, waves travel for long distances in the direction in which the wind was

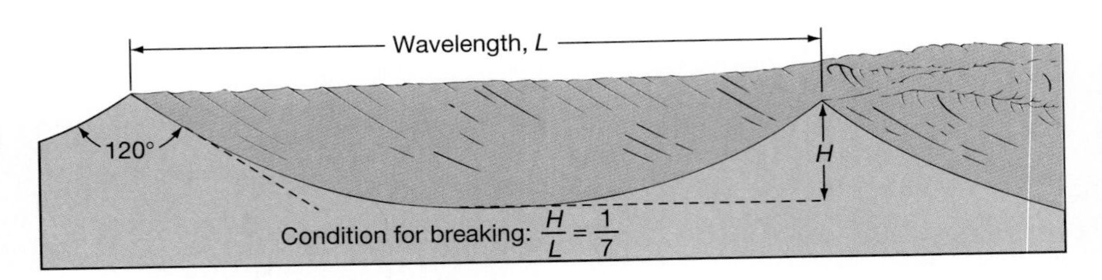

Figure 9–10
Larger waves tend to be more sharp-crested than a simple sine wave. There are limits to which a wave can grow. Waves commonly break when the angle of the crest is less than 120° or the ratio of the wave height to wave length *(H/L)* is 1/7.

blowing when they formed. Because wind speed and wind direction are rarely constant for long, waves formed in a storm "fan out" from their point of origin as they move (this type of wave travel is called angular dispersion). As a result of this fanning out, the wave energy is spread over a larger area, causing a reduction in wave height.

At the same time the waves are fanning out, they are also dispersing by wavelength. Longer waves travel faster than shorter waves. As a result, complex waves of varying wavelengths formed in the generating area are sorted through time as they move away from the storm area, the long waves preceding the shorter waves. Consequently, the first waves to reach a coast from a distant large storm are those having the longest periods. Island-dwellers, sensitive to the normal wave period on their coasts, may be warned of approaching hurricanes by the arrival of such abnormally long waves which travel faster than the storm itself.

Waves travel great distances in swell, crossing entire ocean basins before encountering a coastline. Storms in the North Atlantic, for example, form waves that dissipate their energy in the surf on North African coasts, 3,000 kilometers away. On extremely calm summer days, very-long-period swell strikes the southern coast of England after traveling about 10,000 kilometers from South Atlantic storms. Similarly, waves from Antarctic storms strike the Alaskan coast, more than 10,000 kilometers away.

Wave Height

When we look out over the open ocean, there are waves of many different heights. The scene can be described statistically, however, because in any complex of waves, there is a nearly constant relationship between waves of various heights. One useful index for measuring wave size is **significant height**—the average height of the highest one-third of the waves present (Table 9–1). In general we find that the most frequent waves found in any wave complex are about half as high and that the average wave height is about 60 percent of the significant height. The highest

TABLE 9–1
Wave-Height Characteristics

Wave	*Relative Height*
Most frequent waves	0.50
Average waves	0.61
Significant (average of highest one-third)	1.00
Highest 10 percent of waves	1.29

Source: U.S. Naval Oceanographic Office, 1958, p. 730.

Figure 9–11
A wave 34 meters high was measured by the USS *Ramapo* in the Pacific on February 7, 1933—the largest wave ever measured reliably. The length of the vessel was known, and the angle of its slope as it rode up the wave was measured. With a line of sight from wave crest to horizon, the depth of the trough was calculated.

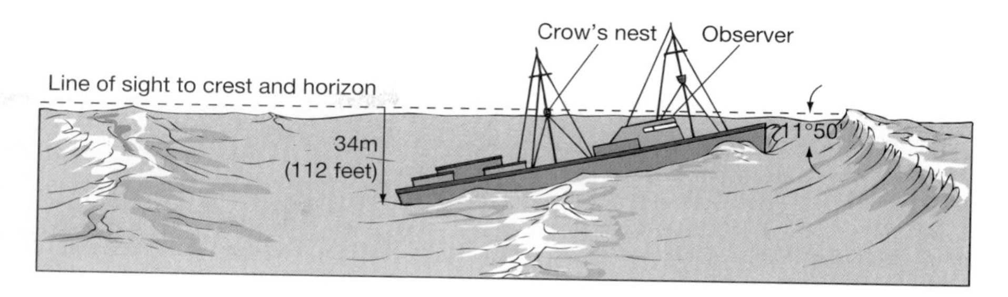

10 percent of the waves average about 1.29 times the significant height. Thus, given the wave height for part of a wave spectrum, we can predict other parts of the spectrum.

For years, the only data we had on wave height came from ships at sea. An observer on a moving ship has no fixed reference points to use in making estimates, however, and therefore does not provide the most reliable information. Today, we no longer have to depend on such subjective observations of on-site data, because satellites now provide nearly instantaneous coverage of wave conditions throughout the ocean.

Existing observations indicate that about one-half of the waves in the ocean are 2 meters or less in height. Only about 10 to 15 percent exceed 6 meters in height, even in such notoriously stormy areas as the North Atlantic or in the strong winds of the Roaring Forties in the southern hemisphere.

There are reliable reports from observers on ships of waves up to 15 meters high in the North and South Atlantic and the southern Indian Ocean. It appears that, in these ocean regions, winds rarely blow from one direction long enough to produce waves that are significantly higher.

The Pacific holds the records for giant waves. The largest deep-water wave that has been reliably measured was in the North Pacific on February 7, 1933. The U.S. Navy tanker USS *Ramapo* encountered a prolonged storm that had winds with an unobstructed fetch of many thousands of kilometers. The ship, steaming in the direction of wave travel, was relatively stable, and her officers measured one wave at least 34 meters high (Fig. 9–11). The wave period was clocked at 14.8 seconds and the wave speed at 102 kilometers per hour, somewhat faster than the theoretically predicted wave speed.

Rogue Waves

Waves interact with currents, winds, and other waves, but these interactions usually cancel each other, so that wind-generated waves rarely become large enough to threaten ships. Unusually high waves can form when two or more large waves combine. Furthermore, wave-current interactions are sometime strong enough to form enormous waves that can sink even the largest ships. These extraordinarily large waves are called **rogue waves.** Off the coast of South Africa, for instance, large waves generated in the Southern Ocean arrive as swell and locally combine with waves from other storms to become even larger. The Agulhas Current moving from the north and west interacts with these very big waves, causing them to steepen and become shorter. Some waves become so steep that their forward face is close to breaking, leaving a deep hole in front of the wave. Ships have been known to sail into such holes, break up, and sink. In the midst of many other large waves, rogue waves can often be seen only as the ship encounters them. At least some of the many ship disappearances that occur each year are attributable to rogue waves. Indeed, such waves may be the reason some ships are lost each year in the Bermuda Triangle or near the *Kuroshio* off Japan.

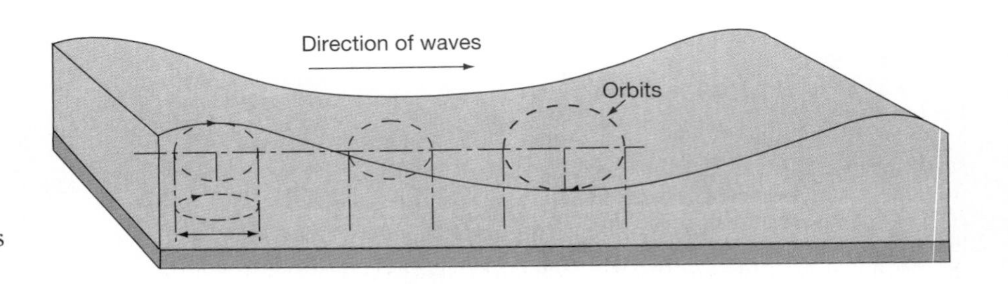

Figure 9–12
In shallow-water waves, wave-induced particle motions are affected by the bottom. Orbits are flattened so that the travel path for any given water particle is elliptical rather than circular.

Waves in Shallow Water

So far we have been discussing deep-water waves, which are unaffected by interactions with the bottom. Where water depths are less than $L/20$, however, the wave-induced motion of water parcels is affected by the presence of the bottom (Fig. 9–12). Waves in which this happens are shallow-water waves. At the water surface, the orbit of water parcels in a shallow-water wave may be only slightly deformed, usually forming an ellipse (a flattened circle) having its long axis parallel to the bottom. Near the bottom, wave action is felt as the water parcels move back and forth. Vertical movement is prevented by the nearness of the bottom.

Sometimes we observe movements of water parcels in shallow waters as waves pass over them—for example, where attached plants move with the water. The speed *(C)* of shallow-water waves, in meters per second, can be calculated by multiplying the square root of the depth *(D)*, in meters, by 3.1. Thus, $C = 3.1\sqrt{D}$). In shallow water, waves are slowed down until the individual wave speed equals the group speed.

We learned above that a wave becomes unstable when its steepness $(H/L) = 1/7$. Another rule of thumb for determining instability is that a wave becomes unstable when its height is about eight-tenths of the water depth. At this point, the wave begins to collapse and is known as a **breaker.** Although some waves are destroyed by opposing winds or cancel each other while still in deep water, most reach the shallow waters along coastlines and become breakers.

Breakers dissipate enormous amounts of energy. A single wave 1.2 meters high, having a 10-second period and striking the length of the U.S. Pacific Coast would release 37,000 megawatts of energy. (The largest power generator in New York City produces 1,000 megawatts of electricity.) Most of this energy is in the form of heat, which is not detectable because of water's high heat capacity. Also, extensive mixing in the surf disperses the heat throughout the nearshore waters.

As waves move from deep water to shallow water, wavelength and speed decrease, while wave period remains constant. At the same time, the wave height first decreases slightly and then, as the wave slows and its kinetic energy is converted to potential energy, wave height increases rapidly as water depth decreases. This series of events is illustrated in Figure 9–13.

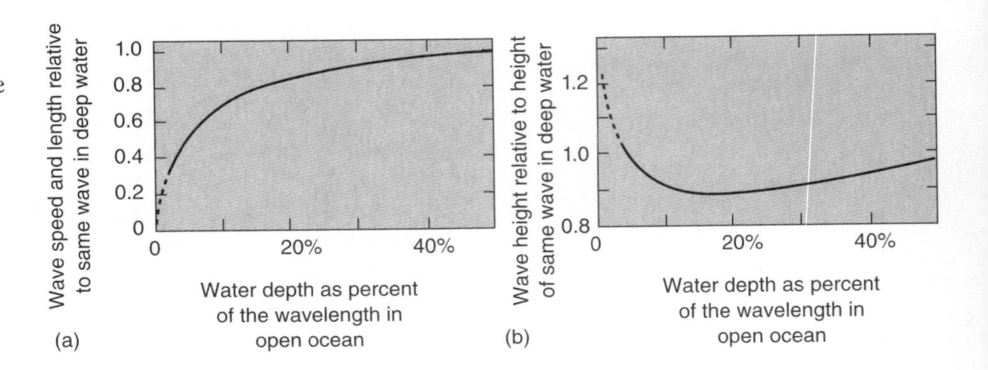

Figure 9–13
Waves change as they enter shallow water. (a) Speed and wavelength decrease as the water shallows; (b) wave height first decreases and then increases as the water shallows.

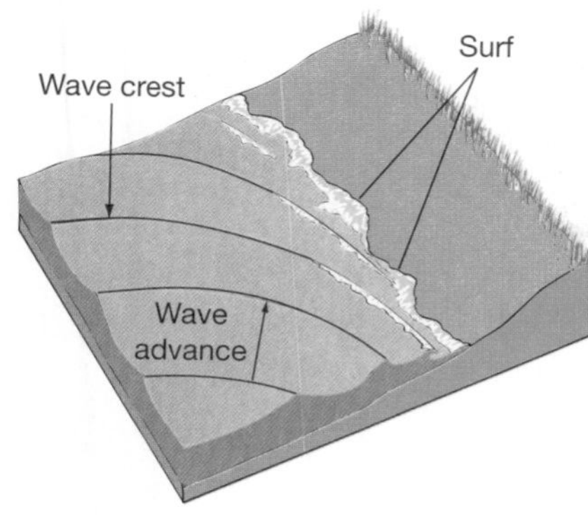

Figure 9–14
Refraction of a uniform wave train advancing at an angle toward a straight coastline over a gently sloping, uniform bottom. Note the bend in the wave crests as they approach the beach. Such waves cause a longshore current moving to the left (in other words, moving in the same direction as the waves) near the beach.

The direction of approach changes as waves enter shallow water. We commonly see breakers nearly parallel to the coastline when they reach the beach, even though they may approach the coast from many different directions. This process, known as **wave refraction,** occurs because the part of the wave still in deeper water moves faster than the part that has entered shallower water. The result is that the crest line rotates to become more parallel to the bottom depth contours, as shown in Figure 9–14.

In the simple case just discussed, the ocean bottom slopes uniformly away from the beach. Obviously this is not always the case; over irregular bottoms, wave refraction is more complicated. Submarine ridges and canyons, for example, cause wave refraction such that the wave energy is concentrated on headlands and spread out over bays, as shown in Figure 9–15. Consequently, headlands erode more rapidly than bays. The eroded material is usually deposited in adjacent bays, eventually smoothing the coastline. Eventually, the crest lines become parallel with the depth contours.

The more complicated the bottom topography, the more complicated the wave patterns on the shore (Fig. 9–16).

Surf

The belt of nearly continuous breaking waves along a shore or over a submerged bank or bar is known as **surf** (Fig. 9–17) and is a mix of breakers. It forms as different types of waves approach a shore and interact with the shallow bottom. Breaker height depends on the height and steepness of the waves offshore and, to a certain extent, on the offshore bottom topography. Breakers may be only a few centimeters high on a lake or protected ocean beach or many meters high on an open beach. Lighthouse keepers report spectacular breakers at their exposed locations. For instance, Minot's Lighthouse, 30 meters tall and standing on a ledge on the southern side of Massachusetts Bay, is often engulfed by spray from breaking waves. The glass in the lighthouse of Tillamook Rock, Oregon, 49 meters above the

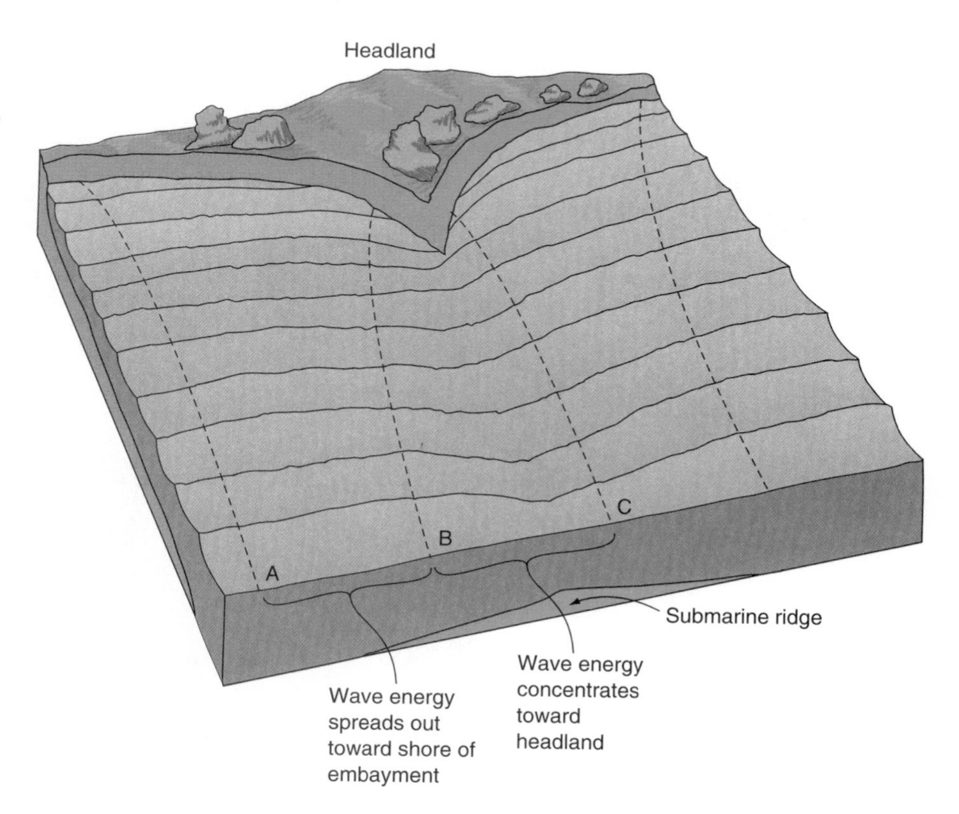

Figure 9–15
Wave refraction causes equal amounts of wave energy to be spread over different surface areas. The amount of wave energy contained between lines A and B is the same as the amount contained between lines B and C. The A/B energy is spread out over the entire bay area, whereas the B/C energy is concentrated in the narrow area in front of the protruding point of land, called a headland. Such unequal distribution of wave energy causes the headland to erode and sediment to be deposited on the shore of the bay. Over time, the result of this differential erosion is a straight shore line.

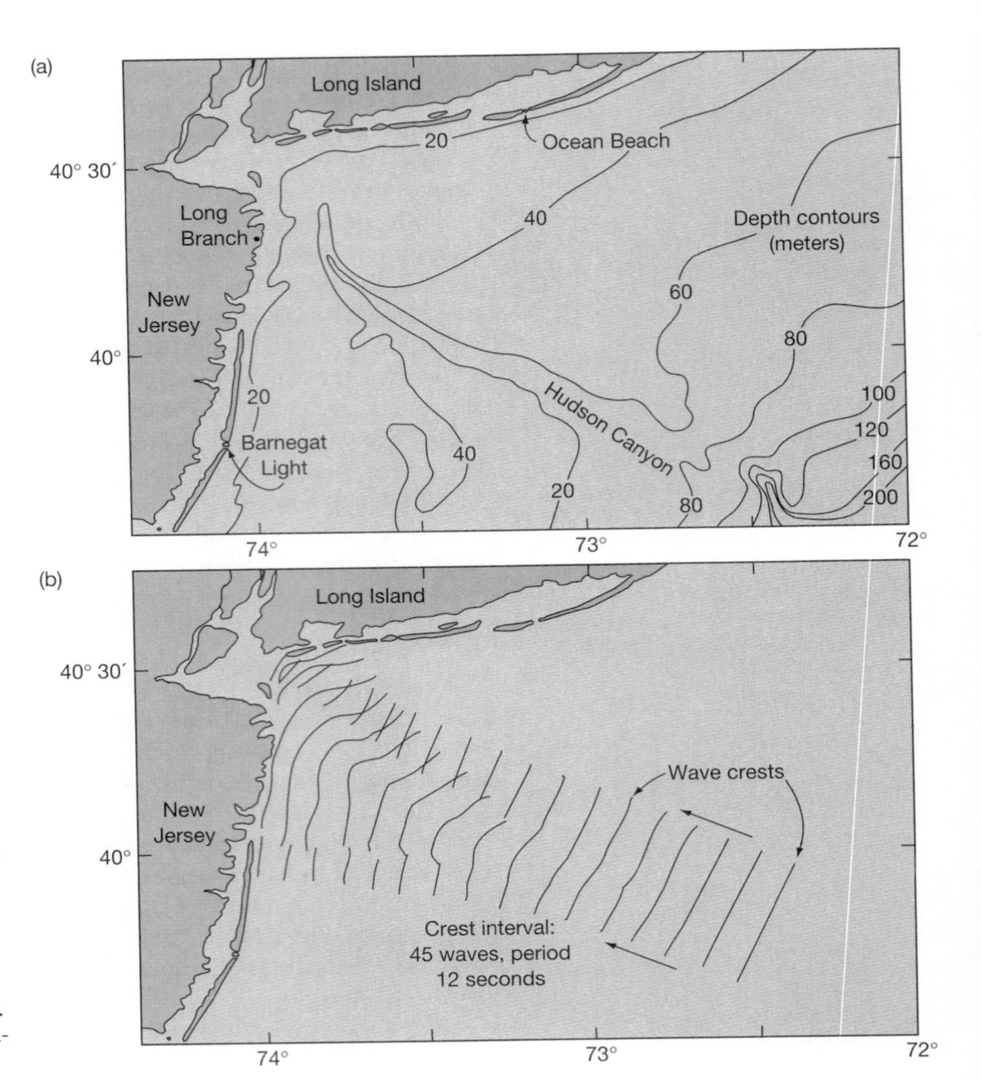

Figure 9–16
Submarine topography causes complicated wave patterns near the entrance to New York Harbor. (a) Hudson Canyon is a salient feature in the waters directly off the harbor entrance. (b) The presence of Hudson Canyon focuses uniform wave trains on the harbor entrance. (After W. J. Pierson, G. Neumann, and R. W. James, *Practical Methods for Observing and Forecasting Ocean Waves by Means of Wave Spectra and Statistics*. U.S. Naval Oceanographic Office, Publication 603. Washington, D.C., 1955.)

Figure 9–17
Breakers in winter surf on Clatsop Spit, near Astoria, Oregon. (David Weintraub/Photo Researchers, Inc.)

Figure 9–18
Lighthouses are normally built on headlands where wave energy is focused. Here we see a lighthouse at Point Fairy, Australia engulfed by spray from breaking waves. (K. Stepnell/Bruce Coleman, Inc.)

sea, has often been struck by waves. We have no observations of the waves that cause such surf (Fig. 9–18).

Surf breakers 14 meters high have twice damaged a breakwater at Wick Bay, Scotland, moving blocks weighing as much as 2,600 tons. Surf breakers about 20 meters high have been reported at the entrances to San Francisco Bay and the Columbia River when onshore gales were blowing. Surf at a river mouth or harbor is likely to be especially high when the incoming waves encounter a current moving in the opposite direction. Under these conditions, ships must wait, often for days, before they can safely enter (or leave) the harbor.

Tsunamis

Tsunamis, or seismic sea waves, are large waves caused by sudden movements of the ocean bottom resulting from earthquakes or volcanic eruptions. They have very long periods and therefore behave like shallow-water waves, even when passing through the deepest parts of the ocean. An earthquake in the Aleutian Islands on April 1, 1946, for example, caused a tsunami having a 15-minute period and a wavelength of 150 kilometers. Even when this wave was passing through water as deep as 4,300 meters, its speed, about 800 kilometers per hour, was controlled by the water depth. In deep water, the crests of this tsunami were estimated to be only about a half meter high, virtually undetectable to ships.

Tsunamis eventually encounter a coast, often with catastrophic results. For instance, upon reaching the Hawaiian Islands, the waves from the 1946 Aleutian earthquake were driven ashore as a rapidly moving wall of water, in a few places up to 6 meters high. Where the water was funneled into a valley, the waves formed enormous breakers. More than 150 people were killed, and property damage was extensive.

Often tsunamis reach their greatest heights near their source. In the 1946 tsunami, the wave was highest in the Aleutian Islands, the location of the earthquake. A concrete lighthouse and a radio tower 33 meters above the sea were destroyed at Scotch Cap, Alaska.

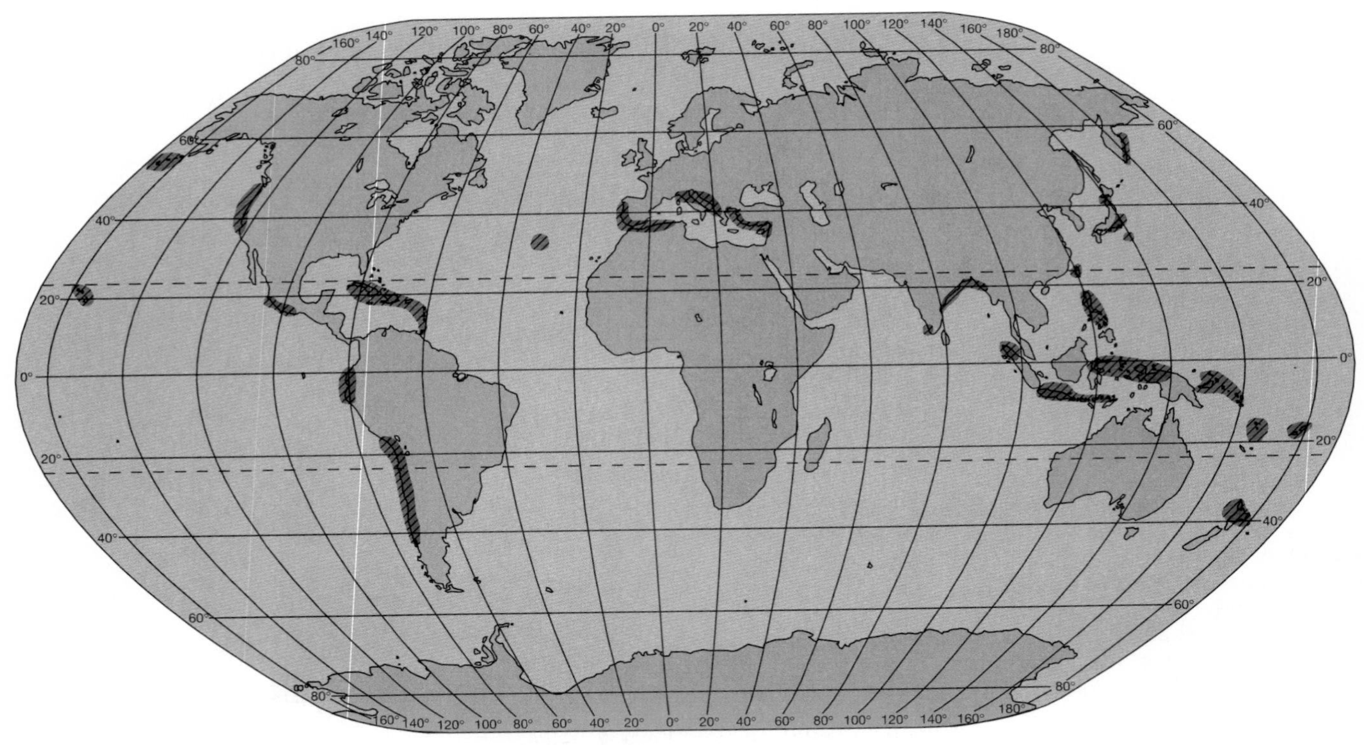

Figure 9–19
Areas that experience tsunamis.

Tsunamis can be extremely destructive. In Japan, the Hoei Tokaido–Nankaido tsunami of 1707 killed 30,000 people and destroyed 8,000 homes. Today, even though Japanese warning systems predict tsunamis generated by nearby earthquakes and extensive barriers have been built to protect especially vulnerable locations, people are still killed by these waves. The deaths often occur because there is too little time between the earthquake and the tsunami striking the coast. People in remote coastal villages often cannot be warned at all. Areas farther from the tsunami-generating areas can be better warned and have time to take precautions.

Because of the frequent occurrence of tsunamis around the Pacific (Fig. 9–19), an international system headquartered in Hawaii issues warnings when large earthquakes that might cause destructive tsunamis occur. These warnings have greatly reduced the number of deaths from tsunamis. For example, a 1957 tsunami killed no one in Hawaii, even though water levels were higher than in 1946.

Standing Waves

So far we have been talking about progressive waves, those that move across a water surface. Another type, **standing (or stationary) waves,** are characteristic of basins ranging in size from familiar teacups and bathtubs to bays and ocean basins. Standing waves can be caused by winds, changes in atmospheric pressure, or tides.

A simple standing wave can be made by tilting a round-bottomed dish of water and then setting it flat on a table. The water surface will then tilt first toward one side and then toward the other. Because of this sloshing motion, standing waves are also known as **seiches** (pronounced *saysh*), a French word meaning sloshes. The water surface at the edge of the dish moves vertically, but along a line, usually in the middle of the dish, the surface does not move. Instead, it acts as a hinge about which the rest of the water surface tilts. This stationary

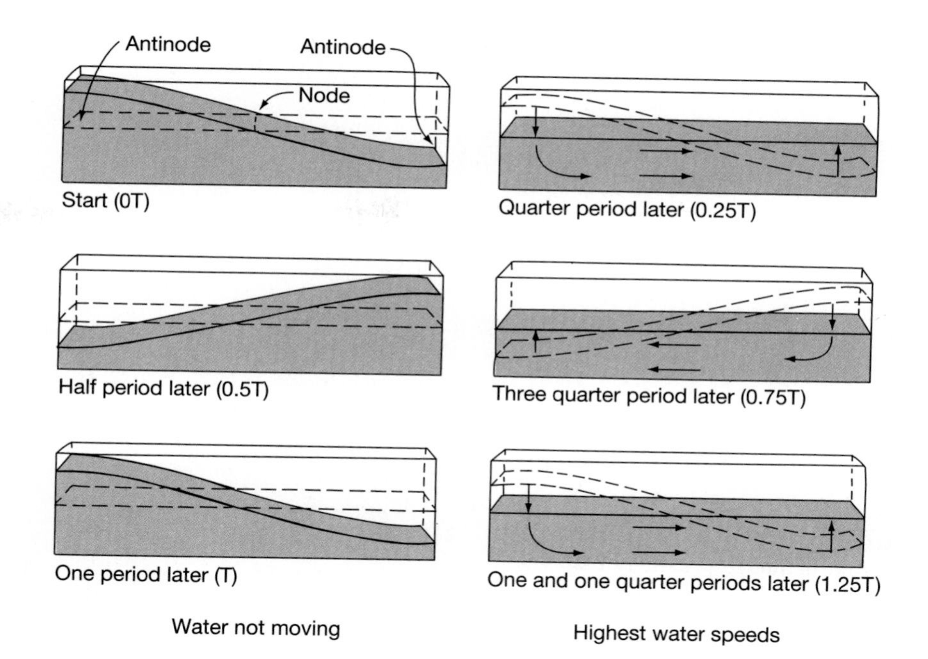

Figure 9–20
A simple standing wave containing one node, shown at quarter-period intervals.

line (or sometimes a point) is called a **node** (Fig. 9–20). The parts of the surface showing the greatest changes in elevation are called **antinodes.** Complex stationary waves have several nodes and several antinodes.

In a seiche, maximum horizontal water movements occur at the node, where the water surface remains horizontal. At the instant when the water surface is tilted most, there is no water motion. This start/stop action makes standing waves distinctly different from progressive waves. In the latter, the water is always moving continuously in orbits as the wave passes through; in the former, the water flows for a distinct period, stops, and then reverses its direction. Also, the wave form alternately appears and disappears in a standing wave, as the sequence in Figure 9–20 shows. The water movements are mostly horizontal rather than the circular or nearly circular and continuous orbits associated with progressive waves.

Like progressive waves, standing waves are modified by their surroundings. They are reflected by the sides of the basin, especially if they are steep and smooth. Their energy is absorbed by shallow or irregular bottoms and by complicated shorelines. The wavelength of a seiche is determined by the size of the basin, which, in the case of large ocean basins, may be thousands of kilometers. Standing waves in large ocean basins are influenced by the Coriolis effect. The resulting wave, instead of simply sloshing back and forth, goes around the edges of the basin in a rotary motion.

Storm Surges

Storm surges are elevations of the sea surface caused by storm winds, usually extratropical cyclones and sometimes hurricanes. A storm surge is a large wave that moves with the storm that caused it.

A storm surge can cause flooding of low-lying coastal areas, especially if the storm that caused it is slow-moving. It is caused by a combination of events offshore and the effects of local winds. A large storm is accompanied by a sudden decrease in atmospheric pressure which causes the sea level to rise considerably. Although this may occur many kilometers from land, the winds and large storm-generated waves drive this mass of elevated sea water to shore. Where the winds

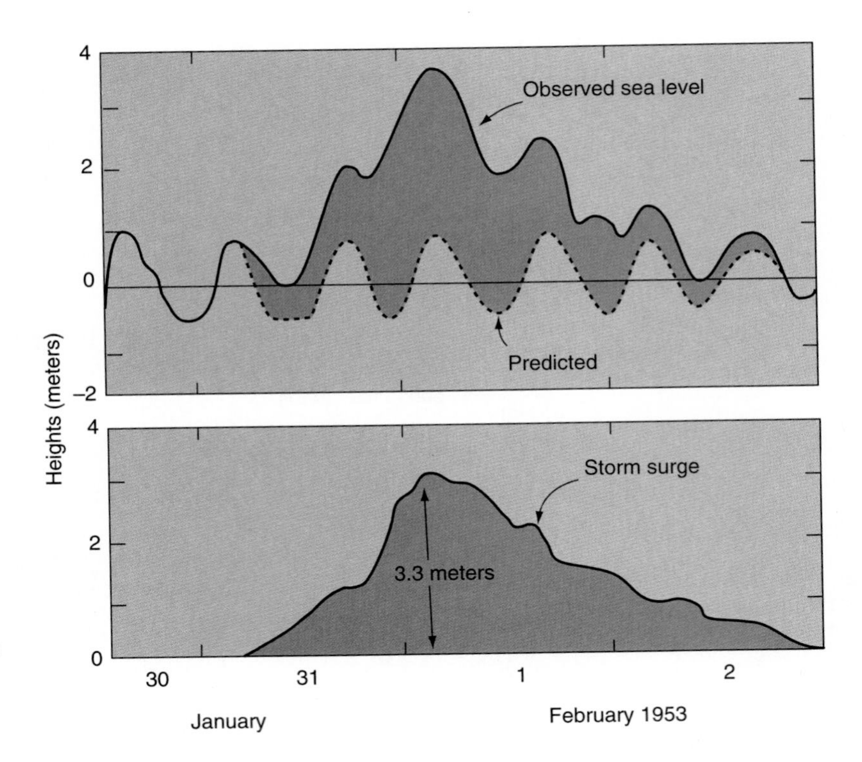

Figure 9–21
Storm surge of January–February 1953 on the Dutch coast. Sea-level changes resulting from the surge were estimated by subtracting predicted tide levels from observed sea levels. (After P. Groen, *The Waters of the Sea*. London: D. Van Nostrand, 1967.)

are onshore when a hurricane hits land, the storm surge will be much higher than where the winds in the same storm are offshore.

A few hours ahead of the storm's arrival, a gradual and slight lowering of sea level, the **forerunner,** is observed along a wide stretch of coastline. When the storm center passes onto land, it causes a sharp rise in water level called the **surge.** This surge usually lasts two to five hours; increases in sea level of 3 to 4 meters have been observed, usually slightly offset from the storm's center. Combined with the very high waves generated by the storm, surges can be extremely destructive. A storm surge is an example of a forced wave.

Following the storm, sea level continues to rise and fall as storm-caused oscillations pass. These oscillations are more-or-less free surface waves and have been termed the **wake** of the storm, like the wake left by the passage of a ship through the water. These resurgences can be quite dangerous, particularly because they are often not expected once the storm has subsided.

Storm surges can be predicted, based on wind speeds and direction, fetch, water depth, and shape of the shoreline. Other factors—such as currents, astronomical tides, and seiches set up by storms—complicate the calculations. Better observations of large storms and the availability of more powerful computers have combined to permit more accurate predictions of storm surges.

Storm surges have caused catastrophic flooding many times. In 1900, Galveston, Texas was destroyed and about 6,000 people were killed by a storm surge resulting from a hurricane. In 1969, Hurricane Camille, one of the strongest recorded storms to hit the Gulf Coast, caused $1.4 billion in damage; even with advance warning, it killed 256 people. A disastrous storm surge occurred in 1876 on the Bay of Bengal in the northern Indian Ocean; 100,000 people were killed. In 1970, another storm surge hit the same area, killing an estimated half million people.

With better warnings, the loss of life can be greatly reduced, although property damage remains high. In Bangladesh in 1994, for instance, a storm surge comparable to the one of 1876 in the Bay of Bengal killed only a few hundred people because an excellent warning system permitted evacuation of the affected coastal areas. In this 1994 event, most of the lives were lost in nearby Burma, where people did not receive adequate warnings.

January 31–February 1, 1953, a northwest gale blew across the North Sea from Scotland to the Netherlands. Its fetch, 900 kilometers, caused sea levels to rise more than 3 meters (Fig. 9–21). As the resulting storm surge combined with high tides and strong waves, waters broke through protective dikes and dunes, flooding low-lying areas on the Dutch and English coasts. Thousands drowned. Better forecasting techniques and improved coastal defenses now protect these areas. In particular, 1.3 billion dollars was spent to build a flood defense system in the Thames River to protect London. The funnel-shape of the Thames estuary makes it especially vulnerable to storm surges caused by storms from the east. Massive flood defense structures were also built on the Dutch coast to protect it against future storm surges.

Internal Waves

Besides surface waves, there are **internal waves,** which occur beneath the water surface and are therefore not easily observed. Internal waves, shown in Figure 9–22, occur along interfaces between layers of different densities. They can be detected either by measuring temperature or salinity changes at a given depth as internal waves pass, or by studying the effects they have on the ocean surface. For example, slicks, areas of smooth ocean surface similar to those seen in Langmuir cells, may form as relatively shallow internal waves pass under the surface layer. These slicks are the result of the convergence of the orbital motions set up by the internal waves (Fig. 9–23).

Internal waves act just like surface waves. However, there may be only a small density difference at the interface where internal waves form, rather than the large difference at the ocean surface interface between air and water. As a result, internal waves usually have greater amplitudes than surface waves. Internal wave heights of up to 40 meters have been observed. Internal waves generally move much more slowly and have less energy than surface waves. Internal waves can break, just like surface waves. In the process, they can cause mixing between the waters above and below the interface.

The astronauts on the space shuttle flights have observed the surface indications of internal waves in a number of regions. Figure 9–24 shows the surface

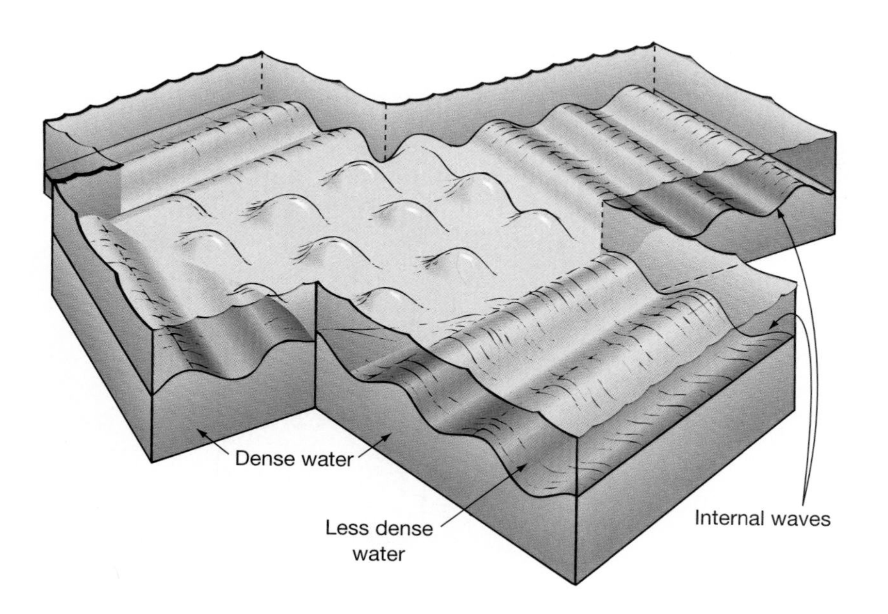

Figure 9–22
Simple internal waves interact at the interfaces between water layers of different densities to form complex waves.

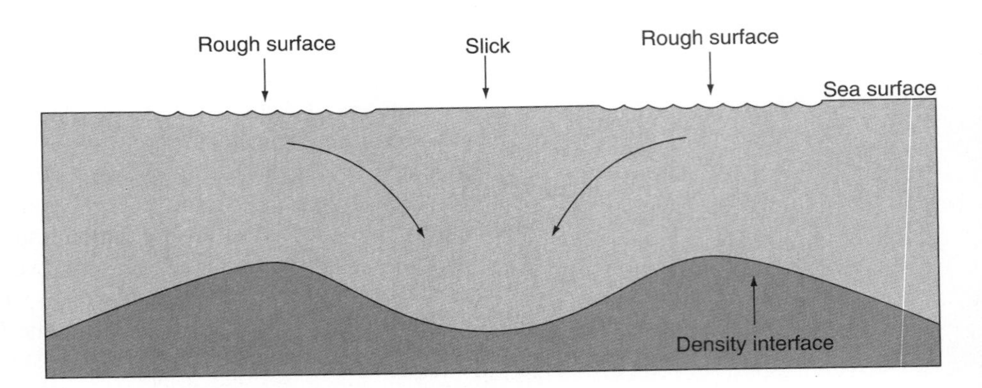

Figure 9–23
Areas of slick on the surface indicate the presence of internal waves below the surface layer. They form as the result of convergent water set in motion by the passage of the internal wave.

effects of the internal waves generated by tidal currents flowing through the Strait of Gibraltar, at the entrance to the Mediterranean Sea.

Internal waves can have significant consequences. Submarine warfare tactics have long taken advantage of the acoustic properties of density interfaces which deflect the sound pulses used to detect submarines. In 1963 the nuclear submarine *USS Thresher* was mysteriously lost with all hands. An internal wave was the suspected cause. The submarine, cruising in the pycnocline layer, was sud-

Figure 9–24
When tidal currents flow into and out of the Mediterranean, they generate internal waves. Surface manifestations of these internal waves can be seen in the Atlantic Ocean in this photograph taken from the space shuttle *Challenger.* (Courtesy NASA.)

denly moved downwards to depths below its pressure resistance capabilities. The hull imploded and all of the crew was killed as the submarine sank.

Waves on Beaches

Waves dominate beach processes. For example, wave-induced currents and turbulence stir up and transport sediments on beaches and in shallow waters. Large amounts of sand are transported while suspended in the water; relatively little is transported along the bottom.

Beaches change seasonally, primarily because of changing wave regimes. During periods of low, long-period swell from distant storms (usually in summer), sand is moved from water to land, usually making the beach taller and wider (Fig. 9–25). Longshore bars migrate shoreward, the troughs separating them from the mainland fill in, and a new berm (nearly horizontal part of the beach) forms, usually at a level lower than the preceding one.

During periods of high, choppy waves from nearby storms (mainly in winter), beaches are cut back (Fig. 9–25). The beach foreshore becomes more gently sloping, and a beach scarp (marked slope cut by waves) forms. Strong currents caused by waves develop deep channels. Bars develop because of the offshore movement of sand from areas seaward of the breakers. Most of the sand removed from the beach is deposited nearby in the offshore zone, to be moved back onto the beach during the next period of smaller waves.

If a wave were to approach land with its crest perfectly parallel to the shore, as in Figure 9–26a, all the energy of the wave would be directly perpendicular to the shore and sand would wash straight up. However, even though refraction causes crests to be nearly parallel to the coast, the process is rarely complete, and so most waves approach the shore obliquely (Fig. 9–27). Because of this oblique angle of approach, some of the wave energy acts parallel to the coast (Fig. 9–27a), and this energy causes **longshore currents** that move in the same direction as the waves.

Longshore currents are strongest between surf zone and beach and when the waves approach the shore at a 45° angle. This situation rarely happens, however; instead, the crests of most waves are within 20° of being parallel to the beach when they strike the shoreline (Fig. 9–27).

Wave-induced sediment movements can be surprisingly fast—up to 25 meters per hour and 1 kilometer per day. A more typical rate would be 5 to 10 meters per day, however.

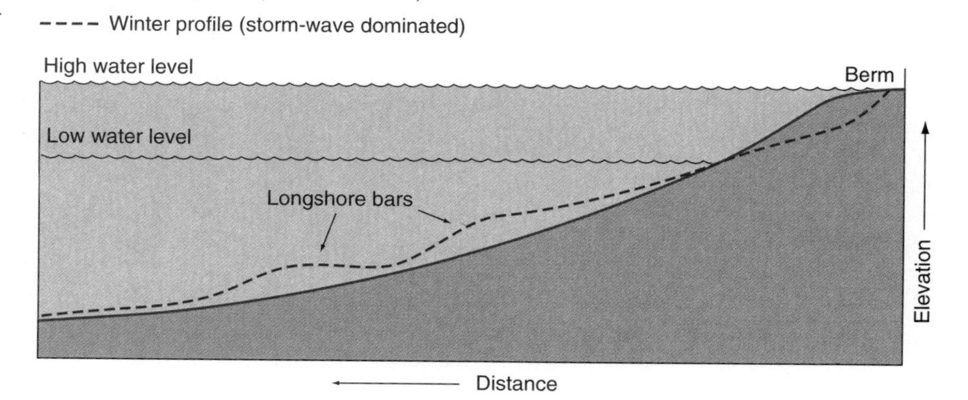

Figure 9–25
Beach profiles change seasonally. In summer, swell from distant storms striking the beach moves sand up onto the beach, making it higher and broader. In winter, storm waves from nearby storms erode the beach. Sand is moved offshore and deposited in longshore bars that parallel the beach.

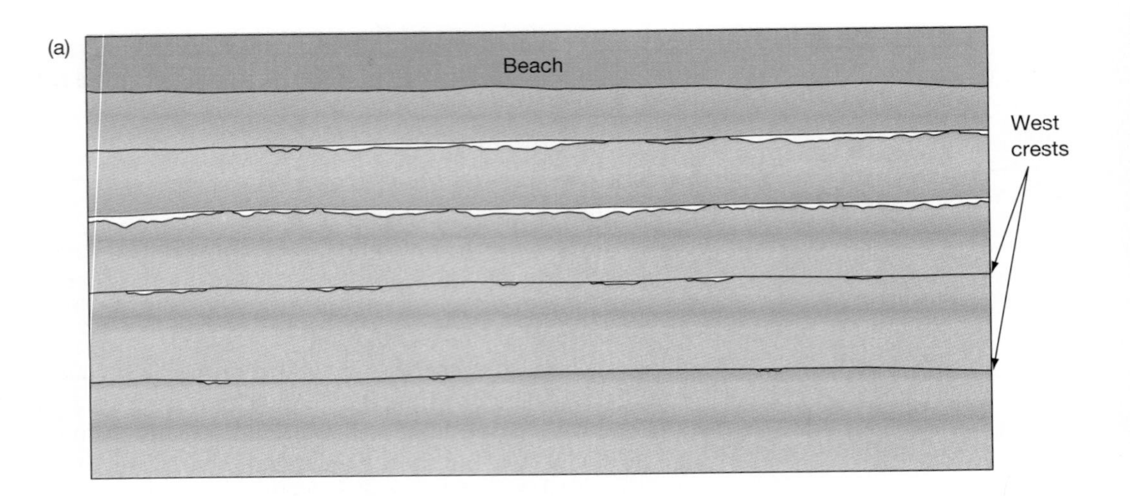

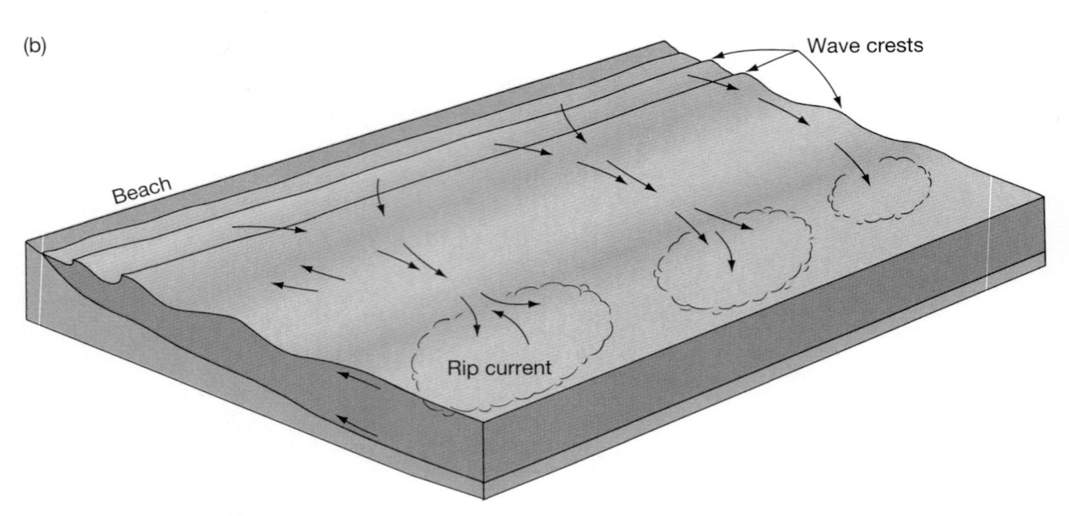

Figure 9–26
(a) Waves approach a beach directly. (b) Waves approaching a beach directly transport sand onto the beach. Water carried toward the beach by the waves returns seaward, forming rip currents.

When a wave breaks, there is substantial water movement. Water at the surface and along the bottom moves toward the beach, carrying with it materials floating or dragged along the bottom, which is why a beach acts as a convergence zone, collecting all sorts of debris as well as sediment.

After moving toward the beach in breaking waves, water tends to flow parallel to the beach for short distances, until it forms a narrow stream of return flow through the breaker zone. This narrow stream is called a **rip current** (Fig. 9–26b). In rip currents the most rapid flow has speeds of up to 1 meter per second until it reaches a distance of a few hundred meters from the coast.

Seaward of the breaker zone, the current becomes more diffuse and spreads out, forming a head. At this point, the water is caught up in the general flow back toward the beach. Rip currents and their associated water movements form a cell-like nearshore circulation system within the breaker zone.

Waves striking certain beaches cause **beach cusps,** which are uniformly spaced, tapering ridges separated by small rounded depressions as illustrated in Fig. 9–28. The regular spacing ranges from less than 1 meter to several tens of meters and is related to wave height, with higher waves associated with wider spacing.

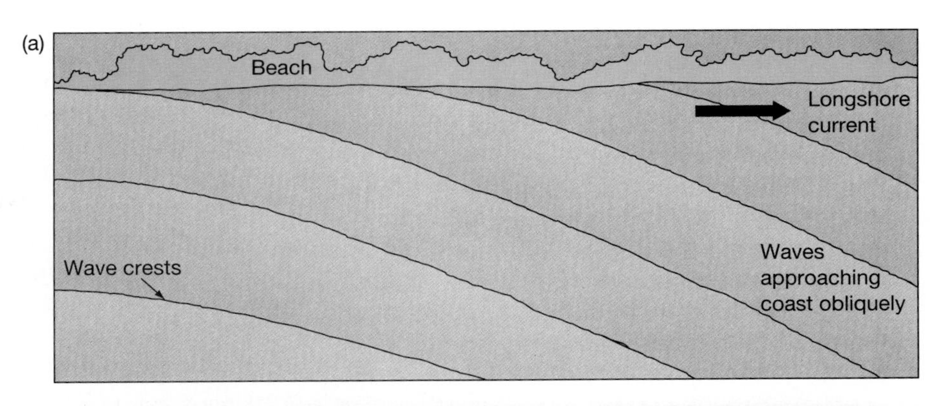

Figure 9–27
(a) Waves approaching a beach are refracted as they enter shallow water. Striking the beach at an angle, they cause a longshore current that moves parallel to the beach in the surf zone. (b) On Angaur Island, Belau in the Pacific, waves strike the reefs at an angle. This type of oblique approach causes longshore currents. (Douglas Faulkner/Photo Researchers, Inc.)

Figure 9–28
Beach cusps at Big Beach, near Makena, Maui, Hawaii. (Carl Purcell/Photo Researchers, Inc.)

Waves and Offshore Structures

The successful prediction of waves and their effects on large structures has multibillion-dollar payoffs for the oil industry. As more oil and gas is found under the seafloor in coastal waters around the world, building structures adequate to survive even hundred-year hurricanes is increasingly important. Since the oil industry first ventured into coastal waters in the 1940s, the industry has learned much about waves. Some lessons were learned the hard way—from disastrous loss of structures in storms.

The benefit of these lessons was seen after Hurricane Andrew hit the Gulf of Mexico in 1991. Of 3,800 offshore platforms, only 249 were damaged and fewer than 500 barrels of oil were spilled. Most of the damaged platforms were older ones. More recent platforms have more legs and are also stronger. Futhermore, their platforms are higher above the water surface. Thus, most storm waves can pass under them without hitting the platforms or their structures.

In addition to offshore drilling platforms for the oil and gas industry, offshore structures of all kinds are increasingly common in Japan where land is scarce. As coastal populations increase worldwide, such structures will become more common. They must be built to withstand the forces of oceanic waves.

Summary

Ocean waves are disturbances of water surfaces caused by energy from winds, earthquakes, or volcanic explosions. In deep-water waves, which are ones in which the water depth exceeds half of the wavelength, water parcels move in circular orbits as waves pass, but then return to their original location. Deep-water waves are unaffected by the bottom. Shallow-water waves, where water depths are less than one-twentieth the wavelength, are affected by the bottom; in these waves, water-parcel motions are essentially parallel to the bottom. Free waves move independently away from their source. Forced waves remain under the influence of the force that caused them.

Waves are classified by size. The smallest are capillary waves, with wavelengths less than 1.7 centimeters and wave periods less than 0.1 second. The restoring force for capillary waves is surface tension. Larger waves are most common, having periods of up to 5 seconds. Gravity is the dominant restoring force.

Most waves are generated by winds blowing across the ocean surface. Where waves are forming, the ocean surface is chaotic and is called a sea. Ripples form first and then grow into larger waves as winds continue to put energy into the water surface. Outside the wave-generating area, waves separate themselves according to size, forming swell, where waves are regular and smooth-crested.

Longer waves travel faster than short ones. Most ocean waves are less than 6 meters high. The highest waves occur in the Pacific and in the South Atlantic.

Waves are altered when they enter shallow water. They change direction by refraction, which occurs when water-parcel orbits drag on the bottom, moving most slowly in shallow water and fastest in deep water. As the waters get shallower, waves eventually become unstable and break, forming breakers.

Large ocean waves, called tsunamis or seismic sea waves, can be formed by sudden disturbances such as large movements of the ocean bottom due to earthquakes or volcanic explosions.

Standing waves, or seiches, occur in basins when wave energy is reflected from the sides; the waveforms do not move, but the water surface tilts in a regular manner.

Very strong, prolonged winds associated with storms can cause large, relatively slow-moving waves called storm surges; such waves can cause flooding of low-lying coastal areas.

Under-sea waves form at density interfaces. These internal waves act like surface waves, but are much larger and move more slowly.

Waves dominate beach processes. Beaches change seasonally because of changing wave regimes. They are cut back during stormy weather and built up during calm weather. Longshore currents result from waves striking the beach obliquely. Rip currents are the return flows of water carried onto the beach by breaking waves.

Key Terms

waves
progressive waves
wave train
crest
trough
wave height (H)
wavelength (L)
wave period (T)
wave frequency ($1/T$)
wave speed (C)
sine wave
deep-water waves
shallow-water waves
dispersion
restoring forces
forced waves
free waves
capillary waves
gravity waves
potential energy
kinetic energy
sea
fetch
whitecaps
fully developed sea
limit of stability
swell
significant height
rogue waves
breaker
wave refraction
surf
tsunamis (seismic sea waves)
standing waves (seiches)
node
antinode
storm surge
forerunner
surge
wake
internal waves
longshore current
rip current
beach cusp

Study Questions

1. Draw an ideal wave and label its parts.
2. Explain the differences between shallow-water waves and deep-water waves.
3. List the forces that form waves in the ocean. Where is each force most important in forming waves in the open ocean?
4. Explain the differences between gravity waves and capillary waves.
5. Define sea and swell. Describe how each forms.
6. What limits the maximum size of wind-generated waves in the ocean?
7. Describe what a tsunami is and how it forms.
8. What areas are most likely to experience tsunamis? Discuss the relationship between these source areas and plate-tectonic processes.
9. Draw a diagram of a simple standing wave in a basin. How do standing waves differ from progressive waves?
10. Describe a storm surge and explain what causes it.
11. Why are storm surges more likely from an extratropical storm than from a hurricane?
12. Describe rip currents and the forces that cause them.
13. What are rogue waves? Where do they form? Why are they important to ships?
14. Describe how complex waves can be modeled by combining different sine waves.
15. Describe the differences between internal waves and surface waves.
16. How does the topography of a continental shelf affect waves on nearby beaches?
17. Discuss the difference between free and forced waves.
18. Describe some devices used to extract energy from waves.
19. *[critical thinking]* Why are mathematical techniques so important in studying waves? What assumptions are involved?

Selected References

BASCOM, W., *Waves and Beaches: The Dynamics of the Ocean Surface*, rev. ed. Garden City, N.Y.: Doubleday Anchor Books, 1980. Elementary.

CLANCEY, E. P., *The Tides*. Garden City, N.Y.: Doubleday, 1968. Nontechnical.

MYLES, D., *The Great Waves*. London: Robert Hale, 1986. Discusses catastrophic waves.

RUSSELL, R. C. H., AND D. M. MACMILLAN, *Waves and Tides*. London: Hutchinson, 1954. Elementary.

Flipper

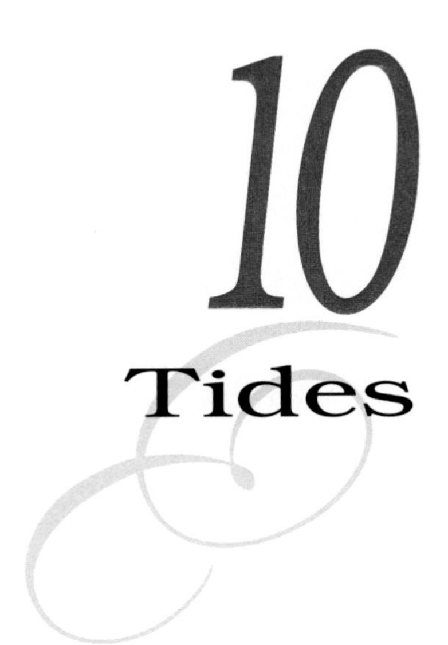

10 Tides

OBJECTIVES

Your objectives as you study this chapter are to understand:

- The general features and causes of ocean tides
- How tides behave in open-ocean basins and in coastal oceans
- The effects of tidal currents, especially in coastal waters

The Bay of Fundy, on the Atlantic coast of Canada, has the world's highest tides—up to 16 meters twice a day. They leave fishing boats and ships stranded on the bottom at low tide.
(Ned Haines/Photo Researchers, Inc.)

Tides, "the ocean's pulse," are the periodic change in sea level relative to the land along a coast. Their effects are most noticeable in coastal oceans, where the periodic rise and fall of the sea surface alternately submerges and exposes shallow ocean bottoms, influencing plant and animal life and behavior. They are the largest waves in the ocean.

Tides are easily observed and measured. All we require is a pole attached to a post or stuck firmly in the bottom of the ocean near the shore. At intervals—perhaps hourly—we record the water level on the pole. The height of the water surface plotted at each interval of time produces a **tidal curve.**

More elaborate installations are needed for continuous tidal observations. A simple mechanical tidal station is shown in Figure 10–1a. A basin with a restricted intake connects to the ocean. Thus, the water level in the basin corresponds to the undisturbed sea level outside but is not disturbed by waves. A float on the water surface in the basin is connected to a pen that plots the tidal curve on a clock-driven, paper-covered drum.

Modern tide gauges work automatically. Observations are recorded electronically for computer processing and also transmitted by satellite to a central recording station (Fig. 10–1b). Instruments on Earth-orbiting satellites can now directly measure the vertical movements of the sea surface caused by the tides. This is especially valuable in the open ocean where, until recently, direct tidal measurements were not feasible without the satellite instruments.

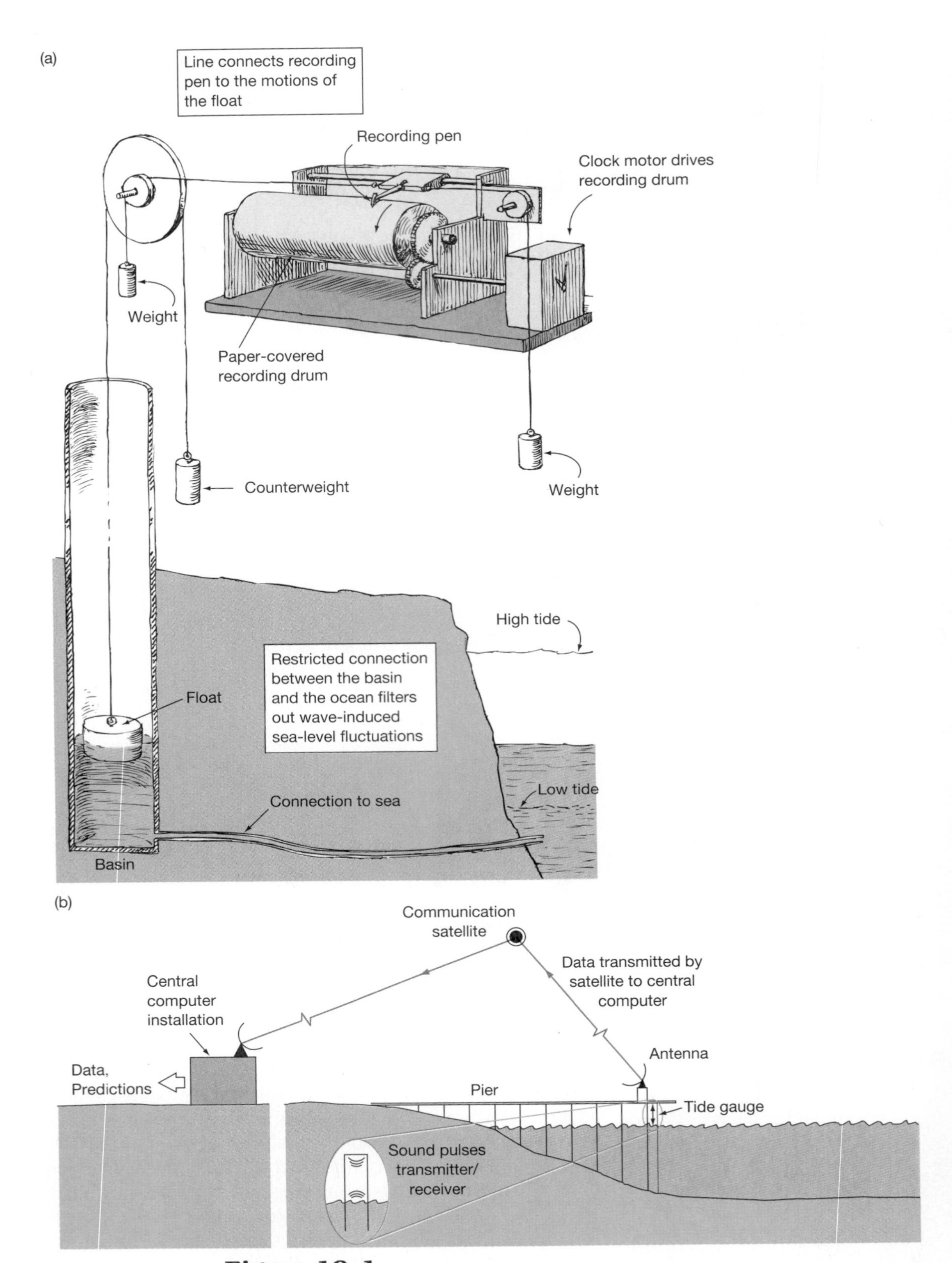

Figure 10–1
(a) A simple mechanical tide gauge. A clock rotates a paper-covered cylinder on which a pencil connected to a float draws a line indicating the changes in sea level. (b) Modern tide gauges use sound to determine sea level. A sound pulse from a transmitter is reflected off the sea surface and its travel time back to the transmitter is measured. The signal is processed at an installation on shore and also transmitted by satellite to a central computer for further processing and data storage.

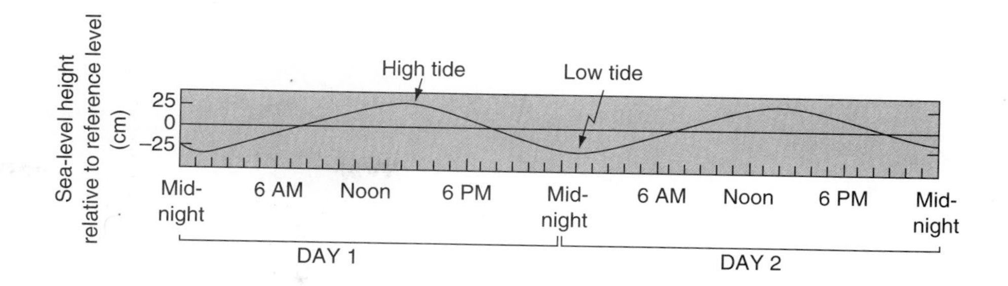

Figure 10–2
Sea-level curve for a daily tide at Pensacola, Florida on the Gulf of Mexico.

Types of Tides

There are three types of tides in the ocean, differentiated by the number of high tides and low tides per day and by their relative heights. A few ocean areas, such as parts of the Gulf of Mexico, have only one high tide and one low tide each day (Fig. 10–2). These are called either **daily tides** or **diurnal tides.**

The second type of tide is found along most coasts: two high tides and two low tides per **tidal day**—24 hours, 50 minutes. (This length of time corresponds to the time between successive passes of the Moon over a given point on Earth.) The time between two successive high tides or two successive low tides is known as the **tidal period.**

When the two high and two low tides are of approximately equal heights, they are called either **semidaily** or **semidiurnal tides** (Fig. 10–3). Such tides are relatively easy to predict at any particular location, because high tides tend to occur at a known time after the Moon has crossed the meridian for that location. Tidal predictions for ports having semidaily tides have been made for centuries, based on their obvious relationship to the lunar cycle.

In the third type of tide, the tidal curves again show two high and two low tides per tidal day, but the two highs are usually of different heights and the same is true of the two lows. These are called **mixed tides** and are shown in Figure 10–4. The higher of the two high tides is called the **higher high water (HHW);** the lower is called the **lower high water (LHW).** There are similar terms for the low tides—**lower low water (LLW)** and **higher low water (HLW).** Mixed tides are more difficult to predict than semidaily tides, mainly because the timing of high-tide and low-tide levels is not simply related to the Moon's passage over a particular location.

From a record of only a few days' length, it is possible to determine the type of tide for any harbor (Fig. 10–5). We can, for example, measure the **tidal range** (the difference between the highest and lowest tide levels) and the **daily inequality** (the difference between the heights of successive high or low tides). Because tides change from week to week, however, we need a record several weeks long to

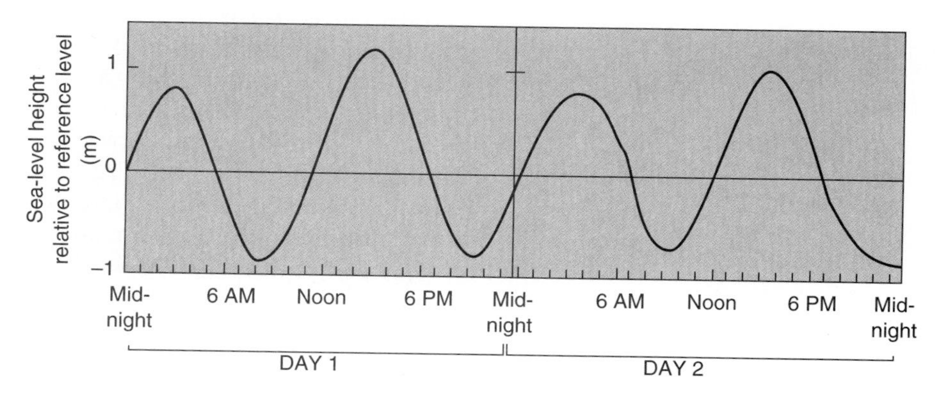

Figure 10–3
Sea-level curve for a semidaily tide at New York Harbor.

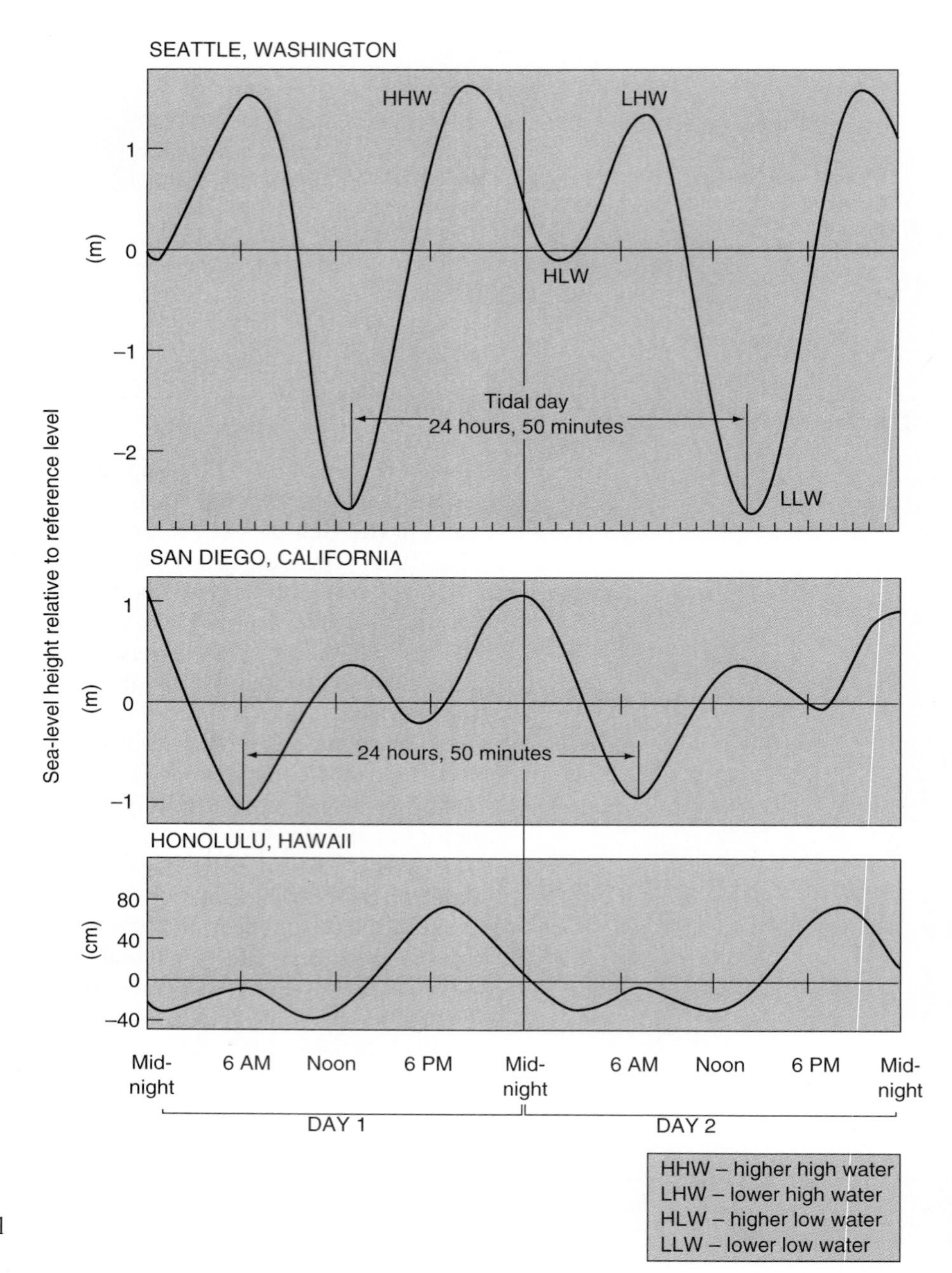

Figure 10–4
Sea-level curves for mixed tides at several Pacific ports.

see a pattern in the changes of tidal range. **Spring tides**—those having the greatest tidal range—occur during full and new moons, the two times when Earth, Sun, and Moon are all in a line (Fig. 10–6a). (The name *spring* is derived from the Saxon word *sprungen,* meaning a strong, active movement. It has nothing to do with the season of the same name.) The spring tidal range is larger than the **mean tidal range** (the difference between the long-term mean high and mean low tides). During the first and third lunar quarters, because the Moon, Sun, and Earth are aligned as shown in Figure 10–6b, the tidal range usually is least; these are called **neap tides.** (From an old Scandinavian word meaning "hardly enough.") As Figure 10–7 shows, there is substantial variation in tides during a month. Other, less striking variations occur over several years.

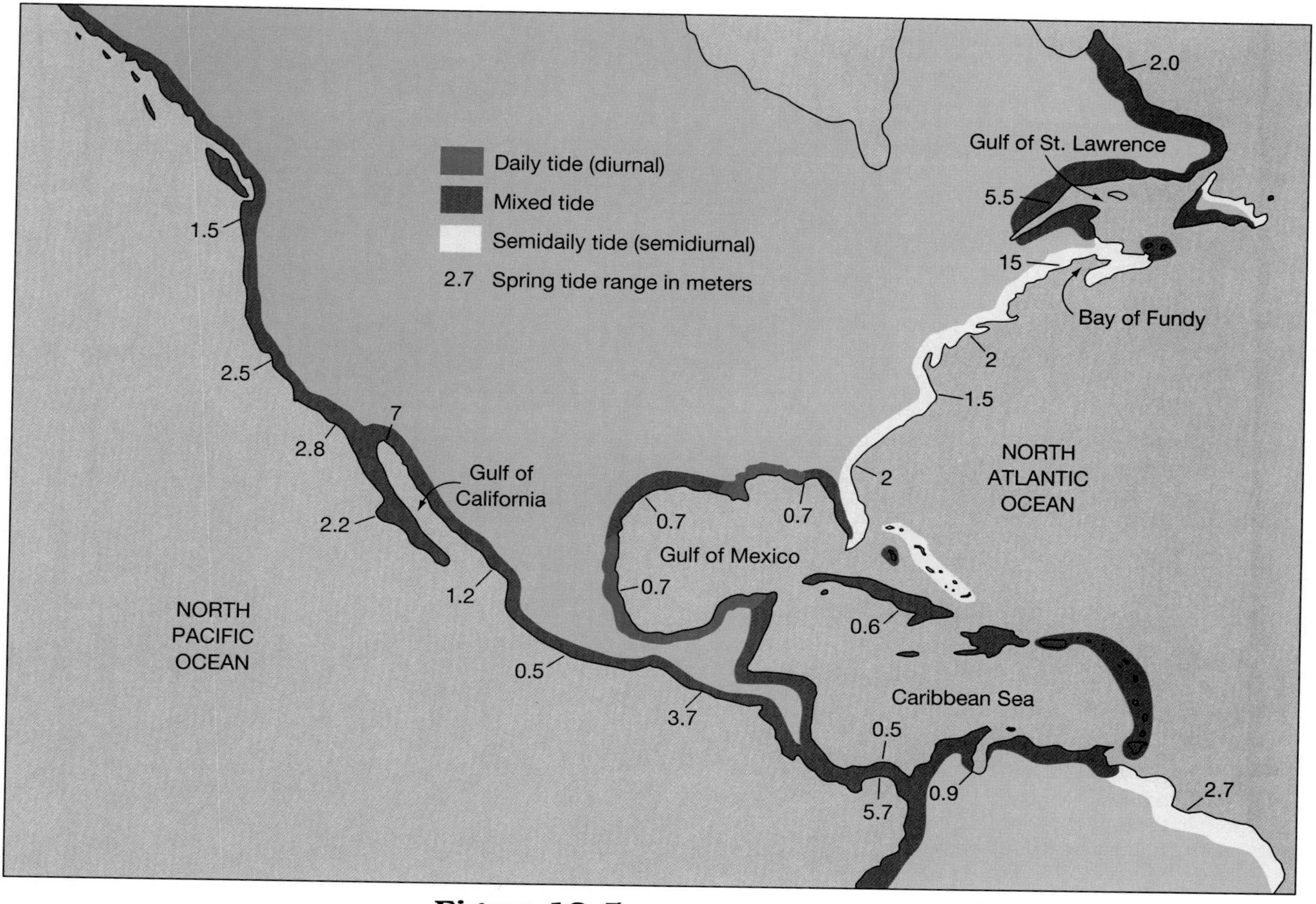

Figure 10–5
Types of tides and spring tidal ranges (in meters) at North American and Central American coasts. (U.S. Naval Oceanographic Office, 1968.)

Figure 10–6
Relative positions of Sun, Moon, and Earth during (a) spring and (b) neap tides.

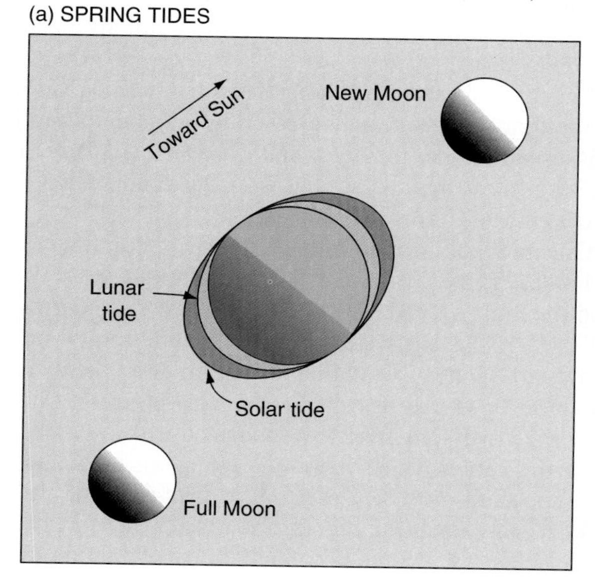

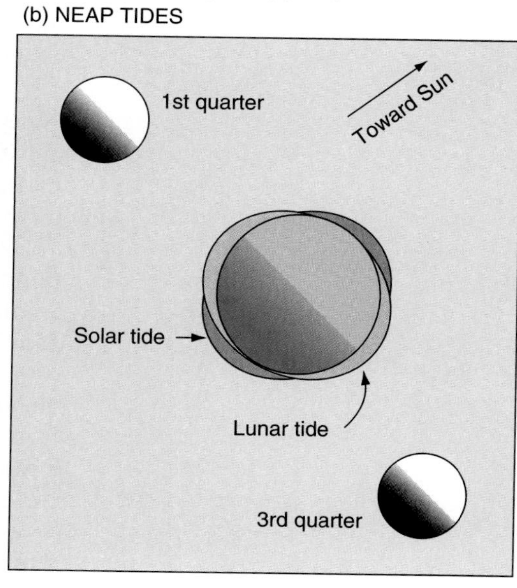

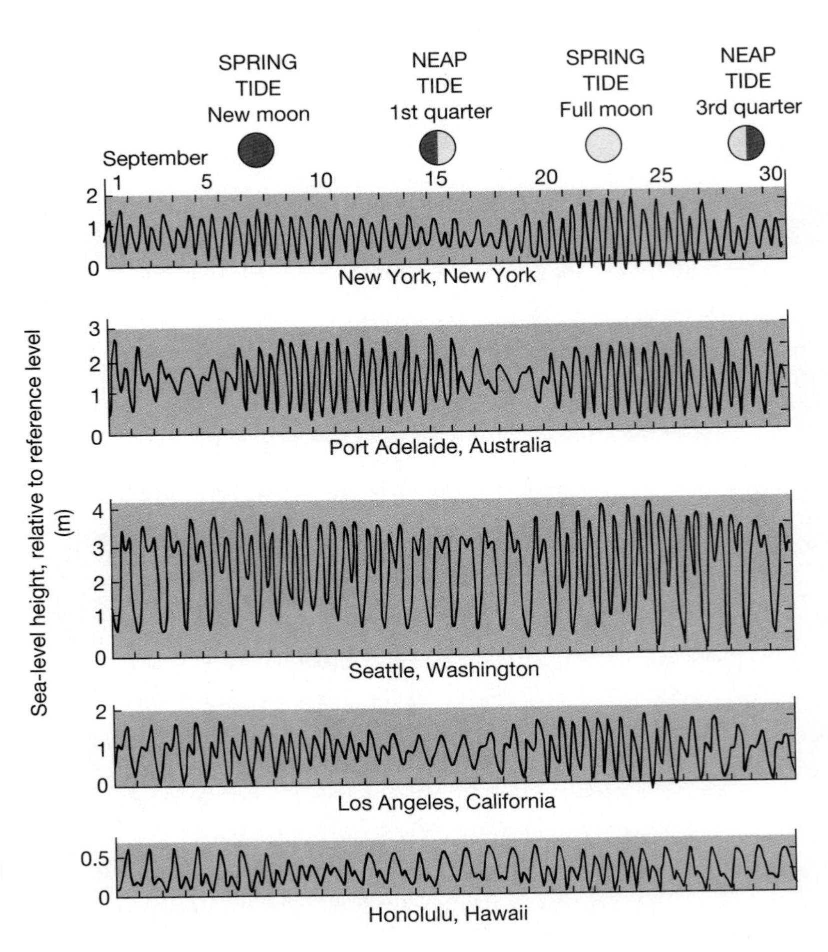

Figure 10–7
Tidal variations during a month. Note the variations in the time of spring and neap tides relative to the new and full moons.

Tide-Generating Forces

Ocean tides are caused by the gravitational attractions of the Sun and Moon acting on the ocean. Because the Earth–Moon distance is so much smaller than the Earth–Sun distance, the Moon has a greater tidal effect on the world ocean. For this reason, we will focus on the lunar tides, but all the basic points also apply to weaker solar tides.

The English scientist Isaac Newton (1642–1727) laid the foundation for understanding and accurately predicting tides. He began by making several simplifying assumptions, mainly a static ocean completely covering a nonrotating Earth. In other words, he started with the simplest case and then progressed to more complicated, more realistic cases.

Gravitational attraction pulls Earth and the Moon toward each other (Fig. 10–8a). These two bodies exert an influence on each other as they rotate around the Sun as a pair. The center of mass of the Earth–Moon system is labeled *M* in Figure 10–8a; the system revolves around this point. (If Earth and the Moon were the same size, *M* would be located midway between them. The mass of the Moon, however, is only about 1/82 the mass of Earth. Consequently, the center of mass of the Earth–Moon system is located within Earth, about 4,700 kilometers from Earth's center. This situation is analogous to an adult and a small child on a seesaw. The adult must sit closer to the pivot than the child to achieve balance.)

There are small but significant unbalanced forces in the Earth–Moon system. Consider first the Moon's gravitational attraction on Earth's surface and on the water covering that surface. In general, the force of gravity exerted by one object on another is inversely proportional to the square of the distance separating

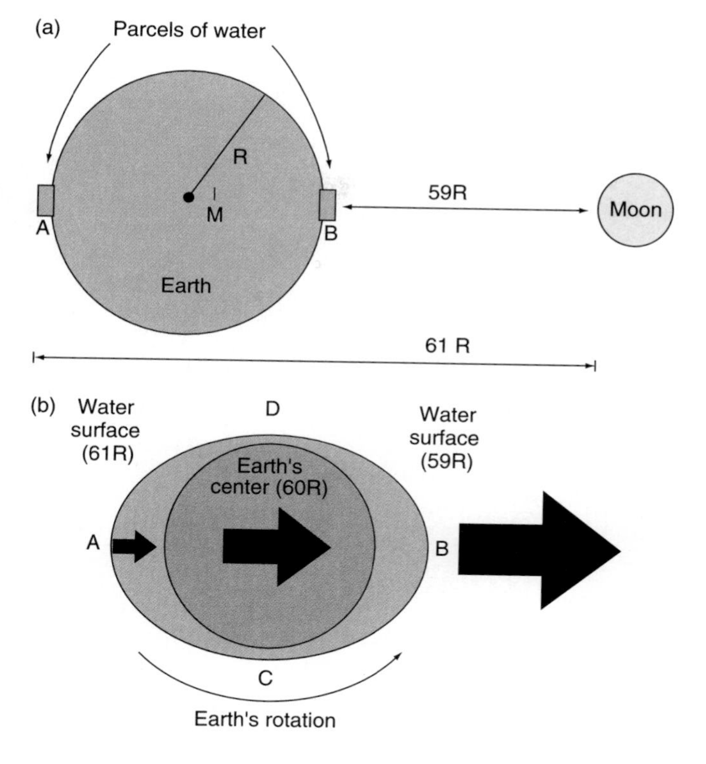

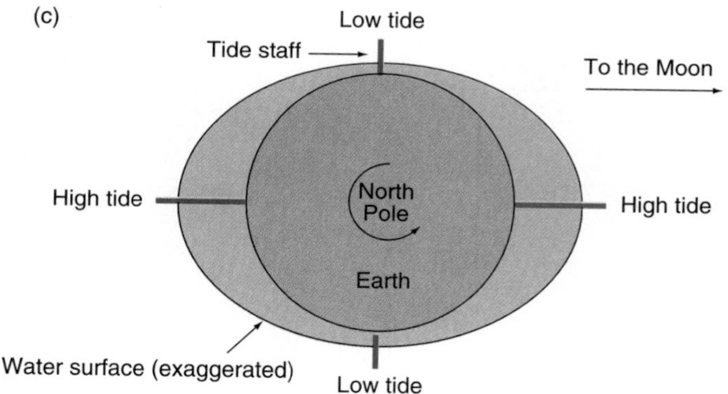

Figure 10–8
(a) Earth and Moon revolve around a common center, *M*. A bit of water nearest the Moon *(B)* is 59 Earth radii *(59R)* from the Moon, while a bit of water on the opposite side *(A)* of the Earth is 61 Earth radii from the Moon *(61R)*. (b) The differences in the gravitational attraction of the Moon on opposite sides of the Earth deform the water surface, forming bulges at *A* and *B*. (c) As Earth rotates within the deformed water surface, any location will experience two high tides and two low tides each tidal day.

the two. (In other words, the force decreases as the distance increases. For example, doubling the separation between Earth and the Moon reduces the force of gravity to one-fourth the original value.) Now consider two parcels of water on Earth's surface, one on the side of Earth closest to the moon (labeled point *B* in Fig. 10–8a) and the other on the side farthest from the moon (point *A* in Fig. 10–8a). The parcel of water at *B* is only 59 Earth radii away from the Moon, but the parcel at *A* is 61 Earth radii away. Therefore, the Moon's gravitational pull is greater at *B* than at *A*.

Because of the gravitational force exerted by the Moon on Earth and on its ocean, there is a bulge in the ocean at both *A* and at *B*. Because ocean water is piled up to create these bulges, the water depth is less at points *C* and *D* in Figure 10–8b, and it is these two bulges and two low points that give us the tides. It is easy to see how the gravitational force exerted by the Moon on the water at *B* pulls the water away from Earth's surface, but not so easy to see why there is a water bulge at *A*. The water bulges at *A* because, in addition to pulling on that water, the Moon is also pulling on the solid Earth beneath that water. The three arrows of different sizes in Figure 10–8b represent the force exerted by the Moon at three locations. The lunar force exerted on the near-side water is the strongest (longest arrow), the

BOX 10–1
Energy from Tides

Figure B10–1–1
Areas where tidal heights exceed 5 meters.

Tides are used to generate electrical power in a few favorable coastal locations, primarily in France, Russia, Canada, and China. Three factors limit the usefulness of tidal power: tidal ranges, topography, and timing. Even with special turbines designed to work on the ebb *and* flood tides, the tidal range must exceed 5 meters to be useful. Such ranges are rare (Fig. B10–1–1), and the range for most coasts is only about 2 meters. Furthermore, many potential sites having sufficiently large tidal range are in remote areas. It would be expensive to transmit any power generated in these remote areas to urban industrial centers.

Topography is a limiting factor because most tidal-power schemes involve one or more dams (Fig. B10–1–2). In a typical system, gates in the dam are opened when the tide is high and then closed, keeping the water behind the dam at high-tide level. When the water level outside the basin has dropped sufficiently, the dam gates are opened and water in the basin allowed to flow out through turbines, thereby generating power.

Finally, there is a timing problem. Tidal power generation is tied to the tidal cycle, which usually does not coincide with periods of peak power demand. The tidal power plant on the Rance estuary in France, for instance, produces about four times as much power during spring tides as during neap tides. Several ideas have been advanced to solve this problem. One is a network of power lines that allow the electricity to be used somewhere, regardless of when it is generated. Another uses multiple dams to store water at high levels, so that one basin serves as a reservoir and another as a collector. Finally, there is the option of generating electricity and storing it in some way for later use.

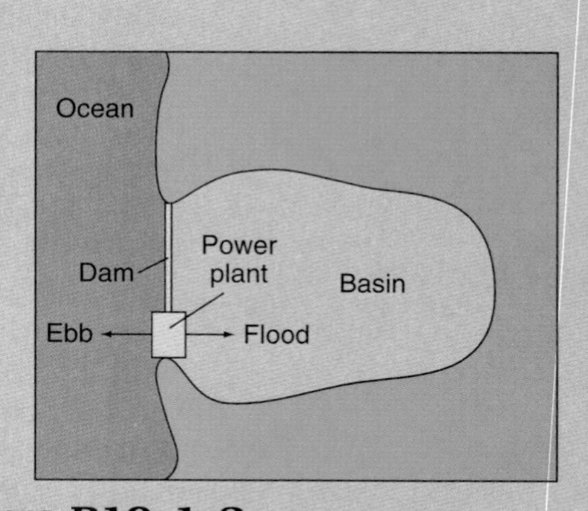

Figure B10–1–2
A simple tidal-power installation. The water enters the basin on the flood tides and exits on ebb tides, each time flowing through the turbines of the power plant to generate electricity.

lunar force exerted on the solid Earth at B is a bit weaker, the force on the solid Earth at A is weaker still, and the force on the water at A is weakest of all (shortest arrow).

Equilibrium-Tide Method

Having considered the lunar tide-generating forces (remember, the Sun also exerts such force on Earth, about half the lunar forces), we can now see how two approaches are used to predict tides. The first and simplest is **equilibrium-tide theory,** so named because it assumes static (or equilibrium) conditions, by which we mean that the water of the ocean does not move; the only thing moving in this model is the planet rotating under its watery covering (Fig. 10–8c). This method predicts the tidal bulges that occur on the side of Earth nearest the Moon and on the side opposite the Moon, as we saw in Figure 10–8c. As a result of the varying strength of the gravitational attraction of the Moon over Earth's surface, the ocean forms a football-shaped envelope over our imaginary, nonrotating, water-covered Earth.

Because Earth rotates inside its deformed watery covering, the equilibrium-tide approach explains successive high and low tides (Fig. 10–8c). When the Moon is in the plane of Earth's equator, as in the left part of Figure 10–9, the tidal bulges are centered on the equator. In Figure 10–8c, a tide gauge on the equator registers high tide when that point is directly under the Moon. After Earth rotates 90°, the tide gauge registers low tide; in other words, the tide gauge is now midway between the two tidal bulges. After Earth rotates another 90°, the tide gauge is directly opposite the Moon and registers high tide.

This simplified equilibrium tide model explains semidaily tides having two equal high tides and two equal low tides per tidal day. Remember that Earth rotates inside its deformed water cover. The more or less football-shaped deformed water surface, corresponding to the equilibrium tide, remains fixed in space, its position determined by the location of the Moon in our simplified case.

The Moon, however, does not maintain a fixed position relative to Earth. Rather it moves, over the course of a month, from 28.5° north of the equator to 28.5° south of the equator. As the Moon changes its position, so does the orientation of the tide-generating forces and the position of the tidal bulge, as Figure 10–9b shows.

A similar model could be made for the Sun-Earth system to determine the Sun's influence on the tides. There are, however, several important differences. First, the greater distance of the Sun from Earth (approximately 23,000 Earth radii) is only partially compensated for by its greater mass (330,000 Earth masses). Thus, the tide-generating effect of the Sun is only about 47 percent that of the Moon. Second, solar tides have a period of about 12 hours rather than the 12-hour, 25-minute period of lunar tides. Finally, the Sun's position relative to Earth's equator also changes—from 23.5° north to 23.5° south of the equator—but this

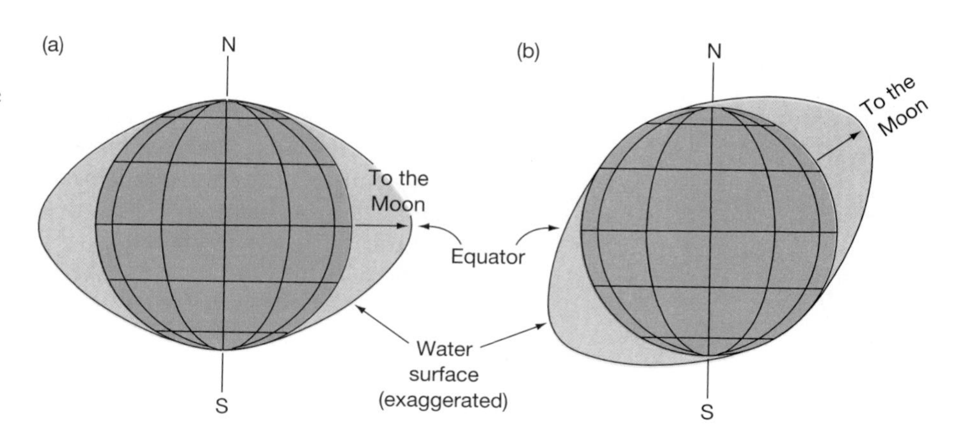

Figure 10–9
The tide-producing forces acting on the ocean surface (a) when the Moon is in the plane of Earth's equator and (b) when the Moon is in a plane north of the equator (right). The tide-producing forces shift orientations as the Moon's position changes.

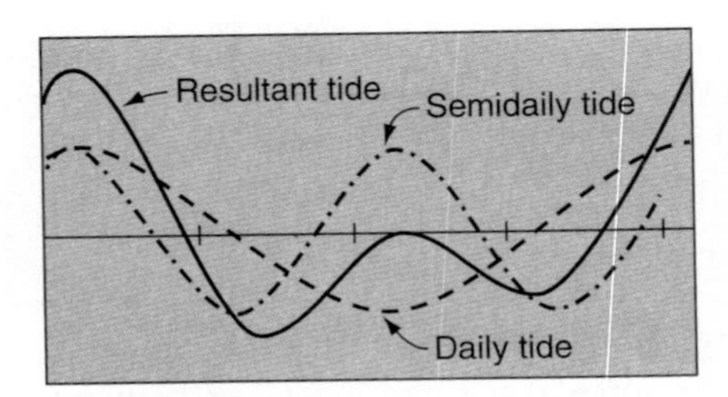

Figure 10–10
A daily and semidaily tide combine to produce a mixed tide.

change of position takes place over a full year, in contrast to the Moon's monthly changes in position relative to Earth.

The equilibrium method also explains spring and neap tides. Spring tides occur every 2 weeks, within a few days of the new and full moons (Fig. 10–6). During the time of the full and new moons, the force exerted on Earth by the Moon and the force exerted by the Sun act together, causing the largest tidal bulges and hence the greatest tidal ranges (Fig. 10–7). During the first and third quarters of the Moon, the lunar force acts at right angles to the solar force, and so the two partially counteract each other. The result is the smallest tidal range—the neap tides.

Dynamic Tidal Method

Since Newton's time, scientists have investigated tides by considering how the ocean responds to tide-generating forces in a model in which the waters of the ocean do not move—in other words, using the static or equilibrium model just discussed. A more realistic approach involves a dynamic rather than a static ocean. It also involves the consideration of the effects on tides of the size and shape of ocean basins and of the Coriolis effect.

The dynamic method treats tides as waves that can be separated mathematically into several components; each component can be treated individually. (We learned about this approach when we discussed waves in the last chapter.) The tide-generating forces exerted by the Sun and the Moon can also be resolved into **tidal constituents,** called **partial tides.** Each of these constituents of the tides can be related to the effects of the Sun or Moon and the interactions among them. As many as 62 tidal constituents are used to make tidal predictions. Four principal constituents account for about 70 percent of the tidal range.

The tide for any location thus can be predicted by combining partial tides (Fig. 10–10), just as complicated waves in a sea can be reconstructed by combining several simple wave trains. When predicting tides using the dynamic model, the first step is to analyze tidal curves mathematically. Then tidal predictions are made by combining partial tides, using computers. Today's more powerful computers permit the use of more sophisticated tidal models, requiring fewer simplifying assumptions, incorporating more tidal constituents, and resulting in more accurate predictions.

Tides are very-long-period waves. Neglecting Earth's curvature, we see that the two water bulges are the crests of a simple wave and the intervening low areas are its troughs. Because of their immense size relative to the ocean basins, tides behave as shallow-water waves. Their wavelength, one-half of Earth's circumference, is about 20,000 kilometers. The ocean basins average about 4 kilometers deep; thus, $D/L = 4/20{,}000$. This depth-to-wavelength ratio is much smaller than the limit of 1/20 for shallow-water waves. If tides were free waves, they would move at a speed of about 200 meters per second (720 kilometers per hour). In order to "keep up with the Moon," however, the tides need to move around Earth in 24 hours, 50 minutes, meaning they must move at a speed of 1,600 kilometers per hour at the equator.

For tide waves to move fast enough at the equator to keep up with the Moon, the ocean would need to be about 22 kilometers deep. Because it is much shallower, the tidal bulges move as forced waves the speed of which is determined by the movements of the Moon. The position of tidal bulges relative to the Moon is determined by a balance between the attraction of the Sun and Moon and frictional effects of the ocean bottom.

Because the tides behave as shallow-water waves, they are felt on the bottom of the deepest ocean basins. Twice a day, the deep-ocean bottom experiences strong currents. As we see in later chapters dealing with bottom-dwelling animals, these currents stir up and move particles. Some of these particles are captured by organisms that are equipped to filter their food from currents.

Open-Ocean Tides

Because the continents separate the world ocean into three north–south basins, it is impossible for tides to move east–west across Earth as forced waves. Only around Antarctica can tides move all the way around Earth. This restriction imposed by the continents adds to the complexity of open-ocean tides. In addition, because the waters in an open-ocean tide move over substantial distances for long times, the tide is modified by the Coriolis effect. The result is a swirling motion such as we might get if we swirled water in a round-bottomed cup. Let us see why such a wave forms and how it behaves.

To see how open-ocean tides behave, we examine the tide in the Atlantic Ocean (Fig. 10–11). Behavior here is typical of all three oceans. The southern basin is open to its north and south and acts like a channel. The northern basin, however, is nearly closed off at its northern end. The tide in the Atlantic has some characteristics of a progressive wave and some characteristics of a standing wave.

After entering the South Atlantic from the Southern Ocean, the tide moves northward. In the South Atlantic, its course is relatively simple. As it progresses northward, the wave from the Antarctic tide interacts with the independent tide generated in the Atlantic.

Each ocean basin has a distinct tide. Unlike tides in the South Atlantic basin, tides in the North Atlantic are altered by interactions with the complicated coastline. A standing wave is set up—not a simple wave (as discussed in Chapter 9), but rather a swirling, Coriolis-influenced wave in which the water surface rotates around a point called an amphidromic point. As the tide continues to move northward into the North Atlantic basin, the water is deflected to the right because of the Coriolis effect, causing a tilted water surface, as shown in Figure 10–12a. As

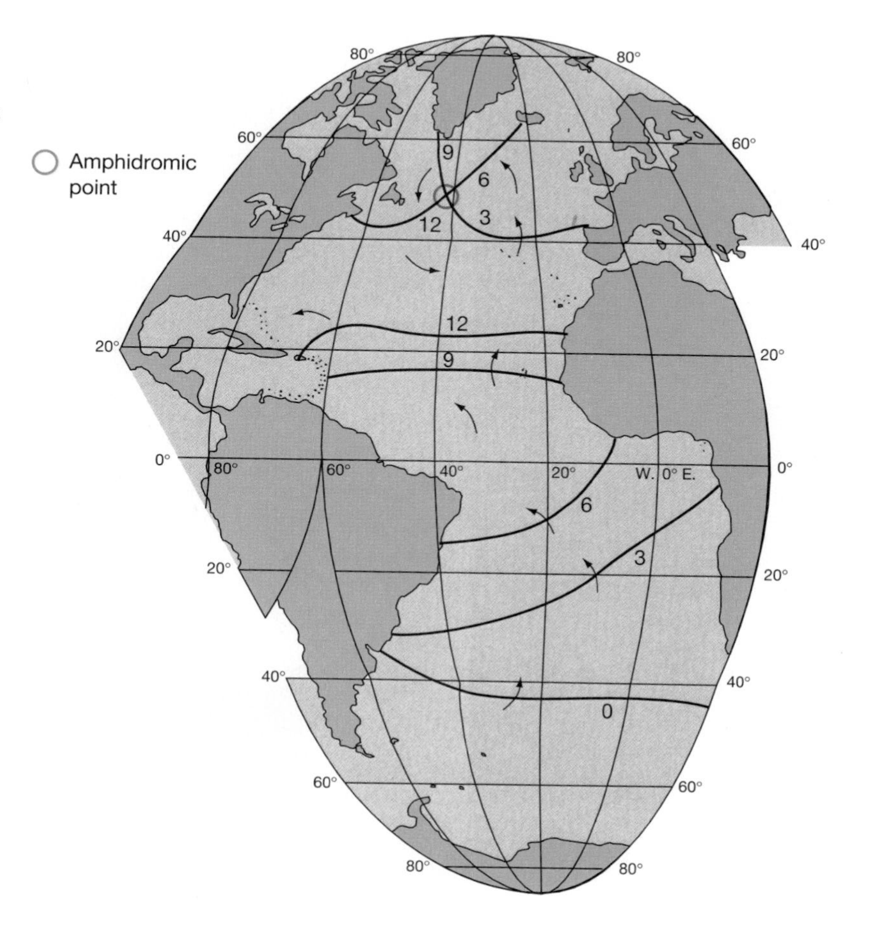

Figure 10–11
Locations of the high water of the principal lunar partial tide—the major tidal component caused by the Moon. Numbers on the dark lines indicate the number of hours since the Moon crossed the longitude of Greenwich (near London). (After A. Defant, *Ebb and Flow*. Ann Arbor: University of Michigan Press, 1958.)

the wave continues to rotate in the basin, the water surface tilts in the opposite sense (Fig. 10–12b). The tilted wave surface rotates around the basin, once during each wave period. Near the center of such a system, called an **amphidrome system** (from the Greek *amphi* "around" and *dromas* "running"), is a point where the water level does not change, the **amphidromic point.**

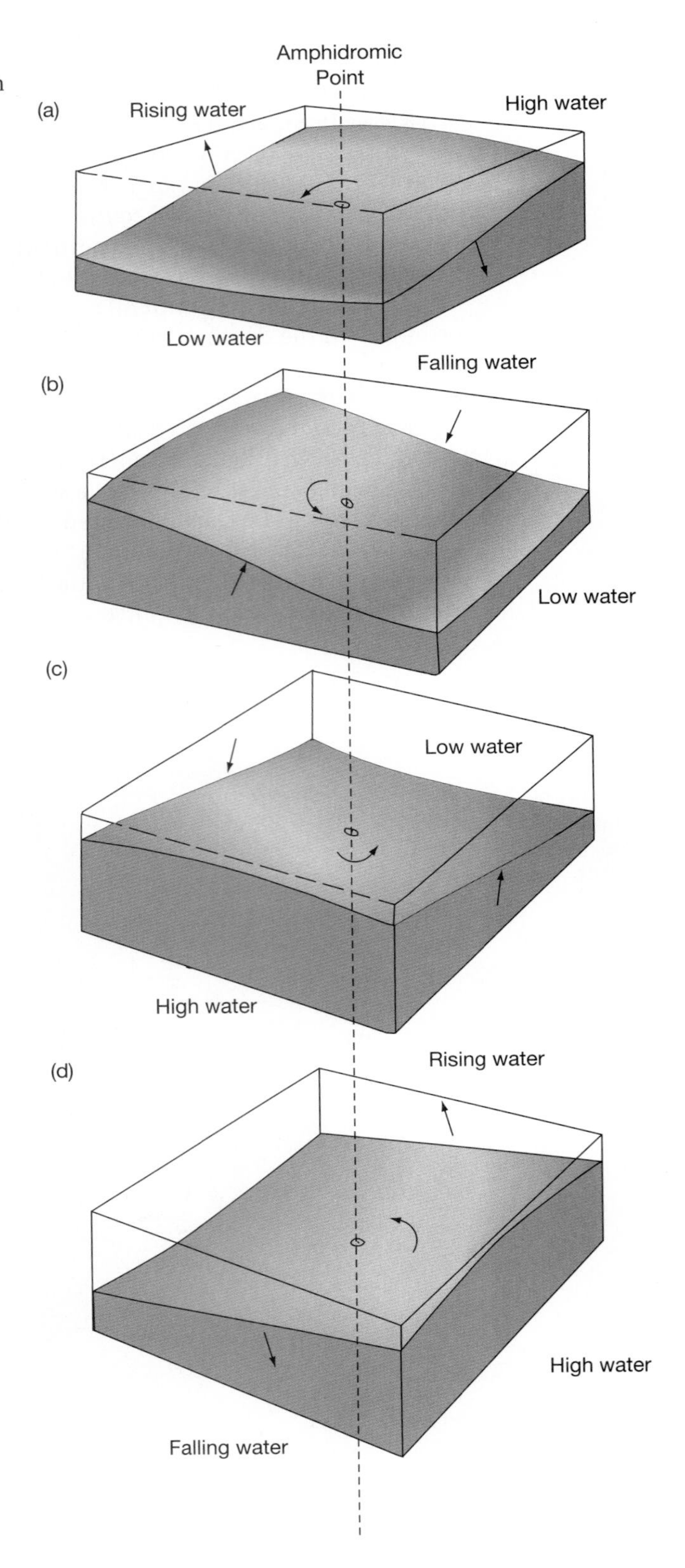

Figure 10–12
Simple amphidromic system in a Northern Hemisphere basin.

Figure 10–11 shows the movement of the high water associated with the principal lunar partial tide (the strongest of the tidal constituents) in the Atlantic. We see a large amphidrome system in the North Atlantic. Other, smaller amphidrome systems (not shown on the map) occur in the English Channel and the North Sea. Sites near an amphidromic point for one partial tide are, by definition, not affected by that partial tide but may be affected by other tidal constituents. For example, the water at the amphidromic point for a semidaily tide may well experience a daily tide.

Every basin has a natural period for standing waves determined by its dimensions. If that natural period is near 12 hours, the standing-wave component of the tide is well developed. If the period of the basin is much greater or less than 12 hours, the standing-wave component is not well developed. Because of their natural periods, ocean basins respond more readily to certain constituents of the tide-generating forces than to others. The Gulf of Mexico, for example, has a natural period of about 24 hours. Therefore, it responds more to daily tidal constituents than to semidaily constituents, and, as a result, much of the Gulf has a daily tide (Fig. 10–5). The Atlantic Ocean, on the other hand, responds readily to the semidaily constituent of the tide-generating forces. It tends to have a semidaily tide. Because of their size and shape, the Caribbean Sea and the Pacific and Indian oceans respond to both daily and semidaily forces. These basins have mixed tides.

The tide in a bay, harbor, or sea is greatly influenced by the magnitude of the ocean tide at its mouth. It is also influenced by the natural period of the basin. A tide wave advancing through a bay (or harbor or sea) is reflected by the coast of the basin. Where the natural period of the basin is near the tidal period, it is possible to set up a large standing wave, giving rise to exceptional tidal ranges. One example is the Bay of Fundy on the Canadian Atlantic coast, the natural period of which is apparently about 12 hours. Spring tides in the inner part of the bay have ranges of 15 meters (Fig. 10–13) because of an especially favorable situation. Large tidal ranges also occur on the northwestern coast of France (about 13 meters) and at the head of the Gulf of California (about 7 meters).

A third factor influencing the tide in any bay, harbor, or sea is the cross section of the opening through which the tide wave must pass. The small tidal range (less than 0.6 meter) of the Mediterranean Sea, for example, results from the small opening through the narrow Strait of Gibraltar into the Atlantic, a narrowness that inhibits exchange of water during each tidal cycle. The small tidal range in several marginal seas of the Pacific can be similarly explained.

Tidal Currents

Tidal currents are horizontal water movements caused by tides. Relationships between tides and tidal currents are not always simple. Some seacoasts have tides but no tidal currents, and a few have tidal currents but no tides.

Let us begin by examining currents caused by a tide that is a simple progressive wave. As we have seen, tides are shallow-water waves because of their extreme length and the relative shallowness of the ocean basins. In such a tide, the crest of the wave is high tide and the trough is low tide. Orbital motions of water caused by the tide are ellipses—greatly flattened circles, each with its long axis parallel to the ocean bottom. In other words, most of the water motions associated with the tides consist of horizontal motions, with little vertical motion involved.

The familiar **reversing tidal currents,** in which water flows in one direction for a while and then reverses to flow in the opposite direction, occur in restricted waters, such as harbors. Such tidal currents can be compared with the water movement associated with the passage of a progressive wave (Fig. 10–14). As the wave crest moves into the harbor, the water flows in; this corresponds to the **flood**

Figure 10–13
(a) High tide and (b) low tide at Halls Harbor, Nova Scotia, on the Bay of Fundy. (Russ Kinne/Comstock.)

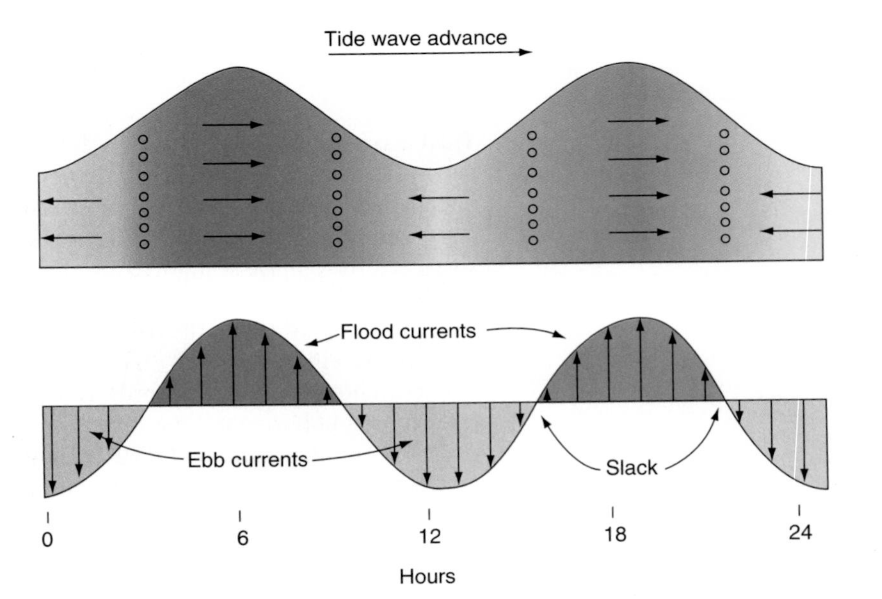

Figure 10–14
Relationship between tides and tidal currents in an idealized progressive-wave tide.

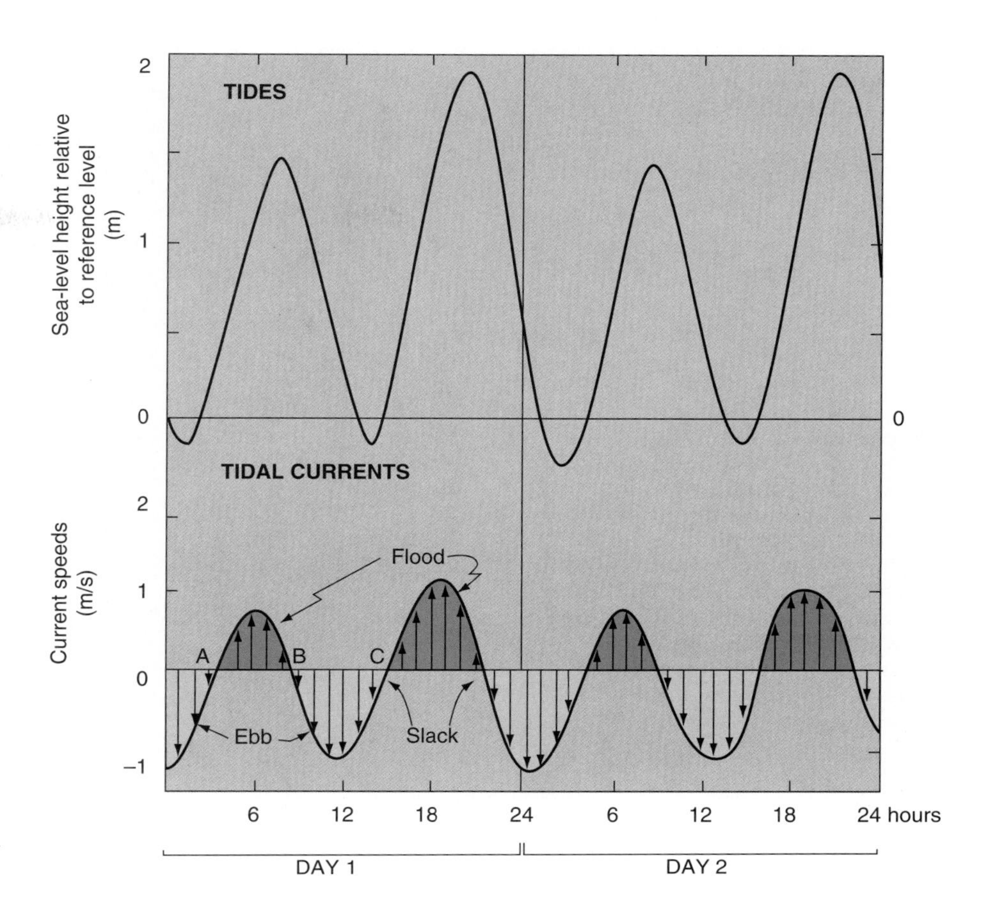

Figure 10–15
Tides and tidal currents over two days in New York Harbor.

current. As the wave trough moves into the harbor, the water flows out; this corresponds to an **ebb current.** Each time the current changes direction, a period of no current, known as **slack water,** intervenes.

Unfortunately, tides and tidal currents are rarely that simple. Because progressive waves are reflected by the shore, the tide observed in most coastal areas consists of several progressive waves moving in different directions, as well as the standing-wave component. Other complicating factors are frictional effects and nontidal currents, such as river discharges. As a result, there is no simple relationship between high or low tides and the times of slack water or maximum currents. Just like the tides, tidal current predictions depend on the analysis of tidal current records collected over a long period of time.

Tidal currents over two days in New York Harbor are illustrated in Figure 10–15. Beginning with slack water before high tide (labeled *A* in the bottom graph), the flood current increases until it reaches a maximum and then decreases again until it is slack water, about an hour after high tide *(B).* After this slack, the current ebbs as the tide falls. The current again reaches a maximum and then decreases until about two hours after low tide, when it is slack water again *(C)* and the cycle begins again.

Elsewhere in New York Harbor, slightly different tidal current patterns may be observed. Tidal currents near shore, for example, are usually weaker than in the middle of the channel because of friction along the channel walls. Tugboat captains often take advantage of these nearshore tidal currents to avoid bucking strong midchannel currents.

Tidal currents are altered by winds and by river runoff. The latter, for example, can prolong and strengthen ebb currents because more water must move out of the harbor on the ebb than comes in on the flood. Also, different tides have different tidal currents. In areas with large daily inequalities between successive high

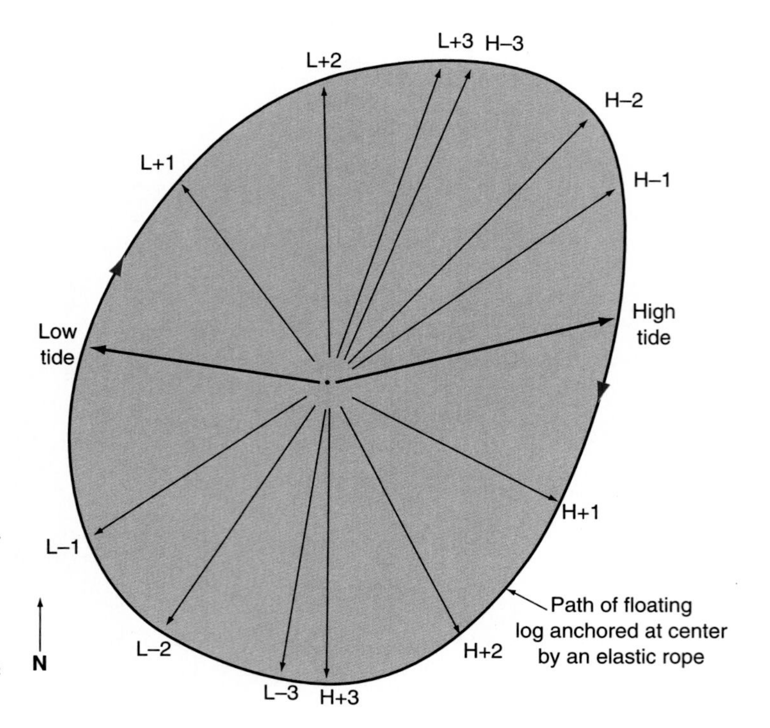

Figure 10–16
Average tidal currents during one tidal period at Nantucket Shoals off the Massachusetts coast. The arrows indicate the direction and speed (shown by arrow length) of the tidal currents for each hour during a tidal period. The outer ellipse represents the movements of a log (tied by an elastic line) in the water in the absence of winds or other currents. The values at each arrowhead are hours before (–) or after (+) a tide.

(or low) tides, there may be days with continuous ebb (or flood) currents that change in strength during the day.

The strength of a tidal current depends on the volume of water that must flow through an opening and on the size of the opening. Thus, tidal range alone is not sufficient information to predict the strength of a tidal current. The large tidal range in the Gulf of Maine, for instance, is accompanied by weak tidal currents because the opening to the gulf is large. Conversely, Nantucket Sound has strong tidal currents but a small tidal range. At a given location and on a given day, however, tidal current strength is generally proportional to tidal range for that day. A spring tide, for instance, is usually accompanied by stronger tidal currents than a neap tide. In general, tidal currents are the strongest currents in coastal regions.

In the open ocean, **rotary tidal currents** exhibit patterns quite different from the reversing tidal currents in coastal areas. Rotary tidal currents continually change direction. To understand the movement of rotary currents, picture a log anchored to the ocean bottom by an elastic tether near Nantucket Shoals (off the southeast coast of Massachusetts). During a simple semidaily tide, this log would move clockwise in an elliptical path, returning to its initial point after 12 hours, 25 minutes, as indicated in Figure 10–16. (The absence of wind and nontidal currents is assumed.)

In mixed-tide areas where the inequality is substantial, the current motion is more complicated because there are two ellipses (Fig. 10–17). A log placed in such a current moves through two elliptical paths and returns to its original position after 24 hours, 50 minutes (again assuming no winds and no nontidal currents).

Instead of the slack-water periods of coastal areas, the open ocean has periods in which the rotary current is at a minimum. In the Northern Hemisphere, the current usually changes direction in a clockwise sense. Open-ocean tidal currents are generally weaker, typically about 30 centimeters per second (slightly more than 1 kilometer per hour), than their coastal counterparts.

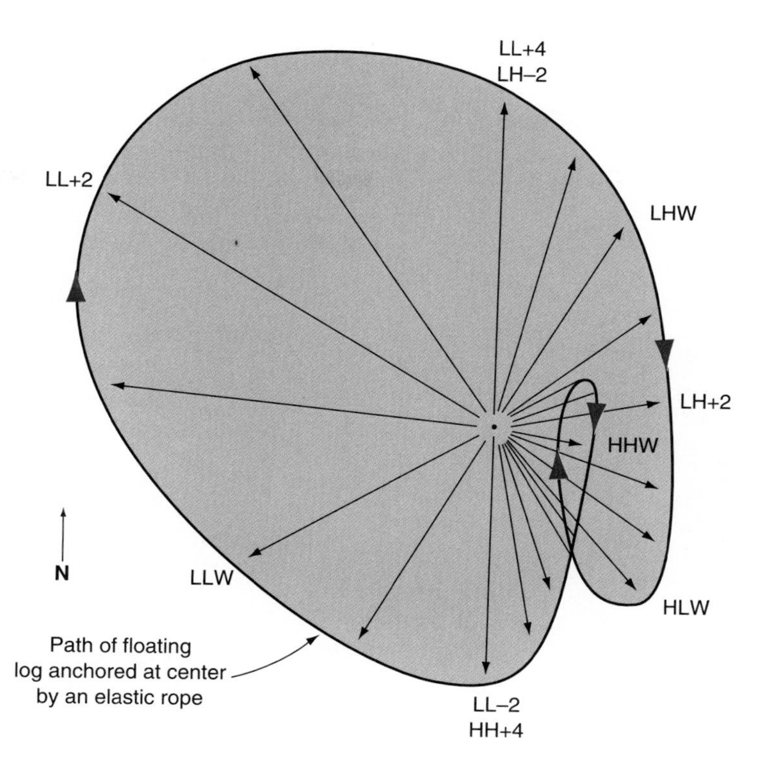

Figure 10–17
Average tidal currents during one tidal day at the entrance to San Francisco Bay. The radiating arrows show the direction and speed (shown by arrow length) of tidal currents for each hour of the tidal day. The outer ellipse represents the movements of a log (tied by an elastic line) in the water in the absence of winds or other currents. The labels at the arrowheads represent hours before (–) or after (+) a tide. The complex shape of the curve is caused by the mixed tides at this location.

Coastal-Ocean Tides

Tides in coastal areas affect bays, river mouths, and harbors. The tide enters a bay as a progressive wave and moves inland. In many bays the wave is gradually damped (reduced in height) by friction and opposing river flow. In some cases, however, the tide wave reaches the end of the bay and is reflected back.

To see how water moves in a tidal embayment let us look at a typical example—Chesapeake Bay on the Atlantic coast of the United States. This area has relatively simple tidal currents associated with a tide that behaves primarily like a progressive wave. To follow the changing tidal currents, we begin when the waters are slack at the entrance, as shown in Figure 10–18a. From just inside the entrance, in the southern part of the bay, up to a second slack-water area, about midway up the bay, tidal currents are ebbing. In the northern part of the bay, above the upper slack area, the currents are flooding. What we learn from this pattern is that, in general, slack water separates sections of ebb currents from sections of flood currents.

Two hours later, both slack-water areas have moved into the bay about 120 kilometers, as shown in Figure 10–18b. By this time, tidal currents are flooding at the bay entrance, and the pattern we observed in (a) has been displaced northward. Four hours later, the southernmost slack-water area has reached the entrance to the Potomac River and flood-tidal currents at the mouth of the bay (at Cape Henry) have weakened somewhat, as shown in Figure 10–18c. A final look at Chesapeake Bay, six hours after slack water before flood shows that the area at the bay mouth is again experiencing slack water, but this time it is slack water before ebb. The arrows shown in the initial picture would now be reversed, meaning that flood currents take the place of ebb currents.

Because of the narrowness of Chesapeake Bay, tidal currents flow generally along the bay long axis. Because there is little tidal flow across most of the bay, current patterns are complicated by crossflow only in the wider parts.

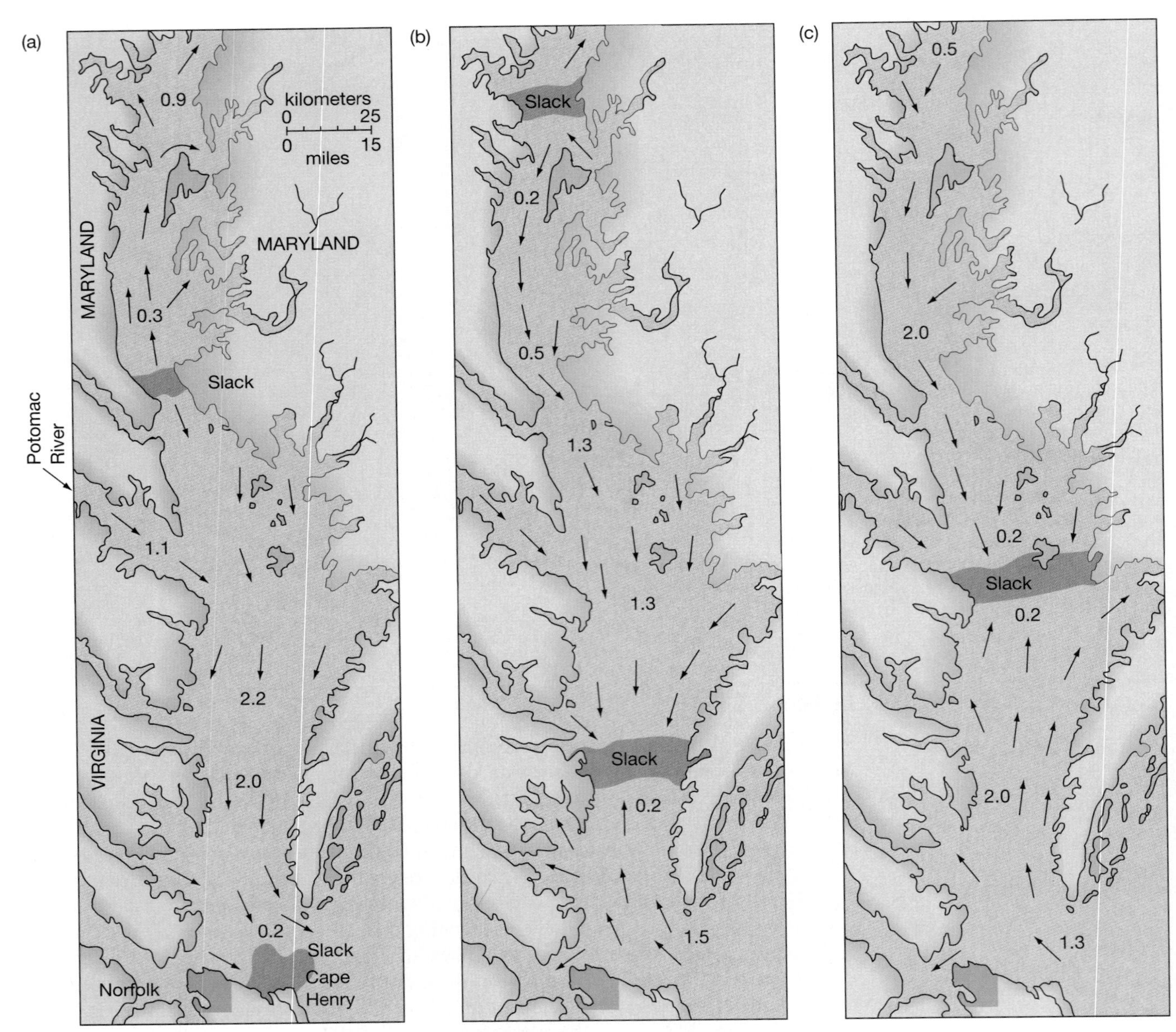

Figure 10–18
Tidal currents in Chesapeake Bay illustrate the behavior of a progressive-wave tide in an estuary. (a) Slack water before flood begins at the mouth of the bay; (b) 2 hours after flood begins; and (c) 4 hours after flood begins. Arrows indicate current direction, and the numbers are maximum current speeds in kilometers per hour during spring tides. (U.S. National Ocean Service.)

In Long Island Sound the tide behaves not like a progressive wave, but like a standing wave. Here the tidal currents are nearly the same over most of the sound at any time. For instance, slack water occurs nearly simultaneously over the entire sound, and flood or ebb currents do, too, although current strength varies from one area to another (Fig. 10–19). The area near the western end of the sound is an exception to the same-all-over description. Here, an area of slack water separates the small area of opposing tidal currents that have advanced into Long Island Sound from the tidal system in New York Harbor, which connects with the sound through narrow channels.

The standing-wave tide in Long Island Sound results in part from the natural period of the sound, which is apparently about 6 hours.

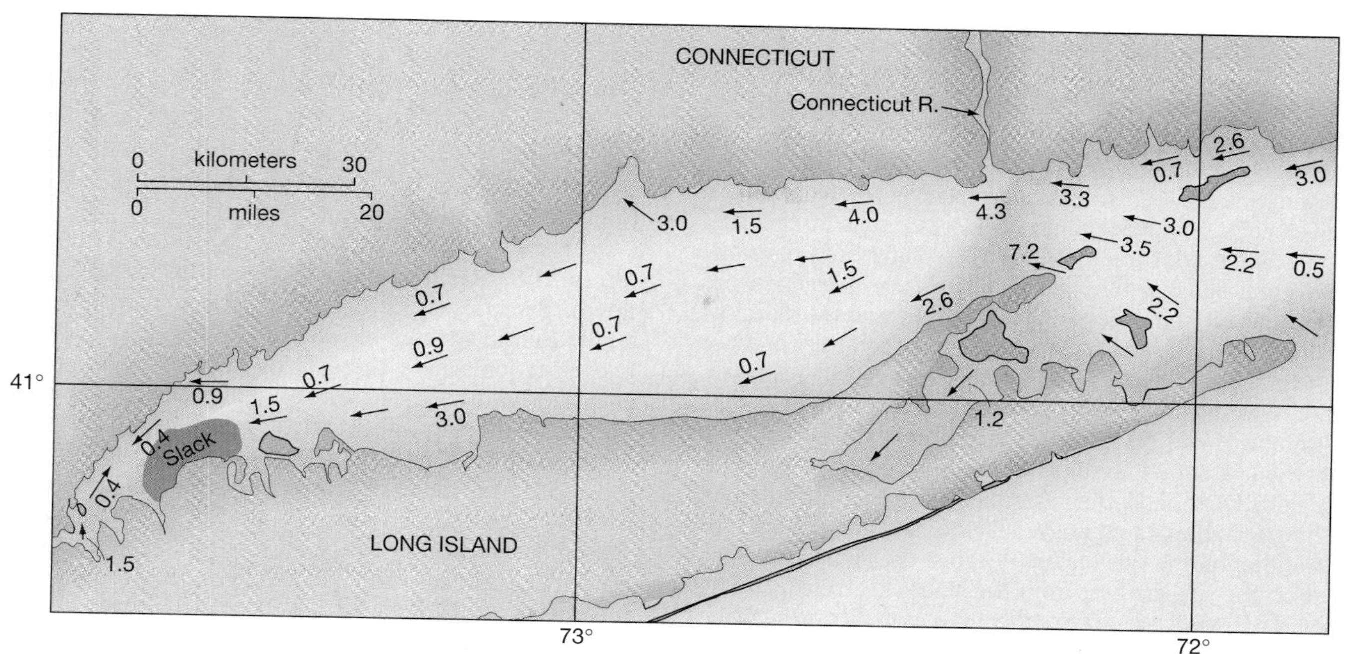

Figure 10–19
Tidal currents in Long Island Sound show a standing-wave tide during flood at the eastern entrance to the sound. The arrows indicate current direction, and the numbers are current speed in kilometers per hour. Note that the tide is flooding everywhere within the sound, in contrast to more complicated tidal current patterns in Chesapeake Bay. (U.S. National Ocean Service.)

Observing Open-Ocean Tides

Direct observations of ocean tides have long been restricted to coastal waters. As we have seen, the rise and fall of sea level is easily measured by even simple instruments. Because of their importance to shipping, tidal records have been kept for many decades in most ports and even longer in some of the most important ports, such as London, New York, or Amsterdam. But direct observations of tides away from the coast was impossible. Only recently have pressure gauges been installed by researchers on relatively shallow coastal-ocean bottoms where the effects of the tides can be measured.

Earth-orbiting satellites offer opportunities to observe tides in the open ocean. Radar altimeters on these satellites can measure the height of the sea level beneath the satellite's orbit. When enough observations are available it will be possible to detect tidal effects. Such observations must extend over many months to permit the effects of waves and storms to be eliminated from the small changes in sea levels due to open-ocean tides.

Studies of open-ocean tides will be greatly expanded by satellite observations. Extensive observations of the ocean made during the Cold War for military purposes are not yet available to ocean scientists. For example, in the 1980s, the U.S. Navy made accurate maps of the shape of the ocean surface to account for local gravity effects, thus enabling accurate aiming of intercontinental missiles launched from submarines. Some of the earliest data from such satellite surveillance have been recently released and more are expected. Such data will provide oceanographers the first ever direct measurements of open-ocean tides.

Summary

Tides are the periodic rise and fall of the sea surface due to the attraction of the Sun and Moon. The tidal period is the time between successive high (or low) tides. The tidal day is the time between successive transits of the Moon over a given point on Earth and is equal to 24 hours, 50 minutes.

There are three types of tides. Daily tides have one high and one low tide per day. Semidaily tides have two nearly equal high tides and two nearly equal low tides per tidal day. Mixed tides have two unequal high tides and two unequal low tides per tidal day.

Two approaches have been used to explain how the Sun and Moon cause tides. The simplest, the equilibrium method, assumes a static ocean completely covering a smooth Earth, and considers only tide-generating forces. The gravitational pull exerted by the Moon on the ocean is strongest on the side of Earth nearest the Moon and weakest on the side farthest from the Moon. This inequality in the strength of the gravitational forces deforms the water surface of the world ocean into bulges on the near and far sides. The Earth rotating once a day within this football-shaped shell of water thus rotates through two areas of high water and two areas of low water.

Tides are very long waves (wavelength half the Earth's circumference), the location of which is determined by the Moon's position with respect to Earth. In other words, tides are forced waves, unlike wind waves, which are free waves.

The dynamic approach includes the responses of the ocean to the different shapes of the ocean basins. This method treats tides as very long waves. Tidal observations are simplified mathematically into constituents that are used to make predictions. Each constituent can be related to observed effects of the Sun and the Moon. Tidal predictions are calculated using these techniques.

Each basin responds differently to tide-generating forces. The natural periods of some basins favor standing waves. Because of the Coriolis effect, such waves form amphidromic systems in which the wave rotates about a fixed location where there is no tidal effect. In nearly enclosed basins, factors controlling tidal effects include the range of ocean tides at the entrance to the basin, the basin's natural period, and the size of the ocean entrance.

Tidal currents are horizontal water movements caused by tides. Such currents may be viewed as water movements caused by the passage of progressive shallow-water waves. In coastal areas we have reversing tidal currents. In the flood stage, the water flows toward shore; this is the crest of the progressive wave reaching shore. In the ebb stage of a reversing current, the water flows away from shore; this is the trough of the progressive wave reaching the shore. For a short time between flood current and ebb current, as the water is changing direction, there is no water movement, and such water is said to be slack. Open-ocean tidal currents flow continuously, with particles moving in elliptical orbits; these are called rotary currents.

Key Terms

tides
tidal curve
daily (diurnal) tides
tidal day
tidal period
semidaily (semidiurnal) tides
mixed tides
higher high water
lower high water
higher low water
lower low water
tidal range
daily inequality
spring tides
mean tidal range
neap tides
equilibrium-tide theory
tidal constituents
partial tides
amphidrome system
amphidromic point
tidal current
reversing tidal currents
flood current
ebb current
slack water
rotary tidal currents

Study Questions

1. Diagram the three types of tidal curves and label the principal features of each.
2. Describe the equilibrium theory of tidal generation. What are the major assumptions of this theory? What features of tides does the theory fail to explain?
3. Explain the dynamic theory of tidal generation. How does this theory differ from the equilibrium theory?
4. Diagram the simplest relationships between tides and tidal currents.
5. Explain the difference between rotary and reversing tidal currents. Draw a diagram showing water flows in each.
6. Discuss the difference between progressive-wave tides and standing-wave tides.
7. Explain why the tide acts like a forced shallow-water wave in the deep ocean.
8. Diagram the relationship between tide stage and tidal currents in an estuary.
9. Diagram an installation for obtaining power from tides. Contrast this with an installation for obtaining power from waves.
10. Describe how tides are measured.
11. Describe the contributions Newton made to understanding tides.

12. Explain the differences between how the Sun and Moon affect tides on Earth.
13. Where are tidal power stations feasible?
14. Why is it more difficult to predict tidal levels in an estuary than on the open coast?
15. *[critical thinking]* Explain how tides behave like very long waves.
16. *[critical thinking]* Why are there so few tidal-power plants?

Selected References

BRIN, A., *Energy and the Oceans*. Surrey, England: Westbury House, 1981. Discusses tidal energy.

DEFANT, A., *Ebb and Flow*. Ann Arbor: University of Michigan Press, 1958. Descriptive and mathematical treatment.

RUSSELL, R. C. H., AND D. M. MACMILLAN, *Waves and Tides*. London: Hutchinson, 1951. Elementary.

ΑΡΓΥΡΟΙ Η.Μ.2

11 Oceanic Life and Ecosystems

OBJECTIVES

Your objectives as you study this chapter are to understand:

- The characteristics of marine environments capable of supporting life
- The principal features of oceanic ecosystems
- Primary and secondary production in the ocean
- How organisms affect seawater
- What controls primary production and its distribution in the ocean

In many coastal areas, the rich biological productivity of the ocean supports economically important fisheries. Here a Greek fishing trawler heads towards Mykonos.

Different oceanic features and processes control the abundance and distribution of life in the ocean. Some oceanic processes are similar to those we know on land, while others are quite different. In this chapter, we examine the ocean first as an environment capable of supporting life and then as an enormous biochemical system in which some of the energy stored in sunlight is incorporated into living matter. In Chapters 12 and 13, we focus on different parts of these oceanic systems to see how energy is used by individual organisms and by communities of organisms. In Chapter 14, we discuss the distinctive features of coastal oceans and how these environments and the organisms that inhabit them have been affected by human activities.

Oceanic Habitats and Life-Styles

The ocean provides a three-dimensional environment for life, from its edges and surface to the deep ocean bottom (and even in the rocks and sediments below the bottom). This three-dimensionality is quite unlike the land environment (Fig. 11–1a), where we land-dwellers are for the most part confined to living in the two dimensions of Earth's surface. Thus, there is much more living space in the ocean.

Most marine plants and animals live near the top, bottom, or sides of the ocean (Fig. 11–1b). The mid-oceanic depths are less populated and also little known, because we have few tools to observe and sample organisms there.

Figure 11–1
(a) Life on land is restricted to a thin zone, no more than a few hundred meters thick. (b) In the ocean, life exists at all depths, up to several kilometers deep. In the open ocean, there are no boundaries for swimming or floating organisms.

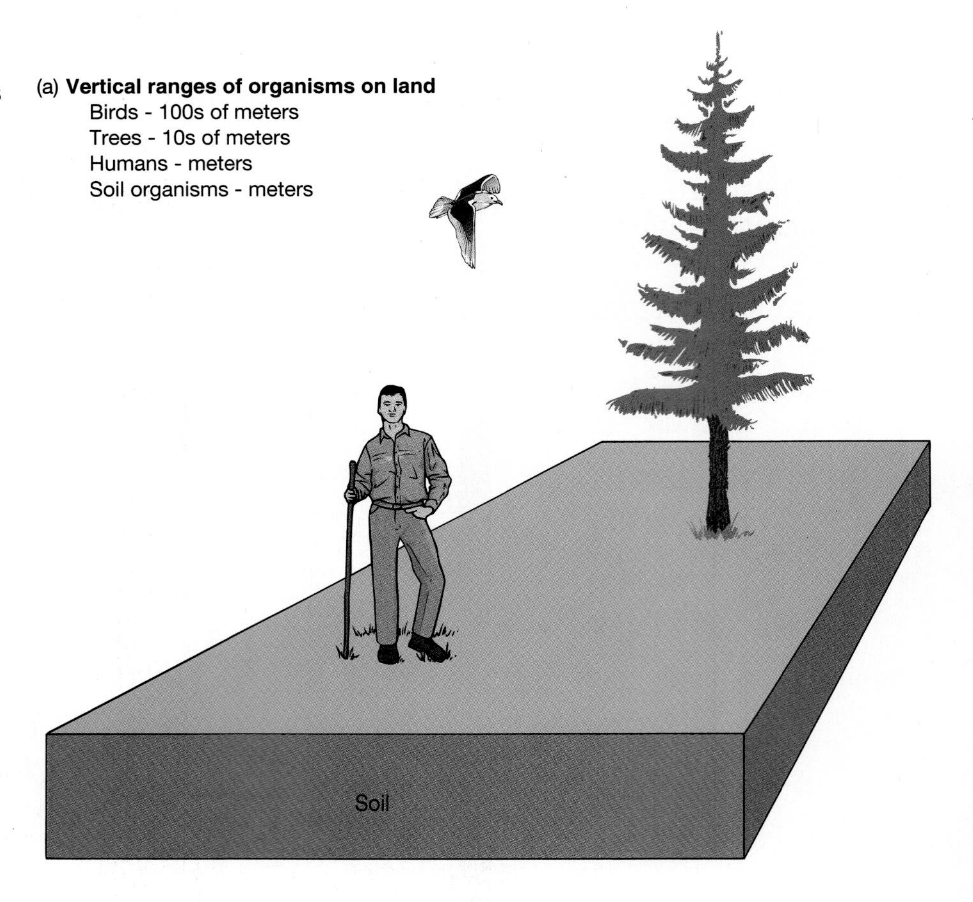

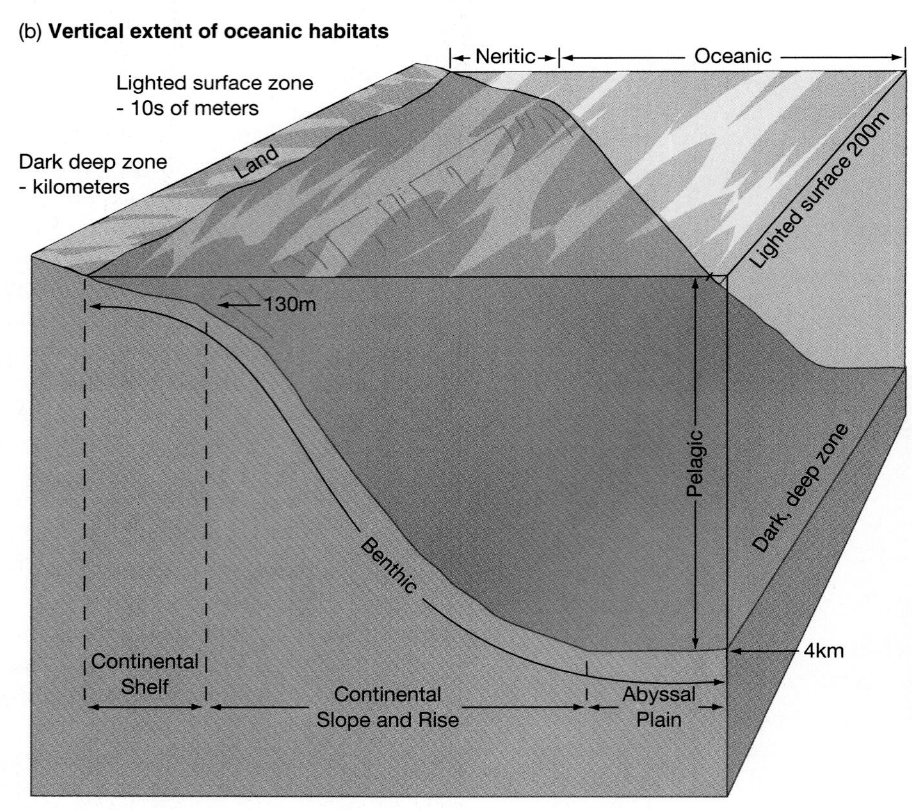

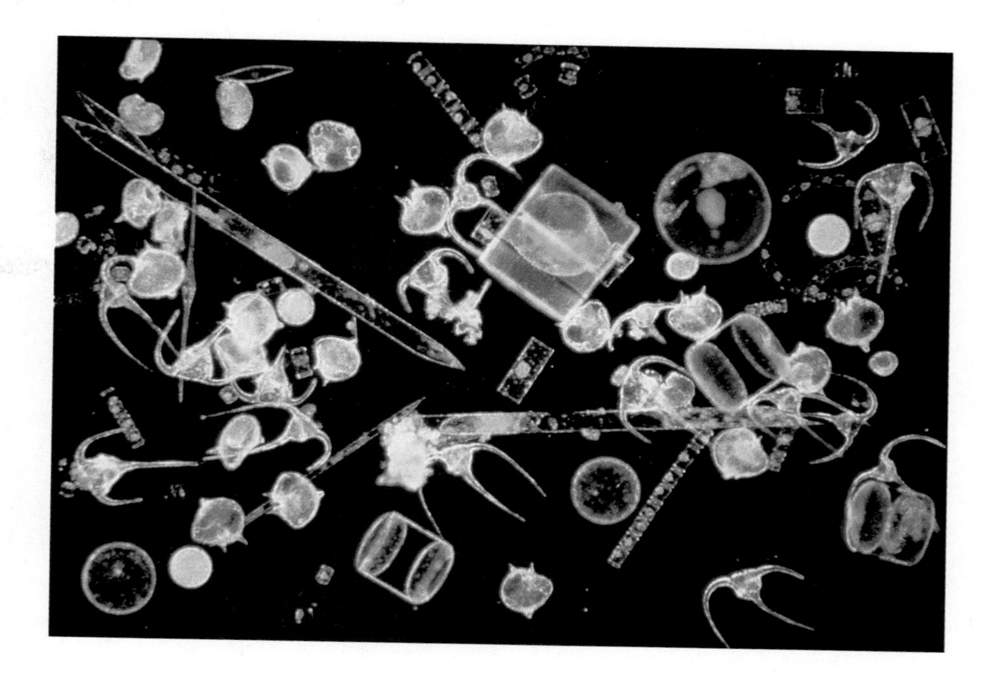

Figure 11–2
A variety of live phytoplankton species from the Atlantic Ocean. They include chain-forming diatoms, large single-celled diatoms (pillbox and needle shaped), and dinoflagellates (with three spines, two of them curved). (Copyright N. T. Nicoll.)

There are three quite different oceanic life-styles: *planktonic* (drifting), *nektonic* (swimming), and *benthic* (attached). As Figure 11–1b shows, **pelagic** organisms live in the waters of the open ocean, **neritic** ones live in coastal waters, and benthic organisms live on the floor of the continental shelf, the slope and the deep ocean.

The drifting organisms, called **plankton,** are weak swimmers at best, buoyant, and easily carried by currents. With a few exceptions, they are small and include **bacteria, phytoplankton** (one-celled plants, Fig. 11–2), and **zooplankton** (planktonic animals, Fig. 11–3). **Nekton** are strong swimmers and include fishes, squid (Fig. 11–4), and whales. **Benthos** include large plants (Fig. 11–5) that grow in shallow waters and bottom-dwelling animals (Fig. 11–6) at all depths.

Figure 11–3
Zooplankton; the most common forms are copepods and their larvae (distinguished by their paired legs and antennae). This picture also includes the larval forms of starfishes and their relatives (transparent with long delicate arms) and the larvae of other common marine organisms. (Copyright N. T. Nicoll.)

Figure 11–4
Squid are fast swimmers and effective predators. They are part of the nekton.

Figure 11–5
Kelp, the largest plants that grow in the ocean, live attached to the shallow ocean bottom and utilize the abundant sunlight available at these depths.

Figure 11–6
Benthic animals compete for living space on the ocean bottom off the California coast.

A word on usage before we go on. *Plankton* refers to a population of organisms; individual members of such a population are called **plankters.** Thus, we say that *phytoplankton* are more abundant in summer than in winter, and only a few *phytoplankters* survive over a winter.

Temperature

Water temperature is the primary factor controlling where organisms live in the ocean. Since many organisms have limited temperature tolerances, the boundaries of their distribution ranges are found where there are marked temperature changes. For example, the region of Cape Hatteras on the U.S. Atlantic coast, where the warm Gulf Stream waters swing offshore, is the location of a distinct change in the types of marine organisms from subtropical to temperate species. The reason is simple: temperature controls the rates of many chemical and biological reactions important to living organisms.

Except for marine mammals and seabirds, marine organisms are **poikilothermic** (cold blooded) which means that their body temperature is controlled by the temperature of their surroundings. The rates of chemical reactions and of most biological processes generally double with each 10°C increase in temperature. Organisms living in the cold waters of polar seas or the deep-ocean live longer and grow more slowly than similar organisms living in tropical waters.

Another factor controlling distributions of marine organisms is their ability to tolerate temperature changes. As we saw in Chapter 7, the largest temperature changes in the ocean occur with depth. Descending through the thermocline, an organism may experience a temperature change as great as 25°C within a few hundred meters. In the surface zone, an organism would have to travel thousands of kilometers from a subtropical gyre to subpolar regions in order to experience a comparable temperature change. Organisms that cannot tolerate large changes in temperature are usually restricted to limited areas; those able to tolerate temperature changes are more widely distributed. Eggs, larvae, and the juvenile stages of most organisms are more sensitive to temperature changes than are the adult forms. Organisms living in tide pools in coastal areas are among the most tolerant of large sudden changes in temperature, able to survive as the shallow waters are cut off by the falling tide and warmed by the sun before the tide rises again.

Below the pycnocline, there is little change in temperature. Thus, as expected, deep-sea organisms are widely distributed.

Marine mammals and seabirds are **homeothermic** (warm blooded). They maintain a constant internal body temperature regardless of the temperature of their surroundings and are therefore largely independent of changes in seawater temperatures. During migrations, these animals experience large changes in temperature without ill effects. The gray whale, for example, migrates annually from the Arctic Ocean to the lagoons of the Mexican Pacific coast to mate and bear young. Antarctic fishes and invertebrates, being cold blooded, require special adaptations to prevent ice forming in their blood.

Salinity

Salinity also affects distribution of marine organisms. Just as with temperature changes, it is in the early life stages that organisms are most sensitive to salinity changes. In fact, the distribution of some organisms is determined by their ability to withstand changes in salinity that their predators cannot tolerate. Oysters are a good example; they thrive in shallow coastal waters because their primary predators—snails and starfish—cannot tolerate the salinity changes there, caused by rains and river discharges. Other animals, such as barnacles and clams, can close

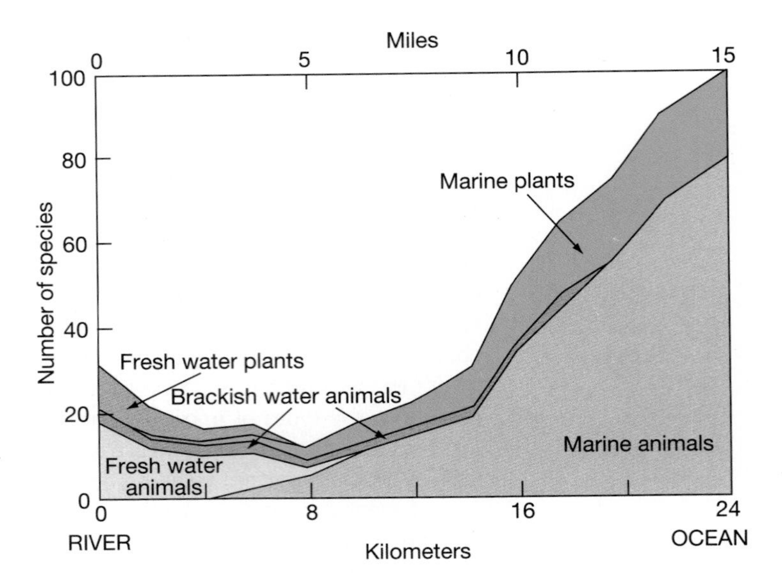

Figure 11–7
The abundance and distribution of plants and animals in the Tees Estuary on the northeastern coast of England is controlled by their tolerances for salt water and changes in its salinity. This is typical of the pattern found in most estuaries. (After W. A. Alexander and others: *Survey of the River Tees.* Water Pollution Research Technical Paper No. 5. London, Her Majesty's Stationery Office, 1935.)

their shells tightly when salinities or other environmental conditions are temporarily unfavorable.

The body fluids of most marine organisms have salt contents similar to that of seawater. These organisms do not need to protect themselves from losing (or gaining) water as land organisms must. Where salinity is variable, as in estuaries or in the coastal ocean, organisms must be able to maintain the salt balance of their blood to survive.

Most marine organisms require fairly high, constant salinities. Thus as one goes from a river, through an estuary with variable salinity, and then into the open ocean, the number of species one finds reflects the ability to cope with changing salinity, as seen in Figure 11–7. Near the river, freshwater forms dominate. As one proceeds seaward, salinities increase slightly and become more variable; here the number of species decreases to a minimum. As one approaches the coastal ocean, the number of species increases markedly as truly marine organisms dominate.

Density

Seawater has a density of about 1.03 grams per cubic centimeter, much higher than the density of air (0.0012 grams per cubic centimeter). The density of seawater is very similar to the density of the soft tissues of organisms; this similarity in density has a profound effect on marine organisms.

The density of seawater supports the bodies of many marine organisms, so that they do not need strong skeletons the way we land-dwellers do. The largest plants in the ocean, kelp (Fig. 11–5), use flotation bladders to keep their flexible, leaf-like structures near the surface to gather sunlight. In contrast, redwoods, the largest land plants, require huge, rigid trunks to support them. Since they do not require skeletons, many marine organisms consist of large delicate jelly-like structures, made mostly of water. We discuss these gelatinous plankton in the next chapter.

Pressure

Except for those fish that have gas-filled swim bladders, marine organisms are little affected by changes in pressure. Pressure increases by approximately one atmosphere (the weight of the atmosphere above you at sea level) for every 10 meters of water depth; thus, the pressure on the deep-ocean floor, at 4 kilometers, is 400 atmospheres. There is some evidence that these extreme pressures may affect metabolic and growth rates, but little is known about the processes involved.

Organisms with swim bladders are extremely sensitive to changes in pressures. When caught in nets and brought aboard ship, deep-sea fishes that have swim bladders usually die as the nets are hauled up, because gases in the bladder expand as pressure decreases.

Humans diving in the ocean must also guard against changes in pressure. The first problem is to keep the lungs from collapsing. Submarines and diving bells protect humans by maintaining the pressure at one atmosphere. Individual divers must prevent the collapse of their lungs by breathing pressurized gases. SCUBA (*s*elf-*c*ontained *u*nderwater *b*reathing *a*pparatus) tanks are commonly used for this purpose.

Breathing gases that are under high pressure can also cause problems. At high pressures, more nitrogen gas is dissolved into the blood and tissues than at surface pressures. This increased level of nitrogen in the blood causes nitrogen narcosis, a condition that essentially intoxicates the diver and is therefore aptly called "rapture of the deep." A person in such a condition is unable to think rationally, and so might become disoriented and swim deeper rather than returning to the surface.

As a diver returns to the surface, the nitrogen dissolved in the blood and tissues must be allowed to come out of solution slowly. If a diver's ascent is too rapid, the nitrogen coming out of solution forms bubbles in the blood stream and tissues, a painful condition called "the bends." The bends can cause paralysis and sometimes even death. For deep diving, helium or hydrogen is sometimes substituted for nitrogen in the breathing gases, since these two gases come out of solution in the blood more readily than does nitrogen.

Air-breathing marine mammals (seals, whales, dolphins) have special adaptations that permit them to make long dives to great depths without experiencing the problems that afflict human divers. First, their blood absorbs and dissolves more oxygen than human blood, and, second, they can tolerate higher carbon dioxide concentrations in their blood. Furthermore, their vascular system adjusts to any lowering of the oxygen level by cutting off blood flow to the extremities, so that blood from the heart flows mostly to the brain. Finally, their lungs collapse during dives, preventing the blood from absorbing large amounts of nitrogen.

Ecosystems

Oceanic plants and animals depend on one another to provide the conditions and materials that make life in the ocean possible. Organisms exchange matter and energy with each other and with the waters around them. An **ecosystem** (Figs. 11–8 and 11–9), whether marine or terrestrial, contains living organisms, the physical environment in which they live, and an energy source (most often sunlight).

Simple or complex, all ecosystems include organisms that produce food, organisms that consume it, and organisms that decompose both producers and consumers once they die. It is in this production–consumption–decomposition cycle that energy moves through an ecosystem, and this cycle is a major theme of this chapter.

The *primary producers* are **autotrophic** ("self-feeding") organisms, usually plants or some types of bacteria. Autotrophs produce organic matter—food—from inorganic substances. They need energy to drive this process. Plants use energy

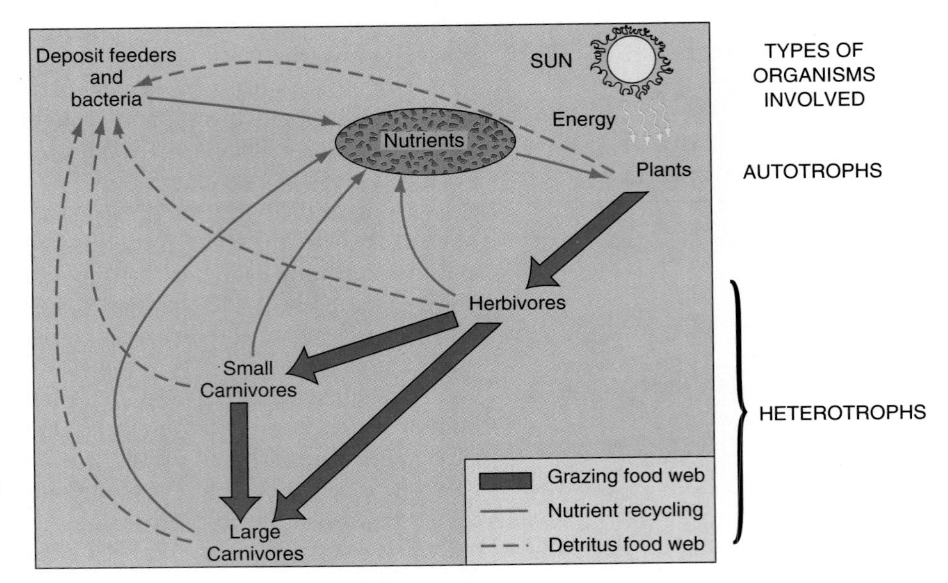

Figure 11–8
A simple ecosystem includes organisms, an energy source, and the environment in which the organisms live. Nutrients are recycled within an ecosystem. Energy is not recycled and flows only one way.

from the Sun in a process called **photosynthesis.** Some autotrophic bacteria use a different source of energy derived from compounds such as methane (a component of natural gas) or hydrogen sulfide, in a process called **chemosynthesis.** (These food-making processes are discussed later in the chapter.)

Heterotrophic ("other-feeding") organisms—animals and most bacteria—eat the organic matter made by the primary producers, thereby obtaining needed energy. Heterotrophs are classified as being *secondary producers, tertiary producers,* and so forth, depending on their position in a food chain. **Decomposers** (primarily bacteria and fungi) break down tissues after an organism's death. This decomposition releases and recycles the chemical constituents originally tied up in tissues or skeletons.

The first part of the ecosystem we consider deals with feeding relationships among organisms. Here we define the concept of **trophic level**—or, more simply put, who eats whom. Plants, the primary producers, are the first trophic level. They are eaten by **herbivores** (secondary producers), the second trophic level. **Carnivores** (tertiary producers), which eat herbivores, are the third trophic level. Higher-level carnivores eat lower-level ones.

Outside this neat arrangement are the **omnivores,** who eat both plants and animals. (Humans are omnivores, for example.) At every stage, **detritus** (organic

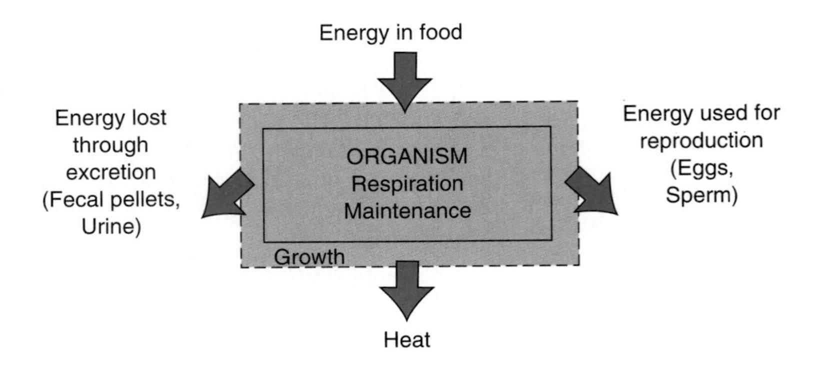

Figure 11–9
Most of the food taken in by an animal is used in maintaining itself and for growth. Some is excreted, and some is used for reproduction.

waste products) is released into the water and then supports another complex of organisms, the **detritivores.**

Food Chains and Food Webs

In a **food chain,** organisms at each trophic level are eaten only by organisms of the next highest trophic level. One example is a *grazing food chain:* primary producers (plants) being eaten by herbivores and the herbivores then being eaten by carnivores.

Detritus is produced at all levels in food chains. The discarded particles sink, thereby supplying food (and energy) to all depths in the ocean and to a great variety of organisms. This supply of organic matter supports the *detritus food chain* of bacteria and deposit feeders which feed on dead organic matter (Fig. 11–8). The dead organic matter is decomposed by bacteria and fungi that are subsequently eaten by microscopic animals. Many of the organisms in detritus food chains are extremely small and poorly studied. As larger organisms feed on the detritus chain, it becomes part of the more familiar grazing chain.

Only a small amount of waste organic matter actually reaches the ocean bottom, and that happens only after a given molecule has been eaten many times. Only the most resistant forms of organic matter with little food value (such as teeth and shells) survive to be buried in sediment deposits. (As we learned in Chapter 5, a small amount of this organic matter in sediment deposits is converted, at high temperatures and under high pressures, to oil and gas.)

A key point to remember is that energy is not recycled. It is a one-way flow, from the sun to heat given off in the process of decomposition at the end of the flow. This heat cannot be reclaimed and recycled the way chemical compounds are. Each time a particle of organic matter is consumed, its energy content is reduced, as Figure 11–9 shows. At each step in a food chain, energy is transferred from one trophic level to the next. Some of this energy is lost at each transfer because organisms use much of the energy they take in for maintenance or reproduction. More energy is lost in excretion (fecal pellets, urine) and as body heat. Only about 10 percent of the energy in the food intake at each trophic level is available for growth. Thus, 1,000 grams of plants support 100 grams of herbivores which in turn could support 10 grams of carnivores.

So far, we have considered only simple food chains, but feeding relationships in the ocean are rarely simple. Most organisms eat more than one type of food. In turn, each organism is eaten by many predators. These more complicated feeding relationships are called **food webs.**

To quantify the amount of organic material in an ecosystem, we use the concept of **biomass,** the amount of plant or animal material (expressed as the weight of organic carbon) per volume of seawater or area of ocean bottom. **Production** expresses how much organic matter is produced in an ecosystem in some specified period of time, while biomass indicates how much is available to be eaten at any one time. For example, an organism may have a high rate of production but quickly use most of the energy in that new organic matter, storing little. A hummingbird, for example, has an extremely high metabolic rate. In that case, production is high but available biomass is low. A whale is the opposite case. It grows slowly for a very long time, so that the biomass is large even though the rate of production may be small.

Another way of expressing this relationship is the **standing crop,** which is the difference between the amount of organic matter produced and the amount consumed (Fig. 11–10). Standing crop is influenced not only by the rate of production, but also by rates of reproduction, by predation, and by the death rates of the organisms in the ecosystem.

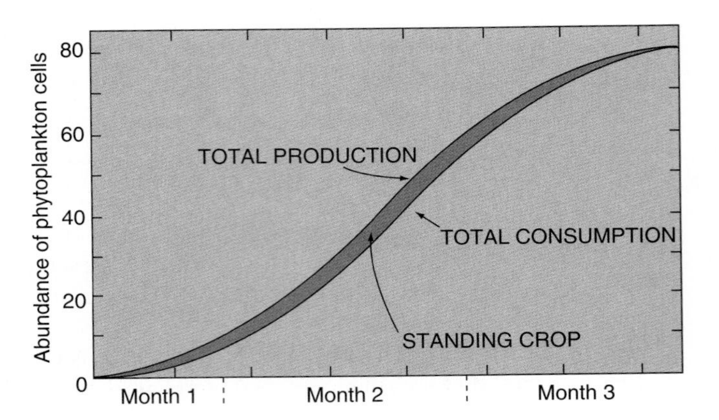

Figure 11–10
The standing crop is only a small part of total plant production during the spring bloom. In this example, only about 3 percent of total production was in the standing crop at any given time. The rest was either eaten by grazing herbivores or lost through other processes, such as sinking.

Primary Production

As we have seen, all ecosystems function by transferring organic material and energy from one trophic level to another in a food chain or a food web. How does this process begin?

All ecosystems are based on **primary production**—the process by which organic material is synthesized from inorganic substances (primarily carbon, nitrogen, and phosphorus) using a source of energy which is stored in the newly made material. Primary production is carried out by photosynthetic plants or chemosynthetic bacteria. In marine ecosystems, most primary production comes from phytoplankton. While the attached marine plants of the coastal areas are more familiar to most of us, they are restricted to well-lit shallow waters, so that their contribution to primary production is small on a global scale.

To understand life in the ocean, it is important to know how much production occurs and where. You could determine this for your yard by collecting and weighing grass clippings after a year. Unfortunately, production of organic matter in the ocean is more difficult to measure. One way of measuring marine primary production is to label the newly formed organic matter in a water sample. Radioactive carbon-14 is commonly used as the labeling agent. It is injected into the water in closed containers that are exposed to light intensities similar to those experienced by organisms living at various depths below the sea surface. After a set time, the water is filtered and the amount of radioactivity incorporated into organic matter is measured in the microscopic organisms retained on the filters. Radioactive carbon not incorporated into tissue passes through the filter and is not counted.

This technique measures *gross primary production,* the quantity of inorganic radioactive carbon made into organic matter. To measure **net primary production** (the amount available to other organisms), one must correct for any carbon used up by the respiration of plants and bacteria in the container (usually 10 to 50 percent). (Any radioactive carbon released by plant respiration would remain in the water and not on the filter.) Such measurements provide only an estimate of production at the time of the observation.

Color sensors on satellites now measure chlorophyll concentrations in surface waters and consecutive satellite images show how these concentrations change with the seasons. Such measurements can be averaged and combined to provide maps of the global distribution of plant growth (Fig. 11–11).

Photosynthesis and Respiration

Phytoplankton and some bacteria synthesize organic matter, using energy from sunlight and nutrients dissolved in near-surface waters. Around the ocean margins, where the waters are shallow enough for the bottom to receive enough sunlight, larger plants, primarily algae, also photosynthesize. Photosynthetic organisms use

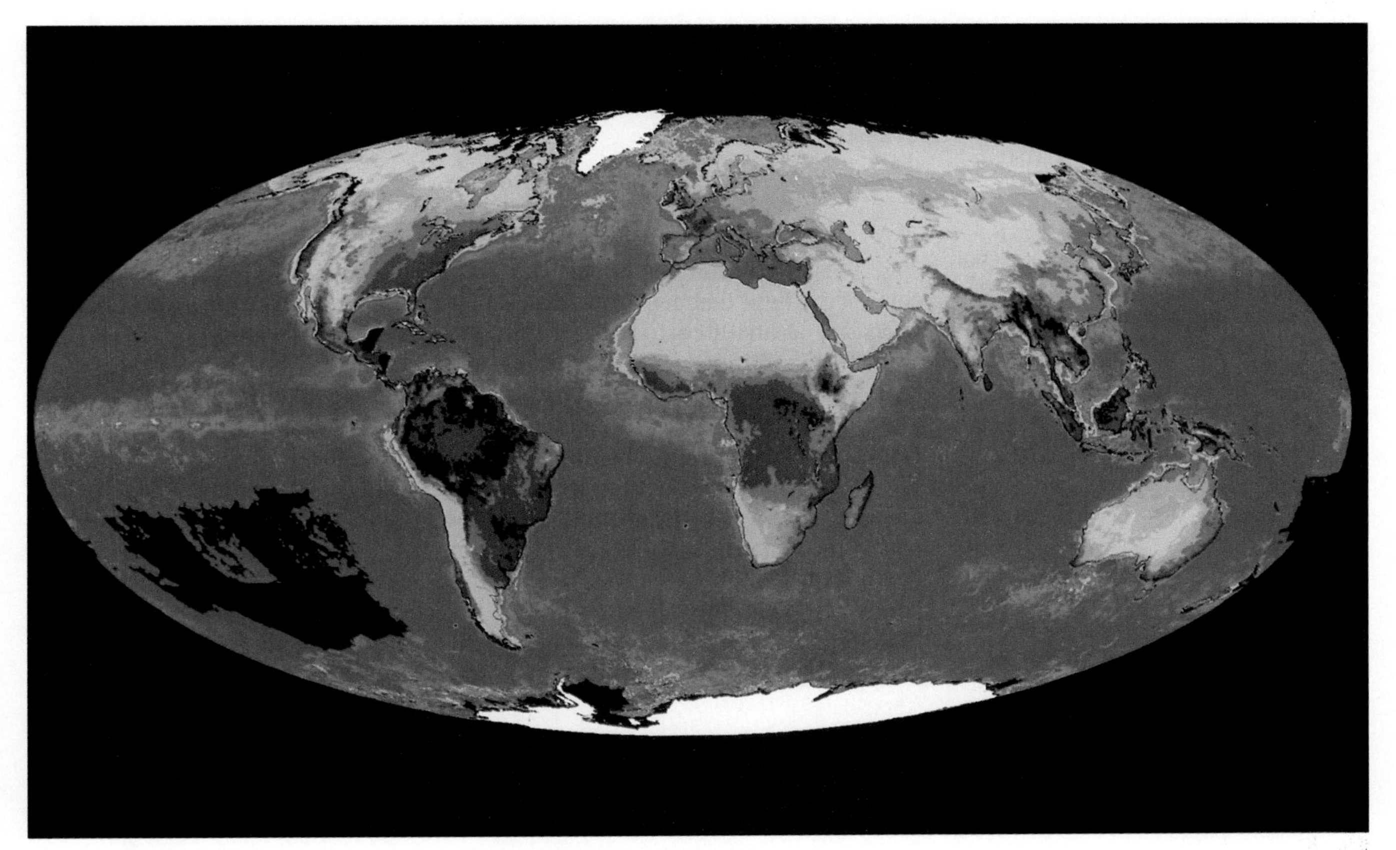

Figure 11–11
Average distribution of primary production in surface waters of the ocean. The highest production levels (red) occur in coastal waters, especially in upwelling zones. The lowest levels (purple) occur in the centers of the gyres in both the northern and southern hemisphere. The most productive land areas are dark green; the least productive are yellow. (Courtesy NASA.)

chlorophyll or some other light-absorbing pigment to produce organic compounds. These pigments capture energy from sunlight, and the plant then uses that energy to combine dissolved carbon dioxide with water, forming *carbohydrates*, which are energy-rich compounds consisting of carbon, hydrogen, and oxygen.

As Figure 11–12 shows, the process is reversed during **respiration,** when oxygen is taken up by an organism, carbohydrates are broken down into simpler substances, and energy is released. Heterotrophic organisms (animals and many bacteria) use respiration to release the energy stored by the autotrophs in substances such as carbohydrates and fats. Plants also respire, using some of the energy captured during photosynthesis for maintenance and reproduction.

Light Limitation

The availability of light controls plant growth and phytoplankton distributions in the ocean. As we learned earlier, most food for marine organisms is photosynthesized by phytoplankton in sunlit, near-surface waters. The **photic zone** is the depth to which sufficient light penetrates for photosynthesis to proceed, no matter how slowly. The bottom of the photic zone in clear open ocean waters occurs between 100 and 200 meters. Below the photic zone there is no photosynthesis; this is called the **aphotic zone.**

Light is absorbed as it passes through water. The clearest open-ocean water is most transparent (absorbs the least amount of light) in the blue-green range of

Figure 11–12
In photosynthesis, plants use the energy from sunlight to manufacture organic matter (carbohydrates) from inorganic constituents (carbon dioxide and water), giving off oxygen in the process. The energy is captured by chlorophyll (or other pigments) and is stored in the carbohydrates. In respiration, the opposite process, both plants and animals combine oxygen and organic matter to release the energy contained in the organic matter, giving off water and carbon dioxide as byproducts.

PHOTOSYNTHESIS
(chlorophyll)

$$6CO_2 + 6H_2O + \text{energy} \rightleftharpoons C_6H_{12}O_6 + 6O_2$$

Carbon dioxide, Water, Carbohydrate, Oxygen

RESPIRATION

colors. As concentrations of particles and dissolved organic matter increase, the color of the light that penetrates deepest into the water shifts from yellowish green in coastal ocean waters to red in the most turbid estuarine waters (Fig. 7–1). Thus, as light gets dimmer with increasing depth, its color also changes. This color change affects plant production because each plant pigment is most efficient with a specific color of light. The combination of pigments found in any type of phytoplankton will determine its optimal depth distribution.

While most production occurs near the surface, respiration is nearly independent of depth. Thus, at some depth the instantaneous rate of photosynthesis equals the instantaneous rate of respiration; this level is called the **compensation depth** (Fig. 11–13). Above this depth, phytoplankters can produce enough food to survive; below it, the rate of respiration exceeds the rate of photosynthesis and an individual cell carried below this depth will rapidly use up its food reserves. At the compensation depth, light intensity is approximately one percent of its value at the ocean surface.

However, we need to take account of the fact that while the rate of respiration is constant with increasing depth, the rate of photosynthesis is not—it is much greater in the near-surface waters. If we consider a population of phytoplankton in a column of water, rather than individual cells, there is a depth at which total photosynthetic gain equals total respiratory loss: this is called the **critical depth** (Fig. 11–13).

To see how the availability of light affects phytoplankton abundances seasonally, we examine the relationships in the mid-latitudes. In winter, low light levels cause the critical depth to be relatively shallow (Fig. 11–14). At the same time, the well-mixed surface zone (of which the lower boundary is the top of the pycnocline) is much deeper than at other times of the year. During winter, surface waters are mixed vertically by winds and waves of winter storms. Therefore, plants are frequently mixed below the critical depth and do not remain in the photic zone long enough for photosynthesis to exceed respiration. Only a few phytoplankters survive over the winter.

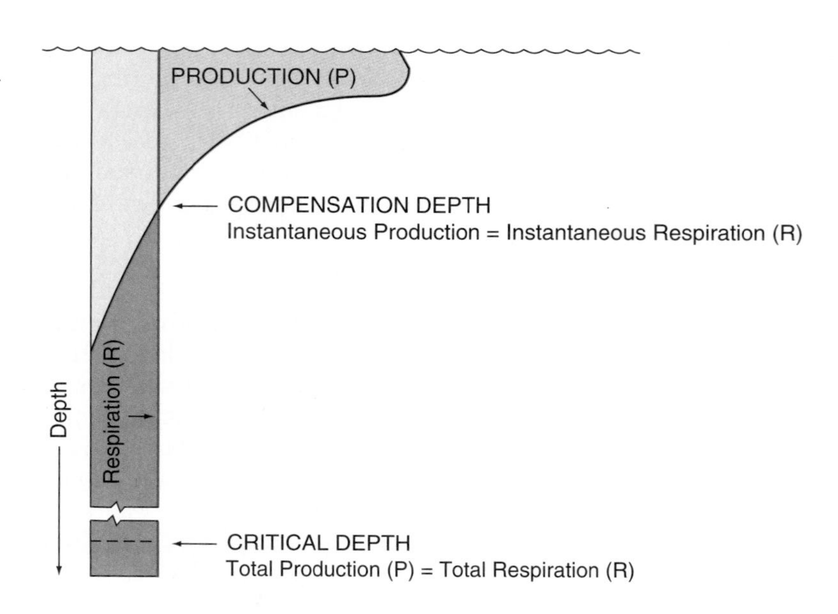

Figure 11–13
Production is highest near the ocean surface and diminishes with depth. Respiration (color) is constant with depth. The compensation depth is where instantaneous production equals instantaneous respiration. At the critical depth, total respiration equals total production.

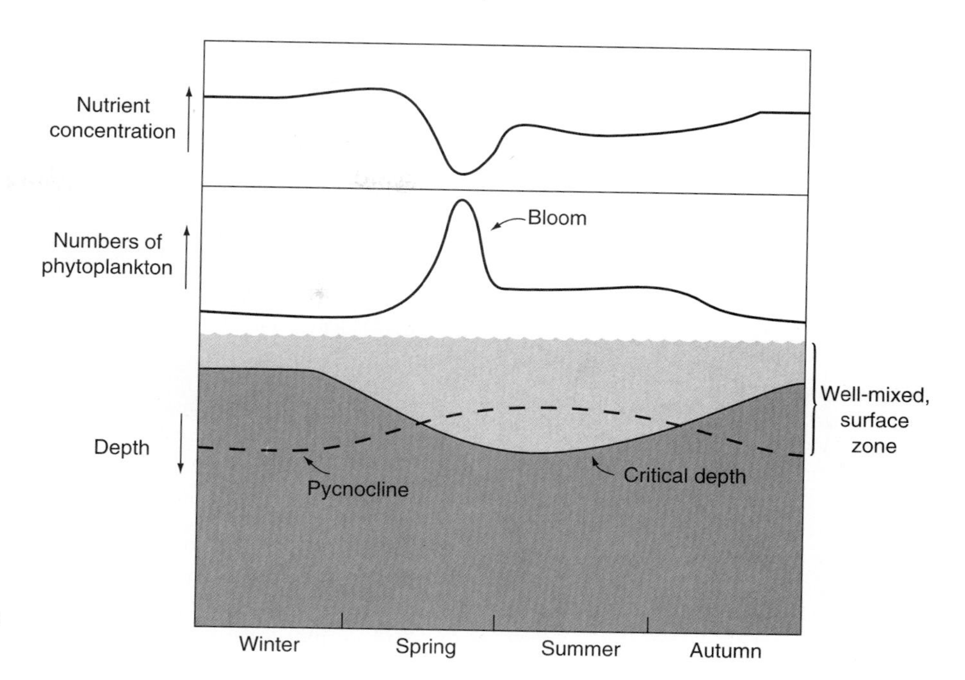

Figure 11–14
In winter, there is little light in surface waters, and the pycnocline is relatively deep. Thus, phytoplankton are readily mixed out of the photic zone, there is little phytoplankton growth, and populations are small. In spring and summer, a shallow pycnocline keeps phytoplankton in the photic zone. A bloom occurs because of the abundance of nutrients built up over the winter in the photic zone and ends when the nutrient supply is exhausted. Continued phytoplankton production after the bloom ends depends on nutrients being recycled from the deep zone into the surface zone.

During spring, light intensities increase, causing the critical depth to deepen. At the same time, surface waters warm and storms abate, so that the mixed surface layer becomes thinner. Consequently, fewer phytoplankters are mixed below the photic zone, and most stay in this zone long enough for photosynthesis to exceed respiration. Therefore, there is a net increase in photosynthesis. This results in a rapid increase in phytoplankton abundance and biomass; such a rapid increase is called a **bloom.** The numbers of many phytoplankton cells can double in a day or two.

The bloom ends when one of two things happens. Either the growing phytoplankton populations use up the available nutrients which they require for photosythesis, or the herbivores eat the phytoplankters, greatly reducing their numbers. As autumn comes, light intensities drop and the critical depth gets shallower. Mixing, due to cooling and wave action, causes a deeper surface zone, and phytoplankton production drops to low winter levels. Diseases may also play a role, but little is known about their effects on phytoplankton populations.

Nutrient Limitation

As we have just seen, the scarcity of essential nutrients often limits population sizes. As plants grow, they take from the surrounding water the elements needed to make their tissues and external shells. Many of these elements—carbon, oxygen, and sulfur to name just three—are available in abundance in seawater, and therefore uptake by plant growth does not change their concentration in the water. Other nutrients, such as nitrogen and phosphorus compounds, are scarce in seawater, and this scarcity limits production for many organisms, even when there is enough sunlight. Scarcity of any one nutrient can limit production, but in many ecosystems two or more nutrients are co-limiting.

As plants and animals die and decompose in the surface zone, about 95 percent of the nutrients contained in their tissues is released into surface waters by decomposition and quickly taken up again by growing plants. The remaining 5 percent is released below the surface zone, because some of the tissues and shells sink before decomposing. Because there is usually too little light for photosynthesis at these lower depths, nutrients released in the aphotic zone accumulate there (Fig. 11–15). These deep-sea nutrients move with subsurface currents and are

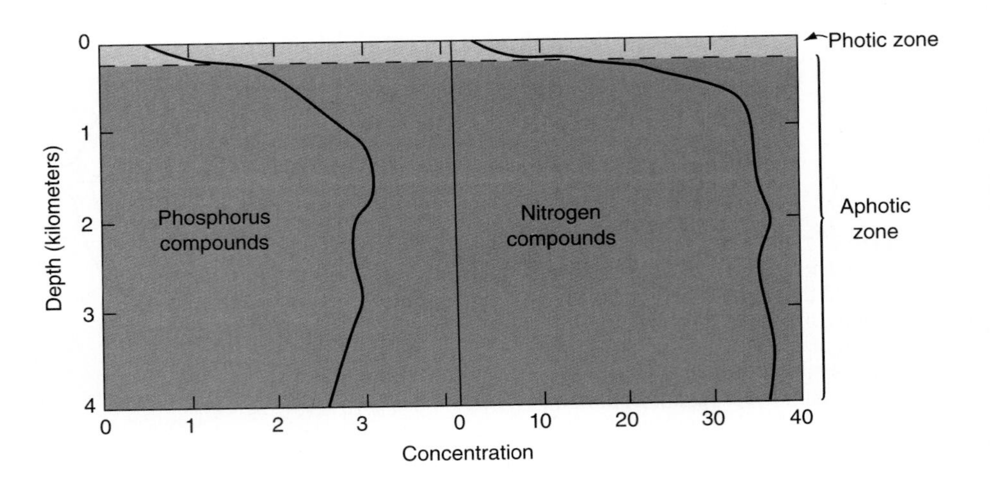

Figure 11–15
In near-surface waters, phytoplankton deplete the water of phosphorus and nitrogen compounds.

returned, often decades or centuries later, to surface waters. This complex of processes is often called the **biological pump.** As we shall learn in Chapter 15, such processes can also remove carbon dioxide from the atmosphere.

In the mid-latitudes, nutrients build up in the surface zone over the winter because, as described above, only a few phytoplankters survive. Once the water warms up in spring, this nutrient stock is quickly depleted by growth of new organisms. In the tropics, where light levels remain high throughout the year, phytoplankton growth continuously removes nutrients from surface waters. There is no opportunity for them to build up seasonally as they do at higher latitudes.

In the mid-latitude example of Figure 11–14, the spring bloom occurs because of the abundance of nutrients present in surface waters when winter ends and light is no longer limiting. The bloom ends in early summer, even though light levels remain high at this season, because the supply of limiting nutrients in the surface layer is now exhausted. Later in the summer, as phytoplankters die and decompose, the nutrients released support additional phytoplankton growth. Small blooms occur over the rest of the summer anytime strong winds pull deep nutrients up into surface waters.

Another factor limiting phytoplankton growth is the presence or absence of organic **trace nutrients** (which we call vitamins when discussing human nutrition). Such substances are essential for the growth of certain plants and animals. Recently upwelled waters that lack these trace nutrients support little plant life, even though they contain high levels of the major nutrients. In addition, some organic compounds produced in seawater in trace amounts by bacteria and marine plants combine with potentially toxic metals, such as copper, thereby detoxifying recently upwelled waters so that plants can grow in them.

Phosphorus and Nitrogen Cycles

Compounds containing either phosphorus or nitrogen are necessary for growth of both plants and animals. Nitrogen is required for the formation of proteins, while phosphorus plays a role in energy transfers and in the formation of cell membranes and genetic materials. Both types of substances are taken up from the water by phytoplankton and subsequently consumed by animals. When plankters die, compounds of both phosphorus and nitrogen are reduced, primarily in near-surface waters. They are also released below the pycnocline because some particles sink before decomposing.

Marine plants can survive short periods of low nutrient concentrations. Whenever these phytoplankters encounter high nutrient concentrations, they store more in their tissues than they need for immediate growth. In this way, phytoplankton populations can grow for several generations by using reserves of stored nutrients. Because of differences in their chemical behavior, phosphorus and nitrogen compounds are released in different ways after an organism's death.

Phytoplankton takes up phosphorus primarily in the form of phosphate. Phosphate is also excreted or released by the decomposition of dead phytoplankters and animals and is rapidly recycled for uptake by other organisms. Phosphate by itself is rarely limiting in the marine environment.

The nitrogen cycle is much more complicated. Dissolved nitrogen is an abundant constituent in seawater, but plants cannot use it in this form. Instead, specialized marine bacteria, called cyanobacteria, must convert dissolved nitrogen to ammonia, in a process called **nitrogen fixation.** Two additional steps involving bacteria are required to convert ammonia into nitrate which can be used by phytoplankters. Marine animals then eat the plants and process their nitrogen-containing compounds, ultimately excreting them in forms that, again, are not readily used by plants. Again, bacteria are required to break down these compounds into nitrate which is then easily recycled to the plants. The complexity of the nitrogen cycle and the time required to complete all of its steps often causes nitrate to be a limiting nutrient for phytoplankton growth.

In addition to the recycling of these compounds, there is a slow resupply of nutrients to the surface zone by waters moving upward through the pycnocline. As these waters rise, they carry nutrients released below the photic zone back into the photic zone. This supply of nutrients balances the nutrients lost when particles fall into the deep ocean before decomposing.

Chemosynthesis

As mentioned earlier in the chapter, some bacteria produce organic matter without sunlight via chemosynthesis (Fig. 11–16). Instead of using sunlight, these bacteria extract energy from hydrogen sulfide, methane, metals, or even hydrogen gas to make new organic matter. Chemosynthesis is common in dark, oxygen-deficient environments, such as in sediment deposits or on the ocean floor near vents discharging sulfides, methane, petroleum, or reduced metals—all energy-rich compounds. Rapidly growing organisms living near hydrothermal vents on the deep-ocean bottom must depend on chemosynthesis for their food because too little food sinks from the surface to reach them. Vents can support food production as long as they remain active.

Chemosynthetic bacteria may live in association with host organisms. In worms, bacteria live in specialized organs; in clams, they live on the gills, and the food they produce is absorbed through the gill surfaces along with dissolved oxygen. These chemosynthetic bacteria thus provide food directly to a host, while waste products from the host provide nutrients to the bacteria. Vent organisms

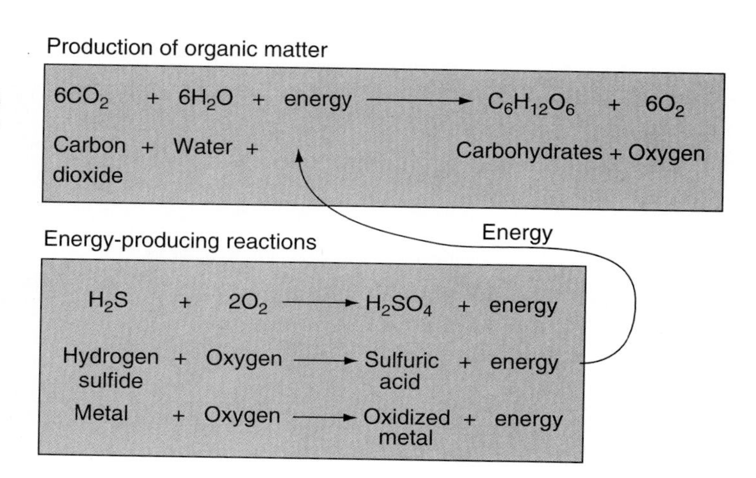

Figure 11–16
Specialized bacteria obtain energy from chemical reactions involving energy-rich compounds to produce organic matter—a process called chemosynthesis.

TABLE 11–1
Estimates of Primary Production and Fish Production

Region	Percentage Area of Total Ocean	Primary Production (millions tons/year)	Fish Production (millions tons/year)
Estuaries	0.5	640	70
Continental Shelves	6.4	3730	356
Upwelling Areas	1.4	1120	116
Open Ocean	91.7	18,920	817
Totals	100.0	24,410	1,359

Source: E. D. Houde and E. S. Rutherford: Recent Trends in Estuarine Fisheries: Prediction of Fish Production and Yield. *Estuaries 16*(2):161–176 (1993).

have special blood proteins to transport sulfides to the bacteria without poisoning themselves. This relationship is an example of *symbiosis*, in which two organisms grow together to their mutual benefit.

Chemosynthetic bacteria also grow in mats on the bottom near the vents, where snails and other grazing organisms feed on them. In addition, filter-feeding organisms ingest bacteria living in the water.

Chemosynthesis occurs widely where there are discharges of methane-bearing or hydrogen-sulfide-bearing waters. Such waters are expelled from sediments subducted at trenches or by ground waters discharging on continental slopes. Communities of chemosynthetic organisms may be common on the deep-ocean bottom. We still know little about them.

Distribution of Production

As Table 11–1 shows, most of the biological production in the ocean occurs in open-ocean waters. However, coastal areas are more productive per unit area, and most of the major fisheries are found in coastal waters, especially in upwelling areas. (In the open ocean, fish are widely dispersed, and harvesting them is not economically efficient.) Coastal ocean food webs are relatively simple, involving few steps (Fig. 11–17) and each step is more efficient than open-ocean food webs in transforming primary production to fish production. Fishing ports are usually close to coastal upwelling areas, making their fish populations easier to exploit.

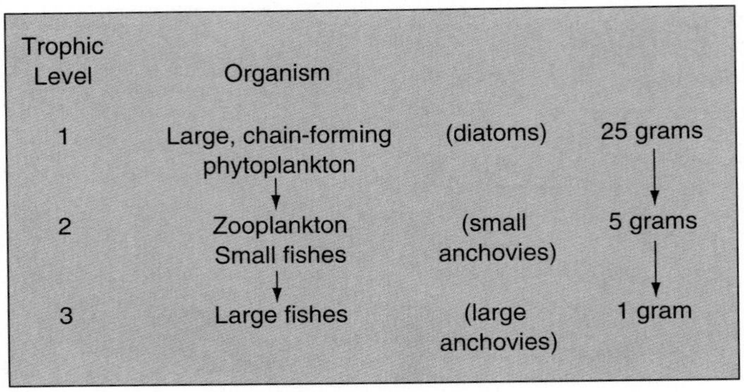

Figure 11–17
A simple food chain in an upwelling area. Note that the food chain is short and that at each level about 20 percent of the energy is transferred to the next trophic level.

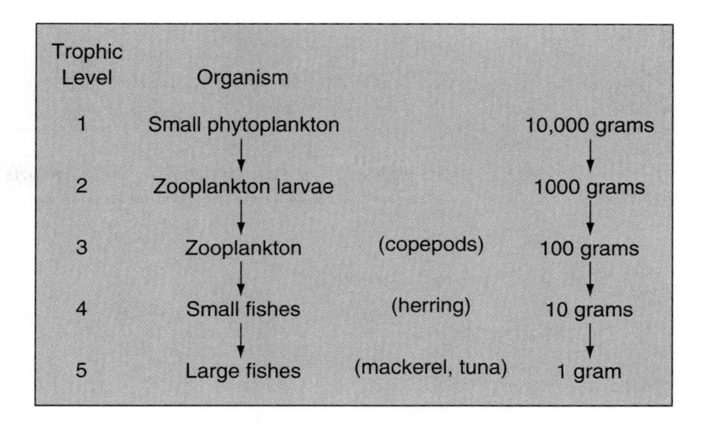

Figure 11–18
A simple, open-ocean food chain showing the amount of primary production necessary to produce one gram of a top predator such as tuna, assuming energy transfers of 10% between trophic levels.

Coastal areas with upwelling are characterized by a large proportion of phytoplankton (principally diatoms) that aggregate either into clumps or into long chains. They can be eaten directly by fishes or by small herbivorous zooplankton. In any case, highly productive areas commonly support short food chains of rapidly growing organisms, where transfer efficiencies may be 20 percent or higher. Little energy is wasted in hunting, and the yield of fish is much higher for the same amount of primary production. About one-third of the world's fish catch comes from estuaries, continental shelves, and upwelling areas (Table 11–1) even though these areas constitute only one-twelfth of the ocean area.

In coastal areas without upwelling, productivities are intermediate between those of the open ocean and those of coastal upwelling areas. Large phytoplankton are often present, especially during blooms. The density of even very small phytoplankton, however, is higher in coastal waters than in the open ocean, and longer food chains undoubtedly exist along with very short ones, depending on local conditions. Energy transfer efficiencies in non-upwelling coastal ocean waters average around 15 percent.

Open-ocean food webs are typically long and complicated, involving many energy transfers. Only the smallest phytoplankters grow in the nutrient-poor, open-ocean waters. Therefore, they can be consumed only by very small herbivores. They, in turn, are eaten by carnivores about ten times as large. In many cases, one or two larger invertebrate animals or fishes form additional links in an open-ocean food chain before it reaches a large carnivore, such as mackerel or tuna. Typically, the top carnivore in such a food chain is a fast-swimming fish that can cover large distances in its search for prey.

Energy is lost each time it is transferred from one trophic level to another (Fig. 11–18). Open-ocean predators grow more slowly and expend more energy hunting for food than coastal-ocean predators. Transfer efficiencies average about 10 percent in open-ocean food chains.

The Effects of Life Processes on the Ocean

So far we have focused on the relationships among marine organisms and between them and their environment. Biological processes can change the physical and chemical characteristics of seawater. For example, intense algal blooms cause surface waters to become quite turbid, thereby increasing the absorption of sunlight in the first few meters. This in turn significantly increases sea surface temperatures. Organisms release organic material that supports other, poorly understood food webs involving groups of microbes. As we have already seen, decomposition can completely deplete deeper ocean layers of their dissolved oxygen.

TABLE 11–2
Relative Abundance of Various Forms of Organic Matter in Seawater

Form	*Relative Abundance (%)*
Dissolved	95
Particulate (nonliving)	5
Phytoplankton	0.1
Zooplankton	0.01
Fishes	0.0001

Dissolved Organic Matter

Most organic material in the ocean is dead, occurring either as dissolved organic compounds in seawater or as small particles mostly in near-surface waters (Table 11–2). Dissolved organic matter remains in the ocean for a very long time—many thousands of years. It is roughly equal in abundance to all living matter on Earth.

There are many sources of dissolved organic matter. Decomposition of dead plant and animal matter is one source. Up to half the mass of a dead organism dissolves in seawater as bacteria decompose it. The remaining tissues are recycled in particulate form, which will be discussed in the next section.

Secretion of organic compounds by living plants is another important source of dissolved organic matter. Usually less than 10 percent of the carbon assimilated during photosynthesis is released to the water as dissolved organic compounds. Under stressful conditions, though, such as unusually high light levels, 50 percent or more of the photosynthesized carbon compounds may be released. Excretion by animals is another source.

Dissolved organic matter is used as a food source in a food chain known as the **microbial loop** (Fig. 11–19). It is based on tiny bacteria less than 0.6 micrometer across—too small to be seen by ordinary light microscopes. These bacteria live floating in seawater and utilize dissolved organic matter, absorbing it directly from their surroundings. They are eaten by flagellates and ciliates, which in turn may be eaten by small zooplankton. So, the microbial loop provides an additional food supply to the grazing food chain by converting dissolved organic matter into particulate organic matter. The details of the linkages and the rates of the processes involved in the microbial loop are still being studied; indeed, its existence was not known to biological oceanographers until very recently.

Particles

In Table 11–2, we see that the second most abundant form of organic matter in the ocean is the non-living particles that result from biological activities. The processes involving particulate organic matter influence the chemical and biological properties of seawater. There are about 10 billion tons of particles in the ocean, mostly in near-surface, coastal-ocean waters. Biological particles in the ocean are relatively large, ranging from one micrometer to one millimeter in diameter. (For comparison, a human hair is about 100 micrometers, or 0.1 millimeter, in diameter.) Up to 70 percent of the particles in the ocean, especially in upwelling areas, are shell fragments, fecal pellets, and pieces of tissue broken and dropped while zooplankton feed. These particles are an important source of food for filter-feeding animals living in the water column and on the bottom. Bacteria on the particles supply dissolved organic matter, as well as needed trace organic compounds. Particles also

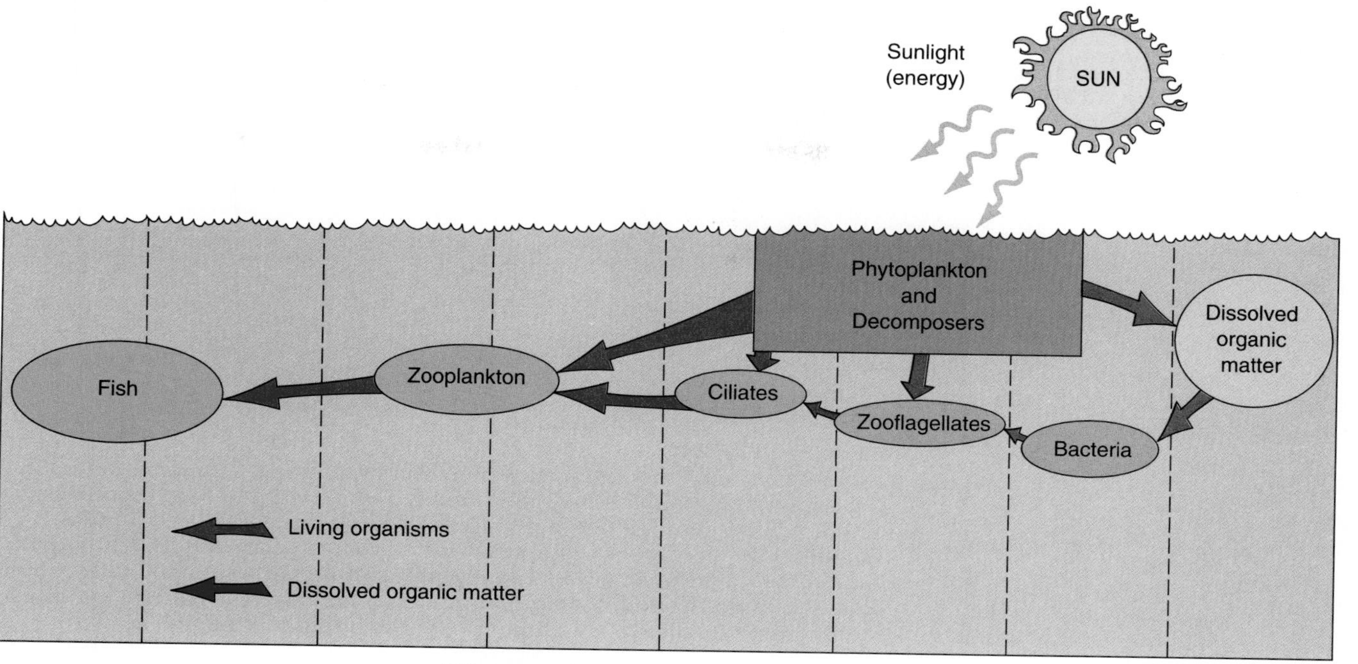

Figure 11–19
Dissolved organic matter, released by phytoplankton and decomposing organisms, is used by bacteria, which are eaten by flagellates and ciliates. These organisms may be eaten by zooplankton, and in this way the dissolved organic matter enters the grazing food chain (represented here by the fish).

help remove toxic metals and organic compounds from the water. Organic carbon constitutes about 25 percent of all oceanic particulate matter.

In areas of high productivity, the many biological particles produced, such as large fecal pellets, are quickly eaten in the upper layers by the abundant filter-feeding organisms. Those which are not consumed sink rapidly to the bottom within a few days in the coastal ocean, or a few weeks in the deep open ocean. They are only slightly altered by biological or chemical processes before reaching the bottom and provide a valuable source of food to bottom-dwelling organisms. (We discuss these in Chapter 13.) Smaller particles sink more slowly and most decompose before reaching bottom.

Another type of large particles are actually aggregates of particles. Some marine zooplankters build large mucous membranes with which they trap particles as food. When these structures become clogged, they are abandoned, forming large particles. They may be so abundant in some areas that, seen in the lights of submersibles, they resemble snow seen in the headlights of an automobile, and are therefore called **marine snow.** These aggregates are colonized by bacteria and other small organisms which use them as a food source.

Particles are usually destroyed in two stages. First, they are broken up, either by physical or by chemical processes (dissolution). Dissolution is especially important for particles smaller than 10 micrometers and for soluble materials such as calcareous or siliceous shells and skeletal parts. These dissolving particles release nutrients, silica, and metals, thus changing the chemical composition of deep-ocean waters and sediment deposits. (This release of nutrients and other constituents back into the water is part of the biological pump which we discussed earlier in this chapter.) Some substances are transported to the bottom by adsorption onto particle surfaces.

The most resistant particles, which are neither eaten nor dissolved before reaching the bottom, add to the marine sediments accumulating on the ocean bottom. The processes of sedimentation were discussed in Chapter 5.

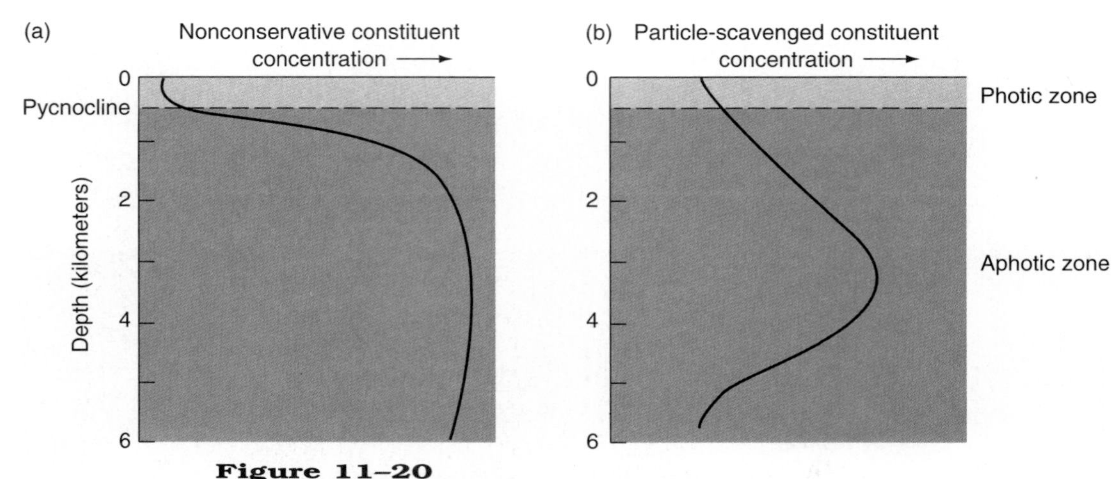

Figure 11–20
The concentration of the constituents dissolved in seawater varies with depth. (a) Some non-conservative elements behave like nutrients. They are removed by phytoplankton production in the near-surface waters and released in the deep ocean when particles decay. This process is part of the biological pump. (b) The concentrations of some nonconservative constituents are controlled by their removal in particles in near-surface waters which decay as they sink. Near the bottom, the constituents are again removed from the water by particles which are abundant in a layer of suspended matter near the bottom.

Trace Elements

Chemical elements that occur in very low concentrations in seawater are usually highly reactive and involved in biological processes. These are the nonconservative constituents we learned about in Chapter 4. The distribution of many of the nonconservative elements involved in marine biological processes is similar to the distribution of nutrients (Fig. 11–20a). In other words, these elements are least abundant in near-surface waters and most abundant in the deep ocean. The explanation is simple: growing organisms remove them from surface waters. Later, when the organisms die, these elements are released to subsurface waters as the organisms decompose. Again, this is part of the set of processes known as the biological pump.

Finally, reactions with particles, including sediment particles, control the distribution of some nonconservative trace elements, such as lead and copper (Fig. 11–20b). Concentrations of trace elements removed by particle interactions are lowest where particles of all sorts are most abundant; i.e., in productive surface waters or near the bottom. Conversely, their concentrations are highest in waters with relatively few particles, such as near-surface open-ocean waters in the centers of gyres, and in the mid-depths of the open ocean.

Dissolved Oxygen

We turn now to how biological processes affect levels of dissolved oxygen in the ocean. The oxygen dissolved in seawater participates in both biological and chemical processes. Most of the oxygen dissolved in seawater comes from the atmosphere through the sea surface. (Photosynthesis in near-surface waters also produces oxygen, but we ignore that source in this discussion.)

Dissolved-oxygen concentrations and nutrient concentrations are inversely related to each other (Fig. 11–21) because primary production consumes nutrients and releases oxygen. Cold water masses that form in Arctic and Antarctic regions are rich in dissolved oxygen because of their low temperatures. They supply dissolved oxygen to the deep ocean. Where such water masses form, dissolved oxygen concentrations are nearly constant throughout the water column, as Figure 11–22 shows (note the curve for Greenland, where the North Atlantic Deep Water forms).

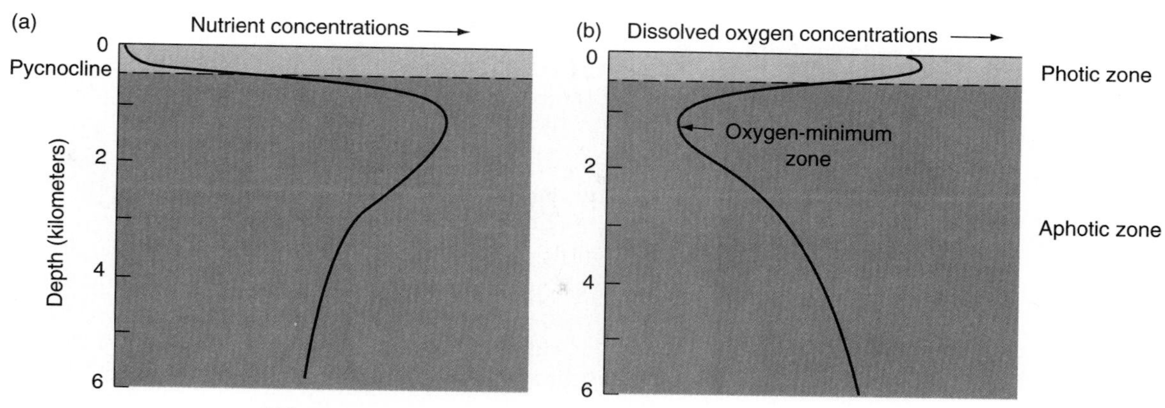

Figure 11–21
Typical variations in nutrient (a) and dissolved oxygen concentrations (b) in the North Pacific. Note that the oxygen minimum zone is near the same depth as the nutrient maximum. The small increase in nutrient concentrations and the corresponding decrease in dissolved oxygen below 2,000 meters is due, in this case, not to biological processes but to subsurface water movements. Note the general relationships: as the dissolved oxygen content of waters decreases due to decomposition below the pycnocline, the nutrient concentrations increase.

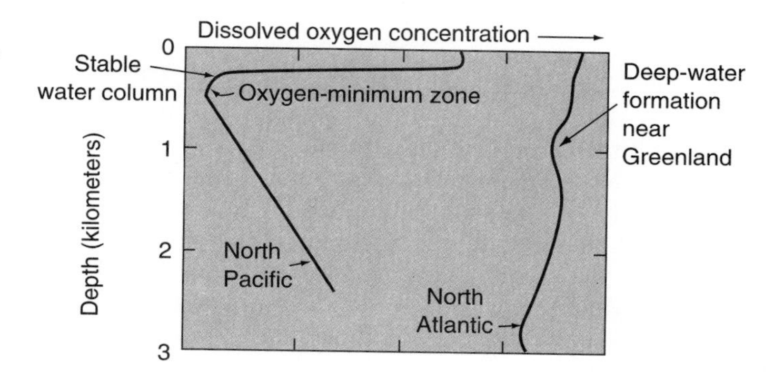

Figure 11–22
Dissolved-oxygen concentrations in a water column in the Pacific and in the North Atlantic south of Greenland. High concentrations near the surface are due to gas exchange with the atmosphere. High concentrations near the bottom are supplied by the sinking of cold surface waters in high latitudes. The North Atlantic near Greenland (right hand curve) is the source of cold, oxygen-rich waters sinking into the deep Atlantic basins. Cold-water masses cannot form and sink in the low salinity surface waters of the North Pacific (left-hand curve). Therefore, the low oxygen minimum zone at mid-depths is caused by consumption of dissolved oxygen by decomposing organic matter sinking to the bottom. Compare to Figure 11–21.

Dissolved oxygen is consumed at all depths in the ocean but is supplied only where the water is exposed to the atmosphere. The longer a water parcel is isolated from the surface, therefore, the lower its dissolved oxygen concentration. Thus, dissolved oxygen concentrations are highest where water masses form (in the Arctic and Southern oceans, for example) and lowest where water masses have been isolated from contact with the atmosphere the longest (in the North Pacific, for example).

Note in Figure 11–22 the low dissolved oxygen concentrations in the mid-depths of the Pacific (between 300 and 500 meters); this is called the **oxygen-minimum zone.** It contains low-oxygen waters that rise from the deep ocean but cannot penetrate the pycnocline and are therefore isolated from the atmosphere. We discussed this as part of the great global conveyer in Chapter 8. These intermediate depths also receive large amounts of organic matter sinking out of the surface zone. As the organic matter decomposes, it uses up more of the dissolved oxygen, contributing to the oxygen minimum.

Oxygen consumption in the deep ocean is slow, mainly because low temperatures reduce the metabolic rates of deep-water organisms. Also, scarcity of food keeps populations sparse. As waters slowly move along the bottom, their dissolved oxygen is used up by bacterial decomposition of organic matter. These patterns of changing dissolved oxygen concentrations are used as tracers to determine the directions and relative speeds of deep-ocean water mass movements, as mentioned in Chapter 8.

Subterranean Ecosystems

Continued exploration of the ocean bottom and of the rocks buried deep below Earth's surface has found enormous quantities of microbes. In fact, the amount of subterranean life probably greatly exceeds the amount of living matter in the ecosystems above Earth's surface. Heat-loving microbes have been discovered in the discharges of waters on the ocean bottom following volcanic eruptions. Vast amounts of organic matter are discharged as the vents begin, indicating that the organisms existed before the eruption. Similar organisms have been found in the waters discharged by hot springs on land in rocks and fluids recovered from deep in Earth's interior. In fact, we do not know how deep in the Earth such organisms can live and the variety of energy sources they may be able to exploit.

An intriguing implication of this find is that many other planets in our solar system, indeed in other planetary systems, may harbor such ecosystems below their surfaces. Future missions to study planets may well incorporate drilling operations to sample below the planet's surface to look for examples of similar subterranean ecosystems.

Summary

Unlike the land environment, the ocean provides a three-dimensional habitat for life. Most organisms are found near its top, bottom, or sides; the vast interior space of the deep ocean is less populated. Organisms exhibit three major life-styles: drifting, or planktonic; swimming, or nektonic; and attached, or benthic.

Temperature and salinity are the principal controls on the distributions of most marine organisms. The densities of seawater and living tissues are nearly equal. Many marine organisms take advantage of this by having delicate, jelly-like structures—they don't need skeletons for support. Pressure is important only for fishes with swim bladders.

An ecosystem includes autotrophic organisms (which manufacture their own food using nutrients and a source of energy), heterotrophic organisms (which must eat autotrophs to get their food), and decomposers (which break down tissues so that essential compounds are recycled to sustain new growth). Trophic levels in an ecosystem describe who eats whom. Plants, the first trophic level, are eaten by herbivores, the second trophic level. Carnivores, the third trophic level, eat herbivores.

The energy contained in organic matter is transferred from one trophic level to the next in a complex series of feeding relationships known as a food web. Energy transfers from one trophic level to another are about 10 percent efficient on average worldwide. Biomass is the quantity of plants or animals per unit volume of seawater or area of sea surface or sea bottom. Primary production is the organic matter produced by plants. Secondary production is the organic matter produced by animals. Standing crop is the amount of organic matter per unit area (or volume) at a given time.

Photosynthesis is the process by which plants use energy from sunlight to make food out of inorganic substances, releasing oxygen in the process. Respiration is the opposite process, in which organisms release energy from their food. In the absence of sunlight, certain bacteria use energy-rich compounds to obtain energy to make food, a process called chemosynthesis.

At the compensation depth, the rate of photosynthesis equals the rate of respiration. Sufficient light is available in the photic zone to sustain photosynthesis. Sufficient light, together with a stable water column and an abundance of nutrients to permit plants to remain in the photic zone, leads to rapid growth of the plants, called a bloom.

Lack of nutrients in seawater (usually compounds of nitrogen and phosphorus) limits plant growth even with enough sunlight. When plants die and decompose, they release nutrients. Many organisms sink below the surface zone and decompose in the pycnocline and deep zones; because there is no photosynthesis at these depths, nutrients from the decomposition build up in deep waters. These deep nutrients are returned to surface waters primarily by upwelling water.

Compounds containing phosphorus and nitrogen are necessary for phytoplankton growth. Bacterial decomposition of organic matter recycles these nutrients. Processes involved in the recycling of phosphorus are relatively simple and rapid; therefore, phosphorus rarely limits phytoplankton production. The nitrogen cycle is slower and more complex, depending upon chemical conversions by specialized bacteria. Scarcity of nitrogen compounds frequently limits phytoplankton production in the ocean.

Biological processes can have an impact on the chemical properties of seawater. Dissolved organic matter is the ocean's largest reservoir of organic carbon. Very small bacteria can use dissolved organic matter directly as a food source. They are eaten by other minute organisms and the organic matter is recycled in what is called the microbial loop. Organic matter also occurs in the form of particles (shell fragments, fecal pellets, and pieces of tissue). Large, rapidly sinking particles are important sources of food to organisms in the water and on the bottom. Small, slowly sinking particles are involved in chemical reactions that affect the composition of seawater. Particles also influence the distributions of trace elements and dissolved oxygen in seawater.

Dissolved oxygen levels are normally inversely proportional to nutrient levels. Changes in dissolved oxygen levels are used to study movements of subsurface waters.

Key Terms

pelagic
neritic
plankton
bacteria
phytoplankton
zooplankton
nekton
benthos
plankters
poikilotherms
homeotherms
ecosystem
autotrophic
photosynthesis
chemosynthesis
heterotroph
decomposers
trophic level
herbivores
carnivores
omnivores
detritus
detritivores
food chain
food web
biomass
production
standing crop
primary production
net primary production
respiration
photic zone
aphotic zone
compensation depth
critical depth
bloom
biological pump
trace nutrients
nitrogen fixation
microbial loop
marine snow
oxygen-minimum zone

Study Questions

1. Describe the types of components that make up an ecosystem.
2. What factors limit production of organic matter in the ocean? In which zone is each factor most important?
3. Describe the processes that control the distributions of phosphorus and nitrogen compounds in the open ocean.
4. Describe photosynthesis.
5. Describe chemosynthesis. Tell how it differs from photosynthesis.
6. Describe the distribution of dissolved oxygen in the three major ocean basins. Explain the differences.
7. Discuss food webs. How does a web dominated by phytoplankton differ from one dominated by bacteria?
8. Explain the different depth distributions of trace elements in the ocean.
9. Describe the processes controlling productivity in upwelling areas.
10. Discuss how undissolved particles can affect the abundances and distributions of trace elements in the ocean.
11. Why is half the world's fish production taken from upwelling zones?
12. Discuss the factors that control the distributions of dissolved oxygen in the open ocean.
13. Compare a typical open-ocean food web with a typical web in an upwelling system.
14. Discuss the difference between primary and secondary production.
15. What is the difference between a food web and a food chain?
16. *[critical thinking]* Discuss the factors limiting the amount of fish that can be caught in the ocean.

Selected References

AUSTIN, B., *Marine Microbiology*. Cambridge, England: Cambridge University Press, 1988.

BANNISTER, K., AND A. CAMPBELL, *Encyclopedia of Aquatic Life*. New York: Facts on File, 1985.

BUCHSBAUM, R., M. BUCHSBAUM, J. PEARSE, AND V. PEARSE, *Animals Without Backbones*, 3d ed. Chicago: University of Chicago Press, 1987. A classic.

CUSHING, D. H., AND J. J. WALSH, *The Ecology of the Seas*. Philadelphia: Saunders College Publishing, 1976. Advanced.

JUMARS, P. A., *Concepts in Biological Oceanography: an Interdisciplinary Primer*. New York: Oxford University Press, 1993. Advanced level text.

LALLI, C. M., AND T. R. PARSONS, *Biological Oceanography: an Introduction*. Oxford: Pergamon Press, 1993. Intermediate level text.

LONGHURST, A., AND D. PAULY, *Ecology of Tropical Oceans*. London: Academic Press, 1987.

NYBAKKEN, J. W., *Marine Biology: An Ecological Approach*, 2d ed. New York: Harper & Row, 1988. Elementary.

PARSONS, T. R., M. TAKAHASHI, AND B. HARGRAVE, *Biological Oceanographic Processes*, 3d ed. Oxford, England: Pergamon Press, 1985. Reviews biochemical processes.

VALIELA, I., *Marine Ecological Processes*. New York: Springer–Verlag, 1984.

OBJECTIVES

Your objectives as you study this chapter are to understand:

- Open-ocean environments and how open-ocean organisms live
- Microscopic floating plants (phytoplankton) and animals (zooplankton)
- Roles of nekton in marine ecosystems
- Mammals and reptiles in the ocean
- Processes controlling the abundance and distribution of nekton
- Life strategies and adaptations of open-ocean organisms

Ctenophores—the comb jellies—are voracious predators on planktonic organisms.
(Courtesy Photo Researchers, Inc.)

12 Open-Ocean Plankton and Nekton

In the open ocean, organisms must either float or swim; otherwise they sink. Water and the organisms in it are constantly in motion, responding to small-scale water movements due to turbulence or to larger eddies and currents. As we learned in Chapter 11, those organisms that cannot swim fast enough and are moved by currents are called *plankton.* Those able to swim against currents, capable of moving independently of the waters around them, are called *nekton.* Plankton includes *phytoplankton* (drifting plants), *zooplankton* (drifting animals), and *bacteria.* In this chapter, we see how such organisms live in the open ocean.

In dealing with open-ocean life, we must consider the space- and time-scales that govern how such organisms live and grow. First, we are dealing with organisms ranging in size from bacteria too small to be seen in ordinary light microscopes, to blue whales, the largest animals that ever lived on Earth. The smallest organisms experience their environment in ways that are unfamiliar to us. For instance, water sticks to the surfaces of microscopic organisms, much like molasses. Thus, the smallest organisms require swimming and food-gathering structures that are quite different from those used by the more familiar large organisms, such as fish.

Size also affects organisms in other ways. One is very obvious—large organisms eat smaller ones. Hence, one survival tactic is to grow as large as possible as rapidly as possible, while expending little energy. Some examples are growing spines or building body tissues consisting largely of water with a minimum of organic matter. We discuss both these tactics in this chapter.

A second aspect of size is that it determines how organisms live. The smallest ones cannot swim well enough to avoid being carried by currents. Many cannot avoid sinking out of the photic zone. More about this later.

We must also consider many different time scales. Bacteria and many of the smallest one-celled organisms reproduce by cell division, sometimes doubling their numbers in a few hours to a few days. Populations of these organisms can expand quite rapidly when conditions (sufficient light, abundant nutrients) are favorable.

The largest organisms, such as whales, can take decades to mature and reproduce. Consequently, they respond much more slowly to changing conditions. For instance, it may take centuries for whale populations to recover from the unlimited whaling of past centuries.

Plankton

Open-ocean plants and animals are mostly microscopic, one-celled plankters (Fig. 12–1). These tiny plants produce food for all other open-ocean organisms. The tiny microscopic animals are the most common forms of life on Earth. Most of them are totally unfamiliar to us land-dwellers.

One simple way in which oceanographers describe planktonic organisms, whether plants, animals, or bacteria, is to classify them by size. Three groupings are commonly used. Those having diameters between 20 and 200 micrometers are **microplankton.** (A human hair is about 100 micrometers in diameter.) Plankters less than 20 micrometers in diameter but larger than 2 micrometers are called **nannoplankton.** They are important in equatorial waters, where they constitute 50 to 80 percent of the standing crop. As a result of their sheer numbers, these tiny planktonic organisms dominate primary production there. The smallest category is the **ultraplankton** (including bacteria), those forms less than 2 micrometers in diameter. These tiny organisms include heterotrophic bacteria, autotrophic cyanobacteria (also called blue-green algae), and the predators that eat them. These are the organisms involved in the microbial loop we discussed in Chapter 11. Table 12–1 summarizes the three size classes of phytoplankton.

One problem unique to planktonic life is sinking, and sinking rates, of course, are related to size. Maintaining the proper depth is critical to an organism's survival, because most open-ocean organisms live at a particular depth that meets their light and temperature needs. Any organism less dense than the surrounding water rises. Any organism denser than the surrounding water sinks, unless it swims or somehow maintains buoyancy. Drag, or resistance to sinking, determines an object's rate of sinking. (You experience drag on your hand when you pull it through the water in a swimming pool.) Drag is greater for large objects than for small ones. By altering its density (by secreting low-density fats into body

Figure 12–1
Phytoplankton take many forms. Some species build chains of individual cells, others have ornate spines. Many of these structural characteristics are adaptations that increase drag and retard sinking. (Courtesy James Yoder, NASA.)

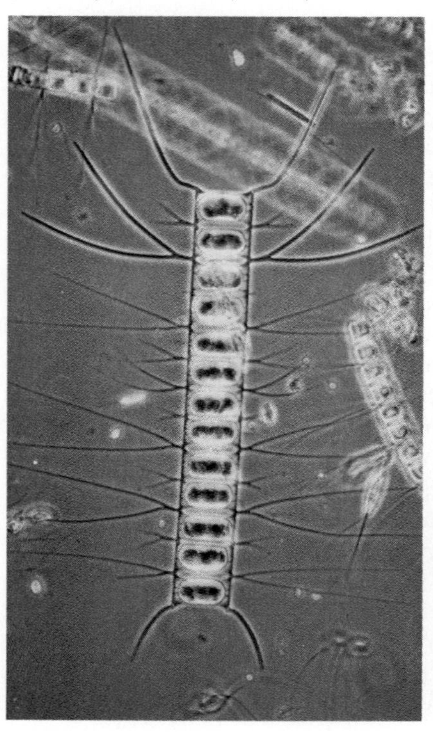

TABLE 12–1
The Three Size-Categories of Phytoplankton

Name	*Diameter (micrometers)*	*Typical Members*
Ultraplankton	less than 2	Heterotrophic bacteria, autotrophic cyanobacteria
Nannoplankton	2–20	Coccoliths, Silicaflagellates
Microplankton	20–200	Diatoms, Radiolaria

cavities), or its size (by growing spines, for example), an organism can change its sinking rate. Also, objects sink more slowly in cold water than in warm water. Organisms with ornate structures or other adaptations to retard sinking are more common in warm waters.

Vertical water movements associated with mixing or upwelling can keep organisms near the sea surface, thus countering sinking. Thus, the larger phytoplankters are more likely to be found where there are strong currents or upwelling. In stiller waters, there is less turbulence to return phytoplankton to the surface layers. Here even the limited swimming ability of many nannoplankters and ultraplankters is an advantage; larger nonswimming forms slowly sink out of the photic layer.

Net plankton is the general term used to refer the types of plankton caught in biologists' nets. Although this term is commonly used, it is not precise as to size. Plankton nets, depending on the size of their mesh, will catch most plankton except the ultraplankton, which will pass through the net. Net plankton has large standing crops, especially at higher latitudes.

Phytoplankton

Relatively few different forms dominate the phytoplankton in the ocean. We begin by considering the most common ones.

Worldwide, **diatoms** (Fig. 12–2a) are the most important primary producers. They are single-celled algae and have hard external skeletons made of silica, in either a pillbox or rod-like shape. Some species have sticky threads or long spines protruding from their bodies and form long chains of individual cells, especially in nutrient-rich waters (Fig. 12–1).

Diatoms grow by division, their cells, encased in glass-like shells, becoming smaller with each generation (Fig. 12–3). When cells decrease to a critical size, both halves of the old shell are discarded. Then the naked cell doubles or triples in size before new shells form. Diatom division rates can exceed one per day. Therefore, populations can increase 500 to 2,000 times over a winter to produce a large "seed crop" from which the next season's diatom crop grows. Diatoms (and many other organisms) can also form resting spores under unfavorable conditions. These spores can remain alive but dormant for long periods, even years. After death, diatom shells dissolve slowly and, as we saw in Chapter 5, contribute to the sediments, especially under highly productive waters.

The second most abundant phytoplankton are **dinoflagellates** (Fig. 12–2b). Even though they are classified as phytoplankton (plants), some dinoflagellates behave like one-celled animals in that they are strictly heterotrophic, and can therefore live on dissolved or particulate organic matter absorbed or ingested from seawater. Other dinoflagellates are strictly autotrophic, carrying out photosynthesis, while still other species are capable of both types of nutrition. Dinoflagellates are capable of weak movements using paired, whip-like flagella. Some have rigid cell walls made of cellulose, while others do not. These single-celled organisms, which cannot easily be classified as either plants or animals, are called **protists.**

Many dinoflagellates can tolerate low nutrient concentrations. Dinoflagellate blooms can exceed those of diatoms, especially in polluted waters with high nitrogen and phosphorus concentrations. This may result from a scarcity of silicon, because low levels of this shell-forming element limit diatom growth but do not affect dinoflagellates, which lack silica shells.

Coccolithophores (Fig. 12–2c), another major group of flagellated phytoplankters, are distinguished by their coatings of tiny calcareous plates. They are important primary producers in the ocean, sometimes forming enormous blooms; they are also major contributors to calcareous sediment deposits. Less common are silicoflagellates (Fig. 12–2d) and other groups of flagellated planktonic organisms.

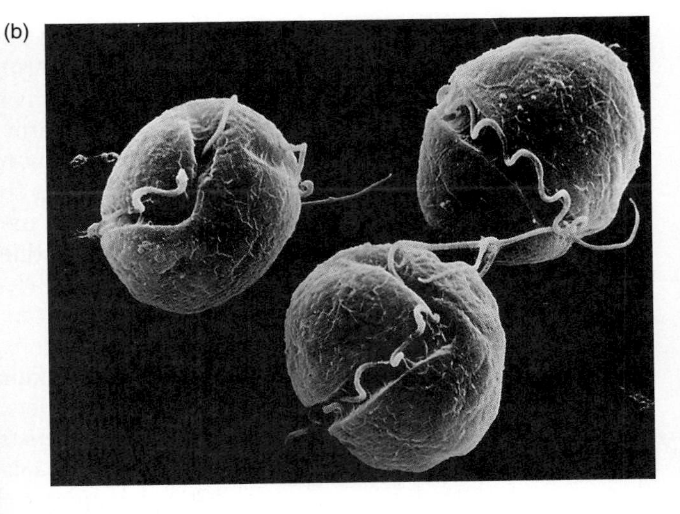

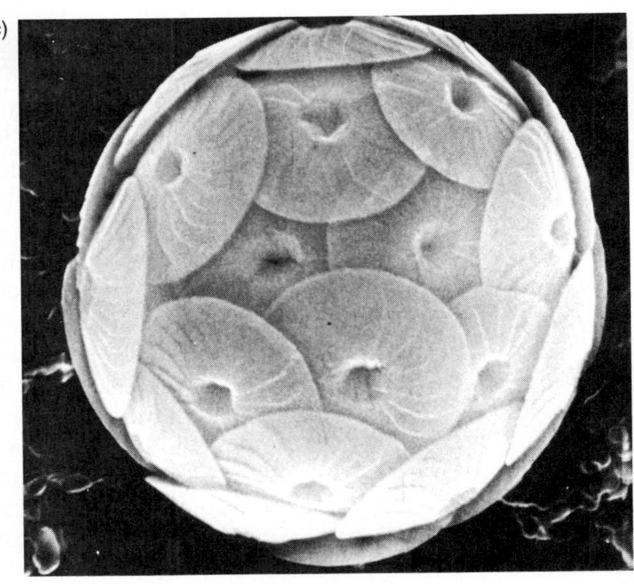

Figure 12–2
Major types of phytoplankton. (a) Diatoms include pill boxed-shaped forms in addition to rod-like shapes and the chains shown in Figure 12–1. (Biophoto Associates/Science Source/ Photo Researchers, Inc.) (b) Dinoflagellates are about 400 micrometers long. (David M. Phillips/ The Population Council/Science Source/ Photo Researchers, Inc.) (c) This coccolithophore measures about 10 micrometers in diameter. Calcareous plates form its shell. (Courtesy Susumu Honjo, WHOI.) (d) Silicoflagellates are about 30 micrometers in diameter, including the spines. (Courtesy J. M. Sieburth and University Park Press, Baltimore.)

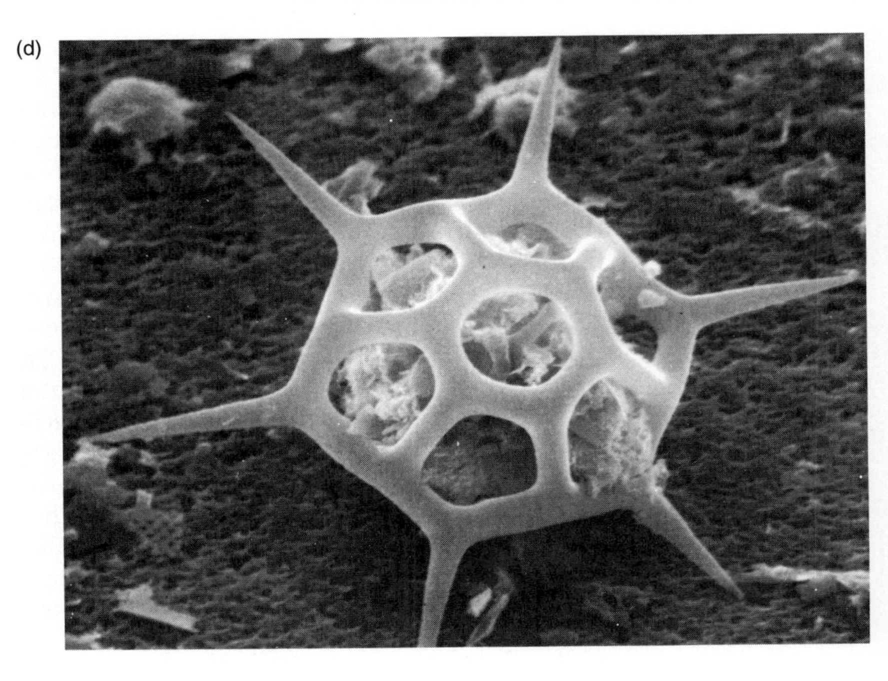

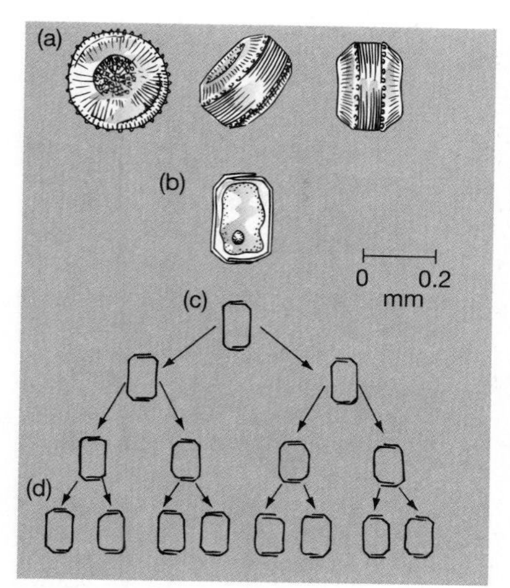

Figure 12–3
Reproduction in a diatom. The box-like shell is made of glass-like silica that contains pores for exchanging nutrients and metabolic wastes with the surrounding water. (a) This cross section of a diatom shows that the outer layer is not continuous; rather, it comprises two halves that fit together like the two parts of a pill box. (b) When a diatom cell grows large enough to divide, the two halves separate, with each half getting roughly half the cell contents. (c) A new silica shell is secreted over the exposed tissues. (d) The cells formed through repeated divisions become progressively smaller.

Finally, a group of bacteria—the **cyanobacteria** (having blue-green pigments)—include some of the smallest types of phytoplankton as well as some of the largest. Some cyanobacteria play a very important role as nitrogen fixers in the ocean.

Small size has several advantages for marine phytoplankton. These organisms rely on diffusion to supply nutrients and remove wastes from their cells. Having a large surface area relative to body volume facilitates the exchange of dissolved substances between cells and the waters around them. To maintain themselves in the photic zone, some phytoplankton can move relative to their surroundings. As we have seen, some of these organisms have flagella to propel them through the water. Phytoplankton lacking flagella sink slowly and depend on turbulence and upwelling to return them to the photic zone. Here, formation of chains, or ornate structures can slow their sinking.

Nannoplankters (having diameters between 2 and 20 micrometers) and the even smaller ultraplankters are the dominant producers and consumers of organic matter in the ocean. As we saw in the last chapter, they account for much of the ocean's productivity.

Many ultraplankters, such as some types of dinoflagellates and bacteria, can live as either autotrophs or heterotrophs. Thus, they are able to survive as heterotrophs for long periods in deeper waters, even if light levels are too low for photosynthesis. Such organisms are often the first to bloom in newly upwelled waters. Indeed, the highly productive upwelling waters, dominated by diatoms, are isolated oases in waters that are otherwise low in productivity.

Minute animals preying on nannoplankton and ultraplankton consume most of the plant matter in the ocean. Surprisingly, there are roughly equal concentrations of organic matter in all size classes of organisms, ranging from bacteria to whales. Because microorganisms process matter and energy much faster than do larger organisms, a large fraction of the total organic matter in the ocean is processed by these abundant but little-known animals.

Zooplankton

Much of our knowledge of the microscopic animals (zooplankton) of the ocean comes from scientists working on problems of fisheries or other commercially exploited organisms, such as whales. Most of these studies have used plankton nets that captured only the larger zooplankters. The smallest zooplankters are poorly understood, although we now recognize that they play important roles in marine ecosystems. In short, we know most about the zooplanktonic organisms which are involved in the food webs that are exploited by humans, and much less about the food webs that are less important to humans.

Lower-trophic-level consumers in marine food webs are mostly zooplankters (Fig. 12–4). Some swim and can actively pursue prey, but most are **suspension** (or **filter**) **feeders,** bearing tiny hairs or mucous surfaces to capture floating food particles. Because these animals usually depend on food particles of a particular size, their distributions depend largely on the availability of food and how the currents have moved them.

Another important factor limiting zooplankton distribution is the narrow temperature range in which a species can reproduce (generally only a few degrees). Adult populations have greater temperature tolerances and so may be carried far out of their breeding range by currents.

Holoplankton

Not all zooplankton remain free-floating throughout their life spans. Those that do, known as **holoplankton,** are generally more important in marine food webs than **meroplankton,** which are the larval stage of pelagic, nektonic, and benthic

Figure 12–4
Mixed zooplankton from the eastern North Atlantic. Copepods are the most common organisms, but this picture includes starfish and sea urchin larvae, as well as a small tunicate. (Copyright N. T. Nicoll.)

organisms. (About 80 percent of shallow-water benthic organisms in the tropics have planktonic larvae.) Holoplankton dominate in the open ocean, whereas meroplankton are more common in shallow coastal-ocean waters.

Many consumers of nannoplankton are single-celled holoplanktonic protists including, for example, the **foraminifera** (Fig. 12–5a) and **radiolaria** (Fig. 12–5b). Foraminifera are single-celled and live nearly everywhere in the ocean—both in the water and on the bottom. As we learned in Chapter 5, they have delicate, porous shells in a large variety of shapes made of calcium carbonate. Thin protoplasm extrusions extend through holes in the shells to capture food particles.

Radiolaria are single-celled amoebae with spherical skeletons made of silica (glass). Protoplasmic strands project in all directions outside the shell as long, sticky filaments. These strands trap tiny phytoplanktonic particles, which are then borne by protoplasm toward the center of the body to be digested. Individuals range from 0.1 to more than 10 millimeters in diameter. Reproduction is by division into many small flagellated cells. As we learned in Chapter 5, both foraminifera and radiolarians are important contributors to deep-sea sediment deposits.

Some types of **crustacea** are the most numerous holoplanktonic organisms, constituting 70 percent or more of the zooplanktonic biomass in the ocean, and one type, the **copepods** (Fig. 12–6), is the most abundant of all. Copepods and another type of crustacean, the shrimp-like **euphausiids** (Fig. 12–7), are most important as herbivores and first-level carnivores in marine food webs. Crustaceans have been called the insects of the sea. In fact, both insects and crustaceans are arthropods ("jointed feet"), animals characterized by segmented bodies and appendages. They have stiff, outer shells made of chitin (a stiff, complex carbohydrate) that serve as skeletons. The appendages are specialized for various functions, such as feeding, moving, sensing, or reproducing.

Copepods are approximately 0.3 to 8 millimeters long and have legs covered with feathery, curved bristles that form a filter chamber in front of the mouth. Copepods occur throughout the ocean; indeed, they are among the most numerous animals on Earth. Depending on temperature and availability of food, large copepods can double their numbers several times in a year. Smaller ones reproduce even more frequently. Most copepods are herbivorous and thus are the primary consumers in marine food webs. Some larger species are carnivores, eating the eggs and planktonic larvae of larger organisms.

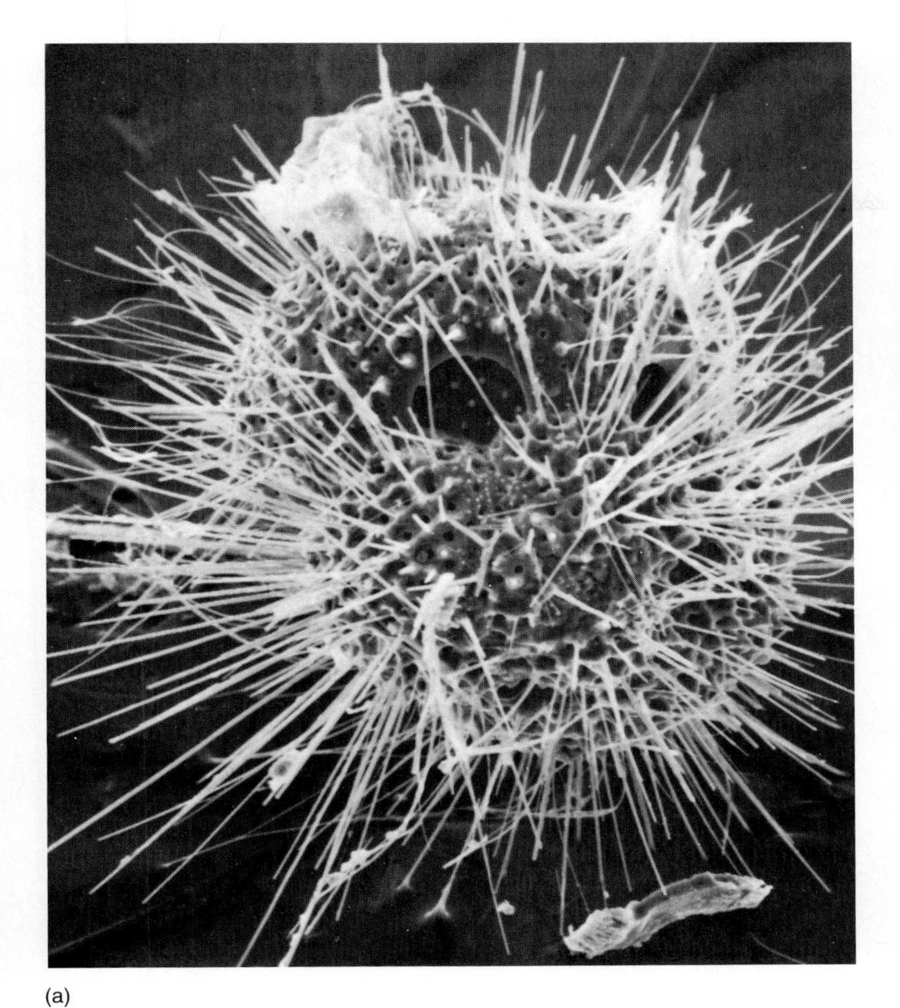

(a)

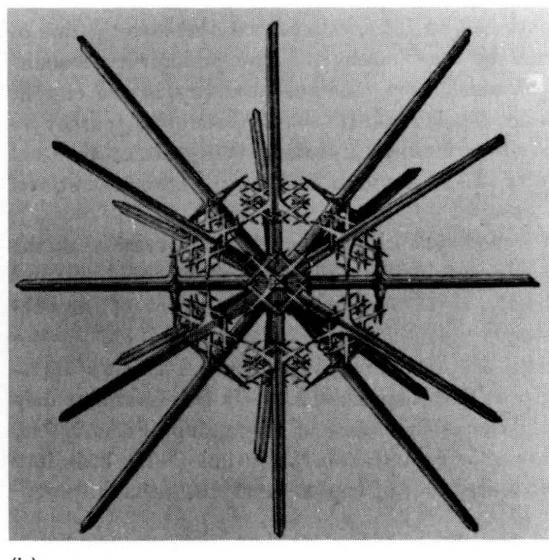

(b)

Figure 12–5
Some open-ocean zooplankton. (a) A foraminiferan. Two apertures, through which food is ingested, are visible just above the center of the organism. Excluding spines, the organism is about 300 micrometers in diameter. (Courtesy J. M. Sieburth and University Park Press, Baltimore.) (b) This delicate shell made by a radiolaria consists of highly soluble minerals that dissolve within a few hours once the organism dies. (Drawn by Ernst Haekel. Report of the Scientific Results, HMS *Challenger*. *Zoology* 18:1887.)

One type of euphausiid, known as **krill** (Fig. 12–7), is important because dense swarms of these animals in the Southern Ocean feed on diatoms, and are, in turn, the chief food of many fishes and of the large whales. Krill are so abundant there that they are fished commercially.

Euphausiids are omnivorous, feeding on phytoplankton, small zooplankton, and detritus. In general, euphausiids are larger than copepods, up to 5 centimeters long. They occur throughout the water column and in all oceans, often in enor-

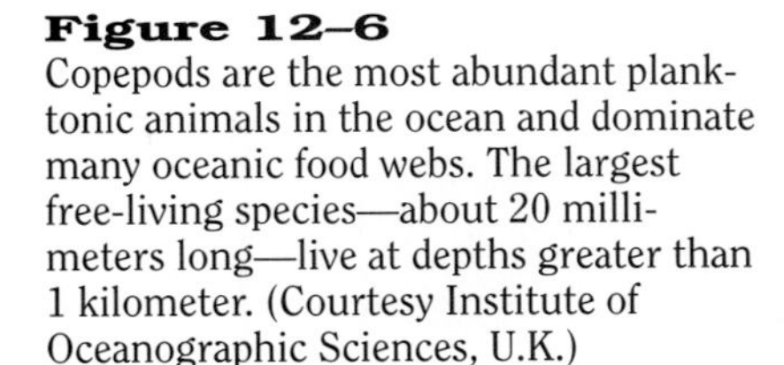

Figure 12–6
Copepods are the most abundant planktonic animals in the ocean and dominate many oceanic food webs. The largest free-living species—about 20 millimeters long—live at depths greater than 1 kilometer. (Courtesy Institute of Oceanographic Sciences, U.K.)

BOX 12–1 Red Tides

Red tides are discolored surface waters formed by intense blooms of planktonic organisms, usually dinoflagellates (Fig. B12–1–1). They occur in nearshore waters and have nothing to do with the tides discussed in Chapter 10. They often give the water a tomato soup or rusty color. Toxins in the dinoflagellate cells cause problems. When filter-feeding organisms eat the dinoflagellates, the eaters retain the toxins in their tissues. These toxins may not

Figure B12–1–1
Red tides are caused by blooms of dinoflagellates. (Jeff Foott/Bruce Coleman, Inc.)

mous swarms. In the Atlantic and Pacific, euphausiids are important as food for fish such as herring and salmon. In Antarctic and other cold waters, euphausiids mature slowly, living up to 2 years. A large baleen whale consumes an average of 850 liters of euphausiids per day.

Besides crustaceans, other types of holoplankton are extremely abundant. **Pteropoda** ("wing-footed") are small pelagic snails that occur in dense swarms in

Figure 12–7
Euphausiids (also known as krill) are large planktonic organisms that are especially abundant around Antarctica. They are the principal food for many Southern Ocean organisms, including the great whales. (Doug Cheeseman/Peter Arnold, Inc.)

immediately harm fish or clams, but they accumulate and become concentrated. Then they can poison people or marine mammals who eat the fish or clams. Thus, fisheries are often closed by health authorities when red tides occur.

Dinoflagellates in a red tide can also release their toxins to surrounding waters and kill marine organisms. Toxins transported by spray can cause lung irritation in people on nearby beaches. Economic damage to coastal communities can run into many millions of dollars, as a result of losses in fisheries and tourism.

Red tides are extreme cases of plankton blooms. They occur when conditions are especially favorable. For example, a sudden increase in surface-water stability (resulting from, say, a heavy rain) keeps planktonic organisms in sunlit surface waters. Then they are no longer limited by scarcity of light, and if nutrients are also abundant, a bloom can follow.

Availability of nutrients can also trigger red tides. Rainfall often contains nitrogen compounds from automobile exhaust. Land runoff contains nitrogen-rich animal wastes and agricultural fertilizers. When nitrogen compounds are mixed into nutrient-limited waters, they can trigger plankton blooms. It is also possible for runoff to provide trace amounts of metals or possibly compounds that detoxify substances in the water (such as copper) that would otherwise inhibit phytoplankton growth.

Dinoflagellates are usually the organisms that cause red tides. These one-celled organisms can photosynthesize in waters that contain few nutrients. Also, they do not require silica, which is needed by diatoms to make their shells. They are widely distributed, especially in nutrient-poor tropical ocean waters. Dinoflagellates can also swim toward the surface to become concentrated in surface waters.

Sustained growth of dinoflagellates usually depletes the nutrients dissolved in the waters. Thus a red tide usually ends after a few days. If there is a continued supply of nutrients—from pollution, for example—a red tide can persist for weeks or months. One source of nutrients is nitrogen-fixing blue–green algae. These organisms take nitrogen gas dissolved in the water and produce ammonia, which is readily used by other organisms. If blue–green algae and dinoflagellates occur together, the combination can support a red tide for many weeks. Eventually, the tide is dispersed by currents and winds, or the weather changes, removing conditions favorable to plankton growth.

Dinoflagellates can form spores, a seedlike stage, when conditions are not favorable. The spores settle out on the bottom and may remain alive but dormant in sediments for years. When conditions are favorable, the spores develop into dinoflagellates and can seed future red tides. Thus, once an area has had a red tide, it is subject to future ones.

all seas (Fig. 12–8). The characteristic snail foot is modified into fins that permit the animals to swim vertically hundreds of meters each day. Pteropods have carbonate shells that are later preserved in pteropod-rich sediments. Generally, pteropods form large, sticky mucous webs which hang in the water and trap phytoplankton, small zooplankton, and detritus. Once the web is full, it is retracted and eaten.

Meroplankton

Meroplankton, the planktonic larval forms of benthic and nektonic animals, are abundant in coastal waters. Nearly 70 percent of all benthic invertebrates and nearly all types of fish are known to have planktonic larval stages. These larvae are an important food source for fishes.

Chapter 13 will discuss the planktonic larvae of benthic organisms (Fig. 12–9) and the subtle processes involved in their recognition of suitable locations in which to settle to the bottom when their planktonic life is completed.

Most fish eggs are released and fertilized in coastal ocean waters. These eggs drift; when sufficiently developed, they hatch and begin to feed. Depending on temperature and species, hatching may be as soon as 1 or 2 days after release. The larval fish then live as meroplankton for several weeks.

Figure 12–8
Most pteropods (sea butterflies) are herbivorous and are extremely abundant in surface waters. (Courtesy Woods Hole Oceanographic Institution.)

Gelatinous Plankton

So far, we have discussed animals that are important members of food chains that support fish populations exploited by humans. Other important zooplankters—called **gelatinous plankton** because they consist of a jellylike substance—are part of food webs not commonly exploited by humans. Jellyfish (also called "medusae") are perhaps the most familiar example of this type of plankton (Fig. 12–10). They have a bell-shaped, two-layered gelatinous body wall surrounding a digestive cavity. The mouth is surrounded by tentacles bearing stinging cells. The tentacles capture particles and move them into the mouth, through which wastes are also eliminated. The animal moves by rhythmic pulsations of its bell.

A group of the gelatinous plankton, related to the true jellyfish, is called **siphonophores** and includes the well-known Portuguese Man-of-War pictured in Figure 12–11. Siphonophores are colonies of medusae and hydroids—individuals that live together and function as one animal, each individual in the colony having

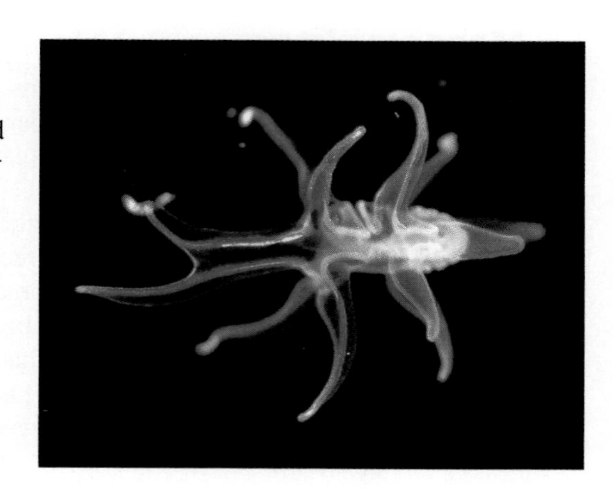

Figure 12–9
The larvae of benthic invertebrates are common in the plankton at certain times of year. This starfish larva will have passed through several stages before it metamorphoses and settles to the bottom as a young starfish. (M. J. Youngbluth/Harbor Branch Oceanographic Institution, Fort Pierce, Florida.)

Figure 12–10
This adult jellyfish *(Pelagia noctiluca)* measures about 10 centimeters across the bell. It occurs worldwide. (Fred Bavendam/Peter Arnold, Inc.)

Figure 12–11
The Portuguese Man-of-War *(Physalia)* has a purple, air-filled bladder that floats at the sea surface. The whole organism consists of a colony of specialized individuals (called polyps), each performing a specific task. Some entrap, paralyze, and engulf their prey, digesting and absorbing it. Other polyps have reproductive functions, producing sexual medusae. (Rod and Moira Borland/Bruce Coleman, Inc.)

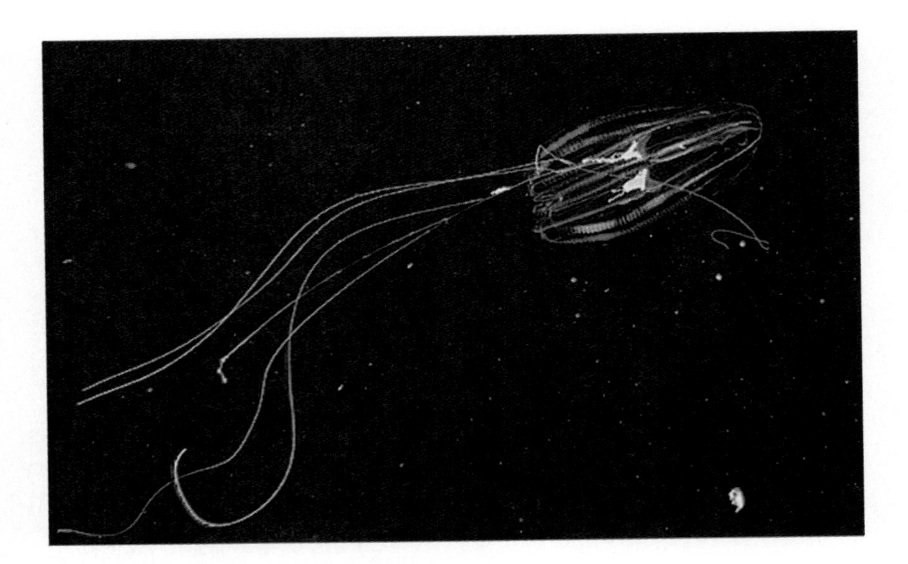

Figure 12–12
Ctenophores are important predators, often occurring in great abundance. (M. J. Youngbluth/Harbor Branch Oceanographic Institution, Fort Pierce, Florida.)

a specialized task. Siphonophores are capable of stunning, capturing, and digesting relatively large fish.

Another group of gelatinous plankton, the **ctenophores** (Fig. 12–12), look something like jellyfish. Small and jelly-like, they are sometimes known as sea walnuts or comb jellies. Some have trailing tentacles for capturing prey. Ctenophores swim smoothly using the beating motion of rows of cilia (hence the common name "comb jellies") rather than the pulsation of the bell used by medusae. Voraciously carnivorous, ctenophores often occur in great numbers and prey on crustaceans and young fishes, greatly reducing their populations.

Some gelatinous plankters, called **tunicates,** are very primitive vertebrates having a notochord, or primitive backbone. One group, the larvaceans, secrete transparent, round gelatinous structures in which the individual, worm-like animal lives. They use their tail movements to pull a current of water through this "house," which contains filters to trap small plankters and bacteria as food. The animal then eats the filter. Another type of tunicate, the salps, has a barrel-shaped gelatinous body and uses a muscular pumping motion to swim and to trap food from the current of water passing through its body (Fig. 12–13). Salps may be solitary, but can form colonies, and often occur in very dense swarms.

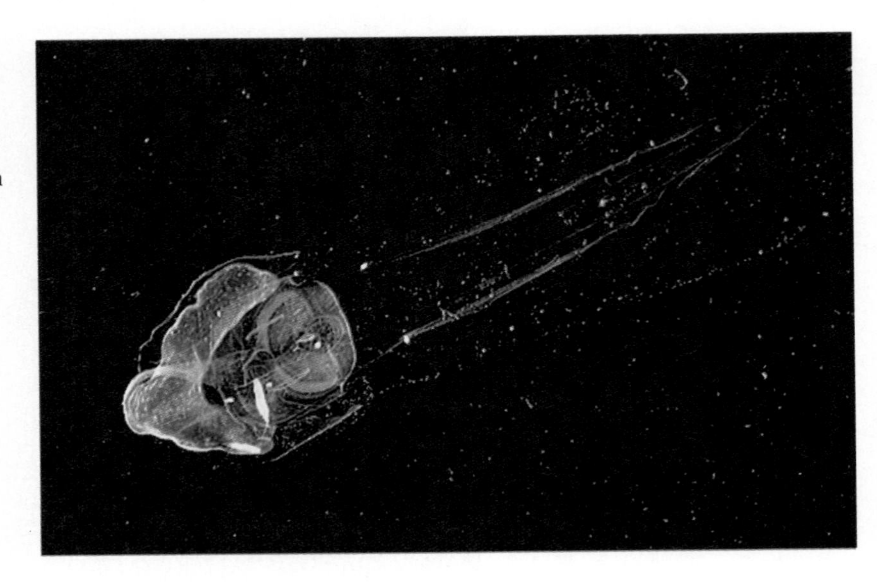

Figure 12–13
Planktonic tunicates form gelatinous structures in which the animals live. The animal pumps water through this "house," and this pumping action brings in food and propels the organism through the water. A filtering system captures particles on a mucous-covered structure which the animal eventually eats. The outline of this transparent structure is seen in this photograph where food particles are concentrated on its surface. (M. J. Youngbluth/Harbor Branch Oceanographic Institution, Fort Pierce, Florida.)

Nekton

As we learned in Chapter 11, marine animals that can swim well enough to move independently of currents are called nekton. This group includes the larger marine animals, such fishes, squids, sharks, marine mammals, and marine reptiles.

The principal characteristic of nekton is their swimming ability. They are relatively large and have body shapes adapted to moving rapidly through the water. Many nektonic animals secrete slimes to reduce drag; others have special ribbing of the skin to reduce drag. (Ribbing of this sort has been used in racing yachts to reduce drag and increase speed.)

Another aspect of drag reduction is body shape. A cone moving through the water pointed end first experiences resistance because of turbulence behind the cone (Fig. 12–14). There is less turbulence when the blunt end is first. The least drag comes from a fusiform body, which is cylindrical, with a blunt end first and a tapered end following (Fig. 12–14c). The fastest fishes have this body shape (Fig. 12–15).

There are other adaptations for speed. Tuna are good examples of fishes adapted for fast swimming. They have characteristic lunate tails (shaped like a crescent moon) and a thin connection to the body (Fig. 12–16). This tail shape increases swimming efficiency.

Fishes

The many different shapes of fishes reflect the varied ways in which they capture food. The typical streamlined shape of many open-ocean fishes shows the value of speed for capturing prey and avoiding enemies. Most fishes can swim rapidly for short periods, about 10 times body length in a second. During migrations, medium-sized fishes, such as salmon and cod, travel hundreds of kilometers in a few days. Large oceanic fishes that feed at high trophic levels, such as blue fin tuna, swim 100 or more kilometers per day for weeks at a time while hunting schools of smaller fish. Estuarine and bottom-dwelling fishes, however, can hide among rocks or plants. They usually have body shapes suitable for concealment rather than speed.

The **demersal fishes,** those that live on or near the ocean floor, are very important to humans; cod is an example. The **flatfishes** (Fig. 12–17), lie on the ocean floor. They have the ability to camouflage themselves to become the color of their surroundings and eat animals living in or on the sediments. The planktonic larvae of the flatfish have one eye on each side of their head. As the larvae mature

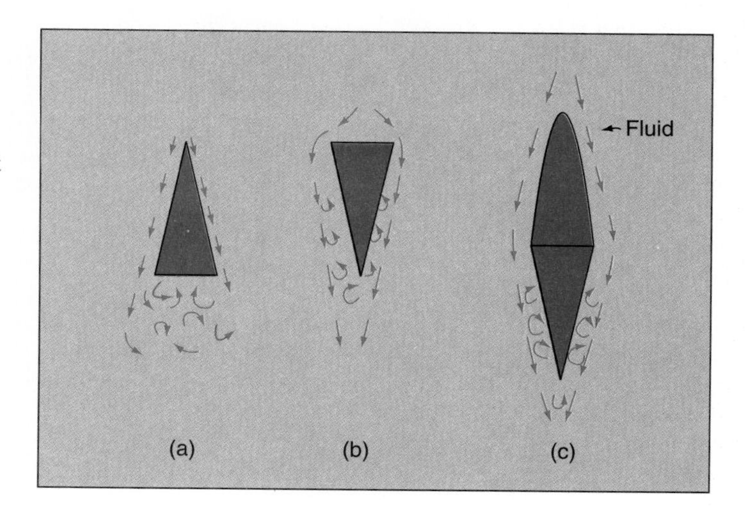

Figure 12–14
(a) A cone pointed into a flowing fluid experiences relatively high drag. (b) A cone with its blunt end facing the direction of flow experiences relatively low drag. (c) A fusiform shape, typical of most fishes, experiences the least drag.

Figure 12–15
Fish have bodies streamlined for speed. This barracuda is among the fastest.

Figure 12–16
The shape of the body and tail fin of the albacore tuna makes it an efficient long-distance swimmer. (Ron and Valerie Taylor/Bruce Coleman, Inc.)

Figure 12–17
A flatfish (in this case, a turbot) lies on the bottom and takes on the color of the bottom sediment as camouflage. Both of the fish's eyes are on the same side of the body. (Courtesy C. Arneson.)

Figure 12–18
Chinook salmon range from southern California to northwestern Alaska. They grow to be 1.5 meters long and are prized as sport fish. (Tom McHugh/Steinhart Aquarium/Photo Researchers, Inc.)

and metamorphose into a demersal stage, one eye migrates so that both eyes are on the upper side of the adult fish.

Mackerel, tuna, and related species are well adapted for strong, continuous swimming in open waters, both near the surface and at depth. Open-ocean species feed at high trophic levels and are voracious carnivores. They, in turn, are heavily fished by humans.

Another group of valuable food fishes are salmon (Fig. 12–18), high-level carnivores, generally confined to fairly cold, ocean waters. They typically spend part or all of their lives in fresh water. Salmon spawn in rivers but attain most of their growth in ocean waters, returning to their original river to spawn and produce the next generation.

Herring (Fig. 12–19) are abundant in coastal ocean waters and in upwelling areas, and these fast-growing fishes support some of the world's largest fisheries. They have been overexploited in many areas, leading to drastic declines in abundance. Overfishing, combined with the effects of *El Niños* in the Pacific, have resulted in spectacular collapses of these fisheries—the California sardine in the 1950s and the Peruvian anchovy in the early 1970s. (Sardines and anchovies are both types of herring.) Herring and their close relatives are important sources of protein worldwide. Indeed, the collapse of the Peruvian fishery in 1972 caused a momentary, worldwide increase in the price of protein feeds for cattle and chicken.

Herring spawn throughout the year in coastal waters. Spawning occurs in the summer and fall in Mid-Atlantic waters. A single female can spawn 50,000 to 700,000 eggs, depending on her size and age; older and larger fishes produce the most eggs. The larval fish enter estuaries when they are about 2.5 centimeters long, spending about eight months there before returning to offshore waters. Herring grow rapidly during their first three years. Adults swim with their mouth open, filtering everything from the water—both plants and animals. One-year-old herring typically weigh about 250 grams, two-year-olds about 500 grams. Then growth slows, so that they weigh only about 1 kilogram as six- or seven-year-old adults. Spawning begins around two years of age.

Environmental factors affect spawning success. If eggs and larvae are carried offshore by currents, the larvae cannot reach the estuaries and wetlands, and they starve to death. On the other hand, if surface currents carry the young toward

Figure 12–19
Herring and related species, such as anchovies, sardines, and menhaden, are important food and commercial fish. (Courtesy C. Arneson.)

shore, they have a much better chance of reaching suitable nursery areas. Those years when large numbers of young survive produce large year-classes that can sustain fisheries for years.

Cod (Fig. 12–20) have supported one of the world's most important fisheries. They were extremely abundant in the high and mid-latitudes of the Northern Hemisphere and were especially important in early American history. Their abundance in the waters off Massachusetts and New England helped feed the first European colonies in the 1600s.

Cod spawn in relatively shallow areas of the continental shelf—Georges Bank, for instance, off the Gulf of Maine. A single female produces up to 10 million eggs, with larger females being the most successful. The larvae grow to 4 centimeters in four months. Individuals mature sexually in about two years, but may live to be more than 15 years old. The largest ones grow to a mass of 90 kilograms.

Like many other valuable fishes, cod have been overfished worldwide. For instance, the cod fisheries off Atlantic Canada and New England were closed after 1992 to protect the few cod that remained. The salmon fishery off British Columbia was closed in 1995 for the same reason. Questions of when and if these stocks will ever recover enough that commercial fishing can resume are hotly debated.

Squid

Squid (Fig. 12–21) are common animals in the ocean. These fast swimmers are successful predators, comparable to sharks. The annual commercial catch is around 1 to 2 million tons and still increasing. Squid are becoming popular on restaurant menus in areas where they have not previously been available.

Figure 12–20
For centuries, the cod fishery supported the economies of New England and Atlantic Canada until its collapse in the 1990s. (Tom McHugh/Photo Researchers, Inc.)

Most squid live at mid-depths in the ocean during the day and migrate to surface waters at night. Many have bioluminescent light organs. Most squid are small, around 10 centimeters to a meter long. We know little about squid behavior because they are so difficult to keep in captivity or to observe in the wild.

The largest is the rarely seen giant squid (up to 23 meters long), which lives at depths between 300 and 600 meters. It is eaten by sperm whales. Giant squid are sometimes brought to the surface by feeding sperm whales, and a few decaying, dead ones have washed ashore. One sperm whale's stomach contained a giant squid 10.5 meters long, weighing 184 kilograms.

Squid swim by filling a muscular mantle cavity with water. They then contract the mantle and forcibly expel the water through a funnel—a form of jet propulsion. The animal controls its direction of movement by moving the funnel. Squid can propel themselves out of the water for 50 meters like flying fish, and have been recovered from ships' decks 3.5 meters above the water.

When threatened, squid eject clouds of ink that can blind their attackers. These ink clouds resemble the squid, fooling the attacker into attacking the cloud and letting the animal escape. Small, fat squid release small, fat clouds; long, thin ones release elongate clouds.

Squid have well-developed eyes that function much like human eyes and are used to locate prey. The eyes can focus over wide distances and adjust to large changes in light intensity. Squid also have long tentacles that they use to catch prey. These tentacles have well-developed senses of touch and smell to assist in

Figure 12–21
Squid are active, fast-swimming predators at all depths in the ocean. This transparent form (about 25 centimeters long) lives at depths around 500 meters. (Tom Smoyer/Harbor Branch Oceanographic Institution, Inc., Fort Pierce, Florida.)

Figure 12–22
Sharks are fast swimmers and effective predators.

locating food. The animals feed primarily on small fishes. They usually feed by swimming backward into schools of small fish and using their tentacles to catch the prey, bringing the food to their mouth where they tear it with their beak, a jaw-like structure.

Sharks

Sharks (Fig. 12–22) occur throughout the ocean. The common image of sharks is of a large, swift predator. Indeed, many sharks are able to kill large prey, including humans on rare occasions. Many sharks are scavengers; like large cats and wolves, they catch vulnerable and diseased animals. They use their rows of teeth to remove large bites of flesh or tear off whole limbs. To a shark, a swimmer may appear to be a disabled animal, which the shark attacks. Most shark attacks on humans occur in shallow, murky waters or at dawn or dusk, when visibility is limited. Groups of sharks also engage in feeding frenzies, stimulated by blood or bits of food in the water. Reasons for these frenzies are not well understood.

Sharks are primitive animals, having skeletons made of cartilage rather than bone. They lack scales but have small, tooth-like plates, called denticles, embedded in their skin making it extremely abrasive. Shark teeth are actually specialized denticles; they occur in rows up to seven deep in the shark's mouth and are easily replaced if lost.

Some large sharks—whale sharks and basking sharks—feed exclusively on plankton. They have large gill slits and swim slowly with their mouths open, trapping plankton in their gills.

Sharks have been hunted for their livers (which are rich in vitamin A), for food, and for sport. The meat is quite tasty when properly prepared to remove the high levels of urea in the blood. In Asia, shark fins are highly prized as an ingredient for soups. Sharks have been overexploited and are now protected in many areas.

Deep-Ocean Nekton

Conditions below the surface zone are markedly different from conditions in the upper waters. Light intensities decrease, and the waters get progressively colder

Figure 12–23
A female angler fish (about 10 centimeters long) taken from a depth of about 1,000 meters. The organ in front of the fish's mouth is bioluminescent, to lure prey to where the fish can seize and swallow it. The fish has a large mouth and so can swallow large objects. Male angler fishes are much smaller and in some species are parasitic, living attached to the females. (Courtesy B. Robison, MBARI.)

with increasing depth below the pycnocline. Both factors influence the types of animals living at any given depth and their life-styles.

The dimly lit deep ocean is home to many bioluminescent animals. For example, angler fish (Fig. 12–23) are common in waters 100 to 500 meters deep. These fishes have distinctive patterns of light-emitting organs, which help them locate and identify potential mates. They also use lighted lures to attract prey. Most deep-ocean animals are black or red, both colors being invisible in the dim light.

Below 1,000 meters, there is no light and waters are uniformly cold, typically less than 4°C. Animals at these depths are widely dispersed, but they occur throughout the ocean. A few suspension-feeders live on detritus sinking out of surface waters, but predation is the norm. Many organisms have special adaptations, such as jaws that disjoint so they can eat animals twice their size. Meals are infrequent in these depths of the ocean; a fish may eat only a few times a year. Food requirements are less because metabolism is slow in the low temperatures and high pressures of the deep ocean.

Marine Mammals

Marine mammals, warm-blooded air-breathers bearing and nursing live young, include the largest animals on Earth. The sea cows are herbivores (e.g., dugongs and manatees; see Fig. 12–24). They graze on aquatic vegetation and spend their lives in shallow coastal waters.

Seals (Fig. 12–25), walruses, and sea lions are all familiar marine mammals because they live in coastal areas. They are pinnipeds (feather-footed), so named because of their distinctive swimming flippers. Although they spend much of their time in the water feeding, they also spend considerable time ashore, where they breed and raise their young in large rookeries. Heavily hunted because of their valuable furs, pinnipeds were reduced almost to extinction in the late nineteenth century. Their numbers have recovered under the protection of international treaties; now their geographic ranges are expanding.

Figure 12–24
Manatees, members of the family known as the sea cows, graze on aquatic vegetation in tropical coastal waters. They are seriously threatened by coastal development and recreational boating. (Doug Perrine/Jeffrey L. Rotman.)

Cetaceans, a group that includes whales (Fig. 12–26), dolphins, and porpoises, are completely oceanic mammals, spending all their life at sea. Some whales are very large. The blue whale, for instance, the largest animal that has ever lived, reaches lengths of 30 meters and weighs 150 metric tons. The largest whales feed on plankton. They swim slowly with the mouth open to take in water and plankton. The tongue acts as a piston to push the water through plates of horny material, called **baleen,** that hang down from the roof of the mouth (Fig. 12–27). These plates filter plankton from the expelled water. The trapped plankton are licked off the baleen and swallowed.

Figure 12–25
A bull fur seal guards its harem and pups. (Tom Bledsoe/Photo Researchers, Inc.)

Figure 12–26
A humpback whale, a medium-sized baleen whale, is typically about 16 meters long and weighs 35 tons. Note the long flippers. It is found in all non-polar oceans. It feeds by swimming with its mouth open through schools of small organisms. (Courtesy NOAA.)

Gray whales, another large species, feed on bottom-living crustaceans. The whales suck up large amounts of mud from the bottom, and the abundant crustaceans are filtered out when the mud and water are expelled through the baleen. These animals leave large scars on the shallow sea floor of their feeding grounds in the Arctic Ocean, between Alaska and Siberia.

Smaller whales, such as porpoises, sperm whales, and killer whales, are carnivores. They are rapid swimmers and use their strong teeth and jaws to capture prey. They feed primarily on fish and squid; killer whales, however, also prey on other whales.

Many marine mammals use echolocation to find prey. The common dolphin, in which echolocation has been most studied, emits clicks from its forehead, probably using its blow hole, and detects its prey using the echoes. Low-frequency clicks are used to locate distant objects, and high-frequency ones are used to

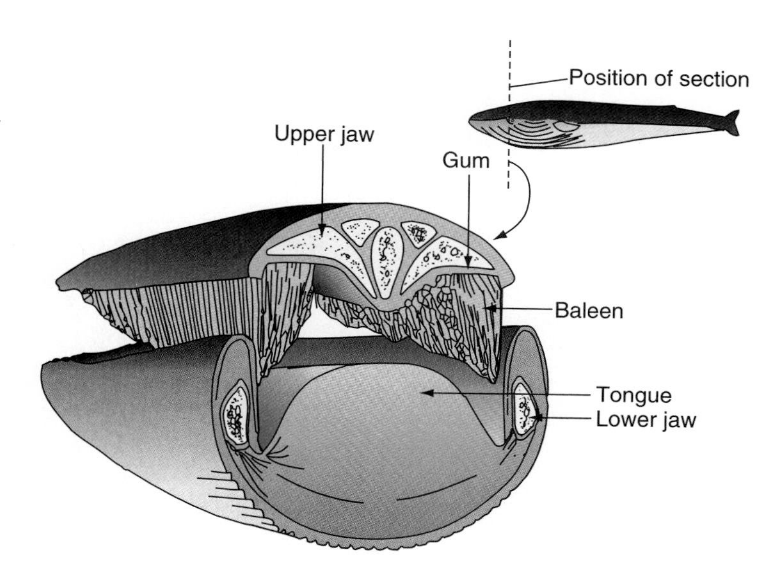

Figure 12–27
The baleen whales take water into their mouths and use the tongue like a piston to force the water out through the feathery plates of baleen, trapping planktonic prey.

discriminate among objects by determining size and shape. Dolphin clicks can be heard by human ears. Sperm whales can locate squid, their primary food, using low-frequency scanning clicks. Whales also produce a variety of other sounds, apparently for communication. Some can communicate across entire ocean basins.

Whales are powerful swimmers, their streamlined body permitting them to swim efficiently. They have an unusual skin structure that deforms in response to water pressure; this deformation reduces turbulence, saving energy as the animal swims.

Whales migrate thousands of kilometers. Many breed and bear their young in the warm low-latitude waters but migrate to exploit seasonal food sources, such as rich plankton blooms in the high latitudes.

Seabirds

Among the most distinctive sights on the ocean are seabirds, adapted to living near or over the ocean. Seabirds have a high metabolic rate and must feed on energy-rich fatty foods, usually schooling fishes. Shorebirds, nearshore foraging ducks, and wading birds feed on benthic organisms.

Wading birds have long legs that permit them to obtain food by wading in wetlands (e.g., herons) or on beaches (shorebirds). The shape of their bills permits a species to either sift food out of sediments (flamingos); or capture their food with long pointed bills (herons). Pelicans (Fig. 12–28), ducks, cormorants, loons, grebes, puffins (Fig. 12–29), and most gulls and terns live along the shores and feed in shallow nearshore waters.

Figure 12–28
Brown pelicans are fish-eating birds that live in coastal regions. (Spencer Grant/ Photo Researchers, Inc.)

Figure 12–29
Puffins are common seabirds all around the North Atlantic.

Because they are carnivores, feeding at high trophic levels in food webs, seabirds accumulate and concentrate pollutants in their tissues. Thus, they are sensitive indicators of environmental conditions. During the 1950s, birds in some coastal ocean areas and in the North American Great Lakes provided early indications of adverse effects of pollutants, especially chlorinated hydrocarbons, such as the pesticide DDT. These compounds affected the birds' eggs, making the shells thin and fragile. Many fish-eating seabird populations in North America were decimated until DDT was banned there. Unfortunately, these compounds are still manufactured and used in many parts of the world, thereby threatening local seabird populations.

Although their food comes from the ocean, seabirds must go ashore to breed and to raise their young. No bird has evolved a truly marine life-style, totally independent of land. Eggs are hatched at 40°C—much higher than ocean-water temperatures—and so must be kept warm by heat from the parents' bodies. Furthermore, the chicks require feeding and protection until they can fly. All this requires nesting ashore.

Oceanic birds—auks, albatrosses, petrels, penguins (Fig. 12–30), and gannets—come closest to being completely independent of the land. For example,

Figure 12–30
Emperor penguins with young chicks on the coast of the Weddell Sea. (Courtesy B. Stonehouse/Scott Polar Research Institute.)

after learning to fly, a young albatross may not set foot on land for years. These open-ocean birds generally lay only one egg during a breeding season, and their chicks have a long adolescence. Adults live long lives, typically decades.

Since most seabirds must return to land during breeding, most have retained their ability to fly. Thus, seabirds cannot optimize their bodies or wings for maximum swimming efficiency. This restriction limits the depths to which they can dive and the length of time they can stay under water.

One seabird that has completely lost the ability to fly because it is so well adapted for life at sea is the penguin. Emperor penguins (Fig. 12–30), the largest ones, can swim under water as well as seals can. These birds dive to depths of 250 meters, swim at speeds of 10 kilometers per hour, and remain submerged for nearly 20 minutes to catch fast-swimming fish and squid.

Seabirds undertake long annual migrations in response to food availability and suitable weather for breeding. For instance, terns and red knots spend summers in the Canadian Arctic and winters off South America and Patagonia. Likewise, shearwaters travel from a volcanic island in the cool, temperate South Atlantic to comparable climates off eastern North America. The timing of these migrations coincides with seasonal food availability. For example, many shorebirds stop in Delaware Bay on their spring flight to South America. There they fatten up for the long flight by feeding on the abundant eggs of horseshoe crabs, which wash up on the beaches.

In addition to responding to pollutants, seabirds are also affected by other human activities. For example, near elimination of whales from Southern Ocean waters has made more planktonic krill available for seabirds and other organisms. In the North Sea, approximately one-third of the diet of seabirds comes from the unwanted fishes and wastes thrown off fishing vessels. In winter, food from this source makes up about half the diet. Sea gulls are abundant in areas where they can feed on garbage. They prey on other seabirds, eating their eggs and chicks.

Marine Reptiles

Few reptiles live in the ocean. The best known are sea turtles (Fig. 12–31), which live in the ocean but nest on land. These turtles graze on sea grasses and rooted aquatic plants growing in shallow waters around Caribbean islands and in southern Florida. Like cows, they have bacteria in their intestines that digest cellulose in the grasses and provide needed nitrogen. Each female turtle lays about a hundred eggs, which she buries in the beach before returning to the sea, leaving the eggs to hatch in the warm sand. As the newly hatched turtles cross the beach on their way to the ocean (Fig. 12–32), they are preyed on by birds and other animals, including humans. Once in the ocean, they must escape predatory fishes. Mature turtles make long migrations between feeding and breeding areas. Turtles have been hunted nearly to extinction because of their valuable meat and shells, which are used for making combs and jewelry.

In the tropical Indian and Pacific oceans, there are sea snakes (Fig. 12–33). These are truly oceanic creatures that reproduce at sea, bearing live young. Their bodies are flattened, which improves their swimming efficiency, and their nostrils can be closed while swimming. Sea snakes are extremely poisonous. They are quite timid and have a small mouth, and so they pose relatively little threat to humans.

In the Galapagos Islands of the equatorial Pacific, there are marine iguanas, a type of lizard (Fig. 12–34). They live on cliffs along the shore and dive into the water at low tide. Using a flattened tail to swim, they feed on seaweed.

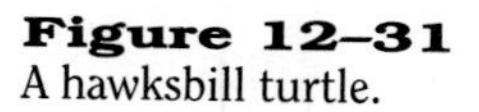

Figure 12–31
A hawksbill turtle.

Figure 12–32
Young loggerhead turtles (8 hours old) cross the beach to reach the ocean. (M. Reardon/ Photo Researchers, Inc.)

Figure 12–33
Sea snakes are truly marine animals and are among the most poisonous of all snakes. (A. Power/Bruce Coleman, Inc.)

Figure 12–34
Marine iguana in a tidal pool on the Galapagos Islands.

Open-Ocean Life Strategies

Open-ocean organisms must cope with conditions that are often radically different from what land dwellers experience. We have already seen that they take many different forms. As we shall now see, they have also evolved many different strategies for dealing with the major tasks of living organisms—feeding, survival, and reproduction. We shall also consider migrations that marine animals undertake as part of their survival tactics.

Feeding Strategies

The feeding strategies used by zooplankton vary with depth (Fig. 12–35). The reason for this is obvious: the abundances of different kinds of food are greatly different at various depths and at different times of day. Thus, vertical migration is part of the feeding pattern of many open-ocean organisms (Fig. 12–36). Even many plankton, although weak swimmers, can move vertically hundreds of meters during a day.

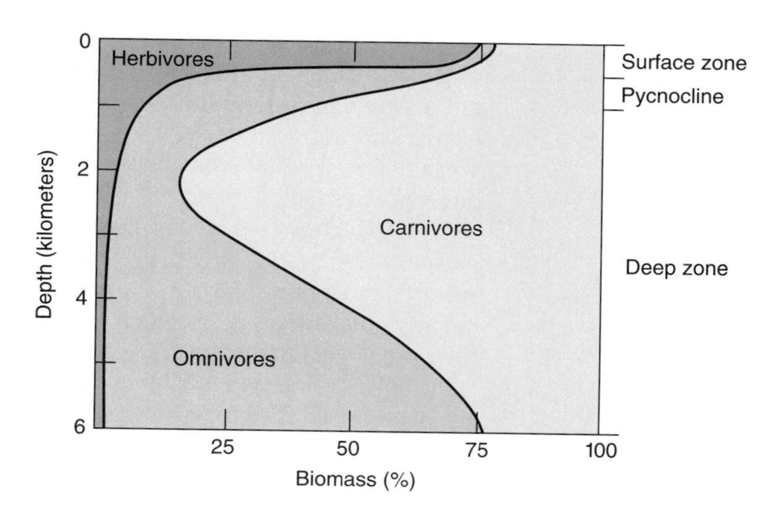

Figure 12–35
The abundance of planktonic organisms using a given feeding strategy varies with depth. Herbivores are most abundant near the surface, where plants are most abundant. Omnivores are most abundant at the greatest depths.

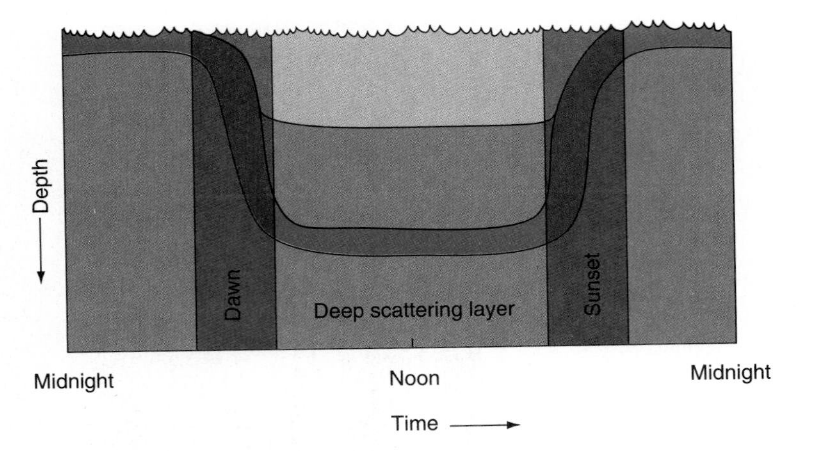

Figure 12–36
Zooplankton migrate to near-surface waters at night to feed. During the day, they return to the safety of the dark aphotic zone, where they are more difficult to see.

During the day, animals stay in the dark water of the pycnocline and the deep zone. At night, they swim to the surface zone to feed. As day brightens, they return to the safety of the dark, where predators who hunt by sight cannot readily find them. Some animals, especially the larger ones, migrate vertically several hundred meters each day.

Herbivorous zooplankters dominate near-surface waters, where phytoplankton are abundant, but their numbers decrease markedly below the photic zone. Carnivorous zooplankters are most abundant in the mid-depths, where they can feed on surface-zone herbivores. Many of the vertically migrating zooplankton are carnivores.

Omnivorous zooplankters occur throughout the ocean but are especially common in deeper waters. Their survival depends on being able to eat anything that sinks out of the surface zone, including nonliving organic detritus.

Many planktonic organisms are filter feeders, that is, they feed by removing small food particles from the water. Some organisms make nets then create water currents to flow through these nets so they can catch any particles suspended in the water. Baleen whales are examples of filter-feeding organisms.

Filter feeding by zooplankton was long thought to involve simply moving an appendage with a filter through water to collect particles, much as one catches butterflies in a net. The "net" was thought to be the hairlike structures (called "setae") on the organism's mouthparts and other appendages. Particles thus captured were then moved into the mouth and ingested.

Observation of filter-feeding zooplankters through microscopes and cameras showed a markedly different picture, however. Because of the small size of the organisms involved, water's viscosity is important, and to these tiny feeders, the water seems as viscous as molasses would to a human. Therefore filter-feeding is not a sweeping process but rather is much more selective, rather like a human hand removing desirable particles from molasses. Apparently, zooplankters can sense the presence of individual particles, probably by chemical means much like our sense of smell.

Many filter feeders can also pursue and capture relatively large prey by grasping their food with their feeding apparatus. For instance, some benthic sponges can capture larger organisms, which they eventually consume. Consequently, filter feeders can switch from one type of feeding to another and select their prey to obtain the quality of food they want as well as the quantity.

Jellyfish and siphonophores paralyze their prey with stinging cells that consist of barbs attached to poison sacs (Fig. 12–37). Dangling tentacles entangle the food and sweep it toward the mouth, inside the bell. (The Portuguese Man-of-War shown in Figure 12–11 is one of the organisms that feed in this way.) Some forms swim upward and then sink with tentacles extended, trapping prey beneath.

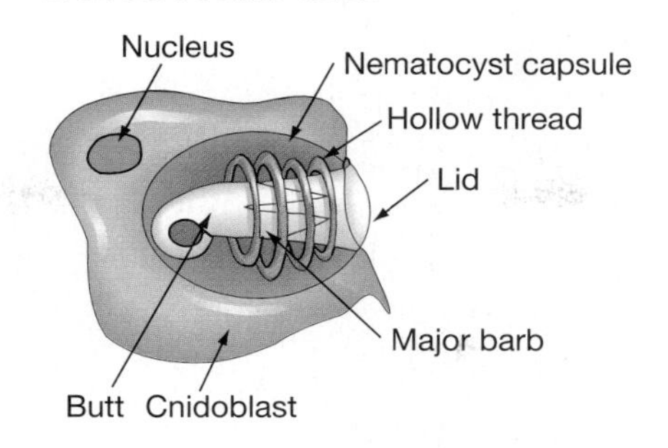

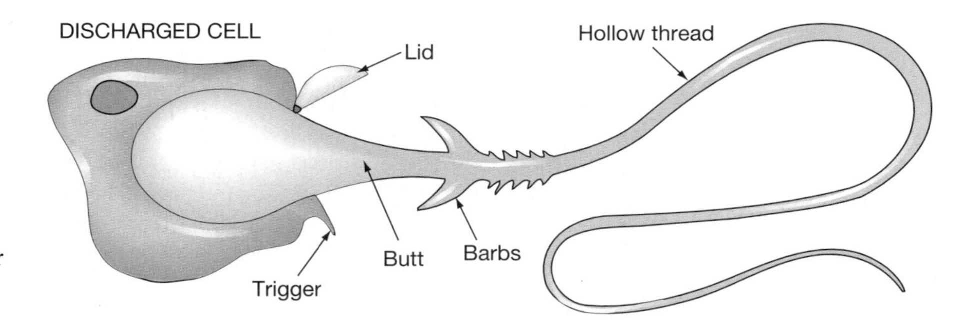

Figure 12–37
The stinging cells of jellyfish, and of their relatives the anemones and corals, are used to poison their prey.

Ctenophores feed in a variety of ways. Some grasp their prey with long tentacles; others have large mouths and simply engulf their prey.

As we have already seen, suspension feeding involves the use of mucous nets and structures. Many organisms, especially gelatinous plankton such as tunicates, secrete mucous structures that trap particles of all sizes from the water. The organism then eats the captured particles by consuming the entire net, as we saw when discussing salps and related organisms.

Defensive Adaptations and Strategies

All organisms seek to survive, by avoiding being eaten by predators. Planktonic organisms exhibit several defensive adaptations and strategies. Small size is perhaps the most common adaptation, having many obvious advantages for plankton. A second is transparency. Transparent or translucent organisms are difficult for predators to see. This adaptation is especially common among gelatinous plankton.

Zooplankton are also color-adapted, camouflaged, to blend into the background. Organisms near the surface are often bluish to match the color of the sea surface. Deeper in the ocean, where light is very dim, animals are often deep red (Fig. 12–38). (Remember that red light is absorbed very near the surface, so that red organisms are invisible.) Zooplankton that live in the deepest ocean are gray or black.

Schooling (Fig. 12–39) is a common defensive strategy among open-ocean fishes and some zooplankters, such as krill. Similar behavior among small zooplankton is called swarming.

Schooling is common among some fishes, such as menhaden, herring, and mackerel. Usually, the fish in a school are of a single species; often, the school consists of individuals of similar size. Schools may be enormous, covering several square kilometers. Fish in a school constantly change position but keep a nearly constant distance from those around them. Schooling fish have wide-angle

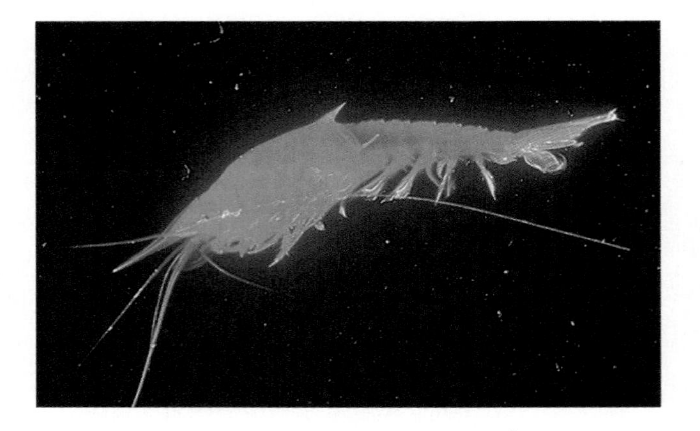

Figure 12–38
A dark red shrimp lives in the dim light around 1,000 meters below the surface. This color is common among deep-sea organisms and is apparently difficult for predators to see. (Courtesy B. Robison, MBARI.)

vision so they sense changes in their neighbors' position and direction. Such schooling is thought to provide protection to individuals by confusing predators. It may also help keep the reproductive members of a population together.

Another defensive strategy is larger body size. This is one of the advantages that gelatinous organisms have. They contain large amounts of water in their tissues, so that a large organism may contain only a small amount of organic matter. Another way to increase body size is to grow spines, like the foraminiferan and radiolarian shown in Figure 12–5. A spiny organism is more difficult to swallow than a spineless one.

Many fishes use countershading, a color pattern in which the upper part of the body is dark, making it difficult to see them from above. The lower part of the body is lighter-colored, so that predators looking up toward the brightly lit surface layer cannot see the countershaded fish so easily. Bioluminescent (light-producing) organs on the lower bodies of some fish living in the darker depths produce a similar effect (Fig. 12–40).

Finally, some open-ocean organisms use toxins, which they make or accumulate, as defenses. Few fishes are truly venomous (see Box 12–3) but many have toxic secretions on their skins, including parrot fish or wrasses. Many of these fishes are brightly colored to warn potential predators to avoid them.

Figure 12–39
Schooling provides protection for individuals by confusing predators.

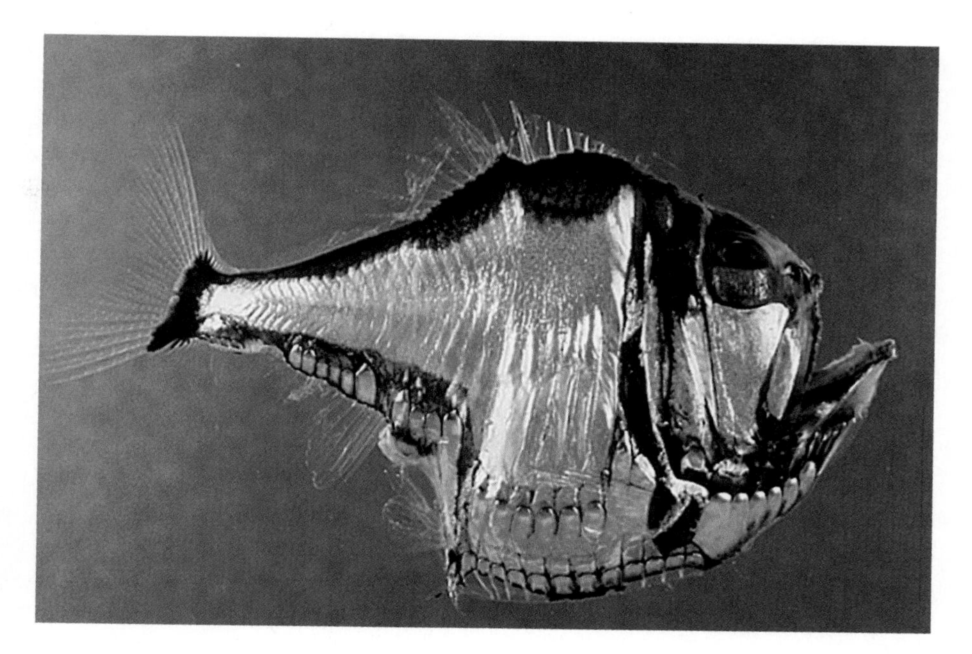

Figure 12–40
Hatchet fish are common in the mid-depths of the ocean. Bioluminescent organs on the bottom of the fish provide light to obscure the shadow the fish would otherwise make in the dim light coming down from the surface. (Courtesy B. Robison, MBARI.)

Reproductive Strategies

Two quite different reproductive strategies are used by marine organisms. One is suited to sparsely inhabited but resource-rich environments, where rapid population growth is advantageous. The other is appropriate for resource-poor environments, where slow population growth is more desirable.

The first strategy (opportunistic) is used by zooplankton that develop rapidly to take advantage of unstable environments or variable resources that other organisms cannot exploit. These opportunistic organisms mature early, have large numbers of small offspring (or eggs) and provide no parental protection. Much of the energy from food goes into producing eggs and sperm. Each female lays millions of eggs. Each male discharges sperm into the water where the eggs are fertilized and later develop. You can think of this as the "many–small" reproductive strategy. This is the strategy used by most plankton and nekton.

The opportunistic strategy has some major drawbacks. The young are small and therefore easy prey for many other larger organisms, including adults of their own kind. The young must essentially grow their way through the food chain. (See Box 2–2.) Such small organisms are also carried by currents. They may well be carried out of areas most suitable for survival and growth. This involuntary migration leads to large year-to-year differences in population sizes. Because of these drawbacks, enormous numbers of eggs and sperm must be produced to ensure that enough larvae survive to reach adulthood and maintain the population.

In fish, the larger females produce the most eggs and, thus, are critical to spawning success for a population. Fish populations are especially sensitive to overfishing, since the largest fishes are caught preferentially.

The other reproductive strategy involves large, slow-growing animals that take a long time to reach sexual maturity and care for their young for some period of time (the duration of which varies from one species to another). These organisms have few, relatively large offspring and protect them against predators. In this strategy, most of the energy from food goes into growth. Think of this as the "few–large," or nurturing, reproductive strategy. Seabirds and marine mammals use this strategy. Sharks and some fishes bear small numbers of young alive. A few fishes protect their young in their mouths or special pouches.

BOX 12–2
Fish and Food Chains

Fish offer a good example of how an organism grows through a food web as it develops from egg, through larval stages, into adult form. Most fish eggs are released and fertilized (Fig. B12–2–1a) while floating freely in the water. Hatching (Fig. B12–2–1b) can occur within a few days, depending on temperature. During this time eggs are eaten by many organisms, including fish of the same species and possibly even the fish that laid and fertilized the eggs. The larval fish then live as meroplankton for weeks.

As larvae (Fig. B12–2–1c), the fish are part of the plankton, moved by currents. They depend on currents to carry them to areas where there is enough food for them to survive after they have used up their yolk sack, the food that came from the egg. Many juvenile fish also depend on currents to carry them into areas where there is shelter to hide from predators. These areas are often called nursery grounds.

As fish larvae grow, they are able to swim better and eventually become part of the nekton (Fig. B12–2–1d), since they can swim against currents. As they grow, their food changes and they are in turn are eaten by larger animals (Fig. B12–2–2). Some of the largest fish, such as tuna, grow large enough to become top predators; that is, fish that are too large for other fish to eat. Thus, fish grow through the food chain as they develop, and also change from plankton to nekton.

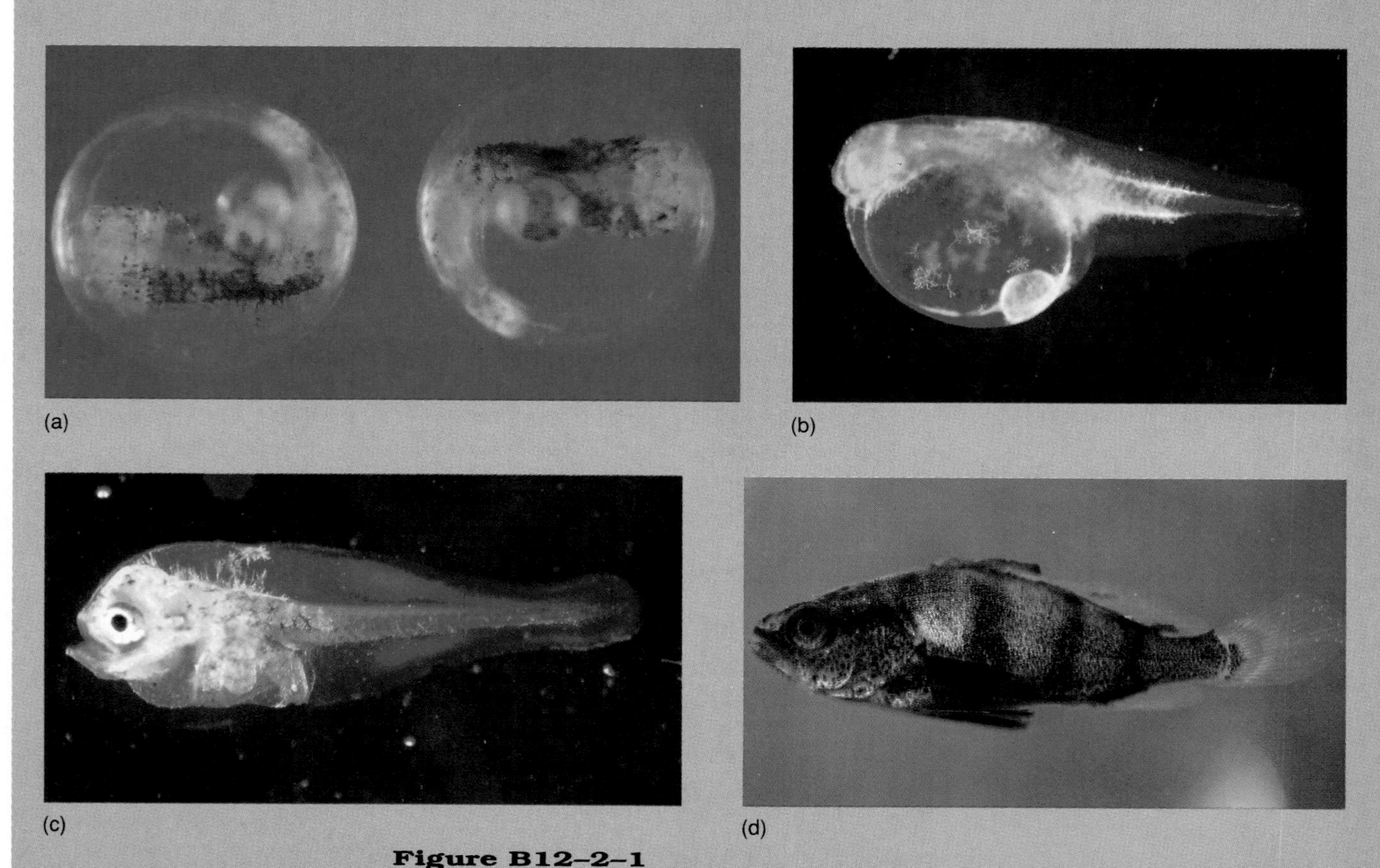

Figure B12–2–1
Development of the white seabass (also known as croaker) is an example of the development of an organism from a planktonic to a nektonic form. It illustrates how an organism can grow its way through a food web by increasing its size. (a) The fish begins as an egg and (b) hatches as a larval fish, seen here with its yolk sac still attached. (c) As it continues to grow, the larval fish matures and grows in size and its swimming ability increases. (d) As a mature fish, it is part of the nekton. (Courtesy H. G. Moser, National Marine Fisheries Service.)

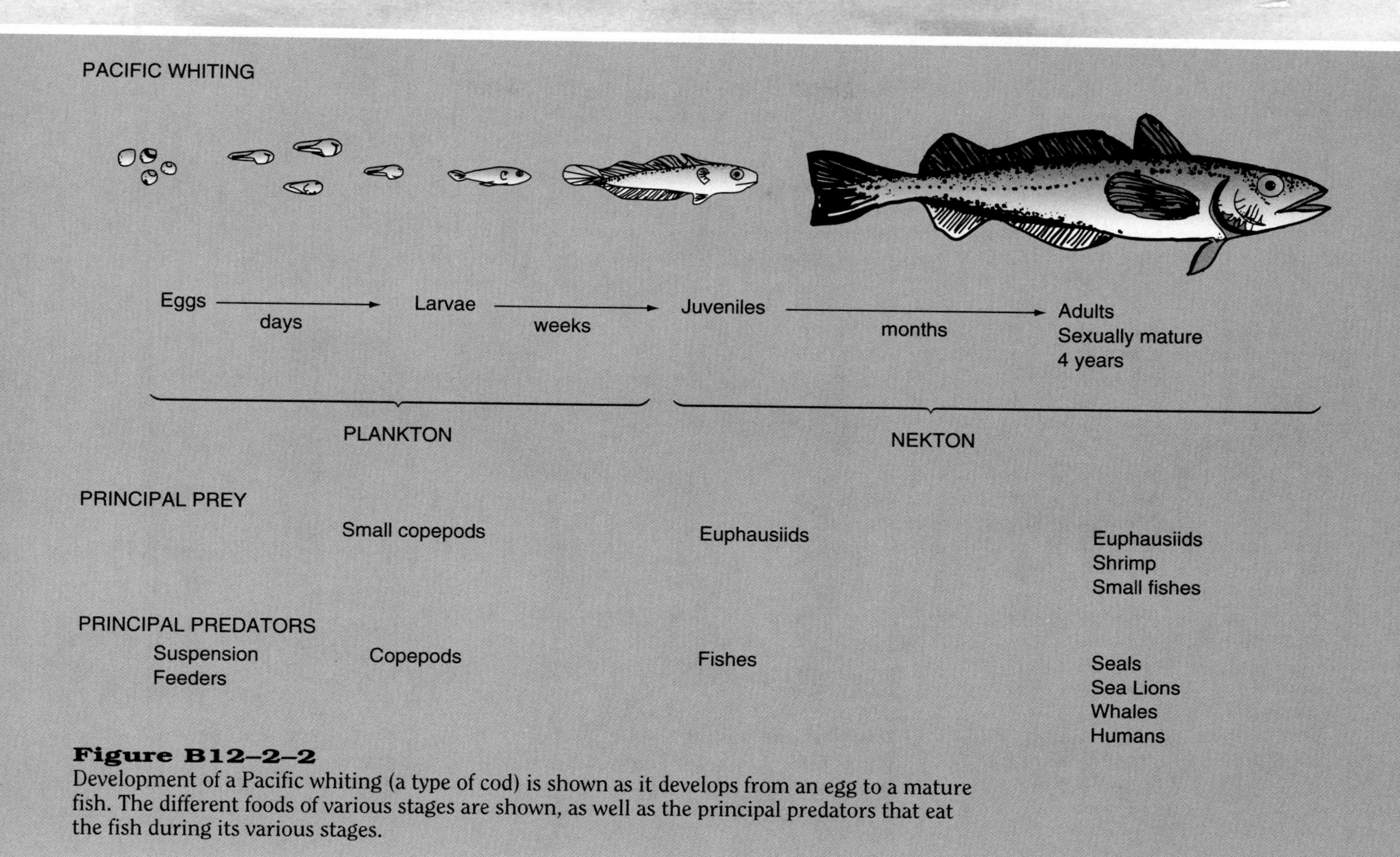

Figure B12–2–2
Development of a Pacific whiting (a type of cod) is shown as it develops from an egg to a mature fish. The different foods of various stages are shown, as well as the principal predators that eat the fish during its various stages.

Migrations

Many nektonic animals have long migrations. The longest known is the 20,000-kilometer annual migration of the California gray whale from the lagoons on Mexico's Pacific coast to the Bering Sea and nearby Arctic Ocean continental shelves. Many migrations permit marine animals to take advantage of seasonally abundant foods.

Many fish migrate as part of their life cycle. Salmon are called **anadromous fish** (from the Greek words *ana,* "up," and *dromos,* "running"), which means they spawn in fresh water, remain there for about a year, and then migrate to the ocean. As adults, they live in the open ocean before returning five or six years later to their home streams to spawn. These migrations cover many thousands of kilometers. Salmon return to the stream where they hatched, at the same time of year, generation after generation.

Mackerel are another migratory fish, only this time part of the migration is in the vertical direction. They leave surface waters about October and aggregate near the ocean floor until January. During this period, they eat crustaceans and small fishes. In January, they move back to the surface in schools, and then in April migrate to spawning grounds near the edge of the continental shelf. Gradually moving closer to land, they feed on plankton, especially copepods. All this time, the mackerel are moving in very large schools. From June to July, they form

BOX 12–3
Dangerous Marine Animals

More and more people are using the coastal ocean for recreation. SCUBA diving, especially, puts humans at risk of encountering harmful organisms. Some jellyfish are extremely poisonous, and the risk from shark attacks is well known. Divers exploring reefs may encounter moray eels, which live in crevices and can attack and bite if threatened. Stepping or brushing against a sea urchin can result in skin punctures from the spines, and, to make matters worse, the spines can break off in the wound. Locals in urchin-prone areas recommend colorful home remedies to get rid of the spines.

Sting rays are common in warm, shallow seas and pose a threat to people walking in the surf zone. The tails of sting rays have spines that can puncture and break off; their venom has no known antidote and can cause prolonged illnesses in humans. These animals burrow into sediment to feed on organisms living there; thus, they are often not seen until encountered. To avoid encounters, push your feet along the bottom as you walk or probe the bottom with a stick before you step.

Lionfish (Fig. B12–3–1) and stone fish, both found in the Indian and South Pacific oceans, have extremely strong venom in their spines. An antidote for stonefish venom is kept available at Australian public beaches.

Perhaps the greatest risk comes from eating tropical-reef or shore fishes that contain ciguatera. This form of food poisoning is a serious problem in the West Indies and in central and southern Pacific islands. It causes tingling of the lips, tongue, and throat, followed by numbness. Recovery is slow and may take months or years. About 7 percent of those who ingest ciguatera die. The best approach is to avoid eating large, long-lived fishes, such as barracuda and grouper during their reproductive seasons. Most especially, follow the advice of local residents regarding which fishes are safe and which are not.

Figure B12–3–1
Contact with the long and venomous spines of the lionfish can result in very painful wounds.

smaller schools and move close to the shore, changing their diet from plankton to small fishes in inshore bays. In the fall, they again form larger schools and seek deeper waters.

North Atlantic eels are **catadromous fish** (the opposite of anadromous, from the Greek word *cata,* "down"). They spawn in the Sargasso Sea in the North Atlantic, live there as planktonic larvae for up to three years, then migrate to the coastal waters of either North America or Europe (Fig. 12–41). There they metamorphose into adult eels and move to the fresh water of rivers for eight to twelve years. After that, they return as adults to the Sargasso Sea to spawn.

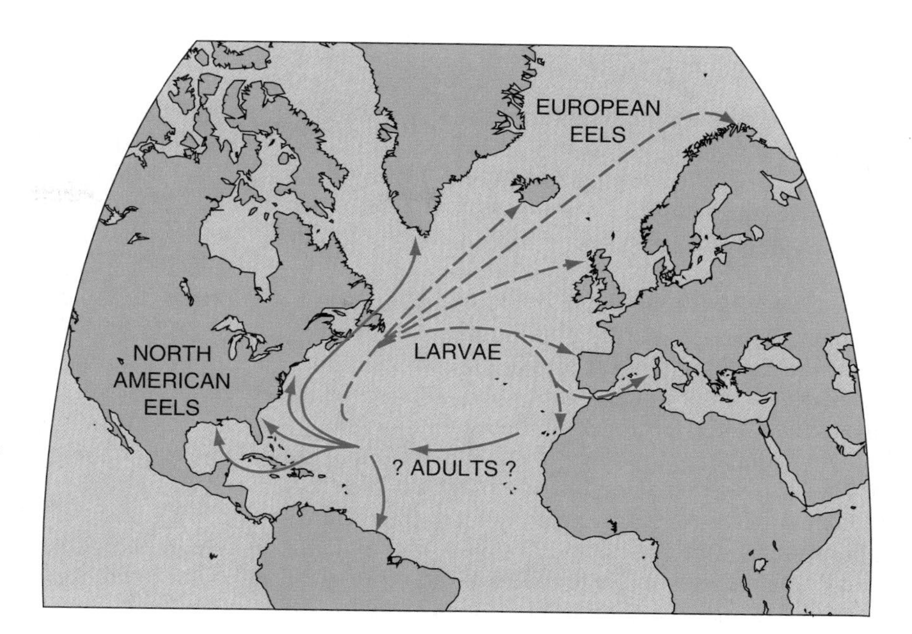

Figure 12–41
Migration routes of eel larvae from the Sargasso Sea to Europe and North America. Note the similarity between migration paths of larvae and the path of the Gulf Stream. Migration paths of adults returning to the Sargasso Sea are not known, but likely involve subsurface currents.

There are many unanswered questions about how eels migrate from the rivers in which they spend their adulthood back to their native Sargasso Sea. One hypothesis is that the eels—both those who matured in European rivers and those who matured in North American rivers—orient themselves at the fronts associated with the Subtropical Convergence. From there the eels are thought to navigate using the Sun as a compass and Earth's magnetic field, in this way approaching the vicinity of where they were born.

Threatened Basking Sharks

Basking sharks are gentle three-ton giants that feed on plankton, much like whales. Denying our usual image of sharks as fierce predators, these ten-meter-long animals pose no threat to humans. The little-known basking sharks receive their name from their normal feeding position, floating at the surface. They are the second largest fish, after whale sharks.

In the 1940s, thousand of basking sharks were observed in California coastal waters, hundreds in Monterey Bay alone. At that time they were widely hunted for their oily flesh and for their livers, a source of vitamin A. These slow-moving animals are still easy targets for hunters and are often killed by boaters and jet-skis.

Surveys made in the early 1990s found disturbingly few basking sharks off California. Similar decreases were reported in Ireland, Norway, Korea, and Japan. Hunting in the 1940s may be part of the problem. Commercial hunting has since ended, because the products once taken from sharks are now produced synthetically more economically.

Pollution and global climate changes may also be a factor. These long-lived animals may be especially vulnerable to pollution, disease, and environmental changes. They reproduce more like marine mammals than fishes, giving birth to only a few young at a time.

Summary

Open-ocean organisms must either float or swim. Microscopic, passively floating organisms are called plankton: floating plants are called phytoplankton and floating animals are called zooplankton. These microscopic organisms experience conditions that are unfamiliar to us land dwellers.

In open-ocean environments, large animals eat smaller ones. Open-ocean organisms must cope with sinking out of the photic zone. Most planktonic organisms live at a chosen level in the ocean. Since they cannot swim, they may sink out of their preferred level.

Phytoplankton produce most of the food in the ocean. The larger net plankton include diatoms, which are common in nutrient-rich waters of upwelling, coastal, and estuarine systems. Dinoflagellates are almost as common.

Zooplankton feed on phytoplankton and other zooplankton. Some are planktonic throughout their life (called holoplankton). These include foraminifera, radiolaria, and other organisms. Meroplankton are the larval forms of organisms that are planktonic for only part of their life cycle. Examples are fish larvae or the larvae of oysters. As fish grow, they become strong swimmers and therefore nekton; adult oysters are benthonic. Gelatinous plankton include jellyfish, siphonophores, ctenophores, and tunicates. Some gelatinous plankton use mucous nets to capture nannoplankton.

Nekton are animals large enough to swim against currents. Most fish swim rapidly and are active predators. Some can lie camouflaged on the bottom to wait for their prey. Fish are adapted to see in dim light. Squid and sharks are common, effective predators. Deep-ocean nekton are adapted to live at low temperatures with infrequent feedings.

Marine mammals are warm-blooded air-breathers. The whales include the largest animals on Earth. Some feed on plankton. Others take small benthic animals. Still others capture squid and other active nekton.

Seabirds are common over the ocean, feeding on fishes and other animals in near-surface waters. All must return to land to lay eggs and raise their young.

Marine reptiles include sea turtles in shallow waters in the Caribbean and off the coast of southern Florida, sea snakes in the tropical Indian and Pacific oceans, and the marine iguana in the Galapagos Islands.

Feeding strategies are different in the various parts of the open ocean. Herbivores dominate near the surface. Carnivores are most abundant beneath the herbivores. Omnivores dominate the deep ocean.

Open-ocean organisms have several defense strategies and adaptations. They can be small or transparent. Some form schools. Others migrate vertically to feed in near-surface waters at night and spend daylight hours in dark, subsurface waters.

Reproductive strategies depend on the amount of resources available. Where resources are abundant, opportunistic forms reproduce rapidly. Where resources are limited, slow-growing forms can dominate.

Many open-ocean organisms migrate to take advantage of seasonally abundant food supplies.

Key Terms

microplankton
nannoplankton
ultraplankton
net plankton
diatoms
dinoflagellates
protists
coccolithophores
cyanobacteria
suspension (or filter) feeders
holoplankton
meroplankton
formaminifera
radiolaria
crustacea
copepods
euphausiids
krill
pteropoda
gelatinous plankton
siphonophores
ctenophores
tunicates
fusiform body
demersal fishes
flatfishes
baleen
school
anadromous fish
catadromous fish

Study Questions

1. Contrast conditions for life on land and in the ocean.
2. Describe the different oceanic life styles.
3. Discuss the different defensive strategies used by planktonic and nektonic organisms.
4. Describe the different types of zooplankton.
5. Discuss the importance of the different forms of phytoplankton.
6. Discuss the factors causing plankton blooms. What causes them to end?
7. Why do fishes and marine mammals migrate?
8. Contrast the reproductive strategies used by most fishes and those of marine mammals.
9. Contrast the life histories of anadromous and catadromous fishes.

10. Discuss the adaptations necessary for organisms to live in the deep ocean.
11. Why do deep-ocean nekton have either vestigial eyes or no eyes?
12. What defensive strategies do gelatinous plankton use?
13. Why are krill so important to Antarctic food webs?
14. How do squid differ from fishes in their swimming?
15. Explain why oceanic birds must return to land.
16. Discuss the differences in feeding strategies with depth in the ocean.
17. *[critical thinking]* Describe the major factors limiting fish production from the ocean.
18. *[critical thinking]* Which parts of the open-ocean ecosystem are most vulnerable to increased levels of ultraviolet radiation caused by the ozone hole?

Selected References

ELLIS, R., AND J. E. MCCASKER, *Great White Shark*. New York: Harper Collins, 1991. Entertaining, comprehensive.

FRASER, J., *Nature Adrift: The Story of Marine Plankton*. London: G. T. Foulis, 1962. Concise survey of plankton.

HARDY, A., *The Open Sea: Its Natural History*. Boston: Houghton Mifflin, 1965. A classic, beautifully illustrated.

IDYLL, C. P., *The Abyss: The Deep Sea and the Creatures that Live in It*. New York: Thomas Y. Crowell, 1964. Elementary.

MARSHALL, N. B., *Aspects of Deep-Sea Biology*. London: Hutchinson, 1958. Ecology of the deep ocean.

MARSHALL, N. B., *Exploration in the Life of Fishes*. Cambridge, Mass.: Harvard University Press, 1971.

MINASIAN, S. M., K. C. BALCOMB III, AND L. FOSTER: *The World's Whales: The Complete Illustrated Guide*. Washington, D.C.: Smithsonian Books, 1984.

ROBISON, B. H., "Light in the Ocean's Midwaters," *Scientific American*, 273(1):60–65. July 1995.

OBJECTIVES

Your objectives as you study this chapter are to understand:

- The environment in which bottom-dwelling organisms live
- How benthic organisms compete for space and food
- Adaptations for using energy sources other than sunlight
- Oyster reefs and coral reefs
- Trophic relationships among organisms on the deep-ocean bottom and in vent communities
- How oil spills and ocean-dumping of waste disposal affect benthic organisms

Competition for space on the ocean bottom is intense. An orange sponge competes for space with green anemones and other small plants and animals. A yellow nudibranch, or sea slug, predator of the anemones (lower right), is on the sponge.

13 Benthos

The *benthos*—the plants and animals that live in or on the ocean floor—includes a great variety of organisms. Their world is basically two-dimensional the way our familiar terrestrial world is, the main difference being that benthic organisms live in a sea of water rather than a sea of air. In this chapter we focus on how these attached organisms and communities of organisms adapt to that world.

Benthic Life

All benthic organisms live by one of three life strategies: (1) attachment to a firm surface, (2) free movement on the ocean bottom, or (3) burrowing in sediments. These three life-styles correspond to the principal ways benthic organisms obtain food: (1) filtering seawater, (2) preying on other plants or animals, or (3) swallowing and digesting sediment.

A single organism can have more than one life-style. Crabs, clams, or worms, for instance, may take shelter in a sand or rock burrow but emerge to hunt prey or scavenge for detritus. Slow-moving benthic animals that have heavy shells, such as snails and sea urchins, feed on attached organisms or detrital particles. Some permanently attached animals, such as sea anemones and barnacles, are also predators, capturing organisms that swim or float past.

Factors controlling distributions and diversity of benthic life include: light levels (if plants are involved); availability of food; temperature; salinity; and nature and stability of the bottom.

Availability of food and suitable places for organisms to live (called substrates) controls the abundance of benthic organisms. Food is most abundant near land and below upwelling areas. Therefore benthic organisms are generally less abundant with increasing water depths and distance from land. As we shall see, benthic organisms can be locally quite abundant if there is enough food.

Stable environments favor diverse communities of benthic organisms, where many different kinds of plants and animals live together. The quiet waters in nearshore tropical areas, for instance, normally have highly diverse bottom communities. Coral reefs, which we discuss later, are prime examples of extremely diverse communities of benthic organisms.

Benthic organisms are less abundant in unstable environments, where waves or currents frequently disturb the bottom, for instance, or in shallow polar seas, where there is only a short growing season for plants and marked salinity changes (due to melting and freezing of sea ice). These unstable situations create environments where only a few species survive—although there are frequently many individuals of these few species. Such ecosystems are said to have low biological diversity.

Intertidal Zone

The **intertidal zone** (Fig. 13–1), lying between the high-tide water level and the low-tide level, illustrates how these conditions control the distributions and abundances of benthic organisms. The intertidal zone (also called the littoral zone) provides a sequence of living conditions for attached organisms, from nearly always dry (at the highest high tide level) to nearly always submerged (at the lowest low tide level). The plants and animals that live in this intertidal zone have been studied extensively because they are readily accessible and experimentation is possible.

Associated populations in the intertidal zone are often sharply divided. The zone is essentially divided into a series of horizontal bands, each band occupied by a particular, well-adapted assemblage of plants and animals. This zonation is caused by intense competition for living space. Where two populations meet, neither enjoys a marked advantage. Above or below that meeting line, assemblages are determined by differences in an organism's ability to endure exposure to air and large changes in water temperatures and salinities. In addition, the organisms must survive disease and avoid being eaten by predators. Zonation among barnacles and mussels on a rocky coast in Rhode Island is shown in Figure 13–2.

Light requirements govern distributions of benthic plants. Green algae live somewhat above the low tide mark down to depths of about 10 meters. Varieties include bright green sea lettuce, up to 1 meter in length, and delicate, mossy plants only a few centimeters long. Red algae occur worldwide. They are most abundant in temperate and tropical waters and prefer dim light in deep waters or shaded pools. Brown algae flourish in colder waters, although some kinds occur on rocky coasts around the world (Fig. 13–3). Sea grasses grow in wetland areas and in shallow waters where light is plentiful and wave action is not too strong.

Grazing by snails, urchins, and other animals affects the presence of benthic plants. Areas actively grazed may have little attached algae left. If the animals are removed (or killed by an oil spill or increased levels of ultraviolet radiation), the plants can grow luxuriantly. Normally they are kept grazed down.

Succession

The regular process by which first one type of benthic organism and then another invades a bare surface is known as **succession** (Fig. 13–4). It begins with an accu-

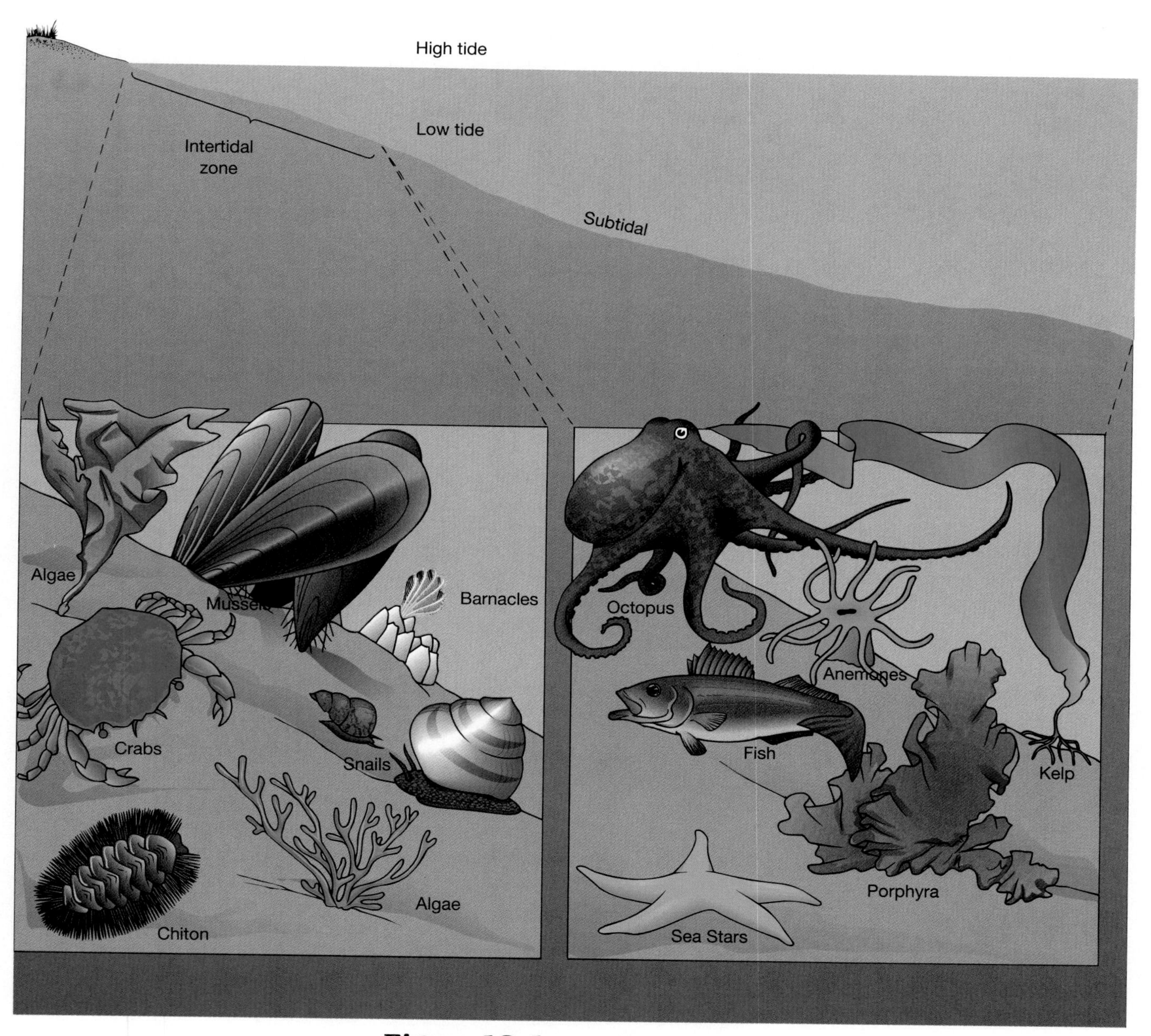

Figure 13–1
The intertidal zone is the area of the shore which lies between the high and low tides. Different assemblages of organisms live at the various levels on a rocky intertidal shoreline, depending on their abilities to withstand drying and wave action.

mulation of dead organic matter. Eventually a bacterial slime covers the organic matter, and then benthic diatoms and protozoans appear. They multiply rapidly, utilizing absorbed organic compounds and products of bacterial decomposition. Hydroids and multicellular algae come next, followed by the settling stages of tunicates, barnacles, mussels, and snails. Eventually, an ecosystem reaches a balanced state and no further colonization occurs. Ecological succession ceases unless the system is disturbed, which causes the process to start afresh.

This process of benthic colonization is familiar to boat owners, because it often occurs on the underside of boats. In such instances it is known as **fouling,** and a great deal of elbow grease is required to remove the benthic community.

Figure 13–2
Barnacles and mussels exhibit zonation on a rocky intertidal shore. The barnacles are higher, being able to withstand more exposure than the mussels. (Michael P. Gadomski/Bruce Coleman, Inc.)

Succession is a process common to all ecosystems when they are disrupted—for example, the sequential replacement of one species, or group of species, by another following a forest fire.

Rocky-Shore Communities

The communities of organisms inhabiting rocky shorelines are especially well known, since they are so easily observed. Such communities extend both above and below the intertidal zone. As we discuss this familiar environment, we will encounter both attached (or sessile) organisms (such as barnacles, anemones, and various seaweeds) and mobile animals (like snails, starfish, and sea urchins).

Rocky shores occur around the world, especially on recently glaciated coasts or where sands and other sediments are removed by wave action. The rocks often form steep faces on cliffs.

Figure 13–3
Rockweed, barnacles, and brown algae grow on rocks in the intertidal zone on the Maine coast. The upper limit of seaweed growth marks the high tide level. (Mary M. Thacher/Photo Researchers, Inc.)

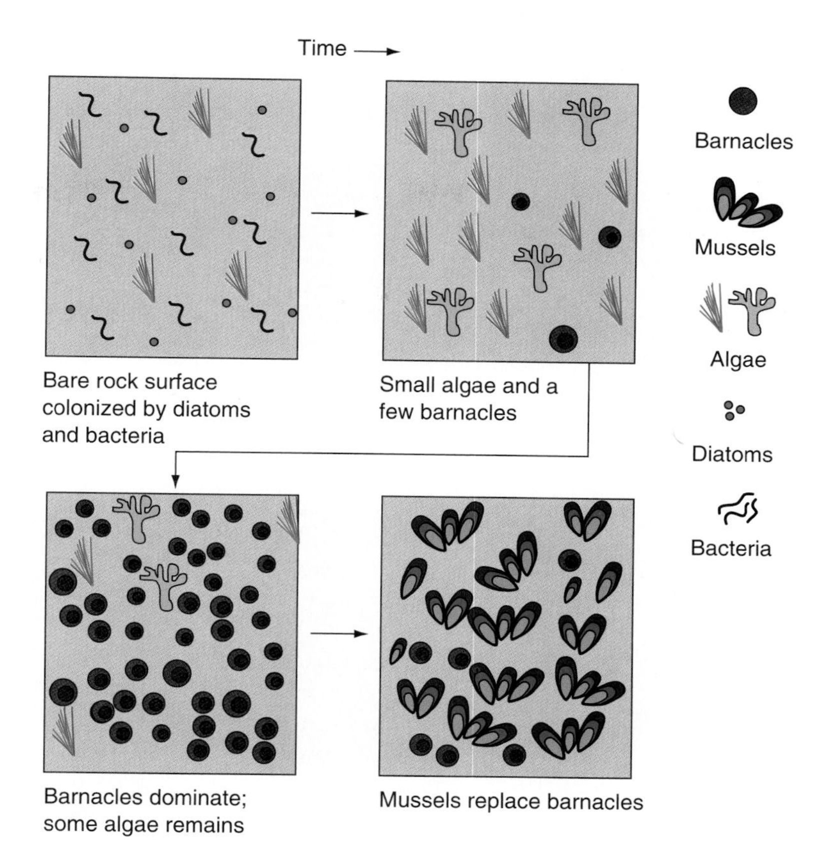

Figure 13–4
Bare rock surfaces on temperate coasts are colonized first by bacteria, then seaweeds, then barnacles, and eventually mussels.

Where wind, sunlight, or waves create an unfavorable environment or where the rock face is too steep, attached seaweeds are sparse or absent, although some species are firmly attached to the rocks so they can exist on coastlines even where wave action is severe. In more protected areas, algae can grow more luxuriantly.

Above the mean high tide mark, seawater covers the rocks only at spring tides and during storms (although spray from breaking waves usually wets these rocks regularly). For organisms living in this zone above the mean high tide level, therefore, resistance to drying is a prime requirement for survival.

On the rocky intertidal zone between high and low tide levels, firm attachment is necessary, especially where there is heavy surf or scouring by strong currents or ice. For this reason, barnacles dominate the upper, more exposed part of this land, which is covered by water for less than half of each tidal day. Barnacle zones in shady, protected areas extend farther up rock faces than do barnacle zones on dry, sunny surfaces. These animals feed when submerged at high tide, filtering small particles from the water. A barnacle shell has a four-part structure that shuts tightly, protecting the animal inside from drying out and dying when exposed at low tide.

Storms also affect plants and animals on rocky coasts. Storm waves can scour rocks and shallow bottoms, leaving nearly bare surfaces for later recolonization. This scouring then is followed by a succession of plants and animals as described above, comparable to the events observed in the fouling of a newly cleaned boat bottom. Ice can have a similar effect on rocky coasts in winter.

Tide pools, which form where the ebbing tide leaves water in depressions between the rocks, contain specialized plants and animals that can cope with the highly variable environment. In protected environments for example, a cave in the intertidal zone, the sheltered conditions permit more delicate organisms to live in such pools. A large variety of plants and animals can live in a small area. Tiny shrimp-like crustaceans (amphipods), swimming worms, starfish, and many kinds

BOX 13–1
Mariculture of Benthic Organisms

In many nations, especially those with rapidly growing populations and poor protein supplies from agriculture, the cultivation of aquatic organisms as a food source is an increasingly important industry. Most aquaculture exploits freshwater fish, but the culture and farming of marine organisms, known as **mariculture,** is becoming more common.

Growing benthic marine organisms commercially is a promising way to increase food production from the sea (Fig. B13–1–1). Because many benthic organisms do not move around, they are readily grown for human consumption. For example, young oysters and clams can be moved from areas where they settled to more favorable areas to fatten for market (Fig. B13–1–2). Another common technique is to enclose oysters or mussels in mesh containers, attach the containers to ropes, and suspend the ropes beneath floats. Suspended and protected this way, the shellfish can mature away from predators. Since both of these species are filter feeders, they can obtain their food supply from the water and can thus be grown even in areas where the bottom may be too muddy for them to live naturally.

Algae are also cultured, especially in Japan. There, a red algal seaweed called *nori* is a major crop. It is used as a protein supplement in foods. The Japanese nori industry employs many tens of thousands of workers and earns several billion dollars per year.

New animals are constantly being investigated as potentially valuable species for aquaculture. An example is the gigantic clam *Tridacna* (Fig. B13–1–3), which is currently being cultured in the tropical Pacific. (This is

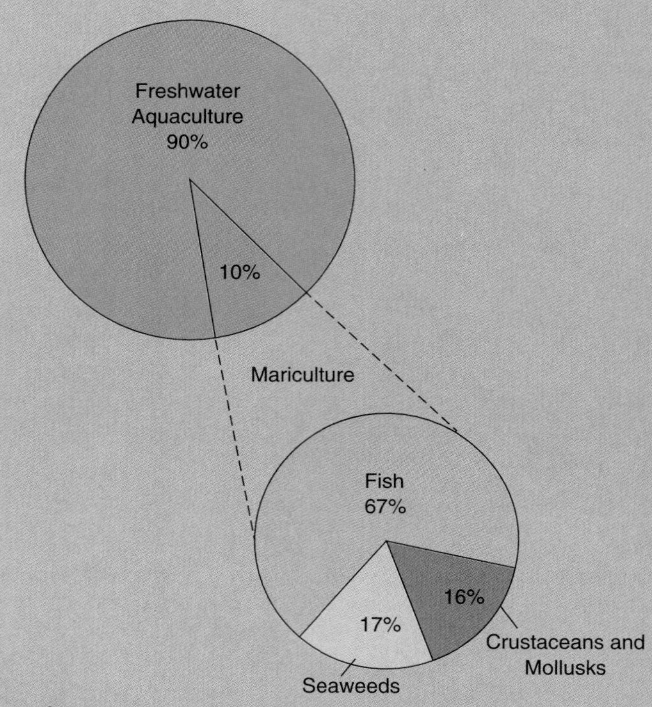

Figure B13–1–1
Mariculture constitutes about 10 percent of the total world aquaculture production. Of this, 67 percent is from the farming of marine fish and 17 percent is accounted for by the culture of seaweeds. The remaining 16 percent involves the culture of invertebrate animals, such as shrimps, mussels, oysters, and abalone.

of snails are common in tide pools, especially where deep crevices or beds of seaweed retain water during low tides (Fig. 13–5). Evaporation, overheating, and oxygen depletion occur on warm, sunny days. Heavy rains can markedly lower salinity within a few minutes. Survival for tide-pool organisms requires the ability to tolerate sudden changes in water temperature, salinity, and dissolved oxygen levels.

The most populous zone on a rocky beach is around and below low tide level. Here starfish and crabs are common, usually hidden in crevices. Small scavenging snails inhabit protected niches containing stagnant water and decaying debris. Sea anemones, sea urchins, sea cucumbers, and mussels (Fig. 13–6) are locally abundant. Hydroids (Fig. 13–7) grow in quiet but not stagnant waters that also favor the nudibranchs, also known as sea slugs, (Fig. 13–8) that feed on them.

Lobsters and crayfish (spiny lobsters) (Fig. 13–9) scavenge on subtidal hard bottoms, both near shore and far out on the sandy and rocky continental shelf. They walk about at night, searching for worms, mollusks, and organic debris. They usually seek shelter under rocks or seaweed during daylight. In autumn, Maine lobsters move offshore to breed, probably to avoid cold shallow waters in winter.

the so-called "killer clam" of underwater adventure movies, a name derived from the fear that a diver could be trapped inside the clam, which can grow to be one meter across. There is no evidence that this has ever happened.)

Tridacna has photosynthetic algae in its brightly colored mantle tissues, which it exposes to sunlight in shallow waters, typically on reefs. The photosynthetic algae provide food to the clam, and it in turn provides nutrients to the algae. The result is a fast-growing animal whose mantle and abductor muscle (used to close the shell) can be eaten. The animal grows to market size within a few years, and in addition to the meat, the large shells are also highly prized because they are used as bird baths or ornaments.

Figure B13–1–2
Oysters grow in containers suspended in the water at an oyster farm in Washington State. (Bruce W. Heineman/The Stock Market.)

Figure B13–1–3
Tridacna, the giant clam of the Pacific, is one of the more recent benthic animals to be used for mariculture. (Lynn Funkhouser/Peter Arnold, Inc.)

The octopus (Fig. 13–10) is a common resident of subtidal, rocky sea floors. Because it hides in crevices during the day and feeds at night, this highly intelligent and shy creature is rarely seen by divers.

Muddy-Bottom and Sandy-Bottom Communities

Many coastlines are sandy, or, where rivers deposit large amounts of sediment or coastal currents are very sluggish, they can be muddy. Typically, such shorelines have only very gentle slopes so that large, flat areas ("mud flats") are exposed by the low tide. In this section we will consider some of the general aspects of life in soft sediments, as well as those features which are specific to muddy and sandy intertidal areas.

The benthic plants and animals that inhabit marshes, sandy beaches, and estuarine shorelines are different from those found in rocky areas. They are seed-bearing, vascular plants with true roots. Sea grasses in marshes contribute plant

Figure 13–5
A rocky intertidal area on the coast of Washington exhibits zonation of attached animals: barnacles, green sea anemones, and starfish in the tidal pool. (Jim Zipp/ Photo Researchers, Inc.)

Figure 13–6
Barnacles and mussels compete for living space on a rock in the intertidal zone at La Jolla, California. Both species are filter feeders and can seal themselves in their shells at low tide to avoid drying out. (Courtesy C. Arneson.)

Figure 13–7
This colonial feeding polyp (called a hydroid) is the benthic asexual stage of the coelenterate *Chrysaora quinquecirrha*. In summer, these jellyfish are abundant in Chesapeake Bay. (Courtesy Michael J. Reber.)

Figure 13–8
Nudibranchs feed on attached benthic organisms. This shellless snail breathes through gill-like projections on its sides. It has few predators because of its unpleasant taste, the result of the accumulation in its body tissues of the stinging cells contained in its prey.

Figure 13–9
Spiny lobsters live in tropical and sub-tropical waters in many parts of the ocean. (Tom McHugh/ Taronga Zoo, Sydney, Australia/Photo Researchers, Inc.)

Figure 13–10
Octopus. The eyes are on the top of the head. Below is a siphon through which water is ejected after passing over the gills. The animal has eight tapered tentacles, or arms, each equipped with a double row of suckers, used to grip surfaces or to hold prey.

debris to sediments and protect deposits from erosion by currents and waves. On open sandy coastlines, however, waves or currents can inhibit growth of rooted plants. These waves and currents can also carry away fine particles and organic debris, leaving behind hard sandy bottoms containing little organic matter. Such environments support many animals adapted to survive on sandy or muddy bottoms.

Infauna—animals that live buried in sediment—are much less conspicuous than surface-dwelling animals, which are called **epifauna.** Many infaunal animals are selective **deposit feeders,** which means they select food particles from among sediment grains (Fig. 13–11). Others pump large quantities of water through filtering devices in their bodies in order to remove edible materials and are referred to as **filter feeders.** Still others are **unselective feeders;** they eat their way through sediments and digest the sediment, extracting their food from the organic detritus contained in it. This latter class includes sea cucumbers and many worms.

Sediment grain size controls the distributions of infaunal animals. Those that feed on filtered plankton and detritus from seawater live in relatively coarse-grained sandy deposits below low-tide level. They require relatively clear, nonturbid waters that do not clog their delicate, mucus-coated filtration devices. Thus, muds, which by definition contain grains much finer than the grains found in sands, are generally not suitable for filter-feeding organisms since their feeding structures clog too quickly. Heavily polluted waters are also not suitable for these filter feeders.

BOX 13–2
Diseases of Benthic Organisms

Entire populations of benthic organisms have been destroyed by diseases; yet, little is known about diseases in the ocean. The reason is easily understood: there are no domesticated marine organisms. The only information we have is our limited observation of how diseases affect commercially harvested organisms or those that live in shallow water.

The best-known diseases in marine animals are those affecting oysters. One of these is known as MSX, or *Multinucleate Sphere X*—an elegant way of saying that no one knows what the disease organism is. What is known is that it is a round, single-celled protozoan that infects oysters in waters having salinities greater than 15. The organism is a parasite that weakens the host oyster. The sick oyster is then more vunerable to other diseases or to predators. Droughts cause MSX to spread, because the lack of water makes estuarine waters saltier than normal. Up to 95 percent of the oyster crop in Delaware Bay was destroyed by the disease when it first appeared in 1957. After that, MSX spread to Chesapeake Bay, where it devastated the oyster industry in the late 1980s.

One way of fighting the disease is to select animals that have lived through an episode. Presumably they are more resistant than those killed. Selectively growing resistant organisms might permit cultured oyster crops to withstand the disease.

Diseases are also an important factor in aquaculture. Keeping large numbers of animals closely confined is an excellent environment for the spread of disease. People raising fish or shellfish must constantly guard against disease. Unfortunately, there are few treatments known for diseases among marine organisms.

Muds are well suited to unselective feeders. The smaller particles in mud usually contain abundant organic matter. They also support bacteria and many other microbes that are food for deposit feeders.

Mud-burrowing animals have specialized breathing structures that are not clogged by sediment grains. Breathing through the skin—as sea stars do, for instance—is not suited to a muddy environment or turbid waters. One widely used method of feeding and breathing uses water pumped through burrows by the animals living there (Fig. 13–12). Dissolved oxygen is supplied for respiration, and sediment particles are swept off the animal's gills as food particles are captured.

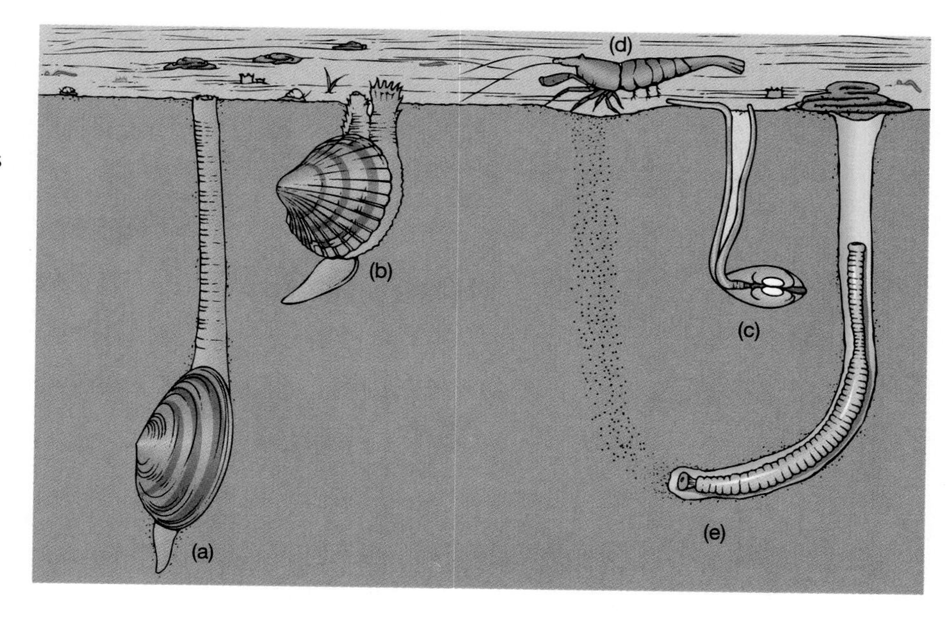

Figure 13–11
In shallow sandy bottoms and tidal flats, a variety of feeding styles are found. The buried clam (a) and cockle (b) on the left are filter feeders, while the clam (c) *Macoma* (right) selects edible particles from the sediment surface. The crayfish (d) is a scavenger. The segmented worm (e) eats sediments and after digesting the organic matter, leaves piles of excreted sediment on the surface near its hole.

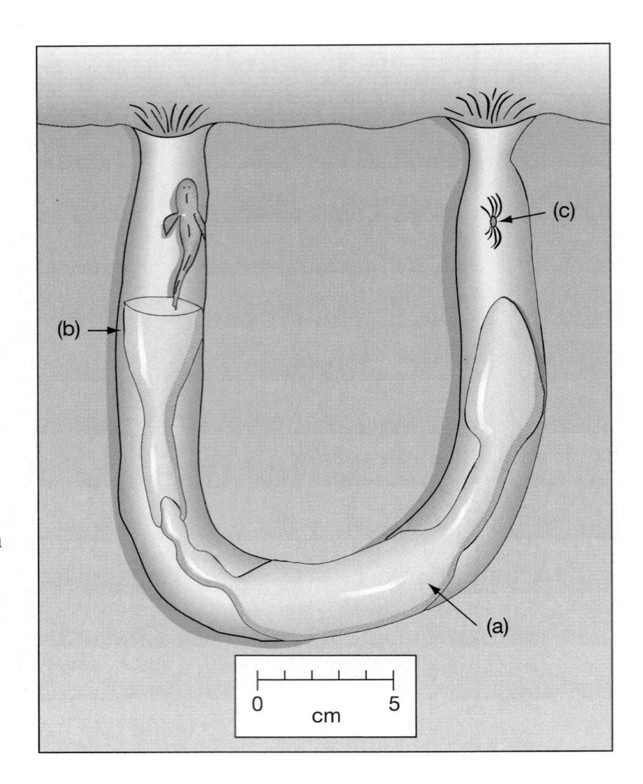

Figure 13–12
Common on Pacific mud flats at or below the low-tide level, the pink echiuroid worm *Urechis cauipo* (a) is called the innkeeper worm, because it usually shares its burrow with other individuals. By scraping away mud with bristles on each side of its body, the worm constructs a deep U-shaped burrow with narrow openings on the ocean floor. Once inside its burrow, the worm secretes a funnel-shaped, fine-meshed mucous net (b) that fits over its head like a collar. As the worm pumps water through the burrow, food particles are trapped in the net. When the funnel becomes clogged, the worm eats it and constructs a new one. Wastes and debris are ejected from the burrow by blasts of water. Also living in the burrow are tiny pea crabs (c), and sometimes small fishes. (After Hedgpeth, 1957.)

In sandy or muddy intertidal areas, deposit feeders generally dominate. Many deposit feeders excrete durable fecal pellets, as we learned in Chapter 5. These pellets bind the sediment particles and thus reduce turbidity at the sediment-water interface. In deep, quiet waters, the bottom may be nearly covered by fecal pellets, and the cover makes the deposits more resistant to erosion, thereby keeping waters clearer.

Below the low-tide level in muddy sediment on continental shelves, there are usually many deposit-feeding bivalves, such as clams (Fig. 13–13). Where waters are too deep for algae and eelgrass (a seed-bearing marine plant, discussed in more detail below) to grow, sediments are often soft and easily eroded, due in part to constant reworking by burrowing organisms. A few tens of clams per square meter can rework sediments to a depth of 2 centimeters once or twice each year. Attached animals are unable to colonize the semi-liquid substrate, and filter feeders are choked by the turbid waters. Thus, the burrowing and reworking of sediments by clams effectively excludes their competitors.

Where deposit feeders are abundant, filter feeders are scarce because the former make the area less habitable for the latter. This is competition by the creation of an undesirable living environment rather than by direct competition for food or living space. On the other hand, oxygen-bearing waters mixed into sediments by burrowing organisms (a process called bioturbation) permit aerobic bacteria, protozoans, and other small benthic animals to live much deeper below the sediment-water interface than is possible in oxygen-deficient muds. Over eons, this burrowing has mixed the sediment layers, thereby destroying much of the detailed record that is preserved in areas where bottom waters and sediments are deficient in dissolved oxygen.

Salt Marshes and Sea-Grass Beds

Salt marshes and sea-grass beds occur along most of the world's nonpolar coastlines. **Salt marshes** are intertidal, which means that plants that grow there are

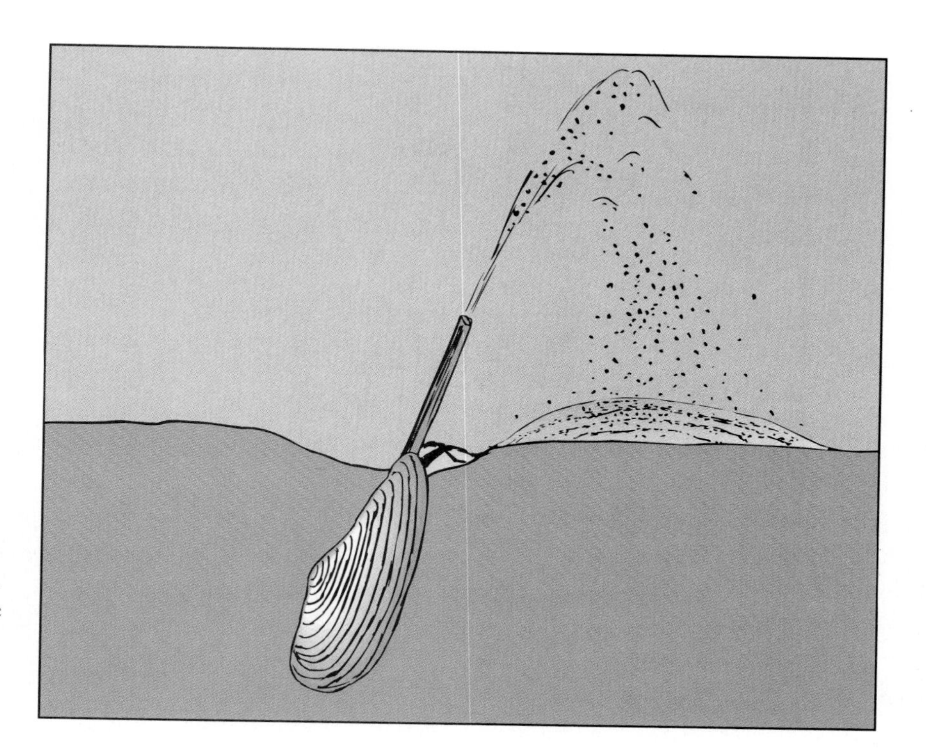

Figure 13–13
Half buried in mud, the clam *Yoldia limatula* uses its feeding appendages to collect particles and bring them to the body cavity where they are digested. Inedible particles are ejected as clouds of loose sediment that deposit nearby, creating mounds of reworked sediment. When resting, *Yoldia* burrows a few centimeters below the surface. (After Rhodes, 1963.)

submerged only some of the time, whereas sea-grass beds are permanently submerged. Both marsh grasses and sea grasses are extremely productive, with a primary production level of 1,500 grams of carbon per square meter per year (comparable to a corn field) being common.

In the absence of strong waves or currents, intertidal sediment deposits, rich in organic detritus, are trapped by marsh grasses, forming wetlands, which we discuss further in Chapter 14. Walking through such an area at low tide, one sees few signs of animal life but the ground is riddled with animal burrows.

Up to half of the organic matter formed in many marshes is carried out of the marsh by tidal currents. Part of this organic matter is consumed on continental shelves, and part is deposited with muds, either on continental shelves or on upper continental slopes. Whether or not a marsh exports organic matter is highly dependent on its connections with the adjacent coastal ocean.

Dense stands of the *sea grass* known as eelgrass grow in shallow waters, as shown in Figure 13–14. Eelgrass has underground stems and ribbon-like leaves as much as a meter long. Specialized communities of plants and animals find food and shelter in eelgrass beds. Leaf surfaces are coated with diatoms, cyanobacteria, protozoans, organic detritus, and small hydroids. Tube-building crustaceans and worms attach themselves to the leaves. Small grazing animals, such as shrimp and snails, scrape off detritus. This cleaning of eelgrass leaves is essential to its continued growth, because if the leaves become too thickly coated with detritus, the plants cannot photosynthesize and thus die.

Sea grasses are more thoroughly marine and live at greater depths than the intertidal marsh grasses, forming expansive underwater meadows. They are limited by the availability of light for photosynthesis. They, too, are highly productive, with productivity levels comparable to those of salt marshes. Some seagrasses can live as deep as 90 meters, if the overlying waters are clear.

In addition to being highly productive, both salt marshes and sea-grass beds provide refuge for young organisms and thus are important nursery beds for many organisms. In addition, they are grazed by sea turtles. Finally, they export large amounts of organic matter to surrounding continental shelf areas.

Figure 13–14
Drawing of an eelgrass community at Woods Hole, Massachusetts. Scallops, shrimps, and crabs live in the eelgrass.

Kelp

Large, brown, benthic algae known as **kelp** (Fig. 13–15) grow on the rocks or hard-sand bottoms in shallow subtidal waters in subtropical to subpolar coastal oceans. In waters cooler than 20°C, kelp grows abundantly wherever it finds suitable substrates. Kelp often forms dense beds or underwater forests, often many kilometers long. In clear waters, kelp grows down to depths of 40 meters.

Kelp alternates between sexual and asexual generations. One form is the conspicuous plants that reach lengths of tens of meters. These large plants are asexual and produce microscopic spores that settle and develop into tiny, filamentous, sexually reproducing plants. These latter plants, about which very little is known, eventually produce the gametes which combine and grow into the familiar large kelp plants. Individual kelp plants can live for several years in areas where they are not destroyed by waves.

How the filamentous sexual kelp plants are distributed in the ocean is poorly understood. The microscopic sexual forms of the kelp are apparently dispersed by currents and are able to remain current-borne for months, perhaps up to a year, thus functioning like seeds in land plants.

Kelp plants have holdfasts that grip rocks or the hard bottom and anchor the plants, much as roots anchor plants on land. Unlike roots, however, holdfasts do not take up nutrients from the bottom. Instead, nutrients are taken up by the leaves, which are called fronds and are kept close to the well-lit surface by gas-filled bladders, which act as floats.

Kelp is highly productive, growing primarily in nutrient-rich waters. Primary production in kelp forests ranges from 500 to 1,500 grams of carbon per square meter per year. This is two to five times the average production by phytoplankton in coastal-ocean waters and rivals the production of the best croplands.

Figure 13–15
Kelp forms large forests on hard bottoms, and this dense plant growth shelters many animals.

Kelp is easily damaged by waves. Often, frond ends are worn away by abrasion, forming fragments. As they are worn away, the fronds are renewed by continued growth. Thus, they are like a conveyor belt of algal tissue. Much of the organic matter produced during the wearing-away processes enters detritus food webs.

Kelp forests offer many different environments to animals in the water or on the bottom. The tops of the tallest plants float near the surface, forming a shaded canopy where animals can hide from predators. Many plants and animals also grow on the bottom among the holdfasts. Such plants must tolerate low light levels. Various animals live in and around the edges of kelp forests, using them as refuges.

Kelp plants are also eaten by fish, sea urchins, and snails (Fig. 13–16). Where sea urchins are especially abundant, they may destroy the kelp forests, leaving areas referred to as urchin barrens. Sea urchins are eaten by sea otters (Fig. 13–17) and lobsters. In the eighteenth and nineteenth centuries, when sea otters were heavily hunted for their furs and nearly became extinct, sea urchin populations expanded and the abundance of kelp beds was greatly reduced. Since 1911, sea otters have been protected from hunting, and their populations have expanded enough to re-exert their control of sea urchin populations. Consequently, kelp beds have recovered over much of their original range. On the east coast of Canada and the northeastern United States, a similar ecological balance exists between kelp, urchins, and lobsters.

Figure 13–16
A snail feeding on kelp fronds. (Courtesy C. Arneson.)

Figure 13–17
Sea otters feed on sea urchins. When the otters were hunted to near extinction, sea urchins became extremely abundant and grazed back the kelp forests. Since sea otters have been protected, they have become more abundant and have reduced the abundance of sea urchins. The kelp forests have subsequently recovered. (Courtesy Kathleen E. Olson.)

Oyster Reefs

Oysters are bivalve mollusks (having two hinged shells) that grow abundantly in coastal oceans and estuaries worldwide. Each mature female produces millions of eggs yearly, most of which are fertilized in the water (Fig. 13–18). Only a few of the planktonic larvae survive (this stage lasts only a few weeks) and settle to the bottom to develop into adults. Larvae select suitable settling areas, preferring oyster shells to any other substrate and apparently favoring live oysters over dead shells. Oysters need one to five years to mature, during which time most of those that settled together as larvae are killed by crowding, starve to death, or are eaten by predators.

Oysters build reefs, rigid, wave-resistant structures (Fig. 13–19). They contain enormous numbers of individuals whose shells are cemented to rocks and to one another. The shells of dead oysters contribute to the reef as well. Tidal flats of partially enclosed bays and river mouths provide advantageous conditions for the growth of these reefs, especially where moving water containing little silt brings fresh supplies of plankton and oxygen and where few predators exist. Oyster reefs were so large in Chesapeake Bay that they were hazards to navigation when the first European colonists arrived, early in the seventeenth century.

Each oyster in a reef pumps many gallons of water each hour. It has been calculated that when the Europeans arrived in the seventeenth century, before oysters were fished on any large scale, oysters were so abundant that they could filter the entire volume of water in the Chesapeake Bay in three to six days. They are now so scarce that this would take more than a year. Plankton and food particles in the water are caught on a mucous net that is moved steadily toward the mouth by ciliary action. Some particles concentrated in this way are swept out of the shell before the oyster can consume them. Nearby filter-feeding animals take advantage of this food supply.

Attached organisms, such as barnacles, mussels, and tube worms, add to the bonding of the material in the reef, thus increasing its stability. Crevices between shells shelter small filter feeders.

Oysters can live in low-salinity waters. In Chesapeake Bay, for instance, large natural beds occur in intermediate-salinity waters (7 to 18). Oysters survive there, but their principal enemies and most diseases cannot. In the absence of their predators, oysters can grow faster and are more prolific in high-salinity waters. If the salinity increases—as a result of drought, say—oysters may be attacked by predators and diseases.

Figure 13–18
Oysters (male on left, female on right) expelling sperm and eggs into water. (Courtesy Michael J. Reber.)

Figure 13–19
Oysters form wave-resistant reefs by cementing their shells together. Spaces between shells shelter many smaller organisms. (Courtesy Michael J. Reber.)

In addition to their value as food, oysters have long been exploited for pearls, which form in the inner portions of oyster shells. Many different kinds of oysters form pearls by coating sand grains or other irritants that get in their shells with layers of the same material that lines the insides of their shells. Commercial gem-quality pearls come primarily from the tropical pearl oyster. In ancient times, pearls came primarily from India and the Arabian Gulf. Now, most pearls are cultivated by inserting objects (made from shell) into the oyster cavity, causing it to form a pearl.

Coral Reefs

Like oysters, corals build structures, again called reefs, that modify their environment. Reef-building corals are colonial animals that build calcareous skeletons (Fig. 13–20). Reefs can be found in tropical and subtropical waters warmer than 18°C. Corals that do not build reefs build flexible, fan-shaped structures (Fig. 13–21) attached to individual rocks. Coral reefs are complex, shallow-water benthic environments that are among the most productive communities in the ocean (Fig. 13–22). As individuals die, others build over and around the dead skeletons, eventually forming massive structures that maintain themselves at the sea surface. Some corals form large branching structures; others build compact forms such as brain coral. Still others grow by encrusting reef surfaces.

Corals reproduce both asexually and sexually; the asexual process (called budding) is the primary means of enlarging coral colonies. In budding, a new individual buds from the side of a parent polyp. Coral polyps apparently take seven to ten years to reach sexual maturity. Sexual reproduction results in free-swimming larvae. When larvae settle and attach after a few days of planktonic existence, they start new colonies.

Figure 13–20
Corals grow in many forms.

Figure 13–21
A red sea fan, a relative of reef-forming corals. The minute animals that formed this beautiful structure capture food from waters flowing past it. (Courtesy C. Arneson.)

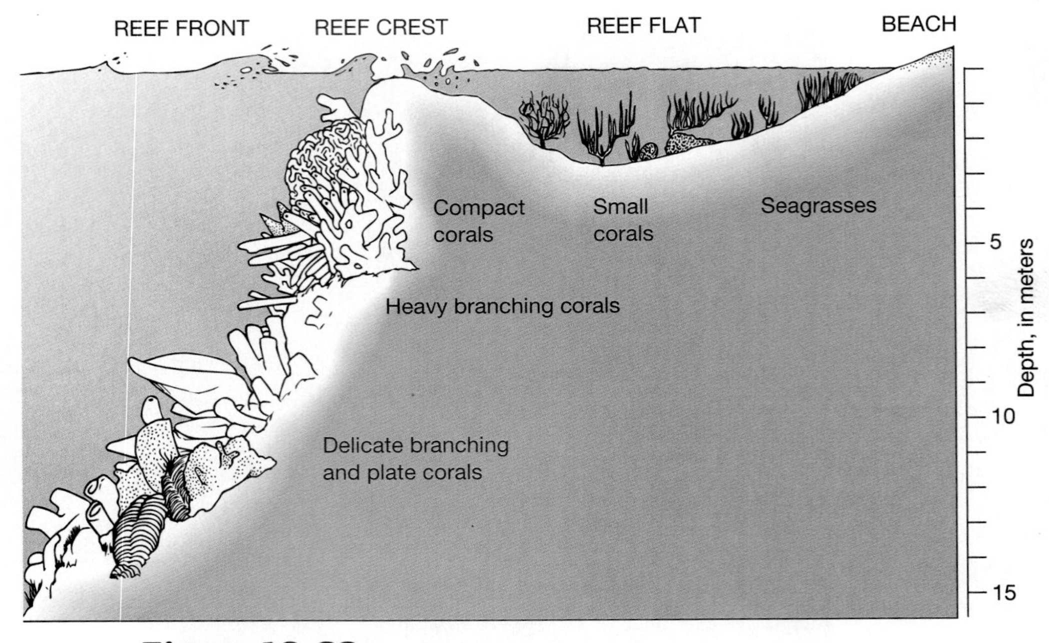

Figure 13–22
Coral reefs are complex ecosystems. Different forms of coral are adapted to life in different areas of the reef; the more robust forms are closer to the surface, where wave action may be heavy at times, especially during storms. More delicate structures grow at greater depths where they are not exposed to wave action at the surface. (Courtesy The World Conservation Union.)

Unicellular algae called zooxanthellae (a type of dinoflagellate) live in the coral tissues of reef-building corals (Fig. 13–23). This is an example of a mutually beneficial relationship, called **symbiosis.** The algae provide oxygen and photosynthetic products to the host coral. In return, the algae receive nutrients and carbon dioxide from the coral, making them more productive Photosynthesis by the algae alters the carbon dioxide concentrations in the coral tissues, greatly increasing the animal's ability to secrete calcium carbonate to make its skeleton. Although corals are truly animals, their symbiotic relationships with the zooxanthellae make them behave ecologically much like plants. When corals lose their zooxanthellae (see Box 13–3) they grow slowly and many die.

Corals are the most conspicuous contributors to the framework of a coral reef (hence the name), but there are other organisms involved. Encrusting, calcareous, red algae are important cement depositors. They bind the coral structures to make the reef more rigid.

Cyanobacteria (unicellular blue-green algae) supply nitrogen compounds to the waters near a reef; furthermore, nutrients are readily recycled in reefs. Currents in and around reefs tend to retain nutrients in the overlying waters. Reefs are also highly diversified in the many kinds of living spaces they provide. Competition for living space rather than lack of food apparently limits the abundance of filter feeders on coral reefs.

Tides periodically expose parts of reef tops. There hardy, soft-bodied algae, barnacles, and coralline algae form rimmed pools a few centimeters above the average sea level. Water splashed into the pools by waves drains slowly, keeping plants and animals in the pools moist. (This is similar to tide pools seen on rocky coasts.) Some encrusting algae form a purplish red ridge, called the algal or lithothamnion ridge, at sea level. This rugged ridge plays an important role in breaking up waves, thus protecting more sensitive parts of the reef.

(a)

Coral reefs are made of coral and algal structures.

(b)

Coral polyps with feeding tentacles extended (right) and retracted (left)

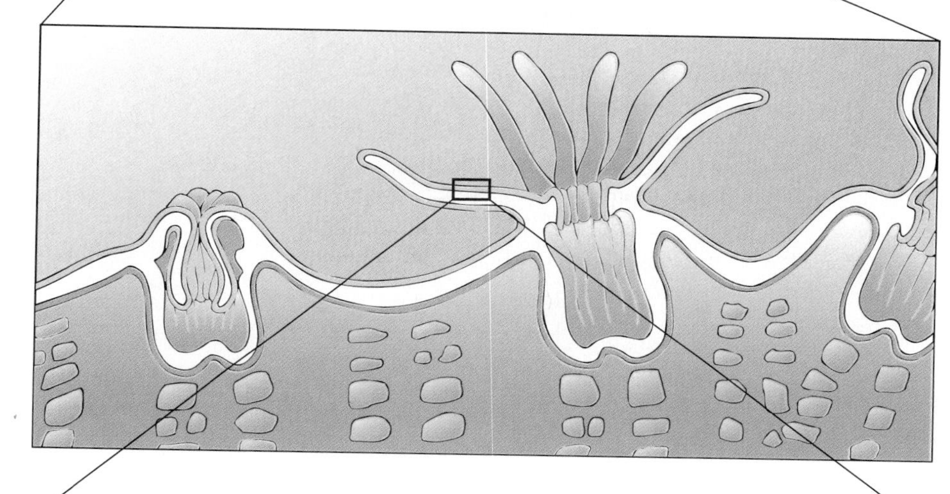

(c)

Outer surface of tentacle has nematocysts (stinging cells) to capture prey. Zooxanthellae are located just below the surface.

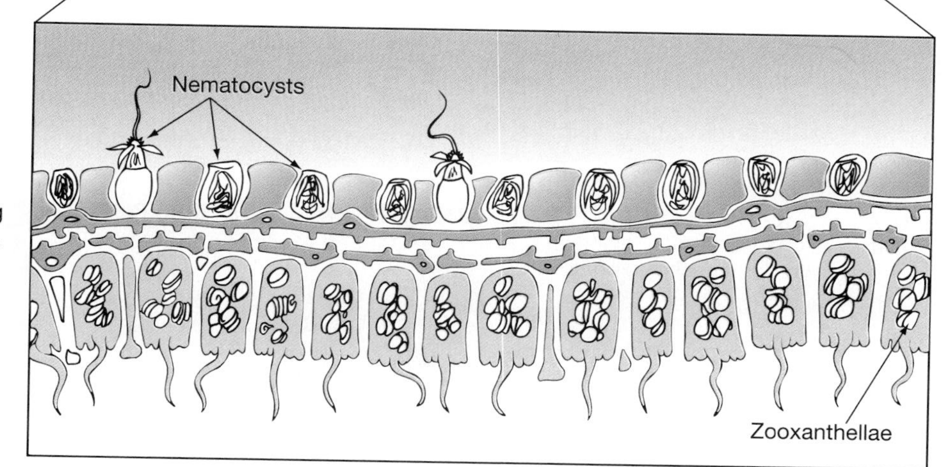

Figure 13–23
(a) A coral reef consists of coral animals and encrusting algal growths. (b) Each individual coral polyp is a hollow, cylindrical animal. The mouth is surrounded by tentacles armed with stinging cells for capturing plankton. During the day these tentacles are folded into the digestive sac. (c) Single-celled algae, called zooxanthellae, that give the coral its green, blue, or brown color are located in the tissue of each polyp. The algal cells can use some of the wastes produced by the polyp. In turn, the algae generate, via photosynthesis, oxygen and organic compounds useful to the polyp. (Courtesy The World Conservation Union.)

BOX 13–3
Coral Bleaching and Global Warming

Repeated and widely scattered occurrences of a phenomenon known as coral bleaching is perhaps the most compelling indication of how global warming affects marine ecosystems.

Corals are unusual animals, in that they contain one-celled algae in their tissues. As mentioned in the text, these symbiotic algae, called zooxanthellae, provide some of the food needed by the corals, and the corals provide nutrients required by the algae. The result is a highly productive ecosystem.

It is these zooxanthellae that give corals their beautiful colors. When the corals are stressed by high temperatures or pollution, they expel their zooxanthellae, and the result is snow-white (bleached) corals. If the bleaching event is short-lived, the corals take up more zooxanthellae and return to their normal growth patterns. If bleaching events are prolonged or repeated frequently, the corals are weakened and may die. Even if they do not die, they are more vulnerable to other stresses such as disease. If too many corals in a reef community die, the reef may be eroded or destroyed.

Starting in 1979, there were widespread reports of coral bleaching. Major coral-bleaching events were reported in 1987 in the Caribbean and in the late summer of 1990 in the Caribbean and near Bermuda, Hawaii, and Japan.

Scientists believe that the most likely explanation is that increases in surface-water temperatures of 1–3°C stressed the corals, which responded by expelling their zooxanthellae. Such elevated temperatures are known to be near the upper tolerance limits for corals.

The probable extinction of one type of coral (*Millepora*) and a reduction in geographic range of another, were caused by the worldwide coral-bleaching events that occurred in the 1980s. These areas were especially hard hit by the El Niño of 1982–1983, one of the strongest ENSO (El Niño–Southern Oscillation) events of the twentieth century.

In the photic zone, reef-building corals are the most conspicuous life forms, but attached plants greatly exceed them in terms of mass of living animal matter. Besides the coralline algae that add calcareous material to the reef framework, filamentous green algae embedded all over its surface manufacture food during the day, using nutrients released by animals and bacteria. At night, coral polyps extend their feeding apparatus (Fig. 13–20a) to capture plankton and detritus from the water.

Many kinds of invertebrates and fish (Fig. 13–24) live in coral reefs, but there are few individuals of any one kind. Some, like sea stars (Fig. 13–25), consume

Figure 13–24
Reef fishes swim near coral growing on a submerged wreck. (Nancy Sefton/Photo Researchers, Inc.)

Figure 13–25
A large crown-of-thorns starfish eats corals, leaving behind bare coral skeletons. Under certain conditions—not well understood—these starfish become extremely abundant and decimate large areas of reefs. (Courtesy C. Arneson.)

coral polyps and algae. Others feed on detritus. Still others—for example, moray eels and sea anemones—prey on animals living on or near the reef. Parrot fish and other browsers graze on the reef surfaces to eat algae and animals living in it.

Because there is always browsing going on their surfaces, coral reefs are constantly changing. An organism having a competitive advantage gains space or position over another, while perhaps losing space to a third. Some coral species even practice chemical warfare. When their tissues touch, one form can release substances that destroy the tissues of the other.

Any disturbance—a severe storm, an infestation of starfish, an oil spill—can set this competition for space in motion by leaving a barren surface. Succession on coral reefs is essentially the same process as succession on rocky shores. First, the most abundant larvae colonize the newly bare surface. Then, as the organisms compete for space with other forms, the composition of that part of the reef may change.

Deep-Ocean Benthos

Soft sediments and deposit-feeding animals dominate much of the deep-ocean bottom (Fig. 13–26), with large filter feeders being rare but conspicuous. Crinoids—a type of deep-sea starfish having the common name sea lily—stand on long stalks above the bottom to filter particles from near-bottom currents. Their arms spread out to trap particles in mucus secretions, and the trapped particles are then swept along ciliated grooves to the mouth. Predatory forms, such as brittle stars, move about on long legs that support their bodies well above the soft-sediment surface.

Filter-feeding sponges, coelenterates, echinoderms (Fig. 13–27), worms, bivalve mollusks, and crustaceans—in fact, all the major groups of animals in the shallow-water benthos—also live on the deep-ocean bottom. Uniform coloration (gray or black among fishes, reddish among crustaceans) and delicacy of structures are typical among organisms living in these dark waters. Deep-sea animals are generally smaller than their shallow-water relatives. The former live much longer and reproduce less frequently. All these differences are adaptations to a scarcity of food.

Most of the detritus that falls from the surface zone is either consumed (usually many times) or decomposed before reaching the deep-ocean floor. Thus, relatively little usable food reaches the deep-ocean floor. Because of the food scarcity, the total biomass of the animals living on the bottom is much lower than the total animal biomass in shallow coastal-ocean waters. Eggs of some deep-ocean benthic forms float to the surface, where food is more plentiful for larvae. The eggs hatch

Figure 13–26
Polychaete worms, relatives of the earthworm, are common in soft sediments. Larvae of these worms are planktonic until they metamorphose into adults. (Courtesy P. I. Blades-Eckelbarger, Harbor Branch Oceanographic Institution.)

there, and the larvae then descend to the bottom. Most deep-water benthic animals produce a few, large eggs and their larval stages are short or nonexistent. Thus, the larvae do not have to go into the photic zone to feed, reducing their chances of being eaten.

In the midst of this scarcity of food, there are infrequent falls of large amounts of it—bodies of sharks and whales. These carcasses attract swimming scavengers, but such events are rare. Thus the ability to sense food from afar—probably by smell—and to move quickly is an important adaptation in this barren environment.

Vent Communities

Dense populations of benthic animals live near hydrothermal vents (Chapter 3), which discharge sulfide compounds (Fig. 13–28). These vent communities contain 10,000 to 100,000 times more living matter than normal deep-ocean benthic communities and are oases of benthic life. Vent communities are also unusual in that

Figure 13–27
A filter feeding sea star anchors to the bottom at 2500 meters depth and extends its long arms into the strongly flowing water to gather suspended particles and plankton. (R. S. Carney/LSU.)

Figure 13–28
The large population of organisms that live near active hydrothermal vents depend on chemosynthetic bacteria for their food. Note the many kinds of animals, including crabs, (lower center) fishes (center), and large worms having blood-red gills. (Courtesy John Edmond/Woods Hole Oceanographic Institution.)

the organisms in them depend on chemosynthetic bacteria for food instead of photosynthesis.

At first glance, hydrothermal vents seem unlikely places to find such rich growths of organisms. In addition to poisonous hydrogen sulfide discharges, the vent waters are extremely hot; temperatures up to 400°C have been recorded. Most vent organisms do not experience such high temperatures, however, because the hot water mixes so rapidly with cold bottom waters.

Scientists know most about the vents and associated organisms that occur on mid-ocean ridges in areas that have recently experienced volcanic eruptions, such as the most active segments of the East Pacific Rise. These relatively shallow vents can be reached by the U.S. Submersible *Alvin*. Benthic communities have also been found around hot-water vents on recently active submarine volcanoes (Fig.13–29). Similar communities have now been found to occur around cold-water seeps on continental margins and in subduction zones where fluids are squeezed out of subducted sediments. These cold-water vents are associated with discharges of methane sulfides or oil and gas. A few active high-temperature vents have been found with no organisms living on them. One example is the young volcanic center Loihi, southeast of the island of Hawaii.

The most conspicuous organisms in vent communities are huge worms (Fig. 13–30), close relatives to much smaller worms that live in sulfide-rich salt-marsh sediments. Vent worms are unusual in that they grow to be nearly 3 meters long and 2 to 3 centimeters in diameter. They have red gill-like organs that they retract into a white plastic-like tube when disturbed. Instead of having a digestive tract, they obtain food from sulfide-oxidizing bacteria that live in specialized organs. Wastes from the worms provide nutrients needed by the bacteria. This is another example of symbiosis, in which two organisms live together to their mutual benefit.

Other organisms inhabiting vents also depend on chemosynthetic bacteria. Clams and mussels have such bacteria on their gills. Clams grow to dinner-plate size in less than two years (Fig. 13–31). (Recall that most deep-ocean organisms are very slow growing.) These organisms have hemoglobin in their blood (like land animals) to transport oxygen. Consequently, their tissues look like raw liver. They also have special enzyme systems to protect them from hydrogen sulfide poisoning.

Dispersal of Vent Organisms

The 1978 discovery of dense benthic communities living around active vents on the East Pacific Rise raised the question of how such organisms find widely

Figure 13–29
A black smoker discharging superheated water and sulfide particles, which discolor the water. Sulfides dissolved in the water supply energy for chemosynthetic bacteria. The bacteria support nearby communities of benthic organisms. (Courtesy Dudley Foster.)

separated vents. The problem was complicated by the fact that these communities can survive only while the vent is active. Each vent field remains active for only a few years to a few decades after a volcanic eruption. Volcanic activity on mid-ocean ridges is episodic, occurring only every few decades. It is obvious that the organism's eggs or larvae must get from an old inactive vent field to a newly formed vent field, perhaps thousands of kilometers away. But how?

Figure 13–30
Worms captured from their habitat near a hydrothermal vent on the deck of a submersible support ship. Such worms can grow to up to 3 meters long, which is about the length of the one being held by the two workers. (Courtesy Jack Donnelly/Woods Hole Oceanographic Institution.)

Figure 13–31
Clams living near ocean-floor vents grow to dinner-plate size in a few years because of the abundant food. These organisms have red flesh because they use hemoglobin to transport oxygen. (Courtesy National Geographic Society.)

The problem of dispersal has been partially solved by subsequent discoveries of similar vent communities on continental shelves and slopes, where organisms have colonized springs discharging methane-bearing waters. Similar communities were found growing around vents where the waters were expelled from subducted sediments. So vent communities are now known to be more common than we originally supposed.

A possible dispersal mechanism has now been discovered. Decomposing whale carcasses in the basins off southern California support vent organisms. These carcasses take years to decompose totally, with the bones taking even longer. As they decompose, the carcasses give off substances that can support dense colonies of vent organisms. There may be many more vent colonies than previously known, and these colonies may not be restricted to just a few localities. These benthic communities can thus serve as steppingstones for organisms, facilitating their dispersal.

Oil Spills and Benthic Communities

Oil spills can occur anywhere oil is produced, transported, or used. Where knowledge of an ecosystem before the spill is scant, it is difficult to determine the effects of the oil spill on the benthos. This lack of knowledge was a limiting factor in understanding the effects of Kuwaiti oil spills in 1991. Conversely, an oil spill from a ruptured oil-storage tank near a major marine laboratory in Panama affected a well-studied ecosystem, permitting scientists to obtain definite information on the effects of the spilled oil on marine ecosystems.

The 1986 Panama spill affected coral reefs, mangrove swamps, and sea-grass beds. These systems are especially sensitive to the effects of oil, as was well known. The new finding was that oil can become associated with sediment deposits where it persists for long periods of time. When these deposits are later disturbed by waves or by periodic flooding, oil is released again, affecting the ecosystem much like a new oil spill. Periodic releases of oil from such contaminated sediments can occur over several years.

After the oil eventually decomposes, it still may take decades to a century for the organisms to recover. Indeed, some parts of an ecosystem may be permanently altered, even destroyed, by an oil spill.

Summary

Benthic organisms are those that live on or near the bottom of the ocean. The three benthic life strategies are (1) attaching to a surface, (2) freely moving on the bottom, and (3) burrowing in sediments. Ways of obtaining food are (1) filtering seawater, (2) preying, or (3) swallowing and digesting sediments.

Stable environments favor diverse communities, and availability of food controls their abundance. Abundance of food usually decreases with increasing water depth and greater horizontal distance from land. Benthic animals are usually most abundant on continental margins and scarce on the deep-ocean floor, far from shore.

Competition for living space and light requirements control which benthic organisms live where. Many organisms must cope with widely changing conditions, such as variable temperatures, salinities, and dissolved oxygen concentrations. For example, organisms on a rocky shoreline form horizontal bands where those adapted to the particular combination of conditions can thrive. These conditions can be disturbed by events such as storms or oil spills, which can cause these distributions to change.

Succession on uninhabited hard surface begins with the growth of bacteria, which form a slime that attracts other organisms and benthic larvae. Other organisms attach, and eventually a community of organisms forms. This process is also known as fouling.

Rocky shores provide living space for attached organisms, but these organisms must cope with changeable living conditions. Kelp are large algae that grow on hard surfaces. They form thick forests that provide shelter for many other organisms. Sea urchins feed on the kelp and in turn are eaten by sea otters.

Sediment deposits—both sands and muds—support many kinds of organisms. Infauna live buried in sediment and usually feed on organic matter in the sediment. Epifauna live on the sediment surface. Some organisms are selective feeders, selecting food particles to be ingested. Unselective feeders simply ingest sediment and digest the usable organic matter. Sediment particle size controls distribution of infauna. Finer-grained sediments usually contain more organic matter and are better suited for unselective feeders. Fine-grained deposits are easily eroded, and the resulting high sediment concentrations are often unfavorable for filter-feeding organisms.

In intertidal wetlands, grasses growing on sediment surfaces stabilize the deposits. Many benthic animals live in the deposits among the plants. Their position in the marshes depends on their requirements for submergence and tolerance for exposure at low tides. Comparable conditions prevail in permanently submerged sea-grass beds.

Kelp are large brown algae that grow subtidally, attached to the bottom. They form highly productive forest-like growths that shelter other organisms. Kelp beds also support large populations of fish, snails, and sea urchins. Sea urchin populations control the distribution of kelp; they are in turn controlled by sea otters that feed on them.

Reefs are wave-resistant structures built by oysters or coral–algal communities. Oysters bind together, forming large structures in which other types of organisms live in the interstices between shells. Oysters thrive in low-salinity waters, which kills their predators (boring snails, starfish). Oysters are subject to various diseases.

Coral reefs grow in warm (>18°C), clear, tropical or subtropical waters of average salinity. Corals are relatively fast growing because their tissues contain symbiotic zooxanthellae, photosynthetic organisms that live in the coral tissues and supply food to the corals. This association also helps the corals secrete their carbonate skeletons. Encrusting calcareous algae bind the coral skeletons together, forming the reef structure. The cavernous reef structure shelters many other kinds of organisms, the skeletons of which contribute to the reef mass.

Benthic organisms also grow on the deep-ocean bottom where food is especially scarce. Many of them feed by filtering the layer of suspended particles just above the bottom. These organisms have evolved special reproductive strategies to cope with the scarcity of food. In the midst of scarcity, there are infrequent falls of large amounts of food, in the form of dead whales or other large animals. These carcasses attract scavengers and also support ephemeral communities of specially adapted organisms, like those found at hydrothermal vents.

Dense growths of specialized benthic organisms grow near the hydrothermal vents on mid-ocean ridges. These include large gutless worms that depend on bacteria living in special organs to provide their food. Other organisms are not harmed by the hydrogen sulfide given off by the vents.

Key Terms

intertidal zone
succession
fouling
tide pool
infauna
epifauna
deposit feeder
filter feeder
unselective feeder
salt marshes
kelp
symbiosis
mariculture

Study Questions

1. Contrast conditions for life in benthic and pelagic environments.
2. Describe the major life-styles for benthic organisms. Give an example of an organism employing each life-style.
3. Contrast epifauna and infauna. Give an example of each.
4. Describe the role of marsh grasses in wetlands.
5. Discuss how the deep-ocean benthic environment differs from the inshore continental shelf environment.
6. Describe some adaptations of deep-ocean benthic organisms to their environment.
7. Describe the unusual features of organisms living around deep-ocean hydrothermal vents.
8. Why are benthic organisms less abundant with increasing distance from land?
9. Describe the succession of organisms on a bare surface in seawater.
10. Describe the different life environments on a rocky coast.
11. Why are relatively few organisms able to live in tide pools?
12. Describe how the type of bottom controls the abundances of filter and suspension feeders.
13. Discuss the symbiosis between algae and corals, and between bacteria and vent worms. How is each member in a symbiotic relationship benefited?
14. *[critical thinking]* How do oyster reefs effect water quality in an estuary?

Selected References

BOADEN, P. J. S., AND R. SEEDS, *An Introduction to Coastal Ecology*. London: Blackie, 1985.

JOHNSON, M. E., AND H. J. SNOOK, *Seashore Animals of the Pacific Coast*. New York: Dover, 1967. Nontechnical.

KAPLAN, E. H., *A Field Guide to the Coral Reefs of the Caribbean and Florida*. Boston: Houghton Mifflin, 1982.

MANN, K. H., *Ecology of Coastal Waters*. Berkeley, CA.: University of California Press, 1982. Intermediate to advanced.

RICKETTS, E. F., J. CALVIN, AND J. W. HEDGPETH (Revised by D. W. Phillips), *Between Pacific Tides*, 5th ed. Stanford, CA: Stanford University Press, 1985. A classic.

STODDART, D. R., AND M. YONGE, *The Northern Great Barrier Reef*. London: Royal Society, 1978.

THORSON, G., *Life in the Sea*. New York: World University Library, McGraw–Hill, 1971. Elementary, emphasizing marine environments.

OBJECTIVES

Your objectives as you study this chapter are to understand:

- The distinctive features of various types of coasts and coastal oceans
- How coasts and coastal oceans have changed through time
- Factors controlling coastal ocean processes
- How specific coastal ocean areas are affected by humans

A composite satellite image of western Europe shows the complexity and variety of coastal environments. Note, for example, the fjords of Norway and the low-lying islands off the Dutch coast.
(Courtesy European Space Agency.)

14 Coasts, Coastal Oceans, and Large Lakes

A **coastal ocean** is the shallow ocean lying between the land on one side and an open-ocean boundary current (Fig. 14–1) on the other. Because of the complex features of coastal oceans—a complexity caused by shallow bottoms and complicated shorelines—the processes that take place in them are highly variable in time and space. Some coastal areas are dominated by large river discharges, others may be ice-covered at times, and many coastal areas are densely populated and seriously affected by human activities. Many parts of the coastal ocean are partially isolated—Hudson Bay, for instance, or the Gulf of California—and these isolated areas have their own distinctive characteristics. Thus, it is difficult to generalize about the processes affecting the coastal ocean as we did for the open ocean.

Although coastal oceans occupy only about 8 percent of the ocean's surface and less than 0.5 percent of its volume, they are especially important to humans. About 90 percent of the world's fish catches come from coastal waters. Adjacent areas on land—called coastal zones—are home to about 60 percent of the world's population, most of its large and rapidly growing cities, the majority of its factories, and the richest agricultural lands. Unfortunately, wastes from these human activities end up in coastal oceans, causing numerous environmental problems.

We begin this chapter by studying the coastal oceans and the characteristic features of these waters. Then we consider the basic processes that affect coastal oceans. Finally, we consider some important coastal areas and the problems each area experiences as a result of human activities. Because very large lakes resemble coastal oceans in many ways, we include them in this chapter.

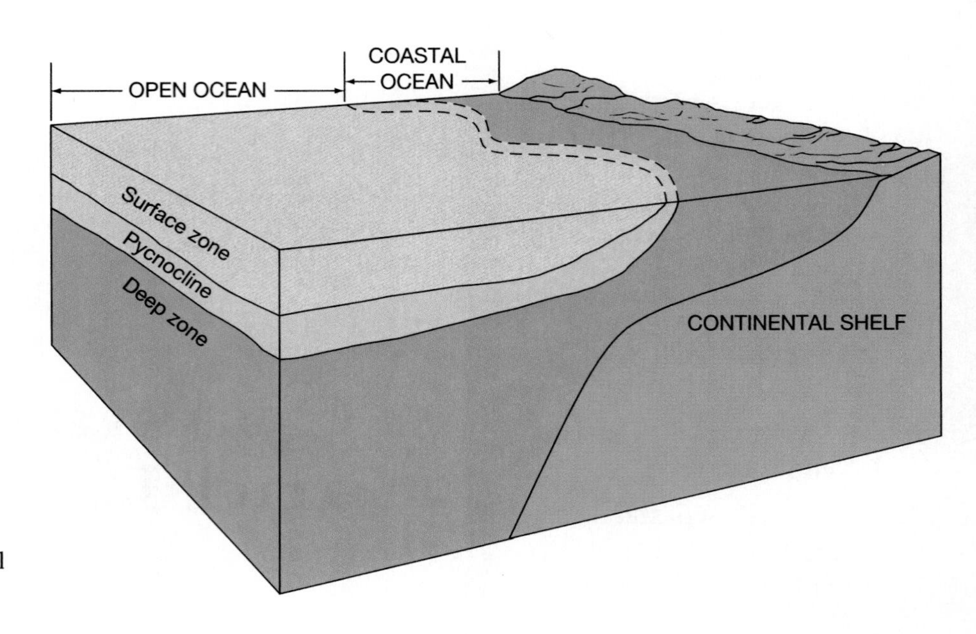

Figure 14–1
A simplified representation of the coastal ocean.

Shorelines

The landward margin of the coastal ocean is the **shoreline** (also called **coastline**); it is the place where land, air, and sea meet (Fig. 14–1). The most dynamic part of the ocean, it is shaped by tides, winds, waves, changing sea level, and human activities. The term **foreshore,** as shown in Figure 14–2, means all the land that extends from the lowest tide level to the highest point on land reached by wave-transported sands. The **backshore** is the landward limit of the beach complex. The influence of the coastal ocean extends much farther inland, however. One commonly used boundary is the 200-meter elevation, which is often used to mark the landward limit of the **coastal zone** which may be many kilometers inland.

Processes affecting shorelines act over periods ranging from a few seconds to thousands of years. Among the most obvious agents of change in the few-seconds category are waves, which break, rush up a beach face, and then quickly retreat. Along exposed open coasts there are few days without waves moving beach materials as they dissipate their energy. Usually these forces act most effectively under extreme conditions, such as storms, which strike most coasts every 10 to 100 years. While these intervals are dwarfed by the billion-year histories of continents, they often profoundly alter coasts, particularly beaches. For instance, the coastlines of southern New England were reshaped within a few hours by a hurricane in 1938. Storm surges (Chapter 9) also cause large changes in the shoreline and the coastal zone. Shorelines are continually altered by waves, erosion, or deposition of sand and gravel (forming beaches) or mud (forming deltas and marshes).

Changing Sea Levels

Advances and retreats of continental glaciers and the resulting sea-level changes shape and reshape coastlines over millions of years. When sea level stood at its lowest, about 18,000 years ago, the shore was near the edge of the continental shelf (Fig. 14–3). When continental glaciers began melting, releasing to the ocean waters that formerly were ice, sea level rose (Fig. 14–4) and the shoreline moved landward across the continental shelf until it reached its present level about 3,000 years ago. Most coastlines are still adjusting to the continuing sea-level rise.

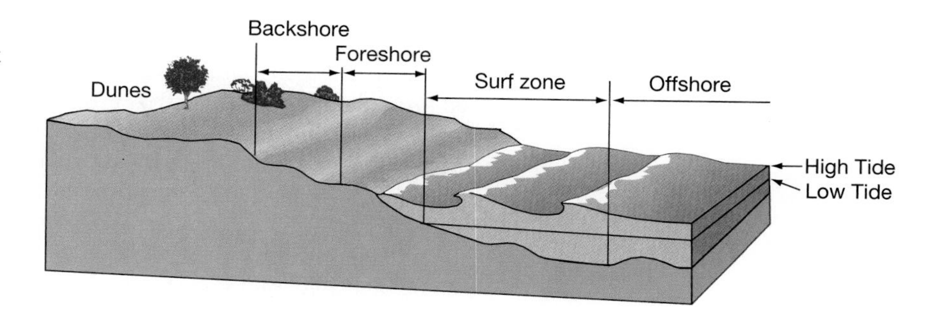

Figure 14–2
Profile of a typical beach and its adjacent coast.

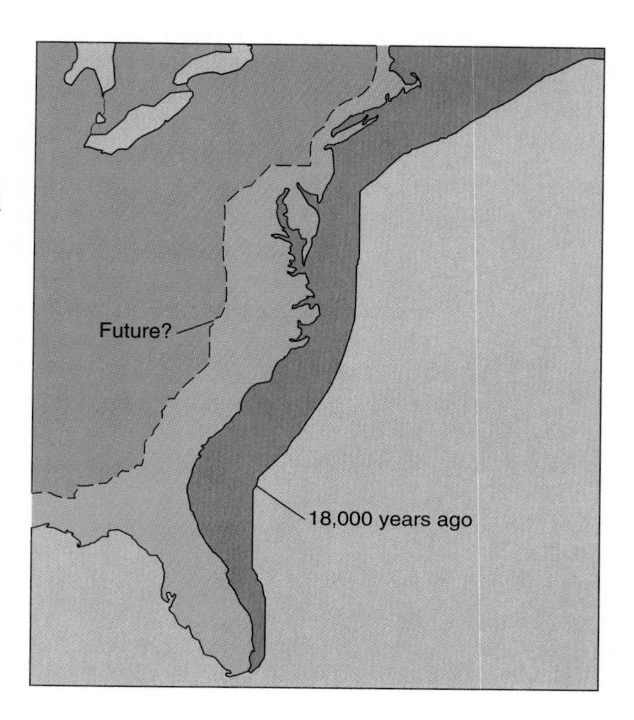

Figure 14–3
When sea level was at its lowest level, before the continental glaciers began melting about 18,000 years ago, the coastline of what is now North America was at the edge of the present continental shelf. Since then, the coastline has moved 200 kilometers to its present position. If all the ice in Antarctica and Greenland were melted, the coastal ocean would cover most of the present coastal plain. (K. O. Emery, U.S.G.S. Prof. Paper.)

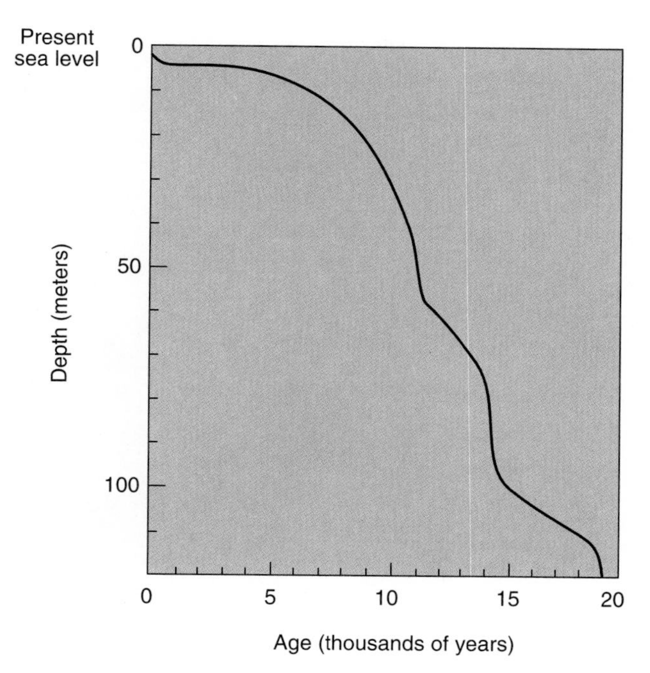

Figure 14–4
Changes in sea level as glaciers have melted during the past 20,000 years. [After R. G. Fairbanks: The age and origin of the Younger Dryas Climate event in Greenland ice cores. *Paleooceanography* 5(6):937 (1990).]

Sea level has not been at its present position long enough to reshape the features along all Earth's land margins, and as a result some of today's coastlines have been formed mainly by terrestrial processes. One example of a terrestrially formed coastline is drowned river valleys, which formed when the sea flooded the valleys of rivers that once cut across a now-submerged coastal plain to reach the ocean (Fig. 14–5). The land making up many of these former river valleys, although mostly submerged today, is little changed: it is now ocean bottom, but was shaped by terrestrial processes.

In glaciated regions, the ancient ice sheets cut deep valleys that have a U-shaped cross section (Fig. 14–6). Now filled by seawater, these former valley bottoms are concealed but have retained their characteristic U-shaped cross sections.

These exceptions notwithstanding, marine processes shape most coastlines. Waves have eroded soft rocks or sediments in the few thousand years since the sea reached its present level. Where bluffs of unconsolidated sand and gravel rise above the water level, they have been cut by waves. The sand and gravel cut from the bluffs are either deposited near the shore or else form small beaches or baymouth bars across bays and inlets. Sand is moved along coasts by longshore currents, forming **barrier islands** and **spits** that separate inlets and bays from the open ocean, as illustrated in Figure 14–7.

Spectacular coastline features are formed by volcanic eruptions. On the island of Hawaii, for instance, lava flows into the sea, enlarging the island (Fig. 14–8). In other areas, volcanoes have been worn down and are now covered by thick limestone caps and growths of corals, forming reefs and atolls (Fig. 14–9).

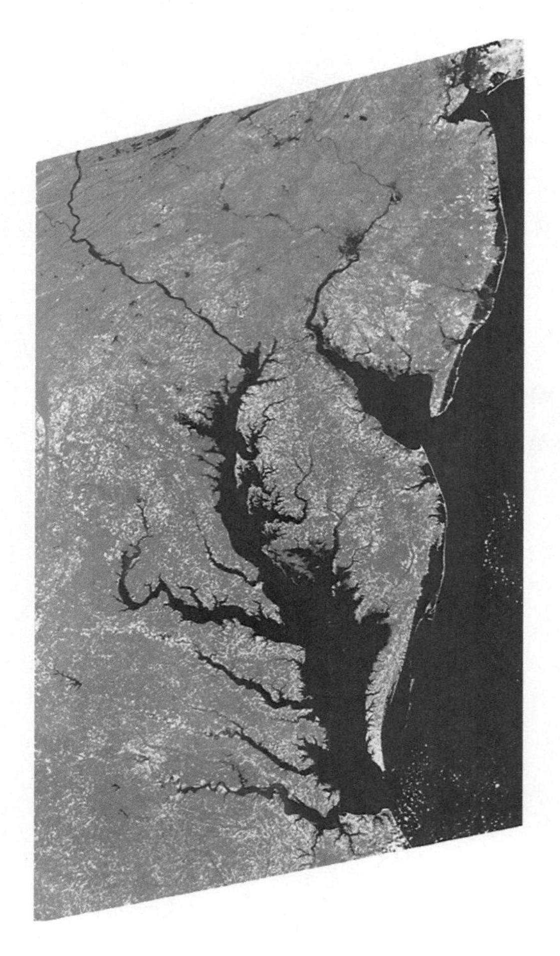

Figure 14–5
The eastern coast of the United States from New York City to southern Virginia is complex. Estuaries formed when rising sea levels flooded river valleys, including Delaware and Chesapeake bays. Such coasts are typical of stable continental margins. (Courtesy NASA.)

Figure 14–6
A U-shaped valley, formed by glacial erosion, has filled with seawater.

Figure 14–7
Barrier islands (light-colored narrow strips) partially isolate Pamlico Sound, North Carolina from the adjacent coastal ocean. (Courtesy NASA.)

(a)

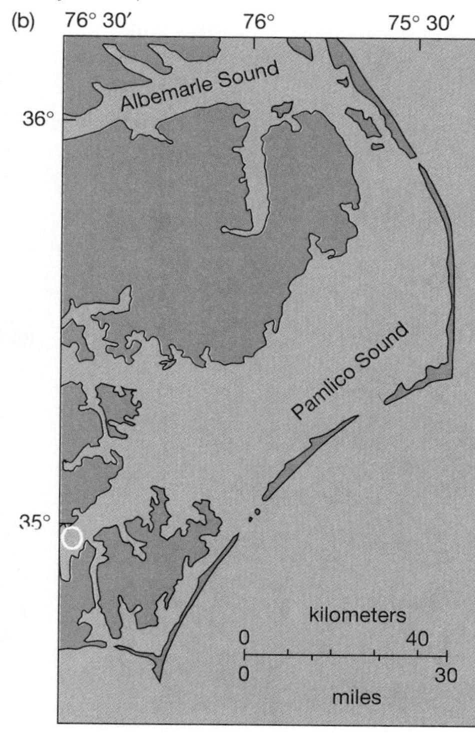

Figure 14–8
Lava flows into the sea from active volcanoes on Hawaii, constantly enlarging the island. (Peter French/Bruce Coleman, Inc.)

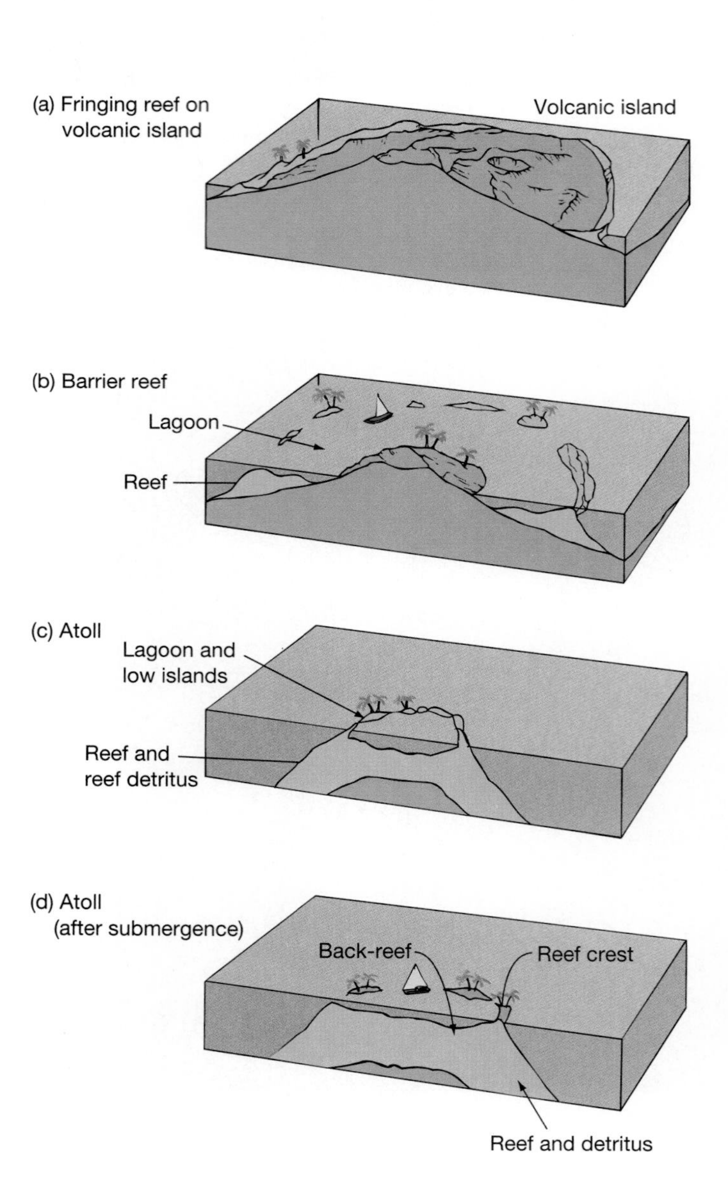

Figure 14–9
Evolution from a volcanic island with a fringing reef to an atoll.

Wetlands

Wetlands are prominent features of the world's coastlines (Fig. 14–10). **Salt marshes** (Fig. 14–11)—also called **wetlands** and one of the benthic habitats we considered in Chapter 13—are low-lying coastal areas that are exposed at low tide, submerged at high tide, and protected from direct wave attack. Such areas accumulate fine-grained sediments (muds and sands) either brought by streams or moved along the coast by currents. The surface of a salt marsh is generally overgrown by salt-tolerant plants and, to a casual observer, looks very much like a grass-covered meadow. If the supply of sediment is adequate, the plants in a salt marsh can trap it and the marsh can maintain its upper surface at sea level for thousands of years, in spite of the opposing processes of sea level rise.

A marsh's size and shape are determined by the depression in which it forms; vertical extent is controlled by the tidal range. The upper vertical limit is generally the highest level to which the spring tides can transport sediment. Most marshes are nearly flat-topped banks of sand or mixtures of sand and silt. The bank tops are commonly exposed at low tide and submerged at high tide. Where these bank tops lack vegetation, they are called **tidal flats.**

Cutting through salt marshes are meandering channels through which seawater enters the marsh as the tide rises and drains as the tide falls. The largest and deepest channels contain water even at low tide. These large channels connect with smaller ones that are often empty at low tide.

In tropical areas, **mangrove swamps** (Fig. 14–12) dominate intertidal wetlands. Mangroves are large, salt-tolerant plants with extensive root systems that form dense thickets. They are important as breeding grounds and provide shelter for the vulnerable juvenile forms of many marine animals. Some organisms are especially adapted to survive in this environment. Mangrove oysters, for instance, attach themselves to roots and branches that are exposed at low tide. The extensive roots of mangroves trap sediment particles and organic debris; eventually mangrove swamps are filled in and replaced by forests of less salt-tolerant trees.

Figure 14–10
Coastlines of the world showing the distribution of intertidal wetlands.

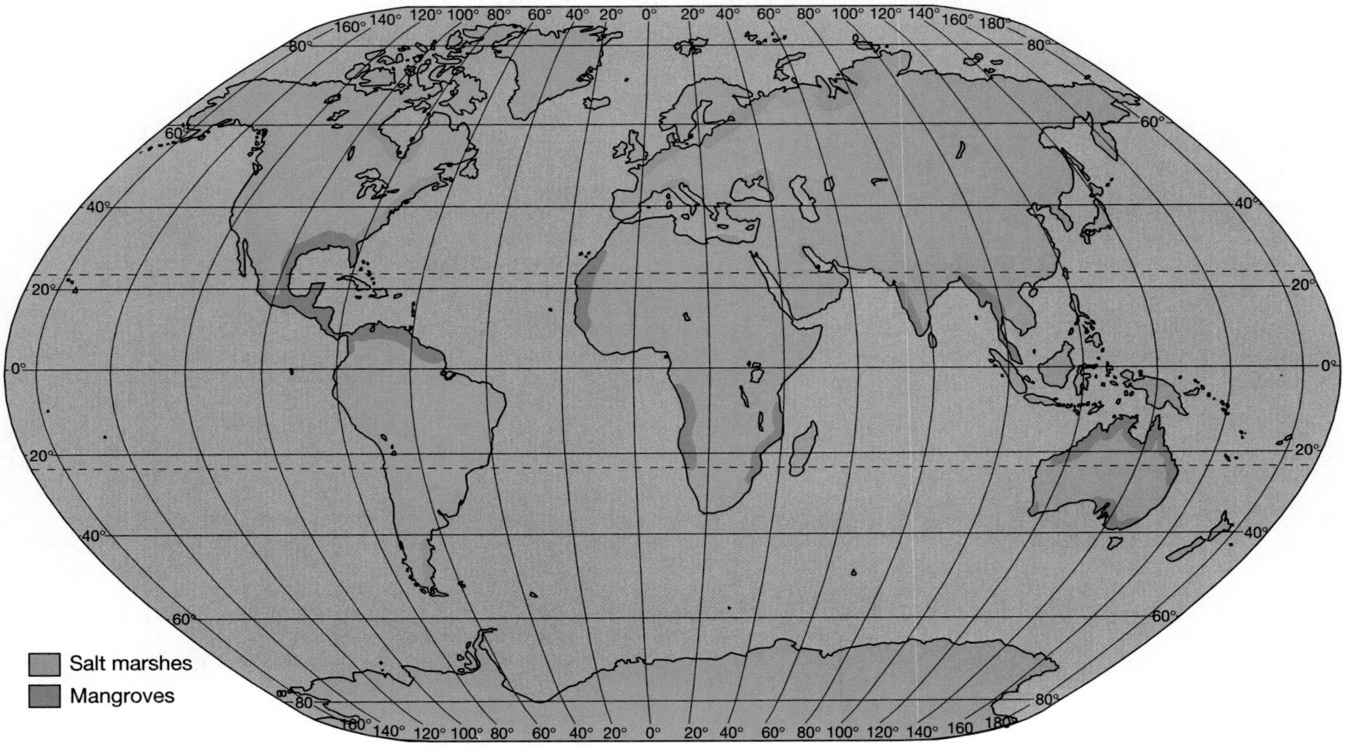

Figure 14–11
A Nova Scotian salt marsh at near high tide.

Figure 14–12
Mangroves at Ten Thousand Islands, Florida. Note the conspicuous root systems projecting above the water surface. Sediments and organic matter are trapped among the roots. (Richard H. Chesher/Photo Researchers, Inc.)

For centuries, humans considered wetlands and mangrove swamps to be wastelands. Large wetland areas were "reclaimed" to provide land for farming or for building industrial developments, airports, waste dumps, or resorts and residences. This misguided practice still continues today but at a greatly reduced pace. Now the role of wetlands in ecosystems is better understood and appreciated. Some "reclaimed" wetlands are being restored to their original state in order to protect the coastline against erosion by waves or to protect the nursery habitats for fish and shellfish populations.

Coastal Currents

Coastal currents (Fig. 14–13) flow parallel to shorelines. These currents are primarily driven by winds and by the fresh water discharged by rivers; they can form, disappear, and/or change directions within a few hours or a few days. Coastal currents are strongest when river discharges are large and winds strong. Winds often control them, especially during storms. For example, surface ocean waters can be blown shoreward and held there by winds, depressing the pycnocline. A sloping sea surface results, creating geostrophic currents paralleling the coastline. (These are essentially the same processes we discussed in Chapter 8.)

As mentioned at the beginning of the chapter, coastal oceans are bounded by open-ocean boundary currents. **Fronts** near the edge of the continental shelf often mark this boundary (Fig. 14–13). Along many coasts, nearshore currents move in a direction opposite that of the offshore boundary currents. When winds diminish or freshwater discharges are low, coastal currents weaken or disappear. Such conditions are especially common in summer (gentle winds, little river discharge); with the onset of winter storms, coastal currents can re-establish themselves within a day or two.

Waters discharged by rivers move slowly across the coastal ocean to mix into the open ocean. To estimate the amount of time required to replace a parcel of water, we use the concept of **freshwater residence time.** For example, in the coastal ocean between Cape Cod in Massachusetts and Cape Hatteras in North Carolina, the lowered salinity of the surface waters indicates that about 2.5 years' worth of freshwater discharged by rivers is present. Other coastal-ocean areas

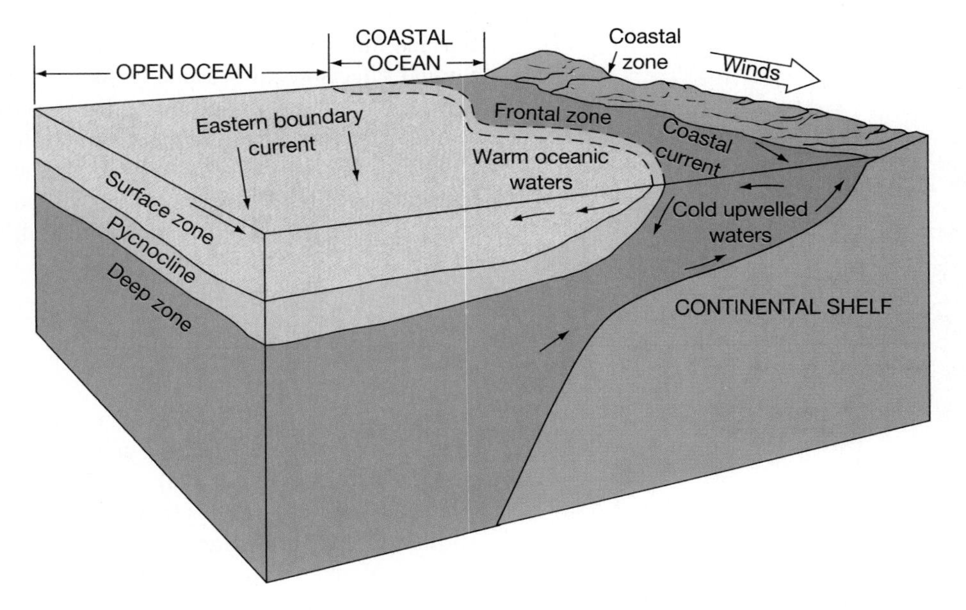

Figure 14–13
Coastal-ocean circulation in a region having an eastern boundary current and wind-induced upwelling. Remember that water moves to the right of the wind in the Northern Hemisphere.

have shorter freshwater residence times—about one month for Narragansett Bay, about three months for the Bay of Fundy (Atlantic Canada) and for Delaware Bay, and about six months for Chesapeake Bay. When we discuss pollution problems in these waters, the freshwater residence times are useful indicators of the time required to flush waterborne pollutants from these systems.

Coastal-Water Temperature and Salinity

In coastal waters, large changes in temperature and salinity occur over small distances and short time periods. Winds coming from continents are much warmer than the ocean surface in summer and much colder (and drier) in winter. In contrast, winds blowing from ocean to land reduce temperature extremes and cause heavy rainfall. (We discussed the processes involved in Chapter 6.)

Because nearshore fresh water discharged by rivers is mixed into the ocean, the lowest salinities in the ocean occur in coastal waters near the mouths of large rivers and near melting sea ice.

Salinity extremes occur where coastal waters are isolated. Where evaporation rates are high—for example, in the eastern Mediterranean and in the northern Red Sea—surface-water salinities are high. In both areas, surface salinities exceed the average for open waters in their latitude.

Extreme temperatures also occur in isolated coastal waters. Surface-water temperatures exceed 40°C in the Arabian Gulf and in the Red Sea during summer, whereas those in the open ocean rarely exceed 30°C. The lowest surface-water temperatures are controlled by seawater's temperature of initial freezing (Chapter 4) at about 20°C. In winter, sea ice forms in high-latitude coastal areas, particularly in shallow bays and lagoons. There, salinities are low due to river discharge and cooling is rapid because of the relatively large surface areas and small water volumes involved. Marked seasonal temperature differences occur in coastal ocean waters, particularly at mid-latitudes. Coastal waters are mixed by storms, with cold waters able to mix all the way to the bottom during winter. Offshore waters never get as cold as continental-shelf waters, because in the open ocean surface cooling is distributed throughout a thicker water column.

Warming and cooling of surface waters (Fig. 14–14) during the day is easily detected, especially in areas where the water is protected from wind and waves. Surface-water temperatures are highest in mid-afternoon and lowest at dawn. Similar effects occur seasonally. As incoming solar radiation increases in spring, surface waters warm, deepening the surface layer; as summer progresses, a pronounced thermocline develops. Conversely, during autumn, surface waters cool by radiation, by evaporation, and by mixing with deeper waters, the latter caused primarily by storms. When winter begins, coastal waters are thoroughly mixed and there is little temperature or salinity difference between surface and bottom waters.

Estuaries and Fjords

Among the important features of coastal oceans are embayments. One of the most conspicuous is the **estuary**—a semienclosed embayment where fresh water discharged by rivers mixes with seawater (Fig. 14–15). The resulting low-salinity waters flow out of estuaries to mix with coastal ocean waters. These plumes of low-salinity waters can be traced for tens to hundreds of kilometers, depending on the amount of river water discharged. On stable continental margins (Fig. 14–5), many estuaries occupy flooded, ancient river valleys. These are called **coastal-plain estuaries.**

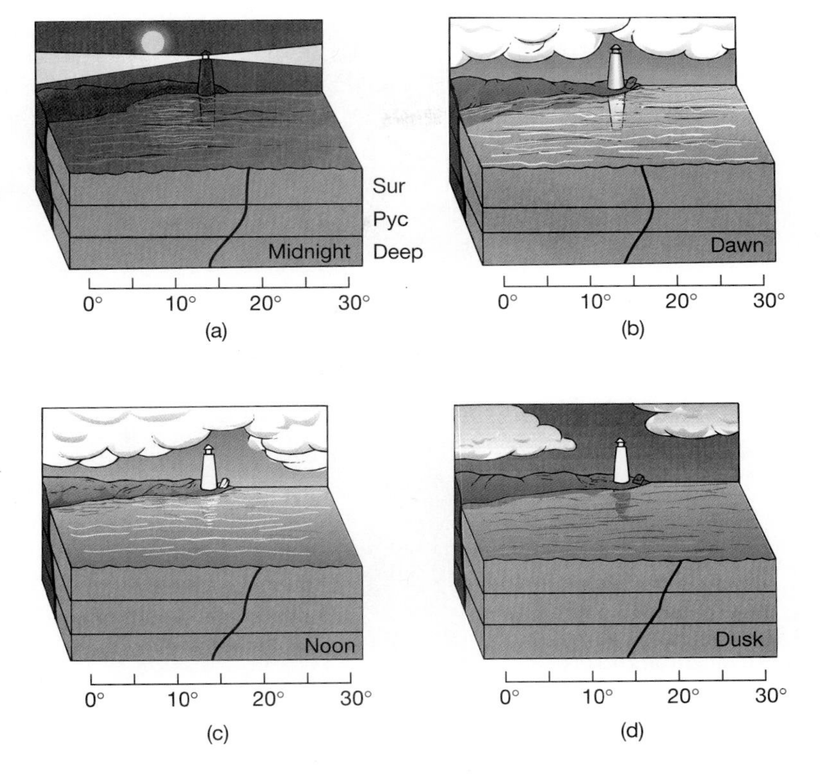

Figure 14–14
Daily variations in surface-water temperatures. At midnight (a), the surface zone is nearly isothermal. The water cools throughout the night, so that by dawn (b) the surface layer is cooler than water in the pycnocline. After the sun rises, surface waters are warmed, and by noon (c) the surface layer is warmer than deeper waters. Warming continues until late afternoon, when surface temperatures are highest (not shown in the illustration). Later, the surface layer begins to cool and is slightly cooler at dusk (d) than at late afternoon. During the night, the surface zone cools until it is again isothermal around midnight. Daily temperature changes are usually confined to the upper 10 meters or less. [After E. C. LaFond: "Factors affecting vertical temperature gradients in the upper layers of the sea." *Science Monthly* 78(4):243 (1954).]

In coastal areas undergoing active mountain building, such as the western coast of North and South America (Fig. 14–16), there are few estuaries. Estuarine systems on unstable (Pacific-type) margins also form where young mountain ranges partially isolate low-lying areas. Where these waterways connect with the coastal ocean, they form **tectonic estuaries.** San Francisco Bay, Puget Sound, and the Strait of Georgia are examples on the Pacific coast of North America. Such tectonic estuaries are also common on Asia's Pacific coast, another place where mountain building is active. Where a mountain glacier flowed into the sea during some past ice age, the flooded previously glaciated valleys, called **fjords,** are usually deep and steep-sided.

Figure 14–15
Mobile Bay, Alabama is a large coastal-plain estuary, which discharges into the Gulf of Mexico. Barrier islands parallel the coastline from Pensacola, Florida (on the right) to Biloxi, Mississippi (on the left). (Courtesy NASA.)

Figure 14–16
The western North American coast is dominated by mountain building and, as a result, has few estuaries. Puget Sound, the Strait of Juan de Fuca, and the Strait of Georgia occupy a large depression that has been modified by recent glaciation. (Courtesy NASA.)

Circulation in Estuaries and Fjords

Estuaries and coastal oceans receiving large amounts of freshwater exhibit distinctive current patterns and processes caused by the mixing of fresh and salt waters. This mixing results in long-term average flows seaward in the surface layers and net landward flows along the bottom. This two-way movement of water is called **estuarine circulation,** even though it is quite common in many coastal oceans. Let us see how it develops.

In an ideal estuary (no friction, no tides, no winds), water flows in from a river at the head of the bay (Fig. 14–17) and, being less dense than the seawater, spreads out over the denser salt water. Once this happens, a pycnocline zone, controlled by the halocline or marked salinity discontinuity, separates the two layers (Fig. 14–18). The subsurface salt water intrusion into such an estuary is wedge-shaped, with its thin end pointed upstream, and so is called a **salt-wedge estuary.**

Simplified salt-wedge stratification is observed only where river flows are large and the tidal range is small, or where large rivers discharge through relatively narrow channels, such as the Mississippi or the Columbia rivers during floods. In these cases, the large river flows are much stronger than the tidal currents, and so the river flows dominate the estuarine circulation. The residence time of fresh water in such salt-wedge estuaries is quite short. A nearly ideal salt-wedge estuary is formed, one in which relatively little salt water is mixed into the upper layer. Nearly fresh water is discharged into the coastal ocean, where it mixes with ocean waters over the continental shelf. The most striking example of this is the Amazon River outflow, where the freshwater is carried many hundreds of kilometers away from the river mouth before mixing with the salt water.

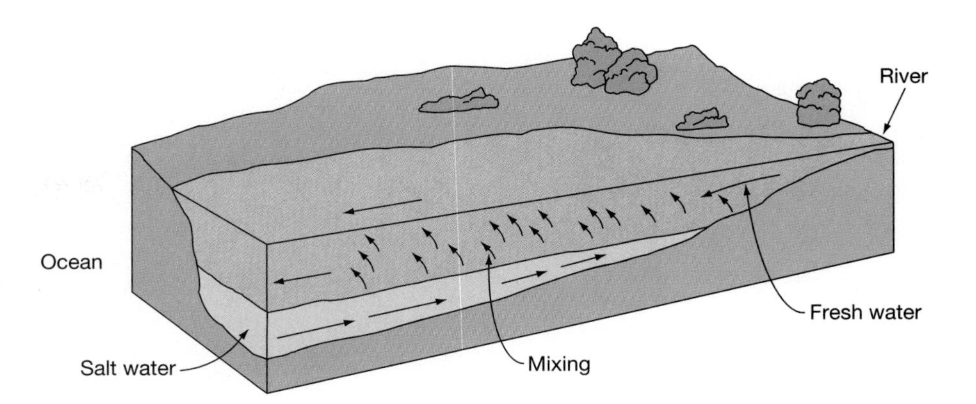

Figure 14–17
A simple salt-wedge estuary, showing the two-layered structure. Salty subsurface waters flow landward, and less saline surface waters flow seaward. [After D. W. Pritchard: "Estuarine circulation patterns." *American Society of Civil Engingeers Proceedings* 81:717 (1955).]

As the amount of fresh water discharged into an estuary diminishes, tidal effects become more important. An estuary may act like a salt-wedge estuary only during floods. During low-flow periods, the same estuary is usually strongly influenced by tides and tidal currents. Then more mixing occurs than in a simple salt-wedge stratification, and the system is called a **moderately stratified estuary** (Fig. 14–19). As we shall see, moderately stratified estuaries have longer freshwater residence times and tend to retain materials discharged to them longer. Consequently they are more readily affected by waste discharges.

The simple situation described above for a salt-wedge estuary is complicated in a moderately stratified condition by several factors. First, friction between seaward-moving fresh water and the underlying seawater causes currents, as shown in Figure 14–20, and, as a result, the wedge shape is not found and the pycnocline is much less pronounced. In this moderately stratified situation, more saline water from below is entrained (dragged up) and mixed with the lower salinity surface layer. Because of these processes, the newly mixed water now contains more salt than it did before; however, it is still less dense than the underlying unmixed water and so cannot re-enter the lower layer. This relatively low salinity layer continues to flow seaward and ocean water flows into the estuary along the bottom to replace that which has been drawn upward and carried seaward.

Each volume of fresh water entering an estuary from a river ultimately mixes with several volumes of seawater, and surface water salinities increase seaward. Landward flows of seawater along the bottom are much greater in volume than river discharges of fresh water into the estuary.

Figure 14–18
Variation in salinity and current speed with depth in a salt-wedge estuary.

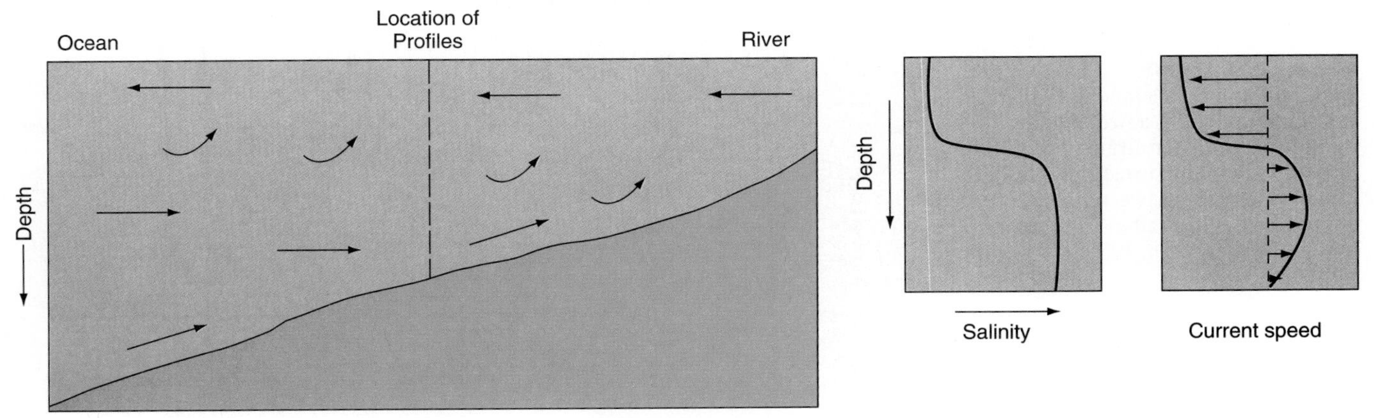

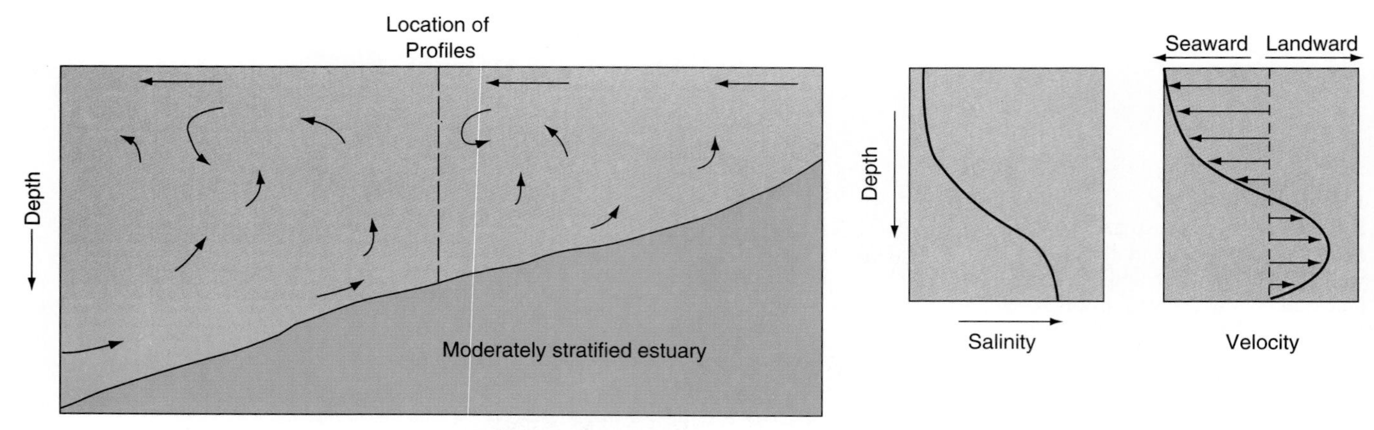

Figure 14–19
Variation in salinity and current speed with depth in a moderately stratified estuary.

Tidal flow into the estuary causes turbulence (irregular, chaotic flow) throughout the water column, and this turbulence increases mixing. Consequently, more salt water is transferred from the subsurface to the surface layer. Some fresh water from the surface also mixes downward, and salinity decreases in a landward direction in both the surface and subsurface layers. As in the salt-wedge estuary, however, there is a net landward flow in the subsurface layer, replacing salt water lost from the system. Seaward-flowing surface layers remove both freshwater and salt.

Consider the volume of water flowing in the two layers. River water has almost no salt as it enters the estuary's surface layer, whereas seawater moving in along the bottom from the ocean typically has a salinity of 33. Mixing equal volumes of fresh and salt water therefore gives a salinity of 16.5. By the time surface-water salinity reaches 30, one volume of river water has mixed with 10 volumes of landward-flowing seawater. In other words, the volume of surface water moving seaward is 11 times the river discharge to the estuary (Fig. 14–21).

Tidal currents dominate most estuaries. Thus, to observe estuarine circulation, one must measure and average currents over several tidal cycles. Where tidal effects are relatively strong, waters in estuaries are less stratified. In estuaries in which river flows are small but tides and tidal currents are large, the waters may be mixed almost completely from top to bottom. These are called **well-mixed estuaries.**

Because of their great depths and irregular bottoms, fjords have more complicated current patterns. Many fjords have a **sill,** or submerged ridge, at the

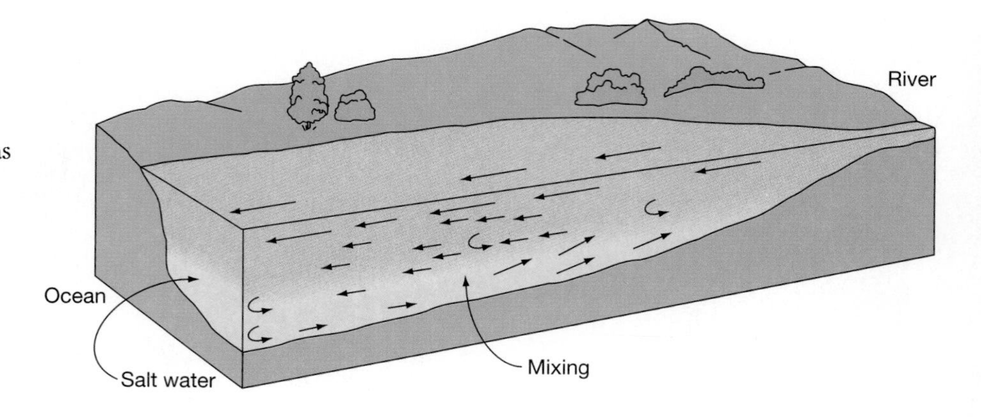

Figure 14–20
Water movement in a moderately stratified estuary. There is a net landward movement in the subsurface layers and a seaward flow in the surface layers. Note that the boundary between layers is not as pronounced as in a salt-wedge estuary. (After D. W. Pritchard, 1955.)

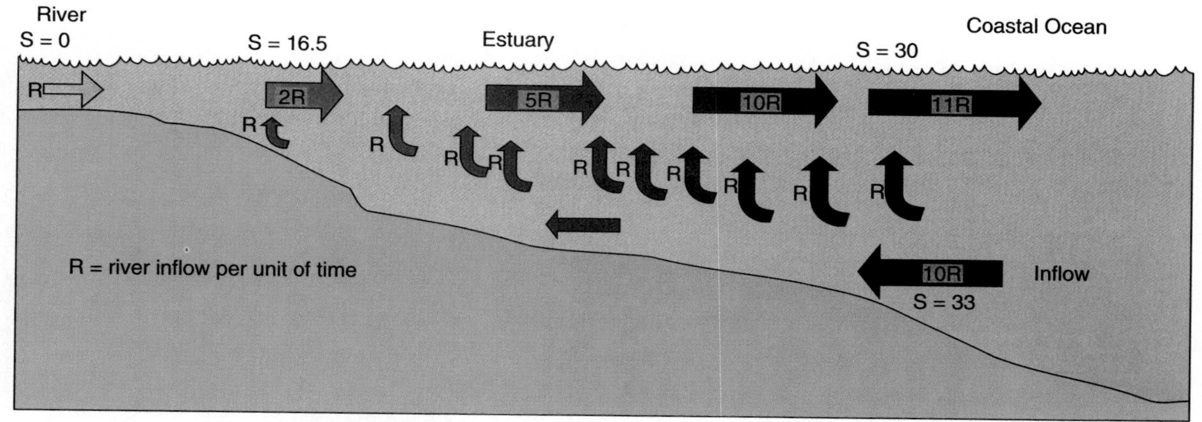

Figure 14–21
In an estuarine system, the volume of seawater drawn upward from below and carried to the ocean is ten times higher than the original volume of the river discharge (R). To replace this outflowing seawater, an equal volume of seawater (10R) moves from the ocean into the estuary. Estuarine circulation thus draws subsurface, high-salinity water into the surface layer; it is essentially an upwelling process.

entrance, left there by the glaciers that cut the fjord. This sill isolates deeper waters from the ocean outside, as can be seen in Figure 14–22. Fresh water flowing into a fjord forms a low-salinity surface layer that moves seaward, in a typical estuarine circulation. This layer usually involves only the waters above the top of the sill, sometimes only a few tens of meters thick.

Although the deepest waters in a fjord may be almost completely isolated from the surface circulation, they are affected by conditions outside the estuary, primarily by density differences. If the water at or slightly above the sill outside the estuary is denser, it will flow into the fjord. There it displaces the deeper waters in the fjord, which then move out. Strong winds can also disturb the deep waters by setting up seiches, which cause the waters to move back and forth. (See Chapter 9 for a discussion of standing waves and seiches.)

Lagoons

Lagoons are shallow embayments in the land along a coast, usually separated from the coastal ocean by offshore barrier islands paralleling the shoreline. Such lagoon–barrier-island systems are common all along the U.S. Gulf coast (Fig. 14–15) and along the U.S. Atlantic coast south of Long Island (Fig. 14–5). They also occur on many other stable continental margins.

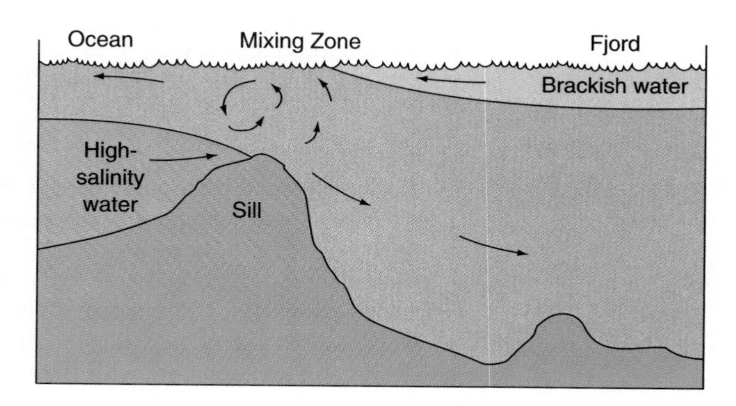

Figure 14–22
Water circulation in a fjord. A water mass forms by mixing surface and subsurface waters while flowing over the sill. [After M. Waldichuk: "Physical Oceanography of the Strait of Georgia, British Columbia." *Journal Fisheries Research Board, Canada* 14(3):321 (1957).]

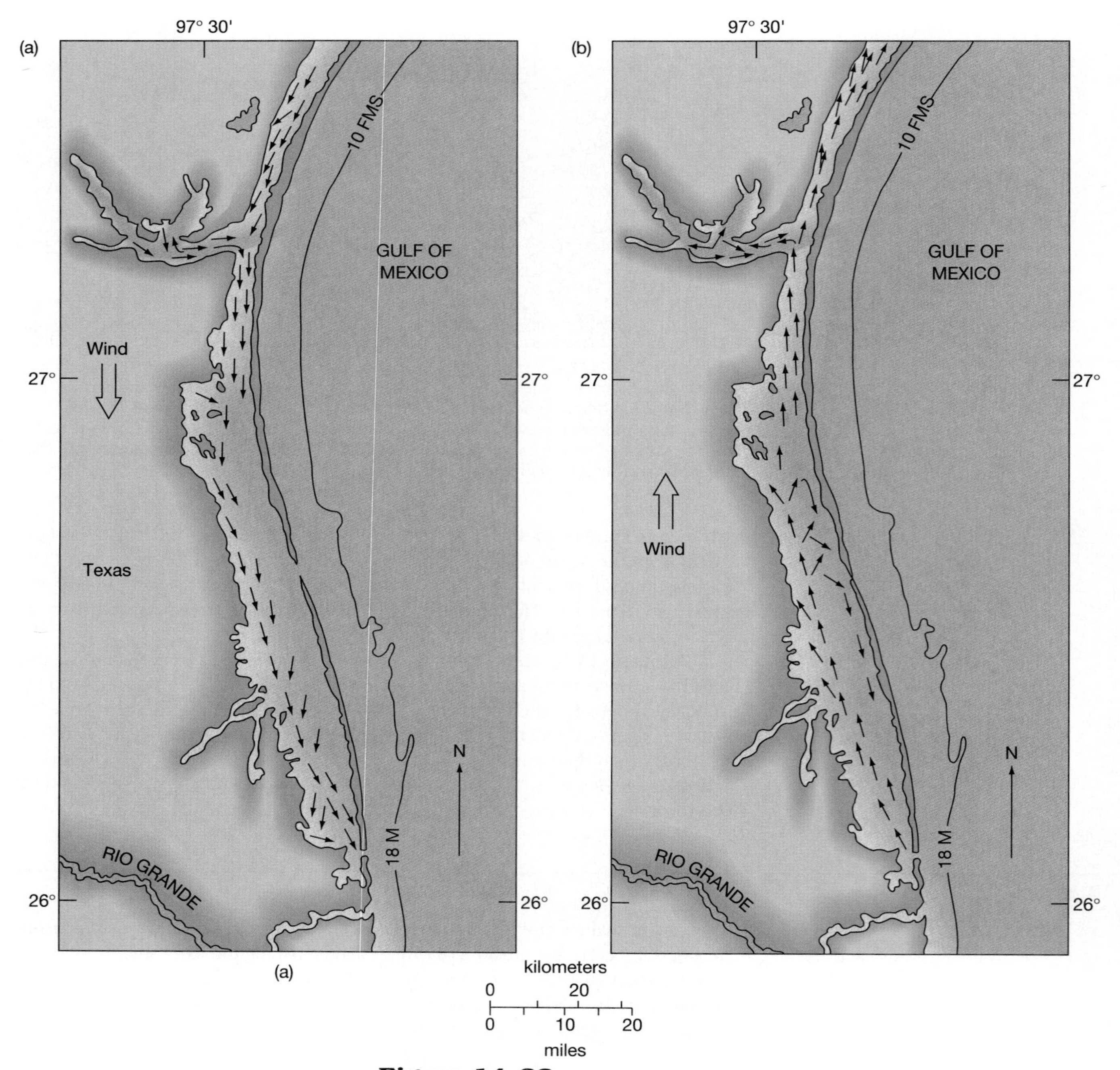

Figure 14–23
Surface currents in Laguna Madre, Texas (*laguna,* Spanish "lagoon"). (a) North wind blowing. (b) South wind blowing. (After G. A. Rusnak, "Sediments of Laguna Madre, Texas," in F. P. Shepard et al. (eds.): *Recent Sediments, Northwest Gulf of Mexico.* Tulsa: American Association of Petroleum Geologists, 1960.)

Narrow inlets cut through barrier islands and permit tidal currents and winds to move water in and out of lagoons (Fig. 14–23). Some lagoons are almost completely cut off from the nearby coastal ocean. Therefore they often have almost no tides and are especially vulnerable to pollution problems. Salinities are usually quite variable, and winds dominate both currents and water levels.

The shallowness of most lagoons means they are strongly influenced by evaporation and precipitation, with the result that both water temperatures and salinities are highly variable. Lagoons on arid coasts often have extensive salt flats and many lagoons have been made into salt pans to produce salt. Where water quality is good, lagoons are often used for mariculture, some having been turned into fish farms.

The shallow bottom of a lagoon also supports bottom-dwelling plants and animals eaten by ducks and other birds. Lagoons can support large populations of birds and other wildlife.

Many lagoons receive large amounts of agricultural and municipal waste carried by the streams that drain into them. In some cases, the nutrients in these waste discharges stimulate phytoplankton production, so that fish production is relatively large. If excessive production is stimulated, the bottom waters may become anoxic because the decomposing organic matter consumes all the dissolved oxygen. The result is that bottom-dwelling organisms are killed off. Excessive waste discharges can also cause red tides. In short, lagoons are extremely sensitive to natural environmental changes and vulnerable to human activities.

Despite their inherent fragility, few lagoons are adequately protected. Most of them have suffered as nearby populations have increased and both industry and agriculture have expanded. Today, lagoons near densely populated areas are at risk worldwide.

Upwelling

We discussed the processes causing upwelling in Chapter 8. Now we examine why coastal upwelling areas—eastern margins of ocean basins, equatorial zones, and around Antarctica—are so rich in marine life.

First we examine coastal upwelling, illustrated in Figure 14–13. As you recall, the pycnocline is nearest the surface in eastern boundary areas and deepest near western boundaries. Upwelling is caused by winds blowing parallel to the coast, resulting in the seaward movement of surface water. (This is the Ekman transport discussed in Chapter 8.) Because the pycnocline is relatively shallow along eastern margins, waters upwelled from depths of 100 to 200 meters come from below the pycnocline (Fig. 14–24). These waters are rich in nutrients, usually low in dissolved oxygen, and colder than surface waters. (In places where the pycnocline is not relatively shallow, upwelled waters come from the surface zone and contain few nutrients. This happens during *El Niños,* as we saw in Chapter 7.)

Landward-flowing upwelling waters receive particles sinking out of the seaward-flowing surface waters above them. The particles decompose in the upwelling waters, and the nutrients from them are quickly recycled into surface waters along with nutrients from below the pycnocline. Also, organisms, larvae, and spores sink into the upwelling waters to seed them. Newly upwelled waters often contain little phytoplankton, but after a day or so, the seed cells recently deposited in the upwelled waters have reproduced. As a result, organisms are 50 to 100 times more abundant in upwelling areas than in nearby nutrient-poor surface waters. In some upwelling areas, the waters contain so many plants and animals that filters on cooling-water intakes for boat motors quickly clog.

Fish and other animals take advantage of this abundance of food. Herbivorous filter-feeders (sardines, anchovies) live in areas of greatest algal production (Zone 2 in Fig. 14–24). Farther downcurrent, in Zone 3, other herbivores (copepods) feed on the algae. Carnivorous fish (mackerel) then feed on the herbivores. The largest zooplankton (euphausiids) live near the edges of continental shelves along with the larger fish (hake) that feed on them.

Little energy is lost in these short, simple food webs. In contrast, typical open-ocean food webs are long and complicated, with much energy lost in the many transfers among trophic levels and total efficiency much lower than in upwelling systems. Whalers learned to exploit the high productivity of upwelling areas by hunting whales downcurrent from them.

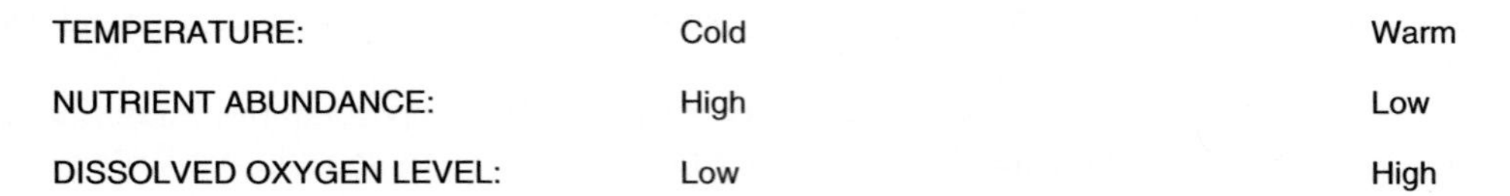

	UPWELLED WATERS	SURFACE WATERS
TEMPERATURE:	Cold	Warm
NUTRIENT ABUNDANCE:	High	Low
DISSOLVED OXYGEN LEVEL:	Low	High

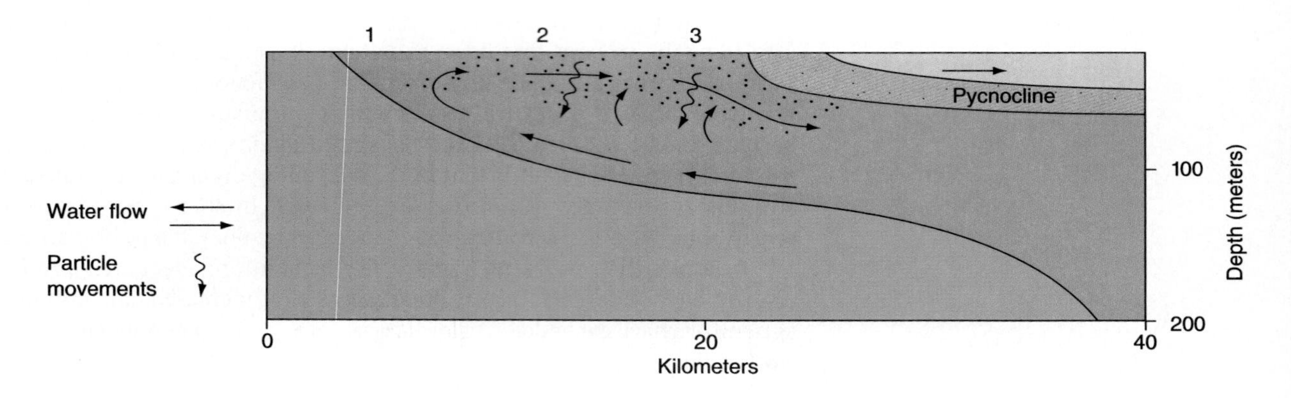

FEATURES OF UPWELLED WATERS

ZONE:	1	2	3
PHYTOPLANKTON			
Abundance:	Low	High	Low
Cell size:	Small	Large	Small
ZOOPLANKTON			
Size:	Absent	Large	Largest
FISHES	Absent	Herbivores (Sardines, Anchovies)	Carnivores (Mackerel, Hake)

Figure 14–24
Schematic representation of the abundance and distribution of nutrients and marine organisms in a coastal upwelling area.

Upwelling in the coastal ocean occurs frequently at capes (Fig. 14–25). The upwelled cold waters and associated large fish populations make these areas preferred locations for fish-eating birds and fishing boats. Until recently, reasons for this association were unknown, but with the help of satellite observations, oceanographers have been able to answer this question. Satellite images of sea-surface temperatures show complex features located near capes; these large complex structures, called jets and squirts, extend seaward from the capes. Not only is upwelling intensified as a result of these capes, but upwelled waters are moved seaward. These jets and squirts form when a current is deflected seaward upon encountering the submerged ridge associated with either a cape or a headland, easily seen off the Peruvian and Northwest African coasts (Fig. 14–25). This deflection causes strong upwelling downcurrent of the ridge (such a ridge is easily detected by the upwelled cold waters).

After the initial seaward deflection caused by the ridge, the currents bend back toward the coast, forming sharp bends. The heads of such features are unstable and form pinched-off eddies that move offshore. (These are eastern boundary currents equivalent to the much larger rings formed by western boundary

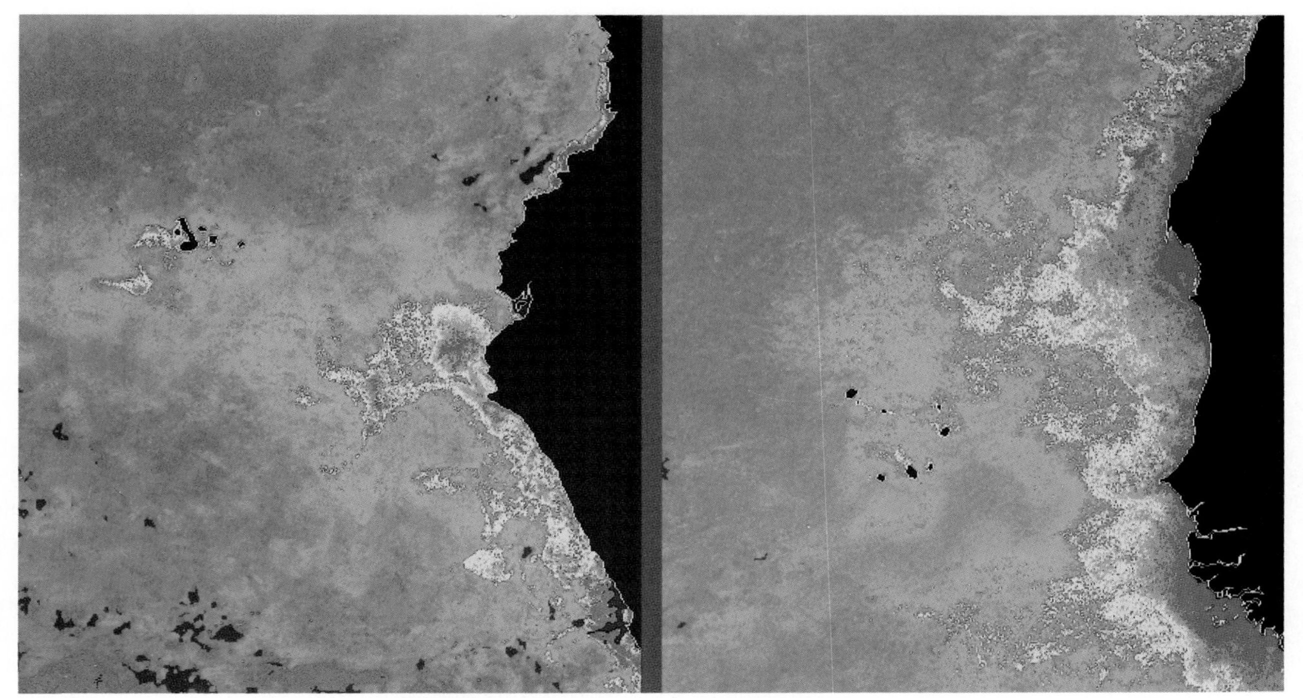

Figure 14–25
Upwelling makes the waters off Peru (left) and northwestern Africa (right) among the most productive in the world. Note the association between the capes and the chlorophyll-rich zones (reds and yellows). Also note that the upwelling is displaced downcurrent from the capes. The currents come from the south (bottom) off Peru and from the north off Africa. (Courtesy NASA.)

currents.) These current patterns persist and move downstream 50 to 150 kilometers, depending on the size and shape of the ridge. In this way, the complexities of shoreline and ocean bottom mold coastal currents into complicated shapes, easily seen by satellites but impossible to detect from ships.

Environmental Problems in Coastal Waters

Sewage, industrial wastes, and agricultural runoff are discharged in large volumes to streams and estuaries, eventually reaching the coastal ocean. Such wastes often cause environmental problems, both in the water where the initial discharge takes place and in the coastal ocean that is its final destination, either killing organisms outright or else rendering them unfit for human consumption. Waste discharges can also affect human health directly through exposure while swimming.

Some disposal systems discharge directly into the coastal ocean. These systems should be designed so that the discharge enters the ocean at a point where currents are swift and wastes are thoroughly mixed with seawater. If a system operates properly, wastes are quickly diluted and their concentrations are low at short distances from the discharge points. Continued movement and mixing by currents dilute the wastes further. Around the United Kingdom, for instance, sewage is discharged to coastal waters where strong tidal currents quickly dilute it and mix it into the nearby coastal ocean.

Another way to dispose of wastes is to discharge them offshore into waters below the permanent thermocline, as is done offshore from Los Angeles and

San Diego. The discharge waters remain at depth and are thus isolated from surface waters, where they might come in contact with people.

Waste disposal causes environmental problems in many coastal areas because estuarine circulation patterns tend to retain wastes near the discharge site rather than mixing and dispersing them.

Treatment methods which remove particles and nutrients from wastes before discharge have improved water quality in rivers, estuaries, and coastal waters. For instance, higher levels of treatment for sewage discharged in the Thames River near London have resulted in improved water quality, so that salmon are now caught in the estuary. Water quality in the upper Potomac River near Washington, D.C., has also improved as a result of increased treatment of sewage discharges.

Many waste materials are either particulate or quickly become attached to particles. These particles usually settle out near the discharge point. Thus, waste deposits near a discharge frequently contain high concentrations of carbon, bacteria, and viruses, as well as oils and waste chemicals. Such materials are also abundant in deposits dredged from urban harbors. Often, this dredged contaminated material is placed in isolated, enclosed areas rather than being dumped back into estuaries or onto continental shelves.

When water circulation near waste deposits on the bottom is sluggish, animals and bacteria in the deposits consume oxygen and deplete the dissolved-oxygen supply of overlying waters. This happened in 1976 over large areas of the continental shelf near New York and New Jersey. When the dissolved oxygen was used up, both benthic organisms and bottom-dwelling fishes died. Hydrogen sulfide, formed by anaerobic bacteria, was widespread in near-bottom waters. Commercially valuable organisms worth millions of dollars died.

In some coastal-ocean areas, depletion of dissolved oxygen in bottom waters occurs naturally because of the high productivity of surface waters and sluggish circulation of bottom waters. For example, such naturally occurring anoxic events occurred in Chesapeake Bay and near the mouth of the Mississippi River before European colonists arrived. Increased waste discharges increase the frequency and severity of such anoxic events, however. Similar processes occur in rivers and lakes.

Bacteria, viruses, and possibly fungi living in contaminated sediment deposits can cause disease in benthic animals. In many urbanized coastal areas, bottom-dwelling fishes and lobsters have large sores. Eventually fins or carapaces are eaten away, a condition called fin rot. Such conditions occur only in the most contaminated areas. Their causes are not known.

Waste disposal in coastal waters often makes shellfish unfit for human consumption. Filter-feeding organisms, such as oysters and clams, concentrate bacteria and viruses as well as various particle-associated materials in their digestive tracts. Once found to be contaminated in this way, these organisms cannot be sold. Although their muscles are not contaminated, these pathogens in the guts of mussels or oysters can cause disease when the whole animal is eaten raw. Large areas of once-productive clam and oyster beds are closed to commercial harvesting because the overlying waters are polluted. Indeed, consumption of raw shellfish is a major cause of gastroenteritis and hepatitis in many developed countries.

A tragic example of human health problems arising from exposure to industrial wastes occurred in Minamata, a Japanese fishing village. Between 1953 and 1960, mercury-containing wastes were discharged into coastal waters near shellfish beds harvested for food. Bacteria changed this mercury into forms that were concentrated in the tissues of shellfish. Persons who ate the contaminated shellfish developed mercury poisoning, which was most serious in pregnant women and children. Some died, and many children developed severe mental disorders. This illness was called Minamata disease.

Chesapeake Bay

Chesapeake Bay (Fig. 14–26) on the U.S. Middle Atlantic coast is a prime example of the environmental problems that humans can cause in a moderately stratified, coastal-plain estuary. The Bay Region, home to nearly 15 million people, forms part of the densely populated metropolitan region extending from Boston on the north to Norfolk on the south.

Chesapeake Bay is relatively shallow, averaging 6.5 meters. Ocean water flows into it primarily through deep, dredged channels. The least dense water flows out, primarily over the shallow portions. The bay is stratified much of the year, especially in summer. When spring tidal currents are especially strong twice each month, bay waters are less stratified but become restratified during the time of the weaker neap tides.

Like other moderately stratified estuaries, Chesapeake Bay retains and recycles nutrients because of its estuarine circulation. Thus, bay waters support a relatively high level of primary productivity, between 400 and 600 grams of carbon per square meter per year.

Chesapeake Bay receives waste discharge from its **watershed,** the entire area from which surface waters drain into the bay (Fig. 14–27). The nutrients contained in this waste stimulate primary production in the surface waters. When the organic matter produced sinks into deeper water, it is decomposed by bacteria that consume dissolved oxygen. The anoxic conditions occur primarily in the deeper channels, during spring and summer. Stratification then prevents surface-layer dissolved oxygen from mixing into deeper water.

Reduction of nutrient discharges by 40 percent from 1980s levels is required by the year 2000 under an agreement among state, local, and federal governments. Reducing nutrient discharges will be difficult, because of the many sources of nitrogen compounds that come to the bay from farms and airborne acid rains. The nitrogen compounds in acid rains come from vehicle emissions (including ships) and from electrical power-generating plants in the airshed, an area about four times as large as the watershed (Fig. 14–27). The **airshed** is the area that

Figure 14–26
The drowned river valleys of the Susquehanna and the Potomac (lower) join many smaller rivers to form the Chesapeake Bay. (Courtesy NASA.)

BOX 14–1 Oil Spills

Among the most dramatic forms of ocean pollution (Table B14–1–1), oil spills are likely to remain a severe problem wherever tankers carry large quantities of oil. Most oil floating on the ocean comes either from spills during normal operations around terminals or from oil discarded into municipal sewage systems. These two nonaccidental sources lead to chronic oil pollution and often to specialized communities of bacteria that decompose the oil.

In addition to these mundane sources of oil pollution, there is also the occasional aberrant event, usually accident or war. When a tanker spills large quantities of oil after running aground, the acute effects can be locally catastrophic. Extremely large amounts of oil were spilled during the Persian Gulf War in 1991, when the retreating Iraqi army destroyed tanks and control structures on the Kuwaiti oil fields. The vast quantities of oil discharged into the gulf left deposits of asphalt on beaches that persisted for years.

Eventually, the marine environment recovers from even the largest oil spills. When the tanker *Exxon Valdez* went aground in March 1989, in Prince William Sound in southern Alaska, large quantities of oil washed up on beaches along the sound and on the Alaskan coast beyond the sound. Exxon spent more than 3 billion dollars and two summers removing the oil from these beaches. Legal battles continued for years to determine the extent of damages and compensation for those involved.

The technology for dealing with large oil spills is primitive, and large spills at sea are virtually impossible to control. Skimming the oil from the surface has only a very limited effect. If the oil is fresh enough, it can be burned, and it can also be dispersed with chemicals. However, both smoke from the burning oil

TABLE B14–1–1
Ten Worst Oil Spills

Location and Date	*Tankers*	*Estimated Spillage (millions of gallons)*
Trinidad July 1979	*Atlantic Empress* *Aegean Captain*	92
South Africa August 1978	*Castillo de Beliver*	77
France March 1978	*Amoco Cadiz*	68
Britain March 1967	*Torrey Canyon*	36
Gulf of Oman December 1972	*Sea Star*	35
Spain May 1976	*Urquiola*	31
N. Pacific February 1977	*Hawaiian Patriot*	30
Sweden March 1970	*Othelo*	18–31
South Africa June 1968	*World Glory*	14
Alaska March 1989	*Exxon Valdez*	11

Source: *The Economist,* August 26, 1989.

contributes significant amounts of airborne wastes to the Chesapeake Bay; it extends as far south as northern Georgia, as far west as the Ohio Valley in the west, and as far north as southern Quebec. The nutrient control regulations were devised to deal with sources lying within the watershed. But it now appears that an effective control scheme must also deal with airborne substances coming from up to 1,000 kilometers outside the watershed.

The bay's waters are also influenced by conditions in the nearby coastal ocean. Another nutrient which stimulates increased primary production is phosphate. Phosphate was eliminated from household detergents in the U.S. and Canada in the 1970s. However, about one third of the present phosphate input to the bay occurs naturally in the inflowing ocean waters. Thus, efforts to further reduce phosphate concentrations must recognize that a large fraction of the bay's phosphate input is effectively beyond human control.

and the dispersing chemicals may themselves cause problems.

Cleanup crews remove oil from a beach (Fig. B14–1–1) by soaking it up with straw or other absorbents and then disposing of the contaminated "sponges." Sometimes the oil-coated beach materials are removed and buried, but this approach can destroy the beach and its marine life. In the *Exxon Valdez* spill, hot water and steam were used to remove loose oil, and booms were placed around beaches to prevent the loosened oil from moving onto other beaches.

Seabirds and marine mammals are seriously affected (Fig. B14–1–2). No one knows how many were killed by the *Exxon Valdez* spill. The techniques for removing the oil often leave the animals weak and unable to survive in the wild. Fishing was not permitted because of the possibility of oil contaminating the seafood. Native peoples living around Prince William Sound who depended on local seafood were especially hard hit.

New techniques now permit workers to clean oil-contaminated beaches via bacterial action rather than physically removing the oil; this technique is called *bioremediation*. The growth of oil-destroying bacteria is enhanced by using fertilizers that provide nitrogen and phosphorus compounds to the bacteria. The result is that the beaches are cleaned without any serious destruction to marine life.

The long-term environmental effects of oil spills are difficult to evaluate, mainly because natural shifts in the environment are often larger than the consequences of the spill or its cleanup. Still, the social, health, and psychological effects of an oil spill can persist for many years.

Figure B14–1–1
Oil coats rocks and beach sands. (Courtesy NOAA.)

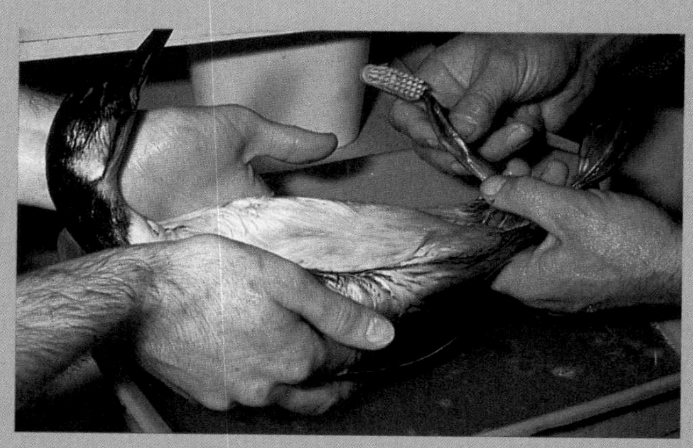

Figure B14–1–2
A sea bird being cleaned of oil. (John Watney/Photo Researchers, Inc.)

Chesapeake Bay supported extremely productive fisheries for oysters, fishes, and crabs until the middle of the twentieth century. Today, little remains of these once-rich fisheries, and the bay shows evidence of the stresses that afflict urbanized estuaries around the world. The living resources of Chesapeake Bay have been decimated by overfishing and by destruction of habitats that sheltered the larvae and juvenile stages of fish and shellfish (crabs).

Oyster reefs, once a prominent feature of the bay, were eliminated by increasingly destructive mechanized harvesting techniques, including the widespread dredging of the bottom of the bay. Since the 1950s, diseases have further reduced oyster populations already weakened by overfishing and habitat destruction. Now most of the organic matter produced in Chesapeake Bay goes not into oysters, fishes, and crabs but into bacteria that cause the anoxia we just discussed. When the bay was first colonized in the early 1600s, the large oyster population

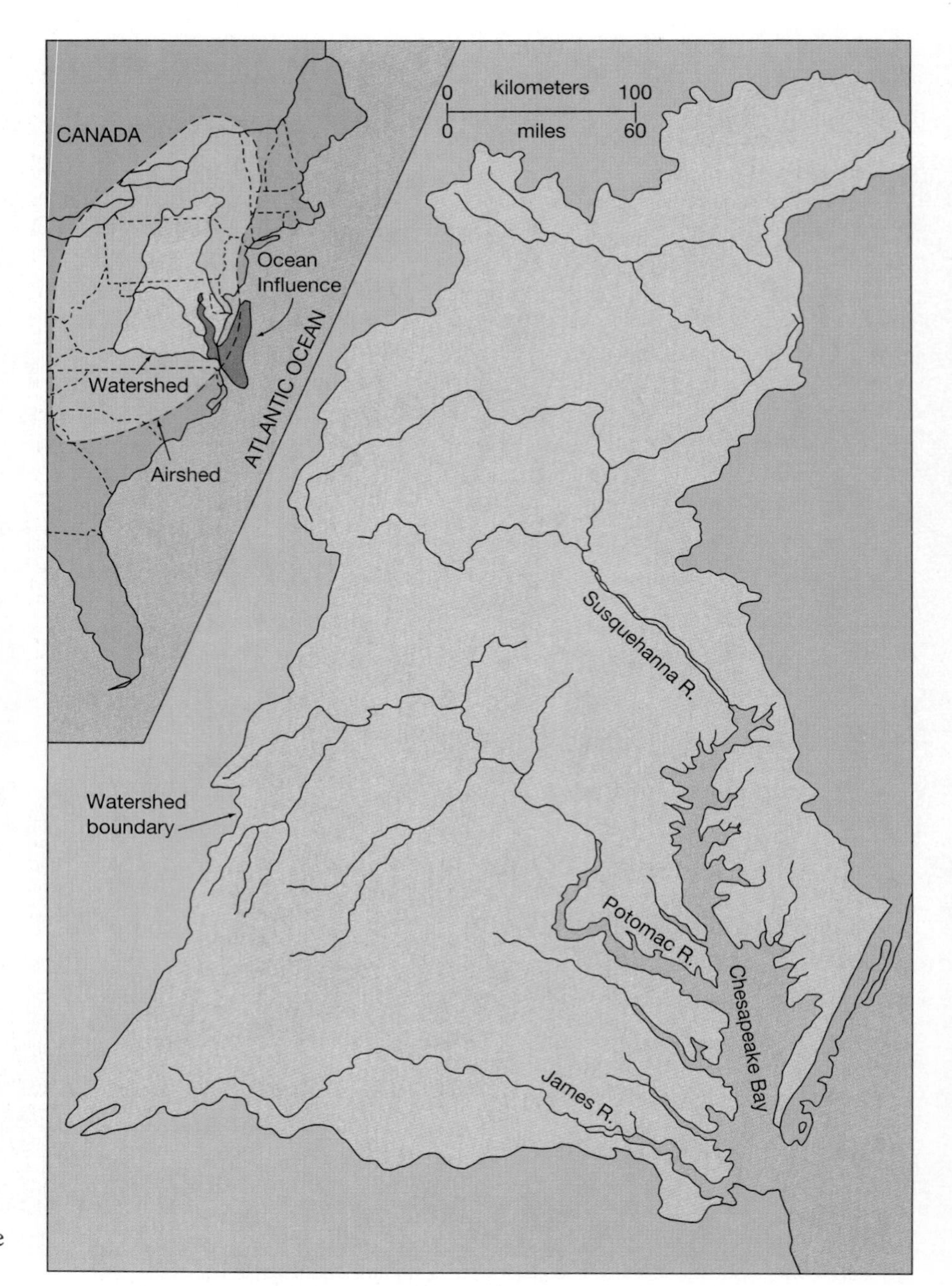

Figure 14–27
All rivers inside the Chesapeake watershed boundary flow into the bay, and all airborne pollutants inside the airshed boundary (red in inset map, upper left) reach the bay. The area of coastal-ocean influence (blue in inset map) is the approximate region supplying seawater to the subsurface flows into the Chesapeake Bay.

filtered the entire bay volume in three to six days. Now filtering takes more than a year, and oyster populations continue to decline, victims of diseases and continued harvesting.

San Francisco Bay

San Francisco Bay (Fig. 14–28), the largest bay on California's coast, is a tectonic estuary. It occupies several basins (Fig. 14–29) formed by movements of crustal blocks along the San Andreas Fault, a major transform fault. The bay formed as sea level rose to its present position over the past 18,000 years, filling the various basins. Most of the bay is quite shallow, with an average water depth of 6 meters. The dredged channels used by ships are the deepest parts of the bay. Its watershed covers 153,000 square kilometers, about 40 percent of the area of California. About

Figure 14–28
San Francisco Bay, California is a series of basins which connect to the Pacific Ocean through the Golden Gate, left center. (Courtesy NASA.)

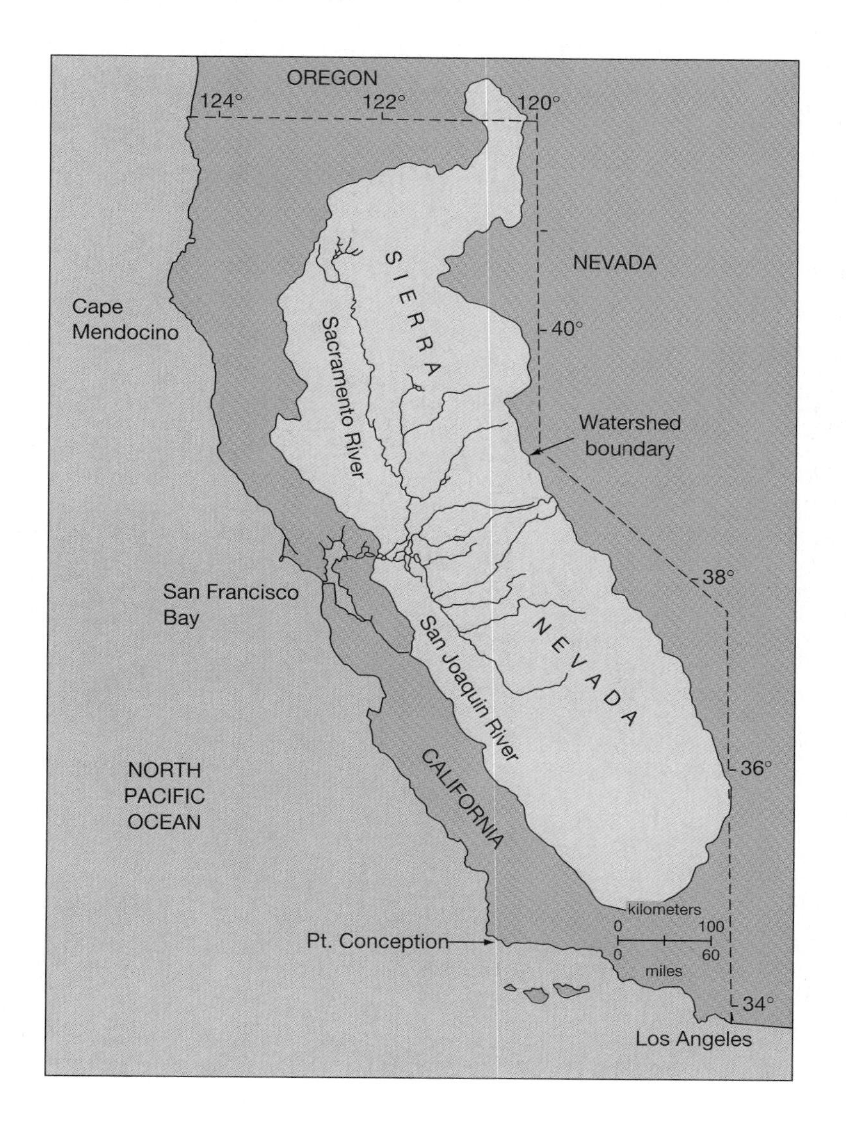

Figure 14–29
San Francisco Bay and its watershed in the Sierra Nevada and Central Valley of California.

90 percent of the fresh water entering the Bay comes from the Sacramento and San Joaquin rivers, which drain the mountains of the Sierra Nevada range and California's Central Valley (Fig. 14–29).

Humans occupied the bay more than 5,000 years ago; the Spanish arrived overland in 1769. The Bay Region was sparsely settled and unknown scientifically during its years of Spanish colonial occupation. European settlement began in earnest with the Gold Rush of 1849. The population around the bay now exceeds 5 million.

We know little about the bay before its alteration by human activities, which began with the Gold Rush. Large volumes of sediment were washed into the streams of the Sierra Nevada as the gold was mined hydraulically. Much of this sediment flowed into the bay before the practice was stopped in 1884. Large amounts of the mining sediments continued flowing into the bay until the mid-1900s. Because of this, and the filling of wetlands for other purposes, the area of the bay has been substantially reduced. Its present shorelines are located approximately where they stood about 5,000 years ago, before the sea level reached its present position. Of the original 2,200 square kilometers of marshlands, only 85 square kilometers remain today; another 75 square kilometers of new wetlands were formed by later sediment deposition on the shallow bay bottom.

A disastrous flood in 1862 filled San Francisco Bay with fresh water, wiping out most native benthic organisms. Arrival of railroads at about that time made it possible to introduce organisms from other areas—shad in 1872 and striped bass in 1879. Both species soon became well established, as did other introduced organisms such as lobsters and oysters.

Now a major problem for San Francisco Bay is the continuing diversion of fresh water from rivers draining into it in order to supply water to southern California. California's problem is that about 70 percent of its fresh water supply lies north of San Francisco, but 80 percent of its water consumption occurs in the drier south. Large amounts of water have been diverted from the Sacramento and San Joaquin rivers, greatly reducing the fresh water supply to the bay and changing its salinities.

These water diversions have had their most marked effects in the Sacramento–San Joaquin Delta, the complex of channels and islands where the two rivers meet (Fig. 14–30). Salinities in the bay have risen by 10 parts per thousand in the past 100 years.

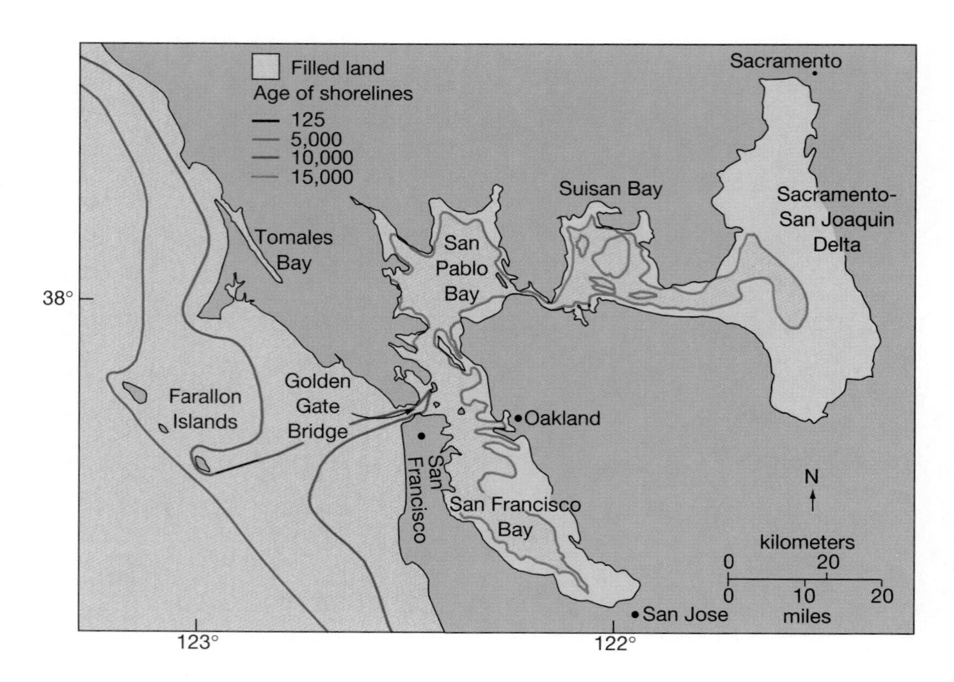

Figure 14–30
San Francisco Bay, San Pablo Bay, and the Sacramento–San Joaquin Delta occupy interconnected basins formed by tectonic activity. Shoreline positions are shown for the last 15,000 years as the sea rose to its present level. In the past 150 years, human activities have filled large shallow parts of the bay so that present shorelines correspond closely to those of 5,000 years ago. (Simplified after B. F. Atwater: "Ancient Processes at the Site of Southern San Francisco Bay," in *San Francisco Bay: The Urban Estuary.* San Francisco: American Association for the Advancement of Science, 1979.)

Wastes discharged to San Francisco Bay have contaminated sediment deposits and organisms. Public health officials have warned against eating large amounts of fish caught in the bay because they contain toxic compounds known to cause cancer or other diseases. The danger is greatest when large, long-lived fish (such as sharks) or bottom-living organisms are consumed. It is less dangerous to eat species such as salmon, which enter the bay only during their breeding season, or short-lived shad which feed only on phytoplankton.

Enclosed and Semi-Enclosed Seas

Human effects on coastal oceans are most easily recognized in enclosed and semi-enclosed seas, which have very long freshwater residence times. Here we examine the Mediterranean and Black seas (Fig. 14–31), both exhibiting signs of large-scale environmental change due to increased nutrient discharges, heavy fishing, and river diversions. As we have already seen, all three factors are very common in coastal ocean areas.

Some of these changes may be reversible with improved environmental protection; others are likely permanent. Let's see what makes these two seas so vulnerable to pollution, why they have responded so differently, and how they might respond to future changes.

First, both seas have limited exchange with the open ocean. Pollutants, including nutrients from sewage and industrial wastes, are not readily removed. In other words, these seas have limited capacities to assimilate wastes without displaying adverse effects.

Secondly, the two areas have large and growing human populations. The Mediterranean basin is expected to have a population of 425 million in the year 2000, double its 1950 level. The most rapidly growing area of the basin is North Africa, where limited financial resources and a very high rate of population growth

Figure 14–31
The Mediterranean and Black seas. Clear, pigment-poor waters are shown in blues and purples. Areas of high primary production along the Mediterranean coast and in the Black Sea are shown by reds, oranges, and yellows. (Dr. Gene Feldman, NASA GSFC/Science Photo/Photo Researchers, Inc.)

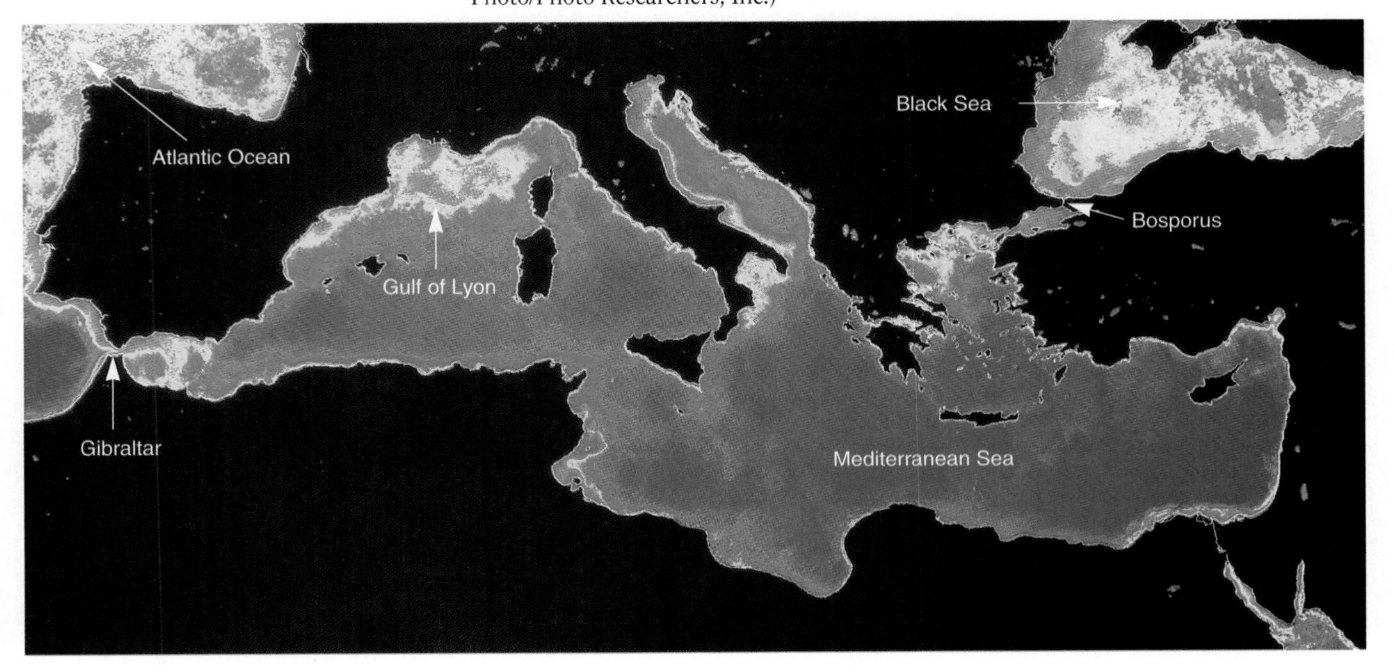

are likely to result in more pollution problems than in the more affluent European sector. The Black Sea has a large and impoverished population, its economy devastated by the collapse of the former Soviet Union in the early 1990s.

The Mediterranean

The Mediterranean is virtually landlocked, connecting with the North Atlantic only through the narrow Strait of Gibraltar. It receives little river discharge from the mostly arid lands around it. Furthermore, the Nile River outflow was essentially shut off in 1966, following completion of the Aswan Dam. Now the only water to reach the Mediterranean Sea from the Nile region is waste discharge from irrigated fields. Other rivers, especially the Po draining northern Italy, also carry heavy loads of industrial wastes as well as agricultural fertilizers and pesticides.

The Mediterranean loses more water from evaporation than it receives from river runoff and from precipitation; this imbalance causes anti-estuarine circulation (Fig. 14–32), in which flow is opposite to that in estuarine circulation; in other words, landward flow in surface water and seaward flow in bottom water. Low-nutrient surface seawater from the North Atlantic enters the Mediterranean Sea through the Strait of Gibraltar. In the Mediterranean itself, denser water masses form by winter cooling in the Gulf of Lyon (south of France) and by evaporation in the eastern Mediterranean. These denser waters flow to the open ocean in a subsurface layer through the Strait of Gibraltar. This anti-estuarine current pattern removes nutrients, thus reducing the Mediterranean's vulnerability to *eutrophication* (high nutrient concentrations, usually due to waste discharges, stimulating primary production).

Reductions in river discharges have affected fish catches. After construction of the Aswan Dam eliminated the Nile River outflow into the eastern Mediterranean, catches of sardine-like fishes were reduced to about one-tenth their former levels. Eliminating the river discharge ended the river-induced upwelling that previously sustained the sardine fishery in these nutrient-poor waters.

As human populations grew in the basin, however, more nutrient-rich wastes reached the coastal waters near the Nile Delta. After 1989, the local fish production in the eastern Mediterranean recovered to nearly three times its historic levels, as greater phytoplankton production more than compensated for the lost upwelled nutrients at the Nile River mouth. Such stimulation of fisheries by waste discharges is fairly common in coastal areas.

Environmental problems have become more severe in partially isolated areas of the Mediterranean such as the Adriatic Sea east of Italy. There the nutrient-rich waters of the Po River stimulate large algal blooms. When these algae wash up on nearby beaches and decay, the odors drive away tourists. Algae decaying in the water cause anoxic conditions and fish kills in the northern Adriatic.

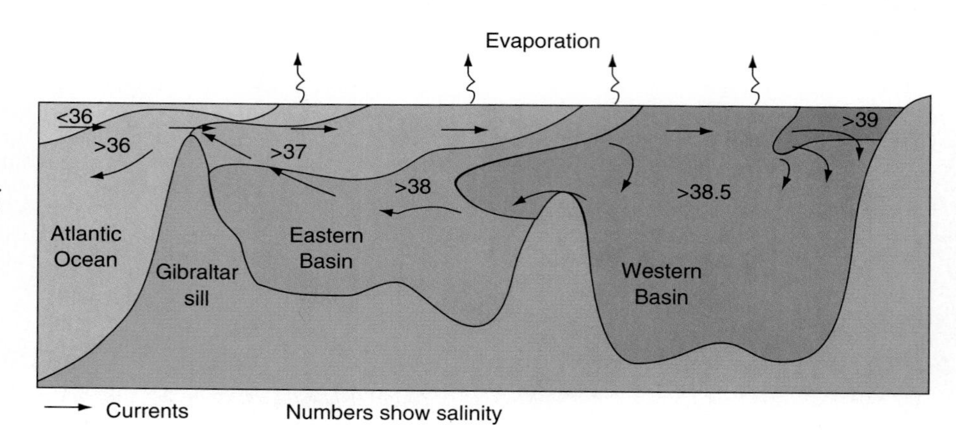

Figure 14–32
Cross section of the Mediterranean Sea, showing the currents. Note that water enters from the Atlantic in the surface layers. After evaporation, its salinity and density increase and the return flow to the Atlantic occurs in the subsurface over the Gibraltar Sill. (After G. Wüst, *Journal Geophysical Research,* 66:3261–3271, 1961.)

These problems have seriously reduced northern Italy's tourist trade and caused large-scale economic losses.

The Black Sea

Because of its long residence time, the Black Sea (Fig. 14–31) is especially vulnerable to human-induced environmental changes and provides the largest-scale example of the collapse of an ecosystem and its fisheries. Because of the waste loads carried by rivers flowing into it, the Black Sea is experiencing severe eutrophication, seen in the high levels of primary productivity shown in Figure 14–31.

This sea has an estuarine circulation, with surface waters flowing out through the narrow Bosporus into the eastern Mediterranean and subsurface waters flowing in from the eastern Mediterranean. Freshwater residence time is estimated at 2,500 years.

Reductions in river discharges have changed salinity distributions in the Sea of Azov, a shallow arm of the Black Sea. Large shallow areas of the Azov now have relatively high salinity waters, a condition that has altered the local ecosystem, making it more favorable for certain new types of benthic organisms. These shallow areas have also experienced anoxia and fish kills due to eutrophication.

Black Sea bottom waters are naturally anoxic below about 100 meters; this condition has persisted for thousands for years. It is the result of the slow rate of exchange of bottom waters and the depletion of dissolved oxygen by decomposition of the organic matter sinking from the surface layer. Now it appears that the top of the anoxic zone is getting shallower due to eutrophication of the surface waters.

In the 1950s, nutrient-rich river discharges to the Black Sea caused local fish production to increase. As the waste discharges continued and increased, however, conditions for fisheries deteriorated, and recent catches clearly demonstrate widespread ecosystem changes.

First, the dolphins and seals that previously fed at the top of the food web disappeared. Next, seasonal predators—bonito and bluefish—that previously migrated into the Black Sea from the Mediterranean no longer did so. The result of the disappearance of these large predators was that smaller plankton-eating pelagic species—mainly anchovies and horse mackerel—dominated the fishery.

In the 1980s, the Black Sea fisheries collapsed because the more saline environment favored jellyfish production. First, the common jellyfish, *Aurelia,* came into the area. These jellyfish fed primarily on plankton but also ate young fish and fish eggs. Later, another jellyfish was introduced by ships discharging ballast water. The new arrival, a ctenophore called *Mnemiopsis leidyi,* eliminated *Aurelia* and continued to prey on fish until their populations finally collapsed. Now little remains of the once-rich Black Sea fisheries.

Ways to restore the Black Sea are being studied and debated. Among the possibilities being considered are reducing nutrient inputs and introducing a new predator to eat *Mnemiopsis.* However, using a newly introduced species to solve a problem often has unexpected and undesirable results. It seems unlikely that the earlier state of the Black Sea can be restored.

The North American Great Lakes

The basins occupied by the Great Lakes (Fig. 14–33) were gouged by continental glaciers during the last Ice Age. These five lakes constitute the largest lake system on Earth, containing about 20 percent of the planet's liquid fresh water supply. Today the lakes' drainage basin (Fig. 14–34) is home to 36 million people and forms the industrial heartland of Canada and the United States.

Figure 14–33
View of the Great Lakes region. Lake Ontario is lower left; Lake Erie, center. Lake Huron is on the right. Lake Michigan is on the horizon, and Lake Superior is beyond the range of this photo taken from the Space Shuttle. (Courtesy NASA.)

Despite their long history of human occupation and utilization, the Great Lakes are in remarkably good shape, mainly because the region around the largest and deepest, Lake Superior, (Fig. 14–35) is relatively unpopulated. Also, the residence time of water in the Great Lakes is only about 230 years. The most polluted lake, Erie, is the shallowest and has the largest human population around it.

Water quality has markedly improved since 1972, the year the United States and Canada began working together to restore the lakes. First, discharges from municipal and industrial sources were reduced. Phosphates from home and commercial laundries were a major problem (phosphates speed up primary productivity), and a mandatory switch to low-phosphate detergents benefited water quality. This change resulted in improved water quality and recovering fish populations in

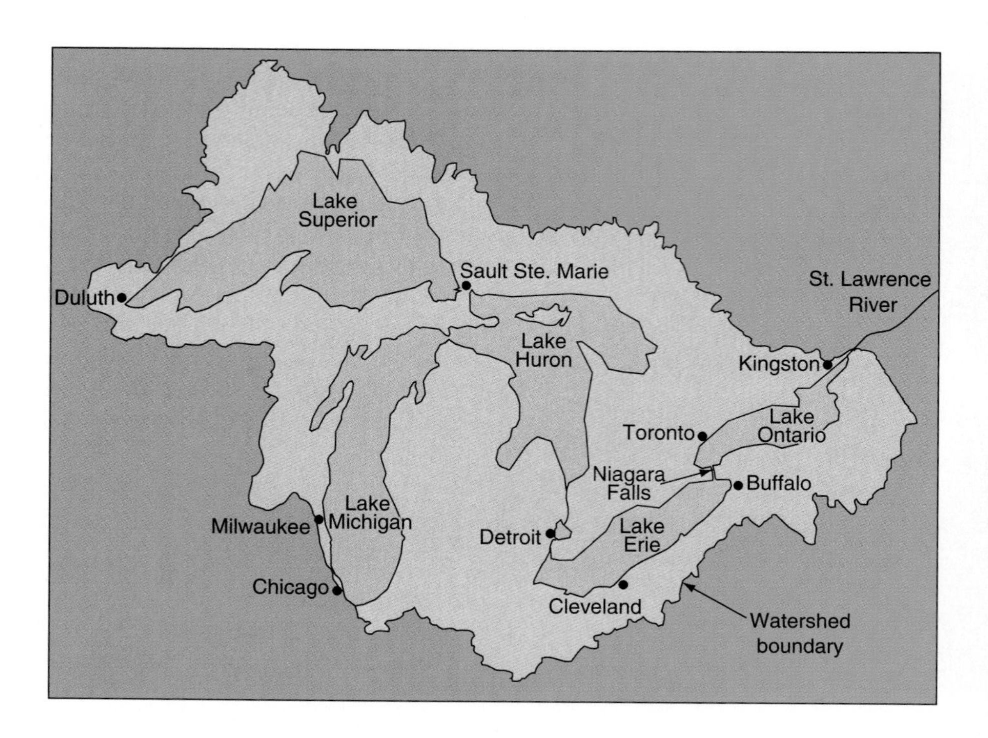

Figure 14–34
The North American Great Lakes and their drainage basin.

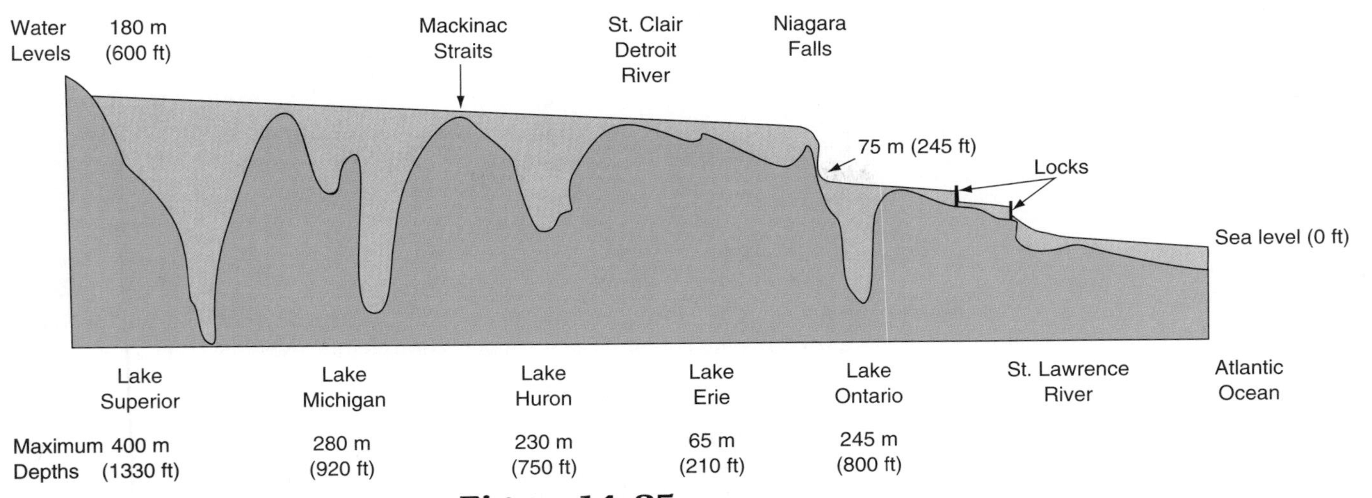

Figure 14–35
Schematic profile through the Great Lakes, showing maximum basin depths and general slope of water surface. (Courtesy International Joint Commission.)

the most seriously degraded areas, especially Lake Erie, which had been called a "dead lake."

Overfishing has seriously degraded ecosystems in the Great Lakes. Many native fish species are gone; of those remaining, several cannot maintain stable populations. As in most overfished populations, the larger and more valuable fishes (lake trout, whitefish) have been replaced by smaller, less desirable ones (smelt, eels, white bass).

Introduced species are another serious problem. Canals connecting adjacent water bodies allowed sea lampreys to enter the Great Lakes and decimate native trout populations. Previously the barrier of the Niagara Falls separating Lake Ontario from the other four lakes had made this impossible. Discharging ballast waters from ships has introduced other new species, such as zebra mussels, which spread unchecked because they have no natural enemies. Today, food webs in all five lakes are dominated by introduced organisms.

There is some good news. Reductions in the amounts of chlorinated hydrocarbon pesticides dumped into the lakes have permitted bird populations to rebuild. Bald eagles were especially hard hit by the most infamous chlorinated hydrocarbon, DDT, which caused birds to lay eggs with very thin shells and drastically reduced the birth rate. The banning of DDT in the United States and Canada has resulted in a strong recovery of the populations of bald eagles and other birds of prey.

The outlook for the Great Lakes is not all rosy, however. A growing threat is loss of habitat due to increasing populations in the Great Lakes basin. Concentrations of dissolved nitrogen compounds are steadily rising despite efforts to control water-borne waste discharges. Acid rain coming from the industrialized airshed of the lakes appears to be a major source of these nitrogen compounds, but the problem has proven difficult to control because cars, trucks, and ships are major sources.

Sediment deposits in the lakes, and especially in the industrialized harbors, contain high concentrations of various toxins. Such deposits can be eroded during floods and re-suspended, releasing their contaminants. Efforts to eliminate such waste sources are difficult, time-consuming, and expensive. Many of these deposits must either be buried where they lie or placed in special structures.

Over time, natural processes will clean up some of these problems. Many toxic chemicals are broken down by bacteria. Some of the deposits will be covered by later accumulations of cleaner materials, taking the contaminated deposits out

Coastal Fisheries

Fisheries in coastal waters around the world have been badly damaged by human activities. One of the most striking is the oyster fishery in Chesapeake Bay. At its peak in 1890, the fishery produced 20 million bushels of oysters, which were shipped by rail all over the continent. By the early 1990s, the oyster fishery in the Bay was so low that only a few oyster boats were active where hundreds were working a few decades ago. Much of the problem arises from overfishing and habitat destruction. Oyster diseases have also decimated the oyster populations, since they first appeared in the region in the 1950s. Some efforts are underway to restore oyster reefs, but little progress is evident.

The story of the return of salmon to the Thames River, which runs through London, England, has had a more favorable outcome. Historically, the river was rich with salmon, but industrial development and pollution had destroyed the salmon runs as early as 1834. After the 1960s, when the British government made extensive efforts to clean up the river, salmon began to return to the river. Some of these fish have grown to adulthood from the young salmon released in the river each spring. Others have migrated from other British rivers.

New fish passages permit the spawning salmon to bypass dams that previously blocked parts of the river to the returning fish. The gravel beds used by spawning salmon have yet to be cleaned of the silt deposits that cover them. When all these efforts are finished, salmon will again be able to breed in the Thames River. Such efforts contribute to the long-term survival of salmon in the North Atlantic.

of contact with the water column. Also natural flushing through the whole lake system will eventually remove some materials, although it takes about 230 years for water to move from the western end of Lake Superior to the Atlantic.

The most compelling case for continued restoration of the Great Lakes comes from concerns about human health. The concentrations of several pollutants are approaching levels where they can cause health problems in humans. At present, pregnant or nursing women are advised to limit their consumption of fish from the Great Lakes.

Summary

Coastal oceans lie on continental margins. They are bounded on one side by exposed land and on the other by open-ocean currents. Shorelines are the most dynamic part of the coastal ocean. They include beaches and wetlands. All shorelines have been affected as sea levels have risen over the past 18,000 years. Some shorelines are formed by terrestrial processes, such as volcanism and glaciation; others are formed by marine processes, such as beach and wetland formation.

Wetlands are low-lying areas usually covered by salt-tolerant plants. In the temperate zones, these plants are usually grasses, forming salt marshes. In tropical areas, salt-tolerant mangrove trees live along the shores and trap sediments.

Coastal currents generally parallel the shore. They are driven by winds and density distributions resulting from river discharges. Waters discharged by rivers are retained in coastal oceans for a length of time called the freshwater residence time. The longer the residence time, the longer waste discharges are retained and the more susceptible a coastal ocean area is to environmental degradation.

Temperature and salinity variations in coastal-ocean waters are greater than in the open ocean. These changes occur daily and seasonally. Salinity changes are greatest near the mouths of large rivers. Highest salinities occur in semi-isolated basins in arid mid-latitude areas.

Estuaries formed during the past 18,000 years as glaciers melted and sea level rose to its present position.

Flooded river valleys formed coastal plain estuaries. Where glacially eroded mountain valleys were flooded, they formed fjords.

Estuarine circulation is a two-way current pattern. Low-salinity surface waters flow seaward, and landward-flowing high-salinity bottom waters come in from the ocean. Separation between flows is sharpest where river flow is large and tidal range small; this is the case in salt-wedge estuaries. Where river flow is small and tidal ranges large, estuaries are well mixed.

Lagoons formed when shallow coastal ocean areas were partially isolated from the ocean by barrier islands. Lagoons are especially vulnerable to human alteration and to pollution from waste discharges.

Upwelling occurs when winds blow surface waters offshore and subsurface waters rise to replace them.

Waste discharge from sewage treatment plants, from industrial sites, and from farmlands are increasingly causing problems in coastal ocean areas. Some pollutants are also carried by winds. Estuarine circulation tends to retain nutrients in coastal ocean areas, and long residence times exacerbate these environmental problems.

Key Terms

coastal ocean
shoreline
coastline
foreshore
backshore
coastal zone
barrier island
spit
salt marsh (wetlands)
tidal flats
mangrove swamp
coastal current
front
freshwater residence times
estuary
coastal plain estuary
tectonic estuary
fjord
estuarine circulation
salt-wedge estuary
moderately stratified estuary
well-mixed estuary
sill
lagoon
watershed
airshed

Study Questions

1. Explain why water temperature and salinity variations are greater in coastal ocean areas than in the open ocean.
2. What causes coastal currents?
3. Describe estuarine circulation, compare it with anti-estuarine circulation, and describe the conditions under which each type forms.
4. Describe how a salt-wedge estuary, a moderately stratified one, and a well-mixed one differ.
5. Describe how an estuary changes into a wetland.
6. Draw a profile of a beach. Label the major features.
7. Discuss the effect of continued sea-level rise on shoreline features.
8. What causes sediment particles to move along a beach?
9. Draw a cross section of a continental margin, and show the relationship to the deep-ocean floor for an Atlantic- and a Pacific-type margin.
10. What is causing the present worldwide rise in sea level?
11. Describe the coastal ocean. How does it differ from the open ocean?
12. Why does upwelling typically occur at capes?
13. Discuss the relationship between upwelling and high chlorophyll contents at capes.
14. Describe the difference between a coastal-plain estuary and a tectonic one, and tell how each formed.
15. Discuss the relationship between salt marshes and mangrove swamps.
16. Where did estuaries occur 18,000 years ago? Where should one look for evidence of their existence?
17. *[critical thinking]* Why are enclosed seas so vulnerable to pollution?
18. *[critical thinking]* Why is the growth in human populations a threat to environmental quality in coastal oceans?
19. *[critical thinking]* How do introduced organisms change coastal ecosystems?

Selected References

CLOERN, J. E., AND F. H. NICHOLS (eds.), *Temporal Dynamics of an Estuary: San Francisco Bay.* Boston: Dr. W. Junk Publishers, 1985.

DAVIS, R. A., JR., *The Evolving Coast.* New York: Scientific American Library, 1994. Modern review of coasts, coastal processes, and coastal problems.

HORTON, T., AND W. E. EICHBAUM, *Turning the Tide: Saving the Chesapeake Bay.* Washington, D.C.: The Island Press, 1991. Well-written explanation of the environmental problems of Chesapeake Bay and some possible solutions.

MCLUSKY, D. S., *The Estuarine Ecosystem,* 2d ed. New York: Chapman & Hall, 1989.

PETERSON, D., D. CAYAN, J. DILEO, M. NOBLE, AND M. DETTINGER, "The Role of Climate in Estuarine Variability." *American Scientist* 83:58–67.

SCHUBEL, J. R., *The Living Chesapeake.* Baltimore: Johns Hopkins University Press, 1981.

OBJECTIVES

Your objectives as you study this chapter are to understand:

- The causes of global climate change
- How global change affects the ocean and ocean processes
- How an ever-enlarging human population affects the ocean and ocean processes
- The significance of marine biodiversity to humans
- The proper management and protection of ocean resources used by all, especially commercial fishing and whaling
- The possibilities of sustainable development on the planet

One of the first expected impacts of global warming is an accelerated rate of melting of the glaciers and ice caps of polar regions, thereby affecting sea level, salinity, and circulation patterns worldwide.

15 The Ocean on a Changing Earth

In all the previous chapters, we dealt mainly with the oceanic parts of the Earth system. In this chapter, we expand our view to look at processes affecting the whole Earth and at how those processes influence the ocean and oceanic life. We close the book with an examination of some possible ways in which we humans at the end of the twentieth century might continue economic development without negatively affecting the options open to future generations.

Global Climate Change

Earth's climate is controlled by incoming solar radiation and by complex interactions among the five subsystems of the Earth system—hydrosphere, atmosphere, biosphere, cryosphere (ice-covered areas), and lithosphere (rocks and soils). During Earth's 4.6-billion-year history, its climate has varied on many different time scales—daily, seasonally, annually, and decadally. All of these variations have been within relatively narrow limits, however, and the main reason for this narrow range is that the planet is covered with water.

The lower limit of Earth's climate range is controlled by water's freezing point. Although some of the water on Earth is frozen at any time, life could not survive if most or all of it were frozen. The upper climate limit is less tightly controlled but is set by interactions between atmosphere and ocean. If surface ocean temperatures rise much above 30°C, the intense thunderstorms that form locally cool both the air and surface waters.

Figure 15–1
The layer of smog above the Los Angeles basin is similar to those found over many cities worldwide. (David R. Frazier/Photo Researchers, Inc.)

That living organisms have been present on Earth during most of its history is compelling evidence that its climate has remained essentially within this narrow temperature range from 0 to 30°C. (Because of the lack of water, temperature ranges on the other planets are much greater than on Earth.)

Since the beginning of the Industrial Revolution (around 1750), human activities have altered Earth's atmosphere on a global scale (Fig. 15–1). The many different gases discharged to the atmosphere have changed how it transmits and absorbs different parts of the incoming solar radiation. These changes affect Earth's climate, because climate is, in part, a function of incoming solar radiation. (We discussed these basic ideas when we discussed the greenhouse effect in Chapter 6.) Several gases are involved, but most attention has focused on carbon dioxide, which is released by the burning of fossil fuels (coal, oil, and natural gas) and by deforestation, the burning or cutting of forests to make way for agriculture or other types of development (Fig. 15–2). The resulting increase in atmospheric carbon dioxide is shown in Figure 15–3.

Figure 15–2
Burning of tropical forests in Amazonia to clear land for use as pasture is a major source of carbon dioxide to the atmosphere. (Shane Moore/Animals Animals/Earth Scenes.)

Gas bubbles trapped in ice cores recovered by drilling deep holes in the Antarctic glacier (Fig. 15–4) show that atmospheric carbon dioxide concentrations have increased and decreased over millennia. Since the late 1950s, direct measurements of atmospheric gases from the top of the inactive volcano, Mauna Loa in Hawaii, (far from major industrial and urban centers) provide direct evidence of the CO_2 buildup. This increase is now about 0.6 percent per year (Fig. 15–5). Similar increases have also occurred for various nitrogen compounds, for sulfur, and for chlorinated hydrocarbons, all of which are released when fossil fuels are burned.

Detecting Climate Change

Climate change is difficult to detect. The initial changes are certain to be small and to occur in a system subject to substantial natural variability. In short, the challenge is to detect a small signal in a lot of noise. It was only in the late 1980s that

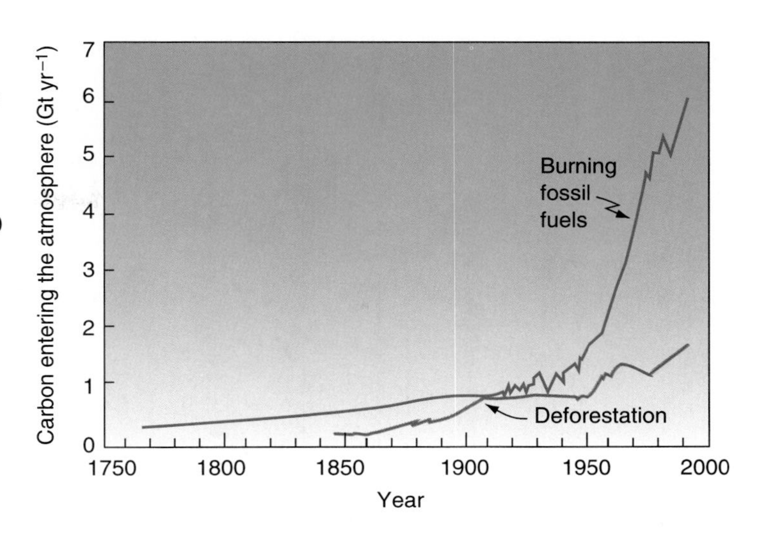

Figure 15–3
Carbon dioxide released to the atmosphere as a result of deforestation and the burning of fossil fuels. Fossil-fuel consumption began to rise sharply once the Industrial Revolution got into full swing in about 1900. The steepening in the deforestation curve starting in about 1950 corresponds to the pressures exerted by an ever-increasing human population in search of new land for building homes, new farmlands, and new sources of cooking fuel. [After B. Moore III and B. H. Braswell Jr.: Planetary Metabolism: Understanding the Carbon Cycle. *Ambio,* 23(1):4(1994).]

Figure 15–4
Gas bubbles in ice cores taken from glaciers in Antarctica and Greenland record changes in the carbon dioxide levels of the atmosphere over the past 200,000 years. (Courtesy A. Fuchs, Physics Institute, University of Bern.)

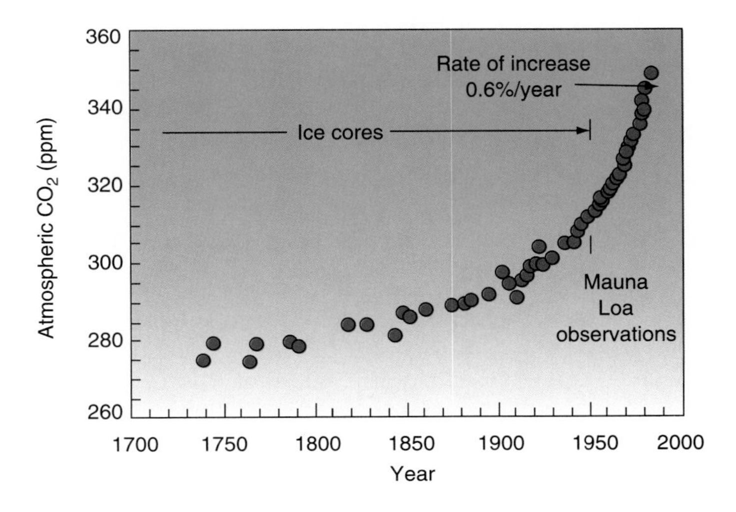

Figure 15–5
Concentrations of carbon dioxide in the atmosphere. The data before 1957 come from gases trapped in ice bubbles recovered in cores drilled in the Antarctic ice. Since 1957, the data come from observations made at an observatory on Mauna Loa, Hawaii. (After Moore and Braswell, *ibid.*)

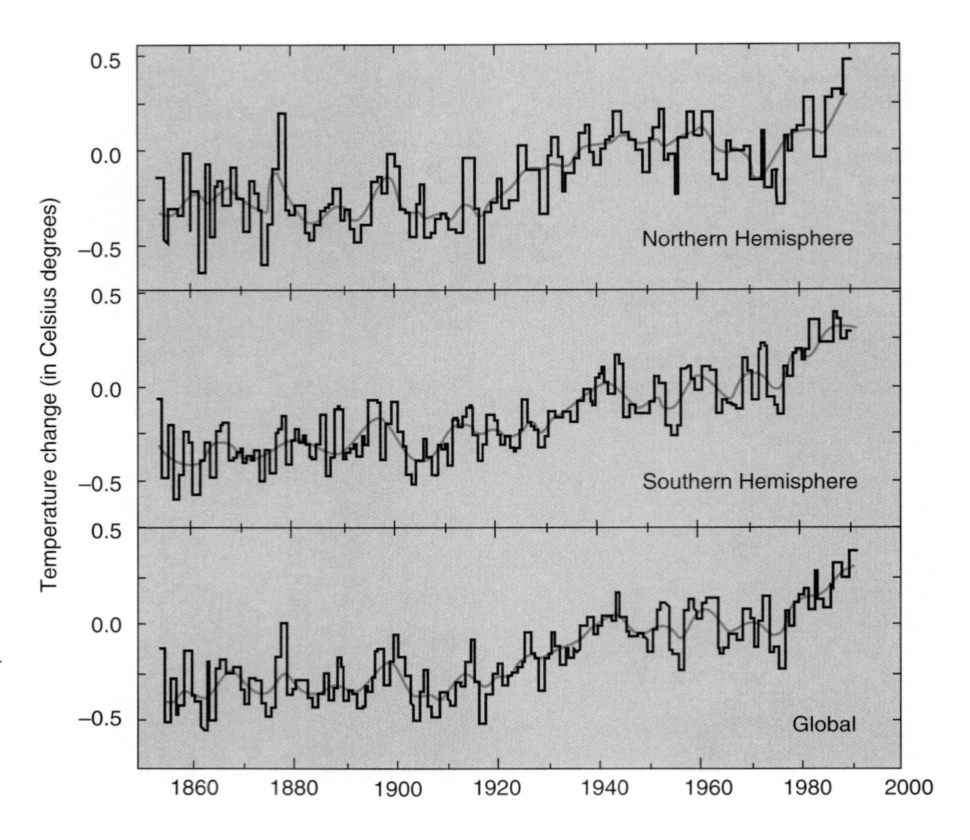

Figure 15–6
Estimates of how annual average temperature has varied from a reference level since the year 1860. Temperatures began rising worldwide about 1920, and this increase has been especially marked in the 1990s.

the increase in global temperatures of approximately 0.5°C could be documented (Fig. 15–6); these increases continue in the 1990s. The small increases and great natural climatic variability leave many people unconvinced that such a change has actually occurred.

Abrupt Climate Change

Ice cores from Greenland and from Antarctica (Fig. 15–4) contain records of Earth's climate for the past 160,000 years and tell us that Earth's climate has changed repeatedly during this interval. These records show that the changes can occur in intervals ranging from a few years to a few decades.

The most obvious recent climate change occurred about 18,000 years ago when the continental ice sheets began melting (Fig. 15–7). Another abrupt climate change occurred between 11,000 and 10,000 years ago. At this time, northern Europe became much colder than it had been. As a result, the glaciers stopped retreating and briefly advanced. The probable cause was the cessation of bottom-water formation in the North Atlantic. A large pulse of meltwater from the melting North American glaciers may have caused a low-salinity surface layer that prevented the bottom-water formation that initiates the global conveyer belt (Chapter 8).

Monitoring Ocean Temperatures

The speed at which sound waves travel through water varies with water temperature and with water pressure, and these changes can be used to measure temperatures in the upper layer of entire ocean basins. The scheme is simple: a powerful underwater loudspeaker is lowered into the water, and the sound emitted by this speaker travels in a natural underwater waveguide (Fig. 15–8). The waveguide is formed by warmer waters above the path of the sound waves and higher-pressure waters below the wave path. The warmer the water, the faster the speed of sound

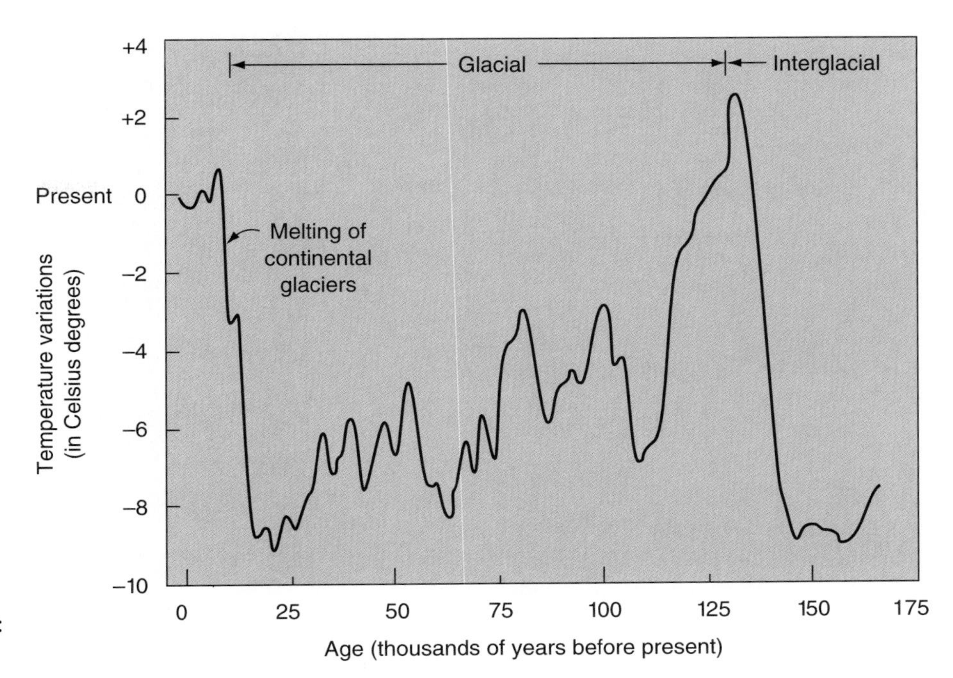

Figure 15–7
An ice core taken from the Russian research station Vostok in Antarctica records changes in surface temperatures over the past 160,000 years. (Source: J. Jouzel and others: Vostok isotopic temperature record. In *A Compendium of Data on Global Change*. Oak Ridge, TN: Oak Ridge National Laboratory, 1994.)

through it, and the more pressure exerted on the water, the faster the speed of sound. Thus, sound travels faster above and below the waveguide than in it. In addition, sound waves hitting either the upper or lower boundary of the guide are reflected back into the waveguide rather than being lost. This reflection greatly improves the range over which the sound pulses can be heard.

Short pulses of coded sound signals are transmitted repeatedly and detected by listening stations in all the major ocean basins. From repeated experiments, it is possible to determine the changes in sound speed over entire ocean basins (Fig. 15–9) and thus, changes in temperatures. These experiments are currently going on; observations over a decade will be required to establish with confidence how the ocean is responding to global warming.

Figure 15–8
Sound pulses are generated by a powerful loudspeaker located in the wave guide. These sounds are detected by microphones in the same layer.

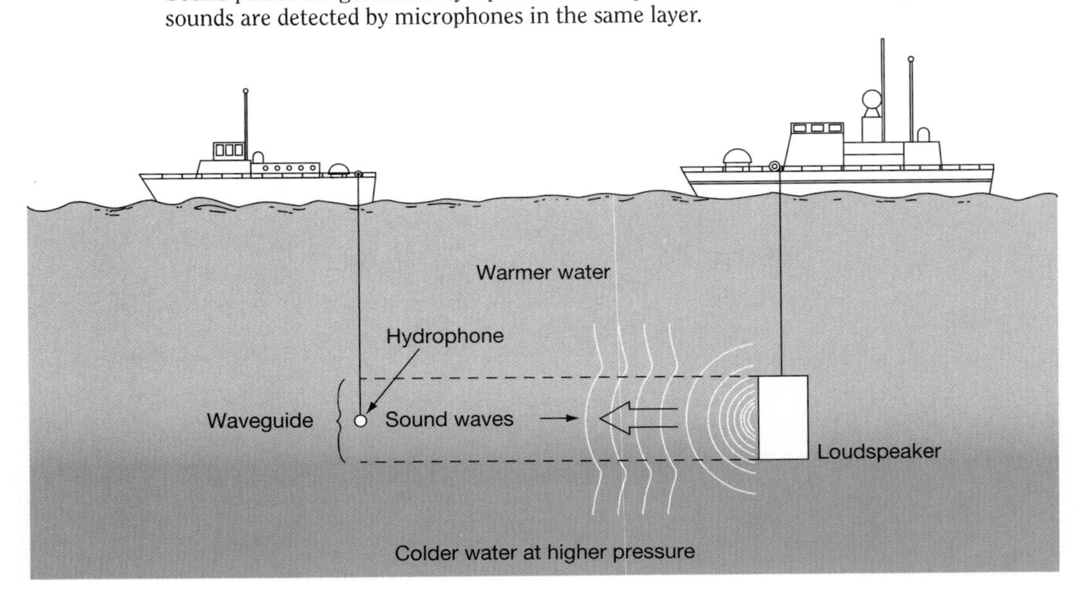

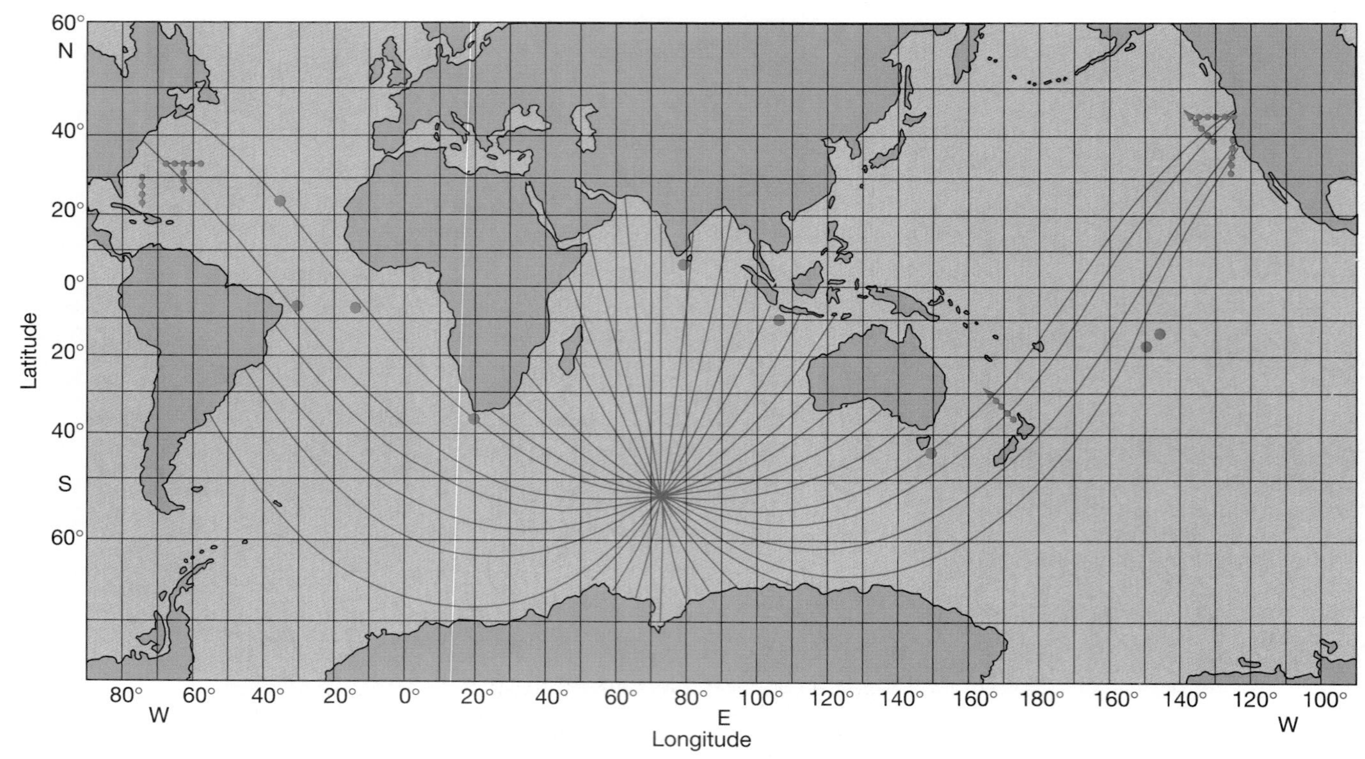

Figure 15–9
In an experiment conducted in January 1990, sound pulses were transmitted from a ship near Heard Island in the southern Indian Ocean. These pulses were detected in all three ocean basins, at the locations indicated by the red dots. Because the speed at which sound waves travel through water varies with water temperature, such pulse transmittals can be used to monitor ocean-water temperatures over long periods and over entire ocean basins. (Courtesy U.S. Office of Naval Research.)

Assessing the Global Carbon Cycle

Carbon dioxide is one of the most important greenhouse gases. Although enormous amounts of carbon dioxide are released when fossil fuels are burned, they are small compared with the amount that moves naturally through the ocean and atmosphere. The challenge to atmosphere and ocean scientists is to account for this relatively small amount of carbon that human activities have released. About one-third of the annual release is unaccounted for (Fig. 15–10). Some of this "missing" carbon may be due to poor estimates of the amount of carbon released by deforestation. Part of it may be stored in forests whose growth rates may have increased due to higher carbon dioxide levels in the atmosphere. A large part of the uncertainty about how much carbon dioxide is "missing" arises from our lack of understanding of the complex processes involving carbon in the ocean and on land. Large-scale field experiments have been conducted to improve our understanding of how carbon is removed from the atmosphere by various oceanic and terrestrial processes.

Climate Predictions

Climate predictions give us a statistical sense of whether the future climate will be wetter or warmer than some specified reference-level value. Our ability to make useful climate predictions is limited to seasonal and year-to-year variability.

The dominant factor in predicting climate is the *El Niño* (warm phase)—*La Niña* (cold phase) cycle. The coupling of the ocean and tropical atmosphere has

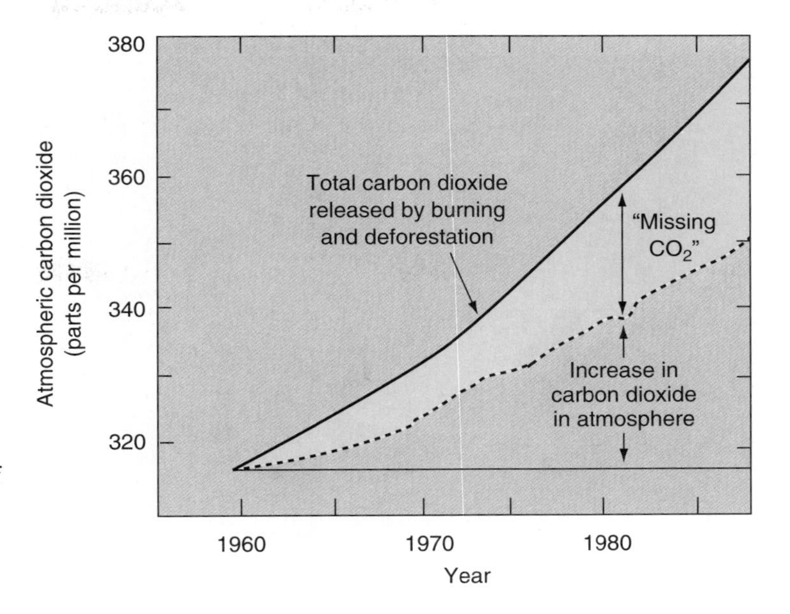

Figure 15–10
A large fraction of the carbon dioxide released by burning fossil fuels and deforestation cannot be accounted for. Some of it may be taken up by forests which have become more productive as a result of increased carbon dioxide levels in the atmosphere.

been extensively studied, as we saw in Chapter 7. Improved understanding of these processes has led to the ability to predict an *El Niño* event up to a year in advance. This early prediction has, in turn, led to improved seasonal forecasts.

This predictive capability has practical applications. Strong *El Niño* years bring very dry conditions to the Australian continent. Government officials can use the prediction to plan their budgets for fighting bush fires. The same *El Niño* brings wetter-than-normal weather to parts of the Americas, where agricultural experts use climate predictions to advise farmers whether to plant corn (dry weather) or rice (wet weather). The ability to make *El Niño* predictions has already helped stabilize the agricultural economies of Peru and the southeastern United States.

Predicting Climate Change

Long-term changes in climate (those that occur over 10 to 100,000 years) are thought to be caused by periodic changes in the amount of incoming solar radiation, changes in atmospheric composition, changes in the configuration of ocean basins resulting from crustal plate movements, or changes in the amounts of gases and particles in the atmosphere after a meteorite impact. All of these possibilities must be studied using mathematical computer models, because it is impossible to observe them directly in a human lifetime.

Two approaches have been employed. One is the equilibrium model, in which calculations are run for extended periods of simulated time to see how the system responds. This approach is useful in providing information about how great the possible change might be. The other approach—called a transient model—uses changeable conditions in the ocean, atmosphere, or some other part of the Earth system. These simulations are more realistic but require even more computer time. Thus, relatively few transient models are run.

All models suffer from uncertainties due to our understanding of the basic processes involved. To fill these gaps in our knowledge, large studies and field experiments are run. An example is the World Ocean Circulation Experiment (WOCE) which is conducting a detailed study of deep-ocean circulation.

Chaos and Climate Prediction

Both weather and climate records have a great deal of short-term variability that we cannot yet explain, but, in theory at least, much of this variability will be

Figure 15–11
Weather forecasters now use a combination of surface measurements of winds and atmospheric conditions and global observations of the atmosphere made from Earth-orbiting satellites. These observations are combined in mathematical models which then predict weather conditions and storm tracks. (Bill Bachman/Photo Researchers, Inc.)

explained as our knowledge of the underlying processes improves. There is another factor, however, that contributes inherent instability to weather and climate: **chaos.** Depending on initial conditions, a system undergoing change may eventually assume any one of many different states. It is impossible to measure conditions accurately enough to predict how such a system will respond to even small disturbances. It is said that the flap of a butterfly's wings in Brazil can lead to a storm in North America.

Weather forecasters have dealt with this problem for decades (Fig. 15–11). Even with global observations provided by Earth-orbiting satellites, forecasts are most reliable over three-day periods. Beyond that, their reliability decreases. Beyond about ten days, it is impossible to predict weather because of the inherent instability—the chaotic nature—of the atmosphere and weather systems.

Severe Tropical Storms and the Greenhouse Effect

Severe tropical storms are the most important natural hazard in tropical and subtropical coastal regions. Strong, hurricane-force winds and their accompanying storm surges damage properties in low-lying areas and kill their inhabitants.

These storms get their energy from warm ocean-surface waters, and one of the most conspicuous effects of the greenhouse effect is to further warm the ocean surface layer. While it is unclear whether sea-surface temperatures will rise much above their present levels, it seems highly likely that warm surface waters will cover much larger areas of the ocean than at present. If water temperatures

BOX 15–1
Global Warming = Global Insurance Problem?

Global warming—if it occurs—will affect many industries and the economies of many countries. Among the first industries to be affected is insurance. Companies around the world insure properties against damage from floods and storms, and in the late 1980s and early 1990s, these companies were threatened financially by the large number of damaging storms that hit heavily populated areas.

The most expensive storm in this period was Hurricane Andrew, which narrowly missed hitting downtown Miami in August 1992; insurance claims cost nearly $20 billion. Before that, three storms had cost the industry nearly $5 billion each: Hurricane Hugo (U.S.) in September 1989 and storms Daria (Europe) in January 1990 and Mirelle (Japan) in September 1991. Because of their losses on these storms, the economic future of the famous Lloyds of London, an insurance exchange, was threatened.

Insurance premiums are based on losses incurred in earlier years, but global warming changes the probabilities of severe storms and where they might strike. Larger areas of warm ocean surface waters could increase the number and severity of hurricanes. Changed wind patterns could also alter the paths of these storms so that they strike areas not previously struck. Even the floods that devastated the midwestern United States in 1993 could be the precursor of new weather patterns that will cause substantial losses for insurers in years to come. Lastly, the continued rise of sea level will likely result in increased erosion of coasts and may increase the damage done by storms.

do indeed rise, stronger and more frequent hurricanes could occur. The more widespread warm waters might lead to more storms forming in areas where hurricanes do not now occur. Furthermore, such strong storms might strike more frequently in areas that now rarely experience hurricanes.

Oceanic Responses to Ozone Holes

As we learned in Chapter 6, ozone is concentrated in a layer high in the stratosphere that absorbs ultraviolet light from the Sun. This ozone layer shields Earth's surface from damaging radiation. In recent decades, scientists have found that this protective shield has been greatly reduced in both the Antarctic and the Arctic.

Since the early 1980s, the amount of ultraviolet radiation reaching the ground in Antarctica each spring has increased—sometimes only 20 percent, other times as much as 200 percent—as a result of the annual formation of the ozone hole (Fig. 15–12). The hole first appears in September, when ozone levels plunge by more than one third, and persists until November (late winter and early spring in the Southern Hemisphere).

The waters surrounding Antarctica are unique in the extent of their ice cover. At the same time, some of these areas are among the most productive in the world. Studies of some important organisms show that they are vulnerable to increased levels of ultraviolet radiation. For instance, algae and bacteria living in the lower layer of the ice, in contact with sea water, are thought to contribute up to half of the annual primary production in the region. Production by these organisms has apparently decreased by as much as 15 percent since the late 1980s, when the effects of the ozone hole were first studied (Fig. 15–13). In the years ahead, decreased production by these communities will have a pronounced effect on the organisms that feed on them (Fig. 15–14). Similar effects are expected in the shallow-water layer formed by fresh water released by melting ice, another highly productive layer. Studies show that marine phytoplankton in the Southern Ocean are already stressed by ultraviolet light.

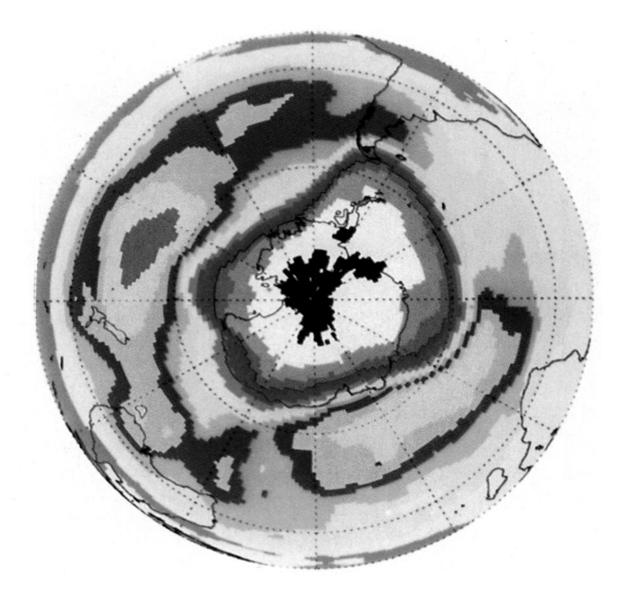

Figure 15–12
Holes in the stratospheric ozone layer over Antarctica form each year. The October 1994 minimum in the thickness of the Antarctic ozone layer is shown by all colors in the area contained within the blue boundary. (Courtesy NASA.)

Precisely how increased levels of ultraviolet light will affect major communities of organisms has not yet been demonstrated, but the problem is under investigation. The deleterious effects are likely to be most serious in one-celled plants and animals (plankton) because these are the organisms most exposed to the radiation. They are responsible for producing the food that feeds the larger animals in the food chain. Larger organisms would be affected in two ways: reduced food supply, and damage to the eyes and skin due to ultraviolet radiation.

At present, the effects of exposure to ultraviolet light are primarily investigated using mathematical models of Antarctic organisms and ecosystems. Using these models, scientists can predict how Antarctic ecosystems are likely to respond to the increased stress resulting from ultraviolet radiation, which is expected to be with us for decades. Of particular concern is the possibility that the loss of some critical organism or group of organisms will inhibit energy transfers through the ecosystems to further endanger already-threatened whale populations.

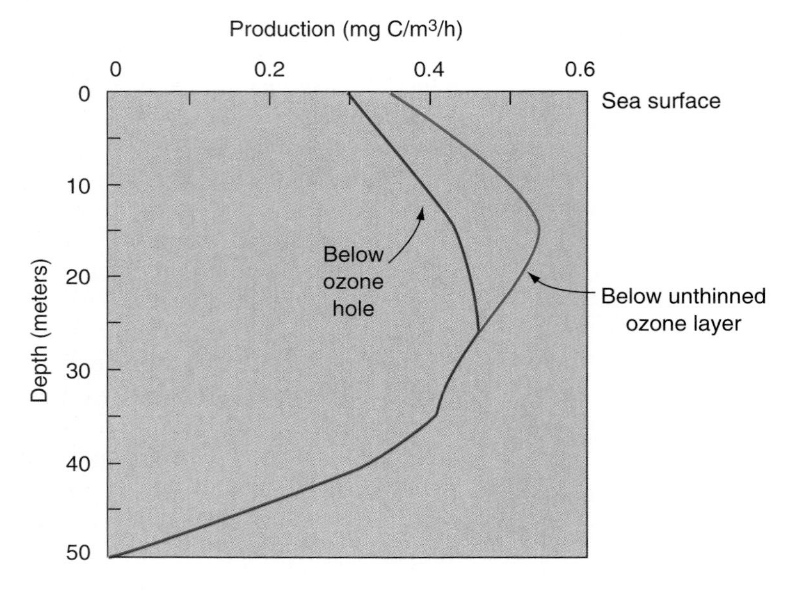

Figure 15–13
Estimates of primary production based on measurements in the ice zone of the Bellingshausen Sea near Antarctica. The left curve represents data taken inside the area exposed by the ozone hole where there is more ultraviolet radiation; the right curve gives estimates of primary productivity taken from ice that lies beneath a part of the stratosphere in which the ozone layer is its usual thickness. (Courtesy Scientific Committee on Problems of the Environment.)

Figure 15–14
Penguins are among the many Antarctic animals that depend on the production of Antarctic phytoplankton.

Protecting the Commons

The interdependence of all the various, interconnected subsystems on Earth highlights the problem of managing and protecting the **commons**—those parts of the planet that fall outside national jurisdictions. These areas—the open ocean, outer space, and Antarctica—are open to uncontrolled exploitation. We have already seen in Chapter 1 that the United Nations Law of the Sea Treaty placed about one-third of the ocean under various national jurisdictions. The rest remains unprotected. Efforts are underway to regulate some types of space activity. And in Antarctica, the Antarctic Treaty places restrictions on how nations may use both the continent and its surrounding waters—the coastal zone of the Southern Ocean.

Since there are no rules to limit use and no one owns the commons, they have been used and misused throughout human history. Commercial whaling destroyed some whale stocks and greatly reduced those that remained. Open-ocean commercial fishing has greatly reduced fish stocks in all the major fisheries. Some may never recover to their levels of productivity. Let us take a closer look at these problems.

Commercial Fisheries

Marine fish and shellfish provide about 5 percent of the total protein and about 10 percent of all animal protein in human diets. These are luxury items (lobster, shrimp, tuna) in diets in most developed countries, but in developing countries, fish are often major sources of protein—80 percent in Bangladesh, for example. Some coastal peoples, especially those living in small island states, get 95 percent of their protein from the sea (Fig. 15–15).

For years, many people thought the ocean could provide unlimited amounts of protein. (A common phrase was "the inexhaustible ocean.") One reason for this optimism was the rapid growth in fish catches between 1950 and 1990 (Fig. 15–16a),

Figure 15–15
Fishing provides an important source of animal protein in many coastal communities, especially on small islands. Here we see a traditional Polynesian net being thrown in Tahiti. (Jack Fields/Photo Researchers, Inc.)

when marine fish production increased fivefold, due primarily to increased fishing efforts and exploitation of newly discovered fish stocks. During this period, the number of fishing boats increased from 2 million to more than 3 million worldwide (Fig. 15–16b). Improved technologies helped commercial fishers find and harvest new fish stocks with greater efficiency. No matter how much commercial fishing expanded, however, the hope was always that, with proper management, the ocean's yield of protein could be virtually unlimited. Indeed, the expectation was that food from the sea would eliminate hunger all over the world.

The first signs that this infinite resource was only a dream came with the collapse of California's sardine fishery in the late 1950s. This was followed by the collapse of the Peruvian anchovy fishery. At its peak in 1972, the Peruvian fishery was the world's largest, producing 12 million tons per year. Since its collapse in the early 1970s, Peru's fish production has varied between 1 to 2 million tons annually, never approaching its former levels.

In 1989, global commercial marine fish and shellfish production was about 86 million tons; it has been decreasing since then. (Estimated maximum production is about 100 million tons.) About two-thirds of the catch is used directly for food. The remainder is used to make fish meal or oil to feed chickens or livestock, or as feed for aquaculture. At least one-quarter of the fish catch is wasted (and unrecorded) because it consists of less commercially valuable species—these fish are simply thrown overboard.

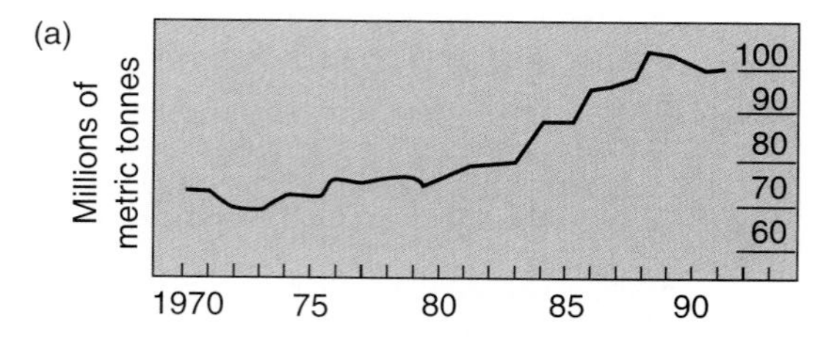

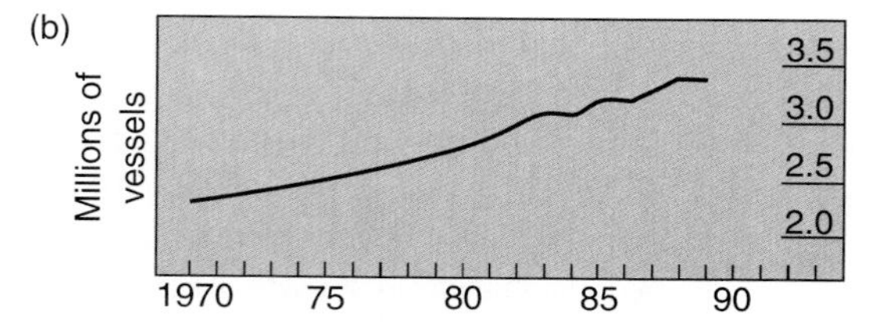

Figure 15–16
(a) Worldwide commercial fish catches increased steadily until 1989, then began to decrease. (b) One cause of the decline is the larger number of commercial fishing vessels currently in service. (Source: *The Economist,* March 19, 1994.)

The prognosis for major increases in fish catches is now poor. In 1990 the United Nations Food and Agricultural Organization reported that one-third of the world's 200 major fish stocks were overexploited. For instance, the Canadian cod fishery on the Grand Banks off Newfoundland has been closed. On the small portion of the Grand Banks outside Canada's Exclusive Economic Zone, however, fishing boats from several countries continue to compete for the disappearing fish stocks.

In 1994, a U.N. conference began to draft a treaty to regulate fishing on the high seas, which heretofore has been totally unregulated. Particular attention was directed to highly migratory fishes (such as tuna and swordfish) and straddling stocks (a fish stock partially in one country's EEZ and partially outside, where it is subject to uncontrolled fishing). These negotiations are likely to last many years.

The outlook for effective future management of fisheries is poor. The present overexploitation of fisheries worldwide is the result of massive subsidies by governments to allow commercial fisheries to buy more boats and to improve their fishing gear. Efforts to implement controls on coastal fisheries have usually been stymied by political pressures. In short, past fisheries management practices have effectively put many fishers out of work by destroying the fish stocks.

Future increases in production of living marine resources are most likely to come from the spread of aquaculture (Fig. 15–17). Already marine plants and animals are produced in large quantities, especially in China, Japan, and southeastern Asia. This industry is spreading to the many coastal communities where traditional fishing is no longer possible owing to overfishing.

Commercial Whaling

Whale hunting began many centuries ago with coastal dwellers using small boats and staying close to shore. By the sixteenth century, commercial whaling was established as European whalers began using large vessels to reach Arctic waters, opening new whaling grounds (Fig. 15–18). One of the richest early whaling grounds for Europeans was near Greenland. There, the *right whale* was hunted until it became nearly extinct. (It was called the right whale because it was easily killed, provided a lot of oil, and floated after death for easier recovery. In other

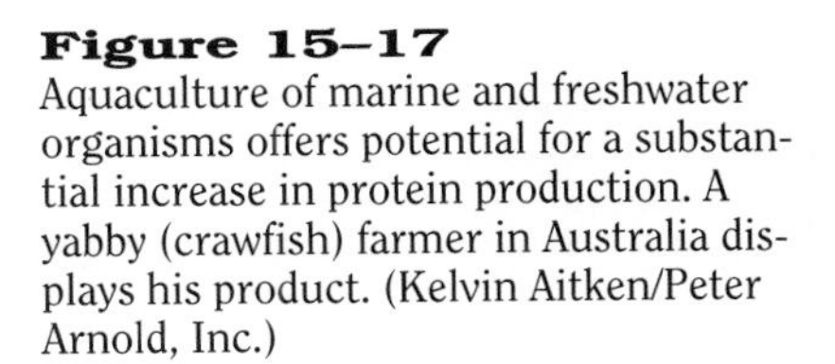

Figure 15–17
Aquaculture of marine and freshwater organisms offers potential for a substantial increase in protein production. A yabby (crawfish) farmer in Australia displays his product. (Kelvin Aitken/Peter Arnold, Inc.)

Figure 15–18
Whaling was an important industry for northern Europeans from the sixteenth to the early twentieth century. (Courtesy Bettmann.)

words, it was just right for the relatively primitive hunting technology of the day.) Whales provided meat and oil for lamps before the days of petroleum. Whalebone (from baleen) was used in dresses and corsets. Sperm whales provided ambergris and spermaceti, oily substances used to make cosmetics and candles. Today, whale oil still has many industrial uses, and whale meat is greatly prized in Japan.

As whale stocks declined, whalers had to go farther to find new ones. In the 1920s, whaling moved into the Southern Ocean around Antarctica. Fast, steam-driven vessels for hunting, factory vessels for processing whales at sea, and explosive harpoons greatly increased whale harvests and decimated whale populations (Fig. 15–19). Extinction of several whales and near extinction of others, including

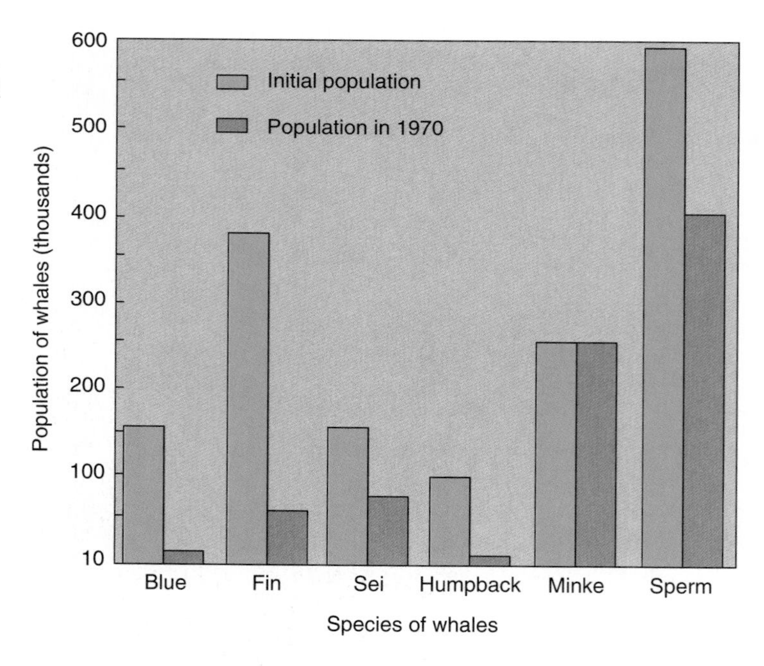

Figure 15–19
Populations of some whale species, showing the effects of unrestricted commercial whaling.

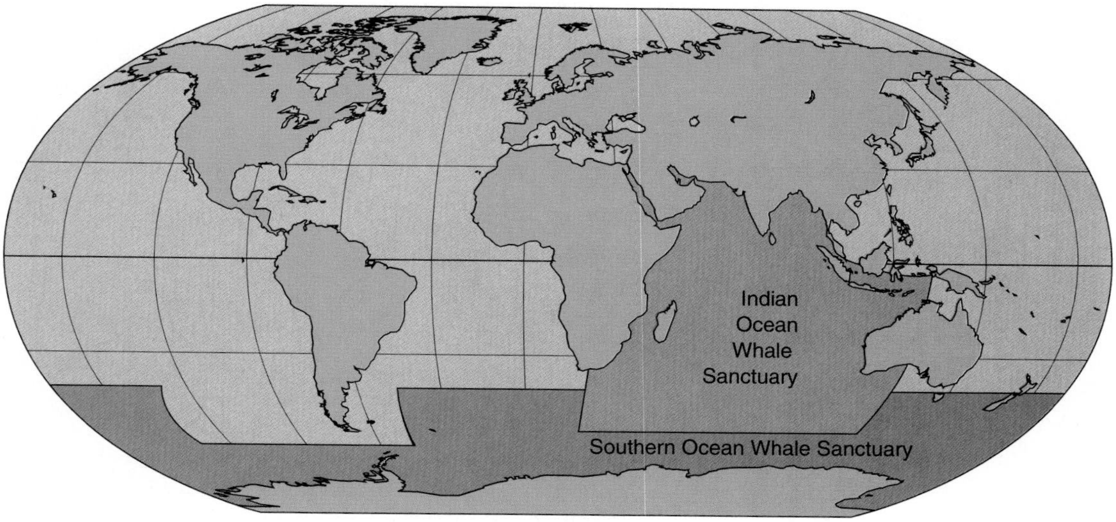

Figure 15–20
Whale sanctuaries in the Indian and Southern oceans. Commercial whaling is prohibited in these regions. (Courtesy U.S. Marine Mammal Commission.)

blue whales, eventually led to their protection. First the large blue whales (up to 150 tons) were taken, until their abundances dropped. Then whalers moved to the next largest species, the fin whale (50 tons). Thus far, the small *Minke whales* (6 tons) have not been taken in great numbers.

Eventually, commercial whaling on the high seas was phased out by the International Whaling Commission during the late 1980s, although some whaling continued for research purposes. A few whaling nations, such as Norway and Japan, have refused to be bound by the moratorium and continue to hunt whales, insisting they do it only for scientific research.

Because most whales are long-lived animals, recovery of their populations will take decades. Some, such as the California gray whale, have recovered enough to be removed from the endangered-species list. Others may never recover.

To promote the recovery of open-ocean whale stocks, two whale sanctuaries have been established: one in 1979 in the Indian Ocean and one in 1994 in the Southern Ocean (Fig. 15–20). All commercial whaling in these sanctuaries is prohibited.

Destruction of Coral Reefs

Coral reefs are extremely vulnerable to destruction by human activities (Fig. 15–21). Many occur in the densely populated coastal areas of some of the world's poorest countries. Thus, these reefs are poorly protected and often destroyed by changes in water quality, construction, and fishing.

A major problem in many coral reef areas is turbidity in the river waters that wash into the coastal ocean adjacent to the reef. The rivers are turbid because the forests through which they run are being cut to obtain valuable tropical woods. Because corals cannot tolerate much silt, they are often killed under these circumstances. Also, many coastal wetlands are destroyed to build tourist or industrial facilities, and such development results in the eventual destruction of nearby coral reefs. Discharges of nutrients in sewage and agricultural runoff can kill corals, as can discharges of pesticides.

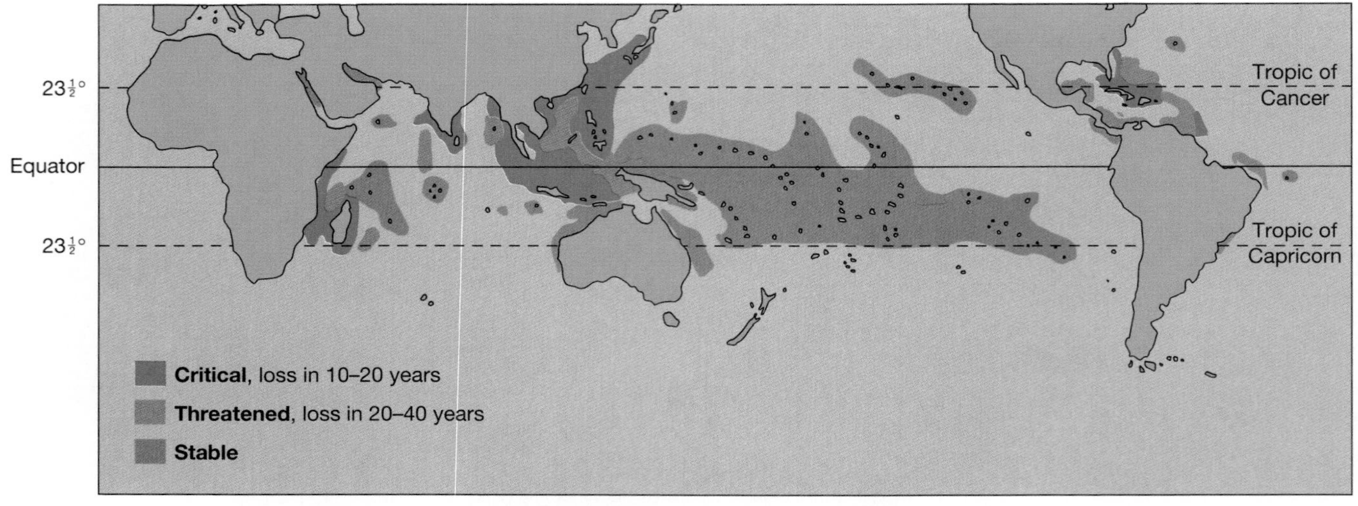

Figure 15–21
Coral reefs in densely populated coastal-ocean areas are threatened by overuse. Only reefs in lightly populated coastal regions (for example, Australia) and on oceanic islands are relatively safe. (Source: C. R. Wilkinson, Australian Institute of Marine Sciences, Townsville, Australia.)

The most obvious destruction of coral reefs comes from some widespread but small-scale fishing activities (Fig. 15–22). One particularly destructive technique uses explosives to stun reef fishes, which are then collected. Unfortunately, large areas of reef are destroyed in the process. Another technique involves breaking corals with sticks or heavy weights to scare fish into nets deployed outside the reef. Finally, poisons such as DDT or pesticides are applied directly to waters over the reef to kill the fish so that they can be caught. All these methods leave dead or dying corals.

Figure 15–22
In many areas, recreational and commercial fishing can dramatically deplete fish and shellfish stocks on coral reefs. (Courtesy Al Grotell.)

In areas of heavy tourism, collection of coral and other organisms can denude a reef. As a result, once-rich coral reefs off Sri Lanka, a large island south of India, no longer exist. Corals are also used to obtain sand for construction and for raw materials to make cement. In some island states, exploitation of coral reefs threatens tourism, which is often the largest single industry.

Biodiversity

Human activities usually superimpose unnatural processes on an ecosystem. These imposed processes can greatly increase the rate of loss of species and may even change the course of evolution. A significant decrease in biodiversity is the result. **Biodiversity**—defined as species richness—is a difficult concept for many of us to understand, mainly because the term *richness* means different things. At the simplest level, biodiversity means many different species. Biodiversity is an index of the number of different species living together in an area, be that area local, regional, or even global. Philosophical concerns about biodiversity have many dimensions. One concern is that human activities are wantonly eliminating species, communities, and even whole ecosystems. This concern is basically a moral one: that it is immoral for humans to reduce Earth's biodiversity and that everyone is diminished when we destroy the richness of life on our planet. Each species, each community, each ecosystem is a unique assemblage, and once any such assemblage is driven to extinction, Earth's biodiversity is irreversibly diminished. A problem in this argument is that biologists do not know the species composition of entire regions, so it may be difficult to know when diversity is being reduced. The system most commonly cited as an example of our lack of knowledge is the tropical rain forest, in the Amazon basin. But the same argument can also be made about the bottom of the deep ocean, coral reefs, and many other communities and ecosystems. The cost for biologists to thoroughly study these systems is extremely high, and the slowness of such activities would doubtlessly result in the loss of many species, communities, and ecosystems as the efforts proceeded.

Another level of concern is utilitarian. Many medicines and other useful substances come from plants and animals. Thus, losing a single species reduces the possibility of discovering substances for human uses. Therefore, the argument goes, humans should seek to retain Earth's biodiversity because it contains unknown resources that may prove to be extremely valuable to us in the future. This type of argument is often used to justify preserving genetic and species biodiversity.

There is also concern about biodiversity on the aesthetic level. With the continuing loss of communities and ecosystems, humans are reducing the richness of our existence. For example, the virtual elimination of tall-grass prairies and the gigantic bison herds that grazed them has left the mid-continental United States much poorer aesthetically. It also virtually destroyed many Plains Indian tribes whose culture was based on hunting bison. European settlers replaced these prairie systems with fields of wheat, corn, and other cultivated crops, making one of the richest agricultural areas in the world. Crop lands are usually used for growing a single species—corn or wheat—replacing much more diverse ecosystems. Tall-grass prairies and bison are now restricted to specially preserved small areas where one must go to experience the environments that once dominated the North American plains. Only about 3 percent of the original terrestrial ecosystems in the contiguous United States remain.

Adding to the confusion over biodiversity is the fact that most species that ever lived became extinct long before humans appeared. Our view of the richness of life forms is only a snapshot of the present state of ecosystems that have continually changed since life first appeared on Earth. Despite our best efforts, extinctions will continue to occur due to natural processes of changing climate, movements of continents, and even catastrophic events like the crashing to Earth, 65 million years ago, of the large meteorite that caused the extinction of dinosaurs. If the dinosaurs had not become extinct, however, mammals most certainly would not have evolved to their present conditions. Following this line of reasoning, one can argue that without the extinction of dinosaurs, humans and human societies would have been impossible.

Exotic Organisms

One result of faster ships and increased ship transport is the unintentional introduction of organisms into new areas. Any species living outside its native habitat is referred to as an **exotic species.** Larval stages of many marine organisms as well as bacteria can survive in the bilges and ballast waters of ships during their oceanic transits. The volumes of water moved in bilge and ballast tanks are quite large. A single coal-carrier takes on a volume of ballast water equal to that of a modest-sized lake; it can take up to three days to fully ballast an empty coal carrier.

When an empty ship reaches port, it pumps out its ballast tanks as it takes on new cargo, and a ship discharging cargo takes on new ballast waters. In this way, larvae and other microorganisms are widely dispersed unless special care is taken to prevent such transfers. The introduction of exotic species in this way is especially likely if there are only small differences in temperatures and salinities between the ports of origin and the final destination. Lacking natural enemies, exotic organisms can expand their range very rapidly and often cause substantial economic losses when they disturb other organisms and other human uses of the affected waters.

One well-known example of such importation of an exotic species is the introduction of zebra mussels (Fig. 15–23) into the North American Great Lakes and their subsequent dispersal throughout nearby lakes and rivers. The first mussels apparently arrived in the ballast waters of a trans-Atlantic freighter in 1985 or 1986. The unwelcome newcomer, native to the Caspian Sea, has no natural ene-

Figure 15–23
Zebra mussels have spread widely in North America after being introduced from the Caspian Sea in 1985 or 1986. (Scott Camazine/Photo Researchers, Inc.)

mies in the Great Lakes except ducks. It first appeared in the waters of Lake St. Clair between Lakes Erie and Huron and has since spread into all five Great Lakes and into the New York State Barge Canal.

Because the thumbnail-sized mussels can survive exposure for several days, they have been transported on boats carried into many inland lakes and rivers. Scientists fear that the mussel will spread throughout most of the United States and southern Canada.

Zebra mussels produce prodigious quantities of larvae that can be transported hundreds of kilometers by currents. An adult female produces up to 50,000 eggs per year. Adults attach themselves to hard surfaces, including pipes, buoys, and piers, where they can obstruct water flows in power plants and water-treatment plants. Since the first appearance of the mussels in North America, it has cost industries billions of dollars to clean water-intake pipes and boat hulls. They can also attach themselves in large numbers on the rocky lake bottoms that fish such as walleyes and lake trout need in order to spawn; the invasion of the mussels makes these rocky areas no longer available to the fish. Thus, the zebra mussel invasion may further damage the local fisheries, which are still recovering from all the problems caused by sea lampreys invading the Great Lakes after the 1959 opening of the St. Lawrence Seaway.

Disease-causing organisms have spread in much the same way, as have red-tide organisms, which have in recent years been dispersed into areas previously free of them. Cholera was spread widely through South America by ballast-water discharges.

One solution is to have ships dump ballast waters in the open sea rather than in port. Out to sea, the transported organisms are unlikely to survive in the saltier water and less likely to be carried into areas where they come into contact with humans or marine resources, such as shellfish. Another control mechanism is to treat ballast waters, thereby killing any organisms in them.

Increasing Human Populations

Predictions are that the world's population will double in the next fifty years, increasing from its 1990 level of 5.3 billion to more than 10 billion by 2050 (Table 15–1). Today, about half the population in industrialized countries lives within 1 kilometer of a coast. In the next few decades, this concentration of population in coastal regions is expected to increase. Thus, much of the expected increase in human populations will take place in coastal areas.

TABLE 15–1
Global Population Growth

Year	*Population (millions)*
1000	400
1650	500
1800	1,000
1930	2,000
1974	4,000
1998 (projected)	6,000
2030 (projected)	9,000

Source: United Nations Population Reference Bureau

BOX 15–2
Plastic Litter

Floating debris, from plastic soda bottles to medical syringes, washes up each day on beaches (Fig. B15–2–1). This debris comes from many sources, primarily ships, ranging from small recreational boats to large cruise liners. New efforts to educate boaters and enforce antilitter laws are reducing plastic pollution.

Plastic nets from commercial fishing are often damaged, leaving pieces to float free in the ocean. These bits of nets continue to catch and kill fish and to ensnare seabirds. Untended plastic lobster traps last for years, trapping and killing lobsters. Plastic lines also persist for many years to wash up on beaches. This source, too, is diminishing as a result of better education of fishers and stricter enforcement of laws against discarding plastic debris.

Discharges from sewage systems continue to supply some of the most troublesome items, such as medical syringes. This plastic garbage, which reaches the ocean through sewers, is perhaps the most difficult to control because of the many sources and because of our inability to prevent debris being thrown into streets and eventually draining into a coastal waterway.

Some plastic debris is dangerous to marine animals. The plastic can rings that hold a six-pack of beer or soda can choke animals. Environmental groups urge us to cut all six rings open before throwing the plastic away. Plastic debris also kills marine animals who mistake it for food. For instance, sea turtles can die after eating plastic bags mistaken for jellyfish. Birds and other animals often die after eating bits of plastic rope that they mistake for worms.

Figure B15–2–1
A variety of plastic litter washes up on even the most remote tropical beaches.

The effects this large population will have on the ocean will be largely confined to coastal lands and coastal ocean areas. The most direct effects will be alteration of coastal land areas to accommodate the expanding population and its attendant industrial activities, roads, airports, and so forth. In addition to filling salt marshes for various uses, offshore islands will be constructed for high-cost facilities such as airports.

Another effect of these large populations will be increased dependence on coastal lands and coastal oceans as dump sites for agricultural, domestic, and industrial wastes. Waste disposal will change ecosystem composition, of course. We shall see these effects first on the top predators—whales, dolphins, seabirds—

but there will also be effects lower down in the food webs. In addition to the environmental havoc it will cause, coastal dumping will inhibit the growth of recreation and tourism as the world's largest industries.

We can estimate the results of future population increases by looking at the effects of population growth over the last few decades. Increasingly, conflicting uses of coastal space will result in elimination or reduction of coastal fisheries. First the fish populations will disappear as a result of over-exploitation and poor resource management. Poor waste-disposal practices will result in increasing pollution problems, such as increased incidences of red tides and of anoxic events in areas that have poor water circulation.

Sustainable Development

Environmental degradation was long thought to be a problem only for rich or developed countries. As developing countries work to achieve the levels of affluence of the more developed countries, they often find that environmental limitations and degradation pose serious problems. This became especially apparent after the end of the Cold War in the early 1990s. The countries of the former Soviet Union and its allies inherited mind-boggling problems of radioactive wastes and oil spills that leaked into rivers and eventually into the Arctic Ocean. Even developed countries have huge toxic waste dumps and massive accumulations of radioactive wastes that pose environmental problems.

Despite this somber outlook, economic development continues to be the primary goal for most developing countries. The problem is how to achieve high levels of economic development and affluence without paying too high an environmental price. An attractive possibility is the concept of **sustainable development,** which means development that improves the lives of today's populations without depleting natural resources or causing environmental problems for future generations to deal with. Such development must occur while respecting environmental constraints. For instance, it requires (1) the use of less material-intensive processes and (2) support and change at many levels of society—industry, local planning boards, national environmental protection/management activities, and international development bankers.

Tapping the creative potential of human populations is essential to sustainable development. The emergence of an information-based economy provides possibilities. Creating wealth by developing new intellectual products, such as software, uses much less material than building automobiles and highways. At the same time, increased affluence is likely to increase the demand for such things as cars.

A major challenge is developing sustainable, nondestructive sources of power to replace fossil fuels and nuclear reactors. Both sources of energy leave their wastes as problems for present and future generations. We discussed several potential energy sources in earlier chapters. For instance, heat from the Earth's interior, or geothermal power, is one possibility. In areas of volcanic activity, there may be useable sources of geothermal energy. Use of energy from this source is growing about 15 percent per year. Present geothermal energy practices can cause local environmental problems, but not on a global scale.

Solar power is another attractive source of energy. Energy from the Sun is already used in many places around the world for heating water in homes. The technology is simple and easily transferred to developing countries. The use of solar energy to generate electricity is also growing. Tapping the Sun's energy still requires developing processes and materials that can compete with fossil fuels, but solar energy is already used in spacecraft and other remote locations.

Burning biomass, such as wood, is already a widespread source of energy in developing countries. The result has been deforestation on a global scale and exposure of humans to the health problems caused by inhaling wood smoke.

BOX 15-3
Model Studies

Models (mathematical expressions of the processes involved) are used to study how various substances behave in the ocean. Models of different complexity have been used. The simplest is a box model of an atmosphere over an ocean, with a mixed layer isolating a deep layer (Fig. B15–3–1a). Gases from the atmosphere move into the deep layer through diffusion. In other words, there are no currents.

The next level of complexity is a model (Fig. B15–3–1b) with the deep ocean waters outcropping on either side of a surface mixed layer. Here, the ocean interacts with the atmosphere through the surface layer and directly through the deep-ocean outcrops in the polar regions. Still more complex models (Fig. B15–3–1c) involve many interconnected boxes that represent the relationships among major oceanic reservoirs with the atmosphere.

Even more complex models consist of a series of boxes (Fig. B15–3–2) in both the ocean and atmosphere. In addition to transfers of materials and energy, such models include chemical and biological reactions. As the models become more complex, they require larger and faster computers for the calculations.

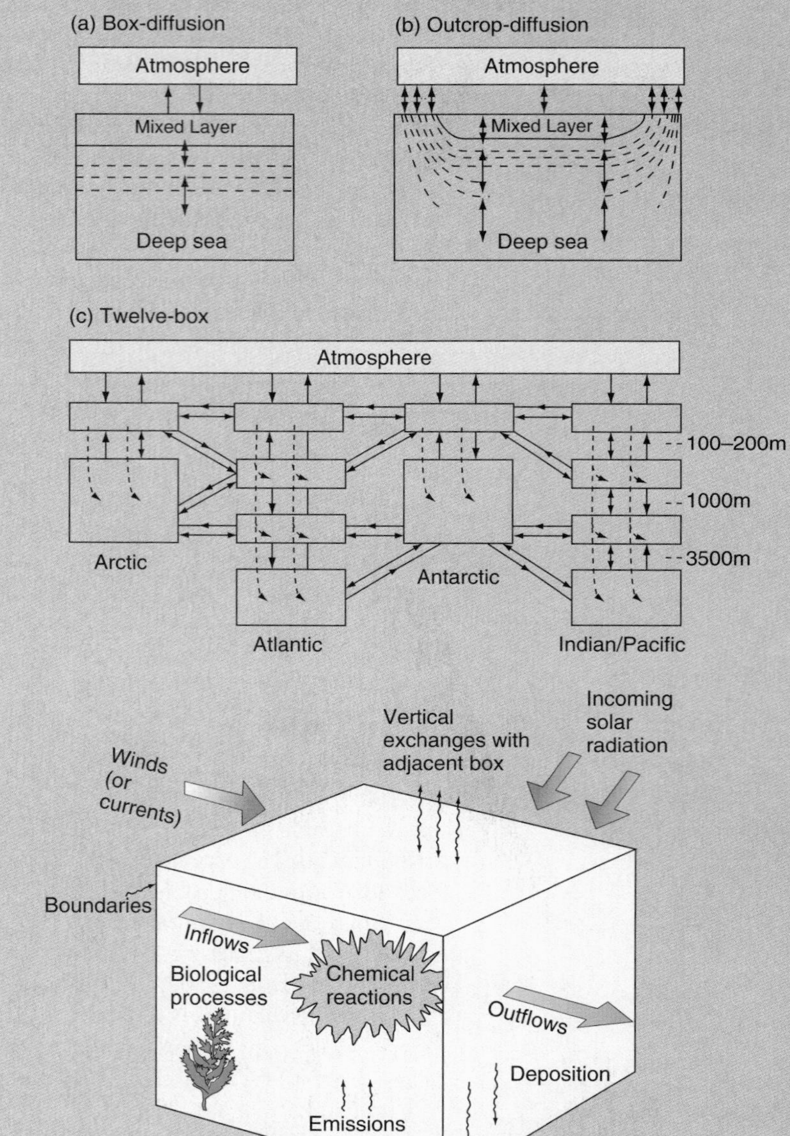

Figure B15–3–1
Different models of increasing complexity are used to study interactions between ocean and atmosphere.

Figure B15–3–2
Conceptual diagram for a component box in a complex-box model used to study oceanic and atmospheric processes. Note that some processes occur within the box, while others involve transfers of energy and matter across the box boundaries.

Cod and Tuna Wars

Competition for the dwindling numbers of many commercially valuable fishes—cod, tuna, swordfish, salmon—has greatly increased the possibility of conflicts on the high seas. The first of these occurred in the 1970s in the so-called "Cod Wars," when British trawlers and Icelandic gunboats confronted each other over the rights to take cod from waters near Iceland.

The Law of the Sea (LOS), an international agreement which went into effect in 1994, ended some of these conflicts. The LOS allows nations to regulate fisheries within 200 miles of their shorelines. Some nations have extended their fisheries jurisdictions even further. Others, such as Canada and Russia, have warned that they will do even more to control what they consider to be illegal fishing of stocks that range beyond their 200-mile limits. Such conflicts have occurred in the Sea of Okhotsk, which Russia has declared off limits to fishing for pollock. Off Newfoundland, both Spanish and Portuguese fishing boats were taken by the Canadians for taking fish in excess of the amounts that they were allotted by international agreements and for using nets that were too small.

Similar conflicts have occurred between British and Spanish boats fishing for tuna near the Azores. Here the problem arose from the ships' using different, incompatible fishing methods. The British use long drift nets that ensnare the fish by their gills as they swim. The Spanish boats, using long lines with baited hooks, were accused of destroying British nets. Gunboats from the two countries were sent to escort their ships.

As commercial fishery stocks are threatened by overfishing, we are likely to see more such "fish wars" in the future.

Despite decades of efforts to develop sources of biomass, such as using crops to make alcohol to power automobiles, this approach still requires enormous governmental subsidies and causes local environmental problems.

Deriving energy from winds is possible in many locations. However, this method of energy production is expensive, causes noise problems, and is considered by many to be unsightly. Further, winds, like tides and waves, provide power that must be used when and where it is generated.

We still lack the ability to store and transport any form of energy easily. One possibility is to make liquid hydrogen fuel to store energy and then transport it the way we transport other gaseous and liquid fuels.

Summary

Earth's climate is controlled by incoming solar radiation and interactions among Earth's various subsystems. Its climate has varied within narrow limits, controlled primarily by the presence of liquid water on Earth's surface. Since the beginning of the Industrial Revolution, carbon dioxide from burning fossil fuels and deforestation has collected in the atmosphere, causing atmospheric temperatures to rise about 0.5°C. Ice cores from Greenland and Antarctica record the increases in atmospheric carbon dioxide concentrations over the past 200,000 years.

Understanding how future carbon dioxide discharges will affect Earth requires a better understanding of its carbon cycle in the ocean, in the atmosphere, and on land. Models of varying complexity are used to calculate how carbon dioxide behaves in the Earth system and how this will affect its climate.

Increased surface water temperatures may affect the frequency and distribution of tropical storms. Discharges of chlorinated hydrocarbons to the atmosphere have affected concentrations of ozone in the stratosphere, resulting in a hole in the ozone layer over Antarctica and the Arctic. The increased ultraviolet radiation reaching Earth's surface as a result of this ozone hole will affect organisms on land and in the surface ocean.

The commons—areas outside national jurisdiction—are especially vulnerable to overexploitation and abuse. Poor management of the world's fish stocks has resulted in overexploitation and a decline in fish production. The prognosis is poor for better protection of fisheries in the future. Future increases in food production from the ocean are likely to come from aquaculture, not from fishing wild stocks.

Uncontrolled harvesting of whales led to the near extinction of many species. These are now protected by an internationally imposed moratorium on whaling, which is respected by all countries except Japan and Norway.

Destruction of coral reefs is a serious problem around the world. Many have been damaged by waste and sediment discharges, and from fishing practices in many countries. Only reefs in lightly populated coastal areas and on oceanic islands are considered stable at this time.

Biodiversity is a serious issue as many species become extinct due to human activities, especially in tropical rain forests and coral reefs. The question has both moral aspects as well as utilitarian ones.

Unintentional introductions of exotic organisms into estuarine and coastal ocean waters is an increasingly serious problem. The organisms are carried in the ballast waters used to maintain stability of large ships. Discharged into new areas with no natural predators, these organisms can quickly multiply to nuisance levels and destroy local ecosystems.

Sustainable economic development, using natural resources today without damaging the systems for future generations' use, is a major challenge for both developing and developed countries. One major problem is to develop energy sources that do not deplete resources for the future.

Key Terms

chaos
commons
biodiversity
exotic species
sustainable development

Study Questions

1. Explain how Earth's surface temperatures are restricted to a narrow range, unlike those on other planets.
2. Discuss why temperature changes associated with changes in Earth's atmosphere have been difficult to detect.
3. Why must mathematical models be used to make predictions about climate?
4. How might a warmer ocean surface affect the frequency and distribution of hurricanes?
5. How might the stratospheric ozone hole affect the ocean?
6. Explain how chlorinated hydrocarbons can cause depletion of the stratospheric ozone.
7. Why are Earth's commons areas so vulnerable to overexploitation?
8. Discuss how government subsidies and fisheries management contributed to the decline of commercial fish catches worldwide and to the collapse of individual major fisheries.
9. How has international regulation of commercial whaling affected whale stocks?
10. How have the development and expansion of tourism affected coral reefs?
11. Describe the problems caused when exotic organisms arrive in coastal ocean and estuaries.
12. *[critical thinking]* Explain how data from ice cores and sediment deposits are used to decipher Earth's climatic history.
13. *[critical thinking]* How could improved forecasts of ENSO events help improve agricultural production?

Selected References

BORGESE, E. M., N. GINSBURG AND J. R. MORGAN: *Ocean Yearbook 11*. Chicago: University of Chicago Press, 1994. Annual summary of marine ecological and economic data.

BROECKER, Wallace S., *How to Build a Habitable Planet*. Palisades, New York: Eldigo Press, 1985. Elementary textbook on the Earth System.

BROWN, L. R., (ed.), *State of the World*. New York: Norton, 1994. Annual global analysis of the world and prospects for developing a sustainable economy.

FIROR, J., *The Changing Atmosphere*. New Haven, CT: Yale University Press, 1990.

GLEIK, J., *Chaos*. New York: Viking, 1987.

GORE, A., *Earth in the Balance.* Boston: Houghton Mifflin, 1992. Discussion of how human activities have affected Earth and its climate.

GRAEDEL, T. E., AND P. J. CRUTZEN, *Atmosphere, Climate, and Change.* New York: Scientific American Library, 1995. Elementary presentation of atmospheric changes and their effect on the Earth system.

MACKENZIE, F. T., AND J. A. MACKENZIE, *Our Changing Planet.* Englewood Cliffs, N. J.: Prentice-Hall, 1995.

OFFICER, C., AND J. PAGE, *Tales of the Earth.* New York: Oxford University Press, 1993. Discussions of major events and processes affecting Earth's climate and weather.

STOLARSKI, R. S., "The Antarctic ozone hole," *Scientific American,* 258(1):30 (1988).

WHITE, R. M., "The great climate debate," *Scientific American* 263(1):36 (1990).

WOODWELL, G. M., AND F. T. MACKENZIE, *Biotic Feedbacks in the Global Climate System.* New York: Oxford University Press, 1994. Comprehensive survey of the effects of global climatic warming on Earth's biota.

WORLD COMMISSION ON ENVIRONMENT AND DEVELOPMENT, *Our Common Future.* New York: Oxford University Press, 1987. Evaluation of the status of environmental and development problems and a framework for solving them without damaging resources needed by future generations.

APPENDIX 1

Conversion Factors

Exponential Notation

It is often necessary to use very large or very small numbers to describe the ocean or make calculations about its processes. To simplify writing such numbers, scientists commonly indicate the number of zeros by **exponential notation,** indicating powers of ten or the number of zeros. Some examples and some common prefixes are

1,000,000,000 = 10^9	(one billion)	
1,000,000 = 10^6	(one million)	
1000 = 10^3	(one thousand)	kilo-
100 = 10^2	(one hundred)	
10 = 10^1	(ten)	
1 = 10^0	(one)	
0.1 = 10^{-1}	(one tenth)	deci-
0.01 = 10^{-2}	(one hundredth)	centi-
0.001 = 10^{-3}	(one thousandth)	milli-
0.000,001 = 10^{-6}	(one millionth)	micro-
0.000,000,001 = 10^{-9}	(one billionth)	nano-

Multiplication

To multiply exponential numbers (powers of ten), add exponents. For example, 10 x 100 = 1000, which is written exponentially as $10^1 \times 10^2 = 10^3$

Division

To divide exponential numbers, subtract the exponent of the divisor from the exponent of the dividend. For example, 100/10 = 10, which is written exponentially as $10^2/10^1 = 10^1$

Units of Measure

Temperature

Conversion formulas

$$°C = \frac{(°F - 32)}{1.8}$$

$$°F = (1.8 \times °C) + 32$$

Conversion Table

°C	°F
–10	14
0	32
10	50
20	68
30	86
40	104
100	212

Length

1 kilometer (km) = 10^3 meters = 0.621 statute mile
= 0.540 nautical mile
1 meter (m) = 10^2 centimeters = 39.4 inches = 3.28 feet
= 1.09 yards = 0.547 fathom
1 centimeter (cm) = 10 millimeters = 0.394 inch
= 10^4 micrometers
1 micrometer (µm) = 10^{-3} millimeter = 0.0000394 inch

Area

1 square centimeter (cm^2) = 0.155 square inch
1 square meter (m^2) = 10.7 square feet
1 square kilometer (km^2) = 0.386 square statute mile
= 0.292 square nautical mile

Volume

1 cubic kilometer (km^3) = 10^9 cubic meters = 10^{15} cubic centimeters = 0.24 cubic statute mile
1 cubic meter (m^3) = 10^6 cubic centimeters = 10^3 liters
= 35.3 cubic feet = 264 U.S. gallons
1 liter (l) = 10^3 cubic centimeters = 1.06 quarts
= 0.264 U.S. gallon
1 cubic centimeter (cm^3) = 0.061 cubic inch

Mass

1 metric ton = 10^6 grams = 2205 pounds
1 kilogram (kg) = 10^3 grams = 2.205 pounds
1 gram (g) = 0.035 ounce

Time

1 day = 8.64×10^4 seconds (mean solar day)
1 year = 8765.8 hours = 3.156×10^7 seconds (mean solar year)

Speed

1 knot (nautical mile per hour) = 1.15 statute miles per hour
= 0.51 meter per second
1 meter per second (m/s) = 2.24 statute miles per hour
= 1.94 knots
1 centimeter per second (cm/s) = 1.97 feet per minute
= 0.033 feet per second

Energy

1 gram-calorie (cal) = 1/860 watt-hour = 1/252 British thermal unit (Btu)

APPENDIX 2

Useful Data about Earth and Its Ocean

TABLE A2–1
Dimensions of Earth

Size and Shape of Earth		
Dimensions	*Miles*	*Kilometers*
Equatorial radius	3963	6378
Polar radius	3950	6357
Average radius	3956	6371
Equatorial circumference	24,902	40,077
Areas of Earth, Land, and Ocean		
	Millions of	
Part of Earth	*Square Miles*	*Square Kilometers*
Land (29.22%)	57.5	149
Ice sheets and glaciers	6	15.6
Oceans and seas (70.78%)	139.4	361
Land + continental shelf	68.5	177.4
Oceans/seas - shelf	128.4	332.6
Total area of Earth	196.9	510.0
Distribution of Land and Water on Earth's Surface		
Hemisphere	*Land (percent)*	*Ocean (percent)*
Northern	39.3	60.7
Southern	19.1	80.9

TABLE A2–2
Heights and Depths of Earth's Surface

Height (Land)	*Feet*	*Meters*	*Depth* (Oceans and Seas)	*Feet*	*Meters*
Greatest height:			Greatest depth:		
Mount Everest	29,028	8848	Mariana Trench	36,200	11,035
Average height	2757	840	Average depth	12,460	3800

TABLE A2–3
Volume, Density, and Mass of Earth and Its Parts

Part of Earth	*Average Thickness or Radius (km)*	*Volume (10^6 km^3)*	*Mean Density (g/cm^3)*	*Mass (10^{24} g)*	*Relative Abundance (%)*
Atmosphere	—	—	—	0.005	0.00008
Oceans and seas	3.8	1370	1.03	1.41	0.023
Glaciers	1.6	25	0.90	0.023	0.0004
Crust					
Continents*	35	6210	2.8	17.39	0.29
Oceanic†	8	2660	2.9	7.71	0.13
Mantle	2881	898,000	4.53	4068	68.1
Core	3473	175,500	10.72	1881	31.5
Whole Earth	6371	1,083,230	5.517	5976	—

*Including continental shelves.
†Excluding continental shelves.

TABLE A2–4
Ocean Provinces

*Ocean**	*Shelf and Slope (percent)*	*Continental Rise (percent)*	*Deep-Ocean Floor (percent)*	*Volcanoes and Volcanic Ridges (percent)*	*Rise and Ridge (percent)*	*Trenches (percent)*
Pacific	13.1	2.7	43.0	2.5	35.9	2.9
Atlantic	19.4	8.5	38.0	2.1	31.2	0.7
Indian	9.1	5.7	49.2	5.4	30.2	0.3
All ocean	15.3	5.3	41.8	3.1	32.7	1.7
Earth's surface	10.8	3.7	29.5	2.2	23.1	1.2

After H. W. Menard and S. M. Smith. 1966. Hypsometry of ocean basin provinces. *J. Geophys. Res.* 71:4305.
*Includes adjacent seas — for example, Arctic Sea included in Atlantic Ocean.

TABLE A2–5
Surface and Drainage Areas of Ocean Basins (in millions of square kilometers) and Their Average Depths (in meters)

*Ocean**	*Ocean Area (millions of square kilometers)*	*Land Area Drained† (millions of square kilometers)*	*Ratio of Ocean Area to Drainage Area*	*Average Depth† (meters)*
Pacific	180	19	11	3940
Atlantic	107	69	1.5	3310
Indian	74	13	5.7	3840

From H. W. Menard and S. M. Smith: Hypsometry of Ocean Basin Provinces. *J. Geophys. Res.* 71:4305 (1966).

*Includes adjacent seas. Arctic, Mediterranean, and Black Seas included in the Atlantic Ocean.

†Excludes Antarctica and continental areas with no exterior drainage.

TABLE A2–6
Average Temperatures and Salinity of the Oceans, Excluding Adjacent Seas

	Temperature (°C)	*Salinity (parts per thousand)*
Pacific (total)	3.14	34.60
North Pacific	3.13	34.57
South Pacific	3.50	34.63
Indian (total)	3.88	34.78
Atlantic (total)	3.99	34.92
North Atlantic	5.08	35.09
South Atlantic	3.81	34.84
Southern Ocean*	0.71	34.65
World ocean (total)	3.51	34.72

After L. V. Worthington: The Water Masses of the World Ocean: Some Results of a Fine-Scale Census. In B.A. Warren and C. Wunsch (eds.): *Evolution of Physical Oceanography.* Cambridge, Mass.: MIT Press, 1981.

*Ocean area surrounding Antarctica, south of 55°S.

TABLE A2–7
Water Sources for the Major Ocean Basins (centimeters per year)

Ocean	*Precipitation*	*Runoff from Adjoining Land Areas*	*Evaporation*	*Water Exchange with Other Oceans*
Atlantic	78	20	104	6
Arctic	24	23	12	35
Indian	101	7	138	30
Pacific	121	6	114	13

From M. I. Budyko: *The Heat Balance of the Earth's Surface* (N. A. Stepanova, trans.). Office of Technical Services. Department of Commerce, Washington, D.C., 1958.

TABLE A2–8
Characteristics of Trenches

	Average		
	Depth (kilometers)	*Length (kilometers)*	*Width (kilometers)*
Pacific Ocean			
Kurile-Kamchatka Trench	10.5	2200	120
Japan Trench	8.4	800	100
Bonin Trench	9.8	800	90
Mariana Trench	1.0	2550	70
Philippine Trench	10.5	1400	60
Tonga Trench	10.8	1400	55
Kermadec Trench	10.0	1500	40
Aleutian Trench	7.7	3700	50
Middle America Trench	6.7	2800	40
Peru-Chile Trench	8.1	5900	100
Indian Ocean			
Java Trench	7.5	4500	80
Atlantic Ocean			
South Sandwich Trench	8.4	1450	90

After R. W. Fairbridge: Trenches and Related Deep-Sea Troughs. In R.W. Fairbridge (ed.): *The Encyclopedia of Oceanography*. New York: Reinhold, 1966.

TABLE A2–9
Major Types of Oceanic Phytoplankton

Diatoms
Silica and pectin "pillbox" cell wall, sculptured designs;
Important for coastal ocean productivity; floating and attached forms
Everywhere in surface ocean, especially in colder waters, upwelling areas, even in polar ice; some heterotrophic below photic zone; some form "resting spores" under adverse conditions
Size: 0.01–0.2 mm
Yellow-green or brownish; single cells or chains of cells; radial or bilateral symmetry; many have spines or other flotation devices
Division, splitting of nuclear material; average reduction of one cell-wall thickness at each division; when limiting size is reached, cell contents escape, form new cell

Dinoflagellates
Next to diatoms in productivity; many heterotrophic, ingest particulate food; some have cellulose "armor"; very small open-ocean species are naked
In all seas, and below photic zone; some parasitic; warm-water species very diverse; some have resting stage for protection; sometimes abundant in coastal areas as "red tides"
Size: 0.005–0.1 mm
Usually brownish, one-celled; have two whiplike flagellae for locomotion; many are luminescent
Simple, longitudinal, or oblique divisions; daughter cells achieve size of parent before dividing

Coccolithophores
Covered with calcareous plates, embedded in gelatinous sheath
Mainly in open seas, tropical and semitropical; sometimes abundant near coasts; some heterotrophic forms at depths to 3000 meters
Size: 0.005–0.05 mm
Many flagellated; often round or oval single cells; when present in great numbers, they give the water a milky appearance. Some individuals form cysts from which spores can develop into new individuals

Silicoflagellates
Very small, have silica skeleton; some heterotrophic forms
Widespread in colder seas worldwide, especially in upwelling areas
Size: about 0.05 mm
Single-celled, one or two flagellae; starlike or meshlike skeleton
Simple cell division

Cyanobacteria (also called Blue–Green Algae)
Small, relatively simple cell structure; cell wall of chitin
Mainly inshore, warmer surface waters, tropics
Size: filaments to 0.1 mm or more
Blue-green or red rafts of mottled filaments; can cause a colored "bloom" in water
Simple division of each cell into two

APPENDIX 3

Presenting Data

Scientific data are compiled and presented in various graphic forms. The wide variety and amount of ocean data pose some special problems. In this section we review some common ways of graphically presenting ocean and Earth data—graphs, profiles, maps, diagrams, and false color—to show their uses as well as their limitations.

Graphs

Of the various techniques used to portray scientific data, **graphs** are perhaps the most widely used. A graph permits the general aspects of a relationship between two properties to be seen at a glance. Furthermore, values of various properties can be estimated from a graph.

In constructing a graph, data are often first organized into tables. For example, if we are studying the increase in the boiling point of seawater with salinity changes, we may arrange our data as follows:

Salinity	Boiling Point Increase (°C)
4	0.06
12	0.19
20	0.31
28	0.44
36	0.57

From this table it is obvious that seawater's boiling point rises with increased salinity. To see this more clearly, we can draw a graph, as in Fig. A3–1. Note that salinity is plotted on the x (horizontal) axis, with values increasing to the right. The boiling point increase, in Celsius degrees, is plotted on the y (vertical) axis, with values increasing upward. We plot our experimental points and then draw a line through these points.

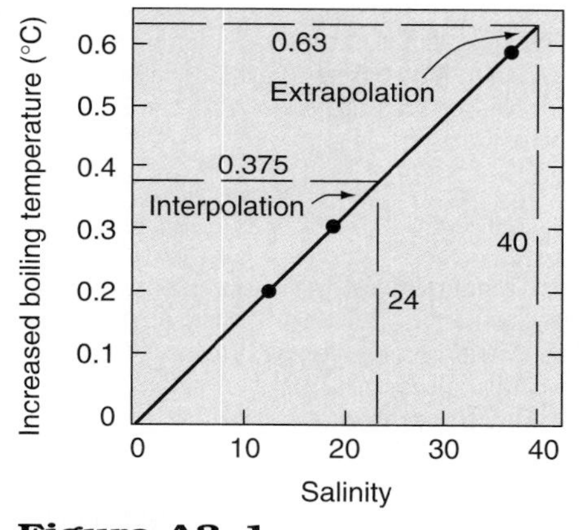

Figure A3–1
Graph showing how the boiling point of water changes with changing salinity. The dots represent experimental data.

Generally, a straight line is the best first estimate. In this case, it is a reasonably good approximation. For many other graphs, we may need to use more complicated curves, but even for complicated curves, a straight line is a reasonable estimate for small portions of the curve.

Note that the graph shows that the boiling point is raised as salinity increases. By using the x and y coordinates for any

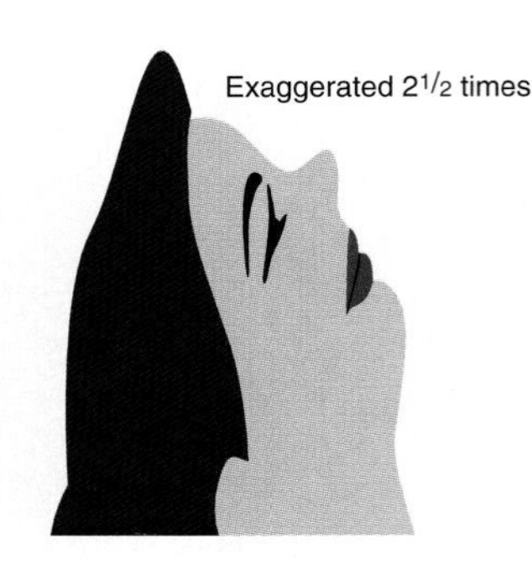

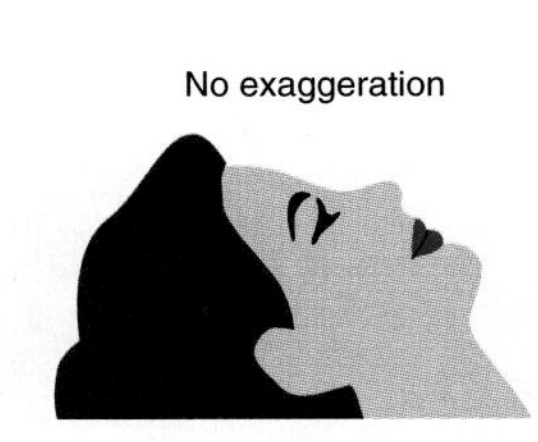

Figure A3–2
Distortion in a profile of a human face caused by different vertical exaggerations.

point along the line, we can estimate the increase in the boiling point for any salinity. When this is done for points that fall inside the experimental data set (which in this case means a salinity between 4 and 36), this process is called **interpolation.** For instance, for seawater having a salinity of 24, we can interpolate a boiling point increase of 0.375°C.

For data points falling outside the experimental data set (in this case, for a salinity less than 4 or more than 40), we use **extrapolation** (or extending) of the curve to estimate the related boiling point increases.

In reading a graph, always determine which property is plotted on each axis. Also, check both the scale intervals and the values at the origins. The appearance of the graph can be changed drastically by changing either. Advertisers, for instance, often use graphs where the scales and origins are chosen to present their data in the most favorable light.

Profiles

Profiles are used to show topography, either of ocean bottom or of land. An ocean-bottom profile can be considered a vertical slice through Earth's surface. Such a profile can be drawn with no distortion—in other words, distances are equal vertically and horizontally. Imagine, however, the problems involved in drawing a 10-centimeter-long profile of the Atlantic Ocean between New York and London, a distance of 5500 kilometers. Because the ocean along this line is only 3.4 kilometers deep at the deepest point, a pencil line would be too thick to portray accurately the maximum relief; thus, such a profile conveys no useful information.

To get around this problem, profiles—including those in this book—are usually distorted in the vertical direction and are therefore said to be **vertically exaggerated.** Profiles showing oceanic features are typically exaggerated by factors of several hundred or several thousand. Consequently, even gently rolling hills look like impossibly rugged mountains. The effect of profile distortion can be seen rather dramatically when it is applied to a human facial profile, as in Fig. A3–2.

Contours and Contour Maps

Various means have been used to portray land forms (topography); the most useful employ contours in contour maps. Contour lines connect points that are at equal elevations or at equal depths. Obviously not all elevations (or depths) can be connected by contour lines. Only certain ones at selected intervals are shown; otherwise, the map would be solid black. The vertical interval represented by successive contour lines is called the **contour interval.**

To interpret a contour map, imagine the shoreline of a lake. The still-water surface is a horizontal plane, touching points of equal elevation along the shore. The shoreline is thus a contour line. If the water surface were controlled and made to fall by regular intervals, it would trace a series of contour lines, forming a contour map on the lake bottom (or hillside). Note that the shorelines formed at different lake levels do not cross one another; neither do contours on maps.

Contours reveal topography. For example, contours that are closed on a map (do not intersect at a boundary) indicate either a hill or a depression. To find out which, look to see if the elevation increases toward the closed contour(s). If so, you are looking at a hill. If the elevation decreases toward the closed contour, it is a depression. Contours around depressions are often marked by **hachures**—short lines on the contour pointing toward the depression. When a contour line crosses a valley or canyon, the contour line forms a V that points upstream.

Contour lines also indicate the steepness of a slope. Closely spaced contour lines indicate a steep slope. A gentle slope is indicated by widely spaced contour lines.

Contours can also be used to depict properties other than elevation or depth. For example, several maps in this book use contours to show distribution of properties, such as ocean surface temperatures and salinities. We would consider them to be temperature (or salinity) hills and valleys. High temperature corresponds to a hill; low temperature, to a valley. Contours may also be used to show the distribution of temperature or salinity in a vertical section of the ocean.

Colors are also used to show distributions of values, such as water temperatures or concentrations of plant pigments. These are sometimes called **false colors,** because they do not show the true colors of the objects involved. Such colored presentations are necessary to illustrate the complex relationships among oceanic phenomena. Neither contours nor various shadings can portray the amount of information that the human eye can extract from a colored presentation.

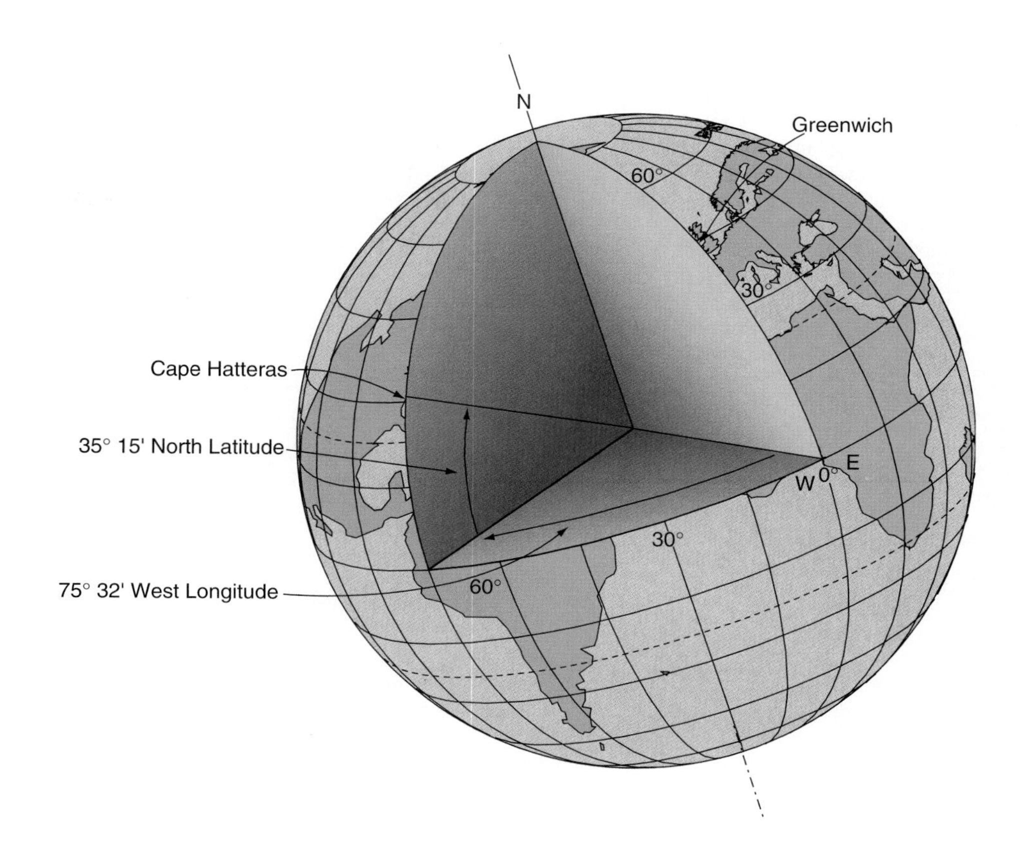

Figure A3–3
Latitude and longitude for Cape Hatteras, North Carolina.

Coordinates—Locations on a Map

A **coordinate** is an address—a means of designating location. The most familiar coordinate system, found in many cities, is the network of regularly spaced, lettered or numbered streets crossing one another, usually at right angles, which enables us to locate a given address. This is an example of a **grid.** As soon as we determine how the streets and avenues are arranged, we can find Fifth Avenue and 42d Street in New York City or 16th and K Streets in Washington, D.C., even though we may never have been in those cities before.

A printed grid is used on large-scale maps to designate locations of features. In making maps of relatively small areas, it is simplest to assume that the world is flat. This works well for areas extending up to 100 miles from a starting point. For larger areas, Earth's curvature must be considered.

Since Earth is a sphere, we must use **spherical coordinates**—a grid fitted to a sphere. A small town or even a state has rather definite starting points for a grid, either the edges or the center. Because a sphere has no edges or corners, however, we must designate those points where our numbering system is to begin.

Distances north or south are easiest to deal with. We can easily identify Earth's geographic poles, where the axis of rotation intersects Earth's surface. Using these points, it is easy to draw a line circling Earth, equally distant from the North and South poles. This is the **equator,** which serves as our starting point for measuring distances north and south. The distance between the equator and either pole is divided into 90 equal parts **(degrees).** The series of grid lines that circle Earth and connect the points that are the same distance from the nearest pole are known as **parallels of latitude.**

To see how this works, imagine Earth with a section cut out, as in Fig. A3–3. Now look at the angle formed by the line connecting any point of interest with Earth's center and the line from Earth's center to a point directly south of that point, on the equator. This angle is a measure of the distance between the chosen point and the equator. The North Pole has a latitude of 90°N, Seattle is approximately 47°N, and Rio de Janeiro is approximately 23°S.

Latitude was easily measured by early mariners. The angle between Polaris (the pole star) and the horizon provides a reasonably accurate measure of latitude. At the equator, the pole star is on the horizon (latitude 0°N). Midway to the North Pole (latitude 45°N), the pole star is 45° above the horizon. At the North Pole, Polaris is directly overhead. Although there is no star directly above the South Pole, the same principle holds, except that a correction is necessary to allow for the displacement from the South Pole of the star used.

Measuring east–west distances on Earth poses the problem: Where do we start? The answer has been to establish an arbitrary starting point—the **prime meridian**—and to indicate distances as east or west of that meridian. Several prime meridians have been used by different nations, but today the Greenwich prime meridian is most commonly used. It passes through the famous observatory at Greenwich (a suburb southeast of London).

Longitude—distance east or west of the prime meridian—is indicated on a map by north–south lines connecting points with equal angular separation from the prime meridian. These lines are called **meridians of longitude,** and they converge at the North and South poles. Longitude in degrees is measured by the size of the angle between the prime meridian and the meridian of longitude passing through the given point, as in Fig. A3–3. Going eastward from the prime meridian, longitude increases until we reach the middle of the Pacific Ocean, when we come to the 180° meridian. Going westward from Greenwich, longitude also increases until we reach the 180° meridian, which represents the juncture between the Eastern and Western hemispheres. Through much of the Pacific Ocean, the 180° meridian is also the location of the **international dateline.** This designation of the

180th meridian as the international dateline—where the "new day" begins—is no accident. Its position in midst of the Pacific avoids the problem of adjacent cities being one day apart in time. This also explains, in part, the choice of the Greenwich meridian as the prime meridian.

Longitude and time are intimately related. Because it takes Earth 24 hours to make one complete turn on its axis of rotation (360°), we calculate that Earth turns 15° per hour. We use this relationship to find our relative position east or west of the prime meridian.

Each meridian of longitude is a **great circle.** If we sliced through Earth along one of the meridians of longitude, our cut would go through the center of Earth. Of the parallels of latitude, only the equator is a great circle. All the other parallels are **small circles,** for a plane (or slice) passing through them would not go through the center of Earth. Great circles are favored routes for ships or aircraft, because a **great-circle route** is the shortest distance between two points on a globe.

To study time and longitude, let us begin at local noon on the prime meridian, when the Sun is directly overhead. One hour later, the Sun is directly over the meridian of a point 15° west of the prime meridian; 2 hours later, it is over a meridian 30° west of the prime meridian. And 12 hours later (midnight), it is over the 180th meridian and the new day begins.

If we have an accurate clock keeping "Greenwich time" (the time on the prime meridian), we can determine our approximate longitude from the time of local noon, when the sun is highest in the sky. Assume that our clock reads 2 P.M. Greenwich time at local noon. The 2-hour difference indicates that our position is 30° from the prime meridian. Because local noon is later than Greenwich noon, we know we are west of Greenwich and our longitude is therefore 30°W. If our local noon occurs at 9:30 A.M. Greenwich time, we are 2.5 hours × 15° per hour = 37.5° east of the prime meridian; our longitude is thus 37.5°E.

Degrees, like hours, are divided into 60 parts known as **minutes.** Each minute is further divided into 60 **seconds.** Consequently, in the last example we would give our position as 37°30′E.

To determine longitude, therefore, a ship need only have accurate time. With modern electronic communications, this poses no problem. For centuries, however, seafarers had no means of keeping accurate time at sea. Not until the 1760s, when the first practical **chronometers**—accurate clocks for use aboard ship—were designed, was it possible for most ships' pilots to determine longitude. Even the Greek astronomer Ptolemy, who made maps with relatively accurate positions north and south, overestimated the length of the Mediterranean by 50 percent, an error not corrected until 1700.

Each map in this book uses latitude and longitude to indicate the positions of the map features. The parallels of latitude may also be used to determine approximate distance on a map. Each degree of latitude equals approximately 60 nautical miles (69 statute miles or 111 kilometers). Each minute of latitude is approximately 1 nautical mile, or 1.85 kilometers. At the equator, each minute of longitude is 1 nautical mile, but at 60° north or south latitude, 1 minute of longitude is only 0.5 nautical mile, and at the poles distance between lines of longitude diminishes to zero.

Maps and Map Projections

A map is a flat representation of Earth's surface. Symbols are used to depict surface features. Because Earth is a sphere, a flat map distorts the shape or size of surface features. The only distortion-free map is a globe, but a globe is not practical for the study of relatively small areas.

In making a map, we would like to make the final product as useful as possible. In general, we would like a map to preserve the following properties of Earth's surface:

Equal area: Each area on the map should be proportional to the area of Earth's surface it represents.

Shape: The general outlines of a large area shown on a map should approximate as nearly as possible the shape of the region portrayed. A map that preserves shape is said to be **conformal.**

Distance: A perfect map would permit distance to be measured accurately between any two points anywhere on the map. Many common maps, such as the Mercator projection, do not accurately portray distances in a simple way.

Direction: Ideally, it would be possible to measure directions accurately anywhere on a map.

No map has all these properties.

A **map projection** takes the grid of latitude and longitude lines from a sphere and converts them to a grid on a flat surface. Sometimes the resulting grid is a simple rectangular one, where longitude and latitude lines intersect at right angles. In other projections, latitude and longitude lines are complex curves that intersect at various angles.

With the network formed from the grid lines, the map is drawn by plotting points in the appropriate spot on the new projection. In this way, the various types of maps are prepared. We shall consider only a few of the many map projections that have been developed to serve specific functions.

The **Mercator projection** is the most familiar. Parallels of latitude and meridians of longitude are straight lines and cross at right angles. The outline shape is a square or rectangle, as shown in Fig. A3–4.

In its simplest form, the Mercator projection can be visualized as being made by a light inside a translucent globe projecting latitude and longitude onto a cylinder surrounding the globe. Even though the cylinder used for the projection is curved, it is easily made into the flat map desired.

The common Mercator projection in our example is most accurate within 15° of the equator and least accurate at the poles. Although shapes are well preserved by this projection, area is distorted, especially near the poles. For example, South America is in reality nine times the size of Greenland, but this fact is not obvious from common Mercator projections. The scale of a Mercator projection changes going away from the equator. In order to avoid serious error, the reader can use the length of a degree of latitude as the scale.

Another property of the Mercator projection useful to mariners is that a course of constant compass direction (a **rhumb line**) is a straight line on this projection. Although a rhumb course is not a great circle and, thus, not the shortest distance between any two points on Earth's surface, it is useful for navigation, because a great circle course requires a constant changing of direction. A rhumb line is slightly longer but easier to navigate.

For world maps, the Mercator projection has distinct limitations; but for relatively small areas, such as navigation charts, it is without equal. Nearly all navigation charts are Mercator projections.

In this book, a projection called a **Hoelzel planisphere** (Fig. A3–5) is used. This modified cylindrical projection shows the continents well, permitting coastal regions to be easily recognized. Like the Mercator, this projection distorts areas near the poles. Instead of converging to a point at the poles, the meridians converge to a line that is only a fraction of the length of the equator.

A special projection (the **interrupted homolosine,** Fig. A3–6) shows the ocean basins without interruptions. In addition, the projection shows area equally. In this projection, the continents are interrupted to show the ocean basins intact.

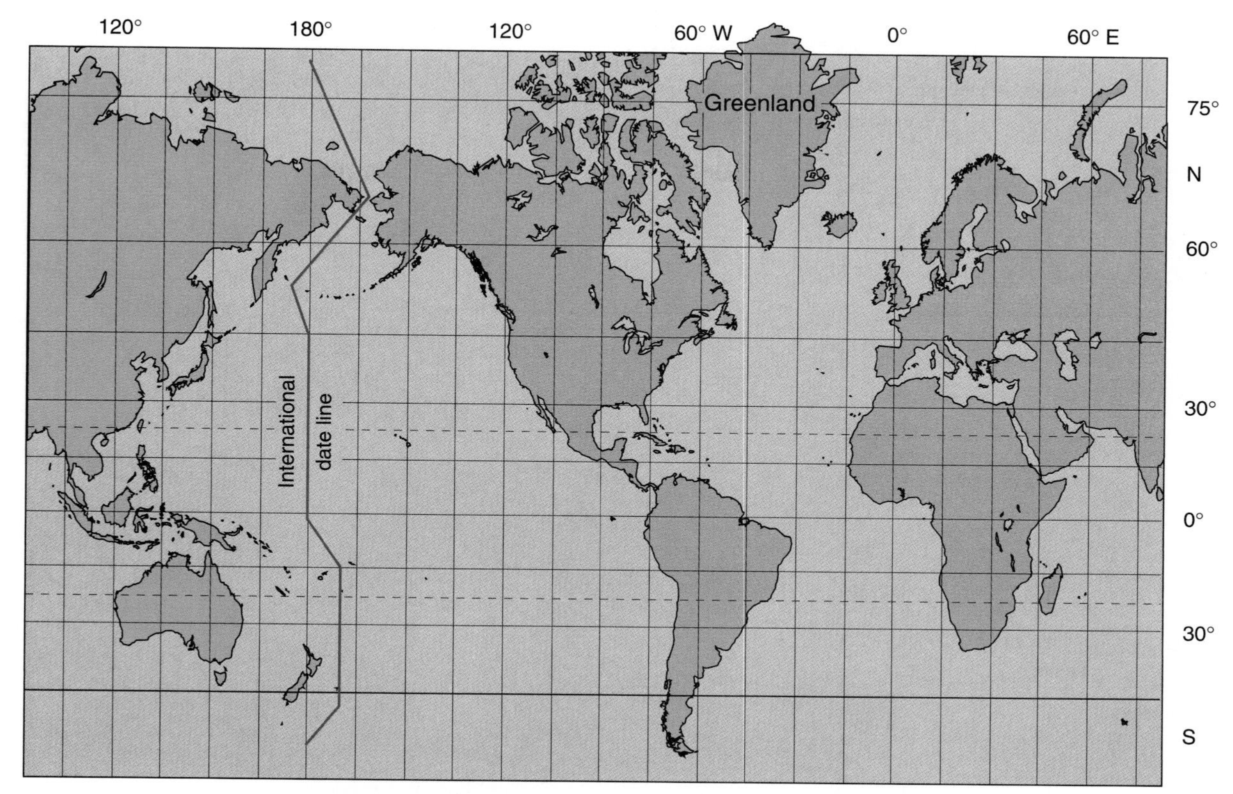

Figure A3–4
Mercator projection. Compare the shape and size of Greenland in this projection with that in Figures A3–5 and A3–6.

Figure A3–5
Hoelzel projection. Note that the meridians of longitude converge to a line shorter than the equator but still not a point.

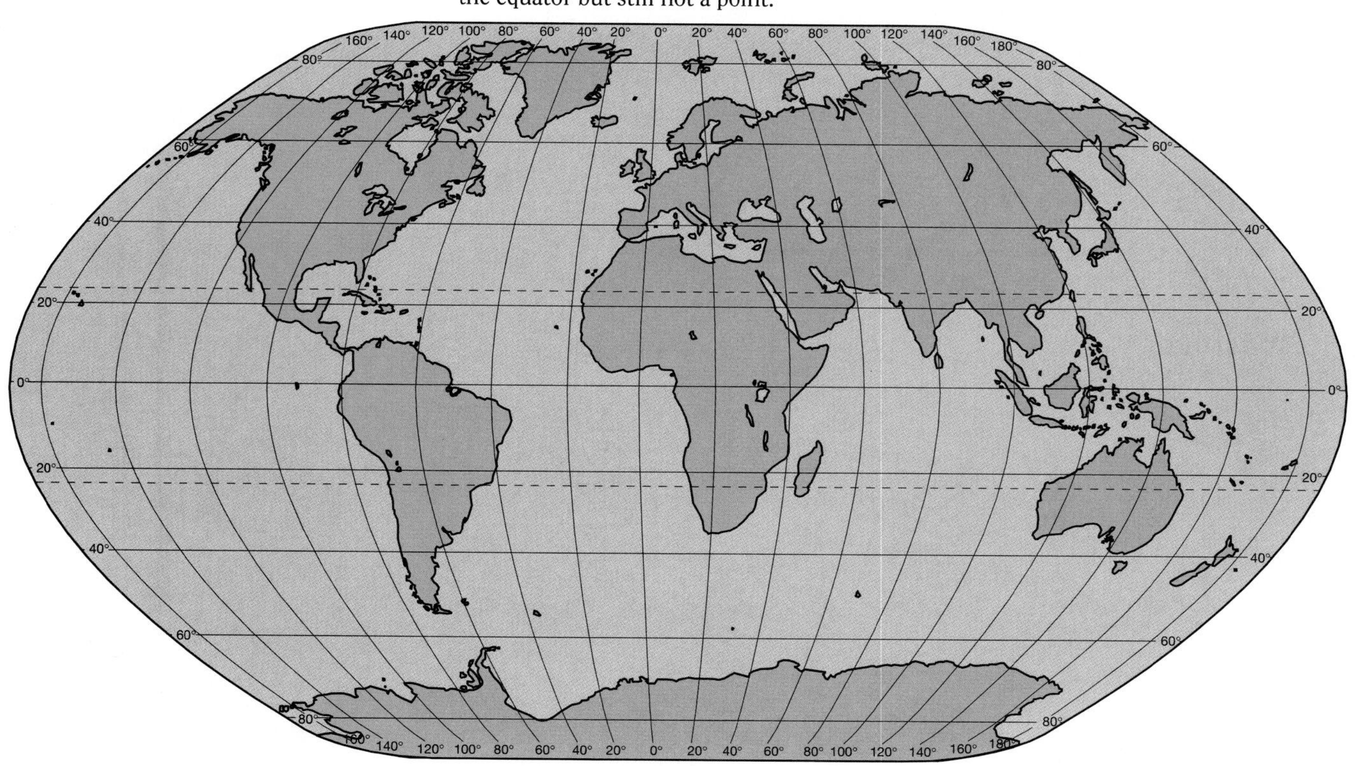

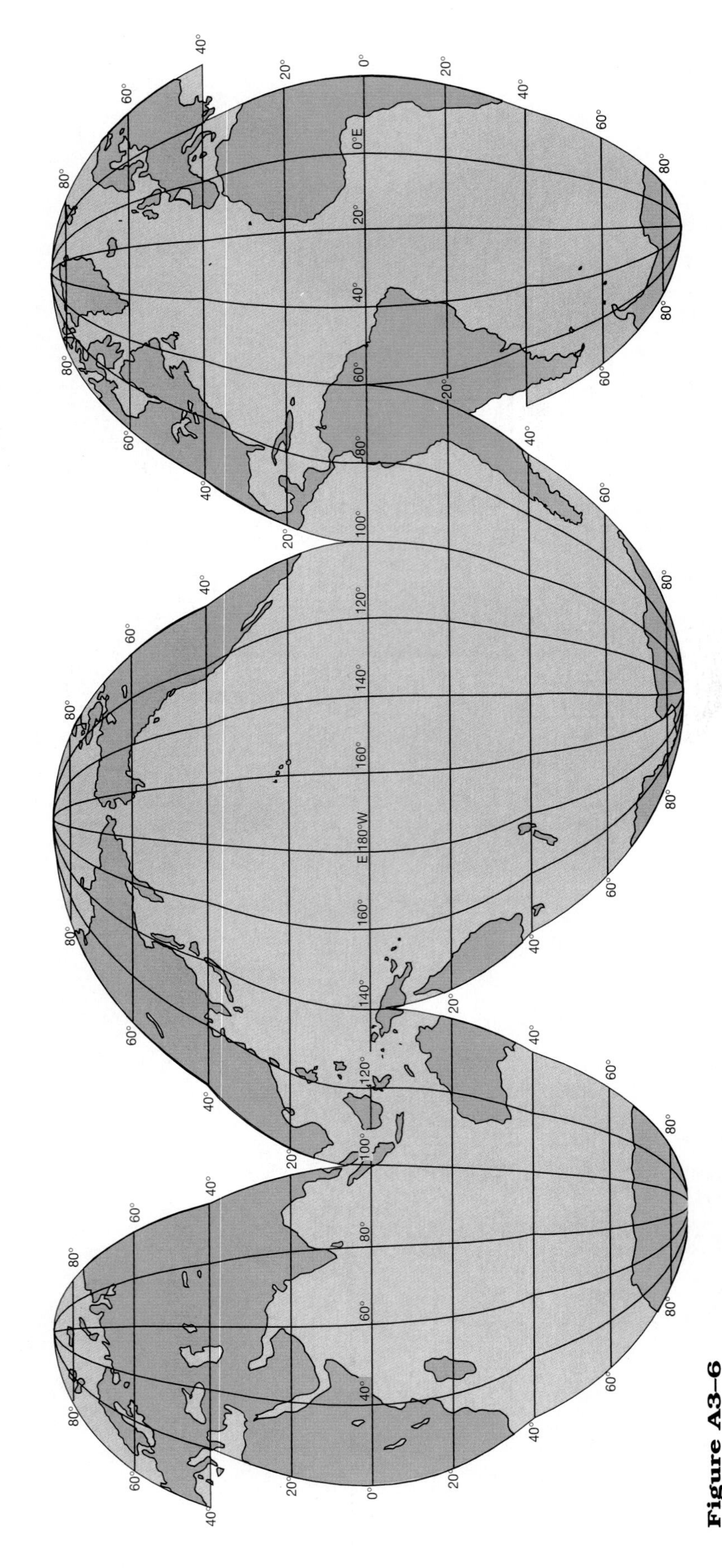

Figure A3–6
Goode's homolosine projection shows the three oceans to best advantage. Note that the meridians of longitude converge to points, thus not distorting the shapes and sizes of high-latitude areas.

APPENDIX 4

Careers in Oceanography

Oceanography as a field of employment dates from World War II. In the 1950s, oceanography emerged as a field of study when American universities began offering courses and degree programs. Initially, these programs trained researchers and scientists needed for the military. In the United States, the rapid growth in federal support hastened the growth of oceanography as a field of science and as an area of employment opportunities.

Later, ocean sciences focused on studies of ocean processes. The emphasis shifted to protecting/regulating living resources (fisheries) and improving our ability to predict global climate change and to understand how such change affects Earth.

Education

Jobs in oceanography normally require university training. The curricula are necessarily broad, because the ocean is involved in almost all aspects of earth sciences; meteorology, geology, and the physical processes controlling ocean currents are some examples. The boundaries between oceanography and its intellectual neighbors are diffuse and increasingly more difficult to define. Thus, students wishing to work in oceanography must take a broad course of study in the physical and life sciences.

Until the 1990s, most universities advised students to take a bachelor's degree in a traditional science field and then follow with a study of oceanography at the graduate level. Because of increasing employment opportunities, many American universities are offering undergraduate programs leading to bachelor's degrees in oceanography.

University Teaching and Research

Teaching and research, primarily at universities, has dominated ocean-science employment in the United States since the 1960s. A significant fraction of all oceanographers have doctoral-level training. With the end of the Cold War, the accompanying reductions in defense spending on research, and slow growth in both universities and the commercial ocean research sector, this area has been growing slowly, if at all. Many doctoral-level oceanographers have taken post-doctoral positions while they await vacancies on faculties or research staffs.

Fisheries

Fisheries offered the first employment opportunities for people interested in the ocean. As early as the late 1800s, fish and whale stocks showed signs of being overexploited. Many countries established groups to study their fisheries and to devise ways to find new fish resources and to protect traditional supplies. Fisheries management remains a major employment area in many countries, and several universities offer programs to train fishery scientists.

Aquaculture

An emerging area of employment is aquaculture—growing aquatic organisms for commercial purposes. As wild fish stocks become depleted by overfishing or environmental changes, the increasing demand for safe fisheries products requires cultivation of both plants and animals. Although there are no domesticated marine organisms, growing marine plants and animals (primarily from wild stocks) is big business around the world, especially in Asia. China and Japan have well-established multibillion-dollar industries. Aquaculture is also expanding in Europe and North America. In the United States, for instance, catfish, trout, and salmon aquaculture are growing rapidly. Demands for cultured products will grow as concerns increase about the contamination of wild populations by diseases and pollution. As aquaculture expands, it will need people trained in related fields of science and in business.

Petroleum Industry

By 1945, oil and gas fields on land had become highly developed, and the petroleum industry began exploring shallow continental margins. Today, oil and gas production in many countries comes almost exclusively from offshore fields. New technologies allow commercial oil and gas production from deeply submerged continental shelves and rises. The most promising areas for finding new oil and gas lie on the continental margins. Thus, many marine scientists and engineers work in the oil and gas industry worldwide. They work on a wide range of problems, such as dealing with the environmental effects of exploration and production and predicting wave and wind effects on offshore structures. This field is likely to continue as an important area of employment for people trained in ocean sciences and engineering.

Environmental Protection, Regulation, and Restoration

Many people trained in the marine sciences work to protect living resources and the environment. Due to their broad educational background, oceanographers find many opportunities in this field. Environmental regulatory agencies employ marine scientists at local, state/provincial, and national levels; oceanographers also work in research and in restoring damaged environments. For example, damaged wetlands can be restored, although doing so has proved difficult. Restoration projects range from planting and growing plants and animals that lived there before reclamation to research to improve our understanding of the complex systems involved.

Emerging Employment Areas

Application of satellite-based remote sensing for ocean research and monitoring offers employment opportunities for oceanographers. In many countries, oceanographers prepare predictions of ocean conditions and other services. For instance, several countries prepare and distribute to fishing boats satellite-data maps that can be used to predict where certain types of fish can be caught. Recreational fishers and companies that supply their needs also use such products.

The enormous amounts of data generated from satellites and computer modeling require specialists in data management. Thus, oceanographers find new employment opportunities in using computers.

In short, oceanographers find work in many of the established sectors of the economy and also in emerging areas such as data handling and remote sensing. The key factors are broad training and flexibility.

Selected References

Environmental Careers Organization, *The Complete Guide to Environmental Careers.* 1994 (ECO also provides job counseling and placement services with governments, nongovernmental organizations, etc.) 68 Harrison Ave., Boston, MA 02111, 617/426-4375

Feibelman, Peter J., *A Ph.D. Is Not Enough.* Reading, MA: Addison-Wesley Publishing Company, 1993. 109 pp. A modern guide to getting and keeping a job in research and education.

Sinderman, Carl J., *The Joy of Science.* New York: Plenum Press, 1985. 256 pp. An insider's view of a career in science.

Sources of Information About Employment Opportunities

Environmental Career Opportunities. Published twice monthly. Available at newsstands and some bookstores. Subscription: 301/986–5545

International Career Employment Opportunities. Published twice monthly. Available at newsstands and some bookstores. Subscription: 804/985–6444

Women's Aquatic Network. P.O. Box 4993, Washington, D.C. 20008. Periodic listing of jobs in Washington, D.C. and surrounding regions. Some national listings.

Professional Associations

Coastal Society. Box 25408, Alexandria, VA 22313–5408, 703/768–1599

Marine Affairs and Policy Association (provides academic network and directory of teaching and research programs, research centers, professionals in the field by region and other related contacts). Center for Study of Marine Policy, Graduate School of Marine Studies, University of Delaware, Newark, DE 19716, 302/831–8086

The Oceanography Society. 4052 Timber Ridge Drive, Virginia Beach, VA 23455, 804/464-0131

APPENDIX **5**

Classification of Marine Organisms*

Biologists categorize organisms using **taxonomic classification** to identify and describe similarities among marine organisms. After basic similarities in external form, internal anatomy, and biochemical characteristics are determined, groups (called **taxa,** or **taxon,** if singular) are assigned Latin names in a rigidly prescribed procedure. Finally, these groups are fitted into a system of increasingly more inclusive categories.

Taxonomic classification is used, among other things, to study evolutionary relationships of organisms. It also shows the many different kinds of organisms that live in the ocean. Indeed, most of the organisms that ever lived on Earth inhabited the ocean. Life has existed in the ocean for at least 3800 million years, compared with only 450 million years on land.

The fundamental unit of taxonomy is the **species,** defined as a group of closely related individuals that can and usually do interbreed. Some 75 million species have appeared since life began on Earth. More than 2 million are living today. In this appendix, we are primarily concerned with major groups of marine organisms, which are classified as follows:

- Kingdom
  - Phylum
    - Subphylum
      - Class
        - Order
          - Family
            - Genus
              - Species

KINGDOM MONERA—dominantly unicellular organisms, lacking nuclear membranes. Nuclear materials occur throughout the cells.

- Phylum Schizophyta—smallest cells, bacteria.
- Phylum Cyanophyta—blue-green algae, contain chlorophyll and other pigments.

KINGDOM PROTISTA—one-celled organisms; nuclear materials confined to nucleus by a membrane.

- Phylum Chrysophyta—golden-brown algae. Includes diatoms, coccolithophores, and silicoflagellates.
- Phylum Pyrrophyta—dinoflagellate algae.
- Phylum Chlorophyta—green algae.
- Phylum Phaeophyta—brown algae.
- Phylum Rhodophyta—red algae.
- Phylum Protozoa—heterotrophs.
  - Class Sarcodina—ameboid, includes foraminiferans and radiolarians.
  - Class Mastigophora—flagellated; includes dinoflagellates.

KINGDOM FUNGI

- Phylum Mycophyta—fungi and lichens.

KINGDOM METAPHYTA—multicellular plants.

- Phylum Tracheophyta—vascular plants with roots, stems, and leaves; separate liquid transport system.
  - Class Angiospermae—flowering plants, with seeds.

---

*After H. V. Thurman and H. H. Webber: *Marine Biology.* Columbus, Ohio: Charles E. Merrill, 1984.

KINGDOM METAZOA—multicellular animals.
Phylum Porifera—sponges.
Class Calcarea—calcium carbonate spicules.
Class Hexactinellida—glass sponges.
Phylum Cnidaria—radially symmetrical, polyp (benthic) and medusa (planktonic) stages.
Class Hydrozoa—polyp colonies, includes Portuguese man-of-war.
Class Scyphozoa—true jellyfish.
Class Anthozoa—corals and anemones.
Phylum Ctenophora—planktonic comb jellies; eight-sided radial symmetry with secondary bilateral symmetry.
Phylum Platyhelminthes—flatworms, bilateral symmetry.
Phylum Nemertea—ribbon worms, benthic and pelagic.
Phylum Nematoda—roundworms, free-living benthic; mostly meiofauna.
Phylum Rotifera—ciliated, unsegmented.
Phylum Bryozoa—moss animals, benthic, branching or encrusting.
Phylum Brachiopoda—lamp shells, benthic bivalves.
Phylum Phoronida—horseshoe worms, shallow-water benthos.
Phylum Sipuncula—peanut worms, benthic.
Phylum Echiura—spoon worms, benthic.
Phylum Pogonophora—tube-dwelling, gutless worms; absorb organic matter through skin.
Phylum Tardigrada—marine meiofauna, which inhabit the water filling the spaces between sediment particles or sand grains.
Phylum Mollusca—soft bodies; possess muscular foot and mantle; usually secrete calcium carbonate shells.
Class Polyplacophora—chitins, oval, flattened body covered by eight overlapping plates.
Class Gastropoda—snails and related forms; many with spiral shell.
Class Bivalvia—clams, mussels, oysters, and scallops; mostly filter feeding.
Class Aplacophora—tusk shells; benthic; feed on meiofauna.
Class Cephalopoda—octopus, squid, and cuttlefish; possess no external shell (except Nautilus).
Phylum Annelida—segmented worms, mostly benthic.
Phylum Arthropoda—joint-legged, segmented bodies; covered by exoskeleton.
Subphylum Crustacea—calcareous exoskeletons; two pairs of antennae; includes copepods, ostracods, barnacles, shrimp, lobsters, and crabs.
Phylum Chaetognatha—arrow worms; mostly planktonic.
Phylum Hemichordata—acorn worms and pterobranchs; primitive nerve chord; benthic.
Phylum Echinodermata—spiny skinned; secondary radial symmetry; water vascular system; benthic.
Class Asteroidea—starfishes, flattened body with five or more rays; tube feet used for locomotion.
Class Ophiuroidea—brittle stars, basket stars; central disc with slender rays; tube feet used for feeding.
Class Echinoidea—sea urchins, sand dollars; calcium carbonate tests.
Class Holothuroidea—sea cucumbers; soft bodies with radial symmetry.
Class Crinoidea—sea lilies; cup-shaped body attached to bottom by jointed stalk or appendages.
Phylum Chordata—notochord; nerve chord and gills or gill slits.
Subphylum Vertebrata—internal skeleton; spinal column; brain.
Class Agnatha—lampreys and hagfishes; most primitive vertebrates; cartilaginous skeletons; no jaws; no scales.
Class Chondrichthyes—sharks, skates and rays; cartilaginous skeletons.
Class Osteichthyes—bony fishes; covered gill openings; swim bladder common.
Class Reptilia—snakes, turtles, lizards, and alligators.
Class Aves—birds.
Class Mammalia—warm-blooded; hair; mammary glands; bear live young.

APPENDIX 6

People in Oceanography

A

Agassiz, Alexander (1835–1910) Swiss-born biologist who studied corals; lived and worked mostly in the United States.

Agassiz, Jean-Louis Rodolphe (1807–1873) Swiss-born natural historian; father of Alexander Agassiz; worked on glaciers, fishes, and other subjects; founded Harvard University's Museum of Comparative Zoology.

Albert I. (1848–1922) Prince of Monaco, supported oceanographic work; founded Oceanographic Museum of Monaco.

Aristotle (384–322 B.C.E.) Greek philosopher and natural scientist; greatly influenced development of science and natural observations.

B

Bede (673–735) English theologian and historian, wrote about the effect of the Moon on ocean tides. Also known as *The Venerable Bede.*

Brendan, Saint (485?–577?) Irish monk who first crossed the Atlantic in a curragh, a skin boat with a wicker frame.

Bullard, Edward (1907–1987) English geophysicist who made major contributions to understanding of plate tectonics and movements of continents.

C

Cook, James (1728–1779) British navigator and explorer of the Pacific.

Cousteau, Jacques-Yves (1910–) developed SCUBA; popularized underwater science. Was Director of the Oceanographic Museum of Monaco for many years.

D

Darwin, Charles Robert (1809–1882) English naturalist who formulated theory of evolution by natural selection; proposed theory of coral reef development.

Drake, Sir Francis (1541–1596) English explorer and buccaneer; first Englishman to circumnavigate the Earth.

E

Ekman, Vagn Walfrid (1874–1954) Swedish physicist and oceanographer; developed theory of wind-driven currents, including upwelling.

Ericsson, Leif (980?–1001) Norse explorer who founded colony in North America in Newfoundland.

Ewing, William Maurice (1906–1974) American geophysicist; founded Lamont-Doherty Earth Observatory.

F

Forbes, Edward (1815–1854) English naturalist; proposed theory of Azoic Zone, postulating that there was no life below about 600 m or 2000 ft).

Franklin, Benjamin (1706–1790) American politician and naturalist who first mapped Gulf Stream.

G

Gama, Vasco da (1460–1524) Portuguese navigator who first opened trade route to India, establishing basis for Portuguese empire.

Grotius, Hugo (1583–1645) Dutch lawyer and philosopher who formulated the legal doctrine of mare liberum (freedom of the seas), which dominated legal thinking about ocean jurisdiction until modified by the U.N. Law of the Sea Conference.

H

Henry the Navigator (1394–1460) Supported early development of Portuguese exploration.

Hess, Harry Hammond (1906–1969) American geologist who was early supporter of plate tectonic theory in North America.

Hudson, Henry (1565–1611) English navigator/explorer who explored North America for Dutch East India Company.

Humboldt, Fredrich Heinrich Alexander von (1769-1859) German explorer, naturalist, and geographer who used personal fortune to support his work; explored Spanish colonies of South America.

M

Magellan, Ferdinand (1480?–1521) Portuguese navigator, whose voyage in 1520–1521 was the first to circumnavigate the world.

Maury, Matthew Fontaine (1806–1874) American oceanographer who first systematically collected data from ships and mapped winds and surface currents.

Meinesz, Vening (1887–1966) Dutch geophysicist who first used submarines to study Earth's gravity at sea.

Mercator, Gerardus (1512–1594) Flemish map-maker who made a global map in 1538; devised map projection which bears his name.

Murray, John (1841–1914) Canadian-born oceanographer who participated in the *Challenger* Expedition.

N

Nansen, Fridtjof (1861–1930) Norwegian oceanographer, explorer, and humanitarian; made extensive exploration in Arctic.

P

Ptolemy, Claudius (90?–168) Greek astronomer, map-maker; produced eight-volume geography book and a map of the then-known world.

R

Ross, James Clark (1800–1862) British explorer who discovered the Ross Sea near Antarctica.

T

Thomson, Charles Wyville (1830–1882) English naturalist who organized *Challenger* Expedition.

W

Wegener, Alfred (1880–1930) German geophysicist/meteorologist who first advanced concept of continental drift, a precursor to plate tectonic theory.

Wilson, J. Tuzo (1908–1993) Canadian geophysicist who developed the theoretical basis for many processes involved in plate tectonic movements.

APPENDIX 7

Geologic Time Scale

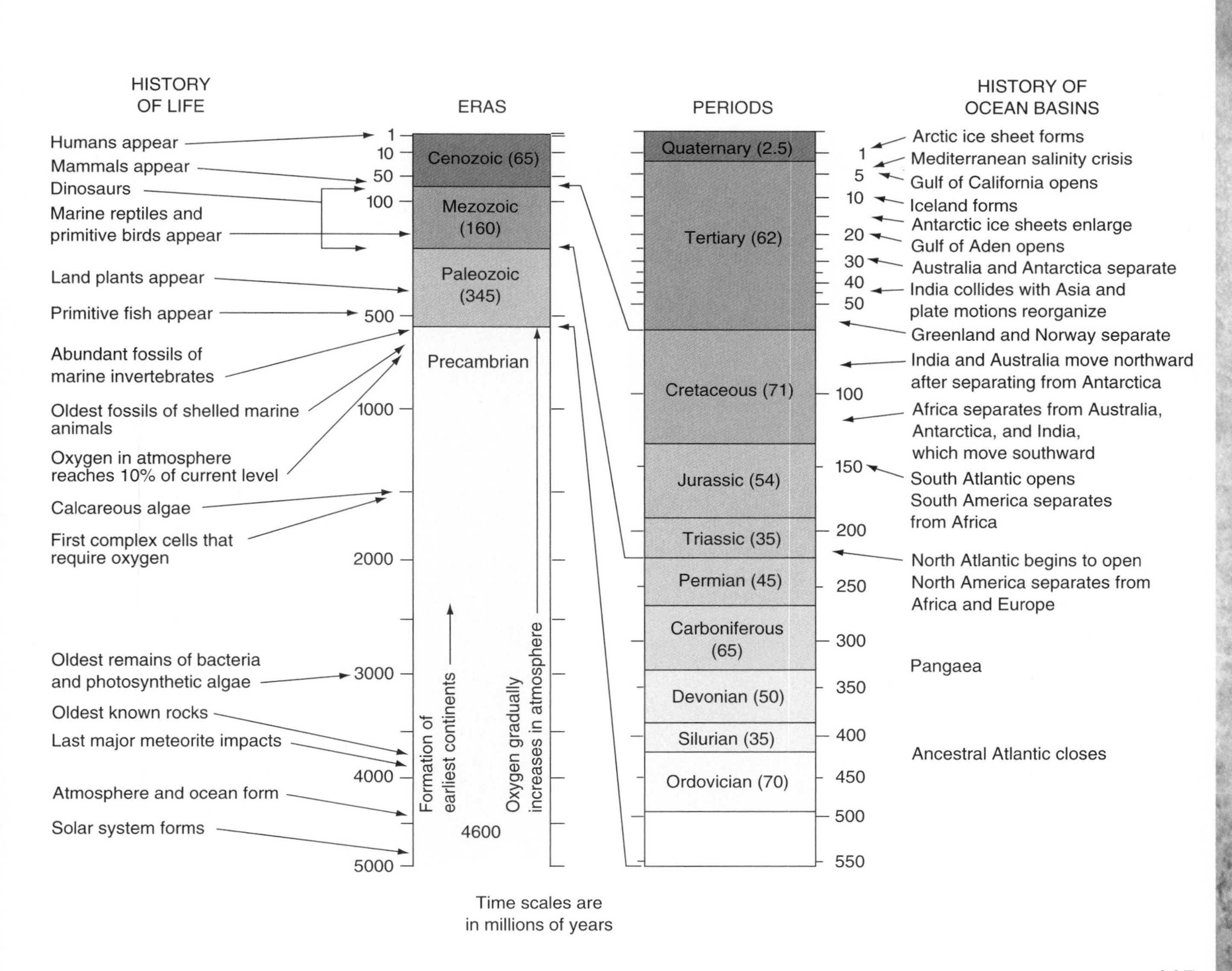

Glossary

A

abyssal pertaining to the great depths of the ocean, generally below four kilometers.

acid solution in which hydrogen ion concentration exceeds hydroxyls.

acoustic tomography technique using changes in sound velocity between acoustic transmitters and receivers in the ocean to obtain three-dimensional pictures of water-mass distributions and their movements.

active margin margin of lithospheric plate where tectonic activity is occurring; also called a Pacific-type margin.

aerobic requires oxygen.

airshed area from which materials are transported by winds to an area or a water body.

algae marine or freshwater plants, including phytoplankton and seaweeds.

algal ridge elevated margin of a windward reef, built by calcareous algae.

alternation of generations mode of development, characteristic of many coelenterates, in which a sexually reproducing generation gives rise (by union of egg and sperm) to an asexually reproducing form, from which new individuals arise by budding or simple division of the "parent" animal.

altimeter satellite-borne radar that measures distances between the spacecraft and the ocean surface. Ocean surface topography, calculated from the distance, indicates the current patterns; roughness of the ocean surface indicates average wave height.

amphidromic point center of an amphidromic system; a nodal or no-tide point around which a standing-wave crest rotates once each tidal period.

amphidrome system a tidal system, usually in an enclosed basin, where the tidal wave crest moves around a fixed point during each tidal period.

anadromous fishes that spend most of their lives as adults in salt waters but return to freshwater streams to lay their eggs; salmon are an example.

anaerobic condition in which there is no dissolved oxygen. Organisms that depend on the presence of oxygen cannot survive. Anaerobic bacteria can live under these conditions.

andesite volcanic rock intermediate in composition between granite and basalt, associated with partial melting of crust and mantle during subduction.

anglerfish deep-sea fishes; a fin attached to the head functions as a lure to attract other organisms.

anion negatively charged atom or radical.

antinode that part of a standing wave where the vertical motion is greatest and the horizontal velocities are least.

aphotic zone where light is insufficient for photosynthesis.

aquaculture cultivation or propagation of water-dwelling organisms.

arrow worm See *chaetognath.*

arthropods animals with a segmented external skeleton of chitin or plates of calcium carbonate and jointed appendages—for example, a crab or an insect.

asthenosphere upper zone of Earth's mantle, extending from the base of the lithosphere to about 250 kilometers beneath continents and ocean basins; relatively weak, probably partially molten.

Atlantic-type margin edge of a continent that is no longer tectonically active; usually has thick sediment deposits.

atmosphere gaseous outer shell of Earth.

atoll ring-shaped organic reef that encloses a lagoon in which there is no preexisting land and that is surrounded by the open sea.

autotroph an organism that manufactures its food from inorganic compounds, such as dissolved CO_2, H_2O, using energy from sunlight (photosynthesizers) or from energy-rich compounds (chemosynthesizers).

B

backarc spreading center spreading center in a backarc basin, associated with a volcanic arc; crust is formed by volcanic eruptions.

backshore part of a beach that is usually dry, being reached only by the highest tides; a narrow strip of relatively flat coast bordering the sea.

bacteria unicellular organisms with no distinct cell nucleus.

baleen (whalebone) horny material growing down from the upper jaw of large plankton-feeding (baleen) whales which forms a strainer or filtering organ consisting of numerous plates with fringed edges.

bank large elevation of the seafloor; a submerged plateau.

bar offshore ridge or mound that is submerged (at least at high tide).

barrier reef that is separated from the land by a lagoon; usually connected to the sea through passes (openings) in the reef.

barrier beach bar parallel to the shore whose crest rises above high water.

barrier island portion of a barrier beach lying between two inlets.

basalt fine-grained igneous rock, black or greenish black, rich in iron, magnesium, and calcium.

base solution in which hydroxyl concentration exceeds hydrogen ion concentration.

baymouth bar extending partially or entirely across the mouth of a bay.

beach seaward limit of the shore (limits are marked approximately by the highest and lowest water levels); usually consists of sand or gravel deposits.

beach cusp low mounds of beach material separated by troughs spaced at more or less regular intervals along the beach face.

beach drift See *littoral drift.*

bed load See *load.*

bends painful condition caused by bubbles of nitrogen that come out of solution in diver's blood; results from improper decompression.

benthic that portion of the marine environment inhabited by marine organisms that live permanently in or on the bottom.

benthos bottom-dwelling marine organisms.

berm low, nearly horizontal portion of a beach (backshore), having an abrupt fall and formed by the deposit of material by wave action. It marks the limit of ordinary high tides and waves.

berm crest seaward limit of a berm.

Big Bang event in which the universe formed about 15 billion years ago.

bight concavity in a coastline; a large, open bay having headlands on either end.

biodiversity the assemblage of different species found in any ecosystem, or on the whole planet.

biogenous (biogenic) sediment deposits containing at least 30 percent by volume skeletal remains (shells, bones, teeth) of organisms.

biological pump the combination of biological processes in the photic zone which remove carbon and other constituents from the water and incorporate them into tissues and skeletons. Dead organisms and fecal pellets sink into the deep zone, transporting these substances which are released through decomposition and dissolution. This depletes the constituents in the surface zone and enriches them in the deep zone. They return to the surface zone as a result of upwelling.

bioluminescence production of light by organisms; results from chemical reactions within certain cells or organs or in some form of secretion.

biomass amount (weight) of living matter per unit of surface area (or volume) of either water or bottom.

bioremediation using biological processes to remedy environmental problems.

biosphere the portion of Earth's surface that contains all organic matter, both living and dead.

biotechnology the manipulation of organisms (often genetically modified bacteria) to obtain useful products or change the characteristics of other organisms.

bird-foot delta delta formed by the outgrowth of pairs of natural levees, formed by the distributaries, making a digitate or bird-foot form.

bivalves mollusks, generally sessile or burrowing into soft sediment, rock, wood, or other materials. Individuals possess a hinged shell and a hatchet-shaped foot, which is sometimes used in digging. Includes clams, oysters, and mussels.

black smoker hydrothermal vent on ocean bottom discharging, at high temperatures, waters that are colored by sulfide minerals.

bloom See *plankton bloom.*

bore See *tidal bore.*

bottom water water mass at the deepest part of the ocean.

boundary currents northward- or southward-directed surface currents that flow parallel to continental margins; caused by deflection of the prevailing eastward- and westward-flowing currents by continents.

brackish water low salinity water, typically with a salinity between 5 and 25.

breaker an unstable wave breaking on the shore, over a reef, etc.

brine water containing more dissolved salt than ordinary ocean water. Brines can be produced by evaporating or freezing seawater.

buffered solution a solution that resists change in pH despite small additions of either acid or base.

buoyancy upward force exerted on a parcel of fluid, caused by density differences within a gravitational field between the water parcel and the waters surrounding it.

C

calcareous consisting of calcium carbonate, like limestone.

calcareous algae marine plants that form a hard external covering of calcium compounds.

calorie amount of heat required to raise the temperature of 1 gram of water by 1°C (defined on basis of water's *specific heat*).

canyon (submarine) See *submarine canyon.*

capillary wave (also called *ripple*) a wave in which the primary restoring force is surface tension; water waves with wavelength less than 1.7 centimeters are considered to be capillary waves.

carbohydrate compound made of carbon, oxygen, and hydrogen; starches and sugars are examples.

carbonate compensation depth (CCD) depth at which all carbonate particles dissolve in the deep ocean; sediment deposits below the CCD have less than 5 percent carbonate.

carnivore animal that eats other animals.

catadromous fish fishes that spawn at sea but mature in fresh water; eels are an example.

cation positively charged ion.

Celsius temperature temperature based on a scale in which water freezes at 0° and boils at 100° (at standard atmospheric pressure); also called *centigrade temperature.*

cephalopods benthic or swimming mollusks having a large head, large eyes, and a circle of arms or tentacles around the mouth.

cetacean the hairless marine mammals; includes whales, porpoises, and dolphins.

chaetognath small, elongate, transparent worm-like animals. Planktonic in all seas from the surface to great depths.

change of state change in the physical form of a substance, as when a liquid changes to a solid due to cooling.

chaos controlled randomness where the state of the system is critically dependent on starting conditions. This limits predictability.

chemosynthesis carbon dioxide fixation (primary production) by certain bacteria in the absence of sunlight, using compounds such as ammonia, methane, reduced iron, hydrogen, or sulfur.

chitin complex nitrogenous carbohydrate compound, forming the skeletal substance in arthropods.

chloride atom or chlorine in solution, bearing a single negative charge.

chlorophyll green pigments that occur chiefly in bodies called chloroplasts and that carry out photosynthesis using energy from sunlight.

cilia hairlike protrusions of cells, which beat rhythmically and propel cells or produce currents in water.

clay inorganic silicate material where the grains have diameters smaller than 0.004 millimeter (or 4 micrometers).

climate meteorological conditions of a place or region, usually averaged over 30 years, in contrast to *weather,* which is the state of the atmosphere at a particular time.

clupeoids fishes of the herring family, including sardines, anchovy, pilchard, and menhaden.

coastal currents currents paralleling the shore, seaward of the surf zone. These currents may be caused by tides or winds or by distributions of mass in coastal waters, associated with river discharges.

coastal ocean shallow portion of the ocean, generally situated over continental shelves.

coastal plain low-lying continental plain, adjacent to the ocean and extending inward to the first major change in terrain features.

coastal plain estuary estuary in coastal plain.

coastal zone the coastal land area, extending inland to an elevation of a few hundred meters, which is influenced by and itself influences the coastal ocean.

coccolithophores microscopic, planktonic brown algae. The cells are surrounded by an envelope on which small calcareous disks or rings (coccoliths) are embedded.

coelenterates animals possessing two cell layers and a digestive cavity with only one opening. This opening is surrounded by tentacles containing stinging cells. Some are sessile, some pelagic (medusae), and some undergo alternation of generations.

cold-blooded animals See *poikilotherms.*

color scanner a radiometer that measures the visible and near-infrared radiation from the ocean surface. These measurements yield ocean color, from which chlorophyll concentrations and locations of turbid water masses can be determined.

commons area where resources are available to all claimants but is not subject to the control of any person or government; examples include outer space, the open ocean, and Antarctica.

community an ecological unit consisting of the micro-organisms, plants, and animals that inhabit a particular area.

compaction decrease in volume or thickness of a sediment deposit under load through closer packing of constituent particles; accompanied by decrease in porosity, increase in density, and squeezing out of water.

compensation depth (carbonate) depth at which carbonate produced in water column is totally dissolved; no carbonate deposition occurs below this depth.

compensation depth (oxygen) the depth at which oxygen production by photosynthesis equals consumption by plant respiration during a 24-hour period.

condensation change of state in which gas becomes liquid.

conduction transfer of energy through matter by internal particle or molecular motions.

conductivity ability of a liquid or solution to conduct an electrical current.

conformal projection map projection in which the angles around any point are correctly represented.

conservative property property whose values do not change in a particular, specified series of events or processes; for example, salinity of seawater. The concentrations of salt are not affected by the presence or activity of organisms but rather by diffusion and currents.

continental climate climate characterized by cold winters and warm summers, where the prevailing winds come from large land areas.

continental crust thickened part of the crust forming continental blocks; consists primarily of granitic rocks.

continental margin zone separating the land from the deep-sea bottom; generally consists of a continental shelf, slope, and rise.

continental rise gentle slope with a generally smooth surface, rising toward the foot of the continental slope.

continental shelf seafloor adjacent to a continent, extending from the low-water line to the change in slope, usually at about 180 meters' depth, where continental shelf and continental slope join.

continental slope a declivity from the outer edge of the continental shelf, extending from the break in slope to the deep-sea floor.

contour line line on a chart connecting points of equal value above or below a reference value; used to portray elevation, temperature, salinity, or other values.

convection vertical movements of air, water, or other Earth materials.

convergence area or zone where flow regimes come together or converge, usually resulting in sinking of surface waters.

convergence zone band along which crustal area is lost. Colliding edges of crustal plates may be thickened, folded, or underthrust, or one plate may be subducted and destroyed.

convergent plate boundary zone where lithospheric plates converge.

copepods minute shrimp-like crustaceans; most are between about 0.5 and 10 millimeters in length.

coral hard, calcareous skeletons of sessile, colonial coelenterate animals, or the stony solidified mass of many such skeletons; also, the entire animal, a compound polyp that produces the skeleton.

coral reef association of bottom-living, and attached, calcareous, shelled, marine invertebrates forming fringing reefs, barrier reefs, or atolls.

coralline algae red algae (bushy or encrusting) that deposit calcium carbonate either on branches or as a crust on the substrate; can develop massive encrustations on coral reefs.

core vertical, cylindrical sample of sediments; also, the central zone of Earth.

Coriolis effect apparent force acting on moving particles resulting from Earth's rotation. It causes moving particles to be deflected to the right in the Northern Hemisphere and to the left in the Southern Hemisphere; the deflection is proportional to the speed and latitude of the moving particle. Particle speed is unchanged by the apparent deflection.

cosmogenous sediments particles derived from outer space.

cotidal points points along a coast that experience high tide at the same time.

countershading protective coloration of animals, where lower parts (in shadow) are normally light-colored; upper parts, in the light are usually dark.

covalent bond chemical bond formed by two elements sharing electrons; water molecules are formed by covalent bonds.

critical depth depth above which the net effective plant production occurs in the water column; total production equals total respiration.

crust outer shell of solid Earth. Beneath the oceans, the outermost layer of crust comprises sediment deposits and basaltic rocks. Its lower limit is the Mohorovičić discontinuity.

crustaceans arthropods that breathe by means of gills or similar structures. The body is commonly covered by a hard shell or crust. The group includes barnacles, crabs, shrimps, and lobsters.

cryosphere portion of Earth's surface covered by ice.

ctenophores spherical, pear-shaped, or cylindrical animals of jelly-like consistency ranging from less than 2 centimeters to about 1 meter in length. The outer surface of the body bears eight rows of comb-like structures.

current ellipse graphic representation of a rotary current, in which current speed and direction at different hours of the tide cycle are represented by vectors joined at one point. A line joining the extremities of the radius vectors forms a curve roughly approximating an ellipse.

current rose graphic representation of currents, utilizing arrows to show the direction toward which the prevailing current flows and a percentage to show the frequency of any given flow.

cyanobacteria one-celled algae which are capable of fixing atmospheric nitrogen gas into more complex forms usable to plants.

cyclone a weather system characterized by a relatively low surface air pressure compared with the surrounding air; same as a "low". Surface winds blow counterclockwise and spiral inward in the Northern Hemisphere.

D

daily (diurnal) inequality difference in heights and durations of two successive high waters or of two successive low waters each day; also, the difference in speed and direction of the two flood currents or the two ebb currents each day.

daily (diurnal) tide tide having only one high water and one low water each tidal day.

debris flow rapid downslope movement of unconsolidated debris.

debris line line near the limit of storm-wave uprush, marking the landward limit of debris deposits.

decomposers heterotrophic and chemoautotrophic organisms (chiefly bacteria and fungi) that break down nonliving matter, absorb some of the decomposition products, and release compounds used by other organisms to produce more organic matter (in primary production.)

decompression release from pressure that a diver experiences when returning to the water surface. Decompression that is too rapid can cause bends.

deep-ocean floor ocean floor deeper than 1000 meters; beyond the continental margin.

deep scattering layer stratified population of organisms in ocean waters that causes scattering of sound as recorded on an echo sounder. Such layers may be from 50 to 200 meters thick. They occur less than 200 meters below the ocean surface at night and several hundred meters below the surface during the day.

deep-sea mud fine-grained deposit on the deep-ocean floor.

deep-water waves water waves whose depth is greater than one-half the average wavelength.

deep zone waters below the base of the pycnocline zone.

delta alluvial deposit formed at the mouth of a stream, tidal inlet, or river.

demersal fishes fishes living on or near the bottom.

density mass per unit volume of a substance, usually expressed in grams per cubic centimeter. In the centimeter-gram-second system, density equals specific gravity.

density current flow of one current through, under, or over another; it retains its unmixed identity from the surrounding water because of density differences.

deposit feeding removal of edible material from sediment or detritus, either by ingesting material unselectively and excreting the unusable portion or by selectively ingesting discrete particles.

depth of no motion depth at which water is assumed to be motionless, used as a reference surface for computing geostrophic currents.

desalination production of fresh water from seawater or brine.

detritivore animal that eats detritus.

detritus loose material produced by rock disintegration. Organic detritus consists of decomposition or disintegration products or dead organisms, including fecal material.

detritus food chain a food chain based solely on dead organic matter.

diatoms microscopic phytoplankton organisms possessing walls of overlapping halves (valves) impregnated with silica.

diffraction bending of a wave around an obstacle.

diffusion transfer of material (e.g., salt) or property (e.g., temperature) by eddies or molecular movement. Diffusion causes spreading or scattering of matter under the influence of a concentration gradient, with movement from the stronger to the weaker solution.

dinoflagellates microscopic or minute organisms that possess characteristics of plants (chlorophyll and cellulose plates) and animals (ingestion of food).

discontinuity abrupt change in a property, such as salinity or temperature, at a line or surface.

dispersion separation of a complex wave into its component parts. Longer component parts of the wave travel faster than shorter ones.

distributary outflowing branch of a river, usually on deltas.

diurnal daily, especially pertaining to actions that are completed within approximately 24 hours and that recur every 24 hours.

divergence horizontal movements of water or air in different directions from a common zone.

divergence zone region along which crustal plates move apart and new lithospheric material solidifies from rising volcanic magma.

divergent boundary area from which lithospheric plates move away; example is a mid-ocean ridge.

divergent margin growing edge of a lithospheric plate; also called a *spreading center.*

doldrums belt of light, variable winds near the equator; an area of low atmospheric pressure.

downwelling area of downward-moving water; a convergence.

drag resistance to a body moving through a fluid.

drift net fishing net suspended in the water vertically so that drifting or swimming animals are trapped or entangled in the mesh.

dune mound or ridge of sand moved by winds.

E

Earth System Science study of Earth as a whole; includes study of Earth's subsystems (hydrosphere, atmosphere, lithosphere, and biosphere) and their interactions.

earthquake rapid movements (horizontally or vertically) of rocks, releasing energy stored during plate movements.

eastern boundary current broad, shallow, slow-moving current on the eastern side of an ocean basin.

ebb current tidal current directed away from shore or down a tidal stream.

echinoderms principally benthic marine animals having calcareous plates with projecting spines forming a rigid or articulated skeleton or plates and spines embedded in the skin. They have radially symmetrical, usually five-rayed, bodies. They include starfish, sea urchins, crinoids, and sea cucumbers.

echiuroids unsegmented, burrowing marine worms.

echo sounding determination of water depth by measuring time intervals between emission of a sonic signal and the return of its echo from the bottom. The instrument used for this purpose is called an echo sounder.

ecliptic plane of Earth's orbit as it revolves around the Sun.

ecological efficiency ratio of the efficiency with which energy is transferred from one trophic level to the next.

ecology study of organisms' relations to one another and to their environment.

ecosystem ecological unit including organisms and the nonliving environment, each influencing the properties of the other and both necessary for maintenance of life.

eddy current of air, water, or any fluid, often on the side of a main current, especially one moving in a circle; in extreme cases, a whirlpool.

edge waves waves in the surf zone that propagate along the beach.

eelgrass seed-bearing, grasslike marine plant that grows chiefly in sand or mud–sand bottoms; most abundant in temperate waters less than 10 meters deep.

Ekman spiral representation of currents resulting from a steady wind blowing across an ocean having unlimited depth and extent and uniform viscosity. The surface layer moves 45° to the right of the wind direction in the Northern Hemisphere; water at successive depths drifts in directions more to the right until, at some depth, the water moves in a direction opposite to the wind. Speed decreases with depth throughout the spiral. The net water transport is 90° to the right of the wind in the Northern Hemisphere.

El Niño warm surface waters offshore from Peru occuring around Christmas; effects felt worldwide, especially in the tropics.

encrusting algae See *coralline algae, red algae.*

epicenter point on Earth's surface directly above an earthquake focus.

epifauna animals that live at the water-substrate interface, attached to the bottom or moving freely over it.

epiphytes plants that grow attached to other plants.

equal-area projection map projection in which equal areas on Earth's surface are represented by equal areas on the map.

equatorial region areas lying within a few degrees of latitude north or south of the equator.

equilibrium tide hypothetical semidaily tide caused by gravitational attraction of the Sun and Moon on a frictionless, nonrotating, entirely water-covered Earth.

equinox a day which occurs twice a year (once in spring and once in autumn) when day and night are equal in length

estuarine circulation characteristic circulation in an estuary; flow is seaward at surface, landward at depth.

estuary a semi-enclosed, tidal body of saline water, with free connection to the sea.

euphausiids shrimp-like, planktonic crustaceans, common in oceanic and coastal waters, especially in colder waters.

euphotic zone See *photic zone.*

evaporites salt deposits left behind by evaporation of seawater.

Exclusive Economic Zone (EEZ) a coastal-ocean zone extending 370 km from the coastline, which is under the jurisdiction of the coastal state for fisheries and resource exploration and exploitation.

exotic terrain fragments of continental masses, or sometimes seafloor, that have accreted to other continents.

exotic species species introduced into an area by human activities.

extratropical cyclone powerful storm that develops on a front, outside the tropics; can reach hurricane strength.

F

fan gently sloping, fan-shaped feature located near the lower end of a canyon.

fast ice ice frozen to the shore or to the bottom, therefore stationary.

fathom a unit of water depth, 6 feet or 1.83 meters. It was originally derived from the distance between the hands of a large man with his arms outstretched.

fault fracture or fracture zone in rock, along which one side has moved relative to the other.

fauna animal population of a location, region, or period.

fecal pellets pellets of organic matter voided by marine animals, usually ovoids less than 1 mm long.

fetch ocean area where waves are generated by a wind having a constant direction and speed, also called generating area; also, the length of the fetch area, measured in the direction of the wind in which the seas are generated.

filter feeding filtering or trapping edible particles from seawater; a feeding mode typical of many zooplankters and other marine organisms of limited mobility.

fjord narrow, deep, steep-walled inlet, formed either by the submergence of a glaciated mountainous coast or by entrance of the ocean into a deeply excavated glacial trough after the glacier melts.

flagellum whip-like bit of protoplasm that propels a motile cell.

flatfishes demersal fish that has a flattened shape, permitting it to lie camouflaged on the bottom.

flocculation process of aggregation into small lumps, especially with regard to soils and colloids.

floe sea ice—either as a single unbroken piece or as many individual pieces—covering an area of water.

flood current tidal current associated with the increase in the height of a tide. Flood currents generally set toward the shore or in the direction of the tide progression.

food chain simplification of a food web.

food web food relationships in an ecosystem, including its production, consumption, and decomposition, and the energy relationships among organisms involved in the cycle.

foraminifera one-celled animals (protists) that secrete carbonate shells.

forced wave wave generated and maintained by a continuous force, in contrast to a free wave, which continues to exist after the generating force has ceased to act.

forereef upper seaward face of a reef, extending above the lowest point of abundant living coral and coralline algae to the reef crest. This zone commonly includes a shelf, bench, or terrace that slopes to 15–30 meters, as well as the living, wave-breaking face of the reef.

forerunner low, long-period swell that commonly precedes the main swell from a distant storm.

foreshore See *low tide terrace.*

foul to attach to, or come to lie on, the surface of submerged human-made or introduced objects, as barnacles on the hull of a ship or silt on a stationary object.

fractals self-similar systems.

fracture zone elongate zone of unusually irregular topography of the ocean floor characterized by seamounts, steep-sided or asymmetrical ridges, troughs, or long, steep slopes.

frazil ice hexagonal spicules of newly formed ice.

free wave any wave not acted on by external forces except for the initial force that created it.

fringing reef reef attached directly to the shore of an island or continental landmass. Its outer margin is submerged and often consists of algal limestone, coral rock, and living coral.

front marked change in properties of water or air.

fully developed sea maximum height to which ocean waves can be generated by a given wind blowing over sufficient fetch, regardless of duration.

fungi plant-like organisms that have no chlorophyll; live by decomposing organic matter.

G

Gaia hypothesis theory that physical and chemical conditions on Earth's surface have been controlled by the presence of life.

gas one of the states of matter.

gastropods mollusks that possess a distinct head (generally with eyes), tentacles, and a broad, flat foot; usually enclosed in a spiral shell.

gelatinous plankton planktonic organism made up of jellylike tissues.

geostrophic current current resulting from the balance between gravitational forces and the Coriolis effect.

geostrophic winds winds that blow parallel to the isolines for atmospheric pressure; the winds are influenced by the differences in atmospheric pressure and by Earth's rotation.

geothermal power power derived from heat energy coming from Earth's interior.

giant squids large cephalopods (length may be 15 meters or more) that inhabit mid-depths in oceanic regions but may come to the surface at night.

gill a delicate, thin-walled structure, often an extension of the body wall, used to exchange gases with the water and sometimes for excreting wastes.

glassworm See *chaetognath.*

glacial-marine sediment high-latitude, deep-ocean sediments transported from the land by glaciers or icebergs.

glacier mass of freshwater ice, formed by recrystallization of compacted snow, flowing slowly from an area of accumulation to areas where snow or ice is removed.

global conveyer belt integrated system of surface and deep-ocean currents that move waters from the polar regions throughout the ocean and return them to polar regions where the deep-ocean water masses form.

Gondwanaland southern portion of Pangaea, during the last supercontinent cycle.

graded bedding type of stratification in which each stratum displays a gradation in grain size from coarse below to fine above.

granite crystalline, igneous rock consisting of alkali feldspar and quartz. Granitic is a textural term applied to coarse- and medium-grained igneous rocks.

gravel loose sediment with particles ranging in size from 2 to 256 millimeters.

gravity anomaly disturbance of Earth's normal gravity field.

gravity wave wave whose velocity of propagation is controlled primarily by gravity. Water waves longer than 1.7 centimeters are considered gravity waves.

grazing food chain food chain in which animals feed directly on plants; contrasts with a detritus food chain.

great circle intersection of the surface of a sphere and a plane through its center—for example, meridians of longitude and the equator are great circles on Earth's surface.

greenhouse gas atmospheric gas (water vapor, carbon dioxide, methane) that absorb energy radiated from Earth's surface.

greenhouse effect warming of Earth's surface caused by penetration of the atmosphere by comparatively short-wavelength solar radiation that is largely absorbed near and at Earth's surface, whereas the relatively long-wavelength radiation emitted by Earth is partially absorbed by water vapor, carbon dioxide, and dust in the atmosphere, thus warming the lower atmosphere.

groin low, artificial, dam-like structure of durable material placed so that it extends seaward from the land; used to slow littoral drift on beaches.

gross primary production the total amount of carbon fixed by plants in the process of photosynthesis.

group velocity velocity with which a wave group travels. In deep water, it is equal to one-half the individual wave velocity.

guyot flat-topped submarine mountain or seamount.

gyre circular or spiral form, usually applied to a very large, semiclosed current system in an open-ocean basin.

H

habitat place or site occupied by a specific plant or animal.

Hadley cell semiclosed system of vertical motions in the tropical atmosphere. Warm, moist air rises in equatorial regions, flows to mid-latitudes (30° N, 30° S), where it sinks, and returns along the ocean surface to the equatorial zone as the trade winds.

half-life time required for the decay of one-half the atoms of a radioactive substance.

halocline water layer with large vertical changes in salinity.

heat budget accounting for the amount of the Sun's heat received on Earth during any one year as equaling the amount lost by radiation and reflection.

heat capacity amount of heat required to raise the temperature of a substance by a given amount.

herbivore animal that feeds only on plants.

heterotroph an organism that utilizes organic compounds for food.

high area of high atmospheric pressure.

high water upper limit of the surface water level reached by the rising tide; also called high tide.

higher high water higher of the two high waters of any tidal day. The single high water occurring daily in a diurnal tide is considered a higher high water.

higher low water higher of two low waters occurring during a tidal day.

holdfast root-like structures that anchor seaweeds to the substrate.

holoplankton organisms whose life cycle is spent in the plankton.

homeotherms so-called "warm-blooded animals;" the birds and mammals whose body temperatures must be physiologically controlled.

horse latitudes belts of atmospheric pressure between about 30° and 40°, both north and south of the equator.

hot spot area of persistent volcanic activity.

hurricane large cyclonic storm, usually of tropical origin, containing winds of 120 kilometers per hour or higher.

hydrocarbons organic compounds composed only of carbon and hydrogen

hydrogen bond relatively weak bond formed between adjacent molecules in liquid water, resulting from the mutual attractions of hydrogen and nearby atoms.

hydrogenous sediment particles precipitated from solution in water, such as manganese and phosphorite nodules.

hydroids colonial polyp form of coelenterate animals that exhibit alternation of generations. They are attached, often branching, and gives rise to the pelagic, medusa form by asexual budding.

hydrosphere water portion of Earth, as distinguished from the solid part and the gaseous outer atmosphere. It consists of liquid water in sedimentary rocks, rivers, lakes, and oceans, as well as ice in sea ice and continental ice sheets.

hydrothermal circulation movement of water through crustal rocks caused by heating of water by recently formed volcanic rocks.

hydrothermal vent associated with high-temperature ground water; for example, alteration or precipitation of minerals and mineral ores.

hypothesis tentative assumption made to test consequences.

I

iceberg large mass of detached freshwater ice floating in the sea or stranded in shallow water.

ice shelf thick freshwater ice formation with a fairly level surface, formed along a polar coast and in shallow bays and inlets, where it is fastened to the shore and often rests on the bottom.

igneous rock formed by solidification of magma.

infauna animals who live in soft sediments.

inner core the central solid portion of Earth's core.

insolation solar radiation received at Earth's surface; also, the rate at which direct solar radiation is incident upon a unit horizontal surface at any point on or above Earth's surface.

instability property of a system where any disturbance grows larger instead of diminishing, so that the system never returns to the original steady state; usually refers to the vertical displacements of water parcels.

interface surface separating two substances of different properties (such as different densities, salinities, or temperatures); for example, the air–sea interface or the water–sediment interface.

internal wave wave that occurs within a fluid whose density changes with depth, either abruptly at a sharp surface of discontinuity (an interface) or gradually.

intertidal zone (littoral zone) zone between mean high-water and mean low-water levels.

intertropical convergence zone (ITCZ) area toward which the trade winds blow; lies near the Equator.

invertebrate animal lacking a backbone.

ion electrically charged atom or molecule.

ionic bond linkage between two atoms, with a separation of electric charge on the two atoms; a linkage formed by the transfer or shift of electrons from one atom to another.

iron-manganese nodules See *manganese nodules.*

island-arc system a group of islands, usually with a curving, arch-like pattern and convex toward the open ocean, with a deep trench or trough on the convex side and usually enclosing a deep-sea basin on the concave side. Generally associated with volcanoes and subduction zones.

isobar a line on a weather map that joins points of equal atmospheric pressure.

isostasy balance of portions of Earth's crust, which rise or subside until their masses are in equilibrium relationship, "floating" on the denser plastic mantle below.

isotherm line connecting points of equal temperature.

isothermal of the same temperature.

isotope nuclides having the same number of protons in their nuclei (and hence belonging to the same element) but differing in the number of neutrons (and therefore in mass number or energy content); also, a radionuclide or a preparation of an element with special isotopic composition, used principally as an isotopic tracer.

J

jellyfish See *medusa.*

jet stream high-altitude swift air current.

jetty a structure built to influence tidal currents, maintain channel depths, or protect the entrance to a harbor or river.

K

kelp large brown marine algae.

kinetic energy energy of motion.

krill small shrimp-like crustaceans; euphausiids.

L

lagoon shallow body of water, generally separated from the open ocean by a barrier beach.

lamina sediment or sedimentary-rock layer less than 1 centimeter thick, visually separable from the material above and below.

laminar flow flow in which fluids move smoothly in streamlines, in parallel layers or sheets; a nonturbulent flow.

land breeze wind blowing toward the sea; caused by unequal heating (and cooling) of land and water.

langmuir cells cellular circulation with alternate left- and right-hand helical vortices, having axes in the direction of the wind; set up in the surface layer of a water body by winds exceeding 3.5 meters per second.

lantern fishes (myctophids) small oceanic fishes that normally live at depths between a few hundred and a few thousand meters. Myctophids characteristically have numerous small light organs on the sides of the body. Many undergo diurnal vertical migration.

larvae immature form of animal.

latent heat heat released or absorbed per unit mass by a system undergoing a reversible change of state at a constant temperature and pressure.

latitude angular distance north or south of the equator.

Laurasia northern portion of Pangaea.

Law of the Sea treaty drafted by the United Nation's Law of the Sea Conference that developed a new legal regime for the ocean and its resources.

layer of no motion See *depth of no motion.*

lead areas of open water between ice floes.

leeward being in, or facing, the direction toward which the wind is blowing; opposite to windward.

levee embankment bordering one or both sides of a sea channel or delta distributary.

lidar laser radar, in which light pulses are from a powerful laser and the elapsed time between transmission of the pulse and return of its reflection is used to determine the distance to the object; returned pulses can also be analyzed to determine the composition of the target and concentrations of different elements; used to study atmospheric gas composition and also to determine water depths in coastal oceans.

limit of stability limit of steepness that a wave can sustain without breaking.

lithogenous sediment sediment composed primarily of mineral grains.

lithosphere outer, solid portion of Earth; includes the crust and part of the upper mantle.

littoral See *intertidal zone.*

littoral drift sand moved parallel to the shore by wave and current action.

longitude angular distance east or west of the prime meridian.

longshore bar See *bar.*

longshore current current located in the surf zone, moving generally parallel to the shoreline; usually generated by waves breaking at an angle with the shoreline.

low tide terrace zone between the ordinary high- and low-water marks; daily traversed by the oscillating water line as the tides rise and fall. This area, together with the vertical scarp that often occurs at its upper limit, is sometimes called the *foreshore;* it ends at the highest point of normal wave uprush.

low area of low atmospheric pressure.

low water lowest limit of the surface-water level reached by the lowering tide; also called *low tide.*

lower high water lower of two high tides occurring during a tidal day.

lower low water lower of two low tides occurring during a tidal day.

lunar tide part of the tide caused by the gravitational attraction of the Moon, as distinguished from that part caused by the gravitational attraction of the Sun.

lysocline depth separating well preserved carbonate shells (foraminifera, etc.) from poorly preserved forms deposited at greater depths.

M

magma mobile, usually molten rock material; capable of intrusion and extrusion; forms igneous rocks when it solidifies.

magnetic anomaly local variation in Earth's magnetic field that differs from the average value. A local value that is greater than the average is called a *positive anomaly.*

magnetometer instrument used to measure intensity and direction of local component of Earth's magnetic field.

maintenance the amount of energy used by an organisms in maintaining its body tissues. Does not include growth or reproduction processes.

manganese nodules concretionary lumps of manganese and iron; found on the deep-ocean floor.

mangrove tropical salt-tolerant trees that grow along shorelines.

mantle the bulk of Earth, between crust and core, from about 40 to 3500 kilometers depth. Also, the tough, protective membrane possessed by all mollusks, within which water circulates.

mantle convection vertical movements in Earth's mantle, driven by heat escaping from the core.

map projection method of representing part or all of the surface of a sphere, such as Earth, on a plane surface.

marginal ocean basin coastal ocean basin that is partially isolated from the open ocean, some by island arcs (Sea of Japan, Bering Sea), others by land (Gulf of Mexico, Hudson Bay).

marginal sea semi-enclosed body of water adjacent to, widely open to, and connected with the ocean at the water surface, but bounded at depth by submarine ridges.

mariculture See *aquaculture.*

marine snow particles of organic detritus and living forms. Sinking of these particles, especially in dense concentration, looks like a snowfall when viewed underwater.

maritime climate characterized by relatively little seasonal change—warm, moist winters, cool summers; result of prevailing winds blowing from ocean to land.

marsh area of wet land. Intertidal flat land periodically flooded by saltwater is called a *salt marsh.*

maximum sustainable yield maximum yield of a fishery that can be sustained without depleting stocks.

meander turn or winding of a current that may become detached from the main stream; sinuous curve in a current.

mean sea level average height of the sea surface for all stages of the tide over a 19-year period, usually determined from hourly readings of tidal height.

mean tidal range difference in height between mean high water and mean low water, measured in feet or meters.

medusa (jellyfish) free-swimming coelenterates having a disk- or bell-shaped body of jellylike consistency. Many have long tentacles with stinging cells.

Mercator projection conformal map projection.

meridian (of longitude) great circle passing through the North and South Poles. It connects points with an equal angular separation from the prime meridian.

meroplankton organisms whose early developmental stages occur in the floating state; adults are benthic.

mesosphere upper portion of the atmosphere, above the stratosphere.

microbenthos one-celled plants, animals, and bacteria living in or on the surface of bottom sediments.

microbial loop food chain based on bacteria, which are eaten by tiny flagellates and zooplankton.

microcontinent isolated fragment of continental crust; forms oceanic plateau.

microplankton plankton between 2 and 20 microns in diameter.

mid-ocean ridge volcanic mountain range extending the length of an ocean basin and roughly paralleling the continental margins; area of oceanic crustal formation.

Milankovitch cycle periodic changes in the amount of incoming solar radiation caused by changes in Earth's orbit around the Sun.

mixed layer near-surface waters down to the pycnocline, where waters show little change in temperature or salinity with depth.

mixed tide type of tide in which a diurnal wave produces large inequalities in heights and/or durations of successive high and/or low waters. This term applies to the tides intermediate to those predominantly semidaily and those predominantly daily.

model system of data, inferences, and relationships, presented as a description of a process or entity.

moderately stratified estuary estuary where mixing has greatly modified the sharp interface between the subsurface and surface water layers that is characteristic of a salt-wedge estuary.

Mohorovičić discontinuity (Moho) sharp discontinuity between Earth's crust and mantle.

monsoons seasonal winds (derived from the Arabic *mausim,* meaning season), first applied to the winds over the Arabian Sea, which blow for six months from the northeast and the remaining six months from the southwest; subsequently extended to similar seasonal winds in other parts of the world.

mucous nets structure made of mucus used to capture food particles; the net and attached particles are then eaten by the organism.

mud detrital material consisting mostly of silt and clay-sized particles (less than 0.06 millimeter) but often containing varying amounts of sand and/or organic materials. It is also a general term applied to any fine-grained sediment whose particle size distribution is unknown.

N

nannoplankton plankton whose length is less than 50 micrometers.

natural frequency characteristic frequency (number of vibrations or oscillations per unit of time) of a body controlled by its physical characteristics (dimensions, density, etc.).

nauplius limb-bearing early larval stage of many crustaceans.

neap tide lowest range of the tide, occurring near the times of the first and last quarters of the Moon.

nearshore that part of a beach between the shoreline and the line of breaking waves.

negative feedback process of mechanism that diminishes or subtracts from an initial disturbance of a system.

nekton active swimmers, pelagic animals such as most adult squids, fishes, and marine mammals.

neritic ocean environment shallower than 200 meters.

net plankton plankton caught by nets.

net primary production total amount of organic matter produced by photosynthesis minus the amount consumed by the photosynthetic organisms in their respiratory processes.

neuston organisms that inhabit the water surface.

new production that fraction of primary production which is supported by nutrients upwelled from the deep ocean rather than being recycled within the photic zone.

niche specific role of an organism in a community and its position in an ecosystem.

nitrogen fixation conversion of atmospheric nitrogen to oxides usable in primary food production.

nitrogen narcosis the "rapture of the deep" resulting from breathing air under pressure.

nodal line line in an oscillating area along which there is little or no rise and fall of the tide.

nodal point no-tide point in an amphidromic region.

node part of a standing wave where the vertical motion is least and the horizontal velocities are greatest.

nonconservative constituent property whose values change in the course of a particular specified series of events or processes; for example, those properties of seawater, such as nutrient or dissolved oxygen concentrations, that are affected by biological or chemical processes.

nonrenewable resource a resource that is not replenished at a rate comparable to its rate of consumption.

nuclide species of atom characterized by the constitution of its nucleus. The nuclear constitution is specified by the number of protons, number of neutrons, and energy content; or, alternatively, by the atomic number, mass number, and atomic mass.

nudibranchs (sea slugs) gastropods in which adults have no shell.

nutrient inorganic or organic compounds or ions necessary for the nutrition of primary producers.

O

ocean basin ocean floor that is more than about 2000 meters below sea level.

oceanic crust basaltic material that lies under the ocean basins.

oceanic (mid-ocean) rise continuous ocean-bottom province that rises above the deep-ocean floor; area of crustal generation.

oceanography scientific study of the ocean.

offshore flow air or water movements directed away from the shoreline.

omnivore organism that eats anything, both plant and animal.

onshore flow water or air movements toward the shoreline

ooze fine-grained, deep-ocean sediment containing at least 30 percent (by volume) undissolved sand- or silt-sized, calcareous or siliceous skeletal remains of small marine organisms, the remainder usually being clay-sized material.

open ocean part of the ocean that is seaward of the approximate edges of the continental shelves, usually more than 2 kilometers deep.

open-ocean current current in the open ocean, outside the coastal ocean.

osmosis movements of water across a semi-permeable membrane, such as a cell wall; water moves from low salinity to higher salinities.

outer core outer, molten portion of Earth's core.

overturning convective overturning of waters, usually refers to lakes.

oxidation loss of hydrogen or electrons; opposite of reduction.

oxygen-minimum zone a layer of water, typically between 500 and 1000 meters, where dissolved oxygen concentrations are lower than in the waters above or below.

ozone layer the portion of the stratosphere where most of Earth's ozone occurs.

P

pack ice a rough, solid mass of broken sea-ice floes forming an obstruction to navigation.

Pacific-type margin See *active margin.*

pancake ice small, pancake-shaped pieces of newly formed sea ice.

Pangaea large continent that split apart about 200 million years ago to form the present continents.

Panthalassia name given to Earth's ocean during times when all continents were collected together, forming a supercontinent.

paradigm a world-view.

parasitism relationship in which the parasite harms the host from which it takes nutrition.

partial tide one of the harmonic components comprising the tide at any point. The periods of the partial tides are derived from various combinations of the angular velocities of Earth, Sun, and Moon, relative to each other.

partially mixed estuary an estuary in which tidal mixing is more influential than river discharge and vertical mixing results in the more-saline bottom layer moving landward with seaward moving surface waters.

passive continental margin a continental margin in plate interior; also called an *Atlantic-type margin.*

patch reef isolated coral growths in lagoons of barriers and atolls; ranging from several kilometers across to small coral pillars or even mushroom-shaped growths consisting of a single colony.

pelagic the environment of the open ocean, not including the bottom or the coastal zone.

pelagic sediments deep-ocean sediment deposits that have accumulated particle-by-particle.

photic zone (euphotic zone) near-surface layer of water that receives ample sunlight for photosynthesis to exceed respiration.

photosynthesis manufacture of carbohydrates and other compounds from carbon dioxide and water in the presence of chlorophyll by utilizing light energy and releasing oxygen.

pH scale logarithmic scale of the concentration of hydrogen ions in a solution; a measure of the acidity or alkalinity of a solution.

phytoplankton single-celled photosynthetic organisms.

pillow lava a characteristic form of lava extruded underwater, where the molten rock forms lumpy, rounded forms, resembling pillows.

pinniped fin-footed, carnivorous, fur-bearing marine mammals; includes seals, sea lions, and walruses.

plankter a single planktonic organism.

plankton passively drifting or weakly swimming organisms.

plankton bloom unusually high concentration of plankton (usually phytoplankton) in an area, caused either by an explosive or gradual multiplication of organisms.

plate rigid unit of lithosphere that moves as a unit.

plate tectonics theory of lithospheric plate movement caused by mantle convection.

poikilotherms the so-called "cold-blooded animals," whose body temperatures vary with that of the environment. Includes all invertebrates and fish.

polar molecule molecule where the electrical charges are separated so that the molecule reacts to electrical charges.

polder land area reclaimed from the shallow ocean bottom, usually separated from the ocean by dikes.

polychaetes segmented marine worms; some are tube builders, others burrow, still others live on the sediment surface or are free-swimming.

polynya persistently ice-free area surrounded by sea ice.

polyp individual sessile coelenterate (jellyfish, sea anemones, and corals). May be a stage in the reproductive cycle of these organisms. See also *hydroids.*

positive feedback process or mechanism that adds to or reinforces an initial disturbance of a system.

potential energy energy in a particle as a result of its position; this energy is released when a particle moves to a lower energy position, such as rolling down a hill.

predator animal that preys on other organisms.

pressure ridge ridges formed when sea-ice floes are pushed together.

prevailing wind systems (planetary winds) large, relatively constant wind systems that result from Earth's shape, inclination, revolution, and rotation, examples are the northeast and southeast trade winds, the westerlies, and the polar easterlies.

primary production amount of organic matter synthesized by organisms from inorganic substances per unit of time in a defined volume of water; also called *gross primary production.*

prime meridian meridian of longitude (0°), used as the reference for measurements of longitude, located on the meridian of Greenwich, England, a suburb of London.

production See *primary productivity.*

progressive wave wave that is manifested by progressive movements of waveforms.

protista single-celled organisms whose nucleus is surrounded by a membrane. Includes the ciliates, protozoa, dinoflagellates and red, brown, and green phytoplankton.

protozoa microscopic, one-celled animals.

pseudopod an extension of protoplasm that can be projected or withdrawn by the animal for capturing food or for locomotion; characteristic of some protozoans.

pteropods free-swimming gastropods in which the foot is modified into fins; both shelled and nonshelled forms exist. In some shallow oceanic areas, accumulated shells of these organisms form sediment deposits called *pteropod oozes.*

pycnocline zone a depth zone in which density increases markedly with depth.

R

radioactivity spontaneous breakdown of an atomic nucleus, giving off energy and often particles.

radioisotope radioactive isotope of an element.

radiolarians single-celled planktonic protozoans possessing a skeleton of siliceous spicules and radiating, threadlike pseudopodia.

radiometer a device for measuring the intensity of radiation from the sea surface beneath a spacecraft or an airplane. Measurements in the infrared bands yield sea surface temperatures, and the visible bands yield color.

radionuclide synonym for *radioactive nuclide.*

red algae reddish, filamentous, membranous, encrusting, or complexly branched plants in which the color is imparted by the predominance of a red pigment.

red clay brown-to-red deep-sea deposit. It is the most finely divided clay material, derived from the land and transported by ocean currents and winds.

red giant stage in the development of a star after it has consumed all of its hydrogen.

red tide red or reddish brown discoloration of surface waters most frequently in coastal regions, caused by concentrations of microscopic organisms, particularly dinoflagellates.

reduction gain in electrons or in hydrogen; opposite of oxidation.

reef off-shore wave-resistant rock which is a hazard to navigation.

reef flat (of a coral reef) flat expanse of dead reef rock, partly or entirely dry at low tide. Shallow pools, potholes, gullies, and patches of coral debris and sand are features of the reef flat. It is divisible into inner and outer portions.

refraction of water waves process by which the direction of a wave moving in shallow water at an angle to the contours is changed, causing the wave crest to bend toward alignment with the underwater contours; also, the bending of wave crests by currents.

relict sediment sediment deposited by processes no longer active.

remote sensing means of collecting data from an area that is far from the observer; an example is an Earth-orbiting satellite observing Earth's surface from space.

renewable resource resource replenished at a rate comparable to its rate of consumption by natural growth or by careful management of the resource.

residence time time required for a flow of material to replace the amount of that material originally present in a given volume. Assuming a steady flow, replacement time can be calculated for any substance, such as salt or water.

respiration oxidation-reduction process by which chemically bound energy in food is transformed into other kinds of energy on which certain processes in all living cells are dependent.

restoring force force in a water surface that acts to restore the surface to its original position after the passage of a wave; surface tension is the restoring force for capillary waves, gravity for most waves, and the Coriolis effect for the largest waves.

reversing current current that periodically changes its direction of flow; ebb and flood currents in tides are one example.

reversing tidal current tidal current that flows alternatively in approximately opposite directions, with a period of slack water at each reversal of direction. Reversing currents occur in rivers and straits where the flow is restricted. When the flow is toward the shore, the current is flooding; when in the opposite direction, it is ebbing.

rift valley narrow trough formed by faulting in a divergence area.

ring body of water separated from surrounding waters by a strong current; formed by a meander of a boundary current.

rip agitation of water caused by the meeting of currents or by a rapid current setting over an irregular bottom; for example, a *tide rip.*

rip current strong current, usually of short duration, flowing seaward from the shore. It usually appears as a visible band of agitated water and is the return movement of water piled up on the shore by incoming waves and wind.

ripple wave controlled to a significant degree by both surface tension and gravity.

rise long, broad elevation that rises gently and generally smoothly from the ocean bottom.

river-induced upwelling upward movements of deeper water that occurs when seawater mixes with fresh water from a river, becoming less dense, and moves toward the surface, becoming less dense.

rogue wave unusually large wave created by interactions between wind waves and currents.

rotary tidal current tidally induced current that flows continually with the direction of flow, changing through all points of the compass during the tidal period. Rotary currents are found where flow directions are not restricted.

S

salinity measure of the quantity of dissolved salts in seawater. Formally defined as the total amount of dissolved solids in seawater in parts per thousand by weight, when all the carbonate has been converted to oxide, all the bromide and iodide have been converted to chloride, and all organic matter is completely oxidized.

salps transparent pelagic tunicates. The body is more or less cylindrical and possesses conspicuous ringlike muscle bands that contract to propel the animal through the water.

salt deposit sea salts deposited when seawater is evaporated.

salt lens small, current-bounded water subsurface mass.

salt marsh see *marsh.*

salt-wedge estuary high-flow circulation pattern with seawater intrusion along the bottom of an estuary, forming a wedge; characterized by a pronounced increase in salinity from surface to bottom.

sand loose material that consists of grains ranging between 0.0625 and 2 millimeters in diameter.

sargassum brown, bushy alga with substantial holdfast (rootlike structure) when attached, and a yellowish brown, greenish yellow, or orange color.

scattering dispersion of light when a beam strikes very small particles suspended in air or water. In light scattering there is no loss of intensity, only a redirection of light rays.

scatterometer microwave radar to measure sea surface roughness beneath the spacecraft or aircraft. Used to determine heights of surface waves and surface-wind velocities.

scavenger animal that feeds on dead organic matter.

schooling large number of one kind of fish or other aquatic animal swimming or feeding together.

science accumulated knowledge built up through use of scientific method.

scour erosion of a sediment bed by waves or currents.

scyphozoans coelenterates in which the polyp or hydroid stage is insignificant but the medusoid stage is well developed. True jellyfish belong to this group.

sea waves generated or sustained by winds within their fetch, as opposed to *swell;* also, a subdivision of an ocean.

sea breeze light wind blowing toward the land, caused by unequal heating (and cooling) of land and water masses.

seafloor spreading process by which lithosphere is generated at mid-ocean ridges. Adjacent lithospheric plates are moved apart as new material forms.

sea grasses permanently submerged beds of marine grasses.

seamount elevation rising 900 meters or more from the ocean bottom.

sea state (state of the sea) numerical or written description of ocean surface roughness.

seawall rock or concrete structure built to protect a coast against wave erosion.

secondary production the amount of organic matter produced by herbivores.

sediment particulate organic and inorganic matter that accumulates in a loose, unconsolidated form. It may be chemically precipitated from solution, secreted by organisms, or transported from land by air, ice, wind, or water and deposited.

sedimentation processes of breakup and separation of particles from the parent rock, their transportation, deposition, and consolidation into another rock.

seiche standing wave in an enclosed or semi-enclosed water body that continues after the cessation of the originating force, which may have been seismic, wind-, or wave-induced.

seismicity distribution, frequency, and magnitude of earthquakes.

seismic tomography technique used to study the Earth's interior, using variations in speed of waves from earthquakes moving though rocks of different temperatures and composition.

semidaily (semidiurnal) tide tide having a period of approximately one-half a tidal day, with two high waters and two low waters each tidal day.

semipermeable membrane membrane through which a solvent, but not certain dissolved or colloidal substances, may pass.

sensible heat portion of energy exchanged between ocean and atmosphere; utilized in changing the temperature of the medium into which it penetrates.

sessile permanently attached by the base of a stalk; not free to move about.

setae small bristles on exoskeletons of arthropods (crabs, lobsters, etc.) or annelid worms.

shallow-water waves waves in water shallower than $L/2$.

shelf break a steepening of the seafloor that marks the edge of the *continental shelf* and the top of the *continental slope.*

shoal submerged bank or bar that is often a hazard to navigation.

shoreline boundary between a body of water and the land at high tide (usually mean high water).

significant wave height (characteristic wave height) average height of the highest one-third of waves of a given wave group.

silica silicon dioxide, a glass-like substance.

silicoflagellate small photosynthetic phytoplankton that secrete a glasslike shell.

sill shallow portion of the ocean floor that partially restricts water flow; may be either at the mouth of an inlet, fjord, or similar structure, or at the edge of an ocean basin, for example, the Bering Sill separates the Pacific and Arctic portions of the Atlantic Ocean.

silt particles between sands and clays in size; 4 to 62 micrometers in diameter.

sine wave trigonometric mathematical function that describes a smooth repeating wave form.

sinking (downwelling) downward movement of surface water, generally caused by converging currents or as a result of a water mass becoming more dense than the surrounding water.

siphonophores colonial coelenterates; many are luminescent and some venomous. Some possess a gas-filled float; others have swimming bell-shaped medusae. All have a collection of both polyp and medusoid individuals, each serving a different purpose and functioning together as a single individual.

sipunculids worm-like marine animals, unsegmented, with the mouth surrounded by tentacles. The anterior (head) end can be withdrawn into the body. They are deposit feeders.

slack water state of a tidal current when velocities are near zero, usually when reversing currents change direction.

slick area of quiescent water surface, usually elongated. Slicks may form patches or web-like nets where ripple activity is greatly reduced.

slump slippage or sliding of a mass of unconsolidated sediment down a submarine or subaqueous slope. Slumps occur frequently at the heads or along the sides of submarine canyons; triggered by earthquakes. The sediment usually moves as a unit initially and may eventually become a turbidity flow.

slurry dense suspension of solids in water.

solar tide partial tide caused solely by the tide-producing forces of the Sun.

solstice time when the Sun is directly over the Tropic of Cancer (summer) or the Tropic of Capricorn (winter).

sorting an indication of the range of particle sizes in a sediment deposit.

Southern Oscillation oscillation in the locations of high- and low-pressure areas in the Southern Hemisphere; associated with *El Niños*.

space scale a particular three-dimensional domain in which objects are located and events happen.

specific gravity ratio of the density of a substance relative to the density of pure water at 4°C; in the centimeter–gram–second system, *density* and *specific gravity* may be used interchangeably.

specific heat quantity of heat required to raise the temperature of 1 gram of a substance by 1°C. The common unit is calories per gram per degree Celsius.

sperm whale See *toothed whales.*

spherule small spherical body.

spicules crystals of newly formed sea ice; also, minute, needle-like or multiradiate calcareous or siliceous bodies in many organisms.

spit small point of land projecting into a body of water.

spreading center growing edge of a lithospheric plate, usually a mid-ocean ridge; also called a *divergent margin.*

spring bloom sudden proliferation of phytoplankton that occurs when the critical depth (as determined by penetration of sunlight) exceeds the depth of the mixed, stable surface layer (as determined by the pycnocline).

spring tide tide of increased range that occurs about every two weeks, when the moon is new or full.

spur-and-groove structure See *forereef.*

stable triple junction triple junction that does not move through time.

stability resistance to overturning or mixing in the water column, resulting from the presence of a positive density gradient: less dense water above denser water.

standing crop biomass of a population present at a specified time.

standing wave type of wave in which the surface of the water oscillates vertically between fixed points, called *nodes,* without progression. The points of maximum vertical rise and fall are called *antinodes.* At the nodes, the underlying water particles exhibit no vertical motion but maximum horizontal motion.

stand of the tide interval at high or low water when there is no appreciable change in the height of the tide; its duration depends on the range of the tide, being longer when the tidal range is small and shorter when the tidal range is large.

steady state absence of change with time.

still-water level level that the sea surface would assume in the absence of wind wave.

stock an interbreeding population of animals.

storm surge (storm wave, storm tide, tidal wave) rise or piling up of water against shore, produced by wind stresses and atmospheric low pressures in a storm.

stratosphere the part of Earth's atmosphere between the troposphere and the mesosphere.

subduction process where material at the surface is drawn down into the interior.

subduction zone inclined plane descending away from a trench, separating a sinking oceanic plate from an overriding plate; usually associated with a *trench* and active volcanoes.

sublimation transition of the solid phase of certain substances into a gas—and vice versa—without passing through the liquid phase.

submarine canyon submarine valley.

subpolar ocean region ocean area lying within the subpolar regions, around 60° N or 60° S.

substrate a surface suitable for settlement by benthic organisms.

subsurface current current usually flowing below the pycnocline, generally at slower speeds and frequently in a different direction from the currents near the surface.

subtropical high one of the semipermanent highs of the subtropical high-pressure belt.

subtropical ocean region ocean area lying within the subtropical regions around 30° N or 30° S.

succession, ecological process of community change whereby communities replace one another.

supernova violent explosive stage in the evolution of a star.

surf collective term for breakers; also, the wave activity in the area between the shoreline and the outermost limit of breakers.

surface-active agent substance, usually in solution, that can markedly change the surface or interfacial properties of the liquid, even when present in minute amounts.

surface tension (surface energy, capillary forces, interfacial tension) phenomenon peculiar to liquid surfaces caused by a strong attraction (toward the interior) acting on molecules at or near the surface in such a way as to reduce the surface area.

surface zone (mixed zone) water above the pycnocline where waves and convection mix the water; results in uniform temperatures and salinities within the mixed zone.

surf zone area between the outermost breaker and the limit of wave uprush.

surge horizontal oscillation of water with comparatively short period accompanying a seiche (see also *storm surge*).

suspension feeding feeding by removing food particles from water.

sustainable development scheme of resource utilization where present resource utilization does not endanger future uses of the resource.

swell waves that have traveled out of their generating area.

swim bladder gas-filled sac lying between the vertebral column and the alimentary tract of certain fishes. It serves a hydrostatic function in most fishes that possess it; in some, it participates in sound production.

symbiosis relationship between two species in which one or both participants benefit and neither is harmed.

synthetic aperture radar radar technique that provides high-resolution images of ocean surface features, such as swell, internal waves, current boundaries, and sea-ice distributions and movements.

T

technology application of science and engineering.

tectonic estuary estuary occupying a basin formed by mountain building.

tektites formed by melting of rock fragments injected into the atmosphere by a meteorite impact.

temperature region region between 30° and 60° in each hemisphere.

terrigenous sediment sediment particles and deposits derived from the land.

tertiary production the amount of organic matter produced by carnivores.

Tethys narrow sea that separated Pangaea into Laurasia (north) and Gondwanaland (south) when the continents formed a supercontinent, before the present spreading cycle began.

theory tentative conclusion, supported by some evidence.

thermocline marked vertical temperature change in a body of water; also a layer in which such a temperature change occurs.

thermohaline circulation circulation induced by differences on water density which is controlled primarily by temperature and salinity.

tidal bore rapid rise of the tide in which the advancing water forms a marked front; occurs in certain shallow estuaries having a large tidal range.

tidal bulge (tidal crest) long-period wave associated with the tide-producing forces of the Moon and Sun; identified with the rising and falling of the tide. The trough located between the two tidal bulges present at any given time on Earth is known as the *tidal trough.*

tidal constituents mathematical components used in the dynamical method of tidal prediction to predict tides.

tidal current current caused by tidal phenomena.

tidal day interval between two successive upper transits of the Moon over a location. A *mean tidal day,* sometimes called a lunar day, is 24 hours, 50 minutes.

tidal flats marshy or muddy areas which are covered and uncovered by the rise and fall of the tide; also called *tidal marshes.* Usually covered by plants.

tidal period elapsed time between successive high or low waters.

tidal range difference in height between consecutive high and low waters.

tidal trough See *tidal bulge.*

tide periodic rise and fall of the ocean and atmosphere, caused by the gravitational attraction of Moon and Sun acting on Earth.

tide curve presentation of the rise and fall of tide; time (in hours or days) is plotted against height of the tide.

tide pool depression—usually water-filled—in the intertidal zone, alternately submerged and exposed by the rise and fall of the tide or wave action.

tide-producing forces slight local difference between the gravitational attraction of two astronomical bodies and the centrifugal force that holds them apart. Gravitational attraction predominates at the surface point nearest to the other body while centrifugal repulsion predominates at the surface point farthest from the other body.

tide rip See *rip.*

tide tables tables that predict the times and heights of tidal phenomena at specified locations.

tide wave long-period gravity wave that has its origin in the tide-producing force; manifests itself in the rising and falling of the tide.

till poorly sorted rock debris, deposited by glaciers.

time scale particular period of time encompassing the duration of an event.

toothed whales dolphins, porpoises, killer whales, and sperm whales.

trace nutrients substances required in minute amounts for plant growth.

tracer substance used to trace water movements; usually one that does not occur naturally.

trade winds wind system in most of the tropics that blows from the subtropical highs toward the equatorial lows. Trade winds are from the northeast in the Northern Hemisphere, from the southeast in the Southern Hemisphere.

transform fault fault along which lithospheric plates move past each other.

trench long, narrow, deep depressions of the ocean floor, typically in a subduction zone.

triple junction intersection of three plates; may involve any combination of *trenches, mid-ocean ridges,* and *transform faults.*

trophic level successive stage of nourishment as represented by links of the food chain. Primary producers constitute the first trophic level, herbivores the second, and carnivores the third and highest trophic level.

tropical region region within a few degrees of latitude of the equator.

tropics equatorial region between Tropic of Cancer (north) and Tropic of Capricorn (south); climate found in the belt close to the equator. Daily variations in temperature exceed seasonal variations; generally high rainfall.

tropopause upper limit of the troposphere.

troposphere portion of the atmosphere next to Earth's surface where temperature generally rapidly decreases with altitude, clouds form, and convection is active.

tsunami (seismic sea wave) long-period sea waves produced by submarine earthquakes, volcanic explosions, or slumps.

tunicates globular or cylindrical, often saclike, animals. Some are sessile; others are planktonic.

turbidite turbidity-current deposit characterized by vertically and horizontally graded bedding.

turbidity current gravity current resulting from a density increase caused by suspended materials.

turbulence irregular motions of air or water; marked departure from a smooth flow.

turbulent flow flow characterized by random velocity fluctuations.

typhoon tropical hurricane in western Pacific.

U

ultraplankton one-celled planktonic organisms, less than 2 microns in diameter.

undersaturation a solution that contains less dissolved matter than the amount that a fully saturated solution could.

unselective feeder organism that feeds on a wide range of particles.

unstable triple junction triple junction that moves through time.

upwelling process by which water rises from a lower to a higher depth, usually caused by divergence.

V

van der Waals forces weak attractive forces between molecules that arise from interactions between the atomic nuclei of one molecule and the electrons of another molecule.

vertical exaggeration exaggeration of the vertical scale relative to the horizontal scale, to permit features to be more readily seen in profile or cross-section.

viscosity internal resistance to flow, a property of fluids that enables them to support certain stresses and thus resist deformation for a finite time.

vorticity whirling or circular motion of a parcel of water around a vertical axis; results from Earth's rotation and vertical shear of water flows.

W

warm-blooded animals See *homeotherms.*

water budget an accounting for the interchanges of water among the land, atmosphere, and ocean.

water cycle circulation of water on and near Earth's surface; also known as hydrologic cycle.

water mass water body usually identified by its temperature and salinity or by some other tracer.

water parcel water mass with a certain temperature and salinity, separated from surrounding waters.

watershed land area from which all waters flow into a body of water such as a river or estuary.

wave disturbance that moves through or over the ocean surface.

wave age state of development of a wind-generated sea surface wave, conveniently expressed by the ratio of wave speed to wind speed. Wind speed is usually measured at about 8 meters above still-water level.

wave energy capacity of a wave to do work. In a deep-water wave, about half the energy is kinetic, associated with water movement, and about half is potential energy, associated with the elevation of water above the still-water level in the crest or its depression below still-water level in the trough.

wave group series of waves in which the wave direction, wavelength, and wave height vary only slightly.

wave height vertical distance between crest and preceding trough in a wave.

wavelength horizontal distance between successive wave crests, measured perpendicular to the crests.

wave period time required for two successive wave crests to pass a fixed point.

wave spectrum distribution of wave energy (square of wave height) with wave frequency (1/period). The square of the wave height is related to the potential energy of the sea surface, so the spectrum can also be called the *energy spectrum.*

wave steepness ratio of wave height to wavelength.

wave train series of waves from the same direction.

wave velocity speed at which individual waveforms advance; also, a vector quantity that specifies the speed and direction with which a wave travels through a medium.

weathering destruction or partial destruction of rock by thermal, chemical, and mechanical processes.

well-mixed estuary estuary where mixing by tidal currents is so complete that there is little or no vertical change in salinity or temperature.

western boundary current strong, narrow, deep currents on western side of ocean basins.

wetland See *marsh.*

whitecaps white froth on crests of waves in a wind.

white smoker discharge of warm waters from the ocean bottom; the waters are colored white by materials that precipitate as they mix with cooler ocean waters.

wind waves waves formed and growing in height under the influence of wind; any wave generated by winds.

windrows rows of floating debris, aligned in the wind direction, formed on the surface of a lake or ocean by Langmuir cells.

windward being in, or facing, the direction from which the wind is blowing; opposite to leeward.

Y

year class organisms of a particular species spawned during a single year or breeding season.

Z

zonation organization of a habitat into bands of distinctive plant and animal associations where conditions for survival are optimal.

zooplankton animal forms of plankton.

zooxanthellae algae, modified dinoflagellates, that live symbiotically in corals or other animals.

Index